7th International Conference on Marine Technology (SENTA 2022)

IOP Conference Series: Earth and Environmental Science
Volume 1166

Surabaya, Indonesia
20 October 2022

Editors:

W. Agoes Praktikto
Daniel M. Rosyid
Mukhtasor
Mahmud Musta'in
Ahmad Fitriadhy

ISBN: 978-1-7138-7498-0
ISSN: 1755-1307

Printed from e-media with permission by:

Curran Associates, Inc.
57 Morehouse Lane
Red Hook, NY 12571

Some format issues inherent in the e-media version may also appear in this print version.

Institute of Physics
Dirac House, Temple Back
Bristol BS1 6BE UK

Phone: 44 1 17 929 7481
Fax: 44 1 17 920 0979

techtracking@iop.org

Additional copies of this publication are available from:

Curran Associates, Inc.
57 Morehouse Lane
Red Hook, NY 12571 USA
Phone: 845-758-0400
Fax: 845-758-2633
Email: curran@proceedings.com
Web: www.proceedings.com

TABLE OF CONTENTS

IOP Conf. Series: Earth and Environmental Science 1166 (2023) 011001 doi:10.1088/1755-1315/1166/1/011001

Conference Proceeding of SENTA 2022

"Sustainable Development and Innovations in Marine Technology for A Post Pandemic Era"

The 7th International Conference on Marine Technology (SENTA 2022) – Faculty of Marine Engineering,

Institut Teknologi Sepuluh Nopember

Surabaya, October 20th 2022

sponsored by

IOP Conf. Series: Earth and Environmental Science 1166 (2023) 011001 doi:10.1088/1755-1315/1166/1/011001

PREFACE
The 7th International Conference on Marine Technology (SENTA 2022)

Assalamualaikum Wr. Wb.

Ladies and Gentlemen,

Since 2001, the Faculty of Marine Technology – Institut Teknologi Sepuluh Nopember (ITS) Surabaya has held the annual SENTA conference, formerly known as Seminar Nasional Teknologi dan Aplikasi Kelautan (National Conference on Marine Technology and Application), which provided a regular forum for researchers and engineers in Marine Technology in Indonesia. Then in 2016, the SENTA conference turned into an international conference, namely International Conference on Marine Technology (SENTA), to expand its research network.

Two years of a pandemic transformed every sector, including the maritime sector. The demand to explore our ocean foster the technology development in marine technology. Embracing the sustainability of our earth and environment, especially our ocean, The 7th International Conference on Marine Technology (SENTA) 2022 brought the theme "Sustainable Development and Innovations in Marine Technology for A Post-Pandemic Era". The conference took place in Institut Teknologi Sepuluh Nopember, Surabaya, on October 20th, 2022, and gathered researchers and professionals in the marine industry.

I would like to extend gratitude to our distinguished keynote speakers, Xabier Agote (President of Albaola Itsas Kultur Faktoria, Spain), Prof. Richard Birmingham (Professor of Small Craft Design, Newcastle University, United Kingdom), and Prof. Dr. Mazlan Abd. Ghaffar (Professor of Faculty of Science and Marine Environment, Universitas Malaysia Terengganu, Malaysia). My deepest gratitude to our authors who submitted and participated in SENTA 2022. We received 99 papers. Of those submitted papers, 81 are presented and accepted for publication. I especially thank the organizing committees and reviewers for their hard work, perseverance, and patience in preparing and organizing this seminar, so it went smoothly and successfully.

Through this conference, let us extend the network and cooperation among all stakeholders of the marine technology sector, especially in Indonesia and the world in general, to build a better maritime world in Indonesia.

Thank you for your attention.

Wassalamualaikum Wr. Wb.

Surabaya, October 20th, 2022

Herman Pratikno, S.T., M.T., Ph.D.

Chairman of SENTA 2022, ITS Surabaya

IOP Conf. Series: Earth and Environmental Science 1166 (2023) 011001 doi:10.1088/1755-1315/1166/1/011001

A. EDITORS

1. Prof. Ir. W. Agoes Praktikto, M.SC., Ph.D.
   Institut Teknologi Sepuluh Nopember | praktikto@oe.its.ac.id

2. Prof. Ir. Daniel M. Rosyid. Ph.D.
   Institut Teknologi Sepuluh Nopember | dmrosyid@oe.its.ac.id

3. Prof. Ir. Mukhtasor, M.Eng., Ph.D.
   Institut Teknologi Sepuluh Nopember | mukhtasor@oe.its.ac.id

4. Prof. Drs. Mahmud Musta'in, M.Sc., Ph.D.
   Institut Teknologi Sepuluh Nopember | mmustain@oe.its.ac.id

5. Assoc. Prof. Ir. Dr. Eng. Ahmad Fitriadhy (Guest Editor)
   Universiti Malaysia Terengganu | a.fitriadhy@umt.edu.my

B. COMMITTEES

ADVISORY BOARD :

1. Prof. Dr. Ir. Mochamad Ashari, M.Eng.
2. Dr. Eng. Trika Pitana, S.T., M.Sc.
3. Dr. Eng. Rudi Waluyo Prastianto, S.T., M.T.
4. Ir. Wasis Dwi Aryawan, M.Sc., Ph.D.
5. Beny Cahyono, S.T., M.T., Ph. D
6. Dr.-Ing. Ir. Setyo Nugroho
7. Herman Pratikno, S.T., M.T., Ph.D.

SCIENTIFIC PROGRAM COMMITTEE :

1.	Prof. Ir. Mukhtasor, M.Eng.,Ph.D.	Department of Ocean Engineering, Institut Teknologi Sepuluh Nopember
2.	Prof. Ir. W. Agoes Pratikto, M.Sc., Ph.D.	Department of Ocean Engineering, Institut Teknologi Sepuluh Nopember
3.	Prof. Ir. Eko Budi Djatmiko, M.Sc., Ph.D.	Department of Ocean Engineering, Institut Teknologi Sepuluh Nopember
4.	Prof. Ir. Daniel M. Rosyid, Ph.D.	Department of Ocean Engineering, Institut Teknologi Sepuluh Nopember
5.	Prof. Drs. Mahmud Musta'in.,M.Sc.,Ph.D.	Department of Ocean Engineering, Institut Teknologi Sepuluh Nopember
6.	Prof. Suntoyo, S.T.,M.Eng., Ph.D.	Department of Ocean Engineering, Institut Teknologi Sepuluh Nopember
7.	Prof. Ir. Sjarief Widjaja, Ph.D.	Department of Naval Architecture, Institut Teknologi Sepuluh Nopember
8.	Prof. Dr. Ir. I Ketut Suastika	Department of Naval Architecture, Institut Teknologi Sepuluh Nopember
9.	Prof. I Ketut Aria Pria Utama, M.Sc., Ph.D.	Department of Naval Architecture, Institut Teknologi Sepuluh Nopember
10.	Prof. Aries Sulisetyono, S.T., M.A.Sc., Ph.D.	Department of Naval Architecture, Institut Teknologi Sepuluh Nopember
11.	Prof. Ir. Achmad Zubaydi, M.Eng., Ph.D.	Department of Naval Architecture, Institut Teknologi Sepuluh Nopember
12.	Prof. Dr. Ir. Heri Supomo, M.Sc.	Department of Naval Architecture, Institut Teknologi Sepuluh Nopember
13.	Prof. Semin, S.T., M.T., Ph.D.	Department of Marine Engineering, Institut Teknologi Sepuluh Nopember
14.	Prof. Dr. I Ketut Buda Artana, ST., M.Sc.	Department of Marine Engineering, Institut Teknologi Sepuluh Nopember
15.	Prof. Ir. Aguk Zuhdi M. Fathallah., M.Eng., Ph.D.	Department of Marine Engineering, Institut Teknologi Sepuluh Nopember
16.	Dr. Ing. Wolfgang Busse	Wismar University of Applied Ssciences, Germany

IOP Conf. Series: Earth and Environmental Science 1166 (2023) 011001 doi:10.1088/1755-1315/1166/1/011001

ORGANIZING COMMITTEE :

General Chair	: Herman Pratikno, S.T., M.T., Ph.D.
Secretary	: Dr. Eng. Shade Rahmawati, S.T., M.T.
Finance Chair	: Dr. Eng. Muhammad Zikra, S.T., M.Sc.
Finance Co-Chair	: Rizky Dwi Putra, S.E.

1. Sufian Imam Wahidi, S.T., M.Sc.
2. Erzad Iskandar Putra, S.T., M.T.
3. Fadilla Indrayuni Prastyasari, ST., M.Sc.
4. Sunarsih, ST, M.Eng., Ph.D.
5. Dr. Dendy Satrio, S.ST.
6. Muhammad Riduwan, S.Kom., M.Kom.
7. Oktaviani Turbaningsih, S.T., M.T.

TECHNICAL COMMITTEE :

1. Ferdita Syalsabila, S.T., M.T.
2. Muhammad Fakhrul Arifin
3. Ridzky Bayu Wicaksono
4. Bhagawanta Atha Adyatma
5. Muhamad Figo Octavianto
6. Salma Arthi Menon
7. Audrey Salsza Maret Roberte
8. Ahmad Qaedi Luthfi
9. Saif Adam Sauqi
10. Mutiara Alifah
11. Adinda Anggraeni Rahmawati Sukarsa
12. Azzura Nurfitriaina
13. Lindah khabibah
14. I Gde Krishnananda Wibawa
15. Zakkiya Al Fakhira Fita Ananda
16. Gabriel Hansel Fabian
17. Advira Nafi Khalida
18. Denny Hartanto
19. Tasya Putri Rhamadani

IOP Conf. Series: Earth and Environmental Science 1166 (2023) 011002 doi:10.1088/1755-1315/1166/1/011002

Peer Review Statement

The 7th International Conference on Marine Technology (SENTA 2022)

All papers published in this volume have been reviewed through processes administered by the Editors. Reviews were conducted by expert referees to the professional and scientific standards expected of a proceedings journal published by IOP Publishing.

Type of peer review	: Single anonymous
Conference submission management system	: Morressier
Number of abstract submission received	: 116 abstracts
Number of paper submissions received	: 99 papers
Number of submissions sent for review	: 96 papers
Number of submissions accepted	: 54 papers
Acceptance Rate (Submissions Accepted/ Abstract Submissions Received × 100)	: 56%
Average number of reviews per paper	: 2
Total number of reviewers involved	: 18 reviewers
Contact person for queries	: Heman Pratikno, S.T., M.T., Ph.D. hermanp@oe.its.ac.id Institut Teknologi Sepuluh Nopember (ITS)

LNG Type C, Cylindrical Weight Design Limitation Based on Shipbuiding Facilities

D A Wibowo[1], C Sasmito[2], M F Manti[1], A Muis[1], Iskendar[1], N I Gutami[1], Z Qonita[3]

[1] The National Research and Innovation Agency of Republic Indonesia, Research Centre for Transportation Technology, PUSPIPTEK Serpong, Banten 15314, Indonesia.
[2] The National Research and Innovation Agency of Republic Indonesia, Research Centre for Hydrodynamic Technology, Sukolilo, Surabaya, East Java 60112, Indonesia.
[3] The National Research and Innovation Agency of Republic Indonesia, Geological Disaster Research Centre, Cisitu, Bandung, West Java 40135, Indonesia

Corresponding author : dimasaldyantowibowo@gmail.com

Abstract. Supporting climate change mitigation and sustainable energy due to its ability to reduce CO_2, Liquefied Natural Gas (LNG) considered as promising alternative fuel. In a particular small market with LNG demand between 2,500 m^3 to 7,500 m^3, utilization of LNG type-C, cylindrical tanks are compatible to use in small-scale LNG carrier. In order to overcome these challenges, Indonesian industry need to design the tank's volume and weight requirements.while also taking into account the capabilities of the shipbuilding facilities. The weight of the tank, which comprises the weight of the outer tank and dished end, are the decision variable in the approach for weight estimation analysis, which carried by using using extrapolation to search optimum diameter and length of the tank The objective function is maximization of the tank capacity. The capability of the lifting equipment in the shipyard is the limitation used in this paper. This research data is based on the shipyard capability of PT. PAL Indonesia, PT. Dok dan Perkapalan Kodja Bahari (Persero), PT. Industri Kapal Indonesia, and PT. Batamec Shipyard. Result obtained that maximum volume of 5,663.369 m^3 LNG type C, weight about 450.671 tons with maximum crane lifting capacity in shipyard are 500 tons

1. Introduction

Up to 2024, Indonesia is anticipated to use more gas as a fuel for electricity production. According to the information in the table 1, taken from the state power company's business plan for the provision of energy in Indonesia for 2021–2030 [1]. In the study, five clusters were simulated for the Eastern Indonesia region. East Nusa Tenggara, West Nusa Tenggara, Maluku, and Papua in eastern Indonesia require LNG ships with sizes ranging from 2,500 m^3 to 7,500 m^3[2]. A small-scale LNG vessel that can accommodate the requirements of a gas power plant that is not too big is required in light of this.

Table 1. Fuel demand forecast for Indonesia's ideal case[1]

No	Type of fuel	Unit	2021	2022	2023	2024
1	Biodisel	Thousand kL	458	330	248	134
2	Biomass	Kilo ton	607	1.508	3.142	4.185
3	Gas	TBTU	365	393	391	405
4	Fuel Oil	Thousand kL	1.883	1.239	909	449
5	Coal	Million ton	116	120	128	135
6	Coal Subtitution	Million ton	-	-	-	-

It was also explained that the need for gas fuel is divided into several clusters and it is planned for central and eastern Indonesia that PLN will use mini-LNG for peak load-bearing plants in large systems in Kalimantan, Sulawesi, Nusa Tenggara, Maluku, and Papua. However, it is possible that mini LNG will be used for other power plants. In the attachment to the decree of the minister of energy and mineral resources of Indonesia number 13K/13/MEM/2020, it is stated that there are 52 gas plants spread throughout Indonesia with a total capacity of 1697 MW, and an indicative gas volume of 148 BBTUD[3]. This need is adjusted to the need for natural gas in PLN's RUPTL 2021-2030. In this study, a study of tank weight restrictions was carried out based on the capacity of the cranes found in several large shipyards in Indonesia.

2. LNG Independent Type C Tanks.

The market for short- and medium-distance transportation of small-scale liquefied gas has grown recently (e.g., domestic trade). Using a separate Type C containment system, which is frequently employed for small-scale liquefied gas cargo transportation, is one of the viable alternatives to address this expanding need [4]. The C-Cylindrical type LNG tank is the subject of this study because it has qualities that make it ideal for small LNG ships that range in size below 30.000 m3 . More than 85% of these newbuilds constructed between 2010 and 2020 come with Type C tanks due to their minimal danger of leakage and structural collapse[5]. This type of tanks do not need a secondary barrier. Depending on the liquefied gas being held, Type C tanks can either be fully or partially pressurized. Compared to type A or type B tanks, this type of tank can also store LNG under higher pressure. According to the Det Norske Veritas-Germanischer Lloyd classification [6], the following components are found in the LNG type C tank :

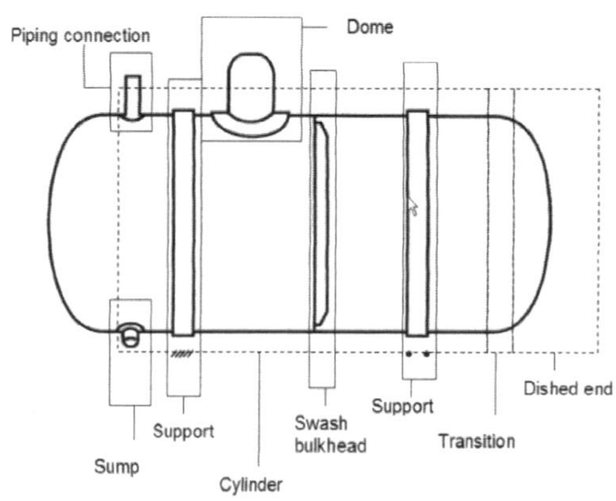

Figure 1. Component division of type C Cylindrical LNG tank [6]

IOP Conf. Series: Earth and Environmental Science 1166 (2023) 012001 doi:10.1088/1755-1315/1166/1/012001

When considering corrosion allowance, the minimum thickness following shell or head formation must not be less than:

- 5 mm for carbon-manganese steels dan nickel steels
- 3 mm for austenitic steels
- 7 mm for aluminium alloys

2.1. Material

Aluminum and stainless steel alloys are used to build the majority of LNG-fueled tanks. They are appropriate for storing LNG at 163 °C because they do not fracture brittlely at such low temperatures[7]. For use in cryogenically welded pressure vessels, ASTM A353 specifies 9% nickel double-normalized and tempered steel plates. The maximum thickness for these steel plates is typically 50 mm. This material can also operate in lower temperature which is essential for cryogenic transportation. In this research ASTM A353 is used for determine weight estimation in LNG tank. Table 2 describes the material properties of ASTM A353. For carbon, carbon-manganese and low-alloy steels the nominal design stress determines using R_{eH} and R_m with temperature up to and including 50°C [8]. R_{eH} stand for minimum upper yield stress that is acceptable at room temperature. If there is no defined yield stress indicated on the stress-strain curve, the 0.2% proof stress must be used. R_m stand for minimum lower yield stress that is determined at room temperature.

Table 2. 9 Nickel Steel Plate in Double-Normalized and Tempered Condition for Pressure Vessels [9]

9 Nickel Steel Plate in Double-Normalized and Tempered Condition for Pressure Vessels	
ASTM Spesification	A 353
Composition, %	
Carbon	0.13 max
Manganese	0.90 max
Silicon	0.15-0.30
Nickel	8.50-9.50
Thickness, in.	2 max
Heat Treatment	Double Normalized and Tempered
Tensile Properties	
Tensile Strength, psi	100,000-120,000
Yield Strength (0.2% Offset), min, psi	75,000
Elongation (2 in.), min, %	20
Charpy Impact (Keyhole-Notch)(ASTM A 300)	
Energy Absorbed, min, ft-lb (Long.)	15 at -320 F
Charpy Impact (V-Notch)(ASTM A 353)	
Energy Absorbed, [a]min, ft-lb	
Longitudinal	35 at -320 F
Transverse	20 at -320 F
Lateral Exspansion, [b]min, in.(Transv.)	.015 at -320 F
a Average of 3 specimen, Supplementary requirement.	
b Required on each specimen.	

IOP Conf. Series: Earth and Environmental Science 1166 (2023) 012001 doi:10.1088/1755-1315/1166/1/012001

3. Method

The weight is calculated by extrapolating linearly from the graph's data in order to determine the ideal diameter. Pressure, corrosion, joint effectiveness, and permissible stress are all taken into account by this procedure. Even with this much information, it is impossible to pinpoint exactly which ratios will result in the lowest total cost because there are many more factors that affect the final cost of a vessel. But choosing the lightest weight is perhaps the most important factor in getting the lowest cost.[10]

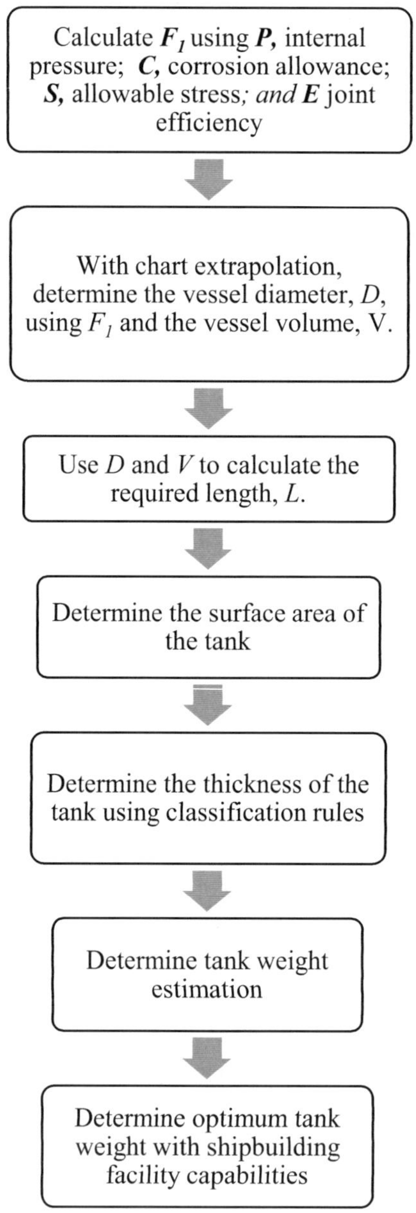

Figure 2. Method use for weight estimation of the tank with shipbuilding capabilities

IOP Conf. Series: Earth and Environmental Science 1166 (2023) 012001 doi:10.1088/1755-1315/1166/1/012001

4. Weight and Volume Estimation

The estimation of weight calculations can be done by using the calculation method. The first calculation method is to calculate $\boldsymbol{F_1}$. After that, then based on figure 2 determine the diameter that will be used to determine the length of ***L***. In the second method, F_2 is calculated and based on figure 3 determines the L/D ratio. After that, the L/D ratio will be used to determine the diameter (***D***). After that, the Volume (***V***) is obtained and used to calculate the Length (***L***). Attached below are the notations and equations used to calculate the weight and volume of the tank. The following is the notation and equation used in the calculation. P set at 10 Bar or 145.038 PSIG

$$F_1 = \frac{P}{C \, x \, S \, x \, E} \tag{1}$$

Where ***P*** is internal pressure in PSIG; ***C*** is corrosion allowance in inch; ***S*** is allowable stress in PSi; ***E*** is joint efficiency.

$$F_1 = \frac{145.038}{0.0393701 \; x \; 18{,}750 \; x \; 1}$$

$$F_1 = 0.1964 \approx 0.2$$

In this study, a variation of the tank volume of 60,000 ft^3; 80,000 ft^3; 100,000 ft^3; 120,000 ft^3, 140,000 ft^3; 160,000 ft^3. Based on the chart from K Abakians Hydrocarbon Proessing and Petroleum Refiner, June 1963[10], extrapolation is made in **Figure 4** to accommodate the calculation of the required volume requirement.

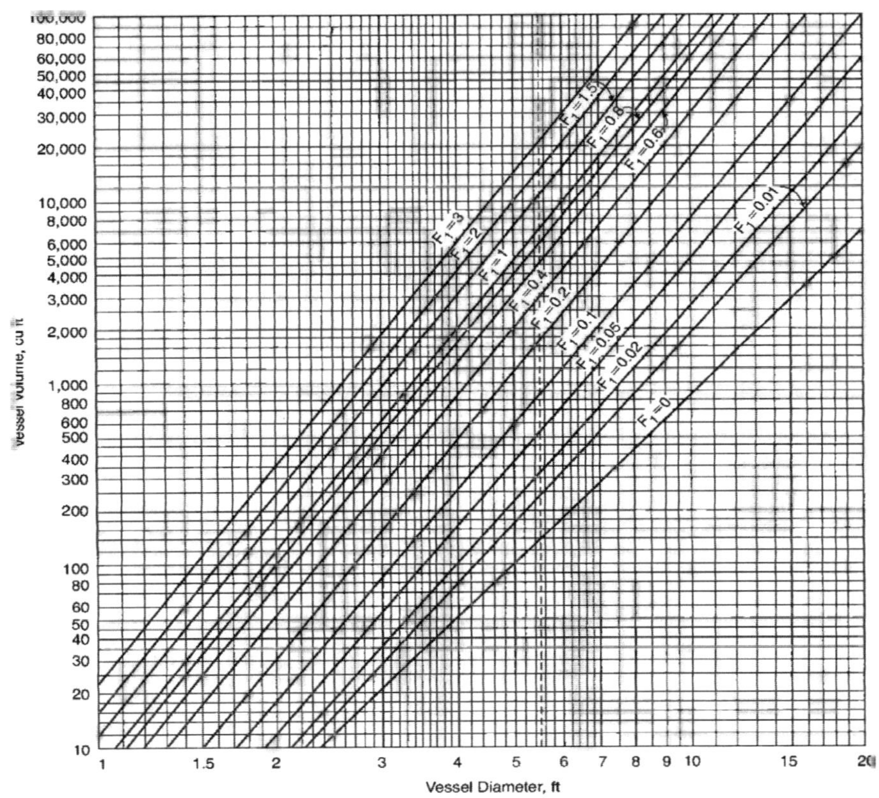

Figure 3. Chart for determining optimum diameter [10]

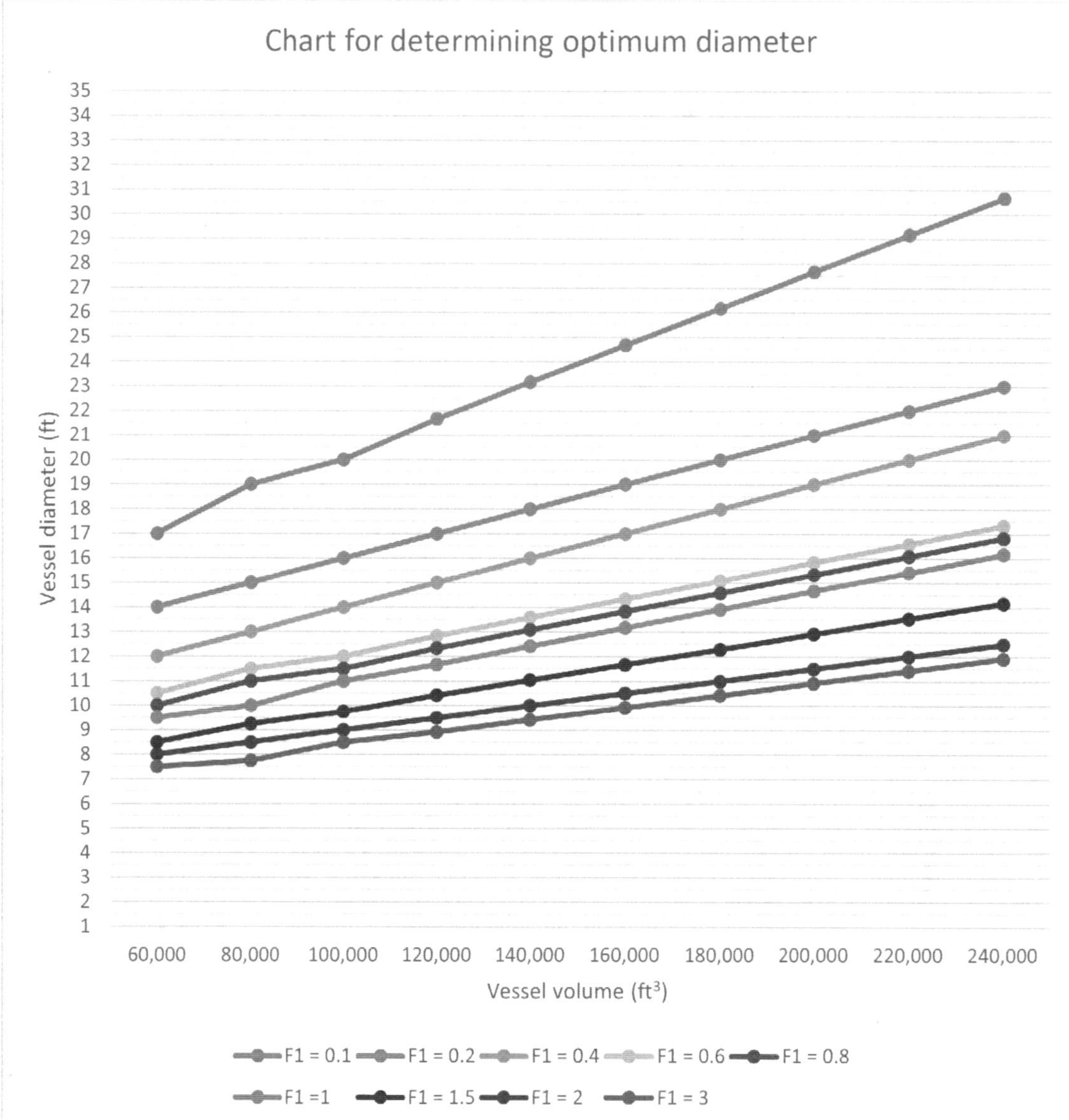

Figure 4. Extrapolation chart for determining optimum diameter

Table 3. Result for $F_1 = 0.2$ based on extrapolation chart for determining optimum diameter

Result for $F_1 = 0.2$			
Vessel volume (ft^3)	Vessel volume (m^3)	Vessel diameter (ft)	Vessel diameter (inch)
60,000	1699.010	14	168
80,000	2265.347	15	180
100,000	2831.684	16	192
120,000	3398.021	17	204
140,000	3964.358	18	216
160,000	4530.695	19	228
180,000	5097.032	20	240
200,000	5663.369	21	252
220,000	6229.706	22	264

Result for $F_1 = 0.2$			
Vessel volume (ft^3)	Vessel volume (m^3)	Vessel diameter (ft)	Vessel diameter (inch)
240,000	6796.043	23	276

The length of the tank can be calculated by obtaining the diameter by using equation (2)

$$L = R\,(\,C_1 + 4K_1\,) \tag{2}$$

Where L is length of the tank in inch; R is radius of the tank in inch; C_1 is constant for ellipsoidal heads = 2,76; K_1 is constant for ellipsoidal heads = 0,5

Table 4. Length of the tank based on vessel diameter.

Vessel volume (ft^3)	Vessel volume (m^3)	Vessel diameter (ft)	Vessel diameter (m)	Length of the tank (inch)	Length of the tank (m)
60,000	1699.010	14	4.267	399.84	10.155
80,000	2265.347	15	4.572	428.4	10.881
100,000	2831.684	16	4.876	456.96	11.606
120,000	3398.021	17	5.181	485.52	12.332
140,000	3964.358	18	5.486	514.08	13.057
160,000	4530.695	19	5.791	542.64	13.783
180,000	5097.032	20	6.094	571.2	14.508
200,000	5663.369	21	6.399	599.76	15.234
220,000	6229.706	22	6.703	628.32	15.959
240,000	6796.043	23	7.008	656.88	16.685

Equations (3), (4) and (5) can be used to calculate the thickness of the tank after establishing its length, diameter and material properties.

Cylindrical shells

$$S \geq \frac{pc\ x\ Do}{20\ x\ \sigma t\ x\ v + pc} + c \tag{3}$$

Where S is the thickness of a cylindrical shells; c is corrosion margin/allowance; D_o is outside diameter; p_c is calculating pressure; v is joint efficiency; σ_t is nominal design stress at calculating temperature[8].

Dished end/Spherical shells

$$S \geq \frac{pc\ x\ Do}{40\ x\ \sigma t\ x\ v + pc} + c \tag{4}$$

Where S is the thickness of a spherical shell; c is corrosion margin/allowance; D_o is outside diameter; p_c is calculating pressure; v is joint efficiency; σ_t is nominal design stress at calculating temperature[8].

Joint efficiency for pressure equipment for LNG shall have a joint efficiency, v=1.0. while for carbon and low-alloy steels, the minimum corrosion allowance c shall be 1.0 mm or 0.0393701 inch, unless other values are specified. Allowable stress determined with divi determined by dividing the yield strength by the safety factor by the IGC Code for LNG type C tanks[11].

Nominal design stress

$$\sigma t = min\{\frac{ReH}{1.7}; \frac{Rm}{2.7}\} \tag{5}$$

Where R_{eH} is specified minimum upper yield stress at room temperature. If the stress-strain curve does not show a defined yield stress, the 0.2% proof stress shall be applied; ***Rm*** is specified minimum tensile strength at room temperature

Following a calculation of the tank's thickness, the tank's thickness result is as follows in table :

Table 5. Thickness of cylindrical shells and dished end/spherical shells

Vessel volume (m^3)	Vessel diameter (m)	Length of the tank (m)	Cylindrical shells thickness (mm)	Dished end/Spherical shells thickness (mm)
1699.010	4.267	10.155	9.344	5.176
2265.347	4.572	10.881	9.940	5.474
2831.684	4.876	11.606	10.535	5.772
3398.021	5.181	12.332	11.131	6.070
3964.358	5.486	13.057	11.728	6.369
4530.695	5.791	13.783	12.324	6.667
5097.032	6.096	14.508	12.920	6.966
5663.369	6.401	15.233	13.517	7.264
6229.706	6.706	15.959	14.113	7.563
6796.043	7.008	16.685	14.708	7.860

Tank surface area $$2\pi \boldsymbol{rT} + 4\pi \boldsymbol{r}^2 \quad (6)$$

Where ***r*** is radius of the tank; ***T*** is length of the tank minus diameter of the tank.

Then equation (6) is used in the computation to arrive at the ratio of the tank's weight to volume, as indicated in the table.

Table 6. Tank volume and weight

Vessel volume (m^3)	Vessel diameter (m)	Length of the tank (m)	Cylindrical shells thickness (mm)	Dished end/Spherical shells thickness (mm)	Total weight of cylindrical shells and dished end of the tank (Ton)
1,699.010	4.267	10.155	9.344	5.176	139.142
2,265.347	4.572	10.881	9.940	5.474	169.78
2,831.684	4.876	11.606	10.535	5.772	204.519
3,398.021	5.181	12.332	11.131	6.070	243.791
3,964.358	5.486	13.057	11.728	6.369	287.756
4,530.695	5.791	13.783	12.324	6.667	336.74
5,097.032	6.096	14.508	12.920	6.966	390.946
5,663.369	6.401	15.233	13.517	7.264	450.671
6,229.706	6.706	15.959	14.113	7.563	516.231
6,796.043	7.008	16.685	14.708	7.860	587.73

According to the calculations, the LNG type C tank with the smallest volume, 1,699.010 m^3, weighed 139,142 tons, while the tank with the biggest volume, 6,796.043 m^3, weighed 587.73 tons.

5. Shipyard crane capacity

The capacity of the crane in the shipyard is a determining factor for the mobilization of type C LNG tanks into mini LNG vessels. So if the type C LNG tank is designed in the form of a single unit, the crane capacity is used as a limiter for the maximum weight capacity of the LNG tank. The following provides data on the availability of material handling particularly crane capacity at several shipyards in Indonesia.

Table 7. PT PAL Indonesia (Persero) Shipyard Facilities [12].

No	Item	Capacity
1	Dry dock	50.000 DWT
2	Dry dock	20.000 DWT
3	Floating dock	2
4	Caisson dock	2
5	Ship lift	1.500 TLC
6	Goliath crane	300 Ton

Table 8. PT. Dok dan Perkapalan Kodja Bahari (Persero) Facilities [13].

No	Item	Capacity
1	Jakarta Shipyard 1	14.500 TLC
2	Jakarta Shipyard 2	9.750 TLC
3	Jakarta Shipyard 3	3.450 TLC
4	Cirebon Shipyard	8.700 TLC
5	Semarang Shipyard	1.500 TLC
6	Banjarmasin Shipyard	4.950 TLC
7	Palembang Shipyard	3.500 TLC
8	Sabang Shipyard	1.000 TLC
9	Batam Shipyard	1.000 TLC
10	Crane	150 Ton

Table 9. PT. Industri Kapal Indonesia (Persero) [14].

No	Item	Capacity
1	Tower crane	60 Ton, 2 units
2	Crawler crane	400 Ton, 1 units
3	Mobile crane	35 Ton, 4 units
4	Gantry Crane	5-15 Ton, 1 units

Table 10. PT. Batamec Shipyard [15].

No	Item	Capacity
1	Overhead Crane	5-30 Ton
2	Gantry Crane	160 Ton
3	Crawler Crane	500 Ton
4	Heavy Transporter	200 Ton
5	Jib Crane	10 – 50 Ton

6. Conclusion

- After calculating using charts that have been extrapolated to find optimum diameter,volume and length, it was found that with the upper limit value of the tank volume from the graph of 240,000 ft^3, with a capacity of approximately 6,796.043 m^3 and weight 587.73 tons, length 16.685 m and diameter 7.008 m.
- Taking into account the capability of the shipyard crane, the maximum weight of the tank that can be lifted is 500 tons and is located at the PT. Batamec Shipyard.
- According to the calculation and maximum crane capacity in PT Batamec Shipyard, LNG tank that can be built without having to add crane facilities at the shipyard are vessel with volume : 1,699.010 m^3; 2,265.347 m^3; 2,831.384 m^3; 3,398.021 m^3; 3,964.358 m^3; 4,530.695 m^3; 5,097.032 m^3; 5,663.369 m^3, with tank weight : 139.142 tons; 169.78 tons ; 204.519 tons; 243.791 tons; 287.756 tons; 336.74 tons; 390.946 tons; 450.671 tons, sequentially
- To accurately calculate the tank weight, a more thorough weight computation is required.

Acknowledgments
Thank you to our colleagues from The National Research and Innovation Agency of Republic Indonesia, Research Centre for Transportation Technology.

References

[1] PLN, "Rencana Usaha Penyediaan Tenaga Listrik (RUPTL) PT PLN (Persero) 2021-2030.," *Rencana Usaha Penyediaan Tenaga List. 2021-2030*, pp. 2019–2028, 2021.
[2] M. A. Satya Dharma Putra, "Jurnal - 213883-desain-rantai-pasok-gas-alam-cair-lng-un," *Tek. ITS*, vol. 5, no. 2, pp. 1–6, 2016.
[3] Menteri ESDM, "Keputusan Menteri No. 13 K/13/MEM/2020," 2020.
[4] A. Bureau of Shipping, "Guidance Notes on Strength Assessment of Independent Type C Tanks 2022," no. January, 2022, [Online]. Available: www.eagle.org
[5] T. M. Germany, "The cutting edge of design.," 2020. https://www.tge-marine.com/wp-content/uploads/January2020.pdf
[6] DNV GL, "CLASS GUIDELINE Liquefied gas carriers with independent cylindrical tanks of type C," no. July, 2016.
[7] W. C. Niu, G. L. Li, Y. L. Ju, and Y. Z. Fu, "Design and analysis of the thermal insulation system for a new independent type B LNG carrier," *Ocean Eng.*, vol. 142, no. November 2016, pp. 51–61, 2017, doi: 10.1016/j.oceaneng.2017.06.067.
[8] DNV-GL, "RULES FOR CLASSIFICATION Ships Part 4 Systems and components Chapter 6 Piping systems," no. July, 2017.
[9] INCO, *NICKEL ALLOY STEEL PLATES, A PRACTICAL GUIDE TO THE USE OF NICKEL-CONTAINING ALLOYS NO 1108*. NICKEL INSTITUTE, 2021. [Online]. Available: https://nickelinstitute.org/media/8d91ba67b9269b2/ni_inco_1108_nickelalloysteelplates.pdf
[10] D. R. Moss and M. Basic, *Appendices - Pressure Vessel Design Manual (Fourth Edition)*. 2013.
[11] J. R. MacGregor, A. Friis, and M. Pavic, "Some aspects in the design of compressed natural gas ships," *Trans. R. Inst. Nav. Archit. Part A Int. J. Marit. Eng.*, vol. 148, no. 3, pp. 35–66, 2006, doi: 10.3940/rina.ijme.2006.a3.8006.
[12] "Fasilitas – PT. PAL Indonesia (Persero)." https://www.pal.co.id/fasilitas/ (accessed Oct. 09, 2022).
[13] "Fasilitas - PT DKB." https://dkb.co.id/fasilitas/ (accessed Oct. 09, 2022).
[14] "IKI Shipyard." https://www.ikishipyard.co.id/app/index/fasilitas (accessed Oct. 09, 2022).
[15] "Facilities." https://www.batamec.com/index.php/id/facilities (accessed Oct. 09, 2022).

Analysis of Fin Stabilizer Base on NACA 0010 in Fast Patrol Boat to Reduce Appendage Resistance on Numerical Method

Cahya Kusuma[1]
[1]Sekolah Tinggi Teknologi Angkatan Laut, Surabaya, Indonesia
{kusumacahya99@gmail.com}

Corresponding author's : kusumacahya99@gmail.com

Abstract. The main function of the fin stabilizer on patrol boats is anti-rolling and providing comfort for ship personnel. However, the installation of the fin stabilizer has a consequence in the form of appendage resistance of the ship. This study analyzes the magnitude of the appendage resistance due to the installation of the NACA 0010 stabilizer fin mounted on a 60m patrol boat at an angle of 0°. Reduction of ship resistance is expected to improve the ship performance which then can optimize the use of fuels and minimize pollution. In the early stages, the bare hull was validated with the results of the model test on the towing tank with a resistance difference of less than 3%. Based on the simulation results, it is known that the average resistance value of Fine Marine Numeca is 1.87% higher than the resistance value of the Maxsurf software. From the results of testing the appendage resistance value with 3 variations of the shape of the NACA 0010 fin, it is known that the appendage resistance value of fin stabilizer 2 is 1.04%, fin stabilizer 1 is 1.26% while fin stabilizer 3 is 1.6%. The appendage resistance of fin stabilizer 2 is smaller than the variation of fin stabilizer 1 and 3 models.

Keywords : Appendages resistance, Fin stabilizer, NACA 0010, Numerical Method, Patrol Boat.

1 Introduction

Fast patrol boats are ships that are very suitable for Indonesia because Indonesia is an archipelagic country which has thousands of islands and shallow seas, especially in western Indonesia.[1] As an archipelagic country, Indonesia has many fast patrol boats, because they can be used as patrol boats as well as combat ships that have high speed [2][3][4][5]. As a fast patrol, the selection of fin stabilizers is very taken into account. The main function of the fin stabilizer is as an anti-rolling for the comfort and safety of passengers [6][7][8] and firing accuracy of ships, especially cannons[3]. However, the fin stabilizer also appendages resistance to the ship, so it is necessary to select the right stabilizer fin so that the value of the appendages resistance caused by the fin stabilizer is as small as possible.

Reduction of ship resistance caused by the fin stabilizer is very important because the reduced ship resistance will increase energy-saving devices. Increased energy-saving devices can reduce the use of ship fuel so that emissions can be reduced and improve the environment, especially the maritime environment. Reducing appendage resistance is very important because of the large number of patrol boats and the very high intensity of their operation.

In this research, four steps will be taken. The first treatment is a bare hull design. Design using numeca fine marine, rhinoceros and maxsurf modeller. The bare hull is calculated as the ship's resistance. The second step is to calculate the appendages resistance with a variety of model 1. The next step is to calculate the appendages resistance with a variety of model 2. The conclusion is obtained by

IOP Conf. Series: Earth and Environmental Science 1166 (2023) 012002 doi:10.1088/1755-1315/1166/1/012002

analyzing all the ship resistance data from each fin stabilizer model and compared to the bare hull. Which fin stab model has the lowest appendage resistance?

2 Basic theory

Research on fin stabilizers is divided into 3 major parts, namely optimization of life force[9][10][11], fin movement control system[12][13][14][15], area and angle and zero speed[16][17]. Several optimization methods have also been used in the selection of fin stabilizers such as the TOPSIS method (a Technique for Order Preference based on Similarity to the Ideal Solution), Analytic Hierarchy Process(AHP) and Entropy Weighting Method (EWM)[18], Adaptive Dynamic Programming (ADP)[19] recurrent neural network (RNN)[20]. All research is done to improve good anti-rolling ability by making the appendages resistance caused by fin stabilizer as small as possible. This study focuses on one fixed angle, namely 0°, with 3 types of shape variations with a fixed area fin stabilizer.

2.1 The design bare hull 3D models

This study uses Fine Marine Numeca simulation software and Maxsurf software with patrol boat objects with the following data shown in table 1 :

Table 1. Patrol Boat Specification

Specification		Value	Unit
Design speed	:	25 - 30	knots
Length over All	:	59.5	metres
Length on Waterline	:	54	metres
Length between Perpendicular	:	53.71	metres
Displacement	:	467	Ton
Draft	:	2.57	metres

In the first stage, drawing the ship's lines plan and then solidifying it with the marine software maxsurf modeller, drawing the hull, and the ship's body in the form of lines and coordinates. After obtaining the bare hull lines plan, it is then running so that it can be seen the Calculate Hydrostatics and the Resistance VS Velocity Curve as shown in Figure 1.

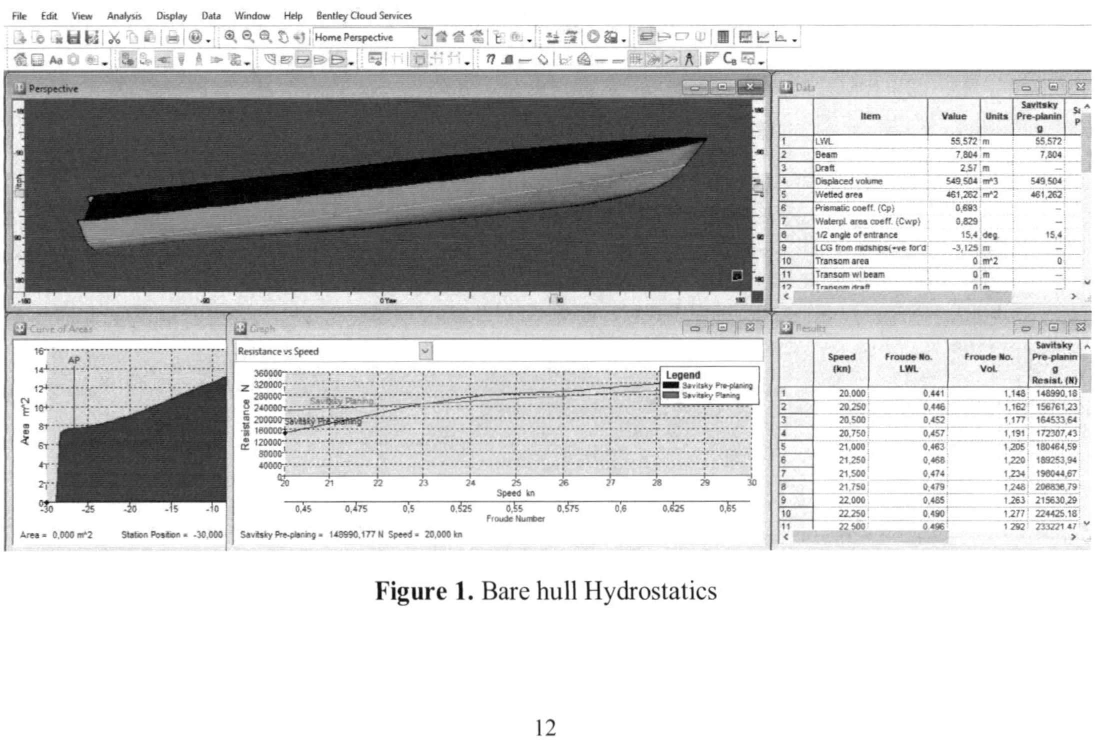

Figure 1. Bare hull Hydrostatics

2.2 NACA 0010

The initial 4-digit, 5-digit, and 4/5-digit NACA airfoil series, were generated using analytical equations that describe the main chamber line of the airfoil as well as the thickness distribution along the airfoil. The next family, including the 6 series, is a more complex form that uses theoretical methods rather than profiling[21]. Before the National Advisory Committee for Aeronautics (NACA) developed this series, airfoil designs were randomly regenerated with no designer guidance except experience with known shapes and experimentation with modifications to those shapes[22].

Methodology began to change in the early 1930s with the publication of the untitled NACA report The Characteristics of 78 Related airfoil section from tests in the variable density wind tunnel[23]. Foil has been standardized by NACA (National Advisory Committee for Airfoil). NACA airfoil is a form of the airfoil which is indicated by a number followed by the word "NACA". The important airfoil geometric properties are camber, thickness-to-chord ratio and chord length of the airfoil's flow configuration. In general, a particular flow configuration is created for an airfoil using the angle of attack and the Reynolds number describing the flow plane for the specified length and thickness of the rope. NACA 0010 is a symmetrical airfoil, the camber for the airfoil is zero while the thickness to chord ratio is ~10%. The thickness and distribution of the mean lines for the 4-digit symmetrical and curved airfoils are expressed in the ratio t/c.

NACA 0010 airfoils are used in the aircraft industry to generate lift over the required wing span during the takeoff, manoeuvring, cruising, and landing conditions required for flight[24][25], gurney flaps[26], propeller aircraft[27], Small-Size Unmanned Air Vehicle[28] and Conceptual Design for a Supersonic Advanced Military Trainer[29]. Airfoil NACA 0010 is also used in stern foil patrol boats [30]

2.3 Model Validation

Model validation is done by comparing the results of the patrol boat design made using the Maxsurf modeler and Numeca software with a resistance test with a maximum error of 3%. Then the bare hull is fitted with a fin stabilizer with NACA 0010 with 3 variations. The results of the model validation are shown in table 2. Based on the validation results in table 2, it is known that the highest error margin value is at a speed of 30 knots with a value of 1.7 per cent, while the lowest value is at a speed of 25 knots with a value of 0.31%.

Table 2. Validation bare hull model test

No	Test Model In Towing Tank			Maxsurf *Resistance*	
	Vs (Knots)	Vs (m/s)	R(kN)	R (kN)	Error (%)
1	25	12.86	284	283.1	0.3
2	26	13.37	295	290.8	-1.4
3	27	13.89	308	303.4	-1.4
4	28	14.40	321	317.3	-1.1
5	29	14.92	334	336.0	0.5
6	30	15.43	346	352.1	1.7

2.4 Variasi model fin stabilizer

The fin stabilizer model is made on Rhinoceros software, based on NACA 0010 shown in table 3. The fin is installed on the starboard side of the ship with the coordinates of the point (x) 24.00 and point (y) 1.1.00 the zero point is at the stern of the ship.

Table 3. Fin Stabilizer Specification

Specification		Value	Unit
Root	:	2.1	metres
Tip	:	1.1	metres
Length Root to Tip	:	0.815	metres
Arge area	:	1.304	Square metres

2.4.1 Design variation 1 Fin stabilizer

At the stage of making this variation of the fin stabilizer model, it refers to NACA 0010. There are 2 sizes of foil geometry with different sizes with trapezoidal shapes. Design variation 1 Fin stabilizer is shown in figure 2.

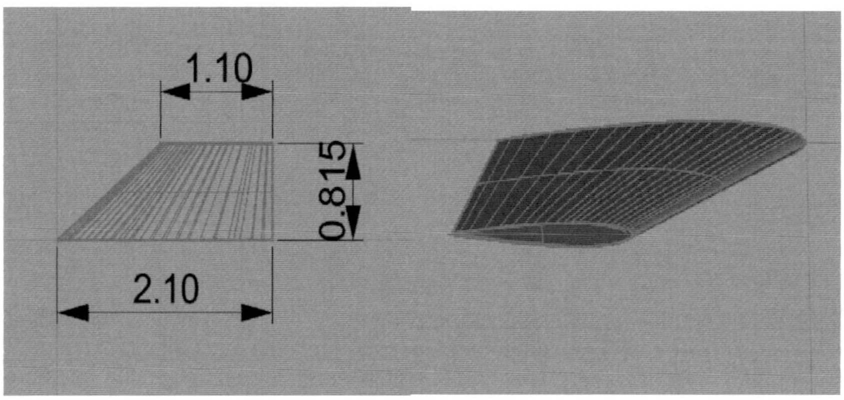

Figure 2. Design variation 1 Fin stabilizer

2.4.2 Design variation 2 Fin stabilizer

Design model fin stabilizer 2 refers to NACA 010 for the same dimensions, the position of the foil measuring 1.1 meters is in the middle of the foil measuring 2.1 m. Design variation 2 Fin stabilizer shown in figure 3.

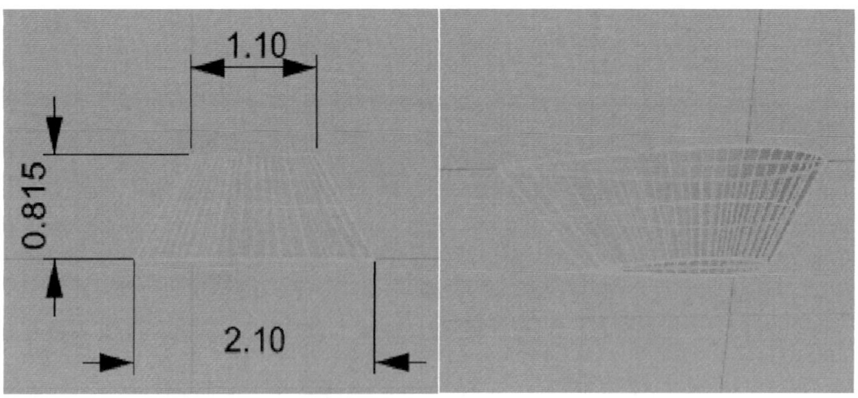

Figure 3. Design variation 2 Fin stabilizer

2.4.3 Design variation 3 Fin stabilizer

Variations in the fin stabilizer 3 design model, the larger trailing edge and leading edge foil aligns with the smaller leading edge foil. Design variation 3 Fin stabilizer shown in figure 4.

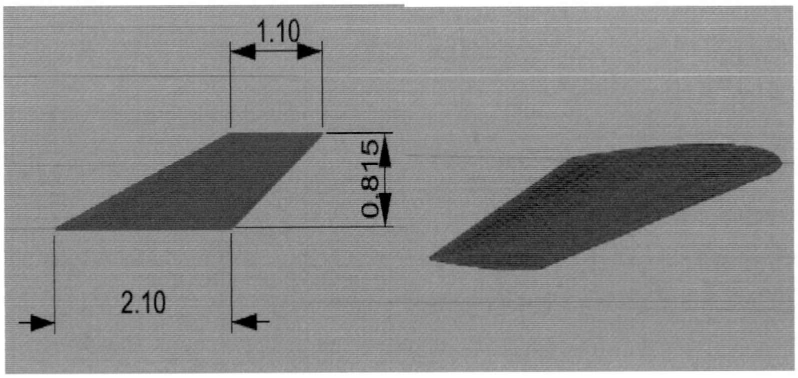

Figure 4. Design variation 3 Fin stabilizer

3 Result and Discussion

3.1 Model variation 1 Fin stabilizer

Model variation 1 fin stabilizer is shown in table 4 and figure 5.

Table 4. Model variation 1 fin stabilizer

No	Velocity Knot	Bare Hull R (kN)	Fin Stabilizer 1 R (kN)	Difference R (kN)	%
1	25	282.0	284.0	2.0	0.7
2	26	292.0	295.2	3.2	1.0
3	27	302.4	307.0	4.6	1.5
4	28	313.0	317.6	4.6	1.4
5	29	324.6	328.8	4.2	1.2
6	30	337.2	343.4	6.2	1.8

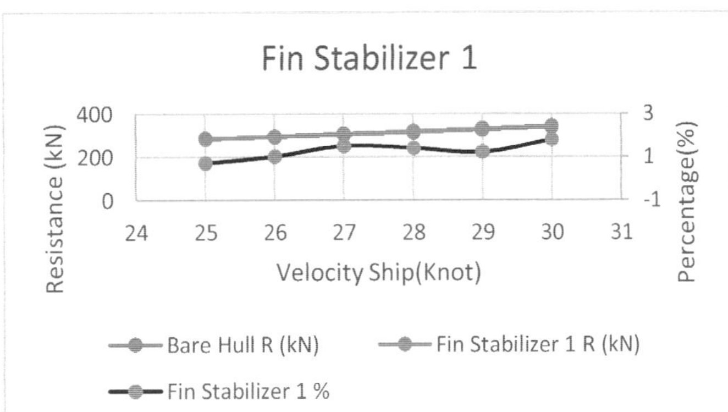

Figure 5. Total resistance fin stabilizer design variation 1.

Variation of fin stabilizer 1 has the highest value at a speed of 30 knots of 343 KN where the difference is 6.2 kN with a percentage value of 1.8 per cent. The value of the lowest resistance difference at a speed of 25 knots is 284, the difference is 2 kN with a percentage of 0.7%.

3.2 Model variation 2 Fin stabilizer

Model variation 1 fin stabilizer is shown in table 5 and figure 6.

Table 5. Model variation 2 fin stabilizer

No	Velocity	Bare Hull	Fin Stabilizer 2	Difference	
		R (kN)	R (kN)	R (kN)	%
1	25	282.0	284.7	2.7	0.95
2	26	292.0	295.2	3.2	1.09
3	27	302.4	305.6	3.2	1.05
4	28	313.0	316.5	3.5	1.10
5	29	324.6	328.6	4.0	1.23
6	30	337.2	340.1	2.9	0.86

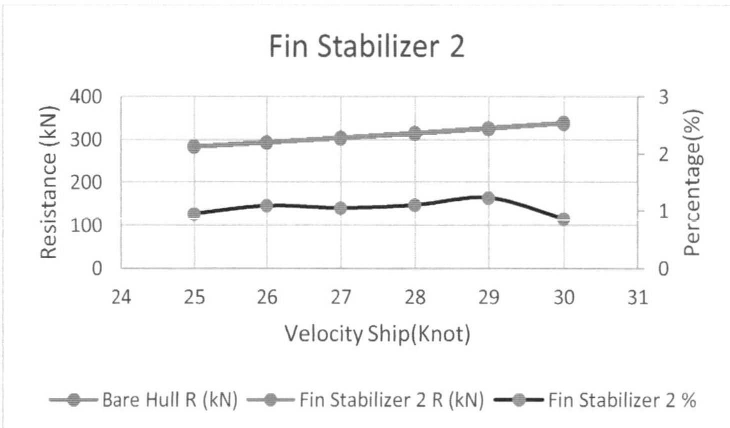

Figure 6. Total resistance fin stabilizer design variation 2.

Based on the data in table 5 and figure 7, the highest value in the variation of fin stabilizer 2 is achieved at a speed of 29 knots of 328.6 KN, there is a difference of 4 KN or 1.23%. The lowest resistance value at a speed of 25 knots is 284.7, the difference is 2.7 KN or 0.95%.

3.3 Model variation 3 Fin stabilizer

Based on the data in table 6 and Figure 8, it can be seen that the highest value of variation in fin stabilizer 3 at a speed of 30 knots is 344.3 kN where the difference is 7.1 KN or 2.1%. The lowest resistance difference value at 26 knots is 295, the difference is 3 kN or 1.02%.

Model variation 1 fin stabilizer is shown in table 6 and figure 7.

Table 6. Model variation 3 fin stabilizer

No	Velocity	Bare Hull R (kN)	Fin Stabilizer 3 R (kN)	Difference R (kN)	%
1	25	282.0	286.0	4.0	1.4
2	26	292.0	295.0	3.0	1.1
3	27	302.4	307.8	5.4	1.7
4	28	313.0	319.0	6.0	1.9
5	29	324.6	331.0	6.4	1.9
6	30	337.2	344.3	7.1	2.1

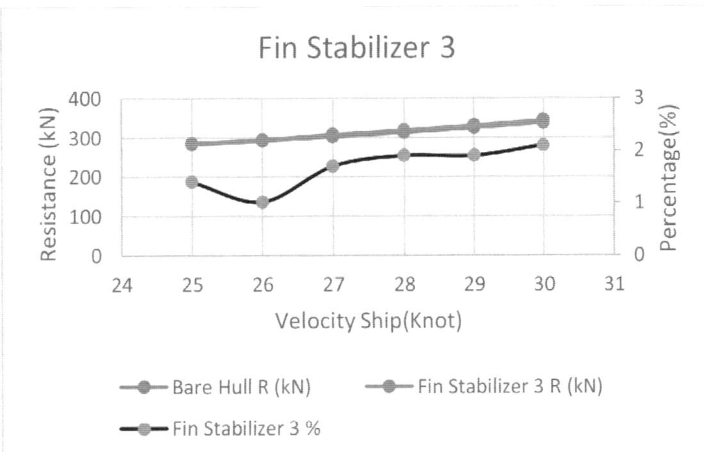

Figure 7. Total resistance fin stabilizer design variation 3.

The results of all resistance due to the addition of a fin stabilizer are shown in Table 7.

Table 7. The results of all resistance by fin stabilizer

No	Velocity	Bare Hull	The model with 3 variations Fin					
			Design 1		Design 2		Design 3	
	m/s	R(kN)	R (kN)	(%)	R (kN)	(%)	R (kN)	(%)
1	25	282.0	284.0	0.7	284.7	0.95	286.0	1.4
2	26	292.0	295.2	1.0	295.2	1.09	295.0	1.0
3	27	302.4	307.0	1.5	305.6	1.05	307.8	1.7
4	28	313.0	317.6	1.4	316.5	1.10	319.0	1.9
5	29	324.6	328.8	1.2	328.6	1.23	331.0	1.9
6	30	337.2	343.4	1.8	340.1	0.86	344.3	2.1

Variation of fin stabilizer 1 has the highest value at a speed of 30 knots of 343 kN with a difference of 6.2 kN or 1.8 per cent. The lowest resistance difference value at a speed of 25 knots is 284, the difference is 2 KN or 0.7%. Fin stabilizer 2 variation has the highest value at a speed of 29 knots of 328.6 kN with a difference of 4 kN or 1.23%. The lowest resistance difference value at a speed of 25 knots is 284.7, the difference is 2.7 KN or 0.95%. The variation of fin stabilizer 3 has the highest value at a speed of 30 knots of 344.3 kN with a difference of 7.1 kN or 2.1%. The lowest resistance difference value at 26 knots speed is 295 kN, the difference is 3 KN or 1.02%.

4 Conclusion

Based on numerical calculation simulation data with Fine Marine Numeca for fast patrol boats with the addition of a NACA 0010 fin stabilizer at a speed of 28 knots, there is an increase in resistance of 4.6 KN or 1.4% in the variation of the fin stabilizer model 1, the addition of resistance 3.5 KN or 1 .1 % in the variation of the fin stabilizer 2 model and the addition of 6 KN resistance or 1.9 % in the variation of the fin 3 models. The simulation results of the NACA 0010 variation of the fin stabilizer model variation of fin stabilizer 2 have a smaller average appendages resistance than the variation of the fin stabilizer model 1 and 3, where fin stabilizer 2 is 1.04%, fin stabilizer 1 is 1.26% while fin stabilizer 3 by 1.6%.

5 References

[1] Marsetio, "Sea Power Indonesia." Unhan, Jakarta, pp. 1–148, 2014.

[2] C. Kusuma, I. M. Ariana, and B. Ali, "Redesign KCR 60m Bow with Axe Bow Type to Reduce Ship Resistance," *IOP Conf. Ser. Earth Environ. Sci.*, vol. 557, no. 1, 2020, doi: 10.1088/1755-1315/557/1/012033.

[3] W. Kusdiana, Semin, I. Made Ariana, C. Kusuma, and B. Ali, "Location Analysis of Patrol Boat Fin Stabilizer Based on Numerical Method," *IOP Conf. Ser. Earth Environ. Sci.*, vol. 972, no. 1, 2022, doi: 10.1088/1755-1315/972/1/012002.

[4] C. Kusuma, I. M. Ariana, W. H. Nugroho, E. B. Djatmiko, A. A. Masroeri, and Sutardi, "Design Propeller Of Fast Missile Boat 60 M By Using Gawn Series," *IOP Conf. Ser. Mater. Sci. Eng.*, vol. 1052, no. 1, p. 012026, 2021, doi: 10.1088/1757-899x/1052/1/012026.

[5] C. Kusuma, I. M. Ariana, and W. H. Nugroho, "An Innovative Proposal for Increasing the Speed of a 60 Metres Fast Patrol Vessel," *Int. Rev. Mech. Eng.*, vol. 15, no. 9, pp. 486–496, 2021, doi: 10.15866/ireme.v15i9.21411.

[6] R. Yao, L. Yu, Q. Fan, and X. Wang, "Experimental and Numerical Resistance Analysis for a Cruise Ship W / O Fin Stabilizers," 2022, doi: https://doi.org/10.3390/jmse10081054.

[7] A. Giallanza and T. Elms, "Interactive roll stabilization comparative analysis for large yacht : gyroscope versus active fins," *Int. J. Interact. Des. Manuf.*, 2019, doi: 10.1007/s12008-019-00618-y.

[8] V. D. Moganti, R. E. Garetno, R. Irvana, and A. Fadillah, "A Study of the effectiveness Fin Stabilizer on Unsada Water Tour Bus to Comfort the Rolling Period to Support Toba Lake Tourism," *IOP Conf. Ser. Earth Environ. Sci.*, vol. 557, no. 1, 2020, doi: 10.1088/1755-1315/557/1/012012.

[9] L. Liang, M. Sun, H. Shi, and T. Luan, "Design and Analyze a New Measuring Lift Device for Fin Stabilizers Using Stiffness Matrix of Euler-Bernoulli Beam," pp. 1–22, 2017, doi: doi: 10.1371/journal.pone.0168972.

[10] S. Della Rosa, S. Maceri, I. M. Viola, and S. Bartesaghi, "Design and optimization of a fin stabilizer using CFD codes and optimization algorithms," *16th Int. Conf. Sh. Shipp. Res.*, 2009.

[11] L. Mao, H. Wang, Y. Li, and H. Yi, "Force model of flapping foil stabilizers based on CFD parameterization," *Ocean Eng.*, vol. 187, no. June, p. 106151, 2019, doi: 10.1016/j.oceaneng.2019.106151.

[12] L. Liang, J. Li, S. Zhang, and Z. Le, "A Novel Real-Time Visual Hardware-in-the-Loop Simulation Platform (RTV-HILSP) for Designing Fin Stabilizer Control System," *IEEE Access*,

vol. 7, pp. 92908–92920, 2019, doi: 10.1109/ACCESS.2019.2927591.

[13] C. Liang and X. Zhang, "Concise and economical control implemented on ship fin stabilizer system based on nonlinear feedback algorithm," *J. Mar. Sci. Technol.*, no. 0123456789, 2020, doi: 10.1007/s00773-020-00723-8.

[14] S. Jiguang, L. Lihua, Z. Songtao, and W. Jiming, "Design and experimental investigation of a GA-based control strategy for a low-speed fin stabilizer," *Ocean Eng.*, vol. 218, no. 145, p. 108234, 2020, doi: 10.1016/j.oceaneng.2020.108234.

[15] Z. Songtao and Z. Peng, "L2-gain based adaptive robust heel/roll reduction control using fin stabilizer during ship turns," *J. Mar. Sci. Eng.*, vol. 9, no. 1, pp. 1–17, 2021, doi: 10.3390/jmse9010089.

[16] Z. Sun, C. Chen, and G. Zhu, "Adaptive Neural Network Control of Zero-Speed Vessel Fin Stabilizer Based on Command Filter," *Appl. Sci.*, vol. 12, no. 2, 2022, doi: 10.3390/app12020754.

[17] Y. Xu *et al.*, "Effect of Sinusoidal Motion of Fin Stabilizer on Ship Roll at Zero Speed," *Ocean. 2019 - Marseille*, pp. 1–4, 2019, doi: 10.1109/oceanse.2019.886715.

[18] Y. Zhao, H. Gao, Z. Lin, Y. Guo, and J. Zhang, "Selection and evaluation of polar cruise fin stabilizer based on combination weighting-TOPSIS method," *Chinese J. Sh. Res.*, vol. 16, no. 5, pp. 121-126+149, 2021, doi: 10.19693/j.issn.1673-3185.02037.

[19] Q. Nguyen, T. Li, J. Ning, and Y. Wu, "Online optimal control for nonlinear fin stabilizer system of marine vessels via time-based ADP algorithm," pp. 264–271, 2020.

[20] C. L. P. Chen and M. Han, "Nonlinear Model Predictive Control for Fin Stabilizer System of Marine Vessels Based on Recurrent Neural Network," pp. 11–16, 2019, doi: https://doi.org/10.1109/ICIST.2019.8836939.

[21] P. Marzocca, "NACA Airfoil Generation Data," *[Online]. Available: http://people.clarkson.edu/.*, pp. 1–6, 1989, doi: 10.1017/CBO9781107415324.004.

[22] R. E. Bilstein, "Orders of Magnitude: A history of the NACA and NASA, 1915-1990," p. 180, 1989.

[23] J. N. Jacobs, K. E. Ward, and R. M. Careas, "The Characteristics of 78 Related Airfoil Sections Sections From Tests In The Variable-Density Wind Tunnel," *Natl. Advis. Commitee Aeronaut.*, 1935, doi: 10.1017/CBO9781107415324.004.

[24] A. Vargas, "Aerodynamic characterization of bio inspired corrugated wings Aerodynamic characterization of bio inspired corrugated wings" MOJ Applied Bionics and Biomechanics, Volume 3 Issue 1 - 2019.

[25] A. Aulnette, "Design Synthesis of a Hybrid Wing Body Transport Aircraft" Department of Aeronautics, Master Thesis Imperial College of Science, Technology and Medicine, South Kensington Campus, London SW7 2AZ, United Kingdom.

[26] T. Nilavarasan and S. Chandel, "Effect of Gurney Flaps on the Aerodynamic Characteristics of NACA 0010 Cascades," 2019, doi: https://doi.org/10.1515/tjj-2019-0013.

[27] P. C. Teixeira and C. E. S. Cesnik, "Propeller Effects on the Response of High-Altitude Long-Endurance Aircraft," pp. 1–15, doi: https://doi.org/10.2514/1.J057575.

[28] A. Pellnäs and J. Sandeberg, "Conceptual Design of a Small - Size Unmanned Air Vehicle Part A : Aerodynamic Performance and," 2021.

[29] R. A. Johansen, "Conceptual Design for a Supersonic Advanced Military Trainer," no. May, 2019, doi: https://doi.org/10.1007/s00773-020-00723-8.

[30] M. F. S. & M. A. Muhammad Arif Budiyanto and Murdianto, "Investigation of the effectiveness of a stern foil on a patrol boat by experiment and simulation a patrol boat by experiment and simulation," vol. 1916, 2020, doi: https://doi.org/10.1007/s00773-020-00723-8.

IOP Conf. Series: Earth and Environmental Science 1166 (2023) 012003 doi:10.1088/1755-1315/1166/1/012003

Resistance on Catamaran Decommissioning Unit: A Comparison of Numerical Analysis (Two Turbulence Models) to Model Test Result

Abdul Ghofur[1], Arifin[1], Erdina Arianti[1], A Bisri[1]and Suwahyu[1]
[1]Research Center for Hydrodynamics Technology, Research Organization for Energy and Manufacture, National Research and Innovation Agency, Sukolilo, Surabaya, Indonesia

E-mail: abdulghofur68000@gmail.com

Abstract. A new concept design Catamaran Decommissioning Unit has been developed by Research Center for Hydrodynamics Technology Indonesia. The catamaran has a unique shape, however in order to achieve resilient and sustainable the decommissioning unit as it has a unique shape, resistance analysis both model testing and numerical calculation must be conducted. The resistance value from the numerical analysis needs to be validated with the results from model testing in the towing tank. A model of a catamaran unit with a scaling factor λ=36 was taken as the object of study. The model test was conducted in a towing tank with service speed varies from 7-15 knots. As a comparison, a numerical analysis using the CFD method was being performed under two different turbulence models (k-ε and k-ω SST) with the same speed regarding the experimental speed parameter. Iin a conclusion, numerical analysis with k-ω SST turbulence model gives a better performance with smaller relative deviation compared to the k-ε turbulence model.

1. Introduction

Indonesia has roughly 22 million barrels of oil and natural gas reserves, with Sumatra, Java Sea, and Natuna waters accounting for 75 percent of total reserves [1]. Most of the oil and gas reserves were found from shallow to mid water depth for about 20-150 meters in depth [2]. As a result, the most convenient and extensively used platform for supporting the oil and gas production phase is a fixed jacket platform. However, 25 percent of the total 500 jacket platforms installed in Indonesian waters were installed in the 1970s, and 30 percent in the 1980s [3]. In addition, 69% of the total installed jackets were being damaged due to a long service life [1] and 72 platforms were reported to be unused [2]. On the other hand, the platform installed on the sea which has surpassed its service life is obligated to be dismantled through decommissioning operation as it has been stated in United Nations Convention on Law of The Sea III 1982 and International Maritime Organisation Resolution A.672(16) [4]. This operation could be executed by cutting the jacket leg near the seabed and move the platform to the shore. Nevertheless, an adequate platform is required to support this process. A huge barge that provides a crane with high lifting capacity is usually carried out.

Platform decommissioning can be accomplished using a variety of technologies, including barges, semi-submersibles, and catamarans. The Iwaki-Oki platform uses a heavy lift vessel and four barges in the decommissioning process. Meanwhile, the platforms that use semi-submersibles in the decommissioning process are Miller and North West Hutton, with the help of a barge to move the jacket from the site to the land [5]. The use of barges and semi-submersibles for decommissioning is inefficient because it takes several units to complete the work. The use of specialized vessels, such as catamarans, which can lift the platform and place it on its own deck before bringing it ashore, is more efficient than the previous type of ship. The Pioneering Spirit is an existing vessel of this type that has successfully decommissioned several platforms, including Brent, Ninian North, Tyra, Valhall QP, and Yme [5]. However, Pioneering Spirit has relatively large dimensions and capacity for platforms in Indonesia, with a large substructure weighing around 2592 tons [6]. As a result, a catamaran with dimensions that correspond to the needs of platforms in Indonesia is required, with a length of 111 meters and a width of 37.8 meters [6].

A catamaran is a platform with two demi hulls positioned abreast of each other [7]. Platform decommissioning activities necessitate structures and infrastructure that are not just sturdy (resilient), but also durable. As a result, the catamaran design must be distinctive enough to be used as a decommissioning unit. Using a catamaran for a decommissioning platform could be an advantage for the operation. The double-hull construction of the platforms provides a wider deck surface which delivers sufficient space for the dismantled platform [2]. Furthermore, a catamaran vessel counted to has higher stability than the monohull type platform. This is due to the fact that the double-hull platform has a greater GM (gravity metacentre) value, resulting in improved heel characteristics[8]. Several studies were conducted regarding the catamaran unit to obtain its characteristics. However, the perseverance in the analysis of resistance in the catamaran platform shall be highlighted. The calculation of resistance has huge significance since it is used to determine an adequate power to operate the vessel [7].

Hull below water lines, resistance prediction for both numerical and experimental results demonstrate good agreement in the range of all given speed conditions within less than 5% variance. However, structures above the waterline (hull, deck, living quarters, crane, and other equipment, etc.) have not yet been considered. So, in the future, the added resistance [9] due to wind and wave should be predicted. Because the ALPO structure has a large windage area, it is required to make a careful estimate of added resistance, both mathematically and through wind tunnel experiments. So that the total resistance could be predicted more accurately.

2. Method

The ALPO unit is a decommissioning vessel in a catamaran form that was designed by BPPT (Agency for the Assessment and Application of Technology) in 2018 and is used as the object of study in the present paper. This vessel is intended to be operated in Indonesian water to support the jacket decommissioning activities. Therefore, the ship dimension and arrangement are specifically adjusted to Indonesian water characteristics. These considerations subsequently led to the ALPO principal dimension written in Table 1 and general arrangement can be seen in Figure 1.

Table 1. ALPO principal dimension.

Parameter	Quantity	Unit
Length waterline (L_{WL})	111.00	M
Breadth moulded (B_{MLD})	37.80	M
Breadth single hull (B_{HULL})	12.00	M
Depth moulded (H_{MLD})	10.50	M
Draft (T)	4.70	M
Displacement (Δ)	8646	Ton
Block coefficient (C_B)	0.582	-
Waterplane coefficient (C_W)	0.590	-
Midship coefficient (C_M)	0.551	-

For the experiment and numerical analysis purposes, the ALPO vessel was built into a model with a scaling factor λ=36. A comparison of principal dimensions between the model and prototype of the ALPO unit is as reported in Table 2.The service speed of the prototype ranging from 7-15 knots was subjected to model in both experimental and numerical study of the model. As a result, the towing tank experiment was conducted in 9 conditions. Additionally, addressing two turbulence models undertaken for the numerical study, 18 CFD models were carried out for the analysis. Froude number used in the analysis model is written in Table 3.

Figure 1. General arrangement ALPO.

Table 2. Principal dimensions of model and prototype.

Parameter	Model	Prototype	Unit
Scale	1:36	1:1	
Length waterline (L_{WL})	111.00	3.08	m
Breadth molded (B_{MLD})	37.80	1.05	m
Breadth single hull (B_{HULL})	12.00	0.33	m
Depth molded (H_{MLD})	10.50	0.29	m
Draft (T)	4.70	0.13	m
Displacement (Δ)	8646	0.181	ton
KG	0.436	15.68	m
LCG (from AP)	1.415	50.94	m

Table 3. Froude number of the model.

Prototype Speed (V_P) (knot)	Model Speed (V_M) (knot)	Froude Number
7	1.16	0.109
8	1.33	0.125
9	1.50	0.140
10	1.66	0.156
11	1.83	0.171
12	2.00	0.187
13	2.16	0.203
14	2.33	0.218
15	2.50	0.234

2.1. Hydrodynamic experiment

The experiment of hydrodynamic to obtain resistance force is performed in Towing Tank (TT) of the Balai Teknologi Hidrodinamika. The ALPO model was built using wood material and covered with fiberglass. A dimensional check through a 3-axis measuring table was conducted before the tests are launched to ensure the accuracy of the model dimension to the design [8]. The experiments were carried out by pulling the model along the towing tank by the carriage and placing a force measuring apparatus to record the occurring resistance along the hull during this process. Addressing the scale factor effect, a turbulence simulator strip with wide of 10 mm which consists of Ø 0.5 mm sand were attached along the model hull section.

2.2. *Numerical analysis set up*

The CFD method is chosen for the numerical analysis approach in this study. As a prior study shows, the problem of wave interference caused by boeth demi hulls in catamaran units may be satisfactorily addressed using CFD methods [10]. A mathematic equation model of RANS was applied to the analysis and the equation is written in Equation (1)[11].

$$\rho\frac{\partial U_i}{\partial t}+\rho U_i\frac{\partial U_i}{\partial x_j}=-\frac{\partial P}{\partial x_i}+\frac{\partial}{\partial x_j}\left(2\mu s_{ij}-\rho\overline{u'_j u'_i}\right) \quad 1)$$

The k-ω SST and k-ε models, which are extensively used to depict turbulence models, were employed as well-established approaches to offer closure to the RANS equations [12]. The turbulence kinetic energy and specific dissipation rate of k-ω SST turbulence model are written in Equation (2) and Equation (3).

$$\frac{\partial k}{\partial t}+U_j\frac{\partial k}{\partial x_j}=P_k-\beta^* k\omega+\frac{\partial}{\partial x_i}\left[\left(v+\sigma_k v_T\right)\frac{\partial k}{\partial x_j}\right] \quad 2)$$

$$\frac{\partial \omega}{\partial t}+U_j\frac{\partial \omega}{\partial x_j}=\alpha S^2-\beta\omega^2+\frac{\partial}{\partial x_i}\left[\left(v+\sigma_\omega v_T\right)\frac{\partial \omega}{\partial x_j}\right]+2\left(1-F_1\right)\sigma\omega^2\frac{1}{\omega}\frac{\partial}{\partial x_i}\frac{\partial \omega}{\partial x_i} \quad 3)$$

Furthermore, in k-ε models, the k stands for turbulent kinetic energy and ε represents dissipation energy, respectively, as indicated in Equation (4) and Equation (5).

$$\frac{\partial}{\partial t}(\rho k)+\frac{\partial}{\partial x_i}(\rho k u_i)=\frac{\partial}{\partial x_j}\left[\left(\mu+\frac{\mu_t}{\sigma_k}\right)\frac{\partial k}{\partial x_j}\right]+P_k+P_b-\rho\in-Y_M+S_k \quad 4)$$

$$\frac{\partial}{\partial t}(\rho\in)+\frac{\partial}{\partial x_i}(\rho\in u_i)=\frac{\partial}{\partial x_j}\left[\left(\mu+\frac{\mu_t}{\sigma_\in}\right)\frac{\partial \in}{\partial x_j}\right]+C_{1\in}\frac{\in}{k}\left(P_k+C_{3\in}P_b\right)-C_{2\in}\rho\frac{\in^2}{k}+S_\in \quad 5)$$

As a computational domain shown in Figure 2, the inlet was placed by 6 times of ship length from the bow. In contrast to it, the outlet region was placed further, which takes 10 ship lengths to prevent the reversed flow that may occur during the analysis. The boundary condition is taken as a no-slip wall element with the fluid flow is heading to the outlet from the inlet.

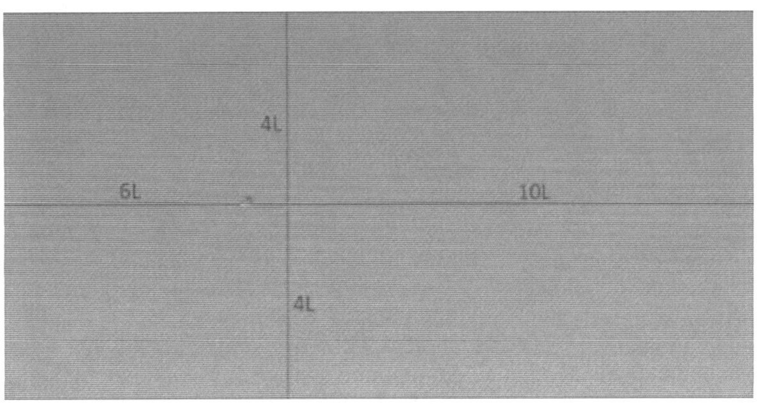

Figure 2. Computational domain of CFD analysis.

2.3. Mesh generation

In FEM based CFD method, any type of element is adequate to perform meshing activity. However, it is claimed that the linear tetrahedral element has performed better solution [10]. Additionally, the element size around the hull was built smaller than other regions as displayed in Figure 3 in order to obtain a more precise solution. Overall, the discretization during meshing resulted in a total of 3,812,313 elements.

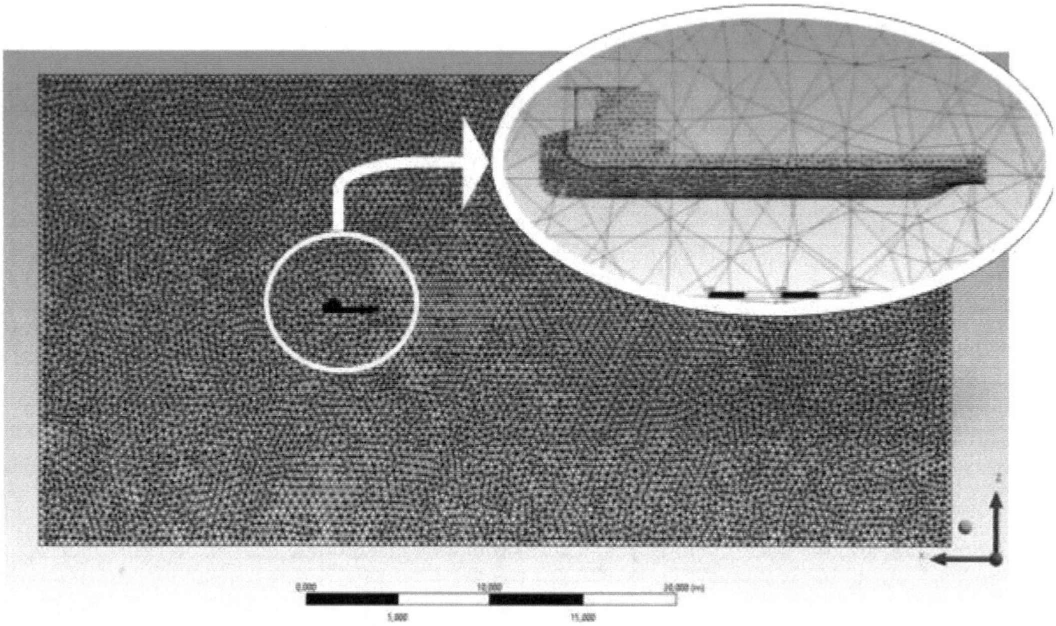

Figure 3. CFD meshing results.

3. Result and discussion

3.1. Experiment result

As required by ITTC-1957, the calculation of total resistance is done by combining the frictional resistance equation with correlation allowance (CA) and form factor (1+k) [13]. The CA value used for this experiment was set to be 0.000785 whereas the form factor was set at 1.198 [8]. Moreover, the system configuration of the model during hydrodynamic testing as depicted in Figure 4.

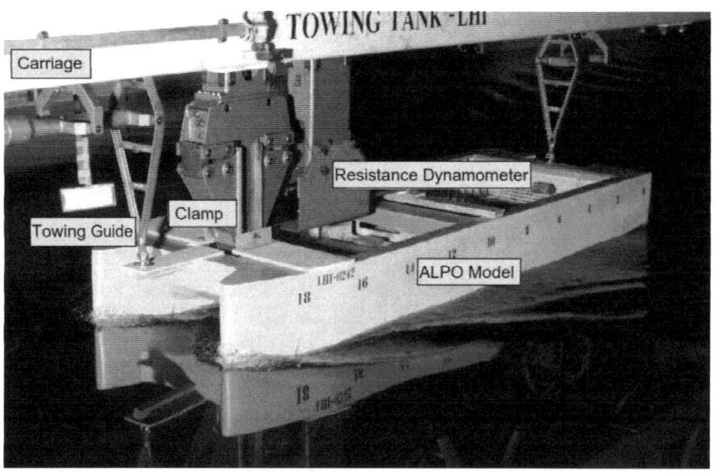

Figure 4. Experiment configuration.

As mentioned in the previous section, the experiment was conducted in 9 different conditions under various Froude numbers. Thus, the resistance result of the experiment is reported in Table 4.

Table 4. Resistance result from hydrodynamic experiment.

Model Speed (V_M) (knot)	Froude Number	Resistance (N)
1.16	0.109	2.37
1.33	0.125	3.06
1.50	0.140	3.87
1.66	0.156	4.89
1.83	0.171	6.14
2.00	0.187	7.58
2.16	0.203	9.43
2.33	0.218	11.58
2.50	0.234	13.54

The result from Table 4 indicates that the value of ship resistance is increasing as the Froude number intensifies. This deduction is linear to the prior studies conducted before [14],[15].

3.2. CFD analysis result

Grid independence study information. Convergence tests are performed to ensure that the number of elements has no effect on the meshing quality. This test is carried out by varying the number of elements in each meshing experiment. Table 5 and Figure 5 shows the convergence test results.

Table 5. Meshingconvergence test results.

Number of Element	Drag (N)
3,702,247	8.2
3,704,247	8.0
3,705,747	7.9
3,707,185	7.8
3,711,233	7.7
3,718,317	7.5
3,732,406	7.4
3,754,156	7.3
3,777,829	7.3
3,812,313	7.3

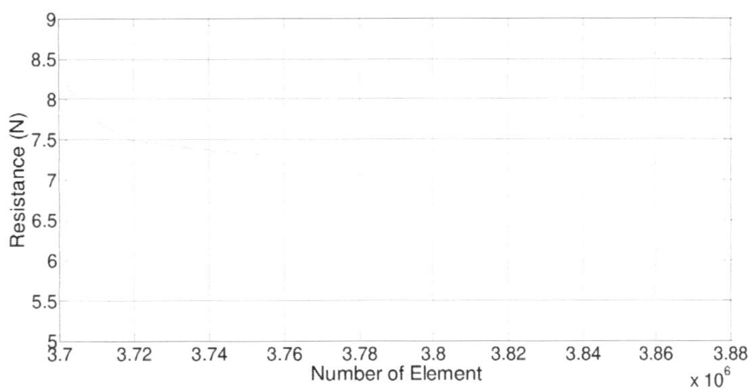

Figure 5. Meshingconvergence test result.

The ALPO model pictures in numerical analysis and towing test could be seen in Figure 6 and Figure 7.

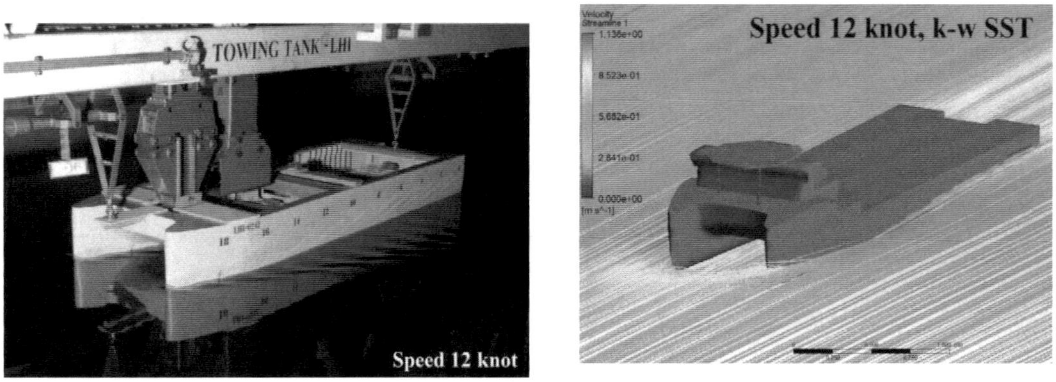

Figure 6. ALPO towing test andnumerical simulation.

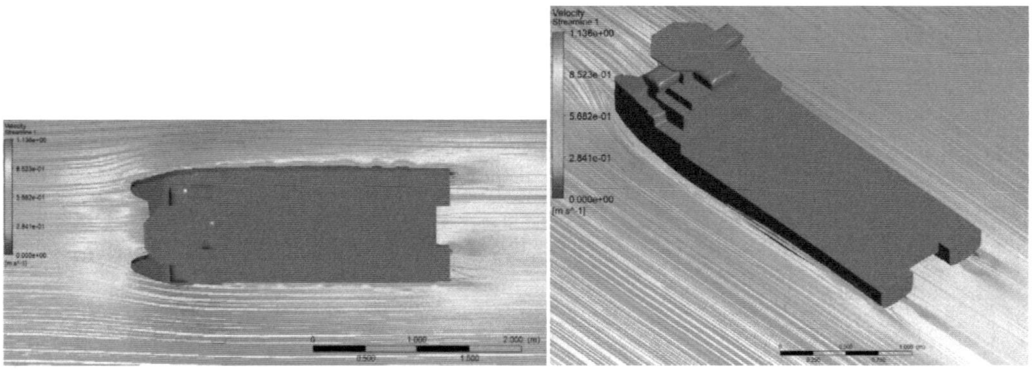

Figure 7. The results of surface flow characteristics at a speed of 12 knots with a turbulence model of k-ω SST.

Table 6. Resistance from CFD analysis.

Froude Number	Resistance (N)	
	k-ε model	k-ω SST model
0.109	2.377	2.360
0.125	3.166	3.020
0.140	4.009	3.821
0.156	4.934	4.825
0.171	6.252	6.001
0.187	7.466	7.304
0.203	9.273	9.140
0.218	11.124	11.318
0.234	12.948	13.097

As it has been highlighted before, 18 CFD model were performed for the analysis. The surface flow at a speed of 12 knots are depicted in Figure 6. Moreover, resistance results from these CFD analyses were written in Table 6.Result from Table indicates that the value of resistance is rising as the Froude number increasing which is in line to the experiment result. However, analysis with k-ω SST model bring in higher resistance result comparing to the analysis in k-ε model.

3.3. Resistance result comparison

The resistance results were compared to experimental data to justify the CFD modeling and analysis. Equation (6) was used to calculate the relative deviation between resistance from the k-ε and k-ω SST models towards the experimental result.

$$RD[\%] = \frac{R_T^{CFD} - R_T^{EXP}}{R_T^{CFD}} 100 \qquad 6)$$

Where R_T^{CFD} is the total resistance calculated numerically through CFD analysis and R_T^{EXP} is total resistance obtained from experiment. The negative deviation as Table 7 shown indicates that the CFD result is underestimating the experiment result [16].

Table 7. Resistance comparison between k-ε and k-ω SST model.

Froude Number	Resistance (N)			% RD*	% RD**
	k-ε*	k-ω SST**	Exp		
0.109	2.377	2.360	2.370	0.295	0.422
0.125	3.166	3.020	3.060	3.464	1.307
0.140	4.009	3.821	3.870	3.592	1.266
0.156	4.934	4.825	4.890	0.900	1.329
0.171	6.252	6.001	6.140	1.824	2.264
0.187	7.466	7.304	7.580	1.504	3.641
0.203	9.273	9.140	9.430	1.665	3.075
0.218	11.124	11.318	11.580	3.938	2.263
0.234	12.948	13.097	13.540	4.372	3.273

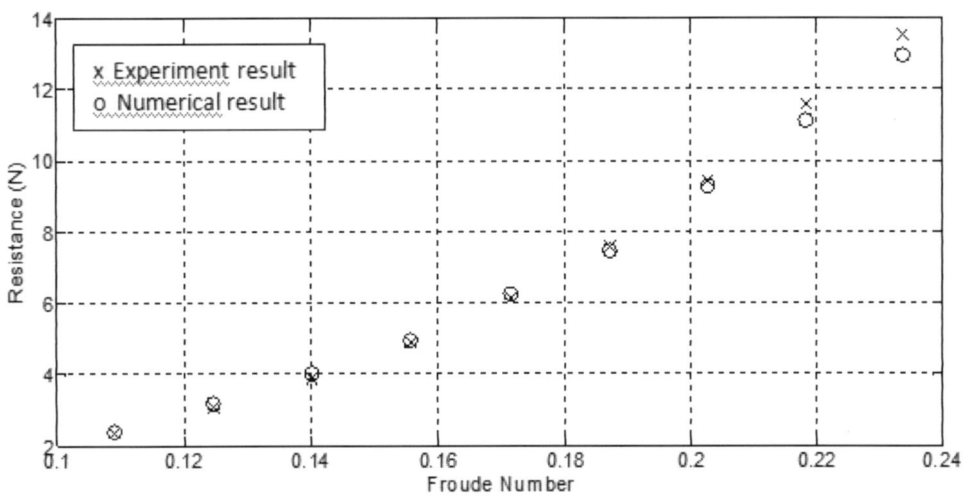

Figure 8. Resistance result from k-ε tubulence model.

In addition, Figure 8and Figure 9 show the comparison of the resistance results from both models to experimental data. Based on the findings, it can be concluded that the CFD model with k-ω SST equation performed better in the analysis and produced a better result than the k-ε model equation.This can be asserted since the maximum relative deviation of resistance calculated by the k-ω SST equation is lower than those calculated by the k-ε equation. Nonetheless, in a low Froude number, both turbulence model equations are thought to perform well. The k-ε equation modeling shows that relative deviation grew by ±0.5% from Froude number 0.109 to 0.160, while relative deviation climbed by ±1% from Froude number 0.160

to 0.234. The CFD model using the k-ω SST equation provides a better result, with an overall relative deviation of less than 4%.

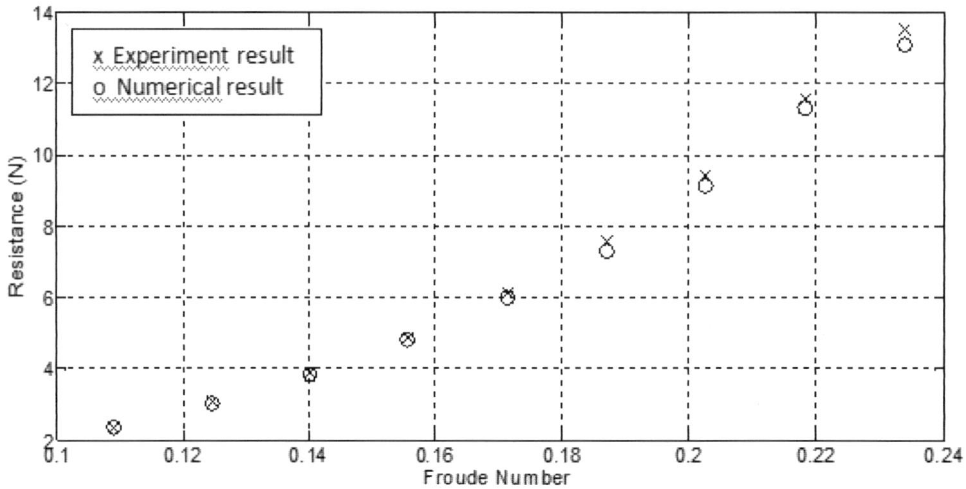

Figure 9. Resistance result from k-ω tubulence model.

4. Conclusion

The focus of this research was to evaluate two methods of experiment and numerical analysis for estimating ALPO's resistance as a catamaran decoommissioning unit. The importance of the research is to determine the best method to calculate resistance as it has significance in settling the propulsion system of a vessel. In general, all the analyses were subjected to a model form with a scaling factor of λ=36. The model was subjected to hydrodynamic testing in nine different conditions, as determined by the Froude number. The experiment's calculations and recommendations are all based on ITTC-1957. Moreover, the numerical analysis is carried out using the CFD approach and differs in two turbulence model equations, the k-ε and the k-ω SST. The results show that when performed at a low Froude number, resistance from CFD analysis produces a good result when compared to the experimental result.This is most likely owing to wake wash interference between two hulls, which caused the equation to be inadequate for the investigation. However, as compared to the k-ε model equation, the k- ω SST produces a better result. Despite this, additional study is needed to enhance the results of the CFD technique approach.in future work, the added resistance due to wind and wave components will be measured both experimentally and mathematically, so that the overall resistance can be anticipated more accurately.

5. References

[1] Wiratno J 2019 *Challenge & Opportunity of Platform Decommissioning in Indonesian Oil & Gas Upstream Business* available at https://kkp.go.id/an-component/media/upload-gambar pendukung/DitJaskel/publikasi-materi-2/marine-energy/Challenge%20%20Opportunity%20 Platform%20Decommissioning%20%20Julius%20Wiratno%20-%20Deputy%20Operation% 20of%20SKK%20Migas.pdf

[2] Ghofur A 2020*Laporan AkhirWahana Angkut Lepas Pantai Pasca Operasi PTRIM BPPT*

[3] O'Sullivan M 2018*IOGP - International Offshore Decommissioning RegulationsVolume 1 – Facilities Abandonment*

[4] IMO 1989 Guidelines and Standards for the Removal of Offshore Installations and Structures on the Continental Shelf and in the Exclusive Economic Zone

[5] Pinheiro M S and Monteiro P R D 2021Decommissioning Plans for Fixed Offshore Platform: A Brief Revision*Asian Journal of Engineering and Technology* **9** 62

[6] Arianti E and Ghofur A 2019 Teknologi Decommissioning Anjungan Lepas Pantai Terpancang Pasca-Operasi *Inovtek Polbeng* **9** 271
[7] Jamaluddin A, Utama I K A P, Aryawan, W D and Widodo B 2012 Experimental Investigations Into the Resistance Components of Symmetrical Catamarans with Variations in Hull Clearances and Staggers *Transactions RINA* **154** *Part B1 International Journal of Small Craft Technology* B33
[8] Ali B 2019 *Laporan Hasil Pengujian Resistance, Paint Smear, Seakeeping dan Stability BTH BPPT*
[9] Kim M., Hizir O., Turan O., Day S. and Incecik A. 2017 Estimation of Added Resistance and Ship Speed Loss in a Seaway*Ocean Engineering* **141**
[10] Salas M, Luco R, Sahoo P K, Browne N and López M 2004 Experimental and CFD Resistance Calculation of a Small Fast Catamaran*Proceeding of International Cnference on High-Performance Vehicles*215
[11] Bustos D S H and Alvarado R J P 2017 Numerical Hull Resistance Calculation of a Catamarán using OpenFOAM*Ship Science & Technology* **11** 29
[12] Terziev M, Tezdogan T, Demirel Y K, Villa D, Mizzi S. and Incecik A 2021Exploring the Effects of Speed and Scale on a Ship's Form Factor using CFD *International Journal of Naval Architecture and Ocean Engineering* **13** 147
[13] ITTC 2017ittc Quality System Manual Recommended Procedures and Guidelines For Resistance Test 7.5-02-02-01
[14] Fitriadhy A, Azmi S A, Mansor N A and Aldin N A 2017 Computational Fluid Dynamics Investigation on Total Resistance Coefficient of a High-Speed "Deep-V" Catamaran in Shallow Water *International Journal of Automotive and Mechanical Engineering* **14** 4369
[15] Nursal R S, Nordin N I, Mohammad M, Hamid M A A, Awang M N and Ariffin A A M 2017 Investigation on Resistance Effects Through Towing Test and CFD Analysis on Catamaran Boat Hull Form *Journal of Engineering and Applied Sciences* **12** 1334
[16] Purnamasari D, Utama I K A P and Suastika I K 2018CFD Simulations to Calculate the Resistance of A 17.500-DWT Tanker *IPTEK Journal of Proceedings Series* **1**

Acknowledgments
The author really appreciates the national research priority program of the Ministry of Research, Technology and Higher Education of the Republic of Indonesia for conducting this research. The author would also like to thank colleagues for their assistance or encouragement, as well as special work by technical staff or other support from the Hydrodynamics Laboratory.

IOP Conf. Series: Earth and Environmental Science 1166 (2023) 012004 doi:10.1088/1755-1315/1166/1/012004

Analysis of technology readiness level of dual fuel harbor tug boat research achievement

K Ismoyo[1]*, Muhajirin[1], Iskendar[1], K Ajidarmo[1], Triyanto[1], A Muis[1]
[1] Transportation Technology Research Centre - National Research and InnovationAgency, Indonesia.

*Email: kuntoismoyo.jat@gmail.com

Abstract. Research and innovation of dual fuel harbor tug boat are carried out to meet eco-friendly vessel. The measure of technological readiness is an indicator of the utilization of the study outcomes that have been carried out. One approach thatcan be used is by mapping technology readiness, assessing the application of research and its implementation activities, lowering the risk of technology failure, and boosting the usage of research and development outcomes. In this paper, an analysis of technological readiness is carried out on the research results of a dual fuel tug boat using a scoring system called techno-meter. The results obtained are that the assessment of the research achievement of the dual fuel tug boat shows the level of technology readiness (TRL) at the 4th level. This shows that the resulting technological research has been validated through testing the scale of the hydrodynamic model which shows the feasibility of the design and maneuverability of the ship in the relevant environment with an achievement score is 80.66%

Keywords. Technological Readiness Level, Research on Harbor Tug Dual Fuel, eco-friendly

1. Introduction

Regarding the assessment and determination of the technological readiness level (TRL), see Regulation of the Minister of Research, Technology, and Higher Education of the Republic of Indonesia Number 42 of 2016 (1). The measurement and determination of TRL aims to know the status of technological readiness, help in mapping technological readiness, evaluate the implementation of research programs or activities, reduce the risk of technological failure, and increase the utilization of research and development outcomes.

One of the research programs carried out by National Research and Innovation Agency is the harbor tug dual fuel ship design program. This research activity is oriented toward efforts to increase the capacity and productivity of the national shipbuilding industry so that it is expected to meet national needs and realize increased capacity and quality of seaport services, as well as increasing the use of integrated and reliable maritime technology and information technology systems.

This paper describes the technology readiness level of the dual fuel harbor tug ship. Design considerations will be discussed in general regarding technological aspects, starting from the design stage, bunkering systems to ship propulsion engines before all the designs are required for the design approval process at the Indonesian Classification Bureau. The qualitative approach taken to assess the technology readiness level as a result of the research is known as the Technology Readiness Level (TRL). This approach is a tool to assess how ready the system or component can be utilized. For this reason, the resulting technology product requires partnerships with vendors and users to determine requirements, provide feedback and evaluate the design and prototype produced. In thefinal process of technological readiness, it is necessary to know aspects related to the technology

IOP Conf. Series: Earth and Environmental Science 1166 (2023) 012004 doi:10.1088/1755-1315/1166/1/012004

that has been tested, such as the cost of implementing activities, operational performance, manufacturing and product launches, and their commercial operations.

One of the efforts that can increase the utilization of technological research results is to generate and document the technology maturity level or technology readiness level measurably produced by the Research Institute in a measurable manner. The qualitative approach taken to assess the level of technological readiness as a result of the research is known as the technology readiness level. This approach is a tool to assess how ready the system or component can be utilized. For this reason, the resulting technology product requires partnerships with vendors and users to determine requirements, provide feedback and evaluate the design and prototype produced. In the final process of technological readiness, it is necessary to know aspects related to the technology that has been tested, such as the cost of implementing activities, operational performance, manufacturing and product launches, and their commercial operations. The level of technology readiness startsfrom the lowest level of TRL-1 to the highest level of readiness of TRL- 9.

This can be done by measuring the TRL of research results to be developed into anindicator of mastery and ability of science and technology which is also useful information to obtain information, implemented and adopted by Users/Prospective Users. The entire assessment process requires documentation and evidence as well as reasons for the eventual readiness assessment. Evidence can include test or application records used to produce the product, technical papers, reports, presentations, and so on. The assessment document describes the technological process and how the process of implementing research and technology development has been carried out. The report must refer to valid sources and these sources as evidence presented in the report on the results of the technology readiness assessment, and whether technology has been carried out in the relevant environment. The assessment report should describe the function of each component, sub-system, and product/system level.

2. Theoretical Review

Technology readiness level is a methodical measurement system that enables the evaluation of a technology's maturity or readiness as well as the comparison of maturity or readiness between various technology kinds. Technology maturity can be viewed as a measure of how prepared/mature a technology is to be used and embraced by potential consumers. According to the notion of "readiness," there may be a distinction between a technology being "ready" and "not ready" for usage, or between several "levels of technology readiness." TRL is a scale from 1 to 9, where each level is tied to the one before it and serves as the foundation for the next, showing the degree of technological readiness. The relationship between the steps of the technology development process is generally carried out with the TRL concept from NASA called TRL meters (2). Figure-1 shows the stages from "basic technology research" to "system testing, launch, and operation" which is a generic sequential step in the context of technology development until it is more ready to be implemented.

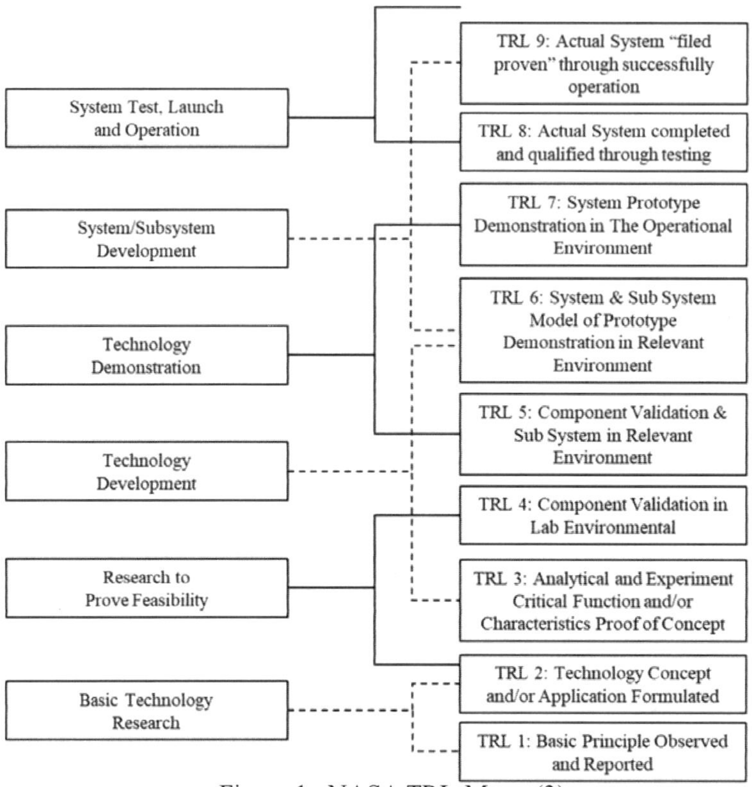

Figure 1. NASA TRL-Meter (3).

The definition of the Basic Technology Readiness Level Framework is shown inTable 1, as follows:

Table 1.The definition of TRL applied in the analysis of theResearch of the dual fuel harbor tugs boat design (3)

Technology Readiness Level (TRL)		
Technology demonstration	9	Actual Product: Tested and proven successful operation and normal commercial service
	8	Full-scale commercial demonstration
	7	Fully functional sub-scale prototype demonstration
Technology development	6	Testing of integrated prototypes in relevant environments
	5	Validate sub-systems in the relevant environment
	4	System validation in a laboratory environment
Basic technology research	3	Proof of concept technology at the component level
	2	Technology/application concept formulation
	1	Observation of basic principles and initial concepts

Based on the TRL framework, a research stage is made, namely the research stage or the stage of determining the concept and design requirements in TRL stages 1-3, then the technology development stage is carried out in TRL 4-6 and then a product demonstration stage in TRL stages 7-9 (3).

In the implementation of the dual fuel tug boat design, research activities begin with the preparation of mission analysis to obtain design requirements from the user in order to determine the design concept or initial design in the TRL 1-3 research implementation range, especially to meet requirements in ship operation, such as speed requirements. ships, propulsion systems and machinery used, navigation systems, loading capacity, loading and unloading systems, maneuvering agility, shipping radius, and so on. Furthermore, the technology development product is carried out at TRL level-6 (4), where product development activities are more focused on component validation in a laboratory environment, where the basic concepts have been tested, and components have been integrated to determine that a technology concept is feasible to operate and then test the ship model in an environment suitable for the ship's operating environment. Then carry out operational trials of the ship prototype at the TRL 7-9 level representing the final stage towards the commercialization stage. The final form or key design of this final product must be through tested and proven through the successful operation.

3. Methodology

The general concept of the technology readiness level measurement method was developedby NASA and was translated contextually and applied in real applications. Realizing this several parties have developed TRL measurement methods and translated them more pragmatically into operational measurements. Quantitative TRL measurement is considered important to provide information as "objectively as possible" and to create a "common sense" of perception (or build understanding) between various parties involved with developed technology.

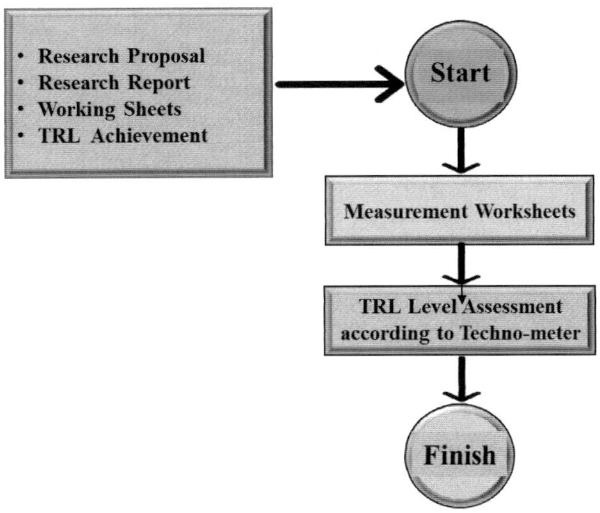

Figure 2. Techno-meter measurement procedure.

Techno-meter-level of technology readiness is a scoring system that is used in the method of analysis of technology readiness for research on dual-fuel tugboats. Techno-meter can represent both the quantitative and qualitative values of research results, serving as both an assessment tool and management tool. Program for technology development because it includes crucial information (in the form of statements) gleaned through surveys or literature reviews to support the management's assessment, allowing for the definition of action items for the process of technology development and its improvement.

Measurement of TRL can be done by assessing the level of technological readinessby using indicators that describe the level of technology readiness. If the achievement of compliance at that level does not meet the level described, it is necessary to review further documents, such as proposal documents, research reports, technology guides and TRL applications, and measurement worksheets.

Furthermore, filling out the worksheet with important points of the research stages, as described in the research document, is measured according to the indicators in the TRL Techno-meter level (4). In the process of filling out the worksheet, it is possible to ask for clarification from the relevant Researcher/Engineer. The achievement value of the indicator is in accordance with the techno-meter guide. The last level that meets the requirements indicates the level of achievement of the technological readiness of theresearch results (TRL level achievement).

4. Discussion

The harbor tug dual fuel ship design research began in 2020, starting with the preparationof mission analysis to obtain the design requirements for the preliminary design. In 2021 a basic design has been produced. Furthermore, the basic design of the ship is tested through verification of the scale test of the hydrodynamic model of the ship. The research stage indicators are described as follows:

4.1 Preliminary design

At the mission analysis stage, the harbor tug dual fuel ship design will produce design requirements as a preliminary design. Furthermore, it can be improved at the basic design and contract design stages through software verification or through hydrodynamic model testing. The results of the preliminary design stage activities resulted in the main size ofthe ship, the outline geometric of the hull, general plan drawings, engine room layout design, and the main structure of ship construction. The shape of the hull is designed in such a way as to meet the criteria for a dual-fuel harbor tug boat, which is capable of serving port activities. The ship has good maneuverability and stability in accordance with the ship's seaworthiness provisions so that it performs as expected. The main size design ofthe ship was obtained with a ship length of 34 m, a width of 11.3 m, and a draft of 4.2 m. Furthermore, the design of the maximum power requirement for ship operations is 2500HP with 2 engines. Based on this information, an engine speed of 800 rpm or 750 rpm is required with a power output of around 2220 kW or 2070 kW.

4.2 Basic design

The basic design of the ship was carried out and has produced the main design drawings of the ship, such as drawings of lines plan and offset, general arrangement, hydrostatic curve, intact stability report (preliminary), speed power prediction, capacity plan, free board calculation, tonnage measurement, lightweight calculation, deadweight calculation, scantling calculation, hull construction, shell expansion, midship section, building profile, and deck plans, hull outfitting & deck machinery, calculation of anode. calculation of equipment number, accommodation, and 3D modeling exterior and technicalspecifications.

4.3 Ship model scale test

To verify the hydrodynamic characteristics of the ship, a ship model scale hydrodynamic test was carried out at the Hydrodynamics Laboratory, in the form of testing the ship's resistance at full load and trim conditions with a speed variation of 8 - 14 knots. The sizeof the ship model is 1:10 scale as shown in the following picture:

From the test results of the harbor tug dual fuel model ship at a speed of 12 knots, the ship's resistance at full load conditions is 128 kN and the effective power is 792 kW. Furthermore, for the ballast load condition, the ship resistance is 123 kN and the effective power is 757 kW (2).

Figure 3. Testing the harbor tug dual fuel ship model.
(Sources: Annual Report Harbor Tug Dual Tug Fuel
PTRIM-BPPT 2021)

Based on the design requirements and basic design of the harbor tug dual fuel ship,a key-plan drawing has been made that has received key-plan design approval from the Indonesian Classification Bureau with detailed key-plan drawings in the form of detailed drawings of machinery outfitting & piping diagram, machinery system, hull outfitting & deck machinery, machinery outfitting & piping diagram (service system), electrical system.

4.4 Dual fuel diesel engine

Referring to the Regulation of the Minister of Transportation, Number 57 of 2015, Chapter VI, Article 38 paragraph (3) that for a tug boat length of 70 - 150 meters, at least 1 unit of tug boat is made with the lowest total power of 2000 HP and the maximum total pulling force. low 24-ton bollard pull(4). Based on these design references. the maximum design power requirement for ship operations is 2500 HP for as many as 2 engines. The enginethat meets these specifications is the NIIGATA 6L 28AHX. - Engine Speed: 800 rpm or 750 rpm - bore: 280 mm - stroke: 390 mm - power output: 2220 kW or 2070 kW - BMEP:
2.31 MPa or 2.3 MPa - mean piston speed: 10.4 m/s or 9.75 m /s - max pressure: 18 MP.

4.5 Refueling system of liquefied natural gas (LNG)

LNG bunkering is the process of giving a ship its own supply of liquefied natural gas for fuel. The primary benefit of using LNG as fuel is the significant reduction in pollution compared to more conventional fuels for ships, such as heavy fuel oil, marine diesel oil (MDO), and marine gas oil (MGO). There is increasing regulatory pressure to reduce emissions brought on by ship traffic. The development of an LNG fueling infrastructure as well as the design of LNG bunkering systems are being driven by the application of higher sulfur content limits on marine bunker fuel. In general, LNG bunkering systems have been widely used, namely land-to-ship, truck-to-ship, and ship-to-ship refueling systems (5). The bunkering system installed on board the tug boat will be the link between the external bunkering installation and the LNG storage tank on board. The bunker piping system is also equipped with a purging system connection.

The bunkering station should be located as far as possible in a place that has adequate natural ventilation and is as close to the tank as possible to minimize the length of the transfer pipe and limit heat exposure in the LNG environment during LNG transfer. Bunkering systems located in enclosed or semi-enclosed spaces must receive special attention, especially with regard to rules for mechanical ventilation and gas-tight separatorsmust be equipped to separate bunkering areas from other areas of the ship. In addition, it must also be equipped with a drip tray and coaming as well as a water curtain system underthe LNG bunker connection flange connection to prevent leakage or LNG droplets directly hitting the ship's hull. The frame of the bunker station including the drip tray is made of stainless steel.

Organizations have established and are developing standards and rules for LNG bunkering (6). For instance, IMO (International Maritime Organization), ISO, EN (CEN- European Committee for Standards), SIGTTO (Society of International Gas Tankers and Terminal Operators), OCIMF (Oil Companies International Marine Forum), and (NFPA (National Fire Protection Association).

4.6 Technology readiness level assessment

The technology readiness level assessment of the results of the dual fuel tug boat research has been carried out by National Research and Innovation Agency. In this regard, information collection and interviews have been carried out with the ResearchImplementation Team of the Transportation Technology Research Centre as the executor of activities and collecting documents for each indicator available at each level. Systematic assessment interviews are carried out by classifying indicators of the results ofresearch implementation every quarter of the implementation of activities, such as quarterly activity implementation reports and preliminary study reports on technology concepts, design requirements, preliminary design, basic design, detailed design, ship key plan design approval process by the Bureau Indonesian Classification, the process of testing ship models in the laboratory and documents for the implementation of related research activities. The TRL level evaluation on the techno meter is then updated to reflect the interview results.

The assessment is continued to the next TRL if the achievement of each TRL meets the lower limit value of 80%; otherwise, it is stopped and the assessment is not continued to the next TRL. The last TRL that is met is the achievement of the level of technological readiness (7), as shown in Table 2 as follows:

Table 2. Observation of research assessment at TRL- 4

No.	Component Indicator	Value Weight	Total Value	Status
1	The scale testing of the ship model in the laboratory has been carried out	5	**86,66%**	**> 80%** (comply)
2	The design requirements that the user expects have been made	5		
3	The test results of the ship model show good performance and can operate	5		
4	The detailed design and key plan of the ship prototype have been done	4		
5	Approval to the key plan design from BKI has been completed	4		
6	The design and integration of all components of the prototype ship have been completed and is feasible to build	3		

Description.: 0=Not Meet; 1=20%; 2=40%; 3=60%; 4=80%; 5=100% (7)

Based on the TRL assessment in Table 2 above shows that the indicators for testing the scale of the ship model in the laboratory, ship design requirements, and the results of testing the ship model are feasible to operate and get an achievement score of 100% (completed), while the detailed design indicators and key-plan design. Design approval from the Indonesian Classification Bureau have not been fully carried out and the integration and feasibility of ship design have only reached a value of 80 % because not all ship components and their integration can be tested in the relevant environment. The measurement of TRL-4 for dual fuel tug boats has reached 86.66% is complied.

The next assessment is technology readiness at the TRL-5 level (see Table 3). The technology status required in TRL-5 includes: The technological element has been examined in the pertinent setting. In order to validate that the functions and properties of the technological components are consistent with the anticipated application, the correctness of the technological components is tested at this level using pertinent environmental tests. In order to test the entire application (at the component, sub-system, or system level) in a simulation or under real-world settings, the fundamental aspects of technology have been combined with supporting features to create a realistic prototype.

To improve the level of readiness of the ongoing technology TRL-5 to a higher TRL stage before completing the research objectives, it is necessary to collaborate with shipyard manufacturers to build ships and conduct demonstration trials of ships at the technology development stage TRL 6-7. By undertaking development steps or consolidating TRL steps together in one "big step".

Table 3. Observation of Research Assessment at TRL-5

No.	Component Indicator	Value Weight	Total Value	Status
1	Preparations for ship production at the shipyard are carried out	1		
2	Market research/Ship Users, as well as manufacturing research in shipyards, have been carried out	2		
3	Approval design for ship production has been approved by BKI	2		
4	The prototype of the ship has been made	0	16,6%	**< 80%** (Not compliant)
5	The ship prototype is ready to be tested (sea trial) in the relevant marine environment	0		
6	The production process has been reviewed by the authorities at the shipyard and by the User and Management	0		

4.7. Partner Cooperation

To accelerate the pace of technology readiness, at the stage of implementing research activities, BRIN has carried out various collaborations with partners so that the research results can be utilized by Prospective Users, among others by PT Pelindo Marine Service (PMS) which is tasked with carrying out and facilitating the implementation of research and optimizing the utilization of converter kits and LNG fuel tanks. PT Lintech Duta Pratama is tasked with implementing and providing support in the fabrication of Mini LNG Tank vessels and as an industrial partner and tasked with building harbor tug vessels and mini LNG vessels as well as developing converter kit products and LNG Tanks (ISO Tanks) to be mass-produced. ITS Surabaya as an educational institution is tasked with carrying out scientific studies from both technical and economic aspects to support the implementation of the national research program including harbor tug dual fuel innovation;Mini LNG ship innovation; and transport vehicle of postoperative offshore platform and has been stated in the cooperation agreement.

5 Conclusion

1) This research has succeeded while evaluating the level of the country's technological readiness research results of the dual fuel tug boat to support policy making in the formulation, implementation, and evaluation of maritime technology research and development programs in Indonesia. The level of technology readiness of the dual fuel tug boat ship has been determined to be TRL-4 after processing and analyzing the data and information obtained. At this level, the technology component has been validated through testing the scale of the ship model in the hydrodynamics laboratory, and the test results show the feasibility ship operates.
2) The formulation of action items must be carried out based on the indicators in TRL-5 that have not been met, namely the implementation of validation and testing of the ship's prototype operation in sea waters so that the validation of functions and characteristics of motion capability ship in accordance with the expected conditions. This will increase the technological readiness of the research results of the ship to the next level, namely TRL-5. A realistic prototype has been created by combining the fundamental components of the technology with the assisting features. In order to get to the actual commercialization stage and the demonstration stage of the ship's prototype functioning, it is also important to project future operations (TRL 7-9).

6 References

1. Ministry of Higher Education. Ministerial Regulation of the Ministry of Research, Technology and Higher Education of the Republic of Indonesia. Directorate of Research and Cvommunity Service, Deputy for Strengthening Research and Development, Ministry of Research and Technology / National Research and Innovation Agency [Internet]. 2016;(879):7. Available from: https://peraturan.bpk.go.id/Home/Download/132670/Permenristekdikti Nomor 42 Tahun 2016.pdf (In Indonesian)
2. Kluk AF, Johnson HC, Mcginnis CP, Ross SL, Sutter HG. T Echnology R Eadiness L Evels H Andbook for. 2008;(September).
3. Mankins JC. Technology readiness assessments: A retrospective. Acta Astronaut [Internet]. 2009;65(9–10):1216–23. Available from: http://dx.doi.org/10.1016/j.actaastro.2009.03.058
4. Ministry of Transportation. Minister of Transportation Regulation number 57 of 2015 concerning Ship Guiding and Towing. 2015. (In Indonesian)
5. Holden D. Liquefied Natural Gas (LNG) Bunkering Study. 2014;1–156.
6. For G. GUIDE FOR. 2018;2017(JANUARY 2017).
7. Suzianti A, Dewi AP, Mubarak A, Shafira ND. Technology readiness level assessment of lithium battery in Indonesia for national electric vehicle program. AIP Conf Proc. 2020;2227(May).

IOP Conf. Series: Earth and Environmental Science 1166 (2023) 012005 doi:10.1088/1755-1315/1166/1/012005

Propeller Performance Analysis on The Modification of Trailing Edge Shape as Anti-Singing

Izzul Fikry[1], I Made Ariana[1] and Cahya Kusuma[2]

[1] Marine Engineering Department, Institut Teknologi Sepuluh Nopember, Surabaya, Indonesia
[2] Mechanical Engineering Department, Sekolah Tinggi Teknologi Angkatan Laut, Surabaya, Indonesia

izzulfikry96@gmail.com, ariana@its.ac.id and kusumacahya99@gmail.com

Abstract. Singing is one of the problems that occurs in the propeller. However, singing will not happen to every propeller. Singing occurs a lot in high-speed propellers and on propellers with wide and thin blade sizes. One of the ships that have high speed and wide propeller is fast patrol boat. To overcome this problem, modifying the trailing edge is one of the most commonly used methods. Sharpening the trailing edge is a way to reduce vortex shedding. However, modifying the trailing edge of the propeller will change the shape of the propeller, which can affect the performance of the propeller. Therefore, this study discussed about the effect of trailing edge modification as anti-singing on propeller performance. Propeller performance analysed with open water test simulation assisted by computational fluid dynamic based software. There are three sharpening model in this study, where the sharpening size in this study refers to Van Lammeren anti-singing design. The size of sharpening are 20 mm, 25 mm, and 30 mm, where 20 mm sharpening is the best modification to overcome singing according to flow analysis. The open water test simulation result shows that the modification of trailing edge as anti-singing can increase the average thrust coefficient 8.96%, average torque coefficient 7.88%, and average open water efficiency 1.29%.

1. Introduction

Singing is a disturbing phenomenon on the propeller. Singing often occurs in propellers with high rotational speeds [1], as well as in propellers with wide and thin blade sizes [2]. Singing can increase airborne noise and underwater noise on ships. Airborne noise can interfere the performance of ship crews, beside that underwater noise can increase the possibility of warships being detected by enemy devices [3][4]. The cause of singing on the propeller is the resonance between the natural frequency of the propeller tip and vortex shedding on the trailing edge [2][3][5][6].

The most common way to deal with singing on a propeller is to sharpen the trailing edge [1]. By sharpening the trailing edge, it is hoped that the separation point will be clearer, and the wake vortices and induced forces are lower than the blunt trailing edge [2]. In addition, sharpening the trailing edge can narrow the vortex street, so that the vortex shedding strength will decrease [7]. Various studies of sharpening the trailing edge have been carried out. Sharpening performed on the back and face trailing edges [6][8][9]. From those research, it is known that sharpening on the back trailing edge can overcome singing [6][8][10].

It has been stated in the previous paragraph that singing occurs a lot in propellers with high rotational speeds, as well as in propellers with wide and thin blade sizes. With those conditions, ships with high speed and wide propeller surfaces certainly have the possibility of singing. One of the ships with that condition is fast patrol boat. However, modify the trailing edge of the propeller can affect the performance of the propeller. Propeller performance may change up to 10% after modification [6]. Therefore, it is very important to ensure changes in propeller performance after modification. Based on that problem, an open water test simulation carried out to determine the performance of the fast patrol boat propeller. The open water test simulation carried out with the help of software based on computational fluid dynamics.

2. Method

2.1. Propeller geometry

The fast patrol boat in this study uses a 5 blade Wageningen B-series propeller, 1.8 meters in diameter, disk area ratio (AE/AO) 1.05, and pitch/diameter ratio (P/D) 1.1. Figure 1 shows the propeller 3D model which modelled with CAD software.

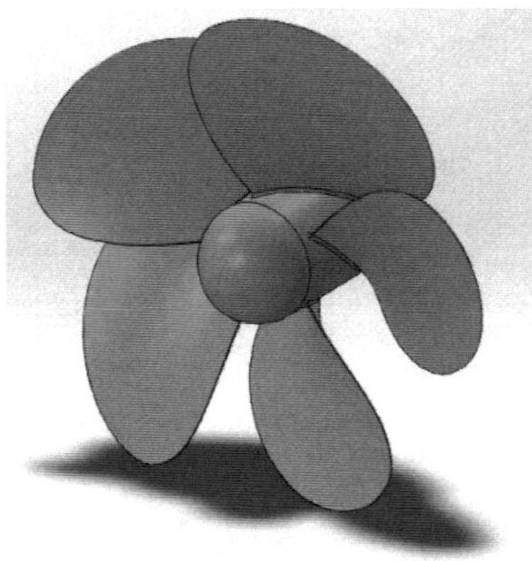

Figure 1. Propeller B5-105 model

2.2. General characteristic of open water test

The open water test in this study carried out with the help of computational fluid dynamic based software. The simulation produce data in the form of thrust (T) and torque (Q). The data processed to get the value of thrust coefficient (KT), torque coefficient (KQ) and efficiency (ηo). The equation (1), (2), and (3) are used to generate the data.

$$K_T = \frac{T}{\rho \, x \, n^2 \, x \, D^4} \tag{1}$$

$$K_Q = \frac{Q}{\rho \, x \, n^2 \, x \, D^5} \tag{2}$$

$$\eta_o = \frac{T \, x \, Va}{2 \, x \, \pi \, x \, n \, x \, Q} \tag{3}$$

To complete the (1), (2), and (3) equation, parameters are needed in the form of fluid density (ρ), propeller rotation speed (n), speed of advance (Va), and propeller diameter (D).

2.3. Propeller modification

B5-105 propeller modified by referring to the design proposed by Van Lammeren which shows in figure 2, where the modifications are performed on the back trailing edge with sizes based on the equation (4) and (5) [1]. The modifications on the trailing edge carried out with variations according to the x and y sizes in table 1. The largest sharpening is 30 mm which obtained from maximum x at equation (4), while the smallest sharpening is obtained from equation (4). The trailing edge sharpened at 0.5 r/R – 1 R as shows in the figure 3.

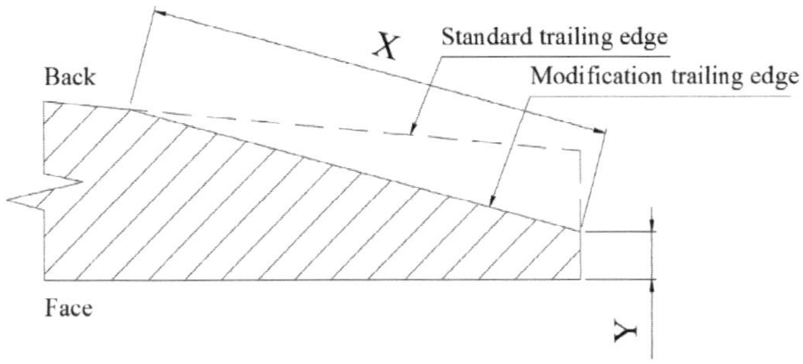

Figure 2. Trailing edge modification geometry

$$x = [20 + 5(D - 2)]|_{max=30\ mm} \quad (4)$$

$$y = 0.1x \text{ mm} \quad (5)$$

Table 1. Variation of trailing edge modification

Variation	X (mm)	Y (mm)
1	20	2
2	25	2.5
3	30	3

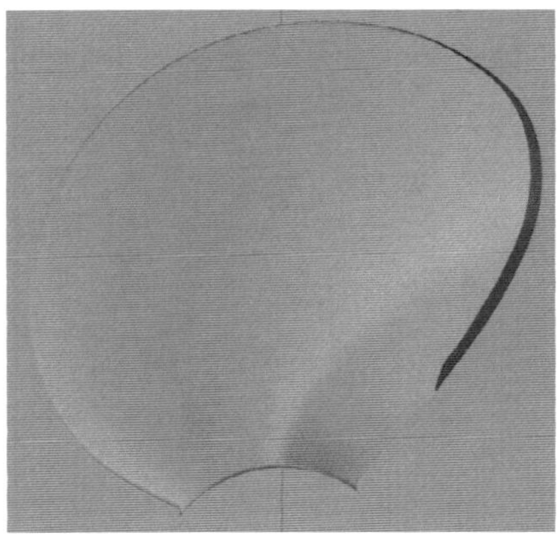

Figure 3. Anti-singing edge position

2.4. Grid Independence Test

Grid independence test is a method which used to generate stable numerical results with a small number of grids [11], so that the numerical results obtained are not much different between a small number of grid and a large number of grid. The grid independence test in this study uses a velocity factor associated with the number of cells. Figure 4 shows the relation between the number of cells and the velocity, where in more than 250000 cells, the resulting velocity does not change much even though the number of cells increases. So that, from the grid independence test it can be concluded that the recommended grid used must be more than 250000 cells.

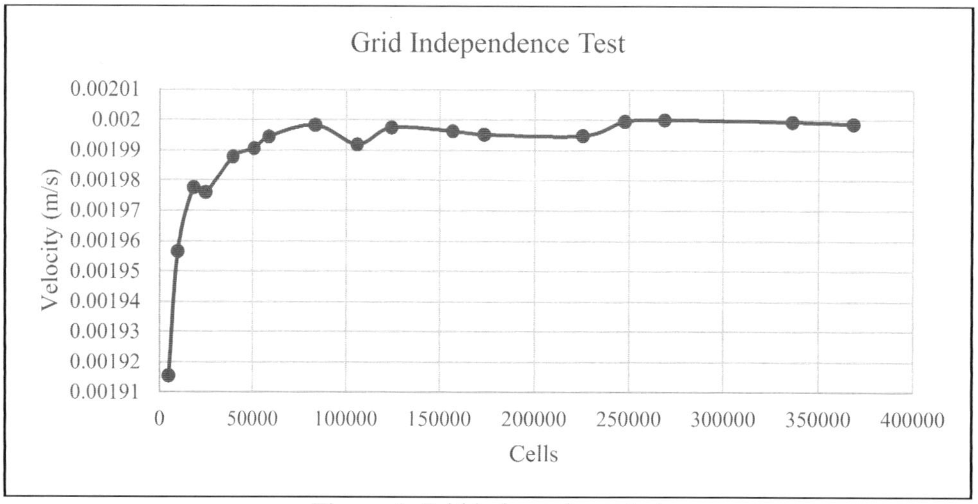

Figure 4. Grid independence test

2.5. Flow analysis

Flow analysis is used to determine the flow that comes out of the trailing edge. Maheswari [12] has conducted a study using flow analysis to analyze the flow on a flat plate that has an anti-singing edge. Flow analysis in this study was carried out on a cross section of 0.7R blade propeller B5-105. The blade cross section will be positioned according to the propeller's 0.7R pitch angle, which is 25.38°, and the flow velocity used is 5.4 m/s. The number of cells used in this flow analysis is 320000 – 325000 cells. Figure 5 shows the cell details in the meshing process.

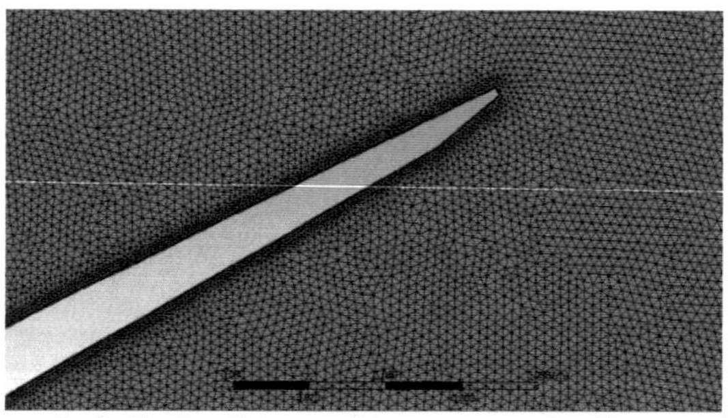

Figure 5. Cells detail in meshing process for flow analysis

Figure 6 shows the flow out of the standard trailing edge, figure 7 shows the flow on the trailing edge with 20 mm sharpening, figure 8 shows the flow on the trailing edge with 25 mm sharpening, while figure 9 shows the flow on the trailing edge with 30 mm sharpening. From figure 6 to 9 shows that the standard trailing edge have strongest vortex shedding and the trailing edge with 20 mm sharpening produces the weakest vortex shedding. So, it can be concluded that the greater the sharpening, the weaker the vortex shedding produced. Vortex shedding in figures 6 to 9 shows the possibility of turbulence generation, where vortex shedding is one of the primary turbulence generation mechanisms [13]. The existence of the turbulence phenomenon can cause noise and cavitation in the propeller [14]. From this statement, it can be seen that by sharpening the trailing edge of the propeller, it can reduce possibility of turbulence that causes noise and cavitation.

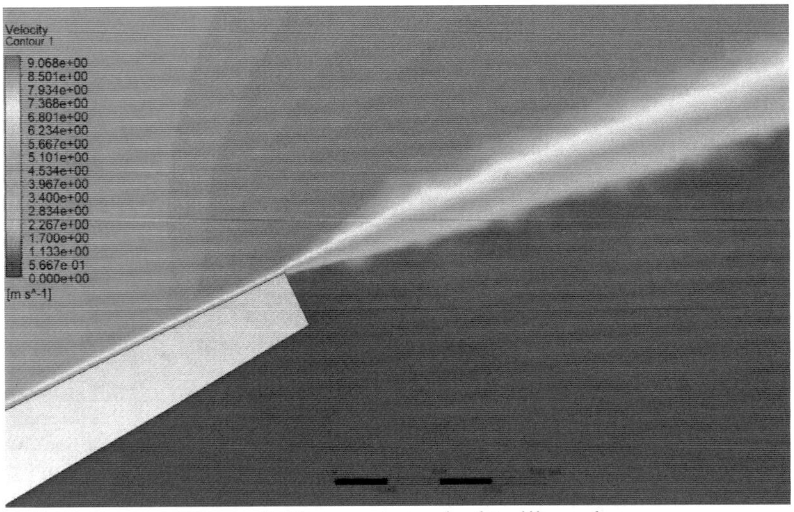

Figure 6. Flow past standard trailing edge

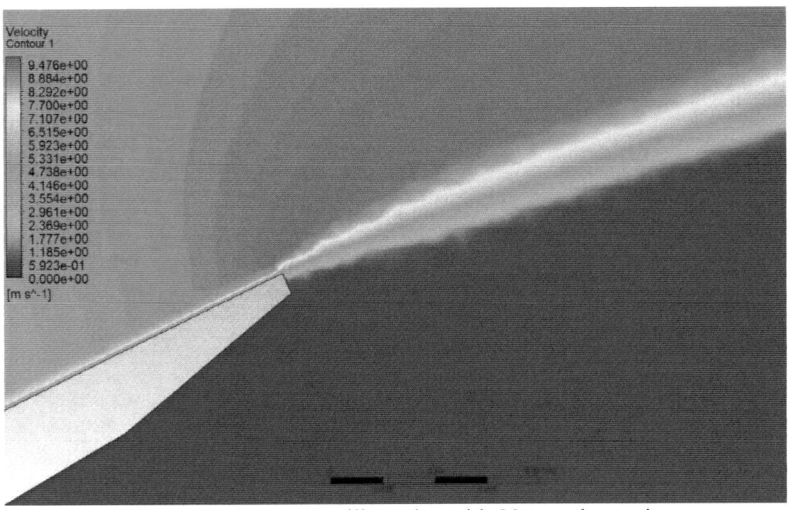

Figure 7. Flow past trailing edge with 20 mm sharpening

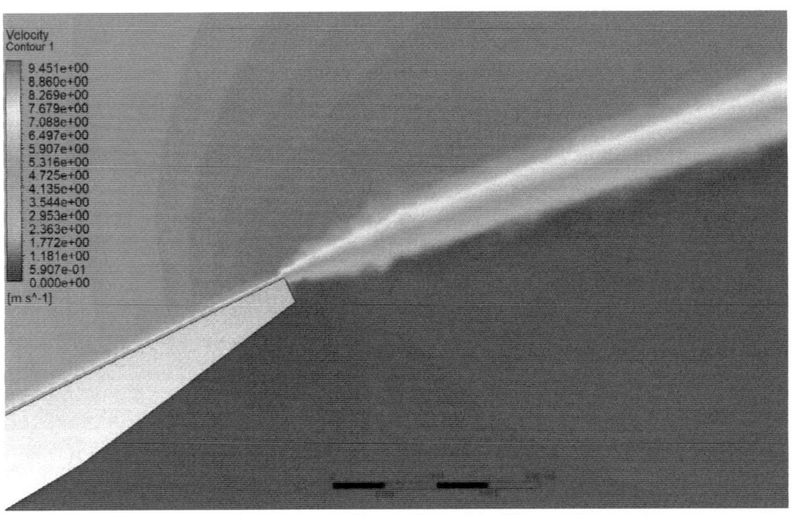

Figure 8. Flow past trailing edge with 25 mm sharpening

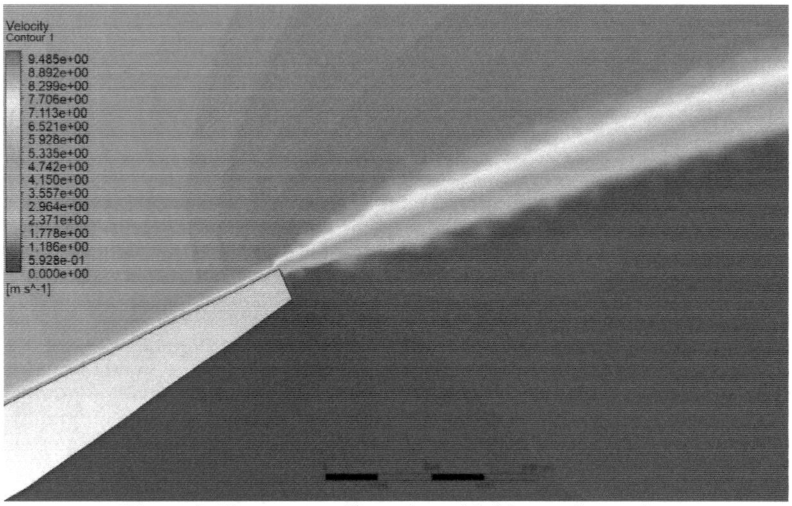

Figure 9. Flow past trailing edge with 30 mm sharpening

2.6. Open water test simulation parameter and meshing

The open water test simulation in this study used Ansys software. The simulation process starts with made a domain. This study uses multi domains, where the domain size is shown in figure 7. The domain for the open water test simulation uses a cylinder with a length 8 times the propeller diameter, with the width 6 times of the diameter of the propeller [15]. The second step was meshing. The curvature of the propeller must give more refinement than another component. In this study used 417399 cells, which figure 8 shows the meshing cell.

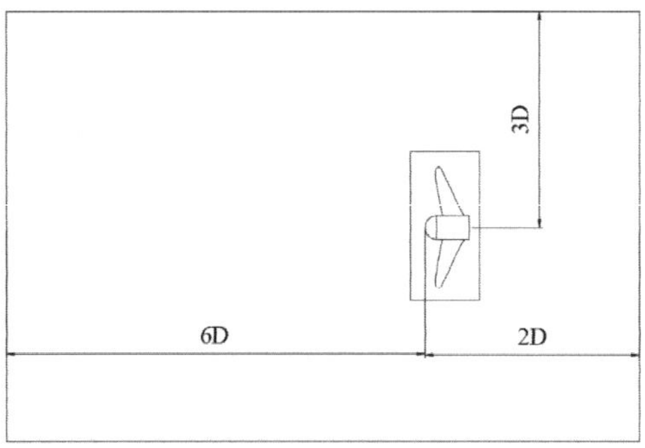

Figure 10. Domain for open water test simulation

Figure 11. Meshing process

At the end of meshing process, the process continues to simulation process. Figure 9 shows the domain which have inlet and outlet channel. In this study there are some parameters used for this simulation.

This study used water fluid which have 998.2 kg/m^3 density, 5.4 until 12.6 m/s speed of advance, and the angular velocity is 600 RPM.

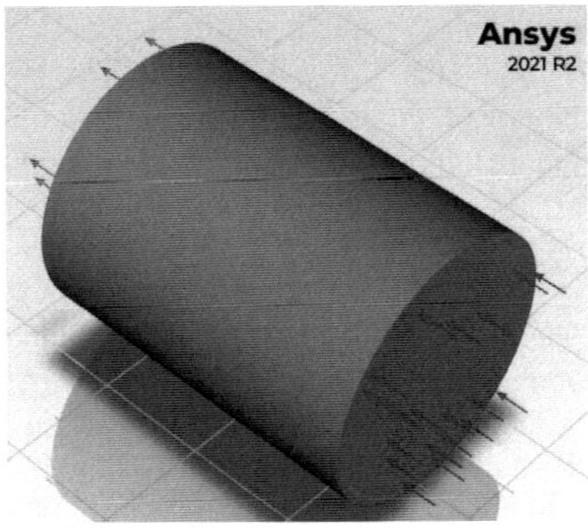

Figure 12. Simulation process

2.7. Validation

Validation is carried out to ensure that the model used in this study is in accordance with the propeller used during the open water test. The validation conducted by comparing the KT and KQ simulation results of the open water test on the standard B5-105 propeller with the results of the B5-105 experimental open water test of the propeller. Kusuma has conducted a B-series polynomial calculation on the B5-105 propeller, the result shows in EFD column at table 2 [16].

Table 2. Comparison between simulation and open water test

J	CFD			EFD			Difference		
	KT	10KQ	ηo	KT	10KQ	ηo	KT	10KQ	ηo
0.300	0.450	0.791	0.272	0.446	0.725	0.294	1%	9%	-8%
0.400	0.401	0.712	0.359	0.397	0.653	0.387	1%	9%	-7%
0.500	0.349	0.629	0.442	0.345	0.576	0.476	1%	9%	-7%
0.600	0.293	0.544	0.514	0.290	0.496	0.558	1%	10%	-8%
0.700	0.235	0.456	0.575	0.233	0.414	0.627	1%	10%	-8%

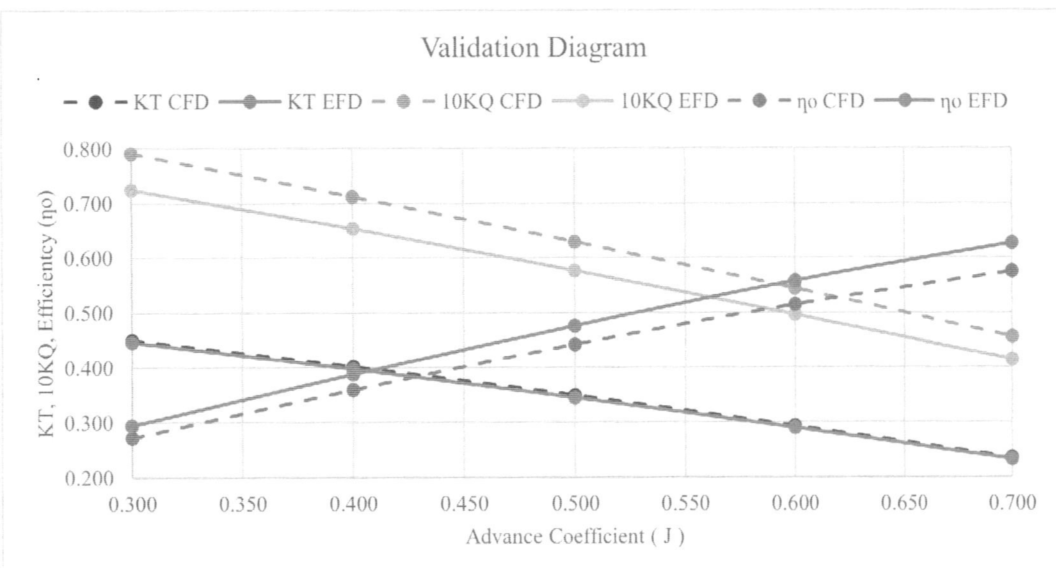

Figure 13. B5-105 validation diagram

Table 2 and figure 10 show the comparison of the results of the open water test with computational fluid dynamic (CFD) simulation to the results of the experimental open water test (EFD) on the B5-105 propeller. The difference in KT values shows stable results, where the average difference in KT values is 1%. Besides that, the comparison of 10KQ results shows an average difference of 9.4%. From the comparison results, it can be concluded that the standard B5-105 propeller model used in the open water test simulation is valid, because the difference between the simulation results and the open water test is less than 10%.

3. Result and Analysis

The open water test simulation generated thrust and torque data. Table 3 shows the data and comparison about B5-105 standard thrust coefficient (KT) and B5-105 modification thrust coefficient. From the table 3, it shows that the sharpening of the trailing edge makes the thrust of the propeller increase, which 20 mm give the highest thrust coefficient increase. The increase in propeller thrust is directly proportional to the flow analysis results, where the weakest vortex shedding also occurs at the 20 mm sharpening. This phenomenon shows that the weaker the vortex shedding, the higher the thrust generated. Oloan [17] also experienced the increase in propeller thrust caused by vortex changes, where the vortex generated by the propeller hub was eliminated by the addition of propeller boss cap fins (PBCF).

Table 3. Comparison between KT B5-105 Standard and Sharpened

J	KT Standard	KT Sharpened					
		30 mm	**Gap**	**25 mm**	**Gap**	**20 mm**	**Gap**
0.300	0.45	0.465	3.33%	0.474	5.37%	0.481	6.91%
0.400	0.40	0.415	3.45%	0.425	5.74%	0.431	7.38%
0.500	0.35	0.362	3.62%	0.372	6.41%	0.379	8.43%
0.600	0.29	0.305	4.01%	0.315	7.39%	0.322	9.82%
0.700	0.24	0.247	4.90%	0.257	9.13%	0.264	12.28%
		Average	**3.86%**	**Average**	**6.81%**	**Average**	**8.96%**

Table 4 shows the comparison between torque coefficient between B5-105 standard and sharpened. From the table 4, it can see that 20 mm sharpening made higher torque coefficient than another sharpening size. Not to different form the thrust coefficient change, the 30 mm sharpening have the smallest impact for the propeller torque.

Table 4. Comparison between 10KQ B5-105 Standard and Sharpened

J	10KQ Standard	10KQ Sharpened					
		30 mm	**Gap**	**25 mm**	**Gap**	**20 mm**	**Gap**
0.300	0.79	0.811	2.50%	0.829	4.81%	0.842	6.45%
0.400	0.71	0.730	2.44%	0.748	5.03%	0.761	6.80%
0.500	0.63	0.644	2.44%	0.663	5.40%	0.677	7.55%
0.600	0.54	0.558	2.55%	0.576	5.96%	0.590	8.53%
0.700	0.46	0.469	2.80%	0.487	6.88%	0.502	10.07%
		Average	**2.54%**	**Average**	**5.62%**	**Average**	**7.88%**

Different from thrust and torque coefficient, the open water efficiency from 30 mm sharpening has the highest increase. The 30 mm sharpening has 1.29% average increase efficiency, 25 mm sharpening have 1.12%, and 20 mm sharpening has 1.00% average increase efficiency which is the smallest increase. The data in table 3 to table 5 are summarized in the diagram in figure 11.

Table 5. Comparison between ηo B5-105 Standard and Sharpened

J	ηo Standard	ηo Sharpened					
		30 mm	Gap	25 mm	Gap	20 mm	Gap
0.300	0.27	0.274	0.81%	0.273	0.53%	0.273	0.43%
0.400	0.36	0.363	0.98%	0.362	0.67%	0.361	0.55%
0.500	0.44	0.447	1.16%	0.446	0.96%	0.446	0.82%
0.600	0.51	0.522	1.43%	0.521	1.35%	0.521	1.18%
0.700	0.58	0.587	2.04%	0.587	2.10%	0.587	2.01%
		Average	**1.29%**	**Average**	**1.12%**	**Average**	**1.00%**

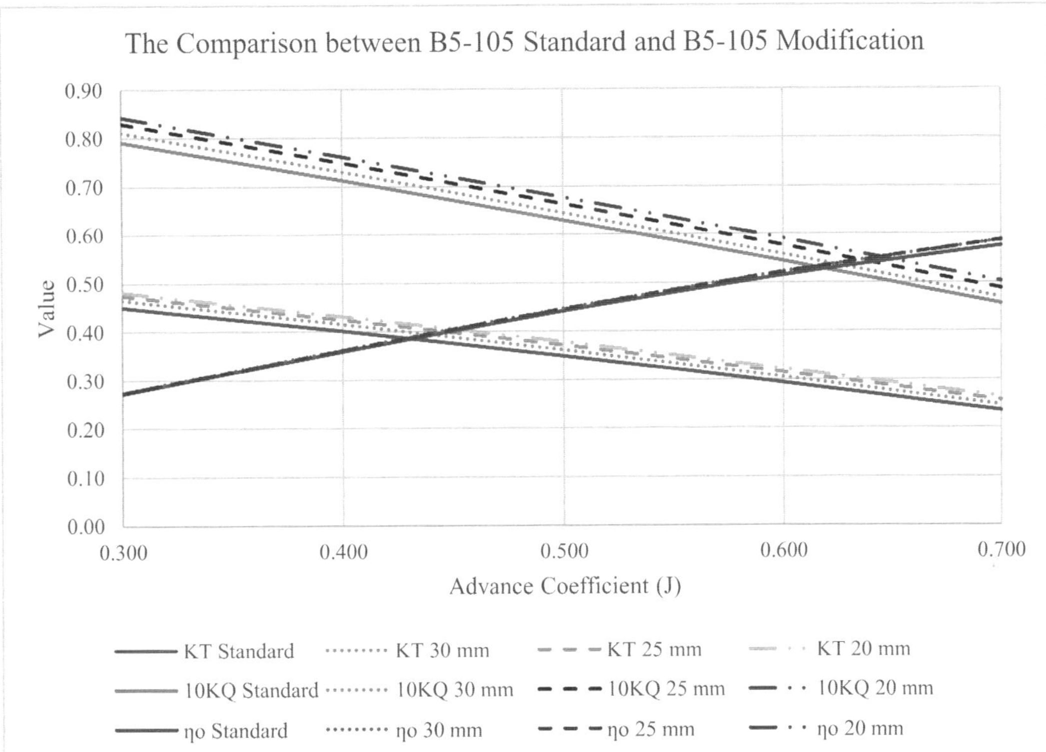

Figure 14. Comparison Open Water Diagram of B5-105 Standard and Modification

4. Conclusion

This study discussed about performance analysis of propeller which have some modification in the trailing edge. The modification conducted in the back of propeller which used to overcome the propeller singing. The anti-singing design in this study refers to Van Lammeren theory. There are 3 sharpening size, 30 mm, 25 mm, and 20 mm. The 20 mm sharpening is the best modification to overcome singing. Trailing edge with 20 mm sharpening has the smallest vortex shedding. Besides that, open water test simulations were carried out to determine changes in propeller performance due to the addition of anti-singing edges. The open water test simulation shows that 30 mm sharpening have the lowest thrust and torque coefficient change with 3.86% average KT and 2.54% average KQ, but it has the highest open water efficiency change with 1.29% average ηo. Meanwhile, the 20 mm sharpening have the highest thrust and torque coefficient change with 8.96% average KT and 7.88% average KQ, but it has the lowest open water efficiency change with 1.00% average ηo. From that statement, it can be concluded that modification of the trailing edge as anti-singing can change the performance of the propeller, which it can increase thrust, torque, and efficiency.

References

[1] Carlton J 1994 *Marine Propellers and Propulsion* (Elsevier)

[2] Fischer R 2008 Singing Propellers—Solutions and Case Histories *Mar. Technol. SNAME News* **45** 221–7

[3] Sastry H R S 1982 Singing Propellers *Def. Sci. J.* **32** 67–73

[4] Work C E 1951 Singing Propellers *J. Am. Soc. Nav. Eng.* **63** 319–31

[5] Park H S, Choi S H and Kim N S 2005 Identification of propeller singing phenomenon through vibration analysis of propeller blade *Proc. Int. Offshore Polar Eng. Conf.* **2005** 591–5

[6] Hasuike N and Yamasaki S 2004 Anti-singing Edge Effect on Propeller Open Characteristics *Pros. Japan Shipbuild. Soc. No. 196*

[7] Yi-mai W 1986 A Study Of Simulation and Elimination of Propeller Singing *Shipboard Acoustics* ed J BUITEN (Dordrecht: Martinus Nijhoff Publishers) pp 75–90

[8] Burril L C 1949 Underwater Propeller Vibration **62**

[9] Heskestad G and Olberts D R 1960 Influence of trailing-edge geometry on hydraulic-turbine-blade vibration resulting from vortex excitation *J. Eng. Gas Turbines Power* **82** 103–9

[10] van de Voorde C B 2018 The singing of ship propellers *Int. Shipbuild. Prog.* **7** 451–5

[11] Lee M, Park G, Park C and Kim C 2020 Improvement of Grid Independence Test for Computational Fluid Dynamics Model of Building Based on Grid Resolution *Adv. Civ. Eng.* **2020**

[12] Maheswari A, Kumar N, Singh V, Saxena V and Kumar Singh M 2019 Numerical simulation of flow past a flat plate with anti-singing trailing edge *IOP Conf. Ser. Mater. Sci. Eng.* **691**

[13] Lage J L 1998 The Fundamental Theory Of Flow Through Permeable Media From Darcy To Turbulence *Transport Phenomena in Porous Media* (Dallas: Elsevier Science Ltd) pp 1–30

[14] Purwana A, Made Ariana I and Wardhana W 2021 Numerical Study on the Cavitation Noise of Marine Skew Propellers *J. Nav. Archit. Mar. Eng.* **18** 97–107

[15] Oloan A F N, Made Ariana I and Baidowi A 2022 Open Water and Performance Analysis of Marine Propeller with PBCF Based CFD Method *IOP Conf. Ser. Earth Environ. Sci.* **972**

[16] Kusuma C, Ariana I M and Nugroho W H 2021 An Innovative Proposal for Increasing the Speed of a 60 Metres Fast Patrol Vessel *Int. Rev. Mech. Eng.* **15** 486–96

[17] Oloan A F N, Ariana I M and Baidowi A 2022 Self propulsion Performance Analysis of Japanese Bulk Carrier with PBCF Based CFD Method *IOP Conf. Ser. Earth Environ. Sci.* **1081**

Dynamic response evaluation due to construction system modification of 155 m barge ship using finite element method

Tuswan Tuswan[1], Wilma Amiruddin[1], Hartono Yudo[1,*], Aditya Rio Prabowo[2]

[1] Department of Naval Architecture, Universitas Diponegoro, Semarang 50275, Indonesia,
[2] Department of Mechanical Engineering, Universitas Sebelas Maret, Surakarta 57126, Indonesia

*Email: hartono.yudo@yahoo.com

Abstract. The application of lightweight sandwich material as a ship structure was an innovative breakthrough that provided an excellent strength-to-weight ratio. In the sandwich application process, the proposed structural design and scantling must be designed according to the equivalent strength requirement with low weight compared to the existing structure. The research aims to evaluate the dynamic response of different construction systems of barge ships due to applying various hybrid sandwich types in the deck, ship hull, and bottom structures using the current Lloyd's Register standard. A total of three proposed construction systems: longitudinal, transverse and mixed framing systems, are investigated under different plate configurations, material types, and scantling sizes. The numerical-based finite element approach is used to evaluate dynamic properties of natural frequency and mode shape using free vibration analysis at low mode numbers. The finite element result from the dynamic characteristic and weight reduction estimation between the existing structure and the proposed construction systems under various plate configurations, material types, and scantling sizes is evaluated to determine the application of lightweight sandwich panels in the barge structure.

Keywords: ship construction, barge, finite element, free vibration

1. Introduction

Shipping has developed highly sophisticated logistics chains that deliver just-in-time parts and goods to manufacturers and consumers to support global economies. Ships are an essential mode of transportation in economic activity, transporting passengers or commodities [1]. Improving competitiveness with other modes of transportation is critical in shipping operations to increase economic value added. One solution for increasing competitiveness is to use lightweight materials for ship structures. Structural design and lightweight optimization aim to increase cargo-carrying capacity in the shipbuilding industry. As a result, many scholars are interested in using lightweight materials as ship structures to increase ship payload, protect the environment, and reduce fuel consumption and CO_2 emissions. Lightweight structures are created by altering existing construction sizes or implementing lightweight materials into potential structural parts. Weight savings should be consequential in applying lightweight material to correct unexpected technical, practical, and financial difficulties during sandwich processing and manufacturing and ensure the proper use of sandwich panels by a shipowner, classification companies, and shipyard management [2].

Many materials have been used in developing ship structures, ranging from traditional materials such as wood and cast iron to materials commonly used today, such as steel, aluminium, and carbon fibre. As technology advances, ship materials must be strong while remaining as light as possible. Generally, this is difficult to achieve because strong material tends to be heavy and lightweight material tends to have a small strength value. Sandwich plates, which combine the configuration of two or more materials consisting of a faceplate and a core layer, are one of the solutions to the previously mentioned problems [3].

In the maritime and construction industries, sandwich panels are currently used on over 35,000 m^2. The overlay approach has found significant applications in ship repair (e.g., ramps and Ro-Ro decks)[4]. Several European Commission-funded projects have achieved this goal and significantly improved knowledge of marine sandwich structures, including the SAND.CORE project [5] provided a best practice guide for sandwich structures in marine applications. Reduced fuel consumption, ease of construction, reduced displacements with consequently increased payload, improved stability and manoeuvrability, reduction in building material volumes and mass with consequent cost, and time savings are all advantages of lightweight applications of ship structures [5].

Sandwich panels can be used to build a wide range of ships. In practice, sandwich materials can be applied to new building schemes or ship repairs. In the sandwich material development stage, each sandwich type has a standard set to meet the standardization of material properties criteria and sandwich material strength. Classification bodies typically issue these standards to ensure high safety and quality standards. All marine sandwich construction rules are based on ensuring that they are equivalent to a single plate or steel-reinforced counterpart. To address the differences in reactions between sandwich and regular structures, classification companies' standards provide dedicated strength requirements and an evaluation technique [6].

In applying sandwich panels for new construction, it is crucial to ensure that the sandwich modelhas the equivalent strength to the existing structure. This study proposes a sandwich barge structure design by varying the construction system based on the Lloyd's Register standard [7]. There are three types of construction system variations will be evaluated, namely transverse, longitudinal, and mixed framing systems. The stiffness comparison between the existing structure and the proposed constructionsystem will be analyzed using FEM by comparing the natural frequency values and shape modes.

2. Schematic Overview of Ship Structural Assessment

Using sandwich materials in ship structures necessitates a strength index calculation to ensure that the newly developed structure has equivalent strength, as Lloyd's Register requires [7]. The advancement of research in practical and scientific areas demonstrates that the application of sandwich material is unique based on the various structural locations and components. Static and dynamic load structural assessments on various structural locations will be discussed. Pioneering research into the use of sandwich material is being conducted in tankers with a double bottom, side shell, main deck, and longitudinal bulkhead. The results show a 2.8% reduction in overall weight and a 20% reduction in workload in the manufacturing process due to the elimination of stiffener use [8]. By varying the stiffener frame spacing, Ismail et al. [9,10] investigated the application of a hybrid sandwich based on steel/polyurethane components in a tanker's side shell and deck. The promising outcome allows for stress and weight loss. The remarkable stress reduction ranges from 20.8 to 27.9%, and the structural weight ranges from 4.2 to 8.8% [2]. A sandwich panel can reduce the need for stiffeners, reducing weight and increasing payload [11].

A thorough investigation using various material types and structural locations yields a new perspective. A structural assessment due to static and dynamic load in the Ferry Ro-Ro ship's car deck is thoroughly investigated [12,13]. Sandwiches with various material types, stiffener configurations, and load types contributed to stress reductions of approximately 14.6 - 15.8% and weight savings of roughly 8.87% - 11.6%. Furthermore, the investigation is carried out at various structural locations on thesame ship. Using sandwich panels on side shells produces promising results, with stress reductions of up to 28.4% and weight savings of approximately 17.1% [14]. The structural assessments resulting fromthe sandwich application on the ramp door are then investigated in terms of static and dynamic behaviour [15,16]. Numerical and experimental methods for assessing the structural integrity of hull sandwich panels have also been developed [17,18].

The preliminary results show promising results in terms of structural strength and weight savings.Its application contributes to 20.75% increase in the strength-to-weight ratio. Furthermore, debonding damages the modal parameter and reduces structural stiffness. Figure 1 depicts the comparative results from three different locations. Several parameter scenarios were studied from the above literature, including the type of sandwich core material, type of loading, thickness configuration, and variations in the number of reinforcements on the ship's strength. From a technical standpoint, using sandwich panels on ship structures increases the structure's strength-to-weight ratio, which varies depending on the structure location, modification configuration, and sandwich material. The use of sandwich plates on the ship allows for a reduction in the amount of bracing attached to the structure. The structure's simplification allows the structure's weight to be reduced while maintaining the structure's strength. Because of the reduced volume of welding work, structural simplification can simplify the manufacturing process. The weight reduction is expected to increase loading capacity and thus improve the operation of sea transportation modes.

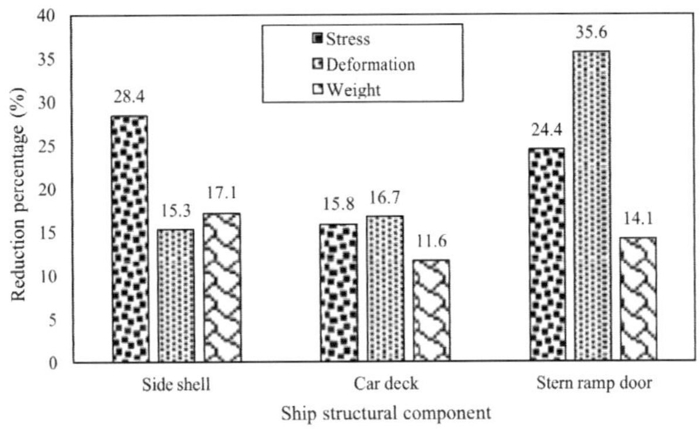

Figure 1. Comparative results of stress, deformation, and weight of three different structure locations.

3. Modelling Description of Application of Sandwich Panel

3.1 Construction System

In general, ship construction is a structural arrangement that strengthens the ship. There are three types of construction on ships: longitudinal, transverse, and mixed. The construction system distributes the load on the transverse members of the ship with the help of the longitudinal structure of the ship. The longitudinal construction system has the main reinforcement installed longitudinally. Based on the regulations of the Indonesian Classification Bureau (BKI) [19], this construction system is recommended on ships with a length of more than 100 m. The transverse construction system is a ship construction system that distributes the load transversely. Transverse construction systems have transverse main reinforcements such as web frames, main frames, deck beams, and strong beams and are generally suitable for ships with a length of less than 90 m. The mixed construction system is a combination of a transverse construction system and a longitudinal construction system that is used simultaneously on the deck and base of the ship operating a longitudinal construction while on the sides using a transverse construction system. Based on BKI, this construction system is recommended for ships with a length of 90 m < L < 100 m [19].

In construction design variations, it is necessary to design a cross and mixed construction systemto determine the profile size to be used. Calculation of the size of the profile according to BKI then it is necessary to calculate the size of the profiles such as web frame, main frame, stinger, deck beam, strong beam, and a floor that will be used in the transverse construction system. The mixed profile construction system used on the deck and bottom of the ship uses the same profile as the longitudinal construction system, and the sides will use the same frame and web frame as the transverse construction system. Before determining the profile size to be used, it is necessary to calculate the value of the load oneach construction under BKI [19].

3.2 Ship Data Particular

A barge is a ship used to transport dry and liquid bulk. In technological developments, an engineeris always required to provide the latest innovations to advance the shipping world. To increase the strength and capacity of the ship's cargo, there are innovations in the material used, namelysandwich material. Sandwich material is a combination of plates separated by a lightweight core material where this combination will provide an efficient ratio of strength and stiffness to weight [20]. The data collected is cargo barge image data in the form of general arrangement and construction profile which can be seen in Figure 2. The main size of the deck cargo barge ship with a size of LOA x B x T of 155 m x 45 m x 6.1 m can be seen in Table 1.

Table 1. The main dimension of a barge.

Main particular	Value
Length Overall (LOA)	155 m
Length Perpendicular (LPP)	154 m
Breadth (B)	45 m
Depth (H)	9.5 m
Draft (T)	6.1 m

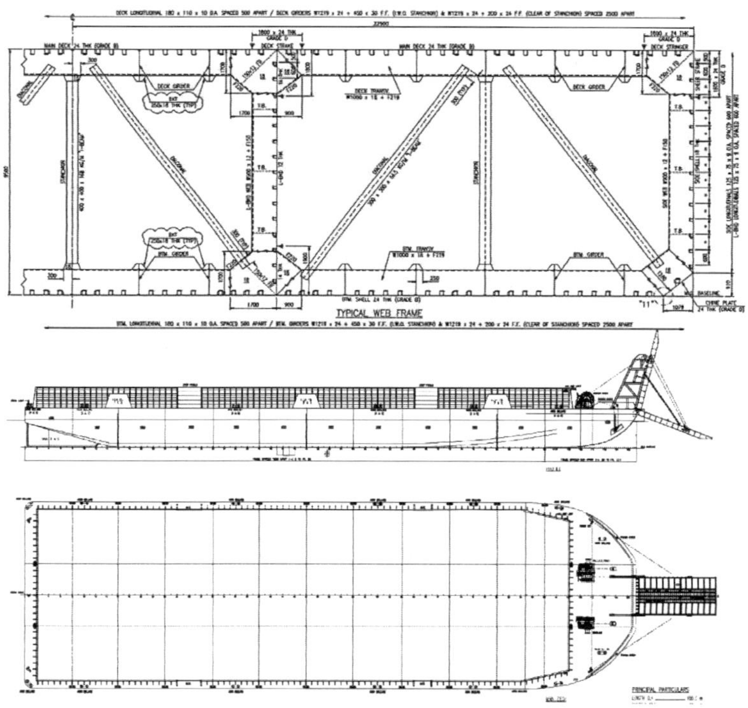

Figure 2. Midship section of 155 m barge.

3.3 The procedure of Sandwich Panel Application Based on Lloyd's Register

Using the regulations issued by Lloyd's Register, modifications to the ship's construction can be done by applying for a sandwich plate so that the stresses can meet the classification criteria. In the first stage, the value of the cross-sectional modulus of the existing ship is calculated. The value of the cross-sectional modulus to the deck and bottom of the vessel must be greater than the minimum modulusvalue specified. Based on the calculations that have been carried out, the value of the cross-sectional modulus of the existing ship is 16,134,463.833 cm^3.

Calculating sandwich plate configuration using Lloyd Register rules for applying sandwich panel construction to ship structure. The sandwich plate calculation procedure is carried out according to the grooves shown in Figure 2. In the work process, the calculation is carried out using the assumption of new building construction, which is in accordance with the provisions contained in the Lloyd Register [7], the sandwich plate has a minimum core thickness of 15 mm, and the minimum thickness of the upper and lower faceplates can be seen in Table 2.

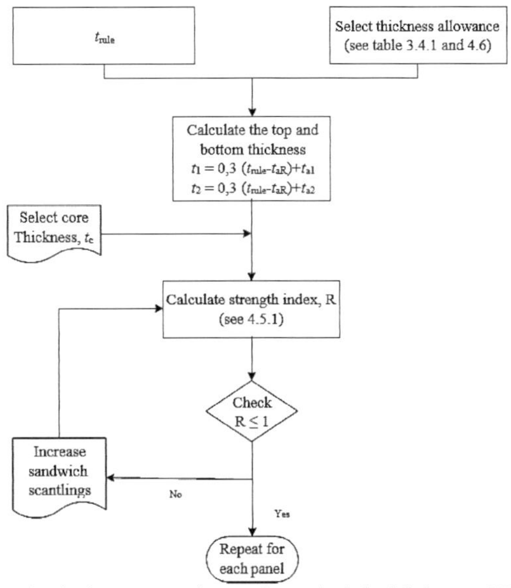

Figure 3. Calculation procedure of sandwich thickness [7].

Table 2. Minimum faceplate thickness.

Items	Minimum thickness in mm	
	New construction	Overlay construction
t_{1MIN}	3.0	3.0
t_{2MIN}	3.0	50% of the as-built thickness

Determination of the thickness of the sandwich plate is calculated according to the equation in Figure 3. To get the thickness of the sandwich plate to be used in numerical analysis. The results of the calculation of the strength index value can be seen in Table 3. For new building configurations, the faceplate thickness value begins by entering the minimum number, which is 3 mm. The sandwich plate configuration used on the bottom and main deck uses a faceplate thickness of 9 mm and a core of 16 mm,while the side shell uses a faceplate and core thickness of 10 mm and 16 mm.

Table 3. Strength index calculation.

Components	t_c (mm)	a (mm)	b (mm)	A_R	$Z\,rule$	s	$PeqR$ (MPa)	Face Plate t_1	t_2	k	d (mm)	R
Bottom plate	16	6000	722	3.96	1581207	500	0.1075	9	9	1	25	0.627
Side shell	16	6000	600	4.47	158120	600	0.0746	10	10	1	26	0.309
Deck plate	16	6000	600	4.47	158120	500	0.1075	9	9	1	25	0.509

3.4 Description of Finite Element Modelling

The ship used as a reference in this research was a barge ship. Three developed models with different construction systems will be compared with the existing model. In this case, the free vibration analyses are carried out within ABAQUS using the linear perturbation load step and the Lanczos iteration method for eigenvalue extraction [21]. Comparisons of dynamic responses between existing and proposed models are used to assess the effects of sandwich panel addition on modal characteristics. To obtain the natural frequencies and mode shapes of the models, the commercial finite element code ABAQUS/Standard is used. The finite element (FE) simulation procedure is divided into three steps: (1) preprocessing, which includes modelling the geometry of the structures, meshing, and assembling the mass and stiffness matrices; (2) a linear perturbation analysis step, and the frequency is extracted using the Lanczos solver. (3) The natural frequencies and mode shapes are obtained using the post-processing module.

The layer-wise solid/shell element can be used to simulate the side hull in the discretization of the finite element model. LR notices that shell elements are used to interpret the faceplate layers, and solid elements are used to interpret the core material [16]. The eight-node quadrilateral shell element (SC8R) was used to simulate the faceplate layers, while the eight-node hexahedral element was used to simulate the core material (3D8I). Shell elements were used to model the stiffeners, which included the mainframe, web frame, side longitudinal, and side stringer (SC8R). Tie constraints modelled contact modelling between all parts of the structure. Meanwhile, the boundary condition should be organized so that it could be the same as the real condition. Pinned support in the rear and front of models, X SYMM (Symmetry in X-axis) in the centre of the model, and free in the side model were applied. The boundary condition is depicted in Figure 4.

Figure 4. Calculation procedure of sandwich thickness [7].

Meshing is a simulation process of dividing components into smaller elements. The meshing process affects the results' accuracy and the simulation's speed. The smaller the size used in the meshingprocess, the longer it will take and get more accurate results. On the other hand, using a larger meshing size will take a shorter time and get less accurate results. Figure 5 shows the results of meshing using astructured mesh type. In this study, the convergence analysis was carried out using a natural frequency variable where the variable must meet the convergence requirements to declare the results valid. To achieve convergence, simulations were carried out on five mesh element sizes starting from 0.5-0.1 m. Convergence has been achieved with the number of mesh elements as much as 741,528 and has a margin of error on the element size 0.1 m of 4.5%. Material properties used for analysis are presented in Table 4. In this study, the barge construction was replaced by a sandwich plate system (SPS) with different thicknesses. The SPS model consists of a steel faceplate and FRPU core material.

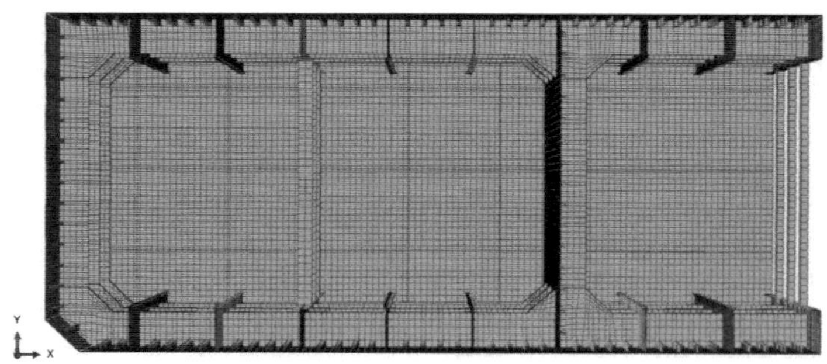

Figure 5. Meshing on existing barge structure.

Table 4. Material properties of faceplate and core sandwich.

Materials	Density (kg/m$^{3)}$	Young Modulus (MPa)	Poisson ratio
Face plate (steel)	7850	206000	0.30
Core (fibreglass reinforced polyurethane/FRPU)	1098	901.95	0.36

4. Result and Discussion

4.1 Comparison of Structural Weight Under Different Framing Systems

Three types of construction system variations are used: longitudinal construction systems, transverse construction systems and mixed construction systems. Each variation will be made into a different model. Variations of the longitudinal construction system use the same construction as the existingship, but there is a change in the structural elements on the deck plate, side and bottom of the ship into sandwich plates. Figure 6a is a 3D model of a ship using a sandwich plate in a longitudinal construction. Then, to model the variation of the transverse construction system, the reinforcing structural elements on the side of the ship were originally composed of longitudinal and transverse sides replaced with frames, web frames and side stringers. Meanwhile, the deck, which originally consisted of longitudinal decks, transverse decks, side girders and centre girders, was replaced with beams, strong beams, side girders, and centre girders. Furthermore, the basic part of the geometry model is replaced by the reinforcing structural elements, which originally consisted of the bottom longitudinal, bottom transverse, side girder, and centre girder to become the floor, side girder and centre girder. Figure 6b is a 3D model ofa geometric structure modelled using a sandwich plate in a transverse construction. In the variation of the mixed construction system made on the geometric structure, the sides have the same reinforcing element structure as the transverse construction. The structure of the reinforcing elements on the sides consists of the main frame, web frame, and side stringer, while the bottom and deck have the same reinforcing element structure as the longitudinal construction system consisting of bottom longitudinal, longitudinal deck, bottom transverse, deck transverse, bottom centre girder. Bottom side girder, deck centre girder, deck side girder and pillar. Figure 6c is a 3D model of the geometric structure of a mixedconstruction system modelled using a sandwich plate.

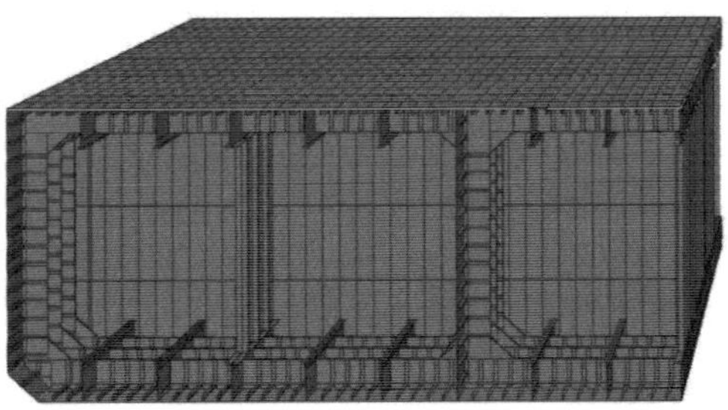

(a)

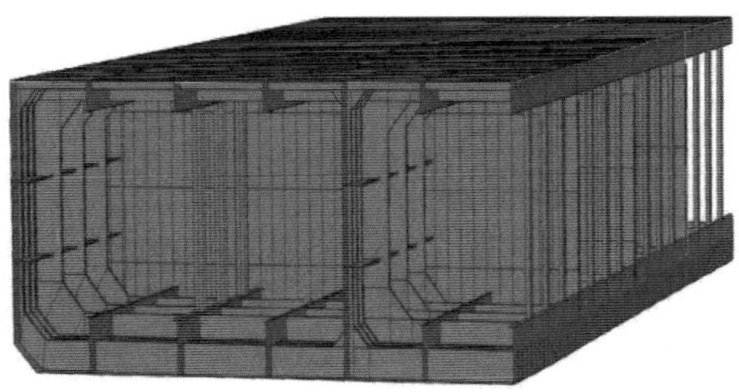

(b)

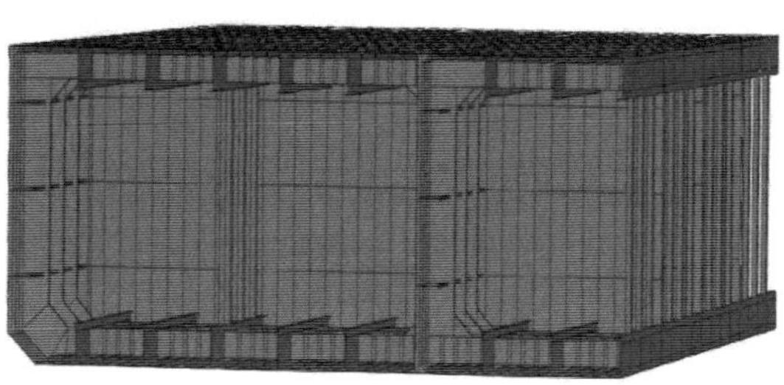

(c)

Figure 6. 3D construction system a) longitudinal framing system, b) transverse framing system, c) mixed framing system.

IOP Conf. Series: Earth and Environmental Science 1166 (2023) 012006 doi:10.1088/1755-1315/1166/1/012006

To compensate for the increased stress and deformation obtained, the application of sandwich plates in place of conventional steel plates reduces the structure's weight which can be used to increase theload carried. Weight is calculated for all structural elements contained in the model by multiplying the volume value by their respective density. The density of steel is 7.85 tons/m^3, while the density of polyurethane core is 1.098 tons/m^3. Weight comparison can be seen in Figure 7 that the sandwich modelwith a longitudinal construction system experienced a weight loss of 5% from the existing ship model. The weight reduction in the longitudinal sandwich model is not significant because the analysis is carried out globally by only converting the ship's steel plates of the deck, bottom, and sides into sandwich plates without changing the construction structure. Furthermore, it can be seen that the transverse ship construction system and the mixed construction system resulted in a lighter weight of 13% and 9% thanthe existing ship model. Both of these can occur due to differences in the use of profile sizes from the longitudinal construction system.

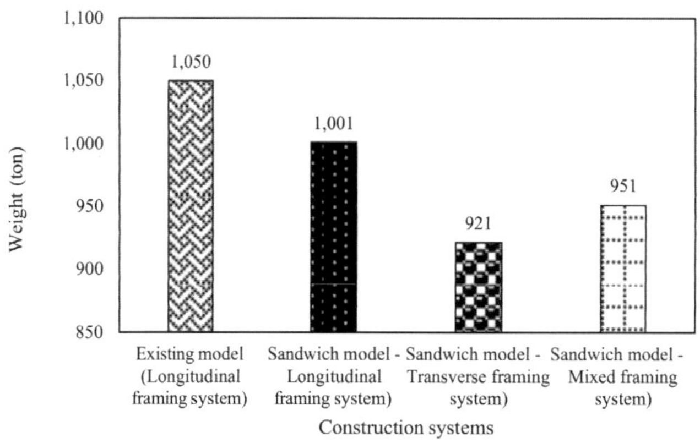

Figure 7. Comparison of structural weight under different framing systems.

4.2 Comparison of Natural Frequency Under Different Framing Systems

In this section, the finite element results are presented to thoroughly observe the comparison of natural frequency due to the addition of the sandwich panel. The result of free vibration analysis is conducted to measure the stiffness of different construction systems. The comparison of the first six natural frequencies for different models is presented in Table 5. It can be found from the result that the highest natural frequency can be found in the existing model before applying the sandwich panel. In contrast, the sandwich with a transverse framing system experiences the lowest natural frequency compared to others. When subjected to certain external forces, these are the frequencies at which the structure will tend to vibrate. These frequencies are determined by the structure's distribution of mass and stiffness. From Figure 8, it can be evaluated that small frequency discrepancies are found in the first three mode numbers. The frequency of all models tends to have similar values. However, the higher differences in natural frequency are found in higher modes (mode numbers 4 to 6).

Table 5. Comparison of the first six natural frequencies under different framing systems.

Mode Number	Natural Frequency (Hz)			
	Existing - Longitudinal System	Sandwich - Longitudinal System	Sandwich - Mixed System	Sandwich - Transverse System
1	2075.27	2065.50	2061.16	2056.50
2	2092.97	2083.85	2082.43	2079.40
3	2096.53	2086.64	2090.33	2088.40
4	2467.99	2401.60	2328.61	2290.70
5	2486.54	2419.70	2348.89	2302.90
6	2489.68	2439.25	2390.25	2345.90

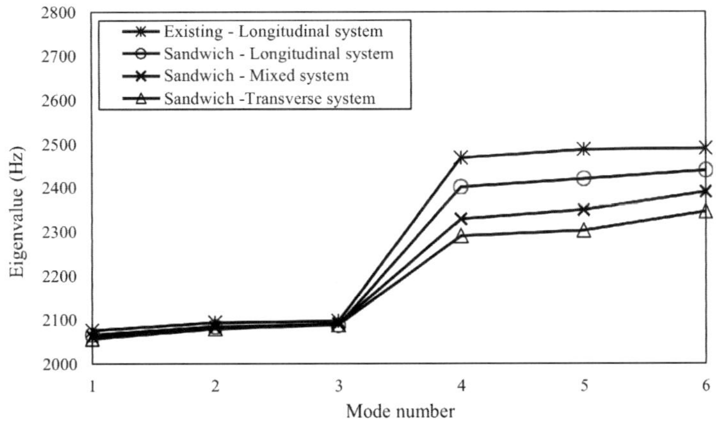

Figure 8. Comparison of natural frequency under different framing systems.

Figure 9 compares the displacement contour of the first mode shape under different framing systems. It can be found that local deformation is located in the same location in the transverse bulkhead. It is caused by the transverse bulkhead plate having the lowest thickness compared to the deck, side and bottom plates.

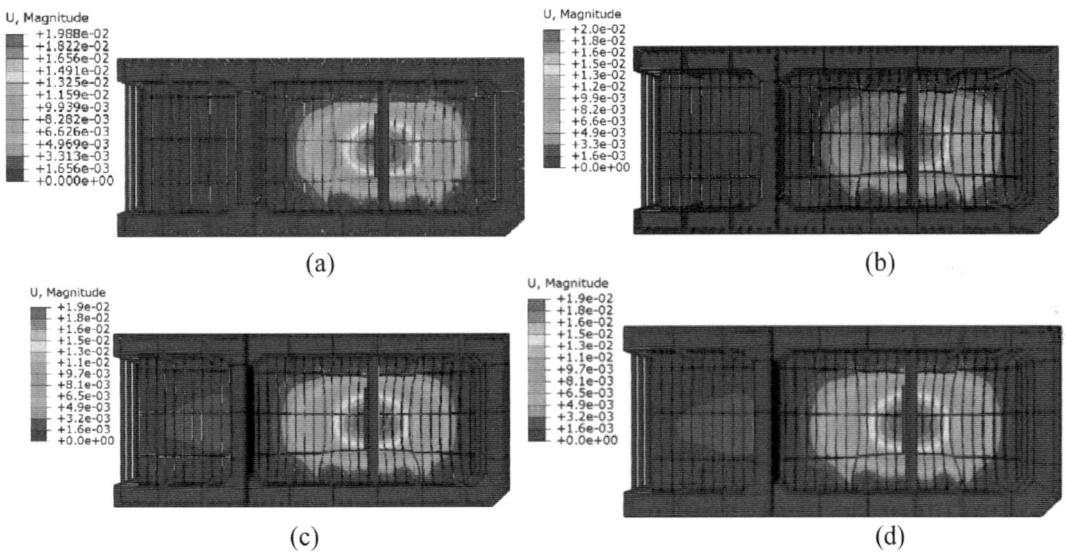

Figure 9. Comparison of displacement on first mode shape under different framing systems.

5. Conclusion

The study is conducted to evaluate the dynamic response of different construction systems of barge ships due to applying sandwich panels in different local structures such as the deck, ship hull, and bottom structure of barge ships using Lloyd's Register standard. A total of three proposed construction systems such as longitudinal, transverse and mixed framing systems, are investigated using free vibration analysis using ABAQUS software. The preliminary study indicates promising sandwich panel application results in a weight saving of about 5-13%. The sandwich panel can reduce the stiffener application to make a simpler construction system. The application of a sandwich panel can reduce the natural frequency. It can be found that the highest natural frequency can be found in the existing model and the sandwich with a transverse framing system experiences the lowest natural frequency compared to others.

Acknowledgements
The research leading to these results has received financial support from "Penelitian Strategis" research scheme, RKAT Faculty of Engineering, Universitas Diponegoro, with contract number 356/UN7.5.3.2/HK/2022.

References

1. Prabowo, AR, Tuswan T, Ridwan R 2021 *Appl. Sci.* **11**(9) 3954
2. Ismail A, Zubaydi A, Piscesa B, Tuswan T, Chandra A R 2021 *J. Eng. Appl. Sci*, **19**(4), 1090-1098.
3. Borsellino C, Calabrese L, Valenza A 2004 *Compos. Sci. Technol.* **64** 1709–1715
4. S. CORe 2013 Best Practice Guide for Sandwich Structures in Marine Applications 279.
5. Noury P, Hayman B, McGeorge D, Weitzenbock J 2002 Lightweight construction for advanced shipbuilding-recent development, *Proceedings of the 37th WEGEMT Summer School 11–15*
6. Palomba G, Epasto G, Crupi V 2021 *Mech. Adv. Mater. Struct.*
7. Llyod's Register 2015 Provisional Rules for The Application of Sandwich Panel Construction to Ship Structure UK
8. Brooking, M., Kennedy, S., 2004. The performance, safety and production benefits of SPS structures for double hull tankers. *Proceedings of the RINA Conference on Double Hull Tankers*
9. Ismail A, Zubaydi A, Piscesa B, Panangian E, Ariesta RC, Tuswan 2020 *IOP Conf. Ser.: Mater. Sci. Eng.* **1034** 012026
10. Ismail A, Zubaydi A, Piscesa B, Panangian E, Ariesta RC, Tuswan 2020 *IOP Conf. Ser.: Mater. Sci. Eng.* **1034** 012027
11. Momcilovic N, Motok M 2009 Estimation of Ship Lightweight Reduction by Means of application of Sandwich Plate System. Faculty of Mechanical Engineering, University of Belgrade, Serbia
12. Tuswan, Zubaydi A, Budipriyanto A, Sujiatanti SH 2018 *SciTePress* **1** 87-96
13. Tuswan, Zubaydi A, Piscesa B, Ismail A 2020 *Open Eng.* **10** 424-433
14. Tuswan, Abdullah K, Zubaydi A, Budipriyanto A 2019 *Mater. Today: Proc.* **13**(1) 109–111
15. Tuswan, Zubaydi A, Piscesa B, Ismail A, Ilham MF 2020 *Procedia Struct. Integr.* **27** 22-29
16. Tuswan, Zubaydi A, Piscesa B, Ismail A, Ariesta RC, Ilham MF, Mualim FI 2020 *J Appl Eng Sci.* **19**(1), 208-216
17. Ariesta RC, Zubaydi A, Ismail A, Al-Syachri MZ 2021 *IOP Conf. Ser.: Mater. Sci. Eng.* ***1034*** 012012
18. Ariesta RC, Zubaydi A, Ismail A, Tuswan T 2021. *Mater. Today: Proc.* **47**(10) 2310-2314
19. BKI 2015 Volume II Rules for Hull – Inland Waterways Edition 2015
20. Mantari JL. Granados, EV 2015 Compos. B: Eng. **69** 317–334.
21. ABAQUS 2016 Abaqus Analysis User Guide, Simulia

IOP Conf. Series: Earth and Environmental Science 1166 (2023) 012007 doi:10.1088/1755-1315/1166/1/012007

The Development of Ship Recycling Facility Cluster in Indonesia

Hariyanto[1], N I Gutami[1], Z Qonita[2], N Shabrina[1], D A Wibowo[1], A Muhtadi[1]

[1] The National Research and Innovation Agency of Republic Indonesia, Research Centre for Transportation Technology, PUSPIPTEK Serpong, Banten 15314, Indonesia
[2] The National Research and Innovation Agency of Republic Indonesia, Geological Disaster Research Centre, Cisitu, Bandung, West Java 40135, Indonesia

Corresponding author: hariyanto1708@gmail.com

Abstract. With a number of 80.849 units ship operating in Indonesia, the Ministry of Transportation estimates that 29,944 units ship of those vessels have been in service for more than 20 years. Furthermore, recycling ships are necessary. Ship Recycling Facility (SRF) should be ecologically friendly following International Maritime Organization (IMO) policy under the Hong Kong International Convention. Docks, cranes, steel cutters, and warehousing are necessary. SRF facilities, along with ecologically suitable hazardous material treatment. If this facility became a cluster, it would operate more effectively. This study aims to provide a different area in Indonesia where a prospective SRF cluster could grow. This research uses Diamond Porter's method to determine a country's comparative economic advantage towards the SRF cluster in Alang City, Gujarat, India, which is the global leader in ship recycling. The factor conditions; related and supporting industries; demand conditions; and firm strategy, structure, and rivalry are all examined in this study. To accommodate ship owners, the SRF business, and the government in carrying out ship recycling, the results of this study are a reference for formulating specific regulations for the SRF cluster. This research found that Kamal and Bojonegara are viable for the SRF cluster.

1. Introduction

As an archipelagic country, Indonesia has a relatively high number of ship fleets. The Directorate General of Sea Transportation, Ministry of the Transportation Republic of Indonesia, reported that until 2019, there were a total of 80,849 units ship Indonesian ships ranging in age from 0 to more than 50 years [1]. With such a large number, some of these ships should be stop operating or recycled because they are no longer safe.

Ships aged over 20 years, tend to no longer be able to operate optimally. Based on the statistics from Det Norske Veritas (DNV), passenger ships and general cargo carriers over 25 years old are no longer guaranteed to be safe or have frequent accidents [2].

As a result, Indonesia's fleet of ships over 20 years old ought to be recycled immediately. The ship fleet's statistics are shown below, which shows that there are 29,994 units ship or 37% of Indonesia's ship fleet are older than 20 years.

Figure 1. Age distribution of Indonesia's total fleet until 2019

1.1. The Hong Kong Convention

Following the International Maritime Organization's (IMO) The Hong Kong Convention regulation when recycling ships, The Hong Kong Convention strives to give management of hazardous materials more consideration. The Hong Kong Convention identifies and measures the potential release of contaminants into any medium and their consequences, focusing on the severe environmental effects of releases. Large-scale releases and the discharge of more dangerous pollutants, which are expected to have the biggest effects, have likely received the majority of attention [3].

1.2. Ship Recycling Facility (SRF) Cluster

Due to the high demand for ship recycling, several SRFs have obtained authorization from the Indonesian government, specifically the Directorate General of Sea Transportation, Ministry of Transportation of the Republic of Indonesia, and are located throughout Indonesia. Facilities for storing and treating waste are required to recycle ships in a safe and environmentally friendly manner in accordance with IMO rules. This waste treatment facility will be more cost-effective if it is in the form of an SRF cluster, which does not yet exist in Indonesia and is the research's goal.

Table 1. Ship Recycling Facility (SRF) Companies in Indonesia

No.	SRF Company	Location
1.	PT. Bahtera Bahari Shipyard, PT. Batamitra Sejahtera, PT. Bumi Natura Indonesia, PT. Citra Shipyard, PT. Karya Sumber Daya, PT. Marcopolo Shipyard, PT. Sentek Indonesia, PT. Kumala Indonesia Shipyard, PT. Batam Expressindo Shipyard, PT. Multi Ocean Shipyard, PT. Harapan Maritim Dunia	Batam, Riau Islands
2.	PT. Marianna Bahagia	Banyuasin, South Sumatera
3.	PT. Arjaya Berkah Marine, PT. Dock Pulo Ampel, PT. Samudera Marine Indonesia, PT. Harapan Teknik Shipyard, PT. Samugara Multikarya Utama, PT. Anugrah Kasih Agape	Java Island
4.	CV. Bahagia Abadi	Hative Besar, Ambon

The SRF cluster is a facility for recycling ships into recycled products that are still valuable. Ship recycling is not the same as ship breaking or ship dismantling. In ship recycling most of the ship's parts, components, and even inventory are reused or recycled. This SRF product is a fundamental material used in the metal industry. As shown in the figure below, the SRF product is 75% metal, 20% machinery, and 5% waste. So the majority of the 95% is in the form of recycled products that are still beneficial, while the remaining 5% is disposed of in the form of waste [4].

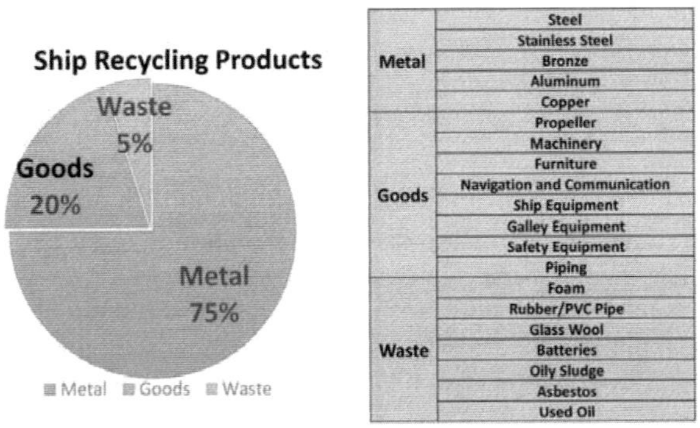

Figure 2. Ship recycling products [5]

Positive economies of scale are provided by clustering in personal and inter-organizational networks (also known as positive externalities) [6]. As a result, the purpose of this research is to make a recommendation, for the SRF in the form of a cluster, resulting in better economy because all industries collaborate. With this economic value, we no longer need to recycle abroad, and we could become scrap steel suppliers, saving the country $400,000,000 in foreign exchange [7].

2. Method

Methodology used in this study is the Porter's Diamond Method, which includes demand conditions; related and supporting industries; factor conditions; and firm strategy, structure, and rivalry.

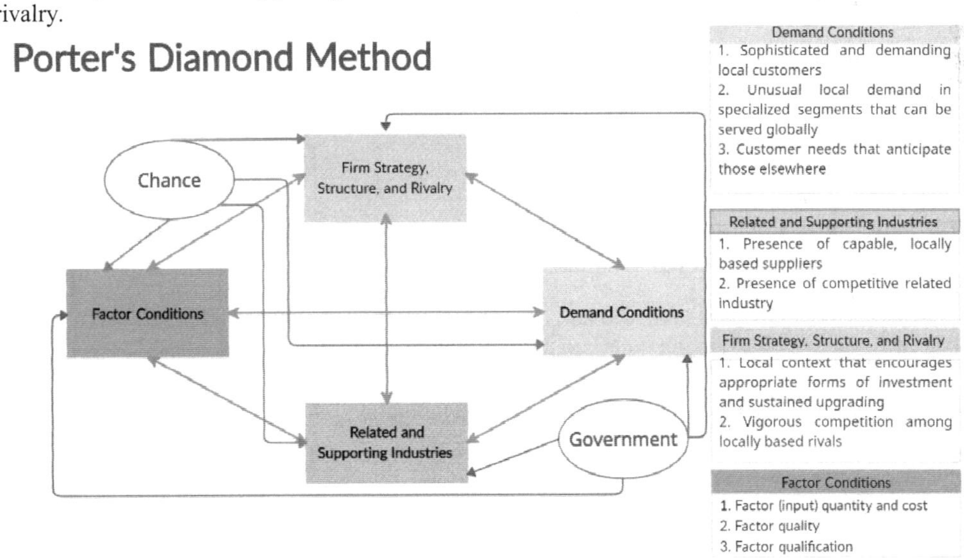

Figure 3. Porter's Diamond Method

2.1. Demand Conditions

Demand conditions in Porter's Diamond Model refer to home market demand for a specific product or service. These are determined by customer demographics, the size of customer segments, and their distribution. High demand forces a company to constantly improve and innovate its product line-up in order to gain a competitive advantage [8].

2.2. Related and Supported Industries

According to Porter's Diamond Theory, the success of one industry is dependent on the success of other industries. Supply chains, exchange of ideas, and resource exchange can all help an organization's market position and create a competitive advantage [9].

2.3. Factor Conditions

In the Diamond Model, actor conditions include the various resources available to a nation or organization. These are classified into two types: basic and advanced. Basic resources include land, natural resources, and unskilled labour, whereas advanced resources include capital, infrastructure, and cutting-edge technology—resources that a nation or organization can develop for themselves. According to Porter, advanced resources are more important than basic resources in establishing a competitive advantage [8].

2.4. Firm Strategy, Structure, and Rivalry

This theory is based on how organizations structure and strategize. Economic, social, and political factors frequently influence strategy and structure. These, together with rivalry, determine an organization's competitive position in the market. Rivalry motivates organizations to improve their productivity, product quality, and efficiency. The greater the rivalry, the greater an organization's desire to distinguish itself [10].

Porter also included the role of government and the chance factor as supporting factors. Porter did not specify whether the nature of these two variables was positive or negative. The government's role is to implement policies and regulations that benefit the growth of domestic industry, thereby indirectly increasing the country's competitiveness. This research aims to look into the feasibility of establishing SRF clusters in various locations throughout Indonesia. By using the Porter's Diamond Method approach, if we do a ship recycling in Indonesia by paying attention to the environment and safety, it has the potential to save the state budget because it can meet the needs of national steel imports, according to the Central Statistics Agency reached US$ 11.96 billion in 2021 [11].

3. Result and Discussion

The following findings and discussion are based on Porter's Diamond Method comparative studies of the situations in Alang, India, with Indonesia.

3.1. Demand Conditions

Because domestic demand for steel is high. Indonesia frequently imports steel. Several steel-importing countries, namely Ukraine and Russia, are at war, which can be used to push companies to seek alternatives from domestic suppliers [12].

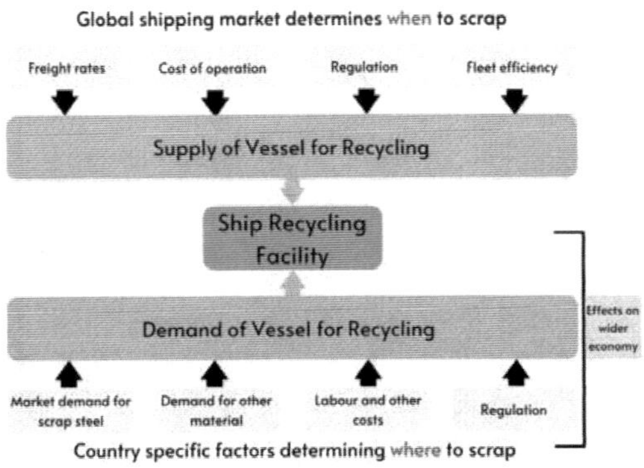

Figure 4. Ship recycling demand and supply [6]

3.1.1. Market Demand

The estimated Industrial Natural Resource Demand for steel per year is 28 million tons of steel [13], 1/3 of which is 9 million scrap steel, and 90% was imported from abroad. The country's 5 million tons of steel shortage will be replaced by recycled products from this SRF cluster. According to UNCTAD data in the Maritime Transport Review 2020 (Figure 5) and 2021 (Figure 6) [14], many ships from other countries are available to be recycled. We can recycle all of these ships in Indonesia, meet domestic steel needs, and even export them by developing this SRF cluster.

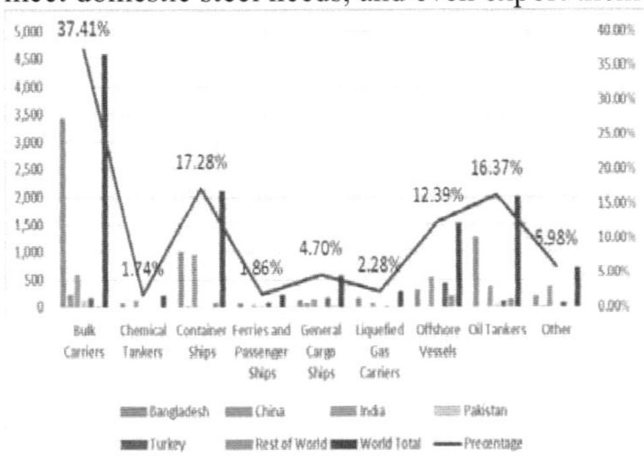

Figure 5. Reported tonnage sold for ship recycling by major vessel type and country of ship recycling, 2019 (Thousand gross tons)

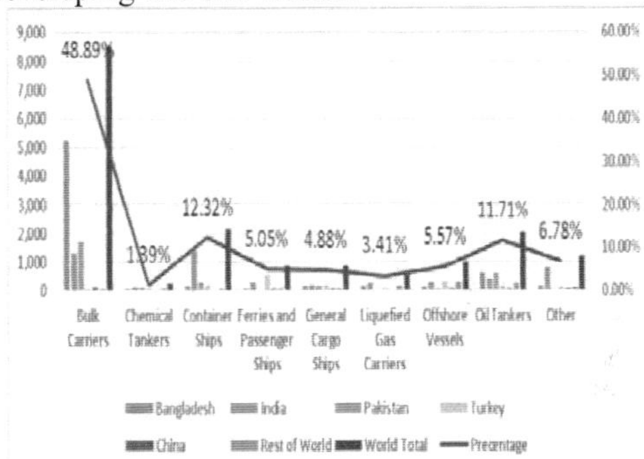

Figure 6. Reported tonnage sold for ship recycling by major vessel type and country of ship recycling, 2020 (Thousand gross tons)

3.1.2. Regulation

- Government of the Republic of Indonesia, Regulation Number 31 of 2021 concerning the Implementation of the Shipping Sector Article 137 paragraph (3);
- Republic of Indonesia, Law No. 17 of 2008 concerning Shipping, Chapter XII, articles 241-242;
- Government of the Republic of Indonesia, Regulation Number 31 of 2020 concerning the Implementation of the Shipping Sector;
- Government of the Republic of Indonesia, Regulation Number 21 of 2010 concerning Maritime Environment Protection, article 2; and
- Minister of Transportation of the Republic of Indonesia, Regulation Number 29 of 2014 concerning Prevention of Pollution of the Maritime Environment, Chapter II Part Two Paragraph 4 Article 51-55.

3.2. Related and Supported Industries

The data shows that the distribution of the steel industry and waste treatment is quite adequate in the area around East Java, as well as in Banten. There are 28 steel companies and 10 waste treatment industries in East Java. [15].

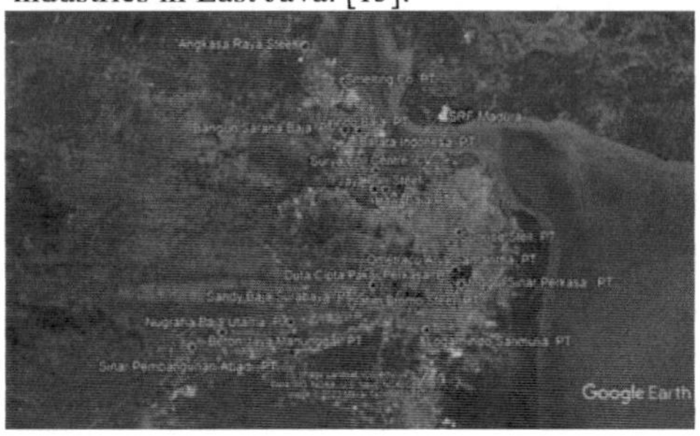

Figure 7. Distribution map of steel industries and companies around the SRF Madura plan with a distance of 25.4 – 65 km

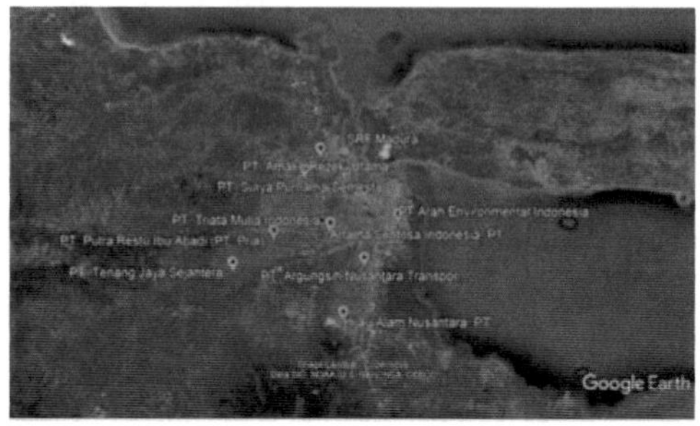

Figure 8. Distribution map of waste management industries and companies around the SRF Madura plan with a distance of 26 – 77 km

Table 2. Steel Industries in East Java [15]

No.	Steel Company	Commodity	Distance (km)	No.	Steel Company	Commodity	Distance (km)
1.	Alim Ampuh Jaya Steel, PT	Spiral Pipe of Steel	37.5	**2.**	Hasil Karya, PT	Steel And Concrete	25.4
3.	Gunawan Dian Jaya Steel Tbk	Steel	31.9	**4.**	Lintech Duta Pratama, PT	Steel Construction Services	37.7
5.	Angkasa Raya Steel, PT	Steel And Iron Pipe	49.7	**6.**	Logamindo Sarimulia, PT	Cast Iron And Steel Components	39.7
7.	Baja Utama Danastri, PT	Stainless Steel Assembly	33.3	**8.**	Nugraha Baja Utama, PT	Plain Reinforcement Steel	63
9.	Bangun Sarana Baja, PT	Steel Fabrication	33.8	**10.**	Ometraco Arya Samantha,	Steel & Plate	35.3
11.	Barata Indonesa, PT	Steel	38.5	**12.**	Sandy Baja Surabaya, PT	Steel	41.3
13.	Beton Jaya Manunggal, PT	Concrete Reinforcement Steel	55.7	**14.**	Sinar Pembangunan Abadi, PT	Steel	65.7
15.	Bukit Baja Anugrah, PT	Steel Pipe	34.8	**16.**	Smelting Co, PT	Steel	45.9
17.	Duta Cipta Pakar Perkasa, PT	Steel Structure	30.3	**18.**	Spindo, PT	Steel Pipe	26.2
19.	Hawi Baja PT	Bendrat Wire	34.5	**20.**	Sunrise Stell, PT	Plated Steel	26.1
21.	Indo Baja, PT	Steel	41	**22.**	Surya Coil Centre, CV	Coil/Steel Cutting Services	32.9
23.	Jatim Bromo Steel, PT	Steel	49.9	**24.**	Tunggal Jaya Steel, PT	Steel	58.6

No.	Steel Company	Commodity	Distance (km)	No.	Steel Company	Commodity	Distance (km)
25.	Jatim Mustika Sarana Steel, PT	Steel Construction Services	34.4	**26.**	Unggul Sinar Perkasa , PT	Steel	36.5
27.	Jatim Taman Steel Mfg Co, PT	Steel Billets	45.5	**28.**	Jaya Pari Steel Corp Ltd, PT	Steel Plate Industry	32.9

Table 3. Waste Treatment Industries in East Java

No.	Waste Treatment Industry	Commodity	Distance (km)	No.	Waste Treatment Industry	Commodity	Distance (km)
1.	Artama Sentosa Indonesia, PT	Waste	55.3	**2.**	PT. Amako Rezeki Utama	Hazardous and Toxic Waste	46.8
3.	PT Arah Environmental Indonesia	Hazardous and Toxic Waste	31.1	**4.**	PT. Surya Purnama Semesta	Waste	56.6
5.	PT. Putra Restu Ibu Abadi (PT. Pria)	Hazardous and Toxic Waste	82.6	**6.**	PT. Tenang Jaya Sejahtera	Waste	79.5
7.	Hijau Alam Nusantara, PT	Hazardous and Toxic Waste	77	**8.**	PT. Ardi Restu Arta	Waste	56.2
9.	PT. Triata Mulia Indonesia	Hazardous and Toxic Waste	26	**10.**	PT. Argungsih Nusantara Transpor	Hazardous and Toxic Waste	53.7

There are 12 steel companies and 10 waste treatment industries in Banten [16]. These facilities demonstrate that the two areas are suitable for SRF cluster development.

Figure 9. Distribution map of steel industries and companies around the Cilegon SRF plan with a distance of 7.2 – 25.2 km

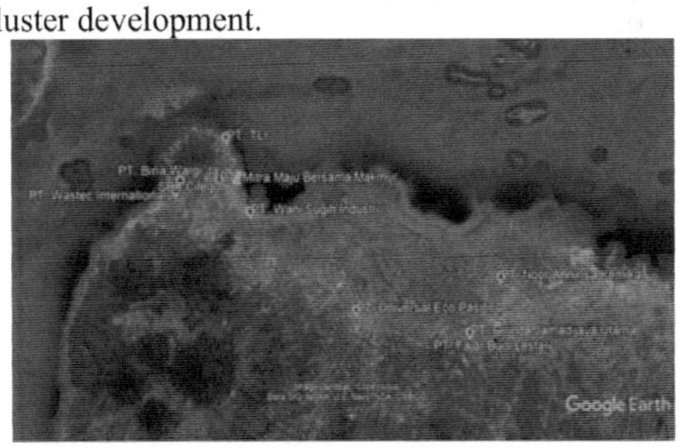

Figure 10. Distribution map of waste management industries and companies around the Cilegon SRF plan with a distance of 3.7– 75.2 km

Table 4. Steel Industries in Banten [16]

No.	Steel Company	Commodity	Distance (km)	No.	Steel Company	Commodity	Distance (km)
1.	Aneka Baja Perkasa Industry	Steel Wire	50.6	**2.**	Krakatau Osaka Steel, PT	Steel	16.6

No.	Steel Company	Commodity	Distance (km)	No.	Steel Company	Commodity	Distance (km)
3.	Cakrawala Panopa Baja, PT	Steel Triangle	23.1	4.	Krakatau Wajatama, PT	Reinforcing steel	11.4
5.	Central Berkat Indonesia, PT	Steel Construction Services	14.8	6.	Multi Fabrindo Gemilang, PT	Steel Construction	15.2
7.	Indofero, PT	Sreel	25.3	8.	Purna Baja Heckety, PT	Scrap Iron Service	15.8
9.	Karunia Berca Indonesia, PT	Steel frame	14.9	10.	Panca Baja Cilegon	Steel distributor	7.2
11.	Karya Logam Agung, PT	Steel Construction	48.2	12.	Steelmaking Plant PT. Krakatau Posco	Steelwork manufacture r	17.7

Table 5. Waste Treatment Industries in Banten

No.	Waste Treatment Industry	Commodity	Distance (km)	No.	Waste Treatment Industry	Commodity	Distance (km)
1.	PT. Mitra Maju Bersama Makmur	Hazardous and Toxic Waste Collector	3.7	2.	PT. TLI (Teknotama Lingkungan Internusa)	Hazardous and Toxic Waste Treatment	8.9
3.	PT. Wani Sugih Industri	Hazardous and Toxic Waste Collector	12.9	4.	PT. Fajar Budi Lestari	Hazardous and Toxic Waste Treatment	71.2
5.	PT. Universal Eco Pasific	Hazardous and Toxic Waste Collector	50.4	6.	PT. Noor Anninsa Kemikal	Hazardous and Toxic Waste Treatment	73.4
7.	PT. Wastec International	Hazardous and Toxic Waste Collector	15.2	8.	PT. Bina Wargi Jaya	Hazardous and Toxic Waste Treatment	12.8
9.	PT. Sinar Laut Biru Logam Perkasa Jaya	Hazardous and Toxic Waste Collector	75.2	10.	PT. Girinda Jamadijaya Utama	Hazardous and Toxic Waste Treatment	67.5

3.3. Factor Conditions

3.3.1. Water Conditions in Alang, Bhavnagar, Gujarat, India

Alang's geography is distinct, with high tidal ranges, a broad continental shelf, a 15-degree slope, and mud-free beaches, making it ideal for ships of any size to become stranded at high tide [17]. The tides are strong in Alang. This allows the ship to go to shore at high tide and then dock when the water recedes. This not only makes work easier, but it can also make collecting valuables and used items easier.

Usually, heavy items will be dropped into the water when the water is high; this aims to minimize damage. Due to the high tidal range, larger ships can come directly to shore. This reduces the total working time on each ship. As can be seen in the following picture, the high tide conditions in Alang can reach 10 m, and the low tide conditions reach 2 m.

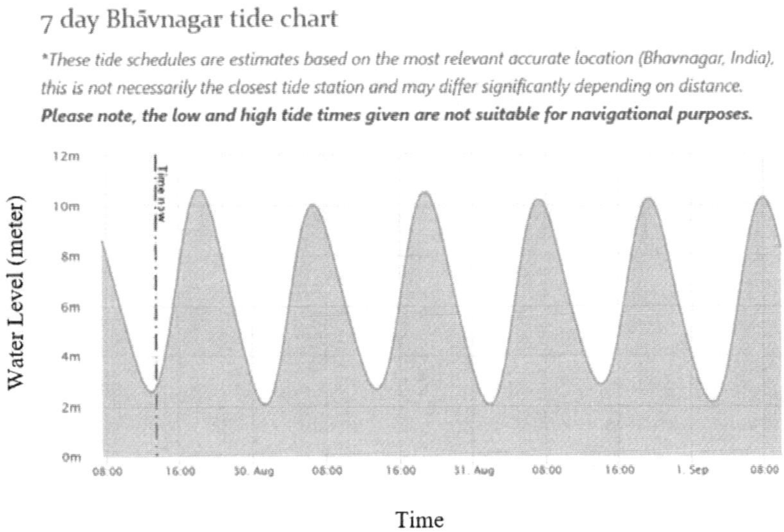

Figure 11. Alang's tide chart [17]

3.3.2. Water Conditions in Kamal, Madura, East Java, Indonesia

Tidal data from the Tidal Model Driver (TMD) Matlab was obtained in 2022 for the Kamal waters. The tidal characteristics of the Kamal waters reveal two high tides and one low tide in a single day. In Kamal waters, the high tide is 1.26 and the low tide is -1.36 m. The ship that can be recycled with these high tides is approximately 5,000 DWT.

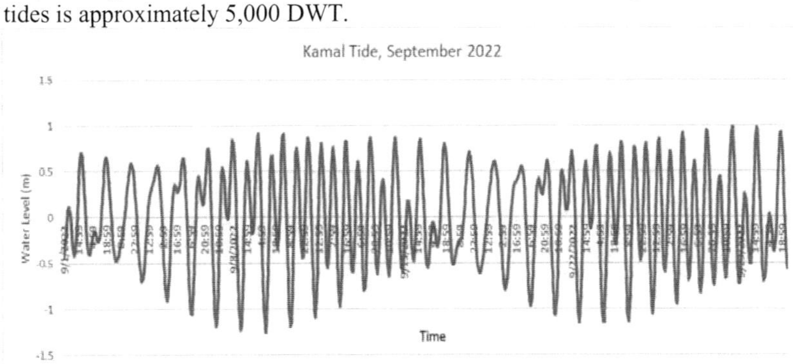

Figure 12. Kamal's tide chart
(Source: Tidal Model Driver Matlab)

3.3.3. Water Conditions di Bojonegara, Banten, Indonesia

Tidal data in Bojonegara water was obtained in 2022 using the Tidal Model Driver (TMD) Matlab. Bojonegara waters have mixed tidal types, which tend to be double daily. The high tide in Bojonegara waters is 0.3497 m, and at low tide, it is -0.4323 m. For the Bojonegara location, the ship that was recycled was 10,000 DWT.

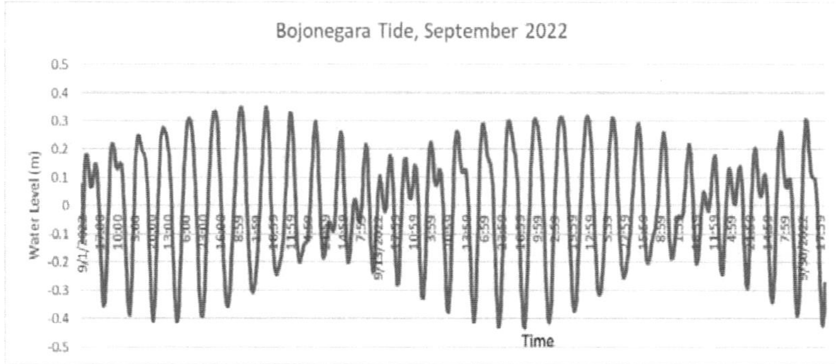

Figure 13. Bojonegara's tide chart
(Source: Tidal Model Driver Matlab)

3. 4. Firm Strategy, Structure, and Rivalry

3.4.1. National Scope

When examining the ship recycling facilities, a strategy is required to highlight certain important Porter Diamond aspects. Kaing Fang et al. state that firm strategy, structure, and rivalry mostly pertain to the structure and management of domestic renewable energy enterprises, in addition to the performance of rivals, while using the porter diamond approach to renewable energy as an example. The industry is more active and more competitive, the more successful the renewable energy company strategy is. The distribution of resources is also more effective the more open and fierce the competition [18]. If we apply it to the SRF context, the SRF will develop more favorably if the rival performs better in the area. The more transparent and intense the recycling market's rivalry, more open and tougher the competition in the recycling facilities, the more efficient the allocation of resources.

Based on field observations in Kamal, Madura conducted by Fariya [19]. It was discovered that East Java's shipbreaking yards converged in the Kamal region. So that shipbreaking facilities flourish in the area. It should be highlighted that these facilities are still only available in small firms that use basic technology and are not subject to strict regulations. However, all of the businesses that are already well-established and have a steady supply of old ships that must be cutdown and sold for scrap.

Things are different in Bojonegara. There, ship breaking activities were conducted in a small ship repair yard. The shipyard, although it has primary business as ship repair yard, does not let the potential opportunity of ship breaking. Bojonegara has several advantages compared to the traditional Kamal ship breaking. In Bojonegara, the shipyards already use sufficient technology, because they are ship repair companies with sufficient facilities and organization. This is reinforced by the results of research proposed by Sunaryo [20].

According to the rivalry principle, it is vital to increase the competitiveness of the neighboring businesses in the Kamal and Bojonegara areas at the same time when building a region where SRF thrives. Inadvertently, it encourages corporate rivalry and fosters an environment of competition and openness in the shipbreaking industry. In Bojonegara, the rivalry already exists because the shipbuilding industry is well established there. Samudra Marine Indonesia, a large shipyard, has already become one of the leading shipyards in Indonesia. Several medium-sized shipyards are based there; Caputra, Krakatau Shipyard, and Dock HTS, many small shipyards and maritime contractors exist there. This small shipyard has the potential to transform into ship recycling facilities.

Table 6. Summary for Porter's Diamond Firm Strategy, Structure, Rivalry

No		Kamal, East Java	Bojonegara, Banten
1	Strategy	Traditional Enterprise and Family Driven Policy	Small to Medium Enterprise, based on Ship Repair Yard Company
2	Structure	Simple Organization	Exist and well managed Organization
3	Rivalry	Small rivalry, all well established with small probability of inter-company conflict	The probability of rivalry between the yard is already exist

3.4.2. Global Scope

Scope of global competition in the SRF industry determined by the price of scrap iron in the country concerned. The data provides in the following table.

Table 7. Price of scrap iron 2022 [21]

Country	Price (US$ / Ton)
Indonesia	US $ 311
Bangladesh	US $ 590
Pakistan	US $ 570
India	US $ 580
Turkey	US $ 260

Table 7 shows that Bangladesh has the highest price at US$ 590, while Turkey has the lowest price at US$ 260. Although prices are competitively tight, the state's regulatory conditions also affect competition. Despite having a high scrap selling price, Bangladesh continues to dominate the global market share. The selling price of scrap in Turkey is the lowest, but many used European ships are in demand. It shows that HKC application is well applied in Turkey. India is currently working to improve its HKC compliance and double its capacity by 2024 in order to gain market share in Europe and Japan.

4. Conclusions

- Demand for scrap steel by the industry is quite high, and Indonesia is currently still importing scrap steel. These imports can be replaced by the growth of the SRF cluster industry.
- The related industry as a supporter of the SRF cluster is the steel industry as a buyer of SRF products centralized in Java. This is supported by adequate road access to transport products from the SRF cluster—from Bojonegara and Kamal—to the steel industry.
- Initial research on conditional factors show that are possible to plan the SRF clusters in Kamal, East Java, and Bojonegara, Banten, particularly for vessels under 10,000 DWT. However, it should be noted that it also possible to develop SRF clusters in several places in Indonesia, such as Batam, Tegal, and Pekalongan.
- SRF competition that occurs still requires significant effort to achieve domestic and global competitiveness.

- The success of the Ship Recycling Cluster in Alang, Gujarat, India serves as a best practice for the development of SRF Clusters in Indonesia.

Acknowledgments
Thank you to our colleagues from The National Research and Innovation Agency of Republic Indonesia, Research Centre for Transportation Technology.

References
[1] Ministry of Transportation Republic of Indonesia Directorate General of Sea Transportation, "Indonesian Fleet Based on Age," 2019. (In Indonesian)

[2] DNV, "'No room for complacency' despite advancement in shipping safety," *13 January, 2022*. https://www.dnv.com/expert-story/maritime-impact/Navigating-new-safety-challenges-after-a-decade-of-progress.html (accessed Sep. 01, 2022).

[3] IMO, "MEPC/63/23/ Annex 5," *Hong Kong Conv.*, vol. **211**, no. March, pp. 1–12, 2012.

[4] E S Ocampo and N N Pereira, "Can ship recycling be a sustainable activity practiced in Brazil?," *J. Clean. Prod.*, vol. **224**, no. March, pp. 981–993, 2019, doi: 10.1016/j.jclepro.2019.03.173.

[5] JICA & Ministry of Industry, "ROLE AND POSITION OF SHIP RECYCLING INDUSTRY WITHIN THE GLOBAL SHIPPING SHIPBUILDING," 2018. (In Indonesian)

[6] H Sornn-Friese *et al.*, *Creating circular economy clusters for sustainable ship recycling in Denmark*. 2021. [Online]. Available: https://www.cbs.dk/files/cbs.dk/red_sustainable_ship_recycling_clusters_in_dk_report.pdf

[7] T Roesdianto, "ACCELERATION OF THE RATIFICATION OF THE HONGKONG CONVENTION IN THE PERSPECTIVE OF THE SHIP FILLING INDUSTRY," 2022. (In Indonesian)

[8] Harappa Education, "Porter's Diamond Model," 2021. https://harappa.education/harappa-diaries/porters-diamond-model/#:~:text=Demand Conditions,customer segments and their distribution. (accessed Sep. 03, 2022).

[9] Ö Öz, "The Competitive Advantage of Nations," *Compet. Advant. Nations Case Turkey*, pp. 1–27, 2019, doi: 10.4324/9780429439087-1.

[10] G M Tasevska, "Macedonian Viticulture," no. 445, 2006.

[11] DataIndonesia.Id, "Indonesia's Iron and Steel Imports Soar in 2021," 2022. https://dataindonesia.id/sektor-riil/detail/impor-besi-dan-baja-indonesia-melonjak-pada-2021 (accessed Sep. 03, 2022). (In Indonesian)

[12] A Products, "PT Gunawan Dianjaya Steel Tbk .," no. 29, pp. 296–297, 2010. (In Indonesian)

[13] Ministry of Industry, "National Industrial Development Master Plan 2015 - 2035," *Rencana Induk Pembang. Ind. Nas. 2015-2035*, pp. 1–98, 2015, [Online]. Available: https://www.kemenperin.go.id/ripin.pdf (In Indonesian)

[14] UNCTAD, *Review of Maritime Report 2021*. 2021. [Online]. Available: http://unctad.org/en/PublicationsLibrary/rmt2015_en.pdf

[15] Ministry of Industry, "Directory of Steel Industry Companies in East Java," *2022*. https://kemenperin.go.id/direktori-perusahaan?what=baja&prov=35 (accessed Sep. 02, 2022). (In Indonesian)

[16] Ministry of Industry, "Directory of Steel Industry Companies in Banten," 2022. https://kemenperin.go.id/direktori-perusahaan?what=baja&prov=36 (accessed Sep. 02, 2022). (In Indonesian)

[17] T T Organization, "Bhāvnagar Tide Times / India," *2022*. https://www.tidetime.org/asia/india/bhvnagar.htm (accessed Sep. 03, 2022).

[18] K Fang, Y Zhou, S Wang, R Ye and S Guo, "Assessing national renewable energy competitiveness of the G20: A revised Porter's Diamond Model," *Renew. Sustain. Energy Rev.*, vol. **93**, pp. 719–731, Oct. 2018, doi: 10.1016/j.rser.2018.05.011.

[19] S Fariya, D Manfaat and K Suastika, "Technical Analysis of the Development of Ship Recycling Yard in Indonesia," *Sci. Technol.*, no. August 2016, pp. 2354–6026, 2018, [Online]. Available: https://www.researchgate.net/publication/307989571

[20] S Sunaryo, E Djatmiko, S Fariya, R Kurt and S Gunbeyaz, "A gap analysis of ship-recycling practices in Indonesia," *Recycling*, vol. 6, no. 3, 2021, doi: 10.3390/recycling6030048.

[21] Go Shipping & Management Inc, "Demolition Market," *2022*. https://www.go-shipping.net/demolition-market (accessed Aug. 26, 2022).

IOP Conf. Series: Earth and Environmental Science 1166 (2023) 012008 doi:10.1088/1755-1315/1166/1/012008

Vulnerability of Ship with a Large Breadth to Draught Ratio Against Excessive Acceleration Criteria

D. Paroka[1], A.H. Muhammad[2], S. Rahman[1], Fuad Mahfud Assidiq[1]

[1]Department of Ocean Engineering, Hasanuddin University, Jl. Poros Malino Km. 6, Bontomarannu, Gowa, South Sulawesi, Indonesia, 92171.
[2]Department of Marine System Engineering, Hasanuddin University, Jl. Poros Malino Km. 6, Bontomarannu, Gowa, South Sulawesi, Indonesia, 92171.

dparoka@eng.unhas.ac.id

Abstract. A ship having a large breadth and small draught may have a large metacentric height. Therefore, the ship could be vulnerable to the excessive acceleration criteria. In order to investigate the vulnerability, an Indonesian ro-ro ferry with breadth to draught ratio of 5.31 was used to apply the criteria. The lateral acceleration and the long-term probability index with limited scatter wave data of International Maritim Organization were calculated for passenger or crew accommodation in full loading condition. The calculation was conducted for variation of metacentric height using the JONSWAP spectrum to modelled irregular wave. The results show that the maximum vertical distance from the roll axis to comply with the criteria level 1 was 8.7% of ship breadth. The metacentric height comply with the criteria level 1 for the passenger or crew accommodation was 1.4 metres corresponding to 8.7% of ship breadth. The maximum wave height comply with the criteria level 2 was 5 metres. These results show that the excessive acceleration criteria can be used to assess stability of a ship with large ratio between breadth and draught. The vertical distance from the roll axis has more significant effect on the lateral acceleration than the metacentric height.

1. Introduction

The International Stability Code (IS Code) was mostly adopted to evaluate intact stability of a ship. A part of the criteria is the weather criterion used to assess stability of a ship have large windage area or a ship operates in light condition [1]. The weather criterion was statistically formulated by using sample ships with a maximum ratio between breadth and draught of 3.5 and the vertical centre of gravity (VCG) divided by draught ranges from 0.7 to 1.5. An overestimate stability can be found if the weather criterion is used to assess stability of a ship with a larger ratio between breadth and draught as well as ratio between VCG and draught [2]. On the other hand, some accidents due to stability occurs to ships comply with the criteria. Therefore, the International Maritime Organization (IMO) developed a new stability criterion with performance-based approach known as the Second Generation Intact Stability Criteria (SGISC). One of dangerous phenomena considered in the SGISC was the excessive acceleration applied to a ship with metacentric height (GM) larger than 8% of ship breadth and the passenger and crew accommodations located with height from the roll axis larger than 70% of ship breadth [3].

Evaluation of ship stability using the SGISC is conducted in three steps. A ship does not comply with the criteria in the first step should be continued to the second step of assessing the ship vulnerability. A direct assessment should be conducted for a ship to be identified to be vulnerable in the second step. For the excessive acceleration failure mode, a simple method was used to calculate lateral acceleration due to roll motion in beam seas for the first level. In the second level, a more complex approach based on the ship performance in irregular beam waves is used to calculate the long-term probability index. Here, the roll damping moment was calculated by using the simplification of Ikeda's method which is statistically developed by using sample ships with a maximum ratio between breadth and draught of 4.5 [4]. For direct assessment, IMO recommend to use a numerical simulation with six-degree of freedom model. On the other hand, operational measurement could be developed for a specific loading condition for a ship that fails in the second vulnerability level or in the direct assessment level. The operational measurement could be operational guidance or operational limitation [5].
A ship having large breadth and small draught tend to have a significant GM because a large metacentric radius. That ship may be vulnerable against the excessive acceleration criterion even the ship has a large vertical centre of gravity and the height of passenger or crew accommodation smaller than 70% of ship breadth. Indications of large lateral acceleration due to a large roll angle had been found in some accidents of Indonesian Ro-Ro ferries such as overthrowing or shifting of a car in vehicle deck [6], [7]. The ratio between breadth and draught of some Indonesian Ro-Ro ferries are larger than 4.5. Therefore, the excessive acceleration criteria may be important in stability evaluation of a ship having large ratio between breadth and draught.
This paper discusses application of the SGISC mainly the excessive acceleration failure mode to a ship with large ratio between breadth and draught. The results can be used as consideration to use the excessive acceleration criteria in stability evaluation of ship with a large breadth and a small draught. On the other hand, this research may also be used to determine operational limitation and operational guidance including guidance to handling payload mainly the car in vehicle deck of ro-ro ferries. The geometric characteristics of ro-ro ferries tends to have a large ratio between breadth and draught [8]. The applicability of the SGISC to a ship with a large breadth and a small draught can be investigate including the method to determine the damping coefficient and the others hydrodynamics coefficients used to calculate the lateral acceleration due to roll motion in beam seas.

2. Vulnerability Criteria Level 1

A ship is assumed to be vulnerable in the level 1 of excessive acceleration failure mode if the following condition is achieved [4]:

$$\varphi \cdot k_L \cdot \left(g + 4\pi^2 \frac{h_r}{T_r{}^2} \right) \leq 4.64 \qquad (1)$$

where φ is the characteristic of roll amplitude, which is determined using the equation (4). g is gravity acceleration, k_L is the factor correspond to roll, pitch and yaw motions. This factor depends on the distance of passenger and crew accommodations from the after perpendicular, h_r is the height of the passenger and crew accommodations from the roll centre of rotation and T_r is the natural period of roll, respectively. The natural roll period is calculated using the equation as follows [1]:

$$T_r = \frac{2 \cdot C \cdot B}{\sqrt{GM}} \qquad (2)$$

where B is the ship breadth and C correspond to the radius gyration coefficient which can be calculated by using the equation (3).

$$C = 0.373 + 0.023 \cdot \left(\frac{B}{d}\right) - 0.043 \cdot \left(\frac{L_{WL}}{100}\right) \qquad (3)$$

Here, d is the ship draught and L_{WL} is the length of waterline. The equation (3) shows that a ship with a large ratio between breadth and draught ratio has a larger value of C. On the other hand, such ship has a large GM. Therefore, the ship may have a small natural roll period. The lateral acceleration of a ship with smaller roll period is larger than the ship with larger roll period.

$$\varphi = \frac{4.43 \cdot r \cdot s}{\sqrt{\delta_\varphi}} \tag{4}$$

Here, r is the effective wave slope coefficient and s is the wave steepness. δ_φ is the non-dimensional logarithmic decrement of roll decay. The equations (5) – (7) show the formula to calculate the non-dimensional logarithmic decrement of roll decay. Here, Cm is the midship coefficient and A_{BK} is the bilge keel area.

$$\delta_\varphi = 0.267 + \frac{2}{3} \cdot \left(100 \frac{A_{BK}}{L \cdot B}\right) \qquad if\ Cm \geq 0.96 \tag{5}$$

$$\delta_\varphi = 0.267 + \frac{1}{3} \cdot \left(1 + 50 \cdot (Cm - 0.94)\right) \cdot \left(100 \frac{A_{BK}}{L \cdot B}\right) \qquad if\ 0.94 < Cm < 0.96 \tag{6}$$

$$\delta_\varphi = 0.267 + \frac{1}{3} \cdot \left(100 \frac{A_{BK}}{L \cdot B}\right) \qquad if\ Cm \leq 0.94 \tag{7}$$

where L is length between perpendiculars of the ship.

The factor corresponding to the simultaneous action of roll, pitch and yaw is calculated based on the longitudinal position of the considered passenger or crew accommodation. The longitudinal position is divided into three parts with the formula for calculating the factor due to simultaneous action of roll, pitch and yaw shown in the equation (8) – (10).

$$k_L = 1.125 - 0.625 \cdot \frac{x}{L} \qquad if\ x < 0.2L \tag{8}$$

$$k_L = 1.0 \qquad if\ 0.2L < x < 0.65L \tag{9}$$

$$k_L = 0.527 + 0.727 \cdot \frac{x}{L} \qquad if\ x > 0.65L \tag{10}$$

where x is the distance of passenger or crew accommodation from the after perpendicular.

3. Vulnerability Criteria Level 2

A ship vulnerable against the criteria level 1 should be tested to the criteria level 2 based on the roll motion characteristic in beam seas. A ship is considered to be vulnerable to the criteria level 2 of the excessive acceleration failure mode if the long-term failure index is larger than 0.00039. The long-term failure index is calculated by using the equation (11) [4].

$$C = \sum_{i=1}^{N} W_i \cdot C_i \tag{11}$$

Here, W_i is the occurrence probability of a pair of significant wave height and mean wave period, C_i is the short-term excessive acceleration correspond to the considered wave condition and N is the wave number. The short-term excessive acceleration failure index for the loading condition and the location under consideration is calculated according to the formula as follows:

$$C_i = exp\left[\frac{-{R_2}^2}{2 \cdot {\sigma_{LAi}}^2}\right] \tag{12}$$

where R_2 is the specified lateral acceleration to be the same as the gravity acceleration. σ_{LAi} is the standard deviation of lateral acceleration corresponding to wave characteristics follow the scatter wave data used to determine the short-term environmental wave condition. This standard deviation is calculated by using the equation (13).

$$\sigma_{LAi}{}^2 = \frac{3}{4}\sum_{j=1}^{N}\left(a_y(\omega_j)\right)^2 \cdot S_{zz}(\omega_j) \cdot \Delta\omega \tag{13}$$

where:

$$a_y(\omega_j) = k_L \cdot \left(g \cdot \sin\varphi_a(\omega_j) + h_r \cdot \omega_j{}^2 \cdot \varphi_a(\omega_j)\right) \tag{14}$$

Here, ω_j is the mid-point of the considered frequency interval, $\Delta\omega$. $S_{zz}(\omega_j)$ is the wave spectrum and $\varphi_a(\omega_j)$ is the roll amplitude at angular frequency, ω_j, in regular beam waves and zero speed, which is calculated by using the following equation:

$$\varphi_a(\omega_j) = \left(\varphi_r(\omega_j)^2 + \varphi_i(\omega_j)^2\right)^{0.5} \tag{15}$$

where $\varphi_r(\omega_j)$ and $\varphi_i(\omega_j)$ are calculated by using the equations (16) – (17) as follow:

$$\varphi_r(\omega_j) = \frac{a(\Delta \cdot g \cdot GM - I_{xx} \cdot \omega_j{}^2) + b \cdot \mu_e \cdot \omega_j}{(\Delta \cdot g \cdot GM - I_{xx} \cdot \omega_j{}^2)^2 + (\mu_e \cdot \omega_j)^2} \tag{16}$$

$$\varphi_i(\omega_j) = \frac{b(\Delta \cdot g \cdot GM - I_{xx} \cdot \omega_j{}^2) - a \cdot \mu_e \cdot \omega_j}{(\Delta \cdot g \cdot GM - I_{xx} \cdot \omega_j{}^2)^2 + (\mu_e \cdot \omega_j)^2} \tag{17}$$

where Δ is the ship displacement and I_{xx} is the roll moment of inertia. a and b are the real and imaginary parts of the roll exciting moment calculated by using Froude-Krylov moment and neglecting the diffraction moment. μ_e is the linearized damping coefficient stochastically calculated according to the linear and quadratic damping coefficients and standard deviation of roll angular velocity in the short-term wave condition. IMO recommended to use the Ikeda's simplified formula to calculate the damping coefficients but the method only valid for a ship with ratio between breadth and draught smaller than 4.5 [9] and the ratio between the VCG and draught from 0.5 to 1.2.

4. Methodology

The excessive acceleration criteria are applied to a ro-ro ferry built and operated in Indonesia with the main dimension and general arrangement are given in Table 1 and Figure 1, respectively.

Table 1. Main dimension of the subject ship

Dimension	**Ship**
Length of the perpendicular (Lbp)	47.25 m
Breadth (B)	13.00 m
Draught (T)	2.45 m
Depth (D)	3.45 m
Linear damping coefficient (α)	0.0329/s
Quadratic damping coefficient (β)	0.7004/rad.
Vertical centre of gravity (VCG)	4.00 m
Windage area (A_L)	432.93 m^2
Height of the centre of windage area from the water surface (C_L)	4.43 m
Down-flooding angle (φ_L)	38.2 deg.

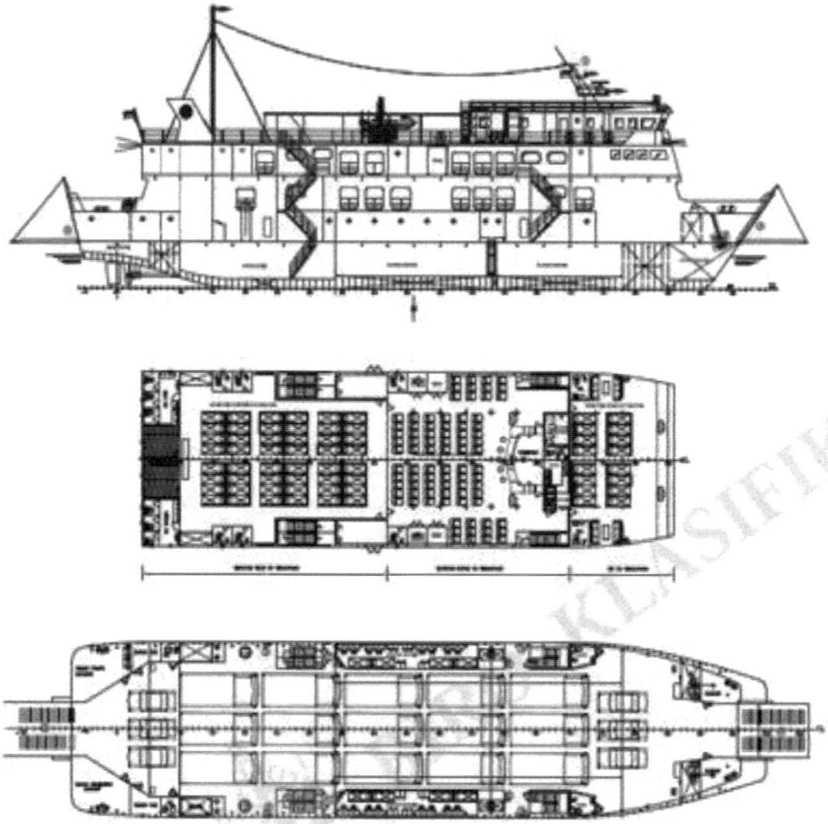

Figure 1. General arrangement

The ship has breadth to draught ratio of 5.31 and the ratio between VCG and draught of 1.63. The ratios are larger than the limitation of method to estimate damping moment of roll given in the IMO guidelines [9]. The passenger accommodations are located in the position of 5.96 metres about the seawater surface, corresponding to 55.27% of ship breadth from the roll axis. The crew accommodation is located with height of 73.73% of ship breadth from the roll axis. These data showed that the vertical distance of the crew accommodation from the roll axis is larger than the vertical distance required to evaluate the stability against the criteria for excessive acceleration failure mode. The GM of the ship is 36.31% of ship breadth, larger than the minimum GM required to apply the excessive acceleration criteria.

The effective wave slope coefficient in the criteria level 1 was determined based on the formula recommended by IMO and the wave steepness was determined based on the natural roll period following the values given in the criteria level 1 of dead ship failure mode. The sine and cosine components of Froude-Krylov roll exciting moment were determined under assumption that the ship is laterally symmetric. The linearized damping coefficient is estimated based on the damping coefficients obtained by simplified Ikeda's method as the recommended method to calculate the lateral acceleration in beam waves.

The irregular wave is modelled by using JONSWAP spectrum with wave parameters consists of significant wave height and mean wave period based of the limited scatter wave data recommended by IMO [10]. The occurrence probability of each pair of significant wave height and mean wave period was given in the scatter wave data. The linearized damping coefficient for the criteria level 2 was calculated based on the linear and quadratic damping coefficients obtained by model experiment. In order to obtain the operational limitation, the calculation is conducted for variation of maximum significant wave height. The simulation is also performed for variation of vertical centre of gravity to determine the maximum GM comply with the criteria level 1 and level 2 corresponding to each maximum significant wave height.

5. Results and Discussion

The lateral acceleration corresponds to the criteria level 1 of excessive acceleration failure mode for variation of height of passenger or crew accommodation from the roll axis is shown in Figure 2. The longitudinal position was 41.0 metres from the after perpendicular. Here, the GM is the same as the GM for full loading condition. The lateral acceleration corresponds to the passenger accommodation is about 6.9 m/s^2 and that for the crew accommodation is about 8.1 m/s^2. Those results mean that the ship should be tested against the criteria level 2 because it does not comply with the criteria level 1.

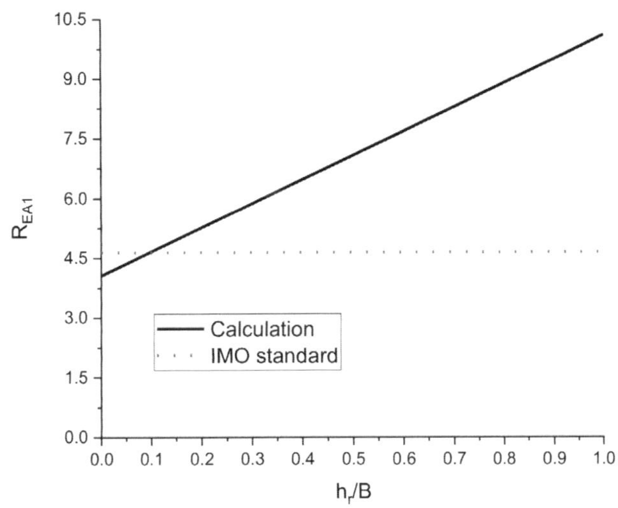

Figure 2. Lateral acceleration of vulnerability criteria level 1

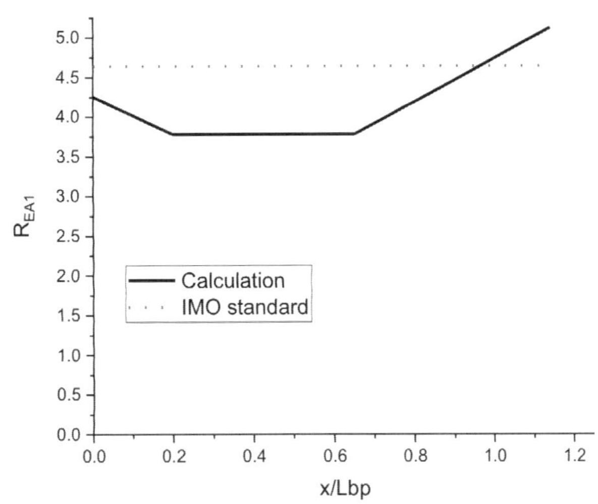

Figure 3. Lateral acceleration based on the vulnerability criteria level 1 corresponding to longitudinal position from the after perpendicular.

Even the height of passenger or crew accommodation from the roll axis is smaller than 70% of breadth, a ship with large ratio between breadth and draught could be vulnerable against the criteria level 1 of excessive acceleration failure mode. A ship with large ratio between breadth and draught tends to have a small natural roll period mainly for small ship because large GM. The roll period of the subject ship is smaller than 6 seconds. Therefore, the wave steepness to calculate the lateral acceleration was 0.1. This wave steepness depends on the operational location of the ship. A smaller wave steepness for the same natural roll period was found for Indonesian seas [11]. On the other hand, the damping coefficient, for the vulnerability criteria level 1 called as non-dimensional logarithmic decrement of roll decay could be underestimate. Some researches corresponding to the effect of bilge keels on the damping moment of a ship with large breadth found a significant effect mainly if the ship has a large vertical centre of gravity [12], [13]. A ship with large lateral acceleration should pay attention on the handling of pay load to avoid shifting which can significantly affect the ship stability.
The effect of longitudinal position of passenger or crew accommodation to the lateral acceleration is shown in Figure 3. Here, the GM was 1.4 metres corresponding to 8.7% of ship breadth and the calculation was conducted with height from the roll axis of 8.36 metres, the same as the height of crew accommodation. The maximum lateral acceleration occurs at the location near the fore perpendicular. The different of lateral acceleration due to the longitudinal position occurs due to simultaneous effect of roll, pitch and yaw motions. The vertical and lateral translations due to pitch and yaw motions increases as the distance from the centre of gravity increases. As results, the lateral acceleration near the after perpendicular and fore perpendicular is larger than that near the midship. The sway motion may also have effect on the lateral acceleration but it was not considered in the formula. The ship with the GM of 1.4 metres complies with the criteria level 1 for maximum height of accommodation from the roll axis of 73.73% of ship breadth when the longitudinal position from the after perpendicular is smaller than 89.74% of ship length.

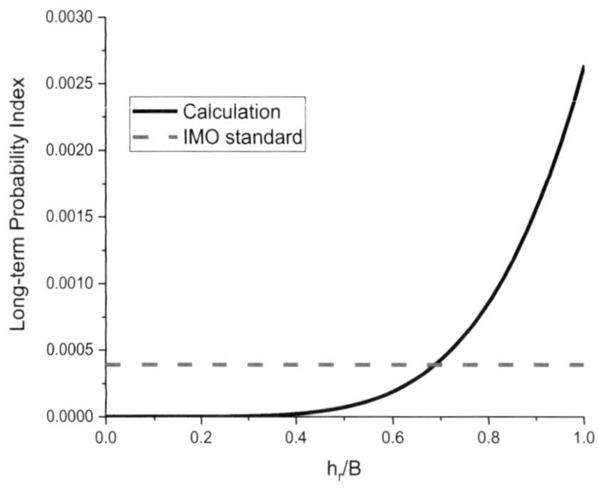

Figure 4. Long-term probability index for variation of height from the roll axis

The long-term probability index corresponding to the criteria level 2 for variation of height of passenger or crew accommodation from the roll axis is shown in Figure 4. This results for full loading condition with longitudinal position of 41 metres from the after perpendicular and the maximum significant wave height of the limited scatter wave data of 5.5 metres. The ship is vulnerable against the criteria level 2 if the height of passenger or crew accommodation from the roll axis is larger than 68.75% of ship breadth. IMO recommend to use the excessive acceleration criteria for a ship with accommodation located with height from the roll axis larger than 70% of ship breadth [3]. However, the GM is larger than the minimum GM recommended by IMO to test against the excessive acceleration failure mode. Effect of longitudinal position of passenger or crew accommodation to the long-term probability index is shown in Figure 5. Here, the vertical distance of passenger or crew accommodation from the roll axis was 68.75% of ship breadth with GM in full loading condition. All passenger and crew accommodations comply with the criteria level 2. The ship does not comply with the criteria when the accommodation for passenger or crew is located in the longitudinal position larger than 86.77% of ship length. This calculation is performed for the maximum wave height is the same as that used in the calculation of vertical distance effect shown in Figure 4.

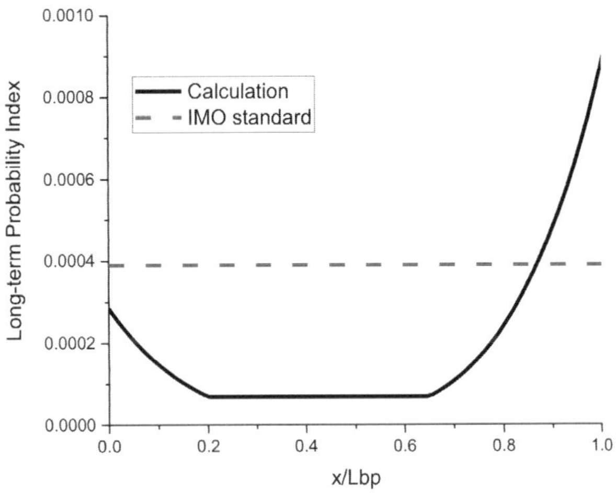

Figure 5. Effect of longitudinal position on long-term probability index

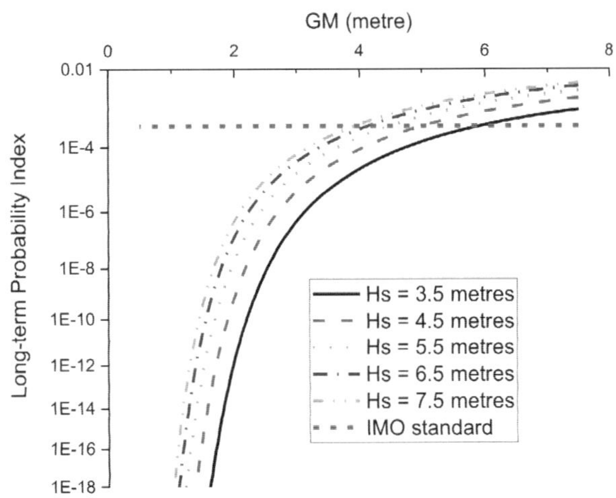

Figure 6. The long-term probability index for the criteria level 2

The long-term probability index corresponding to the criteria level 2 for variation of GM is shown in Figure 6. The calculation was conducted with limited scatter wave data for maximum significant wave height range between 3.5 metres and 7.5 metres. The height from the roll axis and the longitudinal position corresponds to those long-term probability indexes are 8.36 metres and 41.49 metres, respectively. The long-term probability index increases when the GM increases. The ship complies with the criteria level 2 in full loading condition if maximum significant wave height is 5 metres or smaller.

The maximum GM complies with the criteria level 2 decreases when the maximum significant wave height increases. These results means that the subject ship comply with the excessive acceleration failure mode of SGISC. However, an advance investigation corresponding to impact of lateral acceleration to cargo shifting is necessary in order to determine maximum lateral acceleration when the criteria applied to ship load containers or vehicles. An alternative method to estimate the damping coefficients is also important if the criteria is applied to a ship having large breadth and small draught.

6. Conclusion

The excessive acceleration criteria of SGISC had been used to asses stability of a ro-ro ferry with ratio between breadth and draught of 5.31. According to the results and discussions, the conclusion can be formulated as follows:

1. The sample ship does not comply with the criteria level 1 but comply with the criteria level 2 if the maximum significant wave height is smaller than 5 metres.
2. Effect of the height from the roll axis has more significant effect on the criteria level 2 than the GM. Even the GM is larger than 8% of breadth but the height from the roll axis is smaller than 70% of breadth, the sample ship complies with the criteria.
3. The environmental condition of operational location of ship has a significant effect on the application of SGISC. Therefore, an accurate data corresponding to the operational location of the ship is important to apply those criteria.

7. References

[1] IMO (2008). The International Code on Intact Stability (2008 IS CODE). International Maritime Organization, London.

[2] Paroka, D., Asri, S., Rosmani, Hamzah. (2020). "Alternative Assessment of Weather Criterion for Ships with Large breadth to Draught Ratios by Model Experiment: A Case Study on an Indonesian Ro-Ro Ferry". *Int. J. of. Mar. Eng.***162**.

[3] Boccadamo, G., Rosano, G. (2019). "Excessive Acceleration Criterion: Application to Naval Ships". *J. Mar. Sci. Eng.* 2019, **7, 431.**

[4] IMO. (2020). Finalisation of Second Generation Intact Stability Criteria: Report of the Correspondence Group. International Maritime Organization, London.

[5] Pettaco, N., Gualeni, P. (2020). "IMO Second Generation Intact Stability Criteria: General Overview and Focus on Operational Measures". *J. Mar. Sci. Eng.* 2020, **8, 494**.

[6] Tribun Bali. (2022). Available online: https://balitribune.co.id/content/cuaca-butuk-di-selat-bali-truk- terguling-di-dalam-kapal (accessed on 10 February 2022) (in Indonesian).

[7] Detik. (2022). Available online: https://news.detik.com/berita-jawa-timur/d-4549771/ombak-selat-bali- menggila-truk-muat-beras-di-kapal-terguling-timpa-mobil (accessed on 10 February 2022) (in Indonesian).

[8] Kristensen, H.O. (2016). Analysis of Technical Data of Ro-Ro Ships. HOK Marineconsult ApS, Denmark.

[9] IMO (2019). Finalization of Second Generation Intact Stability Criteria: Information Collected by the Correspondence Group on Intact Stability (Part A), SDC 7/INF.2. International Maritime Organization, London.

[10] IMO. (2015). Finalisation of Second Generation Intact Stability Criteria: Information Collected by the Correspondence Group on Intact Stability Regarding Second Generation Intact Stability Criteria. International Maritime Organization, London.

[11] Mappangara, A.S.C, Djalante, A.H, Idrus, M. (2021). "Potential Track in North Maluku as an Alternative Rerouting Ex Ferry of the Bakauheni Merak Track". *J. of Metacentre*, **1, 1**.

[12] Katayama, T., Mastuoka, M., Ikushima, K. (2018). Characteristic of Bilge Keel Roll Damping Component for Shallow Draft, Proceedings of the 13th International Conference on the Stability of Ships and Ocean Vehicles, Kobe, Japan.

[13] Paroka, D., Muhammad, A.H., Rahman, S. Hydrodynamics Factors Correspond to Weather Criterion Applied to an Indonesian Ro-Ro Ferry with Different Weight Distribution. *Int. J. Tech.* 2021, **12, 1**.

Acknowledgments
The data presented is this paper was results of research supported by the National Research and Innovation Agency of Indonesia and Hasanuddin University with the grand number of 1400/UN4.22/PT01.03/2022. The authors extend their great sincere appreciation to those institutions.

Effect of debonding damage on transient dynamic response on stern ramp door of passenger ship 6311 GT

Tuswan Tuswan[1,*], **Wilma Amiruddin**[1], **Hartono Yudo**[1], **Aditya Rio Prabowo**[2], **Haikal Anjasmara**[3]

[1] Department of Naval Architecture, Universitas Diponegoro, Semarang 50275, Indonesia,
[2] Department of Mechanical Engineering, Universitas Sebelas Maret, Surakarta 57126, Indonesia
[3] Department of Naval Architecture, Institut Teknologi Sepuluh Nopember, Surabaya 60111, Indonesia

*Email: tuswan@lecturer.undip.ac.id

Abstract. The sandwich, as an advanced material, increases innovation in the field of ship construction. One of the applications is on the ramp door structure. The characteristics of sandwich material which are light and strong, are the main reasons for the application of this material. However, sandwich material also has drawbacks, one of which is debonding damage, where the bond between the face and core is released. This damage can affect the structure's stiffness, which impacts reducing the strength of the structure on the ramp door. The effect of debonding is identified by performing numerical tests using finite element software. This test aims to identify the ramp door's dynamic response due to the debonding damage's influence. This research uses finite element software with an explicit dynamics method to determine the dynamic response of stress and displacement due to loading on the ramp door. The results of testing with finite element software show that an increase in the ratio of debonding damage causes the amplitude of the dynamic response to increase when subjected to transient loading.

Keywords: debonding, sandwich, dynamic explicit, dynamic response, transient.

1. Introduction

Increasingly advanced technology creates materials that develop from before. Advanced materials or so-called composite materials are developments from previous materials, namely metals, ceramics, and polymers. This composite material is lighter, stronger, and more durable than conventional materials. Much research has led to development of other new materials and their applications for ships. Sandwich Panel is one of them. Sandwich panels have a layered structure composed of faces and cores [1,2]. The face is made of a rigid material, and the core is made of lightweight material. With this structure, sandwich panels can reduce the structure's weight compared to steel materials.

The replacement of conventional steel was replaced by sandwich material, and an analysis of the maximum stress and deformation values was carried out, resulting in values that were not much different and still acceptable by class regulations. Usually, the construction that is replaced with a sandwich construction is a flat part of the construction. This construction provides benefits by reducing the ship's load due to the characteristics of the sandwich material, which is light but has good stiffness and strength [3–5].

However, the use of sandwich material itself is susceptible to damage. Sandwich material damage can occur in the manufacturing process. Damage that may occur during the manufacturing process is debonding. Debonding is a defect where the face material and core material are separated. Debonding can also occur if there is a large difference in thickness and modulus of elasticity between the face and core and a lack of adhesive layer between the face and core [6].

With the material damage, a method is needed to identify the damage. Identifying debonding is quite difficult because the debonding damage is in the core material covered by the face layer. On a small scale, it is possible to check for damage to conventional ship structures using ultrasonic test methods, visual tests, etc. However, in large ship structures, this method is not very practical to be applied. Therefore, we need a method of detecting damage that can be applied to ship structures in general.

One of the methods that can be applied to detect ships, in general, is vibration-based damage identification. This method uses the dynamic response of the structure to identify the existing damage to the sandwich material. Previous studies [7–12] for the effect of debonding damage to the ramp door structure with dynamic response used are natural frequencies. The results of this study indicate that there will be a reduction in the natural frequency for debonding ramp door structures, and the natural frequency value will decrease as the damage ratio increases [13-20].

In this study, an analysis of the dynamic response of the ship's ramp door using steady state and transient vibration analysis will be carried out to find various physical phenomena that occur in the sandwich panel as the ship's ramp door material is debonded and compared with the dynamic response of the ramp door which is not damaged on the sandwich material. So that it can be seen more clearly the effect of debonding so that there is no sandwich material failure under the designed stress.

2. Material and Method

2.1 Ramp Door Model

In this study, ship data and ramp door models will be used as in work [7]. Ship and ramp door data can be seen in Table 1.

Table 1. Vessel main dimension.

LOA (m)	108.20	Gross Tonnage (ton)	6311
LBP (m)	103.90	Net Tonnage (ton)	3058
BMLD (m)	24.30	DWT (ton)	-
HMLD (m)	4.80	Longitudinal Bulkhead	2
LT (ton)	1844	Transversal Bulkhead	17

The construction of the ramp door was changed by applying a sandwich construction. This change process uses the rules issued by the Lloyd's Register class. After being converted into a ramp door model with sandwich construction, several models were made with variations in the number of stiffeners. In addition, models with sandwich construction must also meet the allowable stresses given by the classification. For the reinforcement on the ramp door configuration, use transverse and longitudinal stiffeners where the transverse and longitudinal tensioners are installed with the same distance between the stiffeners. For the ramp door construction size, use the construction dimension listed in Table 2.

Table 2. Ramp door construction dimension.

Model	Plate thickness (mm)	Number of longitudinal stiffeners	Number of transversal Stiffeners
Existing	10	12	16
Sandwich Model	3-15-5	12	12

2.2 Debonding Modelling

Debonding damage made is assumed to be defects that occur due to the manufacturing process. Debonding damage parameter analyzed is the ratio of debonding damage. The damage ratio is the percentage of the debonding damage area compared to the total area of the core material. Figure 1 is an illustration of the debonding damage ratio used.

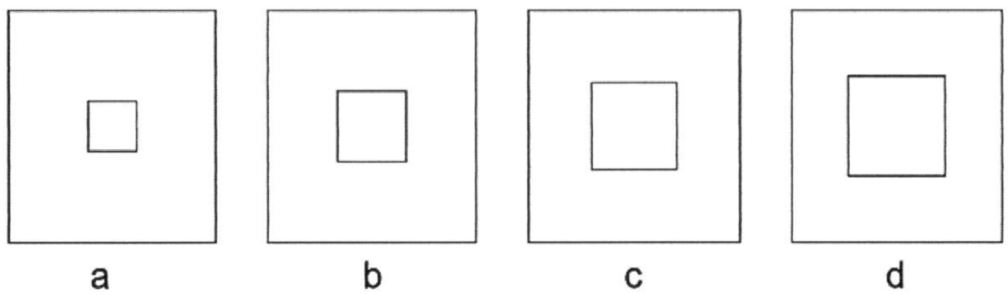

Figure 1. Variation of debonding damage ratio a) 5%, b) 10%, c) 15%, d) 20%.

The modelled debonding damage has a thickness of 5% of the thickest face material in the model, which is 5 mm. The illustration of thickness of the debonding damage is shown in Figure 7. Since the thickness of the thickest faceplate is 5 mm, the thickness of the debonding damage is 0.25 mm.

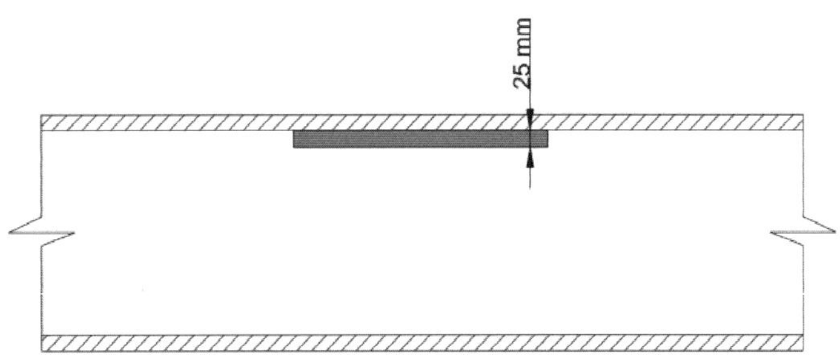

Figure 2. Debonding damage thickness illustration.

The vehicle load used can be seen in Table 3. The maximum load used is 24 tons.

Table 3. Vehicle size and weight data.

Vehicle size	Weight	Car size
Long: 630 cm	Empty weight: 7.2 Ton	Long: 1190 cm
Width: 220 cm	Max. weight: 16.8 Ton	Width: 250 cm
Height: 230 cm		Height: 290 cm
Dimension: 30 CBM		

From the vehicle load used, there is a load distribution for each axle on the vehicle. The load distribution can be seen in Table 4.

Table 4. Vehicle load distribution.

Axis configuration	Axis configuration		Maximum MST					JBI
	Top	Side	I	II	III	IV	V	Max
1.22	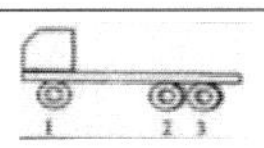	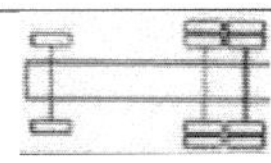	6	9	9	-	-	24

The vehicle load used in the simulation was applied in a short time of about 0.002 s with a total simulation time of 1 s. The amplitude plot can be seen in Figure 3.

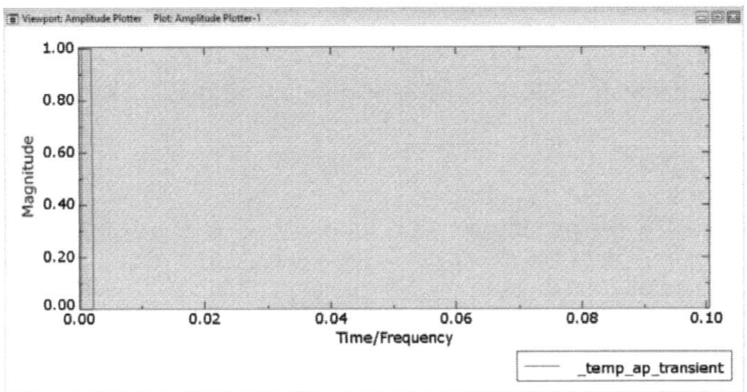

Figure 3. Transient load amplitude.

3. Result and Discussion

To see the effect of transient load, the model is given settings for explicit and transient load analysis. The output is in the form of stress and displacement, which can be seen in Figure 4 and Figure 5. Figure 4 shows the stress that occurs in the 0.1 s time span when the ramp door is given a load in the form of a transient load. It can be seen that the model that has the highest stress is the model with 20% damage, where the stress value that occurs is around 75 MPa and is very high compared to the smaller damage. This shows a pattern where the greater the debonding damage that occurs, the greater the stress the model will receive.

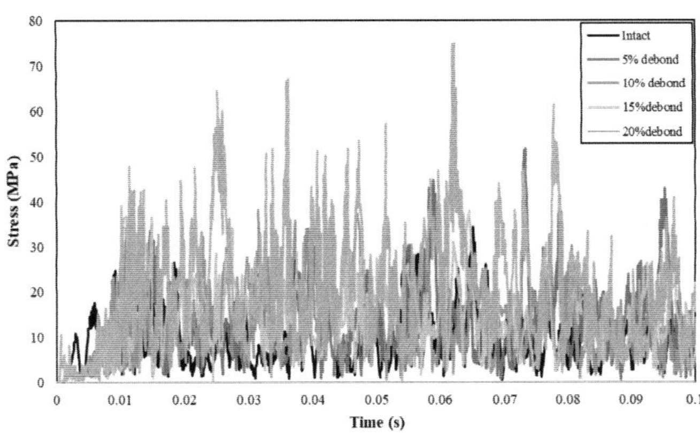

Figure 4. Stress vs. time graph under different damage ratios.

Figure 5 illustrates the displacement that occurs in intact conditions and four damage conditions where each damage increases and the result looks bigger. Displacement at 20% damage of about 16 mm is very different compared to the model without damage, where the displacement value due to the transient is close to 0. The addition of displacement due to this transient load reduces model stiffness.

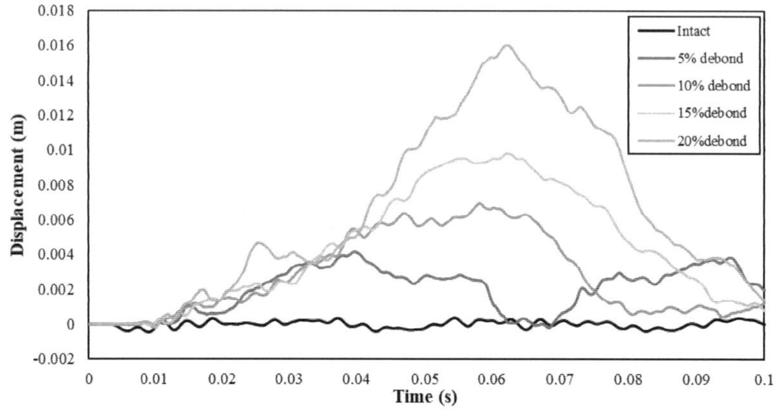

Figure 5. Displacement vs. time graph under different damage ratio

RMS is used again to justify this displacement value to compare the results obtained. Figure 6 shows the trend of the RMS value of the displacement obtained to increase along with the increase in the existing debonding damage.

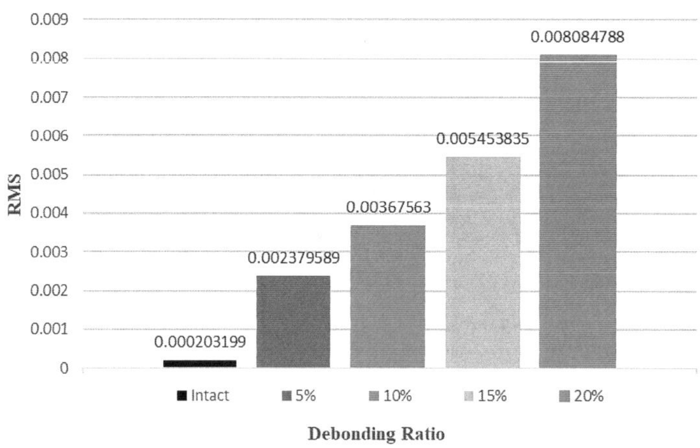

Figure 6. RMS displacement under different damage ratio

From the results described, it can be seen that the greater the debonding damage that occurs in the model, the amplitude of the displacement and stress increases. This additional stress and displacement are undesirable in the industry. The increase in stress will result in the ramp door getting a load that is greater than it should have been planned it will cause other problems in the future. Likewise, the addition of displacement shows that the strength of the ramp door decreases if there is debonding damage. Knowing the pattern that occurs on the ramp door with debonding damage, it is hoped that the debonding problem can be anticipated as early as possible.

4. Conclusions

After the numerical test and analysis, and discussion related to the simulation results were carried out, the following conclusions were obtained:

1. The stress response obtained from transient loading increases with the addition of debonding damage. At 0% damage, the stress experienced by the ramp door is 34.39 MPa and continues to increase with a damage ratio of 20%, experiencing the most significant stress of 74.97 MPa.
2. The displacement response has the same pattern as the stress response, where the greater the debonding damage, the greater the amplitude response is given. At 0% damage, the displacement that occurs is only 3.66×10^{-4} m and continues to increase with 20% damage experiencing a displacement of 1.60×10^{-2} m.
3. The RMS of displacement has increased with the addition of debonding damage. At 0% damage, it has a value of 2.03×10^{-4}. At 5% damage, it has a value of 2.38×10^{-3}. A 10% damage ratio has an RMS value of 3.68×10^{-3}, while a 15% damage ratio has an RMS value of 5.45×10^{-3}, and the largest damage ratio of 20% has an RMS of 8.08×10^{-3}.

Acknowledgements
The research leading to these results has received financial support from "Penelitian Strategis" research scheme, RKAT Faculty of Engineering, Universitas Diponegoro, with contract number 356/UN7.5.3.2/HK/2022.

References

1. Zubaydi A, Budipriyanto A 2020 Material Sandwich: Teori, Desain, dan Aplikasi.
2. Tuswan T, Zubaydi A, Piscesa B, Sari EN, Ismail A 2021 *IOP Conf. Ser. Mater. Sci. Eng.* **1034**(1) 012152
3. Ismail A, Zubaydi A, Piscesa B, Panangian E, Ariesta RC, Tuswan T 2021 *IOP Conf. Ser. Mater. Sci. Eng.* **1034**(1) 012026
4. Ismail A, Zubaydi A, Piscesa B, Panangian E, Ariesta RC, Tuswan T. 2021 *IOP Conf. Ser. Mater. Sci. Eng.* **1034**(1) 012027
5. Ismail A, Zubaydi A, Piscesa B, Tuswan T, Chandra A R 2021 *J. Eng. Appl. Sci*, **19**(4) 1090-1098.
6. Burlayenko VN, Sadowski T. 2010 *Int. J. Non. Linear. Mech.* **45**(10) 959–968
7. Tuswan, Zubaydi A, Piscesa B, Ismail A 2020 *Open Eng.* **10** 424-433
8. Tuswan T, Zubaydi A, Piscesa B, Ismail A, Ilham MF 2020 *Procedia Struct. Integr.* **27** 22–29
9. Tuswan T, Zubaydi A, Piscesa B, Ismail A, Ariesta RC, Prabowo, AR. 2022 *Curved Layer. Struct.* **9**(1) 25-39.
10. Ismail A, Zubaydi A, Piscesa B, Ariesta RC, Tuswan T. 2020 *Open Eng.* **10**(1) 744–752
11. Ariesta RC, Zubaydi A, Ismail A, Tuswan T, Al-Syachri MZ 2021 *IOP Conf. Ser. Mater. Sci. Eng.* **1034**(1) 012012
12. Tuswan T, Zubaydi A, Piscesa B, Ismail A, Ariesta RC, Ilham MF, Mualim FI. 2020 *J. Appl. Eng. Sci.* 1–9
13. Rajput MS 2018 Damage Tolerance of Impacted Composite Sandwich Structures
14. Prabowo, AR, Tuswan T, Ridwan R 2021 *Appl. Sci.* **11**(9) 3954
15. Yulianto AN, Purwanto DB. 2017 Design and Structure Analysis of Ramp Door Using Finite Element Method
16. BKI Consolidated Edition 2021. BKI Rules Hull Vol. II Tahun 2021, vol. II, 2021, [Online]. Available: https://www.bki.co.id/.
17. Burlayenko VN Sadowski T. 2018 *Shock Vib.* **2018**
18. Takeda N, Minakuchi S, Okabe Y. 2007 *J. Solid Mech. Mater. Eng.* **1**(1) 3–17

19. Ariesta RC, Zubaydi A, Ismail A, Tuswan T. 2021 *Mater. Today Proc.* **47**(10) 2310-2314
20. Burlayenko VN, Sadowski T. 2012 *Finite Elem. Anal. Des.* **62** 49–64

Numerical analysis of the effect of a foil-shaped centerbulb on vertical motion and wake on a catamaran sightseeing boat

Kiryanto[1], Parlindungan Manik[1*], Muhammad Luqman Hakim[1], Ahmad Firdhaus[1,2]

[1)] Department of Naval Architecture, Faculty of Engineering, Diponegoro University
[2)] Department of Systems and Naval Mechatronic Engineering, National Cheng Kung University, Taiwan

*parlindungan_manik@live.undip.ac.id

Abstract. Internal and external factors will influence the ship's ability to maneuver. Therefore, proper planning and calculations are needed to produce good ship comfort and performance in designing a vessel. One of them is paying attention to the quality of the movement, which is the ship's response to the influence of the sea, such as waves and wakes which are the comparison between the planned speed and the fluid flow velocity on the propeller under water. In an effort to improve the quality of good motion, this study tries to add variations in geometry and position on the foil-shaped center bulb to see the effect of vertical motion and wake. This study solves the existing problems by analyzing them using CFD software. Based on the results of the numerical simulation. In three different ship speed conditions, it has the best response when the wave heading is 0° on the ship with the addition of a center bulb in the midship area. It can reduce the vertical motion value up to 9.35% against the vessel without a center bulb and by 8.73% at 180° wave heading. So that the shape and placement of the foil-shaped center bulb on the catamaran can improve the quality of the ship's maneuverability.

Keywords: Wake, Ship Motion, Centerbulb, Catamaran

1. Introduction

Ships, as sea transportation, are floating objects that move in the water. When the ship moves, obstacles will naturally occur in the media it passes so that the ship can move at the desired speed. The obstacles experienced by the ship must be overcome by other forces that push the ship alone [1]. The thrust of the ship is generated by the main engine, where the greater the resistance of a ship, the greater the power of the main engine needed by the ship to push. On the other hand, the increase in the main engine power used to propel the ship will significantly impact the amount of fuel consumption which will increase, which means that the costs incurred for its operation will also increase. The amount of ship resistance is influenced by several aspects; where in addition to the depth of water, the most decisive element is the shape of the ship's hull [2]. Changing the shape of the ship's hull from monohull to catamaran hull can also reduce the ship's drag [3]. The shape of the catamaran hull is one technology that can produce good ship performance, especially for tourism cruises with a vast deck and less fuel than monohulls [4].

Catamarans have two hulls, and their name is derived from the Indian Tamil "Kattumaram." There is a bridge between the two hulls of a catamaran, which means that it has a double hull. Compared to a monohull, a catamaran's hull has a slimmer waterline, so there is less resistance. In addition to having less resistance, a catamaran has a greater level of stability than a monohull [5]. An analysis of fishing boats found that the high pitch and roll motion was the problem [6]. Further, Iqbal modified the hull shape of the catamaran by 10%, which improved the boat's seakeeping. [7]

Samuel et al. used a bulbous bow on catamaran fishing vessels to reduce ship resistance. According to the results, ship resistance can be reduced and increased based on the bulbous bows used [8]. As a result of Danisman's optimization of the elliptical center bulb, it has been positioned and shaped accurately using experimental methods and methods, obtaining a reduction in wave resistance of about 13% and a small margin between numerical and experimental results [9] Studies have also relied on the concept of a center bulb to reduce catamaran resistance [9], [10]. Seakeeping can also be enhanced by the application of center bulbs [11]–[13]

Therefore, the purpose of this work is to offer a CFD simulation that incorporates potential flow and diffraction theory in order to forecast the heave and pitch motion performance of a ship catamaran with and without a foil-shaped center bulb system. The simulation focused primarily on taking into account a number of different impacts, including ship velocity and wave direction. The vertical motion and wake of the catamaran will be evaluated after the geometry has been varied and the location of the center bulb has been changed to take the shape of a foil.

2. Methods

In this study, ships obtained from the results of field surveys on one of the traditional catamaran ships will be used with the main ship sizes as follows:

Table 1. Ship Principal Dimension

Item	Value	Unit
Length of Waterline	8.90	m
Length Overall	10.00	m
Draught	0.50	m
Height	0.80	m
Breadth Demihull	1.10	m
Breadth	3.40	m
Service Speed (Vs)	9.00	knot

Ships with catamarans with the specifications in table 1 will later be used as sightseeing boats that are used to support tourism in the Labuan Bajo area. The characteristics of the catamaran in this study have a ratio of S / L = 0.25, which, based on research, is one of the recommended aspect ratios because it has good wave interference between two demihulls[14], [15]. Where the 3D model can be seen in Figure 1 below

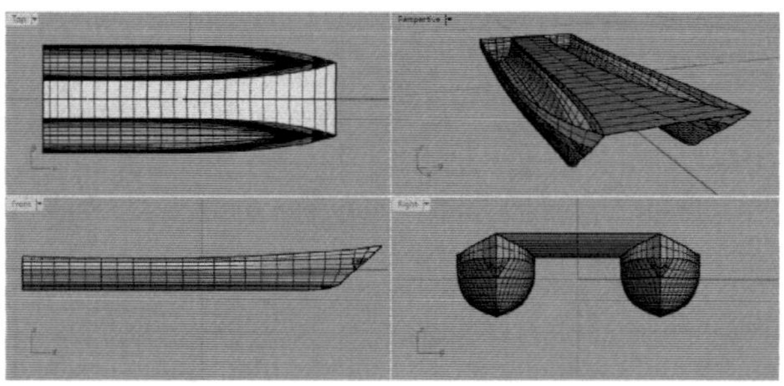

Figure 1. Initial 3D model of the sightseeing catamaran before the addition of the center bulb

The dimensions and sizes of the center bulb are used based on the reference from Danisman as follows in Figure 2. The geometry and size of the center bulb are based on the top and front views. The center bulb used in this study is in the form of a foil based on NACA 0030 and is varied into several configurations and shapes, as shown in Figure 3

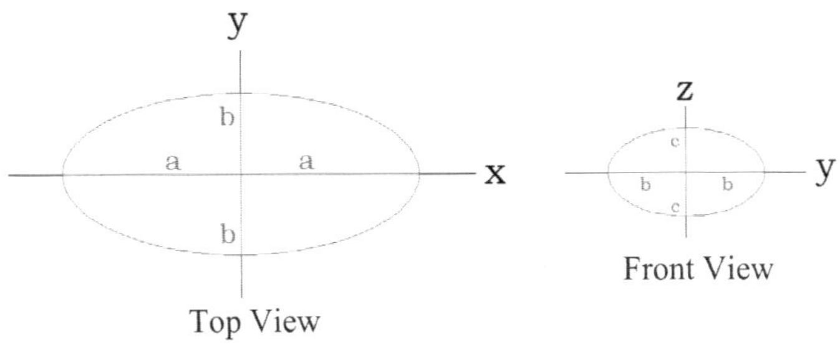

Figure 2. Center bulb geometry with details from the top view is (a1) =1.41m, (b1) = 0.62m, and (c1)=0.34m, while from the front view it is (a2) = 0.5m, (b2) = 0.26m, and (c2) =0.12m

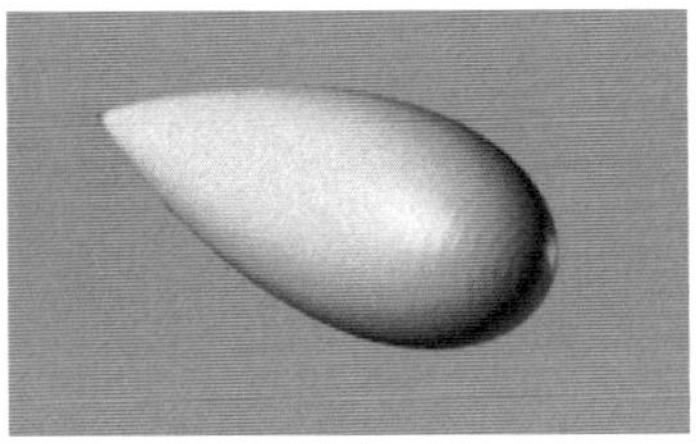

(a)

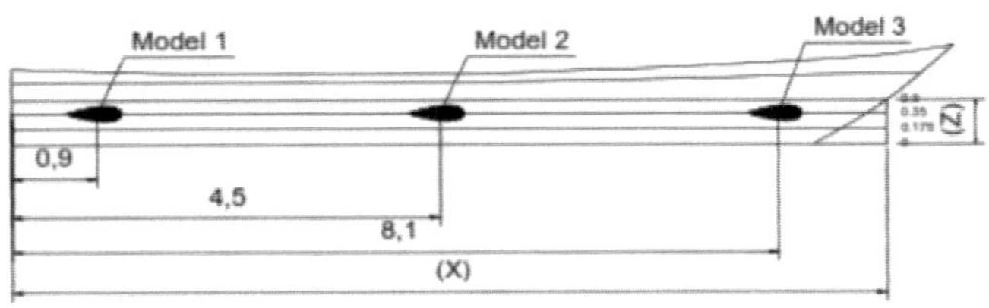

(b)

Figure 3. 3D model of the foil-shaped center bulb (a) and three different configurations in numerical simulation (b)

This study focused on the effect of changing the oval center bulb to a foil-shaped center bulb on the ship's motion and wake. This study was simulated to determine the optimal or smallest value of ship motion and wake and adjusted to the standard seakeeping criteria for fishing vessels.[6], [16]

Table 2. Centerbulb Geometry Variations and Centerbulb Position

Model	Original	1	2	3	4	5	6	7	8	9
Height	-	100%	100%	100%	90%	90%	90%	80%	80%	80%
Width	-	100%	100%	100%	90%	90%	90%	80%	80%	80%
Length	-	100%	100%	100%	115%	115%	115%	130%	130%	130%
Position	-	Aft	Midship	Stern	Aft	Midship	Stern	Aft	Midship	Stern

In this study, calculations were made using numerical simulation. Table 1 displays the formats of the data sets utilized. Maxsurf is used for analyzing hydrostatic properties of the ship, while CFD is used for processing ship motion data. The motion characteristics and performance of the hospital ship catamaran are studied using ANSYS-AQWA in calm seas. The principles of potential flow and diffraction are the theoretical foundations for this numerical program. In this case, we determine the flow rate of an incompressible, viscous, and irrotational fluid. The numerical condition of the test uses CFD software where the significant wave height (hs) used is 1m, where the value is obtained according to the characteristics of the waters in the Labuan Bajo area with an average period (T_{av}) of 3.21 seconds. The observed wave direction is a wave from behind and the front or at an angle of 0° and 180° at three speeds from slow to high (Fr 0.35, 0.5, 0.65). In Figure 4, you can see a model of a catamaran with a center bulb that has been meshed for further numerical simulation calculations on CFD software.

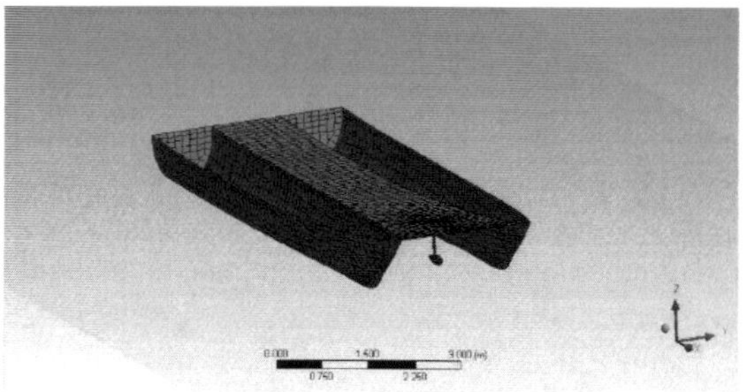

Figure 4. 3D mesh model of a catamaran with the center bulb in pre-process using CFD software

The equation of motion of the ship on the CFD software is expressed as (1)

$$\{m + A\infty\}\ddot{X}(t) + cX(t) + K\ddot{X}(t) = F(t) \qquad (1)$$

In this equation, *m* represents the mass matrix of the structure, *A* represents the mass matrix of the fluid at infinite frequency, and *c* represents the damping matrix that takes into account the linear radiation damping effects. In this case, *K* represents the whole stiffness matrix, while *F(t)* is the external force. [17]

One transfer function that characterizes this frequency-dependent module/raft motion response is known as the Response Amplitude Operator (RAO). These are often not dimensionless with respect to the height of the wave. The theoretical peak power extraction frequencies may be determined with the aid of RAO [18]. RAO is defined by the following equation:

$$RAO = \frac{F_e\omega}{K - \omega^2\big(M + M_a(\omega)\big) + j\omega C(\omega)} \qquad (2)$$

Where $F_e\ \omega$ denotes the force exerted by the excitation wave (including both incident and diffracting forces); The hydrostatic stiffness is denoted by *K*, while the mass of the structure is denoted by *M*. $M_a\ (\omega)$ denotes the mass that was added; Radiation damping is denoted by $C(\omega)$, and the wave frequency is denoted by ω.

Table 3. Numerical parameter for seakeeping calculation

Item	Value
Domain	
Length	22 m
Width	20 m
Height	15 m
Boundary Condition	
Inlet	Velocity inlet
Outlet	Velocity inlet
Bottom	Velocity inlet
Top	Pressure outlet
Wall	Velocity inlet
Ship surface	No-slip wall

The numerical parameters that are used in the computation of seakeeping are detailed in Table 2, which is based on the suggestion of ITTC [19], As can be observed in Fig.4, the formed computational domain has dimensions of -2.3L x 2.7L (22 m) in length, -2.3L y 2.3L (20 m) in width, and -2.3L z 1.1L (15 m) in height. The starting point of the coordinate system is at the same level as the placid sea. It is positioned longitudinally along the after perpendicular and transversely along the ship's centerline, with positive axes pointing toward the bow, port side, and sky. The numerical wave tank has a bottom and four vertical sides, all of which are subject to the velocity inlet boundary condition. Hydrostatic pressure from above is represented by a pressure outlet applied to the upper border. In this scenario, the ship's surface will serve as the no-slip wall boundary condition. It is recommended that there be at least 1.5 ship lengths (or more than one wavelength) of space between the ship's hull and its sidewalls (whichever is greater). The necessary waves may be created in a timely manner before they reach the ship. The ship's wake may be simulated and seen in a big open space located behind the vessel. As can be seen in, the depth of water is set to 15 meters (2.3 liters), which may be thought of as representing an endless depth.

3. Results and Discussion

Based on the diffraction theory's validated potential flow, this section simulated the regular waves sailing in the head. The selected velocities were Fr 0.35, 0.5, and 0.65, and the operating wave characteristics are shown in the previous section. Seakeeping tests have been conducted on steady-state conditions to simulate rolling, heave, and pitch. The calculation of seakeeping in this study is focused on finding the response amplitude operator (RAO) value using a numerical method on ANSYS AQWA, which means the ship's response when receiving waves per 1 m amplitude. This analysis uses seakeeping software in the form of ANSYS AQWA. The resulting data is in the form of a graph presented below.

3.1. Vertical Motion Analysis

The analysis results from the ansys aqwa software produce a graph in the form of Response Amplitude Operator (RAO) for heave and pitch movements, which can be seen in the image below.

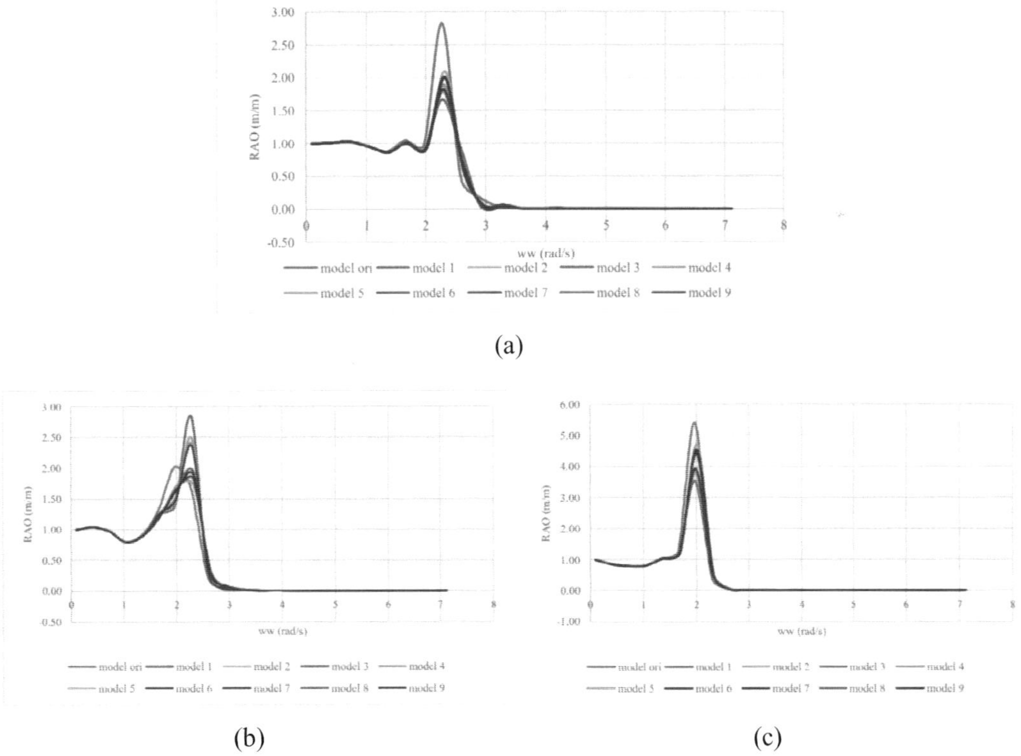

Figure 5. Graph of Response Amplitude Operator (RAO) heaving against wave frequency on catamarans without and with center bulb for nine different conditions with a wave heading 0° conditions at three different speeds, namely (a) Fr 0.35, (b) Fr 0.5, and (c) Fr 0.65

The comparison of RAO response at three different speeds at the same wave heading condition is shown in Fig. 5 (a), (b), and (c), which can be found above. In this situation, the waves are arriving from the same direction as the stern of the ship. Based on the findings shown in graphs 3a, 3b, and 3c, it has been determined that the form of the RAO curve follows a pretty consistent pattern regardless of the speed being considered. This demonstrates that the position of the center bulb in each of the different designs does not have a significant impact on the heaving movement of the ship. The most significant RAO value occurs in vessels with high-speed circumstances at Fr 0.65 (3(c)) without the inclusion of a center bulb with a RAO value of 5.41 m/m at a frequency of 1.97 rad/s. This number indicates that there is no center bulb present. The RAO number that is the lowest when considering catamarans without foil or original versions.

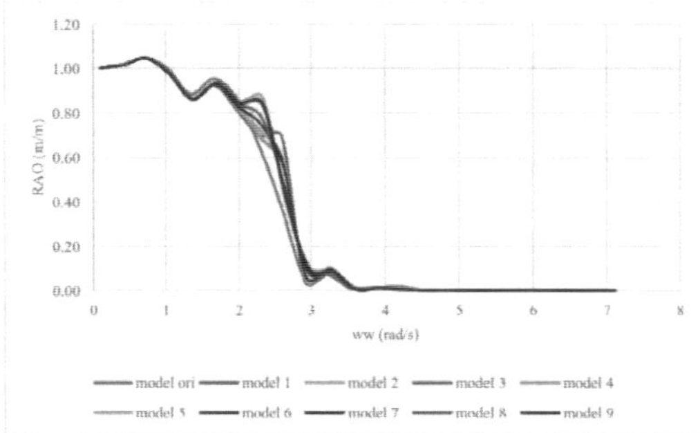

(d)

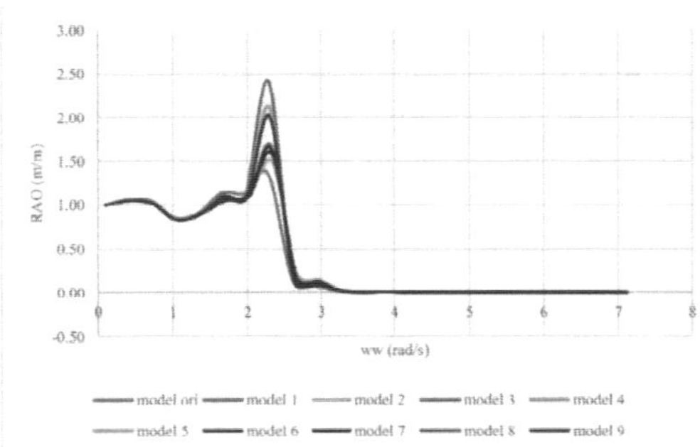

(e)

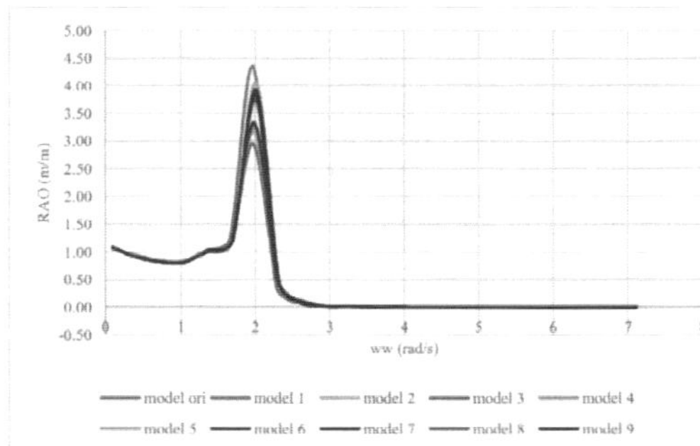

(f)

Figure 6. Graph of Response Amplitude Operator (RAO) heaving against wave frequency on catamarans without and with center bulb for nine different conditions with a wave heading 180° conditions at three different speeds, namely (d) Fr 0.35, (e) Fr 0.5, and (f) Fr 0.65

Observations on the graphs 6(d) to 6(f) indicate that the shape of the RAO curve has a similar trend at three different speeds for the direction of the incoming wave, 180 movements of the ship's motion. This indicates that the location of the center bulb in two different wave directions has no significant impact. Extreme to the ship's heaving motion. The most significant RAO value occurs in boats with a high-speed condition of 6(f) and a center bulb at the bow of the ship with a RAO value of 4.41 m/m at a frequency of 1.97 rad/s, corresponding to the wave direction from the stern of the ship.

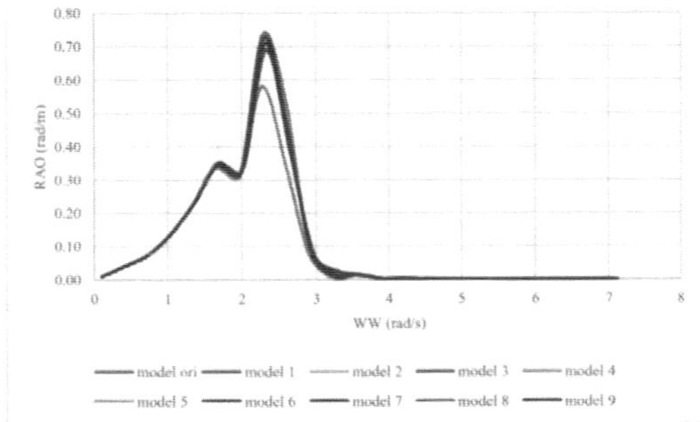

(g)

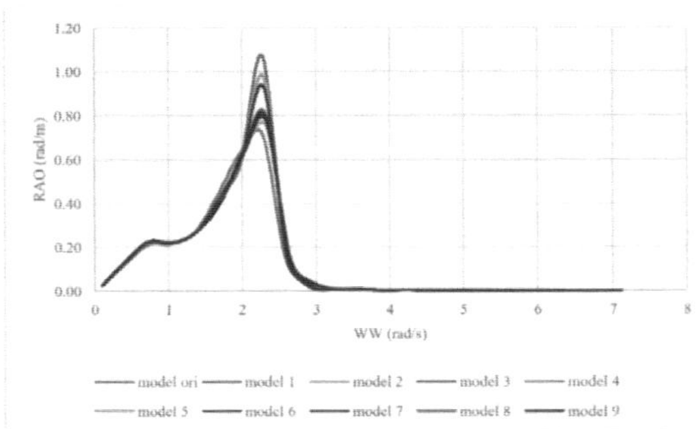

(h)

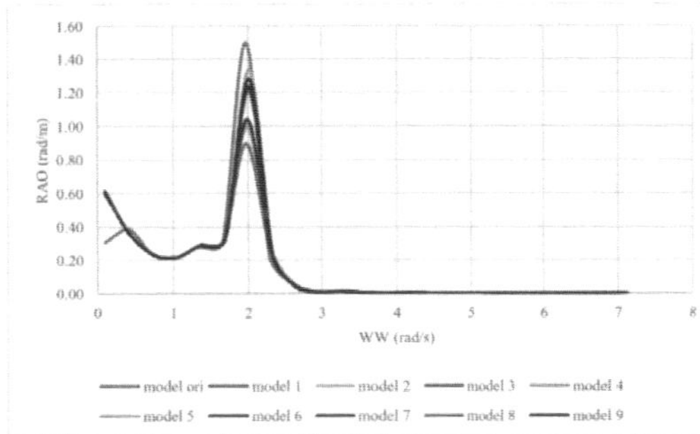

(i)

Figure 7. Graph of Response Amplitude Operator (RAO) pitching against wave frequency on catamarans without and with center bulb for nine different conditions with a wave heading 0° conditions at three different speeds, namely (g) Fr 0.35, (h) Fr 0.5, and (i) Fr 0.65

Figures 7(g) through 7(i) demonstrate pitch RAO at three speeds under a 0° wave incident angle. At Fr 0.35, the original model's max pitch RAO is 0.78 rad/m at 2.34 rad/s. Except for the original model ship, all RAO curves appear the same. RAO curve is left-shifted (towards low frequencies). Original model pitch is worse at 2.60 rad/s. Figure 7(h) shows that at medium speed, model 3 has the highest pitch RAO with 1.15 rad/m at 2.18 rad/s. Each model's RAO curve looks the same, except on the original. RAO peak curve is left-shifted (towards low frequencies). Fr 0.65 in Figure 7 is relatively high (i). Peak pitch RAO is 2.00 rad/m at 1.50 rad/s in the original model. Each model's RAO curve is same.

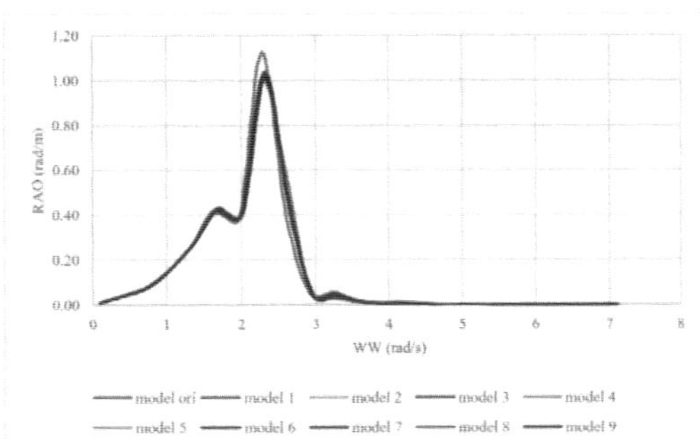

(j)

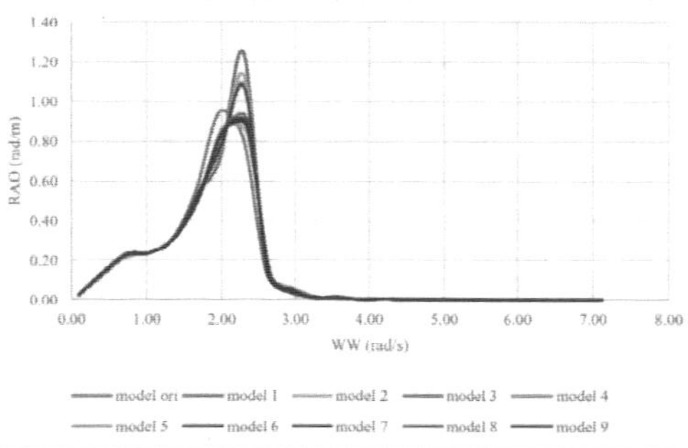

(k)

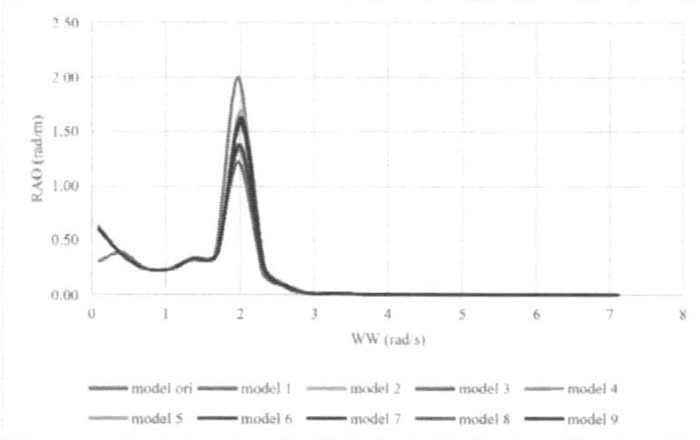

(l)

Figure 8. Graph of Response Amplitude Operator (RAO) pitching against wave frequency on catamarans without and with center bulb for nine different conditions with a wave heading 180° conditions at three different speeds, namely (j) Fr 0.35, (k) Fr 0.5, and (l) Fr 0.65

Figures 8(j) to 8(l) show the RAO of pitch motion at three different speeds under a 180° wave incident angle. At low speed or Fr 0.35, the peak RAO of pitch motion is found in the original model with an RAO value of 1.12 rad/m at a frequency of 2.28 rad/s. And the shape of the RAO curve of each model looks the same, except for the original model ship. The shape of the RAO curve is slightly shifted to the left (towards low frequencies). When at 2.60 rad/s, the most inferior pitch movement is found in the original model. Furthermore, at the medium speed shown in Figure 8(k), It can be seen that the peak RAO of pitch movement is found in model 3 with an RAO value of 1.24 rad/m at a frequency of 2.29 rad/s. The shape of the RAO curve of each model even looks the same, except on the original model ship. The shape of the RAO peak curve is slightly shifted to the left (towards low frequencies). The latter is a relatively high speed, namely Fr 0.65 in Figure 8(l). It can be seen that the peak RAO of pitch movement is found in the original model with an RAO value of 2.00 rad/m at a frequency of 1.97 rad/s. And the shape of the RAO curve of each model looks the same.

Adding a foil-shaped center bulb to the ship can increase or decrease the vertical motion characteristics. This is caused by variations in the geometry and positioning of the center bulb, ship speed, and wave blades. From the overall Fr, at an angle of 0°, model 2 has the best response and can reduce the vertical motion value up to 9.35% against the original model. At a slope of 180°, model 2 can reduce the sheer motion value up to 8.73% compared to the original model. The foil-shaped centerbulb for catamarans has a favorable influence on ship performance, according to past study [20], [21].

3.2. Wake Fraction Analysis

In determining the wake value, the results of the advanced speed on the scaled model are then calculated by the Fn value to find the original advanced speed value. Then from the initial advance speed, the wake/current value is searched using the formula

$$W = 1 - (\frac{V_a}{V_S}) \tag{3}$$

Table 4 shows the comparison of the value of V_a For each variation model.

Table 4. V_a and Wake values at Fr 0.35

Fr	Model	Vs (m/s)	Va (m/s)	Difference (%)	Wake	Difference (%)
0.35	Original	1.0394	0.5088	0.00%	0.5105	0.00%
	1	1.0394	0.4846	-4.75%	0.5338	4.55%
	2	1.0394	0.4932	-3.05%	0.5255	2.93%
	3	1.0394	0.4974	-2.23%	0.5215	2.14%

4	1.0394	0.4916	-3.38%	0.5271	3.24%
5	1.0394	0.4952	-2.66%	0.5236	2.55%
6	1.0394	0.4949	-2.72%	0.5239	2.61%
7	1.0394	0.4937	-2.96%	0.5250	2.84%
8	1.0394	0.5103	0.29%	0.5091	-0.28%
9	1.0394	0.4997	-1.79%	0.5193	1.71%

Table 5. V_a and Wake values at Fr 0.49

Fr	Model	Vs (m/s)	Va (m/s)	Difference (%)	Wake	Difference (%)
0.49	Original	1.455	0.735	0.00%	0.494	0.00%
	1	1.455	0.700	-4.75%	0.518	4.85%
	2	1.455	0.713	-2.93%	0.509	2.99%
	3	1.455	0.718	-2.29%	0.506	2.34%
	4	1.455	0.711	-3.21%	0.511	3.27%
	5	1.455	0.727	-1.08%	0.500	1.11%
	6	1.455	0.715	-2.69%	0.508	2.75%
	7	1.455	0.712	-3.02%	0.510	3.08%
	8	1.455	0.735	0.05%	0.494	-0.05%
	9	1.455	0.720	-2.00%	0.505	2.05%

Table 6. V_a and Wake values at Fr 0.65

Fr	Model	Vs (m/s)	Va (m/s)	Difference (%)	Wake	Difference (%)
0.65	Original	1.930	0.997	0.00%	0.483	0.00%
	1	1.930	0.949	-4.77%	0.507	5.10%
	2	1.930	0.945	-5.26%	0.510	5.62%
	3	1.930	0.968	-2.92%	0.498	3.13%
	4	1.930	0.964	-3.36%	0.500	3.59%
	5	1.930	0.973	-2.37%	0.495	2.53%
	6	1.930	0.977	-1.97%	0.493	2.10%
	7	1.930	0.968	-2.95%	0.498	3.16%
	8	1.930	1.001	0.37%	0.481	-0.39%
	9	1.930	0.979	-1.82%	0.492	1.94%

From table 4 to table 5, model 8 can reduce the wake fraction value and has a better value than the original model at Fr 0.35, 0.49, and 0.65. At Fr 0.35, the optimal 8 model reduces the wake value by 0.28%. Also, at 0.49 Fr, model 8 reduces to 0.05%. At Fr 0.65, model 8 reduces the wake value by 0.39%

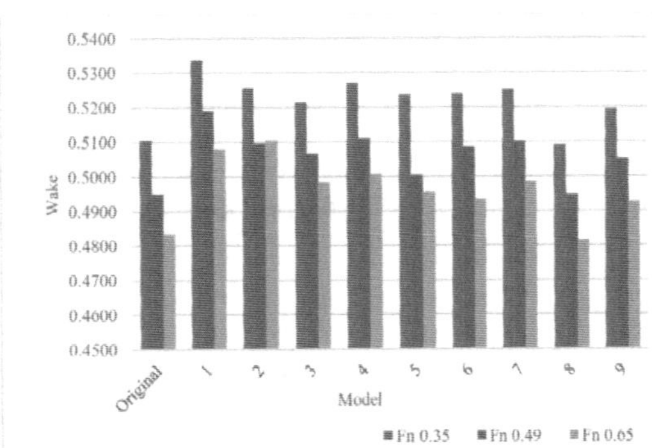

Figure 9. Summary of wake fraction calculation results

In Figure 4.35, it can be concluded that the wake fraction value decreases at Fr 0.35, Fr 0.49, and Fr 0.65 for each model.

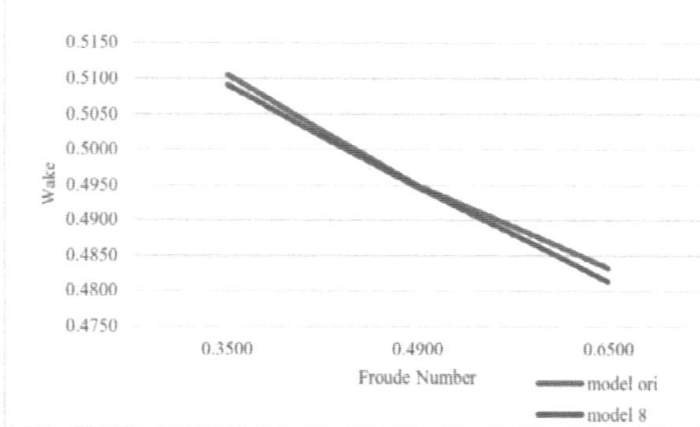

Figure 10. Results of Comparison of Wake Fraction Values of the catamaran without center bulb original model and the catamaran with slightly flattened and more extended center bulb

Figure 10 shows that at Fr 0.35, Fr 0.49, and Fr 0.65, the wake value of the catamaran with a slightly flattened and more extended center bulb (model 8) can reduce the wake value to the original model's wake value.

The wake analysis results show that the model's variation can reduce or increase the value of the wake fraction. At Fr 0.35, Fr 0.49, and Fr 0.65 model 8, the wake fraction value can be reduced to the original model. As seen at Fr 0.35, model 8 can reduce the wake value by 0.28% against the original model. Also, at Fr 0.49, model 8 can reduce the wake value by 0.05% against the original model, and at Fr 0.65, model 8 can reduce the wake value by 0.39% against the original model.

4. Conclusions

Analysis of the addition of a center bulb in the form of a foil on a catamaran using numerical calculations has been carried out and obtained the following results, the addition of a center bulb can provide advantages and disadvantages for vertical motion characteristics. This is caused by variations in the geometry and positioning of the center bulb, ship speed, and the angle of incidence of the waves. In three different ship speed conditions, it has the best response when the wave direction is 0° on the ship with the addition of a center bulb in the midship area. It can reduce the vertical motion value up to 9.35% against the model without a center bulb. Then in the direction of the incoming wave against the ship (180°), there is a decrease in vertical motion value by 8.73%. Meanwhile, for wake analysis on vessels with the addition of a center bulb, the slightly flatter and pointed shape of the center bulb, which is placed in the midship area of the ship, provides an advantage, where at three different speeds, the wake fraction value for ships decreased to 0.40%. The results of numerical calculations in this study still need to be validated using experimental testing to get more convincing results.

References

[1] S. Anggara and I. Utama, 'Analisa CFD Pengaruh Penambahan Appendage pada Lambung Katamaran terhadap Hambatan Viskos', *Jurnal Teknik POMITS*, vol. 1, no. 1, 2012.

[2] T. Suteja, K. Suastika, and Murdijanto, 'Analisa Hambatan Kapal dengan Bulbous Bow dan tanpa Bulbous Bow di Perairan Dangkal', *Undergraduate Thesis of Naval Architecture and Ship Building Engineering*, 2015.

[3] M. Iqbal and Samuel, 'Traditional catamaran hull form configurations that reduce total resistance', *International Journal of Technology*, vol. 8, no. 1, pp. 989–997, 2017, doi: 10.14716/ijtech.v8i1.4161.

[4] J. Tupan and R. B. Luhulima, 'Perbandingan Kapal Monohull, Catamaran, Trimaran Berdasarkan Tinjauan Operasional Penggunaan Bahan Bakar', in *Archipelago Engineering*, 2021. doi: https://doi.org/10.30598/ale.

[5] P. Adi and A. Amiadji, 'Analisa Penerapan Bulbous Bow Pada Kapal Katamaran Untuk Meningkatkan Efisiensi Pemakaian Bahan Bakar', *Jurnal Teknik ITS*, vol. 3, no. 1, 2014, doi: 10.12962/j23373539.v3i1.6033.

[6] M. Tello, S. Ribeiro e Silva, and C. Guedes Soares, 'Seakeeping performance of fishing vessels in irregular waves', *Ocean Engineering*, vol. 38, no. 5, pp. 763–773, 2011, doi: https://doi.org/10.1016/j.oceaneng.2010.12.020.

[7] M. Iqbal and R. Good, 'Optimasi bentuk demihull kapal katamaran untuk meningkatkan kualitas seakeeping', *Kapal: Jurnal Ilmu Pengetahuan dan Teknologi Kelautan*, vol. 12, no. 1, pp. 19–24, 2015.

[8] Samuel, D.-J. Kim, M. Iqbal, A. Bahatmaka, and A. Rio Prabowo, 'Bulbous bow applications on a catamaran fishing vessel for improving performance', *MATEC Web of Conferences*, vol. 159, 2018, [Online]. Available: https://doi.org/10.1051/matecconf/201815902057

[9] D. B. Danışman, 'Reduction of demi-hull wave interference resistance in fast displacement catamarans utilizing an optimized centerbulb concept', *Ocean Engineering*, vol. 91, pp. 227–234, 2014, doi: https://doi.org/10.1016/j.oceaneng.2014.09.018.

[10] G. K. Saha, K. Suzuki, and H. Kai, 'Hydrodynamic optimization of a catamaran hull with large bow and stern bulbs installed on the center plane of the catamaran', *J Mar Sci Technol*, vol. 10, no. 1, pp. 32–40, 2005, doi: 10.1007/s00773-004-0186-6.

[11] D. Bruzzone, A. Grasso, and I. Zotti, 'Nonlinear seakeeping analysis of catamarans with central bulb', in *6th International Conference on High-Performance Marine Vehicles (HIPER'08)*, 2008.

[12] R. Prio Aprijal and M. Iqbal, 'Minimisasi Hambatan Dan Gerak Vertikal Kapal Multihull Catamaran Dengan Centerbulbs', *TEKNIK*, vol. 39, no. 1, pp. 48–54, 2018, doi: 10.14710/teknik.v39n1.10565.

[13] I. Zotti, 'Medium speed catamaran with large central bulbs: experimental investigation on resistance and vertical motion', in *International Conference on Marine Research and Transportation*, 2007, pp. 167–174.

[14] A. S. Sakti and I. K. A. P. Utama, 'Analisis CFD dan Eksperimen Hambatan Lambung Katamaran Asimetris Flat Side Outside dengan Variasi Jarak Demihull', vol. 1, no. 3, pp. 78–83, 2012.

[15] M. Insel and A. F. Molland, 'An Investigation Into Resistance Components of High Speed Displacement Catamarans', *The RINA*, vol. 134, no. January, pp. 1–20, 1992.

[16] M. Iqbal, M. Terziev, T. Tezdogan, and A. Incecik, 'Operability analysis of traditional small fishing boats in Indonesia with different loading conditions', *Ships and Offshore Structures*, pp. 1–20, Aug. 2022, doi: 10.1080/17445302.2022.2107300.

[17] ANSYS, 'Aqwa Theory Manual', in *ANSYS 17.2 - Aqwa Theory Manual*, 2016, pp. 137–158.

[18] B. Bosma and B. Borma, 'On the Design, Modeling, and Testing of Ocean Wave Energy Converters', p. 87, 2013, [Online]. Available: http://ir.library.oregonstate.edu/xmlui/handle/1957/41003

[19] ITTC, 'ITTC-Recommended Procedures and Guidelines Practical Guidelines for Ship CFD Applications', 2011, pp. 2–18.

[20] M. Iqbal, P. Manik, S. Hadi, and A. Kurniawan, 'Pengaruh Posisi Centerbulb Berbentuk Foil Terhadap Komponen Hambatan Kapal Ikan Katamaran MV. Laganbar', *Jurnal Integrasi*, vol. 64, no. 1, pp. 2548–9828, 2020.

[21] M. Iqbal, U. Budiarto, K. Z. Hidayat, H. H. Hernanta, and A. Trimulyono, 'The influence of foil-shaped center bulb geometry into catamaran fishing vessels resistance', *IOP Conf Ser Mater Sci Eng*, vol. 1034, no. 1, p. 012030, Feb. 2021, doi: 10.1088/1757-899x/1034/1/012030.

Design optimization of connector on ship interceptor

O Mursid[1], Samuel[1], S Yulianti[1] and N Huda[2]

[1]Department of Naval Architecture, Diponegoro University, Indonesia
[2]Mechanization Fisheries Study Program, Sorong Polytechnic of Marine and Fisheries, Indonesia

Correspondence author : samuel@ft.undip.ac.id

Abstract. Currently, research on an energy-saving device only focuses on the effective operation of this device. Structural assessment is needed to evaluate the reliability of the interceptor in operation conditions. In this study, 25 mm thick interceptors are evaluated using the finite element method to find an optimized position of the bolt connector. The simulation shows that a smaller size of bolt connector leads to a higher maximum stress of interceptor. Based on the position of the bolt connector, it has a similar pattern in every bolt size. Overall, the best choice to install this device on the boat is between 140 mm and 240 mm of bolt distance. This distance is advised for installing an interceptor in the existing boat structure to avoid weakening the boat structure.

1. Introduction

Interceptor is an ESD (energy saving device) on the ship in a rectangular shape, installed on the ship's stern. Interceptors reduce ship resistance and increase ship performance . Compared to other devices, an interceptor is a controllable device that can change its position depending on the ship's environmental condition and velocity [2].

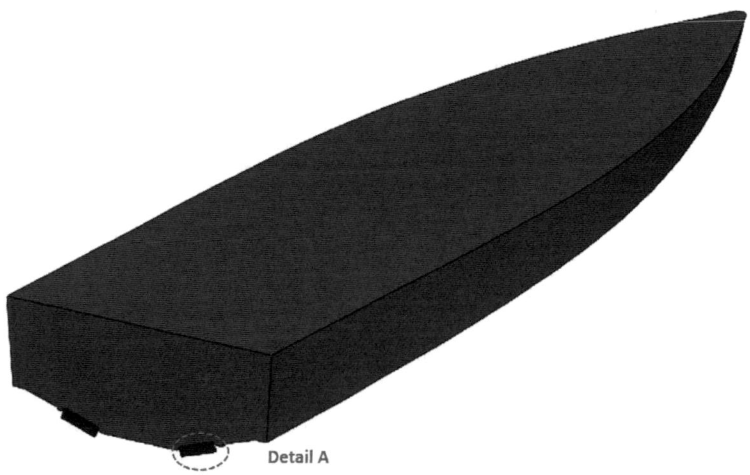

Figure 1. Interceptor Position

Several types of research on the energy-saving device still focus on the ship's resistance and hydrodynamics. On another side, research on the structural strength aspect has little attention. Other energy saving devices have calculated structural strength [3]–[7]. A strength evaluation of the ESD is needed to asses this reliability. Several cases show that additional structures can increase the probability of failure in the existing structure. We need to note that the existing structure is designed without additional devices.

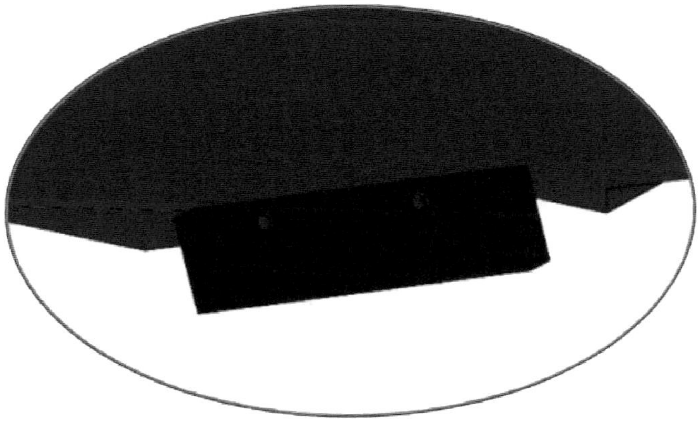

Figure 2. Interceptor Detail

Bolt connection is a common construction method with easy installation and disassembly. Figure 1 shows interceptor installation on the deep-V planing hull ship, which details in figure 2. Based on figure 6, the location of the bolt connection needs to consider the existing boat's structural aspect to minimize the structure's reduction strength. The design of the connection interceptor needs to adapt hull structure where each ship structure has a different construction. This study aims to optimize the location and size of bolt connectors in an interceptor. This study's result is expected to advise on additional devices on the ship.

2. Methods

Variation size and distance of the bolt connecter are used in this study. Three different sizes of bolt connections such as M10, M8, and M6, are used in this study. This study calculates three small bolts to minimize modification to the existing stern ship structure. This bolt is used to connect the interceptor to ship stern construction. The bolt's distance is changed, shown in figure 3 in the "B" notation.

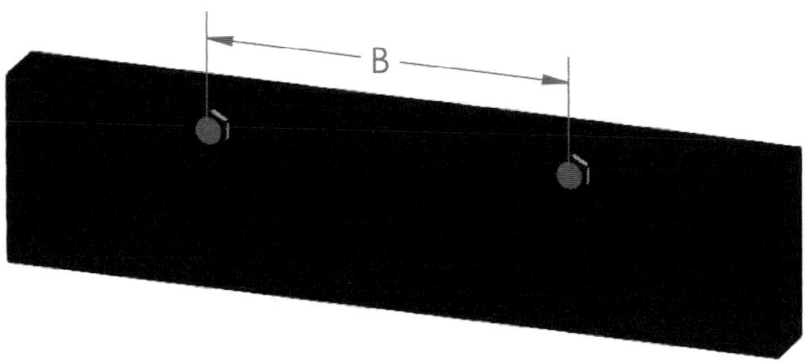

Figure 3. Interceptor detail.

300x70x25mm thick plate aluminium 6031 marine grade is used to simulation. Comparing to steels material aluminium have about three times lighter, it needed to an additional device which do not more effect to shift of centre of gravity. 6031 of aluminium have 62.03 MPa of yield strength.

In this study even load of pressure is applied on interceptor. Finite element modelling is used to evaluate strength of interceptor. On interceptor 75% of front area is exposed to fluids flow that means the rest area of the interceptor are touch to the stern plate of the boat. 25 knots of water flow is applied to pressure load in interceptor.

3. Result and discussion

Based on the simulation, a smaller bolt size can increase the maximum stress on the interceptor. This situation is caused by smaller contact between bolt and interceptor hole. Three different bolt sizes have similar patterns when applying a pressure load. 160 mm of bolt space is the most effective bolt position that divides an interceptor in balance pressure load.

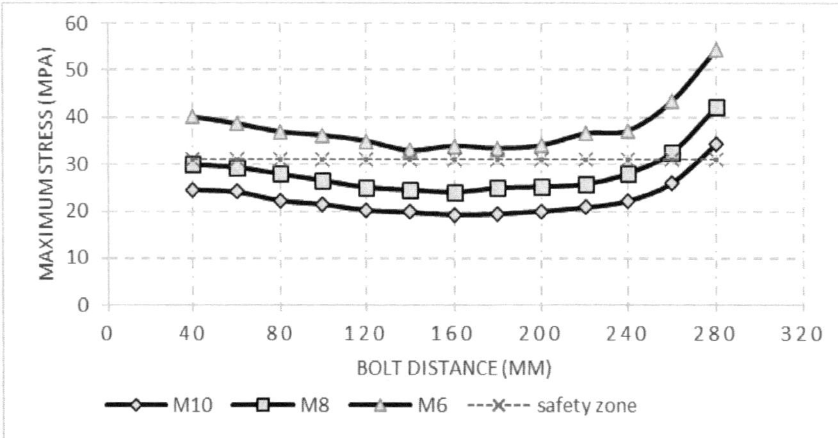

Figure 4. Maximum Stress Comparison

At 25 knots of ship velocity, M6 of the bolt is not recommended because the maximum stress is higher than the safety stress of the Aluminium 6031. The alternative can be affordable in the lower speed of the boat or higher tensile of materials. Overall, applying a narrow and wider space of bolt connection on the interceptor must be avoided to get lower stress. Higher maximum stress can lead to a damaging and less durable interceptor.

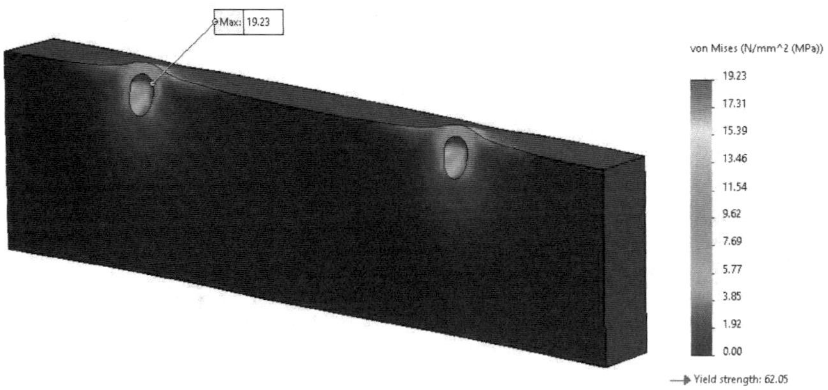

Figure 5. Stress Plot M10

Based on figures 5 – 7, each bolt diameter at 160 mm of space bolt has an exact location of maximum stress in the hole of a bolt. This condition describes that connection structure is the most important in the structural evaluation of the interceptor. In installing an interceptor to an existing ship, awareness of structural aspects is needed, especially in the connection area.

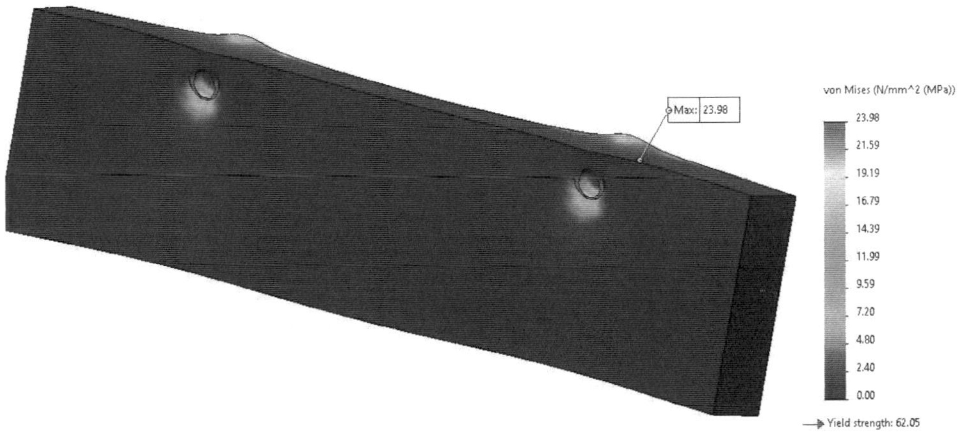

Figure 6. Stress Plot M8

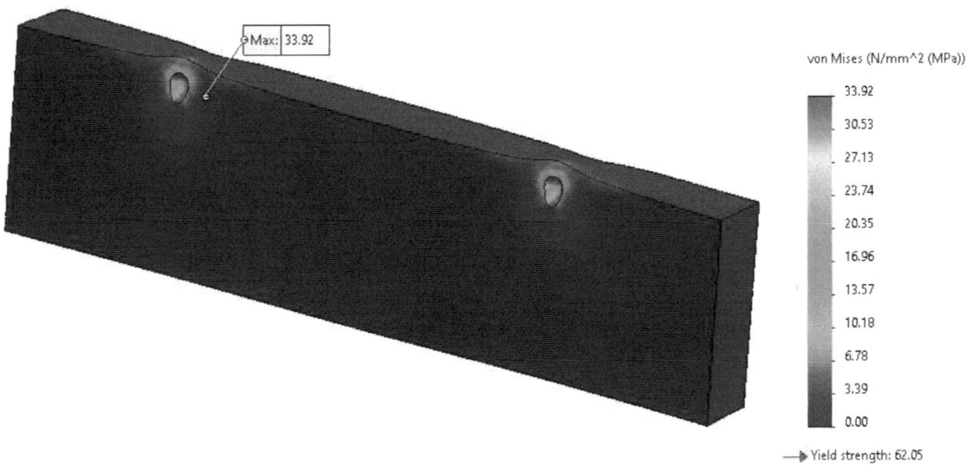

Figure 7. Stress Plot M6

4. Conclusion

The simulation shows that connection size and position can affect the strength of the interceptor. 10 mm of the bolt is the lowest maximum stress compared to other. Spacing of bolts connector has optimized position in 140 up to 220 mm in the interceptor. This simulation can advise development interceptors, especially in applying to existing structures.

Acknowledgments

The Authors would acknowledge to Faculty of Engineering University of Diponegoro for RKAT 2022 number 356/UN7.5.3.2/HK/2022 for funding this research.

5. References

[1] K. Song, C. Guo, J. Gong, P. Li, and L. Wang, "Influence of interceptors, stern flaps, and their combinations on the hydrodynamic performance of a deep-vee ship," *Ocean Engineering*, vol. 170, pp. 306–320, 2018, doi: https://doi.org/10.1016/j.oceaneng.2018.10.048.

[2] C. Gang, W. Wei, and H. Cong, "Experimental study on pitching reduction of ship model with controllable interceptor in wave," 2020.

[3] D. Lee, B.-S. Jang, and H. Kim, "Development of procedure for structural safety assessment of energy saving device subjected to nonlinear hydrodynamic load," *Ocean Engineering*, vol. 116, pp. 165–183, 2016, doi: https://doi.org/10.1016/j.oceaneng.2016.02.038.

[4] H.-B. Ju, B.-S. Jang, D.-B. Lee, H.-J. Kim, and C.-K. Park, "A simplified structural safety assessment of a fin-typed energy saving devices subjected to nonlinear hydrodynamic load," *Ocean Engineering*, vol. 149, pp. 245–259, 2018, doi: https://doi.org/10.1016/j.oceaneng.2017.12.022.

[5] A. Bakica, N. Vladimir, H. Jasak, and E. S. Kim, "Numerical simulations of hydrodynamic loads and structural responses of a Pre-Swirl Stator," *International Journal of Naval Architecture and Ocean Engineering*, vol. 13, pp. 804–816, 2021, doi: https://doi.org/10.1016/j.ijnaoe.2021.09.002.

[6] P. W. Wang, P. K. Liao, C. Y. Hsin, Y. Quéméner, and C. W. Lin, "Ship motions contribution to the fatigue life of a pre-swirl stator," *Proceedings of PRADS2016*, vol. 4, p. 8th, 2016.

[7] J. Jiao, H. Ren, and C. Guedes Soares, "A review of large-scale model at-sea measurements for ship hydrodynamics and structural loads," *Ocean Engineering*, vol. 227, p. 108863, 2021, doi: https://doi.org/10.1016/j.oceaneng.2021.108863.

Sustainable ship health monitoring for side hull structure using the random decrement method

R C Ariesta[1], **A Zubaydi**[1*], **N P L Sukma**[1], **S H Sujiatanti**[1], **H Siswanti**[1]
[1]Department of Naval Architecture, Faculty of Marine Technology, Institut Teknologi Sepuluh Nopember, Surabaya, Indonesia
Email: zubaydi@na.its.ac.id

Abstract. Damage detection of sandwich material used in the ship structure has attracted more awareness to providing diagnostics for Structural Health Monitoring (SHM) systems. This study reported damage identification of sandwich material based on the Random Decrement Method (RDM) using experimental and numerical approaches. The investigation was assumed by modeling the material in two conditions which are undamaged and damaged conditions to analyze the dynamic behaviors of vibration response. The results of the vibration response obtained a similar trend between undamaged and damaged material conditions. Furthermore, a decreased stiffness occurs due to damage to the sandwich plate. Therefore, the random decrement method can be selected as a detection method for identifying damage with fair accuracy and obtaining sustainability for the ship structure.

1. Introduction

Lightweight materials like sandwich materials are introduced as sustainable and advanced and used in different fields such as aerospace and marine industry due to the characteristics of sandwich materials reducing the lightweight by up to 15%, corrosion resistance, heat insulation, and vibration damping [1],[2],[3]. Sandwich materials develop in the shipping industry to replace the existing material in reparation conditions. However, the sandwich material's ease is the complexity of the configurations, depending on the geometry, load, and conditions, which will affect the failure. Therefore, the development of structural health monitoring (SHM) system for sandwich material has been an option and concern this year [4]. Structural health monitoring (SHM) is started by visual inspection to approximate the damage's location. However, the damage in the bonded structures is barely visible or even invisible. Hence, SHM aims to obtain the diagnosis of the condition material [5]. Numerical and experimental analysis was performed on sandwich material applications for structural construction [6],[7]. However, the application in the maritime industry is minimal. Thus, non-destructing testing (NDT) methods are the best choice to detect internal damage in a short time. One of the non-destructing testing (NDT) methods, namely the vibration-based technique, has been developed to extract damaging information, such as the dynamic properties of sandwich material [3], [8], [9]. Several studies have been performed to characterize the structure's dynamic behaviour based on indicators, i.e., natural frequency, frequency response function, frequency-time domain, and mode shape [10], [11].

The present study analysis focuses on the damage identification of sandwich material in the parallel middle body of Tanker 17,500 DWT using the Simplifying and analysing data in real time is the main reason for developing this method. Furthermore, the steps of work are divided into two components: (1) identifying the modal parameter of the material response and (2) applying of random decrement method (RDM). Therefore, the modal parameter can compare the intact and damaged models using RDM. random decrement method. The main reason for using this method is to simplify and analyze data in real-time. This work is divided into two components: (1) identifying the modal parameter of the material response and (2) applying of random decrement method (RDM). Therefore, the modal parameter is able to compare the intact and damaged models using RDM.

2. Method

2.1. Design of sandwich plate

In this study, the sandwich plate composition is an A36 steel plate and polyurethane elastomer with reinforced fiberglass. Configurations of faceplate and core material were used to substitute the existing hull plate on the tanker's side, as shown in Figure 1. First, a minimum thickness of sandwich material must comply with the limitations of hull structure requirements. Then, Llyod Register R explains the minimum thickness calculated with the standard using the strength ratio (R) on each part of the ship with a different value. At last, the configuration thickness requirement defines using Equation (1).

$$R = 0{,}01 \,.A_R \left[0{,}2\,.\frac{b^2}{d(t_1+t_2)} + 11{,}7\left(\frac{b.t_c}{d^2}\right)^{1{,}3}\right] k.p_{eq}.R \qquad (1)$$

Where R is the strength index that the maximum value must be less than one, t_c is the core of the panel represented with a mm. Moreover, t_1 and t_2 are the thickness of the top and bottom plating mm, and b is a breadth of a panel at the shortest edge mm.

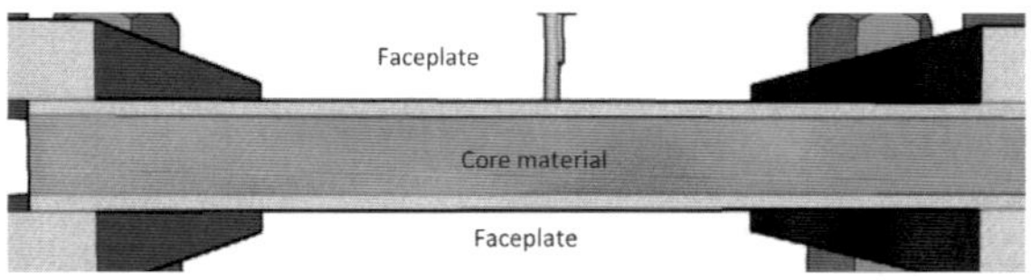

Figure 1. Sandwich material design configuration

2.2. Experimental modal analysis

Experimental Modal Analysis (EMA) is performed in two conditions, intact and damaged conditions but made in a scale model [12]. The following experimental setup, as seen in Figure 2. was clamped and secured with bolts to determine the settle position on both sides of the sandwich material. In this test, the load was applied with hammer force on the surface of the sandwich, and the accelerometer transmitted the vibration response; the natural frequency measurement will focus on the range 500-1500 Hz. Furthermore, data acquisition will record the vibration response. The main dimensions of the specimen used in this experiment are 300 mm x 300 mm, and the faceplate and core material thickness are 4 mm and 20 mm.

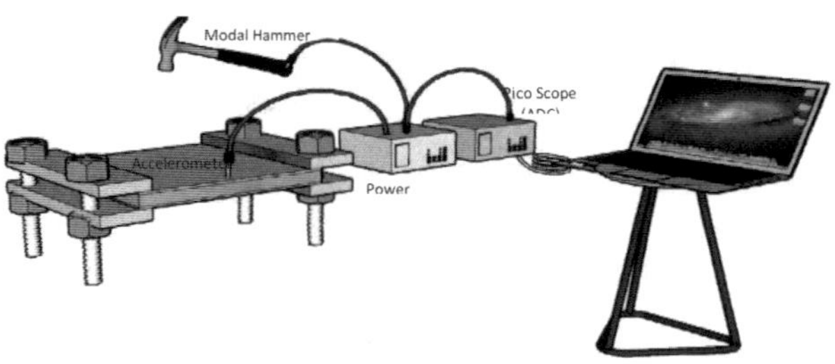

Figure 2. Configuration of Experimental Modal Analysis

Then, PicoScope 6 was generated with Mathcad 14 tools to process data obtained from the acquisition data to get the amount value from vibration amplitude.

2.3. Finite element analysis

Numerical analysis is carried out with finite element analysis (FEA). The analysis settings are adapted to the experimental conditions where both sides are clamped. Discretization of elements defined for each material, shell-type elements applied on faceplates, and solid element types are used for core materials. Contact between elements is bound to assume the existence of adhesive bonds in both materials. The material properties that used for this analysis are summarized in **Table 1.**

Table 1. Materials properties of materials

Characteristics	Faceplate	Core material
Density ρ (kg m-3)	7850	1098
Modulus Young ρ (Pa)	200×10^9	901.95×10^6
Poisson's Ratio	0.3	0.36

A total of two models were generated, as follows in EMA configuration. The faceplate element was defined as a continuum shell element (SC8R), and the solid element (3CD8R) was applied to the core.

2.4. Random Decrement Method

Many methods can identify the vibration response of damage. However, this research aims to investigate applying the random decrement method (RDM). RDM function is to calculate an averaging of specified data random response of the structure determined using a triggering shape with the same initial conditions. The equation of random decrement method can calculate using Equation (2):

$$RD(\tau)\} = \frac{1}{N}\sum_{i=0}^{N}\{x(t_n + \tau)\} \quad (2)$$

N defined a segment number of responses, signal over time represented by x $(t_n+\tau)$ and t simultaneously to verify the initial conditions. To determine the difference between the intact and damage model, the application of the equation takes in the initial vibration response. That will compile with the theory, the stiffness will decrease, and the unbonded sandwich will cause the acceleration to be high because of crack initiation.

3. Result and discussion

3.1. The procedure for selecting bandwith ranges

FRFs models for damaged and undamaged are drawn in Figure 3, which shows that FRFs have a similar pattern, and the magnitude is almost the same for the two conditions. The graph also shows that the highest amplitude of the graph is around the natural frequency of around 500-600 Hz, and it is the first mode. An anti-resonant was investigated in the natural frequency of around 240-300 Hz.

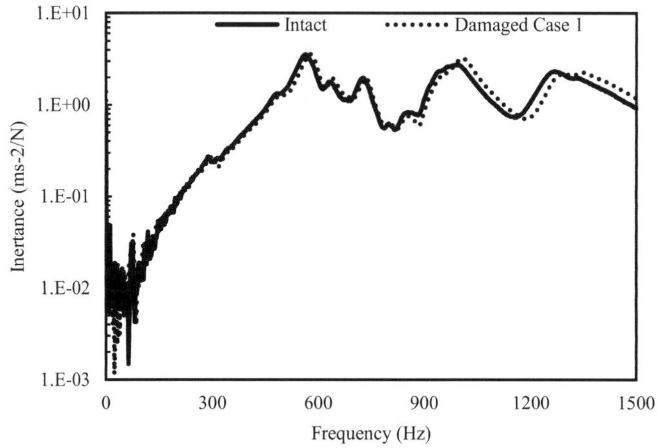

Figure 3. FRFs graph [7]

3.2. Comparison damage evaluation of sandwich material using FEA and EMA

Figure 4 shows the force resulting from the hammer force on the sandwich material. The FEA and EMA test result found that the vibration response has the same trend for the two conditions named intact and damaged. This result was under the research conducted by [13]. Vibration responses of EMA would be smaller than FEA on the same trend data [14]. The most noticeable theory of sandwich material is that the faceplate will change behavior as a component that transfers the load caused by the hammer force to the core material. Moreover, the equipment testing, and the specimen will likely affect the result.

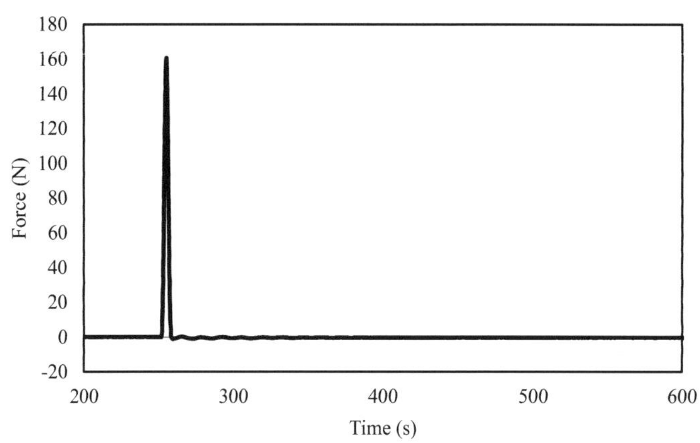

Figure 4. Impact force from hammer

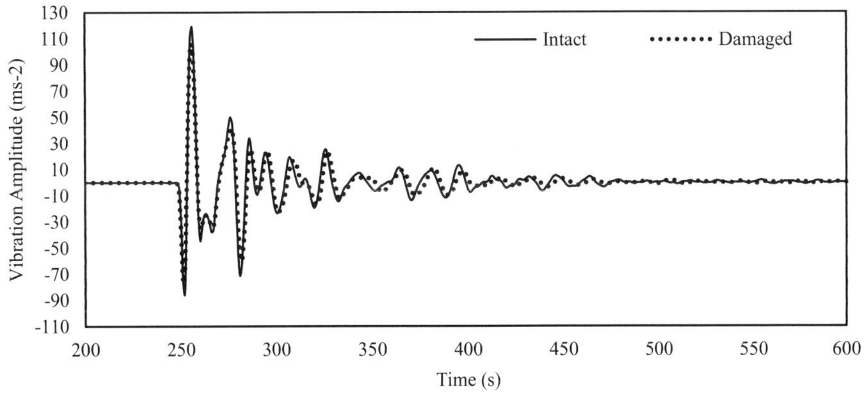

Figure 5. Vibration Response of Experiment Modal Analysis (EMA) [13]

Time interval, also called bandwidth, will affect the signal response. Therefore, it is necessary to do an equation in the time domain. Time domain theory explains that it is divided into several types of waves, including random responses changed in random response filters and made into random signatures. In this study, the results of the filtered response were obtained that it needed to be adjusted from the narrow bandwidth value to the overall bandwidth value extracted into the random signatures response. Then, EMA analysis results are verified using the finite element analysis (FEA) model with time domain adjustments that are present in Figure 6 and Figure 7. Based on the FEA show that results complete in the same trend.

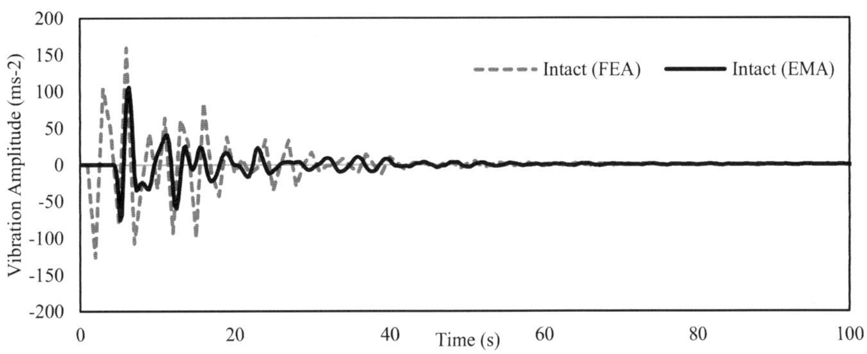

Figure 6. Vibration Response of model intact (FEA) and EMA

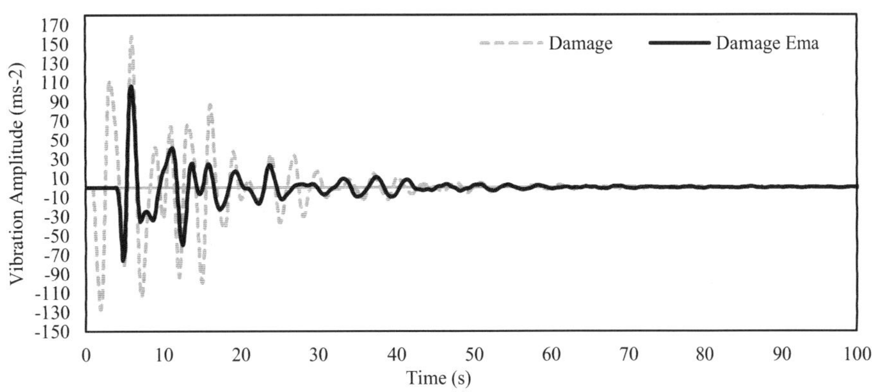

Figure 7. Vibration Response of model damage (FEA) and EMA

3.3. Damage evaluation of sandwich material using Random Decrement Method

Development of damage identification using the Random Decrement Method (RDM) must be implemented immediately. As the next step of FEA and EMA, a study of the sensitivity of the random decrement signature to damage identification has also been carried out. The results show that the random decrement signature has a similar behaviour for intact and damaged models. Where both have the highest peak after being given a load, the amplitude value will move towards zero with the addition of time.

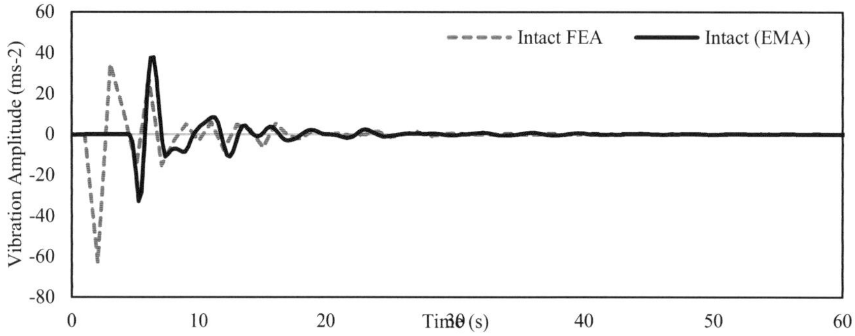

Figure 8. Vibration Response of model damage (FEA) and EMA

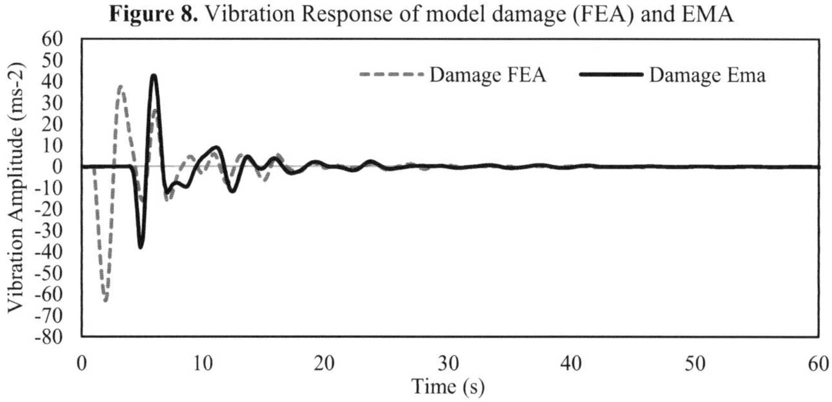

Figure 9. Random Decrement vibration of model Damage (FEA) and EMA

The change of vibration response is caused by damage that occurs. In detail, the random decrement equation results show the average data summarized in Figure 8 and Figure 9. The response between intact material and damage shows a significant difference in the initial conditions. A wave characterizes it in acceleration response at 0 – 10 seconds, which has an almost nearest trend. More clearly, the trend changes that occur are observed by enlarging the time interval to get a clearer picture of the acceleration caused by the sandwich material. Previously, this case had been investigated where there is a deviation when there is a difference in response. More details on the response at time intervals in the narrower range are presented in Figure 10 and Figure 11, where a peak change in the vibration amplitude response can be observed, indicating damage to the structure. Therefore, in the future, this modelling study can develop continuous structural identification to investigate the presence of vibrational damage to ships, especially sandwich materials.

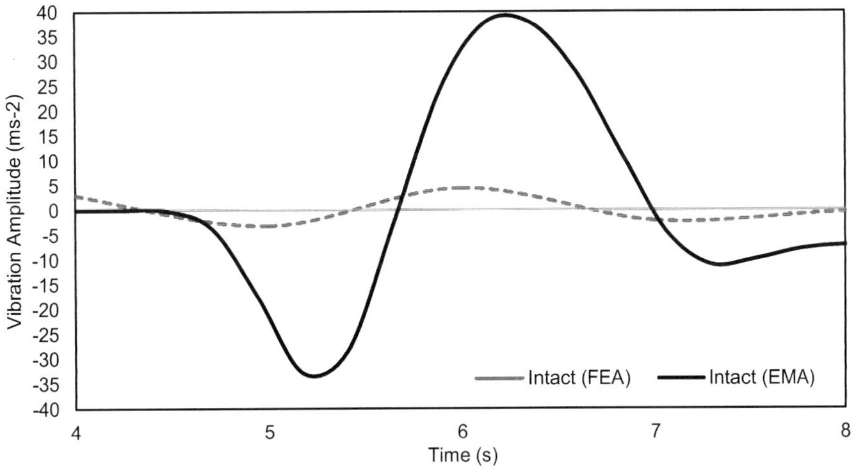

Figure 10. Amplitude of model intact FEA and EMA

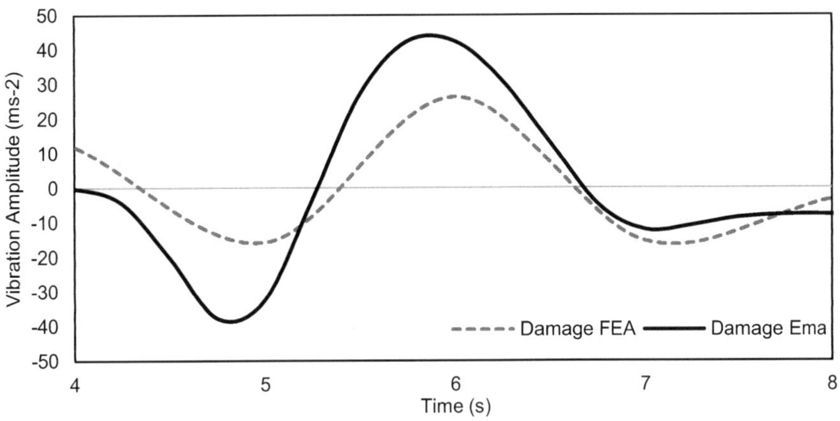

Figure 11. Amplitude of model damage FEA and EMA

4. Conclusion

Based on the analysis results, the occurrence of damage results in a change in the response. The random decrement method was completed for analysis. That represents the average vibration response value that can show and observed when damage occurs, resulting in reduced stiffness; therefore, the vibration response becomes large. This technique can identify the global ship structure in future studies and easily find damage in ship construction. Sustainable health monitoring needed investigation, especially on the old ship structure.

Acknowledgements

This work was supported by Penelitian Magister - Year 2022 with entitled "Pengembangan Teknik Random Decrement untuk Identifikasi Kerusakan Pelat Sandwich pada Struktur Lambung", with Grant/Contract No. 084/E5/PG.02.00.PT/2022, and 1430/PKS/ITS/2022. The support is gratefully acknowledged by the authors.

References

[1] Tuswan T, Abdullah K, Zubaydi A and Budipriyanto A 2019 Finite-element analysis for structural strength assessment of marine sandwich material on ship side-shell structure *Materials Today: Proceedings* vol 13

[2] A. Zubaydi and A. Budipriyanto 2020 *Material Sandwich: Teori, desain, dan aplikasi, 1st ed* (Surabaya: Airlangga University Press)
[3] Ismail A, Zubaydi A, Piscesa B, Ariesta R C and Tuswan 2020 Vibration-based damage identification for ship sandwich plate using finite element method *Open Eng.* **10**
[4] Sha G, Radzieński M, Cao M and Ostachowicz W 2019 A novel method for single and multiple damage detection in beams using relative natural frequency changes *Mech. Syst. Signal Process.* **132**
[5] Katunin A and Przystałka P 2015 Automated wavelet-based damage identification in sandwich structures using modal curvatures *J. Vibroengineering* **17**
[6] Lou J, Wu L, Ma L, Xiong J and Wang B 2014 Effects of local damage on vibration characteristics of composite pyramidal truss core sandwich structure *Compos. Part B Eng.* **62**
[7] Ariesta R C 2022 Identification of Damage Size Effect of Natural Frequency on Sandwich Identifi cation of Damage Size Eff ect of Natural Frequency on Sandwich Material using Free Vibration Analysis Identifi kacija učinka prirodne frekvencije na veličinu oštećenja na sendv
[8] Tuswan T, Zubaydi A, Piscesa B, Ismail A, Ariesta R C, Ilham M F and Mualim F I 2021 Influence of application of sandwich panel on static and dynamic behaviour of ferry ro-ro ramp door *J. Appl. Eng. Sci.* **19**
[9] Budipriyanto A, Haddara M R and Swamidas A S J 2007 Identification of damage on ship's cross stiffened plate panels using vibration response *Ocean Eng.* **34** 709–16
[10] Tuswan T, Diponegoro U, Zubaydi A, Piscesa B and Ismail A 2021 PANEL ON STATIC AND DYNAMIC BEHAVIOUR
[11] Tuswan T, Zubaydi A, Piscesa B, Ismail A, Ariesta R C and Prabowo A R 2022 A numerical evaluation on nonlinear dynamic response of sandwich plates with partially rectangular skin / core debonding *Curved Layer. Struct.* **9** 25–39
[12] Ewins D J 2001 MODAL ANALYSIS, EXPERIMENTAL | Applications *Encyclopedia of Vibration*
[13] Ariesta R C, Zubaydi A, Ismail A and Tuswan T 2021 Damage evaluation of sandwich material on side plate hull using experimental modal analysis *Mater. Today Proc.*
[14] Ariesta R C, Zubaydi A and Ismail A 2021 Identification of damage in a ship hull sandwich plate by natural frequency Identification of damage in a ship hull sandwich plate by natural frequency

Powering Design of Portable Dynamic Positioning Thruster for 150 FT Cable Laying Barge: A Case Study of Untung Jawa-Tanjung Pasir Submarine Power Cable Laying

I Made Ariana[1], Georgius Steven Wiryadinata[1]

[1]Department of Marine Engineering, Institut Teknologi Sepuluh Nopember (ITS), Surabaya, Indonesia

corresponding author's e-mail: georgiussteven@gmail.com

Abstract. Indonesia, an archipelagic nation with most of its electrical power produced by steam power plant. The distribution of electrical power is done through the main island to the smaller island, by means of according to Ministry of Energy and Mineral Resource is done through a submarine cable. In the study case area, Untung Jawa-Tanjung Pasir sea, the most appropriate cable laying platform is a barge. A barge is not equipped with the propulsion system and considering the cable laying operation requirement, barge given condition, environmental disturbances, the barge must be equipped with portable dynamic positioning thruster. An azimuth thruster is the proposed way to overcome this problem. The research done shall determine the barge's necessary engine and propeller to fulfil its need. Using ship manoeuvring equation, the required thrust of the barge for surge, sway, and yaw are 274.36 kN, 289.21 kN, 3292.67 kNm respectively. To produce the required forces and moment, the researcher utilizes Cummins KTA38 (634 kW) diesel engine, Masson Marine gearbox (5.542:1), and Ka4-70 Nozzle 24 with the diameter of 1.6 m and P/D value of 1.06. In the free running condition, the barge will run with the speed of 3.52 m/s or 6.85 knots.

Keywords: Barge, Hydrodynamic coefficients, Ship maneuvering, Submarine Cable Laying, Thruster

1. Introduction

Nowadays, steam power plant become the biggest electrical power source in Indonesia. Indonesia relies on steam power plant as the source of electrical power source for almost 4 decades **[1]**.There is more than 50% of Indonesia's steam power plants located in Java Island. Considering the size of the steam power plant, it must be located at the main island. The previous statement is also supported by the data given by the table above. As we can see from the table above, those 8 steam power plants are located at the main island of Java. Given by the condition explained above, the author concludes that the electrical power distribution of Indonesia flows from the main or larger islands to the smaller islands.

Based on the data of Ministry of Energy and Mineral Resource, submarine cable is one of the electrical power distribution methods **[2]**. The installation of submarine cable called submarine cable laying. Submarine cable laying can be done by using several platform options, for example, cable laying vessel, supply vessels, and barge **[3]**. In determining the utilization of cable platform, the most definitive factor is the sea depth. Cable laying vessel and supply vessels are commonly used to lay cable in the deep ocean or sea, while barge is commonly used for both deep ocean and shallow sea.

As the submarine cable method is used to distribute electrical power from the main or larger island to the smaller island, the marine subsurface geology of the sea is one of the parameters to choose suitable vessel to lay the submarine cable. The sea between Tanjung Pasir and Untung Jawa island is having the maximum seabed depth of approximately 20 m under water surface. Considering the deepest point of the mentioned sea, using cable laying vessel or supply vessels would be an excellent choice, but the overall condition of the marine subsurface geology would not permit the utilization of such cable laying platform. Hence, the most suitable cable laying platform for Untung Jawa and Tanjung Pasir sea is barge.

Different with cable laying and supply vessel, barge is not equipped with propulsion system. Due to the absence of propulsion system, the barge cannot perform the cable laying duty. Due to given condition of the barge, the cable laying company should install additional propulsion system to the barge. Considering the ease of installation, portable propulsion system is proposed.

To install the portable propulsion system on the barge, it is essential for cable laying company to estimate the power requirement of the barge. One of the parameters should be considered to estimate barge's power requirement is the barge speed during laying operation. During the consideration and determination of the barge laying speed, maximum cable tension is the most influential factor to be considered. The barge laying speed is to be designed within the condition where the maximum cable tension is not exceeded.

Besides barge laying speed, environmental disturbances should be considered in designing the barge portable propulsion system. Along the cable laying route which length is 5 km, marine environmental disturbances such as wind, wave and current are influencing the barge motion. If the disturbances are strong enough, the barge could go offtrack, hence the allocated submarine cable length on the barge could be inadequate for the given cable laying route length. Therefore, the portable propulsion system should be able to move the barge forward and maintain the barge on the given cable laying route. In another words, the portable propulsion system should have dynamic positioning capability. In this research the proposed portable propulsion system is azimuth thruster.

Considering all the statements mentioned within the previous paragraph, there are 2 main objectives of this research. The first objective is to determine 150ft barge power requirement for 2 purposes during cable laying operation. Those purposes are dynamic positioning and self-propelling. The second objective is to determine engine and propeller specification of the azimuth thruster based on required power.

2. Methods

Methods represent the processes implemented by the author to gain or to meet the objective of the research. As already explained within the introduction section, there are 2 main objectives of this research. As the consequences, in general, the methods can be divided by 2. The first part is used to determine the 150-ft barge power requirement. The second part is used to translate or convert the power requirement into engine and propeller specification. Figure 1 represents the method implemented by the author to meet this research objective

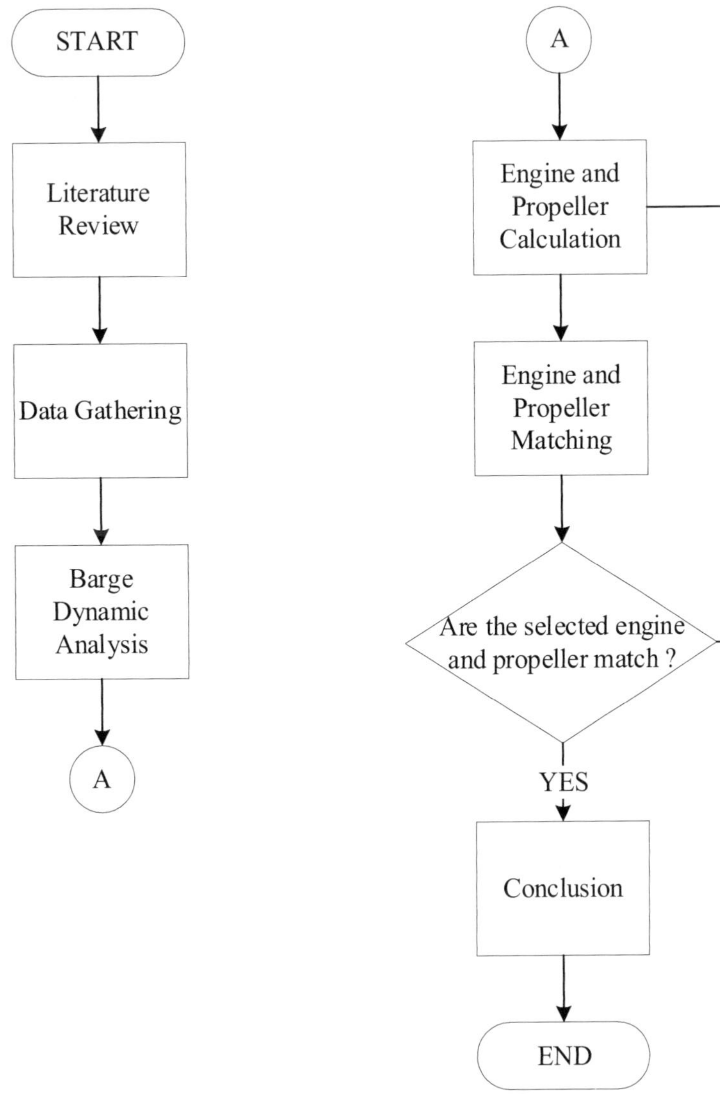

Figure 1. Flowchart of Research Methodology.

2.1. Data Gathering

To fulfill the objective of this research, there are 4 main data required by the author. Those data are 150-ft barge data, list of cable laying equipment, environmental data, and submarine cable data. 150-ft barge data concerns about the principal dimension of the barge such as overall length, breadth, depth, and deadweight tonnage. List of cable laying equipment represents the equipment being loaded onboard to support and to fulfill cable laying operation. List of cable laying equipment are essential for stability analysis where several parameter such as displacement, draft, waterline length, and form coefficient can be acquired. Environmental data required by the author are wind, wave, and current at Untung Jawa and Tanjung Pasir sea. Environmental data are essential to determine the value of environmental disturbance matrix of ship maneuvering equation. The last data, submarine cable specification are essentail for both stability and tension analysis. The specification mainly concerns about the cable weight, diameter and maximum allowable tension.

2.2. *Barge Dynamic Analysis*

The aim of barge dynamic analysis is to determine the value of the thrust matrix which represents the required thrust to perform self-propelling and dynamic positioning. To determine the value of thrust matrix, the author used ship maneuvering equation. In the equation, there are several parameters which value must be acquired. Those parameters are accelerations, hydrodynamic coefficients, velocities, and environmental disturbances. Accelerations and velocities can be acquired through simulation. Hydrodynamic coefficients are partly dependent to velocities. Hydrodynamic coefficients can be acquired by processing the result of stability analysis and velocities simulation. Environmental disturbances can be acquired by processing environmental data and the result of velocities simulation. After all the mentioned parameter values already determined, it will be inputted to the ship maneuvering equation. Hence, the value of thrust matrix can be determined.

2.3. *Engine Propeller Calculation*

Using the principles of moment, the values of forces and moment within thrust matrix are being converted into propeller thrust or T. With the given diameter and nominal rotation, propeller thrust value can be converted into Kt which later on used as the coefficient which value is to be plotted into KQ-Kt-J diagram of the selected propeller. By reading KQ-Kt-J diagram using the determined Kt value, several propeller parameters can be acquired. Those parameters are P/D and KQ. From the acquired KQ value, engine power requirement can be determined.

2.4. *Engine Propeller Matching*

The aim of engine propeller matching is to validate whether the power absorbed by the propeller almost equal to or equal to power produced by the main engine at the designed speed or not. From the earlier statement, the author limits the matching point into some point where propeller consumes 99 percent or more power produced by main engine. If it less than 99 percent, the author will go back to engine propeller calculation to adjust several parameters such as diameter or rotation of the propeller. The author will redo the calculation until the author found the propeller which consumes 99 percent of produced engine power or more.

3. Result and Discussion

3.1. *Barge Specification and Environmental Data*

The process of barge dynamic analysis is initiated by determining load draft, displacement, and center of gravity. It is conducted by inputting list of equipment onboard to Maxsurf Stability. Based on the result obtained by using Maxsurf Stability, the center of gravity (COG) is located 0.008 m behind midship, 0.04 m from B/2 and 2.30 m from barge keel. The displacement and draft obtained from stability analysis result is shown in the table 1

Table 1. 150-ft Barge Principal Dimension Based on Stability Analysis Result.

Parameter	Value	Unit
Displacement	403.9	ton
Draft at FP	0.78	m
Draft at AP	0.78	m
WL Length	42.45	m
Beam max extents on WL	12.19	m
Wetted Area	582.04	m2
Waterpl. Area	517.51	m2
Prismatic coeff. (Cp)	0.98	-
Block coeff. (Cb)	0.98	-

In designing the dynamic positioning thruster, it is essential to concern about the environmental condition of the operating area. This study is taking Untung Jawa and Tanjung Pasir sea as its case area.

Therefore, the environmental data must be taken from the case study area. On the other hand, from the operation point of view, based on **[4]**, cable laying operations considered as weather restricted operations since its planned operation time is less than 72 hours. The environmental data condition is shown in the table 2

Table 2.Environmental Data Scenario.

Environmental Data	Parameter	Value	Unit
Wave	Height	1.25	m
Wind	Speed	22.13	Knots
Current	Speed	36.69	cm/s

3.2. Determination of Barge Velocity and Submarine Cable Tension Analysis

In ship maneuvering equation, there are matrixes are dependent and to be multiplied by the values of maneuvering velocities. Those matrixes are damping and Coriolis matrix. Hence, the determination of barge particular velocity is essential to complete the barge maneuvering equation. Besides for completing the ship maneuvering equation, the values of maneuvering velocities are essential and to check whether the laying speed is suitable from the submarine cable tension point-of-view.

3.2.1. Barge Particular Velocity

This study is using ship maneuvering equation to determine the required thrust of the barge, therefore there are 3 velocities to be determined. Those velocities are surge velocity, sway velocity, and yaw velocity. To determine the mentioned velocities the author is using simulation. Table 3 provides the result of barge maneuvering velocities due to 8 kinds of wave directions

Table 3. Barge Maneuvering Significant Velocities at 1 Knot Forward Speed from 8 Wave Directions.

Wave Direction	Surge Velocity		Sway Velocity		Yaw Velocity	
	Value	Unit	Value	Unit	Value	Unit
0	0.209	m/s	0.1850	m/s	0.3337	°/s
45	0.8508	m/s	1.0126	m/s	0.3092	°/s
90	0.1888	m/s	0.1959	m/s	1.1654	°/s
135	0.3131	m/s	0.4242	m/s	3.9656	°/s
180	0.1848	m/s	0.1940	m/s	0.6066	°/s
225	1.2489	m/s	1.3474	m/s	0.4808	°/s
270	0.2202	m/s	0.2187	m/s	1.0445	°/s
315	0.2846	m/s	0.2931	m/s	3.1687	°/s

Based on the cable laying project document, the maximum pay-out speed of the cable tensioner is 1000 m/h. The value is also equal to 0.28 m/s. In order to maintain the cable angle of attack relative to sea surface, the maximum surge velocity of the barge should be equal to cable tensioner pay-out rate. Therefore, the maximum surge velocity is 0.28 m/s. To choose the representative velocity, there are several steps taken by the author. The first step, the author chose the wave direction which have the closest surge significant velocity compared to the maximum value. The second step is the author calculates the root-mean-square value of surge, sway and yaw velocity and take the value as the representative barge velocity.

Table 4. Barge Representative or Root-Mean-Square Velocity.

Wave	Surge Velocity		Sway Velocity		Yaw Velocity	
Direction	Value	Unit	Value	Unit	Value	Unit
270	0.14	m/s	0.12	m/s	0.79	°/s

3.2.2. Submarine Cable Tension

Tension analysis is done to ensure at the given forward speed and sea depth, the submarine cable tension not exceeding its maximum tension. The maximum cable tension of the existing submarine cable is 120 kN. For the safety consideration, the author reduces the maximum submarine cable tension at operation by 20 percent. Therefore, the maximum cable tension at operation is having the value of 96 kN. Before implementing the tension analysis of the submarine cable, it is essential to determine the value of submarine cable parameter such as horizontal bottom tension, layback length, minimum bending radius (MBR), and catenary length. The determination of the submarine cable parameter will be done in 3 variations of depth 18.92 m,7.29 m, and 2.79 m. The arrangement of cable laying is shown in the figure 2 to figure 4.

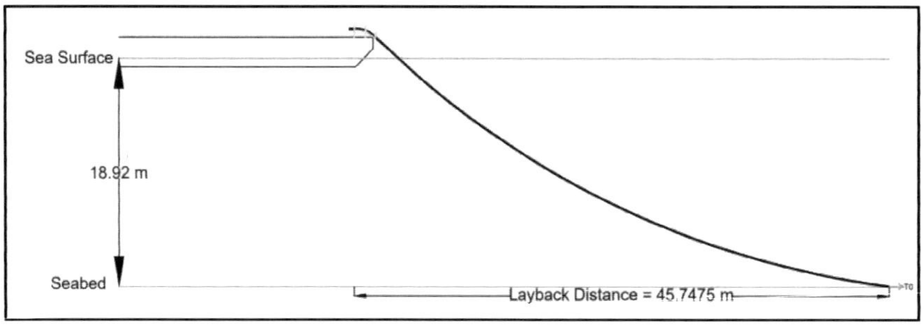

Figure 2.Submarine Cable Parameter at 18.92 Depth.

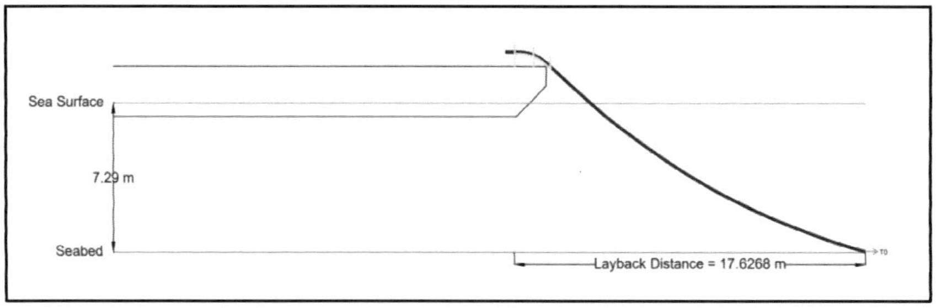

Figure 3. Submarine Cable Parameter at 7.29 Depth.

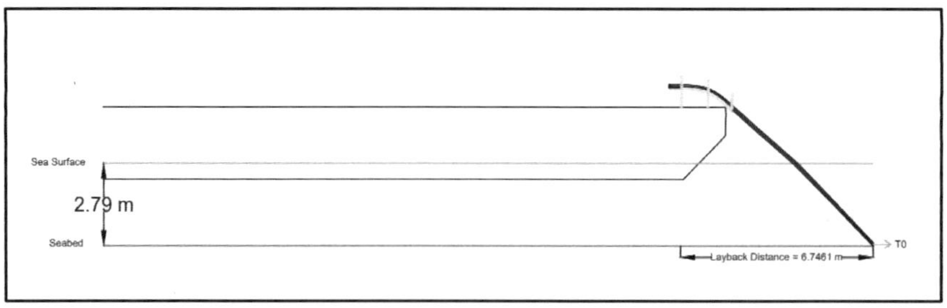

Figure 4.Submarine Cabe Parameter at 2.79 m Depth.

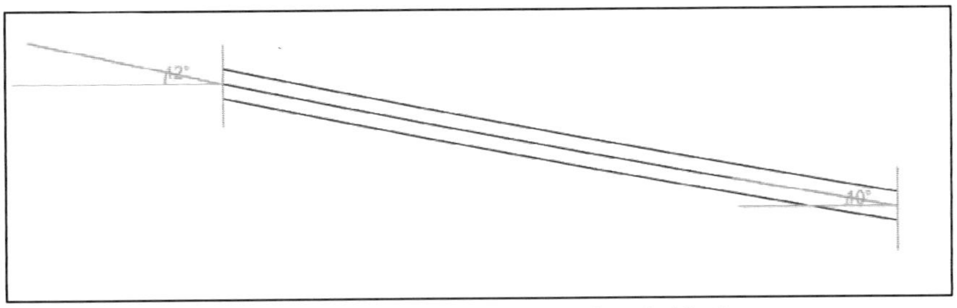

Figure 5.Change of Angle on Submarine Cable Segment.

Figure 5 shows there's a change of angle in cable catenary segment from 10 to 12 degrees. The angle change in cable catenary segment makes the cable tension is getting higher when it is closer to the sea water surface. The next step after determining submarine cable parameters and laying configuration is to determine current relative speed. The formula used to determine current relative speed is taken from [5] as shown by equation (1):

$$V_{rc} = \sqrt{u_{rc}^2 + v_{rc}^2} = \sqrt{(u - u_c)^2 + (v - v_c)^2} \quad (1)$$

From equation (1) the relative current velocity at several current angle of attack is to be presented in table 5

Table 5. Relative Current Speed at Various Current Angle of Attack.

Current Speed (m/s)	Angle of Attack (degrees)	Relative Current Speed (m/s)
	0	0.45
	45	0.53
	90	0.55
	135	0.49
0.3669	180	0.37
	225	0.23
	270	0.20
	315	0.31

Since the speed is proportional to force, therefore the author decides to implement tension analysis at the highest relative current speed, in this case, 0.55 m/s. To determine the submarine cable peak tension, the author used the formula taken from [6] can as shown by equation (2) and (3)

$$T_1 = T_0 \times e^{\int_{\varphi_0}^{\varphi} \frac{\omega \sin\varphi - K_t(V_p - V_{cs}\cos\varphi)^2}{\omega\cos\varphi - k_d V_{cs}^2 \sin^2\varphi} d\varphi} \quad (2)$$

$$T_{n+1} = T_n \times e^{\int_{\varphi_n}^{\varphi_{n+1}} \frac{\omega \sin\varphi - K_t(V_p - V_{cs}\cos\varphi)^2}{\omega\cos\varphi - k_d V_{cs}^2 \sin^2\varphi} d\varphi} \quad (3)$$

The value of submarine cable peak tension will be presented in the table 6

Table 6. Relative Current Speed at Various Current Angle of Attack.

Depth (m)	Submarine Cable Peak Tension (kN)
18.92	15.24

Depth (m)	Submarine Cable Peak Tension (kN)
7.29	5.87
2.79	2.51

3.3. *Barge Particular Acceleration*

The determination of barge particular acceleration is also essential to complete the ship maneuvering equation. Since the scope of this study is concerning the ship maneuvering equation, therefore surge, sway and yaw acceleration must be determined. To determine the barge particular acceleration, the author also uses simulation and implement the same step to determine the barge particular velocity. In the previous section, the author already explained that the representative barge representative velocities were chosen from the 270 degrees wave direction simulation result. Responding to the earlier statement, the author decides to use the acceleration from 270 degrees wave direction as the representative accelerations. Therefore, the barge representative accelerations can be written as shown by the table 7

Table 7. Relative Current Speed at Various Current Angle of Attack.

Wave Direction	Surge Acceleration		Sway Acceleration		Yaw Acceleration	
	Value	Unit	Value	Unit	Value	Unit
270	0.65	m/s^2	0.70	m/s^2	4.71	$°/s^2$

3.4. *Hydrodynamic Coefficients Calculation*

Hydrodynamic coefficients defined as terms representing hydrodynamic forces and moments acting on the vehicle [7]. The calculation and value of hydrodynamic coefficients are essential to complete the ship manoeuvring equation. Hydrodynamic coefficients calculation is done by using several equations from [8], [9], and [10]. Afterwards, several hydrodynamic coefficients should be normalized. The normalization is done by using equations from [8] and [11] .There are two hydrodynamic coefficients are considered, inertia-Coriolis matrix hydrodynamic coefficient and damping matrix coefficient. Table 8 and 9 shows the value of hydrodynamic coefficients after being normalized

Table 8. Value of Inertia and Coriolis Matrix Hydrodynamic Coefficients.

Hydrodynamic Coefficient	Value	Unit
$X_{\dot{u}}$	-14096.11	kg
$Y_{\dot{v}}$	-130771.07	kg
$Y_{\dot{r}}$	1173953.33	kg.m
$N_{\dot{v}}$	632706.84	kg.m
$N_{\dot{r}}$	-20808396.32	$kg.m^2$
I_z	23131789.04	$kg.m^2$

Table 9. Value and Unit of Damping Matrix Hydrodynamic Coefficients.

Hydrodynamic Coefficient	Value	Unit
X_u	-4179.96	kg/s
Y_v	-2155.67	kg/s
Y_r	344.77	kg/s
N_v	-7165.12	kg.m/s
N_r	-410068.61	kg.m^2/s

3.5. Environmental Disturbances and Environmental Matrix

There are 3 types of marine environmental disturbances. Those disturbances are wind, wave, current. Those disturbances are relative to the cable laying barge heading. Environmental disturbance matrix is obtained by calculating the sum of force generated by wave, wind and current towards ship surge, sway, and yaw degree *[12]* as expressed below

$$\tau_{env} = \begin{bmatrix} X_{wave} \\ Y_{wave} \\ N_{wave} \end{bmatrix} + \begin{bmatrix} X_{wind} \\ Y_{wind} \\ N_{wind} \end{bmatrix} + \begin{bmatrix} X_{current} \\ Y_{current} \\ N_{current} \end{bmatrix} \tag{4}$$

Since this study is designing the thruster of barge's dynamic positioning, the author selects the minimum data of each disturbance to get bigger thrust requirement. The summary of the selected disturbances forces and moments are provided by the table 10.

Table 10. Selected Value of Environmental Disturbances of Each Degrees of Freedom.

Degrees of Freedom	Wave		Wind		Current	
	Value	Unit	Value	Unit	Value	Unit
Surge	-1.1519	180°	-2873.1535	90°	-771.0719	90°
Sway	-0.8514	180°	-6122.9603	90°	-2293.8683	90°
Yaw	-0.2937	180°	-92462.8853	90°	-32661.0518	90°

The values in the table 10 used as the input to obtain environmental matrix

$$\tau_{env} = \begin{bmatrix} -3654.3773 \\ -8417.6800 \\ -125124.2308 \end{bmatrix}$$

3.6. Barge Maneuvering Equation and Thrust Determination

In this subsection, the author will determine the value of inertia, damping, and Coriolis-centripetal matrix to obtain the thrust matrix. The detail of the process will be detailed in following sections

3.6.1. Inertia Matrix and Acceleration Matrix

Inertia matrix and acceleration matrix are represented in form of $M\dot{v}$. In ship maneuvering equation, $M\dot{v}$ is represented in form of matrix detailed by equation (5)

$$M\dot{v} = \begin{bmatrix} m - X_{\dot{u}} & 0 & 0 \\ 0 & m - Y_{\dot{v}} & mx_g - Y_{\dot{r}} \\ 0 & mx_g - N_{\dot{v}} & I_z - N_{\dot{r}} \end{bmatrix} \begin{bmatrix} \dot{u} \\ \dot{v} \\ \dot{r} \end{bmatrix} \tag{5}$$

After the value of inertia matrix is determined, the inertia matrix must be multiplied by acceleration matrix and produce value as shown

$$M\dot{v} = \begin{bmatrix} 270777.8801 \\ 279857.7334 \\ 3163928.451 \end{bmatrix}$$

3.6.2. Damping Matrix and Velocity Matrix

Damping and velocity matrix are represented in form of D(v)v. In ship manoeuvring equation, D(v)v is represented in form of matrix detailed by equation (6)

$$D(v)v = \begin{bmatrix} -X_u & 0 & 0 \\ 0 & -Y_v & -Y_r \\ 0 & -N_v & -N_r \end{bmatrix} \begin{bmatrix} u \\ v \\ r \end{bmatrix} \tag{6}$$

The next step is to multiply damping matrix with velocity matrix which result:

$$D(v)v = \begin{bmatrix} 573.4907 \\ 148.0725 \\ 3793.7359 \end{bmatrix}$$

3.6.3. Coriolis Matrix and Velocity Matrix

Coriolis and velocity matrix are represented in form of $C(v)v$. In ship maneuvering equation, $C(v)v$ is represented in form of matrix detailed by equation 7

$$C(v)v = \begin{bmatrix} 0 & 0 & -m\left(x_g r + v\right) + Y_{\dot{v}}\, v + Y_{\dot{r}}\, r \\ 0 & 0 & mu - X_{\dot{u}} u \\ m\left(x_g r + v\right) + Y_{\dot{v}}\, v + Y_{\dot{r}}\, r & -mu + X_{\dot{u}} u & 0 \end{bmatrix} \begin{bmatrix} u \\ v \\ r \end{bmatrix} \tag{7}$$

Then after Coriolis matrix is determined, multiply with velocity matrix. The result of Coriolis matrix and velocity matrix multiplication is:

$$C(v)v = \begin{bmatrix} -647.2319 \\ 785.6822 \\ -172.6317 \end{bmatrix}$$

3.6.4. Thrust Matrix

After the value of inertia, damping, Coriolis, and environmental matrix are already determined, the next step is to determine the value of thrust matrix. The value of thrust matrix represents the required forces and moment must be produced by the engine and propeller. To determine the thrust matrix, the author must determine the sum of inertia, damping and Coriolis matrix after then subtracted by environmental matrix as expressed by equation (8)

$$\tau = M\dot{v} + D(v)v + C(v)v - \tau_{env} \tag{8}$$

Then the result of mathematical equation above is:

$$\begin{bmatrix} \tau_X \\ \tau_Y \\ \tau_N \end{bmatrix} = \begin{bmatrix} 274358.5162 \\ 289209.1681 \\ 3292673.786 \end{bmatrix}$$

The thrust matrix above means that the required surge thrust for 150-ft barge to operate at 1 knot forward speed is 274.36 kN. The required sway thrust for 150-ft barge to operate at 1 knot forward speed is 289.21 kN. The required yaw thrust for 150-ft barge to operate at 1 knot forward speed is 3292.67 kNm

3.7. Propeller Thrust Allocation

The cable laying barge will be equipped with 4 thrusters as shown in figure 6 and considering the yaw moment that could be produced by the propeller, the author decided to make all 4 propellers produce the same or equal thrust value.

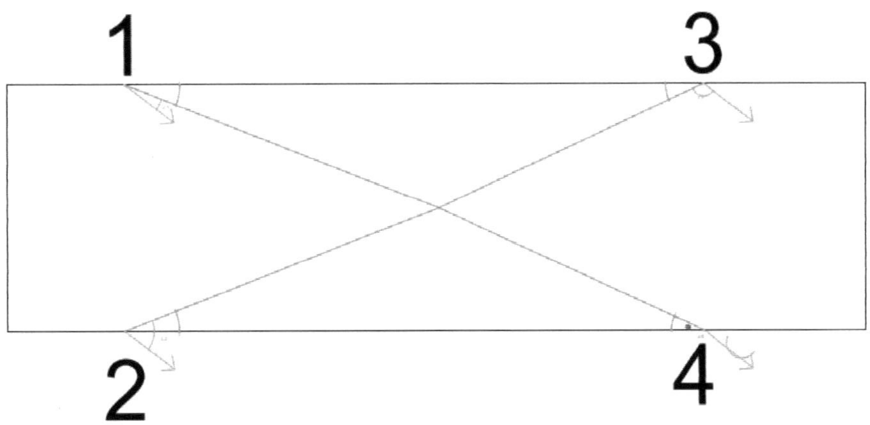

Figure 6. Position of 4 Thruster's Propellers.

The thrust produced by all 4 propellers should produce equal or more than 3292.67 kNm of yaw moment, 274.36 kN of x-axis force and 289.21 kN of y-axis force. In order to determine the thrust value, the author applied physics concept about moment. Based on its basic equation, moment is the result of force times its arm. In this research, the force represents by the thrust and the angle of the thruster while the arm represents by the distance between the center of gravity (COG) to each thruster's propeller. The arm length of each propeller will be detailed in table 11

Table 11. Arm Length of Each Thruster's Propeller.

Thruster Number	Arm Length from COG (m)
1	17.0821
2	17.0528
3	14.8819
4	14.8482

The author varies the thruster angle of attack by 45 degrees to produce required surge force, yaw force, and yaw moment with the minimum thrust value. Besides varies the thruster angle, the author also conducts sensitivity analysis towards surge force, sway force, and yaw moment. The sensitivity analysis shows that using sway force as the basis requires the smallest thrust value as shown in table 12

Table 12. Selected Thruster Angle Scenario from the First Phase.

α_1 (°)	315
α_2 (°)	315
α_3 (°)	90
α_4 (°)	180
T (kN)	109.5261
Fx (kN)	289.2092
M(kNm)	4273.0022

After selected the scenario which requires smallest thrust, the author continues the sensitivity analysis to obtain the smallest thrust possible. After several sensitivity analysis, the author obtained the smallest thrust with the angle configuration detailed in table 13

Table 13. Thruster Angle Configuration of the Taken Thrust Value.

α_1 (°)	314
α_2 (°)	315
α_3 (°)	90
α_4 (°)	175
T (kN)	105.7626
Fx (kN)	274.3907
M(kNm)	4285.0666

3.8. Propeller Selection and Calculation

As already mentioned, and detailed within the introduction section, the proposed portable propulsion system for this research is azimuth thruster. Azimuth thruster is using Ka-series propeller **[13]** . Based on the reference mentioned before, the author decided to use Ka-series propeller for the thruster's propeller.

3.8.1. Determination of Propeller Parameter

Before plotting to the Kq-Kt-J diagram of the selected propeller, the author must calculate and define the value of propeller rotation, propeller diameter, Kt, J, and Va. In this research, the author will use 2 values of propeller rotation 325 and 225 rpm. Propeller diameter value is determined by the clearance of the barge at certain distance from the shore where the cable tension value is 96 kN. In determining the distance, the author utilizes formula detailed by equation (9)

$$Cable\ Tension = \text{Cable weight per m } \times \text{Distance } \times \text{g} \tag{9}$$

With the cable weight value of 29 kg/m and gravity constant of 9.81 m/s^2, the tension of the submarine cable would reach 96 kN if the cable pulled 337.4460 m from the shore. At such distance, the sea depth is 2.62 m, and the barge draft is 0.5 m, therefore the clearance of the barge is 2.12 m. After the clearance has been determined, the author deducts 20 percent of the clearance value and take it as the value of maximum propeller diameter. Hence, the maximum propeller diameter is 1.7 m. In this research, the author will use 2 values of propeller diameter based on the size used by azimuth thruster manufacturers. Based on [14] and [15], the author decided to use 1.6 m and 1.7 m as the value of propeller diameter. The other propeller parameters such as Kt, J and Va required further calculation detailed by the table 14

Table 14. Propeller Parameter.

Parameter	Value				Unit
n	325		225		rpm
D	1.6	1.7	1.6	1.7	m
Kt	0.5366	0.4211	1.1196	0.8785	-
Va	0.5144				m/s
J	0.06	0.06	0.09	0.08	-

3.8.2. Calculation of Required Engine Power

After plotting the value of propeller parameters into KQ-KT-J diagram of Ka4-55 Nozzle 19A, Ka4-70 Nozzle 19A, Ka4-70 Nozzle 22, Ka4-70 Nozzle 24 and Ka5-75 Nozzle 19 A, the author found the value of the required engine BHP as shown by the table 15. The range of required BHP MCR is from 626.10 kW to 951.71 kW

Table 15. Range of Required Engine BHP MCR.

Propeller Diameter (m)	RPM	Propeller Type	P/D	10KQ	Q (Nm)	BHPmcr (kW)
1.6	325	Ka 4-55 Nozzle 19A	1.08	0.5164	16284.4927	678.91
		Ka4-70 Nozzle 19A	1.05	0.5002	15773.6314	657.62
		Ka 4-70 Nozzle 22	1.05	0.4907	15474.0522	645.13
		Ka 4-70 Nozzle 24	1.06	0.4812	15174.4730	632.64
1.7	325	Ka 4-55 Nozzle 19A	0.90	0.3544	15133.0479	630.91
		Ka4-70 Nozzle 19A	0.90	0.3573	15256.8793	636.07
		Ka 4-70 Nozzle 22	0.91	0.3517	15017.7566	626.10
		Ka 4-70 Nozzle 24	0.91	0.3397	14505.3509	604.74
	225	Ka 4-55 Nozzle 19A	1.58	1.1154	22827.6733	951.71
		Ka4-70 Nozzle 19A	-	-	-	-
		Ka 4-70 Nozzle 22	1.50	1.0472	21431.8984	893.51
		Ka 4-70 Nozzle 24	1.51	1.0430	21345.9416	889.93

3.8.3. Validation of Required Engine Power

The purpose of the required engine power validation is to determine the rated propeller rpm and to ensure whether the required engine power is acceptable or not. The validation is done by comparing calculation result of required engine power and maximum input power of azimuth thruster recommended by several manufacturers. Based on [15], maximum power input for azimuth thruster with the propeller diameter 1.6 m is 725 kW. On the other side, based on [14], maximum power input for azimuth thruster with the propeller diameter 1.7 m is 660 kW. The maximum input power for each propeller diameter already determined, hence comparison as detailed in table 16 can be done

Table 16. Range of Required Engine BHP MCR.

Calculation Result			Manufacturers Recommendation		
Propeller Diameter (m)	RPM	Required Power Range (kW)	Propeller Diameter (m)	Maximum Power Input (kW)	Status
1.6	325	632.64-678.91	1.6	725	Acceptable
1.7	325	604.74-630.91	1.7	660	Acceptable
	225	889.93-951.79			Not Acceptable

Table 16 shows that while using 325 rpm as rated propeller rotation, the required engine power matched with manufacturers recommendation. Therefore, the author decided to use 325 rpm as rated propeller rotation.

3.9. Engine Propeller Matching

During the plotting process in Kq-Kt-J diagram of Ka4-70 Nozzle 24 using 325 rpm and 0.0594 of J value, the author found that the required engine BHP is around 632 kW which is near the Cummins engine with the specification detailed in table 17

Table 17. Diesel Engine Specification.

Manufacturer	Cummins
Engine Model	K38-M
No. of Cylinder	V-12
Bore x Stroke	170 x 215 mm
Rated Output	634 kW
Engine Speed	1800 rpm
Dry Weight	4218 kg

To make sure the selected engine match with the power requirement, the author divides the power requirement of the trial phase with 634 kW. It turns out that the value is 99.79%. Due to the value, the author must select the reduction gear which enables the propeller rotates as close as or equal to 325 rpm. To reduce the engine rpm from 1800 rpm to 325 rpm, it requires the reduction gear with the ratio of 5.5385:1. After several selection and filter process the gearbox ratio which enable to reduce propeller rotation from 1800 rpm to almost near 325 rpm is 5.542:1. Using the gearbox with the ratio 5.542:1 the propeller will rotates with the rated rotation of 324.79 rpm or 5.41 rps.

Table 18. Gearbox Specification.

Manufacturer	Masson Marine
Gearbox Model	MM W4400
Ratio	5.542:1
Power/RPM	1.23
Max RPM	1800

With the rated propeller rotation of 324.79 rpm or 5.41 rps, the author should plot the Kt value of 0.5373 with the J value of 0.06. After the plotting and interpolation process, with specified parameter explained before, the Ka4-70 Nozzle 24 is having the P/D value of 1.06 and 10KQ value of 0.4831. With such value, the calculation result shows that the required BHPmcr is 633.92 kW. After BHPmcr value is already determined, the author will ensure that the engine and the propeller is matched by dividing the required BHPmcr with the BHP of the engine. The required engine BHP will consume 99.99% of rated BHP engine. Therefore, the author concludes that the engine and the propeller is already matched.

3.10. Thrust Recalculation and Checking

After the barge is being loaded with the selected engine-gearbox configuration, the center of gravity (COG) of the barge will change. The change of the COG also affecting the length of force arm. In this section, the author will recalculate the thrust to ensure that with the change of force arm, the required yaw moment, surge force, and sway force can be produced by the selected powering configuration. After being loaded with 4 diesel engines and 4 gearboxes, the barge is having the COG and the arm force which detailed by the table 19

Table 19. New COG and Force Arm After Loaded with Diesel Engine and Gearbox.

Parameter	Value
LCG (Longitudinal Center of Gravity)	-0.04 m
TCG (Transversal Center of Gravity)	0.04 m
Force Arm of Thruster Number 1	17.05 m
Force Arm of Thruster Number 2	17.02 m
Force Arm of Thruster Number 3	14.93 m
Force Arm of Thruster Number 4	14.87 m

The selected engine produces 107.59 kN of thrust. The calculation is conducted as previous step before. Starting from determining moment and force produced by the thruster configuration and then recalculating the thrust since the force arm is changing. The next step is the author must make sure that the new value of moment and forces is equal or more than the required moment and forces which values are determined during the hydrodynamic calculation. To ensure whether the new value of moment and forces are adequate, the author decides to compare between the new value and the required value which determined during the hydrodynamic calculation. The comparison will be detailed in table 20

Table 20. New COG and Force Arm After Loaded with Diesel Engine and Gearbox.

Parameter	Moment	Fx	Fy
Required Value	3292.67 kNm	274.36 kN	289.21 kN
New Value	4360.66 kNm	279.14 kN	294.21 kN
Description	The new value is bigger than the required		
Status	Adequate		

3.11. Free Running Speed

After the diesel engine-gearbox configuration being loaded, the next step done by the author is to determine free running speed. Free running speed determines the speed of the barge if all thrusters produce maximum thrust in empty draught condition. To determine the free running speed, the author will Guldhammer-Harvald method and Maxsurf to determine the barge resistance. Before determining the free running speed, the author must determine 2 essential values, EHP which produced by all 4 thrusters, and barge resistance at specific range of speed. The first step taken by the author is to determine the EHP produced by all 4 thrusters. To determine the EHP, the author requires the KQ-KT-J diagram of Ka4-70 Nozzle 24 with P/D value of 1.0636 as shown by figure 7

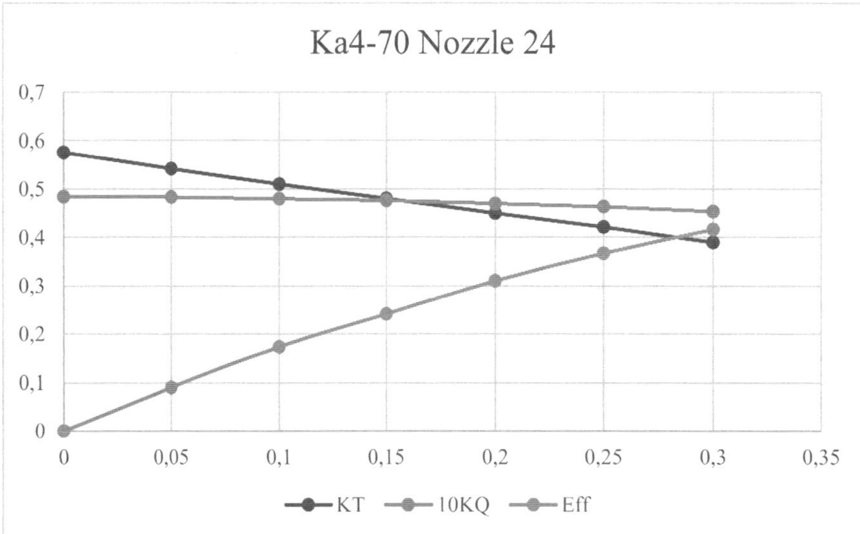

Figure 7. KQ-KT-J Diagram of Ka4-70 Nozzle 24 with P/D Value of 1.06

After reading the diagram above, parameters detailed in the table 21 can be determined

Table 21. Parameter Acquired by Reading KQ-KT-J Diagram of Ka4-70 Nozzle 24 with P/D 1.06.

Parameter	Value
10KQ	0.4832
J	0.04
KT	0.5466
Efficiency	0.08

The most essential parameter acquired by reading the diagram above is the propeller efficiency. The propeller efficiency is one of the parameters required to convert DHP to EHP. Besides, efficiency, there are 2 other parameters essential to convert DHP to EHP. Those parameters are hull efficiency and relative rotative efficiency which value is 1.00 and 1.05 respectively.

After propeller, hull and relative rotative efficiency of the propeller are already determined, the value of EHP can be acquired by implementing the following equation

$$EHP=DHP \times \eta_p \times \eta_{Hull} \times \eta_{RR} \quad (10)$$

$$EHP=517.56\ kW \times 0.08 \times 1.00 \times 1.05 \quad (11)$$

$$EHP=41.95\ kW \quad (12)$$

$$Total\ EHP=4 \times 41.95\ kW \quad (13)$$

$$Total\ EHP=167.81\ kW \quad (14)$$

Based on the detail calculation, 41.95 kW represents the EHP produced by a thruster while 167.81 kW represents the EHP produced by all 4 thrusters. After the total EHP already determined, the next step is to derive an equation from Guldhammer-Harvald method **[16]**. Before deriving the equation, there are some resistances will be neglected due to the absence of speed which are residual resistance due to displacement and residual resistance due to LCB correction.

Guldhammer-Harvald represents the total resistance of the ship by the following equation

$$Rt = 0.5\,(\rho\; Ct\; S + \;\rho_{air}\; CAA\; Area\; Foreship)V^2 \tag{15}$$

While on the other hand, the relationship between EHP and Rt can be written in the form as follows:

$$EHP = Rt\; \times V \tag{16}$$

From calculation, the author acquires free running speed of 4.68 m/s or 9.09 knots. The value of the speed is just the preliminary stage due to negligence of several resistance components. After the preliminary free running speed is already determined, the author runs Maxsurf Resistance to determine the barge resistance from 0 to 4.68 m/s speed range. The result of the Maxsurf Resistance will be shown by the graph at figure 8

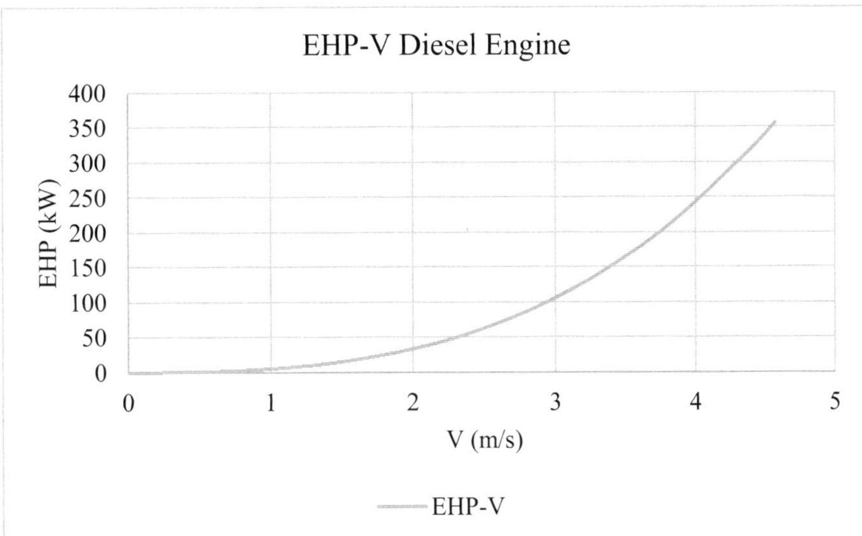

Figure 8. EHP-V Graph of Diesel Engine-Gearbox Configuration.

From the graph detailed by figure 8, the author plots the value of 167.81 kW to find the free running speed. From the graph, the free running speed of diesel engine configuration is 3.52 m/s or 6.85 knots. The acquired value is acceptable because when its value compared to the preliminary value, its value is smaller. The Maxsurf Resistance free running speed value is having smaller value than the preliminary one is because there some resistance components negligence in the preliminary phase.

4. Conclusion

After being loaded with several cable laying equipment which function is to support and to perform cable laying operation at 1 knot laying speed, 150ft-barge is having surge, sway, and yaw velocities of 0.14 m/s, 0.12 m/s, and 0.01 degree/s respectively. In the mentioned operation speed, the author found that the highest submarine cable peak tension is 15.24 kN. Therefore, such laying speed is safe for the submarine cable with the given specification since its allowable maximum tension is being limited to 96 kN. At the same operating speed and wave direction of 270 degrees, the 150ft-barge is having surge, sway, and yaw acceleration of 0.65 m/s^2, 0.7044 m/s^2, 0.0822 rad/s^2 respectively. Utilizing the ship manoeuvring equation, all accelerations, and velocities data and 150ft barge principal dimension and environmental data which has been already processed can be further calculated to acquires the value of thrust matrix. The result of the calculation shows at the mentioned operating speed the 150ft barge requires 274.36 kN of surge force, 289.21 kN of sway force, and 3292.67 kNm of yaw moment.

By implementing the principles of moments, the author converts the values of forces and moment within the thrust matrix into the thrust should be produced by the propulsion system, in this case azimuth thruster. Being loaded with 4 azimuth thrusters, each thruster's propeller must produce nominal thrust of 105.76 kN to fulfil the forces and moment requirement. After several calculation process, to produce the mentioned thrust value, each thruster must be powered with Cummins KTA 38 engine with brake horsepower of 634 kW. Further on, the rotation of the engine must be reduced by Masson Marine MM W4400 gearbox which ratio value is 5.542:1. The power produced by main engine will be consumed by Kaplan propeller with 4 propeller blades, Ae/Ao ratio of 0.7, and nozzle number 24. The selected propeller has the diameter of 1.6 m, and P/D value of 1.06. In the free running condition, where the barge not being loaded with submarine cable, all 4 thrusters will produce effective horsepower of 167.81 kW which enables the barge run at the speed of 3.52 m/s or 6.85 knots.

References

[1] Media Indonesia, "S," Media Indonesia, 13 September 2021. [Online]. Available: https://mediaindonesia.com/ekonomi/432299/saran-bagi-pln-dalam-restrukturisasi-bisnis-pltu. [Accessed 27 July 2022].

[2] Director General of Electrical Power, "Kebijakan Penyediaan Tenaga Listrik Nasional," September 2020. [Online]. Available: https://gatrik.esdm.go.id/assets/uploads/download_index/files/095b0-2009030p-penyediaan-tenaga-listrik-nasional.pdf. [Accessed 27 July 2022].

[3] T. Worzyk, "Cable Laying Vessels," in *Submarine Power Cables: Design, Installation, Repair Environmental Aspects*, Berlin, Springer, 2009, pp. 162-163.

[4] Det Norske Veritas, "Weather Restricted Operations," in *DNV-OS-H101: Marine Operations, General*, Det Norske Veritas, 2011, p. 29.

[5] T. I. Fossen, "Nonlinear DP Model using Current Coefficients," in *Handbook of Marine Craft Hydrodynamics and Motion Control*, Trondheim, John Wiley & Sons Ltd., 2011, p. 155.

[6] N. Yang, D.-S. Jeng and X. Zhou, "Tension Analysis of Submarine Cables During Laying Operations," *The Open Civil Engineering Journal,* vol. 7, pp. 282-291, 2013.

[7] D. Jones, D. Clarke, J. Brayshaw, Barillon and B. Anderson, The Calculation of Hydrodynamic Coefficients for Underwater Vehicles, Victoria: DSTO Platform Sciences Laboratory, 2002.

[8] D. Humphreys and K. Watkinson, Prediction of Acceleration Hydrodynamic Coefficients for Underwater Vehicles from Geometric Parameters, Florida: Naval Coastal Systems Laboratory, 1978.

[9] V. Bertram, "Ship manoeuvring," in *Practical Ship Hydrodynamics*, Oxford, Butterworth-Heinemann, 2000, p. 155.

[10] T. I. Fossen, "Linear Viscous Damping," in *Handbook of Marine Craft Hydrodynamics and Motion Control*, Trondheim, John Wiley & Sons Ltd., 2011, pp. 123-125.

[11] I. Ariana, T. Akbar, M. Prakosa, J. Prananda and W. Tyasayumranani, "Conceptual design of dynamic positioning system for tugboat to improve safer operation in Indonesia, case study: port of Cilacap," *IOP Conference Series: Earth and Environmental Series,* vol. 557, 2020.

[12] A. Alessandri, S. Donnarumma, G. Luria, M. Martelli, S. Vignolo, R. Chiti and L. Sebastiani, "Dynamic Positioning system of a vessel with conventional propulsion configuration: Modelling and simulation," *Maritime Technology and Engineering,* 2015.

[13] S. Utami, I. Ariana, A. Fathallah and I. Asmara, "Optimum Propeller Interaction Analysis of Harbourtug with Azimuth Thruster," in *IOP Conference Series: Materials Science and Engineering*, 2020.

[14] Schottel, Schottel Product Guide, Schottel.

[15] Kongsberg Maritime, Kongsberg Azimuth Thrusters, Kongsberg: Kongsberg Maritime.

[16] S. A. Harvald, Resistance and Propulsion of Ships, Lyngby: John Wiley & Sons, 1983.

Technical and Economical Feasibility Impact on Landing Craft Ship Conversion to Comply with Greenship Requirements

Andreas Mahendri Putra[1], Ketut Buda Artana[1,2]*, I Made Ariana[1,2], Hayy Nur Abdillah[1]

[1] Department of Marine Engineering, Faculty of Marine Technology, Institut Teknologi Sepuluh Nopember, Surabaya, Indonesia
[2] The Center for Excellence Maritime Safety and Marine Installation (PUI KEKAL), Institut Teknologi Sepuluh Nopember, Surabaya, Indonesia

*Corresponding author's e-mail : ketutbuda@its.ac.id

Abstract. This research aims are additional requirements to comply with greenship requirements, analysing technical and economic due to LNG dual fuel conversion. As the technical aspect of using LNG as fuel, this LNG LCT ISO Tank carrier able to utilize LNG as fuel which it will be installed dual fuel genset from MAN B&W 6L23/30DF for sailing and manoeuvring scenario and VOLVO D16 MG HE KC for additional genset that will be operated for cargo handling and at port scenario. While it needs small power requirements, therefore diesel electric propulsion chosen to propel the ship which is need 401 kW for each shaft. For economic aspect the conversion into LNG as fuel has economically valuable which payback period at year eight, ROI at 12,68% and IRR 15%.

1. Introduction

Gas is considered one of the solutions to replace fuel oil because gas has higher calorific value at a cheaper price compared to fuel oil and it is known as clean and green energy. PT.PLN (Persero) prioritizes the planning of peak load power plants that operate using gas (LNG, mini-LNG, and CNG). On the other hand, the minister of Energy and Mineral Resources through Ministerial Decree Number 1750 of 2017 concerning the determination of the allocation and utilization of natural gas for electricity supply by PT.PLN has been determined to provide certainty of gas allocation for all gas-fired power plants by RUPTL 2018-2027 will be increasing.

The government's consistency in supporting the growth of the maritime industrial climate and reducing the dependence of ships on imported products is by procuring LNG carriers through the roadmap of the Agency for The Assessment and Application of Technology (BPPT). Given that environmental concerns and ship emissions have dominated issues in the shipping industry, particularly in relation to the IMO rule that restricts ship fuel emissions so that LNG (Liquid Natural Gas) may be utilized as an environmentally friendly alternative energy.

2. Landing Craft Conversion

There are two options that are considered economically viable and fulfil the emission regulations regulation, those are utilization of exhaust gas treatment systems and LNG as fuel [1].Because of the demand for low-emission ships, the shipping industry has turned to natural gas as an alternative fuel source [2]. In this conversion, the methodology to estimate how is the impact to the environment will be estimated with numerical approach which already done by Banawan, et. al. [3] and EEDI calculation also be used for estimate the impact of technical changes due the LNG dual fuel conversion, landing craft ship which carrying LNG (International Standard Organization) ISO Tank will be converted to dual fuel engine system. The purpose of LCT conversion is to minimalize CO2, NOx and SOx emission. The research object of this research is the existing landing craft vessel design from National Research 2020.

2.1 Ship Dimension and Propulsion

To give clearer perspective of LCT vessel, below are the data of LCT vessel which it will be converted to dual fuel engine system.

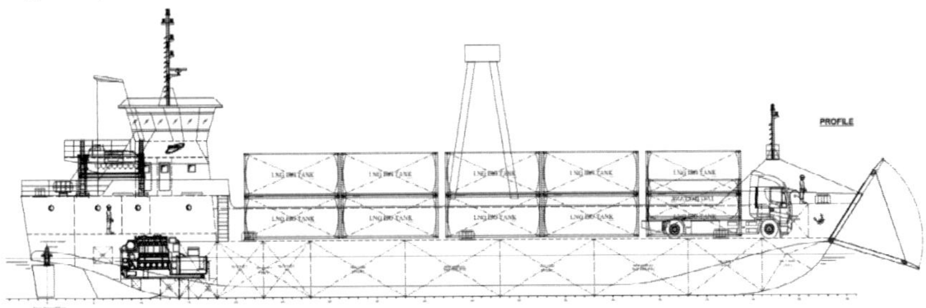

Figure 1. Ship General Arrangement

Table 1. Ship Dimension

Ship Dimension		
Length Overall	53.597	Meter
Length Perpendicular	45,71	Meter
Breadth Max	12,4	Meter
Height	3,5	Meter
Draught	2,5	meter
V service	≥ 9	knot

In this research paper, several activities taken for recalculating the propulsion power needed. Starting with resistance estimation in computational methods and towing test method. So, the result of resistance estimations are show below.

Table 2. Resistance Estimation Results

Method	Result (kN)
Guldhammer-Halvard	41.3
Holtrop Manual Methode	50.17
Maxsurf (Holtrop)	45.56
Towing Test	51.89
Average Resistance	**47.23**

To find the total resistance of LCT ship, there are several resistances that added into the calculation, those are appendages resistance and wind resistance. The total of additional resistance of appendages resistance and also wind resistance adding almost 15% of the average resistance. After finding the total resistance of LCT ship, it can be estimating the power requirement.

IOP Conf. Series: Earth and Environmental Science 1166 (2023) 012014 doi:10.1088/1755-1315/1166/1/012014

Table 3. Power Estimation

Power Result (kW)	
Effective Horse Power	494
Delivered Horse Power	632
Shaft Horse Power	651
Motor Propulsion Input Power @SCR	688
Motor Propulsion Input Power @MCR	810

For efficiency of shaft and motor taken 95%. Therefore, it found the motor propulsion power needed is 810 kW in total. However, initial plan is to use conventional propulsion configuration by using dual fuel main engine, but there is no dual fuel main engine that has small power for this LCT ship. There are other possibility of propulsion configuration, those are diesel electric propulsion or dual output marine gearbox. Based on this scenario, diesel electric propulsion and dual output marine gearbox will be compared below.

Table 4. Propulsion Configuration Comparison

Configuration Type	Advantages	Disadvantages
Diesel Electric Motor	1. Lower fuel consumption 2. Less maintenance 3. Reduce noise 4. Less vibration 5. No limitation of coupled shaft connecting to the engine 6. Less weight	1. Higher acquisition cost 2. Higher efficiency losses 3. Increased weight
Two Output Marine Gearbox	1. High torque	1. More weight 2.Many moving parts

Based on the table above, diesel electric motor configuration have more advantages than two output marine gearbox. Also due to small engine room on this LCT ship design, diesel electric is applicable that comes up with its flexibility.

2.2 Electric Motor Selection

After calculating all the electrical and machinery load which it will affect to the genset selection, the next step is to calculate the generator load. However, the ship condition scenario for cargo handling and at port have small electrical load, so it will using non-DF for its additional genset while the main genset is running with LNG dual fuel.

Table 5. Electrical Load

Electrical Load Scenario	kW
Sailing	1111.2
Manoeuvring	1144.7
Cargo Handling	369.3
At Port	231.6

Then, several gensets can be listed according to power needed for each scenario. Four gensets chosen as load balance calculation and after calculating electric load balance, one gensets can be chosen that has ideal balance for each scenario.

Table 6. Main Genset Load Balance Calculation

Manufacturer	Sailing	Manoeuvring
MAN B&W 5L23/30DF	0.62	0.65
MAN B&W 6L23/30DF	0.78	0.81

Manufacturer	Sailing	Manoeuvring
Wartsila 6L20DF Genset	0.6	0.62
Wartsila 8L20DF Genset	0.9	0.93

Table 7. Aux Genset Load Balance Calculation

Manufacturer	Sailing	Manoeuvring
Cummins K19-Cp	0.62	0.65
KOHLER 350EOZCS	0.78	0.81
VOLVO D16 MG HE KC	0.6	0.62

Below is the table for the genset scenario which using two kinds of genset while the power requirement for cargo handling and at port is small then VOLVO D16 chosen as the additional genset and MAN B&W 6L23/30DF become the dual fuel main genset.

Scenario	Genset Manufacturer
Sailing Manoeuvring	MAN B&W 6L23/30DF
Cargo Handling At Port	VOLVO D16 MG HE KC

2.3 Fuel Oil and Lube Oil Piping & Instrumentation Diagram (P&Id) Design

P&ID drawing will be referring to BKI Classification and IGF Code which it regulates about gas as fuel. As LNG LCT using dual fuel genset, the P&ID of the fuel oil will consist of 3 drawing, those are gas supply drawing, fuel oil drawing and also genset system drawing. The P&Id of those drawing will be attached below

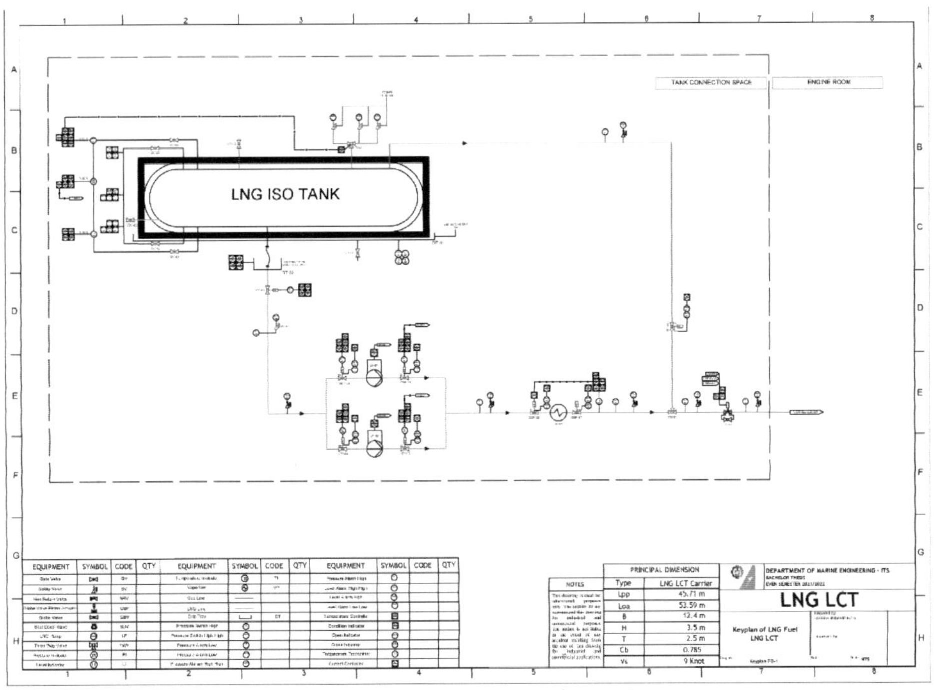

Figure 2. LNG ISO Tank P&ID

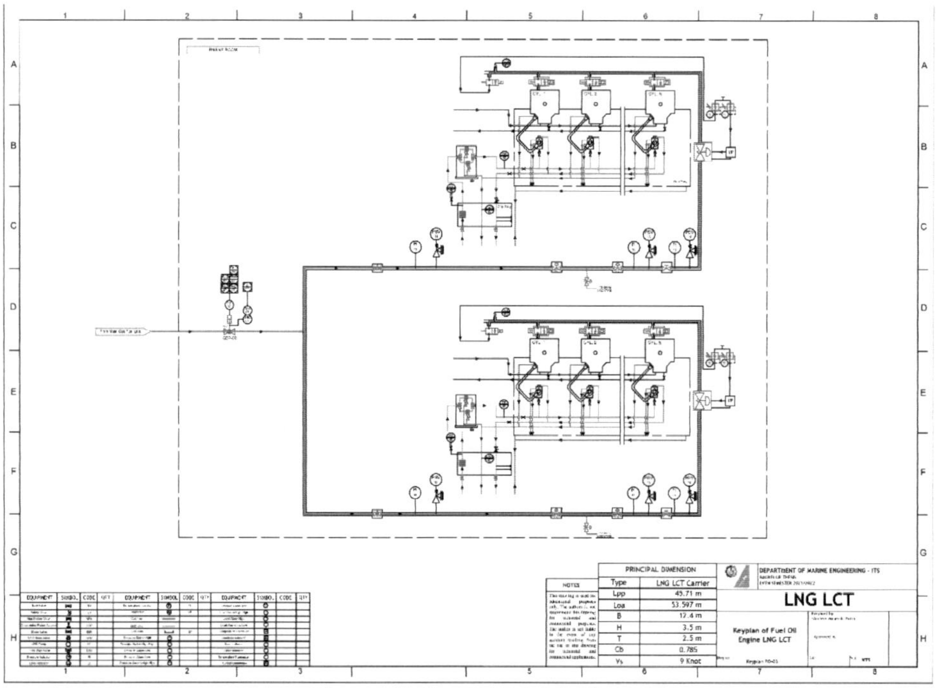

Figure 3. Gas Inlet to Engine

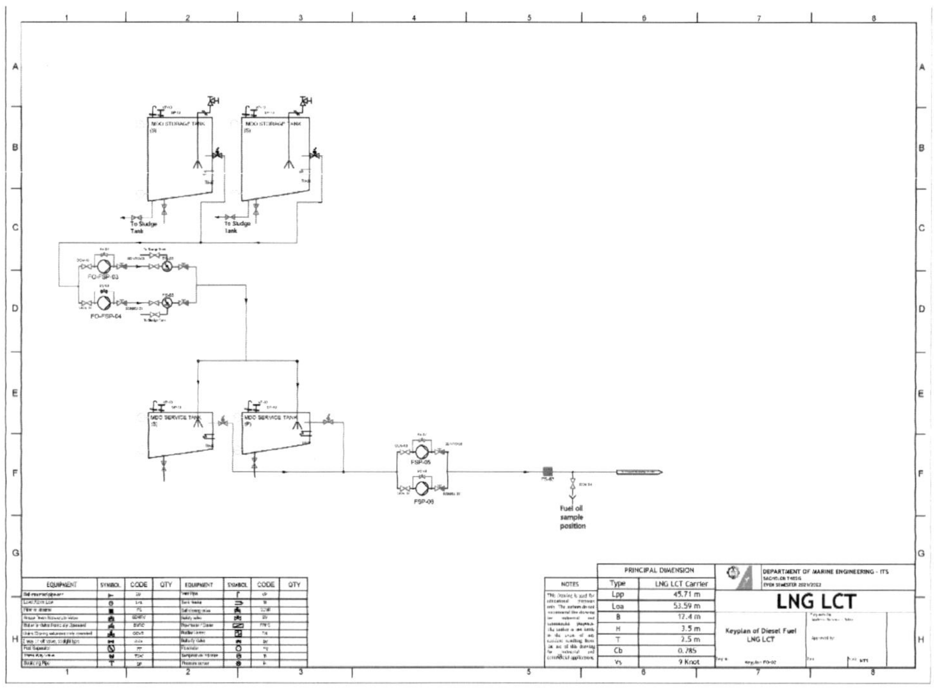

Figure 4. MDO Fuel P&ID

With LNG equipment specification stated below for LNG pump and the vaporizer/

LNG Pump

Manufature	:	Vanzetti
Capacity	:	1.4 litre/min

Pressure	:	20 bar
Power	:	15 kW

Vaporizer

Manufature	:	DOER
Capacity	:	1200 Nm3/h
Weight	:	2100 kg

3. Economic Analysis

The economic analysis using economic feasibility criteria in the form of Net Present Value, Internal Rate Return and Payback Period with the duration project contract for 25 years with 4 kinds of distribution route. To be able to calculate the three criteria, the components calculations such as CAPEX and OPEX have been obtained firstly. Economic feasibility criteria will be compared with the previous ship system which using conventional diesel engine. Therefore, the result can be obtained which one has better economic aspects.

Table 8. CAPEX and OPEX

Distribution Route	CAPEX		OPEX	
Benoa-Lombok Peaker-Benoa	$	10,095,329	$	3,411,163
Benoa-MPP Lombok-Benoa	$	10,095,329	$	3,301,610
Benoa-Gilimanuk-Benoa	$	5,047,664	$	3,267,111
Benoa-Pemaron-Benoa	$	5,047,664	$	1,718,447
Standby	$	5,047,664	$	2,299,113
Total	$	40,381,315	$	13,997,447

It will be in total 2 ship for Lombok peaker route, 2 ship for MPP Lombok route, 2 ship for Gilimanuk route, 1 ship for Pemaron route and 1 ship at standby position. Also, below is the table of gas distribution on each route/year that can be calculated to calculate revenue each year.

Table 9. LNG Delivered to Plant

	ISO Tank/Year	**Delivered Gas (mmBtu)**
Benoa-Lombok Peaker-Benoa	8415	4,102,968
Benoa-MPP Lombok-Benoa	8085	3,942,068
Benoa-Gilimanuk-Benoa	6270	3,057,114
Benoa-Pemaron-Benoa	4620	2,252,610
Standby	-	-
Total	27,390	13,354,761

The selling margin using 1.5 USD/mmBTu, with LNG bunkering price at 13 USD/mmBTU. Then the result of the economic aspect is it got payback at year 8 and almost has same value for the ROI and IRR. It can be concluded that LNG dual fuel has better economic aspect with the price of LNG is cheaper than MDO that affect to the economical aspects, also LNG has bigger calorific value than MDO.

Table 10. Economic Analysis Result

Configuration	Payback (Year)	ROI	IRR
Diesel Engine	9	12.08%	15%
Dual Fuel Genset	8	12.68%	15%

In case that LNG price still increasing in the future, below is the table of sensitivity analysis of LNG bunkering price, this analysis will estimate how is the impact of increasing LNG price into the economic aspect.

Table 11. LNG Bunkering Sensitivity Analysis

Margin 1.5 USD (Dual Fuel Genset)					
No	**LNG Bunkering Price**		**Payback (Year)**	**ROI**	**IRR**
1	$	9.00	6	16.48%	20%
2	$	10.00	6.5	15.53%	19%
3	$	11.00	7	14.58%	17.5%
4	$	12.00	7.5	13.63%	17.5%
5	$	13.00	8	12.68%	15%
6	$	14.00	8.5	11.73%	13%
7	$	15.00	9.5	10.78%	12.5%

Based on the table above, it can be concluded that the minimum bunkering price is $13. So, if bunkering price is higher that $13, diesel electric will not economically viable.

4. Greenship Indicator

4.1 EEDI (Energy Efficiency Design Index)

By using the formula below, EEDI can be calculated by considering impact to environment divided to benefit to society. In this case, conventional diesel engine configuration will be compared to the new conversion of dual fuel genset configuration which using LNG as its fuel and electric motor for its propulsion.

$$EEDI = \frac{((Pme\ x\ SFC\ x\ CF) + (Pae\ x\ SFC\ x\ CF))/(DWTx\ Vs)}{Deadweight\ x\ Speed} \quad (1)$$

4.1.1 Power Main Engine

Although the system is using diesel electric propulsion, the EEDI still can be calculated by using motor power as the main engine as power references. As the formula below, power main engine can be calculated

$$Pme = 0.83\ \text{x} \frac{Motor\ Power}{Electrical\ Efficiency} \quad (2)$$

Motor Power	:	1000 kW
Electrical Efficiency	:	91.3 %
Pme	:	909 kW

However, in this EEDI calculation, the value of EEDI will be compared with non-diesel electric propulsion system which is conventional diesel propulsion. So, the calculation of power main engine become different, as the formula below

$$Pme = 0.75 \text{ x Main Engine MCR} \tag{3}$$

Total Main Engine MCR : 882 kW
Pme : 662 kW

Conventional diesel propulsion will be using Volvo Penta D8-600 which has power 441 kW per engine.

4.1.2 Power Auxiliary Engine (Pae)

Pae is the amount of auxiliary engine power needed to maintain a ship's conventional maximum sea load while also providing power for accommodation, such as main engine pumps, navigational systems, and equipment, but not for propulsion machinery or systems, such as thrusters, cargo pumps, cargo gear, ballast pumps, or maintaining cargo, such as reefers and cargo hold fans.

$$Pae = (0.05 \text{ x Main Engine MCR}) + 0.02 \text{ x Pme} \tag{4}$$

Main engine MCR : 1000 kW
Pme : 909 kW
Pae : 68 kW

Also, the power auxiliary of conventional diesel engine can be obtained as the formula below

$$Pae = (0.05\ x\ Main\ Engine\ MCR) \tag{5}$$

Main engine MCR : 1000 kW
Pae : 44 kW

Then the result of EEDI calculation is the table below

Table 12. EEDI Result

Dual Fuel Genset Configuration				Conventional Diesel Engine Configuration			
P ME	=	909	kW	P ME	=	662	kW
SFC	=	204	g/kWh	SFC	=	207	g/kWh
Cf LNG	=	2.75		Cf MDO	=	3.206	
P AE	=	68	kW	P AE	=	44	kW
SFC	=	204.083	g/kWh	SFC	=	207	g/kWh
Cf LNG	=	2.75		Cf MDO	=	3.206	
DWT	=	900		DWT	=	900	
Vref	=	9	kN	Vref	=	9	kN
EEDI	=	67.71272	gCO2/ton.mile	EEDI	=	65.0533	gCO2/ton.mile

Based on the calculation table above, it can be concluded that the value of diesel engine configuration is bigger than dual fuel genset configuration. Because diesel electric propulsion needs bigger power to power up the motor propulsion which 500 kW, slightly different with the conventional diesel propulsion which the diesel engine available at 441 kW.

4.2 SOx and NOx Calculation on Gas Mode

To measure the decreasing amount of SOx, NOx emission. Below are the formula to calculate the estimation of SOx, NOx emission from dual fuel engine and also diesel engine.

$$E\ gas\ = P\ x\ Lf(0{,}05efd + 0{,}95efg)xT \tag{6}$$

Which :

E = emission quantity (g)

P = eng.power @maximum continuous rating (kW)

efd = emission factor in diesel mode(g/kWh)

efg = emission factor in dual fuel mode (g/kWh)

T = engine running time (T)

With :

Table 13. Emission Factor for LNG Dual Fuel
Source : [3]

Emission Factor Dual Fuel (g/kWh)					
	NOx	SOx	PM	CO2	HC
At cruise	2.16	0	0	0	0
At standby and manouvering	2.16	0	0	0	0

Table 14. Emission Factor 5% Diesel Oil
Source : [3]

Emission Factor 5% Diesel Oil (g/kWh)					
	NOx	SOx	PM	CO2	HC
At cruise	2.59	0.2	0.015	553	0
At standby and manouvering	12.7	4.5	0.9	710	0.6

Calculating the SOx, NOx emission with formula above for every scenario on dual fuel genset and auxiliary genset with respecting the input data below

Table 15. Input Data for Dual Fuel Genset

Dual Fuel Genset Input Data			
P	=	1420	kW
Lf (Scenario)	=	85%	Sailing
		20%	Standby
T average	=	54	hours
Trip/year	=	142	

Table 16. Input Data for Auxiliary Diesel Genset

Diesel Genset Input Data			
P	=	460	kW
Lf (Scenario)	=	85%	Sailing
		20%	Standby
T	=	20	hours

Then the result of calculating the SOx and NOx of LNG dual fuel configuration as below

Table 17. Total Emission LNG Dual Fuel

Emission Quantity Total Dual Fuel Genset + Diesel Oil Genset (ton/year)					
	Nox	SOx	PM	CO2	HC
At cruise	23,310	1,176	235	185,509	157
At standby and manouvering	23,310	1,176	235	185,509	157

4.3 SOx and NOx Calculation on Conventional Diesel Engine

$$E\ diesel = P\ x\ Lf\ x\ efd\ x\ T \tag{7}$$

Which :

E = emission quantity (g)

P = eng.power @maximum continuous rating (kW)

efd = emission factor in diesel mode(g/kWh)

efg = emission factor in dual fuel mode (g/kWh)

T = engine running time (T)

Which it has emission factor

Table 18. Emission Factor for Conventional Diesel Oil

Emission Factor for Diesel Oil (g/kWh)					
	NOx	SOx	PM	CO2	HC
At cruise	10.81	4.1	0.3	645	0.2
At standby and manouvering	12.7	4.5	0.9	710	0.6

Calculating the SOx, NOx emission for every scenario on dual fuel genset and auxiliary genset with respecting the input data below

Table 19. Conventional Diesel Genset Input Data

Diesel Engine Genset Input Data			
P	=	700	kW
Lf	=	85%	Sailing
	=	20%	Standby
Time average	=	54	hours
Trip/year	=	142	

Table 20. Aux Diesel Genset Input Data

Aux Diesel Fuel Genset Input Data			
P	=	150	kW
Lf	=	85%	Sailing
	=	20%	Standby
T	=	20	hours

Then the result of calculating the SOx and NOx of conventional diesel engine configuration as below

Table 21. Total Emission Conventional Diesel Engine

Emission Quantity Conventional Diesel Engine (ton/year)					
	Nox	SOx	PM	CO2	HC
At cruise	53,234	20,191	1,477	3,176,341	985
At standby and manouvering	62,542	22,161	4,432	3,496,438	2,955

As the conclusion, huge decreasing of emission if the ship is using LNG. Based on the calculation below.

Table 22. Emission Reduction per Year

Emission Reduction (ton/year)					
	Nox	SOx	PM	CO2	HC
At cruise	29,925	19,015	1,242	2,990,832	828
At standby and manouvering	39,232	20,985	4,197	3,310,929	2,798

As the composition of LNG is majorly ethane and methane, which low of carbon then it has huge reduction for its CO2 per year that has almost 2 million ton/year.

5. Conclusion

The methodology to analyse environmental impact because of technical changes have been carried out by Banawan, et. al. [3] and EEDI calculation form IMO. As the several regulations which regulate about greenship or environmentally ship design, this LCT vessel already fulfill the installment of several equipment which supporting greenship regulations, such as LNG as fuel. In this case LCT vessel using diesel electric as its propulsion system due to small power requirements, then motor propulsion from Siemens Simotics HV A-compact PLUS was chosen at 500 kW for its power. To generate electricity for motor propulsion, LCT vessel using dual fuel genset, which it has 2 main genset to generate electricity. The electric load analysis was already fulfilling the requirements to choose MAN B&W 6L20/32DF as main genset at 0.78 for sailing scenario and 0.81 for maneuvering scenario, while Volvo D16 MG HE KC as auxiliary generator to generate electricity when the ship is on loading & unloading scenario and at port scenario.

In economic analysis, it resulted quite satisfying result for the LNG LCT conversion into dual fuel LNG. The factor that LNG is used as fuel because its higher calorific value than diesel fuel. Therefore, it resulted better economic feasibility for LNG LCT conversion.

Table 23. Economic Analysis Comparison

Configuration	Payback (Year)	ROI	IRR
Diesel Engine	9	12.08%	15%
Dual Fuel Genset	8	12.68%	15%

As the table above, the conversion to dual fuel LNG has better payback period in 8 years. It is better than the conventional diesel engine which it has payback period in 9 years. However, the ROI value and IRR value almost has the same value for both of them.

6. References

[1] S. Seo *et al.*, "An economic evaluation of operating expenditures for LNG fuel gas supply systems onboard ocean-going ships considering availability," *Ships and Offshore Structures*, vol. 11, no. 2, pp. 213–223, 2016, doi: 10.1080/17445302.2014.984389.

[2] G. Theotokatos *et al.*, "Design of lng storage and feeding system for an open type ferry," *Towards Green Marine Technology and Transport - Proceedings of the 16th International Congress of the International Maritime Association of the Mediterranean, IMAM 2015*, no. June 2016, pp. 473–481, 2015, doi: 10.1201/b18855-63.

[3] A. A. Banawan, M. M. El Gohary, and I. S. Sadek, "Environmental and economical benefits of changing from marine diesel oil to natural-gas fuel for short-voyage high-power passenger ships," *Proceedings of the Institution of Mechanical Engineers Part M: Journal of Engineering for the Maritime Environment*, vol. 224, no. 2, pp. 103–113, 2010, doi: 10.1243/14750902JEME181.

[4] J. L. Kirtley, A. Banerjee, and S. Englebretson, "Motors for Ship Propulsion," *Proceedings of the IEEE*, vol. 103, no. 12, pp. 2320–2332, 2015, doi: 10.1109/JPROC.2015.2487044.

[5] C. Vossen, "Diesel electric propulsion on ZIGMA class corvettes," *2011 IEEE Electric Ship Technologies Symposium, ESTS 2011*, no. February, pp. 288–291, 2011, doi: 10.1109/ESTS.2011.5770883.

[6] C. R. Young, "Diesel-electric ship propulsion," *Journal of the Institution of Electrical Engineers - Part I: General*, vol. 91, no. 46, pp. 383–384, 1944, doi: 10.1049/ji-1.1944.0097.

[7] E. S. D. Poggiali, F. Pilo, G. Pio, T. Iannaccone, M. Carboni, "GUIDELINES FOR EMERGENCY PLANNING OF LNG INFRASTRUCTURES IN PORT AREAS."

IOP Conf. Series: Earth and Environmental Science 1166 (2023) 012015 doi:10.1088/1755-1315/1166/1/012015

The effect of using biodiesel B50 from palm oil on lubricant oil degradation and wear on diesel engine components

A Iswantoro[1*], B Cahyono[1] and C N Ruyan[1]

[1]Department of Marine Engineering, Institut Teknologi Sepuluh Nopember, Surabaya, Indonesia
*adhi.iswantoro@gmail.com

Abstract. Alternative fuels in diesel engines are getting attention due to the current world energy crisis. Biodiesel is a fuel where the main composition from palm oil, it is an alternative fuel may can replace fossil diesel fuels. However, the use of biodiesel has several impacts on the engine if its not maintained. The using biodiesel usually results decrease of lubricating oil viscosity and increases the metal wear rate. This research will present results effect of biodiesel B50 on lubricating oil degradation and metal wear on engine components. The method used is an experiment by testing engines for a long time with the Engine Manufacturer Association (EMA) standard. The use of B50 biodiesel causes significant degradation of lubricating oil than Pertamina Dexlite. The lubricating oil test results that with B50 biodiesel decreased viscosity by 16.17%. Lubricating oil also contains the metallic aluminium 8 ppm, 27% metallic iron, and 1 ppm metallic chromium. Using B50 biodiesel also causes journal bearing worse and significant piston ring gap. In addition, the deposit resulting from the using B50 biodiesel was 30.97% greater in the diesel engine component.

1. Introduction

To reduce consumption of fossil fuel, the Indonesian government obliges the use of biodiesel for diesel engine. Presently, researchers have attracted considerable attention from biodiesel fuel to reduce consumption of fossil fuel and exploring the renewable energy. Biodiesel can be said to be alternative fuel. Biodiesel have some advantages when used, that are can reducing carbon emissions and the engine performance are comparable to diesel oil. But, biodiesel can affect the performance of the diesel engine in some case [1]. In other research, reported that using of biodiesel from palm oil resulted increasing in fuel consumption on engine from 11.93% to 13.48%, and decreasing in engine thermal efficiency from 2.96% to 5.33% [2]. For volumetric efficiency is relatively same. In other research also show about using of biodiesel on ship, resulted that biodiesel causes the engine load, engine power, and the exhaust gas temperature decrease, rather than using High-Speed Diesel (HSD) fuel [3]. Other, using of biodiesel causes the fuel consumption increase, when compared with HSD. In other research about the effect of Karanja oil methyl ester, caused the brake horse power on the 780 HP engine decrease [4]. Other research show that rapeseed biodiesel doesn't generate a significant difference in performance. compounds, so biodiesel is a cleaner fuel and easier to handle than diesel [5]. Another characteristic of biodiesel is that it has a higher cetane number and viscosity and better lubrication properties than diesel.

The mandatory implementation of B50 in Indonesia in early January 2020 is said to be able to save foreign exchange and is also more environmental friendly. However, several complaints emerged from consumers regarding its effect on the engine in its development. In response to this, the Ministry of Energy and Mineral Resources Research and Development Center of Republic of Indonesia described the tests carried out. As a result, there is a decrease in power and fuel consumption of up to 0.87 percent. However, tests of large cars such as trucks over 3.5 tons show an increase in power.

The using biodiesel results the level of metal wears on the engine increase [6]. Using biodiesel based on a mixture of 20% crude oil (KOME20) increases the metal content of iron, aluminum, copper, chromium, nickel, zinc, lead, and magnesium. This is more than using fossil fuels in the lubricating oil after 200 hours of use [7]. After using Pongamia oil biodiesel for 256 hours, the metal content of iron, copper, nickel, lead, chromium, magnesium, zinc, and aluminum in the lubricating oil increased diesel fuel [8].

However, using biodiesel causes more significant degradation of the lubricating oil properties. This can cause a loss in the lifetime of the engine because, in the diesel engine, there are components that move with each other, both linear motion and rotary motion. Movement or contact between elements in a engine will cause friction. If the variance is allowed to occur, it will cause losses such as wear and tear to damage. Therefore, lubrication is required. The main function of engine lubrication is to reduce friction between moving part and other engine components. Also to minimize the risk of debris or damage to the engine. In addition, lubrication can reduce the heat generated due to combustion and minimizes engine power wasted against friction forces.

This manuscript will discusses the effect of using biodiesel B50 from palm oil on the degradation of lubricating oil and wear on engine components. The method use is experiment to find degradation on lubricating oil that have been working for a long time. It is hoped that biodiesel will only slightly affect the characteristics of the lubricating oil compared to the use of diesel oil.

2. Research Method

Literature review was conducted by collecting references on biodiesel and its effects on lubricating oil and engine components. The process of making biodiesel using palm oil includes degumming, esterification and transesterification processes. Biodiesel B50 is a diesel engine fuel that has a mixture of 50% palm oil by volume. In biodiesel itself, microbes can also grow which can trigger problems in the engine. At this stage, preparations are made to be able to carry out the experiment, the preparations are:

2.1. Preparation and engine set-up

The diesel engine used in the experiment and data will be taken regarding the effect of biodiesel use on the degradation of lubricating oil and wear of diesel engine components. The engine used in this experiment is single cylinder diesel engine and four-strokes with maximum power are 5,5 kW @2200 RPM.

2.2. Biodiesel B50 production

To make biodiesel B50, the following steps are carried out:

- The methoxide solution was made by mixing KOH with methanol in ratio of 6:1. For every 1 liter of palm oil, 4 grams of KOH and 166 ml of methanol are needed. [9]
- The sample and methoxide solution were put into the beaker. Mixture then stirred while heated using heater at temperature of 65°C for 1 hour. After heating and stirring, it was allowed to settling for 1 night for separate the glycerine and methyl ester. The methyl ester is on top and the glycerine is on the bottom. [10]
- The methyl ester is put into a beaker for the washing process. Next, distilled water and methyl ester was added to the beaker with a ratio of 1:1. Then 0.5% vinegar was added to 1000 ml of methyl ester. This washing process use bubble method by utilizing air bubbles to bind glycerol which has been bound to methanol with water. This process takes a time, with minimum 8 hours. [10]
- After the solution was bubbled for 8 hours, then the solution was allowed to settling for 1 day so that the solution was separated between the glycerol and the methyl ester.

- The results of the washing in the methyl ester were removed and put into a beaker, then dried by heating at temperature of 80-100°C. This process is needed to reduce water content in the oil. [10]
- After going through the drying process, methyl ester or biodiesel will be produced with a volume reduction of 20%.

2.3. Standard procedure of experiment

Experiments in this research aimed to determine the results of the using biodiesel on lubricating oil and engine component wear. The diesel engine test is the durability of engine with refere to the Engine Manufacture Association (EMA). The duration of the test is about 200 hours, when engine rotation and load was variation, that is: [11]

- Low idle for 30 minutes: engine with no load at 850 rpm the throttle was varied to achieve the manufacturer's recommended curb idle.
- Idle high for 30 minutes: the engine dummy load is 750 watt at 1980 rpm, was set at 25% of maximum torque and the throttle is varied to achieve an engine speed of 90% of rated speed.
- Rated power speed for 60 minutes: the engine was operated at full throttle, with engine dummy load 250 watts at 2200 rpm, load was applied until the engine speed decreases to the manufacturer's specified rated speed.
- Maximum torque and speed for 60 minutes: the engine was operated at full throttle, the engine dummy load is 3000 watts at 1900 rpm, load was applied until the engine speed decreases to the speed of rated torque as described by the manufacturer.

After completing the maximum torque and speed, the steps will be repeated until the time hours accumulates to 200 hours.

2.4. Lubricating oil sampling test

Lubricating oil sample in this research was carried out by taking 100 ml of sample from engine that had been running. Lubricating oil samples were taken at 0 hours (before running), 100 hours, and 200 hours. Laboratory sampling test aims to determine the characteristics or properties of the lubricating oil from engines, that are run using biodiesel B50 and diesel oil. The characteristics of the lubricating oil examined were kinematic viscosity, viscosity index, total base number (TBN), Fourier Transform Infrared Spectroscopy (FTIR) and metal content in the lubricating oil.

3. Result and Discussion

To know the effect of using biodiesel B50 on the combustion process, the mass of the deposit was measured. Measurement of deposit mass is carried out by taking deposits from several parts of the engine components which are then weighed to determine their mass. Deposits are taken on several diesel engine components, namely injectors, pistons, and cylinder head shown in table 1.

Table 1. Deposit mass on diesel engine components.

No.	Engine component	Dexlite fuel (gr)	Biodiesel B50 (gr)
1	Injector	0.03	0.06
2	Cylinder head	1.31	1.68
3	Piston	1.01	1.65
Total mass		2.35	3.39

To know the effect of using biodiesel B50 on the lubricating oil degradation, a lubricating oil test was carried out. The test was carried out by taking lubricating oil sample after experiment and tested on the engine with biodiesel B50 for 200 hours. Furthermore, the lubricating oil sample was tested in the laboratory to know the properties of the lubricating oil. The lubricating oil properties tested included kinematic viscosity at 40°C, kinematic viscosity at 100°C, viscosity index, total base number (TBN) and water content. In addition, the metal content of the lubricating oil was also tested to analyze the metal wear of diesel engine components. The result of laboratory test shown in table 2.

Table 2. Properties of lubricating oil before and after engine running in 200hrs.

Parameters	Method	Unit	Before running	Running with Dexlite	Running with Biodiesel B50
Kinematic viscosity at 40°C	ASTM D7279-20	cSt			
Kinematic viscosity at 100°C	ASTM D7279-20	cSt	14.16	12.06	11.87
Viscosity index	ASTM D2270-10e1	-			
Total base number (TBN)	ASTM D2896-15	mg.KOH/gr	10.06	10.7	9.88
Total acid number (TAN)	ASTM D974-21	mg.KOH/gr			
Water content	ASTM E2412-18	%v	<0.05	<0.05	<0.05
Wear metal:					
- Iron (Fe)	ASTM D5185-18	ppm	2	35	48
- Copper (Cu)	ASTM D5185-18	ppm	<1	2	2
- Aluminium (Al)	ASTM D5185-18	ppm	2	8	8
- Chromium (Cr)	ASTM D5185-18	ppm	<1	4	3

Oil rating or lubricant condition is the overall condition of the lubricant which is affected by oxidation, viscosity, chemical reaction of the lubricant, depletion of additives, etc. All of these things indicate the performance of the lubricant or a change in properties which may make the lubricant unfit for further use. The results that affect the overall condition of the oil include the Physical Test and the FTIR test, can be seen in table 3 and table 4.

Table 3. Physical test of lub oil 15W-40 before and after running

Type of fuel	Kinematic viscosity at 100°C (cSt)			Total base number/TBN (mg.KOH/gr)		
	Before running	100 hrs	200 hrs	Before running	100 hrs	200 hrs
Dexlite fuel	14.16	12.84	12.06	10.06	8.52	10.07
Biodiesel B50	14.16	12.60	11.87	10.06	10.78	9.88

The lubricating oil in the engine was tested using Dexlite fuel show an increase in the TBN value, that is 0.01 mgKOH/gr or 0.09% of the initial TBN value. Meanwhile, the lubricating oil in the engine was tested with Biodiesel B50 show a decrease in the TBN value, that is 0.18 mgKOH/gr or 1.78%. The decrease in the value of TBN is due to its nature as an acid neutralizer that has been used in the combustion process. So that the additive content in lubricating oil which is generally alkaline is reduced.

Table 4. FTIR test of lub oil 15W-40 before and after running

Fuel	FTIR-abs before running (abs/cm)				FTIR-abs 100 hrs (abs/cm)				FTIR-abs 200 hrs (abs/cm)			
	Soot.	Oxi.	Nit.	Sulp.	Soot.	Oxi.	Nit.	Sulp.	Soot.	Oxi.	Nit.	Sulp.
Dexlite fuel	<1	<1	<1	<1	1	1	2	1	1	1	2	1
Biodiesel B50	<1	<1	<1	<1	1	1	3	1	1	1	3	1

Equipment rating is used to see the presence of particles in the lubricant which can indicate component wear. It is usually used to evaluate the performance of the engine and equipment, also can be used to schedule preventive maintenance to increase productivity. The value of equipment rating shown in table 5.

Table 5. Equipment rating of lub oil 15W-40 after running

Fuel	Metal content at 100 hrs (ppm)				Metal content at 200 hrs (ppm)			
	Fe	Cu	Al	Cr	Fe	Cu	Al	Cr
Dexlite fuel	27	1	4	3	35	2	8	4
Biodiesel B50	26	2	5	2	48	2	8	3

The metal content parameters used to see the condition of the equipment are Iron (Fe) is the main component of all equipment, for wear and contamination in the trivector, most of which are caused by this element, the cause of which can be due to friction between materials or the oil is polluted by dirt from the outside. The use of biodiesel B50 affects the wear and tear of ferrous metals in diesel engines. Copper (Cu) it is used as a mixed element because of its malleability, good thermal and electrical conductivity, making it suitable for heat exchanger equipment and bearings. Aluminium (Al) is a material that has high strength with a light weight and is corrosion resistant because there is an oxide layer. Aluminium alloying with certain metals can withstand high temperatures The using of biodiesel has affects to the lubricating oil performance, so that it affects the metal content in the lubricating oil. In the results of the second test, the lubricating oil was tested in the diesel engine fueled by biodiesel B50 after running 200 hours has the same aluminum metal content, namely 8 ppm based on laboratory testing but still within reasonable limits with the maximum value and danger above 18 ppm metal content in the oil lubricant. The wear of chromium (Cr) metal can be seen in the piston ring gap of engines that have been tested with biodiesel fuel B50 and Dexlite.

3.1 Wear analysis of diesel engine composition components fueled by dexlite and biodiesel B50

3.1.1 Piston Rings

The piston ring condition when its at Top Dead Center (TDC) the piston ring gap on the engine tested using Dexlite fuel has an increase of 0.25 mm; 0.10 mm; and 0.05 mm on the compression piston ring number 1, piston ring number 2, and piston ring number 3. Meanwhile, when the piston ring is at the Bottom Dead Center (BDC), the piston ring has increase gap of 0.1 mm; 0.1 mm; and 0.15 mm on the compression piston ring number 1, piston ring number 2, and piston ring number 3.

With using biodiesel B50, the piston ring condition when its at Top Dead Center (TDC) the piston ring gap has an increase of 0.20 mm; 0.20 mm; and 0.05 mm on the compression piston ring number 1, piston ring number 2, and piston ring number 3. Meanwhile, when the piston ring is at the Bottom Dead Center (BDC), the piston ring has increase gap of 0.05 mm; 0.25 mm; and 0.20 mm on the compression piston ring number 1, piston ring number 2, and piston ring number 3. Can be seen in table 6.

Table 6. Gap of piston ring comparasion

Ring number	Dexlite				Biodiesel B50			
	TDC		BDC		TDC		BDC	
	Before running	After running	Before running	After running	Before running	After running	Before running	After running
1	0,5 mm	0,75 mm	0,5 mm	0,6 mm	0,5 mm	0,7 mm	0,5 mm	0,55 mm
2	0,5 mm	0,6 mm	0,4 mm	0,5 mm	0,5 mm	0,7 mm	0,4 mm	0,65 mm
3	0,5 mm	0,55 mm	0,3 mm	0,45 mm	0,5 mm	0,55 mm	0,3 mm	0,5 mm

3.1.2 Engine Journal Bearing

Visually, there are significant differences in the engine journal bearing with biodiesel B50 compared to Dexlite fuel. The journal bearing on the engine with biodiesel B50 has minor damage in the form of a small "coke", while the engine with dexlite fuel only has thin scratches on the surface. The gap of journal bearing can be seen in table 7, that the measurement results of the gap of the journal bearing on the engine with biodiesel B50 has greater size than the journal bearing on the engine with dexlite fuel, namely with a size of 0,055 mm and 0,065 mm with measurements on the edge of the journal bearing surface after running using dexlite and biodiesel B50.

Table 7. Gap of journal bearing

Type of fuel	Gap before running	Gap after running
Dexlite	0,050 mm	0,055 mm
Biodiesel B50	0,050 mm	0,065 mm

4. Conclusion

Based on analysis, it was found that the use of biodiesel B50 caused a greater degradation of the lubricating oil than the use of Dexlite. It is proven by the lubricating oil sample was tested with biodiesel B50 has a decrease in viscosity of 16.17% and from the lubricant data used oil analysis on engine components in attention condition but lubricant is still within tolerance limits, decreased TBN is 1.78%, and in the FTIR test the Nitration value is greater than 1, this nitration number comes from the reaction of the lubricant component with nitrogen oxides than the lubricating oil tested with Dexlite fuel. From this test, the result is use of biodiesel B50 causes greater metal wear in the lubricating oil than use of Dexlite. Proven by the metal content in the lubricating oil sample was tested for each fuel type. Lubricating oil fueled by biodiesel B50 contains the same of aluminium and chopper, 27% greater iron, and 1 ppm more chromium metal in the lubricating oil.

Acknowledgments

Authors very grateful to Marine Power Plant laboratory in Department of Marine Engineering, Institut Teknologi Sepuluh Nopember Surabaya. Also all staff and member of laboratory for support. Thank you very much.

References

[1] Iswantoro. et al 2021 The effect of use of biodiesel B30 from palm oil to degradation of oil lubrication in 1-cylinder diesel engine 4-stroke *IOP Conference Series: Earth and Environmental Science,* **649** 1 1-7
[2] Hanif 2004 Diesel Engine Performance Test using Biodiesel as Alternative Fuel *Journal of R & B,* **4** 1 19-28
[3] Nurhadi I 2015 The Effect of Biodiesel use on Performance and Main Components of the Main Engine KRI Weling-822 *Master thesis report*
[4] Pandey A K & Nandgaonkar M 2010 Experimental Investigation of the Effect of Esterified Karanja Oil Biodiesel on Lubricating Oil and Wear of a 780 HP Military CIDI Engine *SAE Int. J. Fuels Lubr.* **3** 2 273–279
[5] Yuksek L 2009 The Effect and Comparison of Biodiesel-Diesel Fuel on Crankcase Oil, Diesel Engine Performance and Emissions *FME Transactions* **37** 2 91-97
[6] Galbi M & Ishak A 2016 Prediction of Diesel Engine Oil Lubricant Replacement Set Based on Viscosity Change Rate and Total Base Number with Linearity Approach *Journal of Bina Teknika* **12** 1 111-120
[7] Dhar, A., & Agarwal, A. K. 2014. Experimental Investigations of Effect of Karanja Biodiesel on Tribological Properties of Lubricating oil In A Compression Ignition Engine. *ELSEVIER*, 112-119.
[8] Gopal K N & Raj R T 2016 Effect of Pongamia Oil Methyl Ester-Diesel Blend on Lubricating Oil Degradation of Di Compression Ignition Engine *Elsevier* **165** 105-114
[9] Ristianingsih, Y., Hidayah, N., & Sari, F. W. 2015. Pembuatan Biodiesel Dari Crude Palm Oil (CPO) Sebagai Bahan Bakar Alternatif Melalui Proses Transesterifikasi Langsung. *Jurnal Teknologi Agro-Industri*. **2** 1 1-8

[10] Suresh Vellaiyan, K. A. 2016. The Role Of Water-In-Diesel Emulsion And Its Additives On Diesel Engine Performance And Emission Levels: A Retrospective Review. *Alexandria Engineering Journal*, 2643-2472.

[11] Engine Manufacture Association (EMA) Guidelines.

IOP Conf. Series: Earth and Environmental Science 1166 (2023) 012016 doi:10.1088/1755-1315/1166/1/012016

Design of Automatic Underwater Object Detection and Tracking System for Penta Tubular ROV in Blue Economy Environment

R S Widjaja[1*], Y A Hermawan[1], D B Purwanto[1], T Yulianto[1], A S Akintoro[1]

[1] Department of Naval Architecture, Faculty of Marine Technology, Institut Teknologi Sepuluh Nopember, Surabaya, Indonesia

*sjarief@na.its.ac.id

Abstract. The development of underwater vehicles has opportunity to be further improved. One of the most challenging for developing underwater vehicle is developing automatic tracking system for Autonomous Underwater Vehicle (AUV). This research develops AUV with automatic object detection and tracking systems. These systems are developed by using vision-based method. The AUV is designed with Penta Tubular type whereas the automatic object detection system is started from captured image taken by underwater camera in the form of Blue-Green-Red (BGR) color space, which then will go through a masking process. The object resulted by object detection system is then to be used as followed object for automatic tracking system. Meanwhile, Automatic object tracking system is established by using two vertical lines that are used as motion parameters of the AUV. By considering the object position compared with the two vertical lines, the automatic tracking guidance system can be generated. The developed automatic object detection and tracking systems can be used well for AUV, in which verage travel time for forward and backward motion are 14.20 seconds in 0.14 m/s, and 19.82 seconds in 0.1 m/s while average travel time is 15.36 seconds and 13.58 seconds for right and left motion test respectively.

Keywords: Autonomous Underwater Vehicle, AUV, Object Detection, Object Tracking, Automatic Tracking System, Blue Underwater Sustainability

1. Introduction

In carrying out underwater activities, humans use diving equipment such as goggles and oxygen cylinders, the tools that must be used by a diver, while the physical and stamina of humans have limits. Time and quality of view are the weaknesses of humans in the water. Oxygen cylinders that are commonly used also could not be used for last long [1]. A lot of underwater work can be conducted automatically by using an underwater vehicle. Monitoring of environmental conditions, inspection of damage, repair of oil and gas installations, and inspection of pipelines are activities that are often carried out by using underwater vehicle. In general, humans have limited abilities for underwater activities and they can only reach a limited depth. Thus, the underwater environment has many opportunities for the development of underwater vehicles [2].

Underwater robots are a rapidly growing field of study in the autonomous field. Underwater robots are often used in various underwater activities, including exploration, surveillance, and inspection tasks [3]. In such cases, automation of every part of the control process can be an important improvement. It can have a direct impact on reducing the number of errors in work, task execution time, and costs [4]. In this research, the automatic object detection and object tracking system are developed to be implemented on underwater vehicle to support automatic navigation and control system for autonomous underwater vehicle (AUV). This research is expected to be the first step in the development of an underwater assistant to assist in underwater exploration.

2. Overview for Object Detection and Tracking System on AUV

2.1. Autonomous Underwater Vehicle (AUV) Design

Autonomous Underwater Vehicle (AUV) is an undersea unmanned vehicle that used for special purpose i.e. underwater survey, underwater inspection, military, etc. AUV is also known as an unmanned cable robot, which is equipped with various sensors for a specific purpose. In its application, AUV contributes by helping humans in efforts to empower marine exploration in Indonesia, by inspecting underwater pipelines, mapping underwater areas, and for military purposes [5].

In the construction of an underwater vehicle, considering every aspect and element of the vehicle is important to avoid over-designing of the vehicle. The right and appropriate design of AUV including its control system can make operational costs more efficient. The main element that needs to be emphasized in the design is the propulsion system. The propulsion system must have a high thrust ratio. Every component of the propulsion system needs to be designed with care because the greater the motor power used, the heavier the propulsion system on the vehicle. Another aspect is the placement and number of motors used on the rides. Generally, the three motors on the rides will make the rides have forward, backward, up, and down motions; the four motors on the ride will give additional lateral thrust; and the five motors on the ride will make the ride have thrust in all horizontal directions. Placing the motor off the longitudinal axis of the vehicle will allow the vehicle to have a better turning moment [6].

The choice of the type of material in the construction of the vehicle also affects the strength and buoyancy of the vehicle. Therefore, the selection of materials must be done properly. The choice of material depends on the desired depth of operation of the vehicle. Materials commonly used to aid the buoyancy of the rides are glass, acrylic, glass-reinforced plastic, synthetic foam, synthetic rubber, titanium, steel, and aluminum [7].

2.2. AUV Control System

Autonomous underwater vehicle (AUV) uses a closed loop control system, commonly known as a closed control system. This system works by sending feedback in the form of an error signal to the control unit. The error signal is caused by a disturbance that affects the process value, where the error signal is obtained from the difference between the input signal and the output signal. The control unit that receives the feedback will work so that the input given is the same as the desired output [8].

2.3. OpenCV

OpenCV is a collection of libraries that have programming functions for real-time applications of computer vision. Various algorithms in this platform can be used for face recognition, face tracking, artificial intelligence, etc. It supports artificial neural networks and deep learning. The source code has a function to use various programming languages such as C++, Java, Python, JavaScript, and others [9].

2.4. Object Detection System

Object detection basically involves one or more objects in an image, then the system will process the image so that the object can be identified according to the command given. The algorithm will be trained to identify various targeted objects [10].

The RGB (red-green-blue) color model is a color model that is often used in computer vision. The input usually taken by the camera is the RGB color model. This model must be converted to the HSV (Hue, Saturation, and Value) color model to match the threshold value [11].

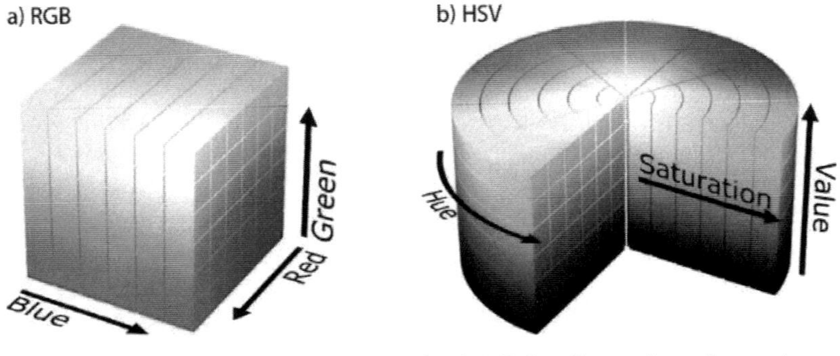

Figure 1. *RGB Color Space and HSV Color Space (google.com)*

The HSV color space can be defined by a cylindrical shape. This color model defines the color space into three components. Hue represents a color type such as red, blue, or yellow that ranges from 0 to 360 degrees. Saturation is the brightness level of a color that ranges from 0 to 100%. Value (intensity) is the strength of the color that ranges from 0 to 100%. The RGB to HSV color conversion process can be described in an equation and diagram as follows [12].

$$H = \begin{cases} 0 + \frac{43 \times |G-B|}{Max(R,G,B) - Min(R,G,B)}, & \text{if Max(R, G, B) = R} \\ 85 + \frac{43 \times |B-R|}{Max(R,G,B) - Min(R,G,B)}, & \text{if Max(R, G, B) = G} \\ 171 + \frac{43 \times |R-G|}{Max(R,G,B) - Min(R,G,B)}, & \text{if Max(R, G, B) = B} \end{cases}$$

$$S = \left[\frac{Max(R,G,B) - Min(R,G,B)}{Max(R,G,B)} \right]$$

$$V = Max(R,G,B)$$

Figure 2. *RGB to HSV Conversion Equation [12]*

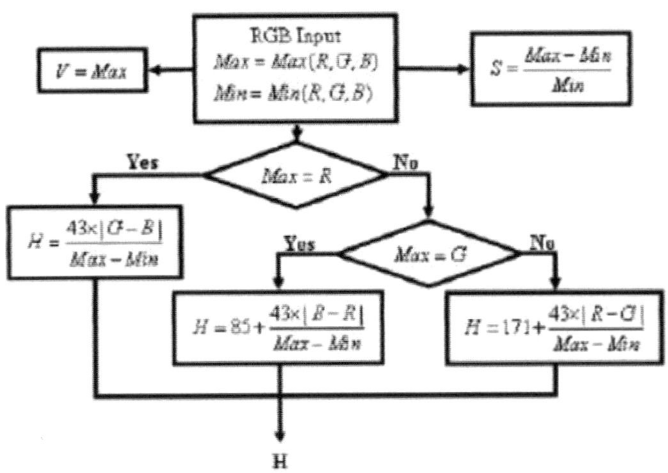

Figure 3. *RGB to HSV Conversion Diagram [12]*

2.5. Object Tracking System

After detecting the object, the object color is matched against the stored HSV value range. If the color matches, then the AUV follows the object, keeping a constant distance from it. The radius and center of the minimum circle are the references used to measure the distance of the object from the camera and the direction in which the object is moving. As the object moves forward, the radius of this circle decreases, and this gives instructions to the AUV to move forward. The AUV moves towards the object until the specified distance is reached. If the object moves to the right, then the center of the circle shifts to the right, and this gives an indication to the AUV to move to the right. In the same way, if the person moves to the left, the AUV moves to the left and tries to keep the center in the standard position. The Euclidian distance between the camera and the object is used to predict the distance. This distance is the length of the line connecting the center of the circle drawn on the object and the camera. This distance will be kept constant to avoid collisions between the AUV and the object [10].

3. Working Process

3.1. Problem Identification and Literature Study

The initial step in working process is to determine the problems must be solved, the opportunities for the development of underwater vehicles, and the many benefits of automation. This research is focused on an idea in the form of implementing object detection and object tracking on autonomous underwater vehicles (AUV). Furthermore, necessary information and references are collected to support the problem for developing object detection and object tracking on an autonomous underwater vehicle (AUV). The literature studies carried out are those related to the determination of design and size, calculation of vehicle characteristics, basic concepts of object detection, and object tracking.

3.2. Data Gathering

To support the computer programming process, object tracking and object detection command scripts are obtained from open source libraries that can be accessed freely. One of the libraries used is OpenCV. The script obtained is modified so that it can function as desired. The compilation of AUV-related journals was used in determining the main measure based on the existing AUV. The data that has been collected is then processed into a decision related to determining the main size and design of the vehicle, designing an automatic control system, and determining the layout of the control system components. Where in the study compared with literature studies that have been obtained.

3.3. Size and Design Selection

Based on the literature, the AUV type with a stylish shape has good maneuverability and is easy to manufacture with PVC pipe as a raw material. The research that was carried out did not use bionic hands or claws to pick up items, which are usually done by box-type ROVs because they require more stability, so there is no strong reason to choose the box-type ROV design. The choice of a stylish penta-tubular body shape aims to make the ROV have good stability. Where the body consists of 1 main pipe and 4 pipes around it as ballast. These four ballast pipes not only function as fixed ballasts but also act as balancers for the main body.

3.4. Design AUV System and Configuration

In designing AUV system, the electronic devices will be installed on AUV must be selected first. The electronic components are determined based on the concept that has been planned according to the desired specifications. The electronic components must be compatible with the other electronic devices to ensure the system can work well. Furthermore, the design of the automatic control program for AUV must be determined to make it easier for making computer control programs that will be developed later.

The automatic control design considers the number of thrusters to be used, the ability of each component in the applied automated system, and other control system considerations. The control computer program must be developed for running the automatic control system. After the computer control program is completed, it is necessary to conduct function test of the automatic control program. This test aims to ensure motor and thruster can work properly according to an input provided by the camera or users.

After electronic devices, control system, and computer control program are ready, the next stage is design arrangement and layout of all components that must be installed on AUV. The arrangement can affect the cable management of the rides, and the stability of the rides, as well as the size of the rides used. In addition, in making the vehicle, the placement of electrical components in the vehicle must be carefully considered. The electrical components must be ensured to be able to enter the vehicle completely, so that the space inside the vehicle must be able to contain all the electrical components.

3.5. Built AUV Prototype and Integrating System

The activity of making a prototype of the vehicle begins with making details of the necessary tools and materials. In the manufacture of the vehicle, the body of the vehicle is made of PVC pipes of various sizes. An additional 3D printed mount is required to connect the ballast tank and the main body of the AUV. Then, the electronic components are determined based on the concept that has been planned according to the desired specifications. The compatibility of each component with the other components must be ensured.

3.6. Trial Phase

At this stage, data are collected from the process of vehicle testing. The data taken in the testing are the ability to detect objects on land by giving 5 different colors to the vehicle to determine its detection ability, the testing of detection ranges in underwater by measuring the maximum distance of the vehicle, speed test on forward and backward motion by running the vehicle on a straight track to determine the speed of the vehicle in following the object, and measuring the travel time of the vehicle in a 180-degree maneuver with variations of right and left rotation. The trial data results are then analyzed to investigated the performance of automatic object detection and object tracking used on AUV.

4. Development AUV with Object Detection and Tracking Systems

4.1. Design of Penta Tubular AUV

The dimensions of AUV used in the research include the size of the components that make up the body of the vehicle and the depiction of the vehicle in the form of 3D images can be described on Table 1 as follows,

Table 1. *Detail Dimension of AUV Component*

No	Name	Dimension	unit	total
1	PVC pipe 6 inch	Ø16.5 x 35	cm	1
2	PVC pipe 2 inch	Ø6 x 35	cm	2
3	PVC pipe $1^{1/2}$ inch	Ø4.8 x 35	cm	2
4	PVC pipe ¾ inch	Ø2.6 x 58	cm	1
5	Pipe Dop	2	inch	4
6	Pipe Dop	$1^{1/2}$	inch	4
7	Clean Out	6	inch	1
8	Tee	$1^{1/2}$ x ¾	inch	4

No	Name	Dimension	unit	total
9	Tee	2 x ¾	inch	4
10	Knee 45°	¾	inch	4
11	Socket (Coupling)	2	inch	6
12	Dome Acrylic	Ø6	inch	1
13	Acrylic thickness 3 mm	16 x 35	cm	1
14	AWG camble Diameter 3 mm	7	m	1
15	Banana Plug	3.5	mm	36
16	3D Print PETG (*Mounting Frame*)	268	gram	2
17	3D Print PLA+ (Vertical *Mounting Thruster*)	45	gram	2
18	3D Print PLA+ (Horizontal *Mounting Thruster*)	51	gram	2
19	3D Print PLA+ (Propeller)	15	gram	6
20	3D Print PLA+ (Mounting Sensor)	151	gram	1
21	Bolt M3	Ø2.58	mm	24
22	Bolt M6	Ø5.14	mm	6

The design of Penta Tubular AUV used in this research is shown on Figure 4.

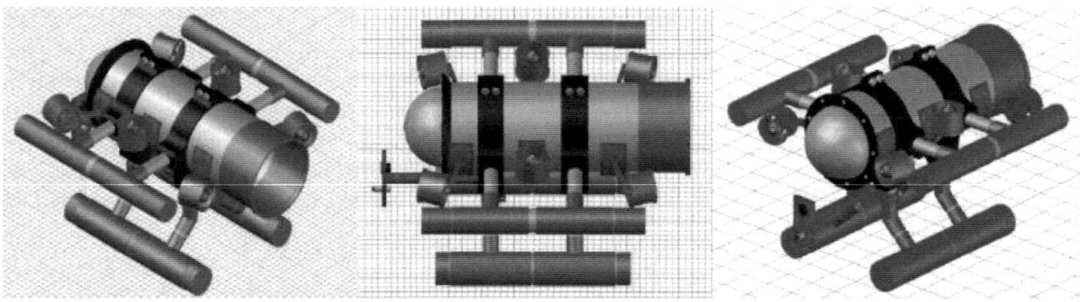

Figure 4. *Three Dimensional Design of Penta Tubular AUV*

4.2. Hydrodynamic Pressure Analysis

Hydrostatic pressure is the pressure caused by the force of the liquid on an object that is in the water at a certain depth. This AUV vehicle is designed to operate at a depth of 1 meter so that the hydrostatic pressure acting on the vehicle can be calculated by Equation (1) as follows:

$$\begin{aligned} P &= P_{atm} + \rho \, x \, g \, x \, h \qquad (1) \\ &= 101325 + 1000 \, x \, 9.8 \, x \, 1 \\ &= 111125 \, N/m^2 \\ &= 1.11125 \, Bar \end{aligned}$$

In the above formula P_{atm} is atmospheric pressure, is the density of water, g is the acceleration due to gravity, and h is the depth of operation. Pressure The pipe used for the main body has a diameter of 6 inches, which is designed to be able to withstand a water pressure of 8 bar based on SNI data. The hydrostatic pressure acting on the vehicle is 1.11125 Bar, so that the hydrostatic pressure on the vehicle at a depth of 1 meter will not have much effect on the strength of the vehicle structure.

4.3. Buoyancy Analysis

The buoyancy analysis is carried out by three condition i.e. without ballast (Condition 1), 6.8 kg ballast (Condition 2), and 10.2 kg ballast (Condition 3) as shown on Table 2. Condition 1 is conducted to check the buoyancy of the vehicle with the weight of the vehicle without the addition of ballast. Condition 2 is providing additional load as ballast in 4 spaces in the ballast tube while Condition 3 is adding additional load as ballast and water in 4 spaces in the ballast tube. Giving the load on the vehicle aims to determine the weight needed to submerge the vehicle. The submerging condition for each condition can be illustrated on Figure 5.

Table 2. *Buoyancy Analysis Condition*

Condition	Prototype weight	Ballast weight	Prototype condition
Condition 1	9.3 kg	0 kg	Floating
Condition 2	9.3 kg	6.8 kg	Floating
Condition 3	9.3 kg	10.2 kg	Sink

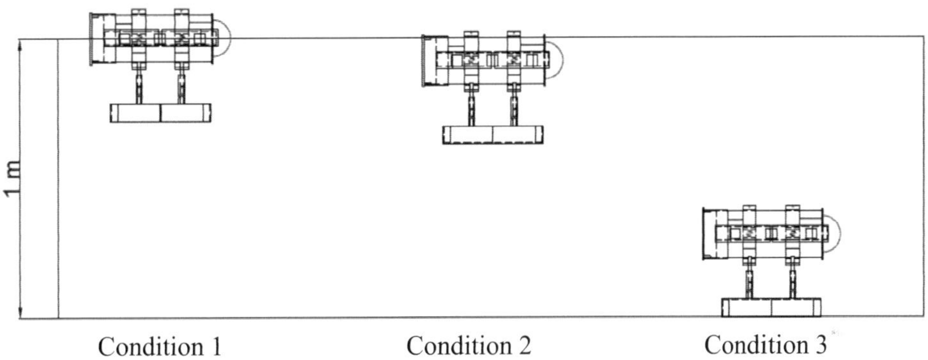

Figure 5. *Illustration for Submerging Condition*

4.4. Power Requirements Analysis

The calculation of power on the AUV is carried out to get an estimate of the power needed to move the AUV according to the planned speed. In determining the speed of the designed AUV, apply Newton's third law, where when the object is in equilibrium, the thrust = resistance. So the object can be said to move when the thrust > resistance. The resistance obtained is used to determine the power requirements of the propulsion motor. Based on the resistance calculation, the resistance value is 12.356 N at a speed of 1 m/s. Thereby, the required motor power can be determined by the following calculations (Equation (2)):

$$\begin{aligned} Power &= R \; x \; Vt \\ &= 12.356 \; x \; 1 \\ &= 12.356 \; Watt \end{aligned} \tag{2}$$

Calculating power with multiplication of resistance (R) by speed (Vt). The speed increases every 0.1 m/s starting from 0 m/s until it reaches the maximum motor power at a speed of 2.8 m/s. Based on the motor specifications, the maximum power of the motor is 150 W, with a safe limit of 140 W. The use of two motors means the maximum power used is 280 W in each movement. The results are presented in the form of a graph shown on Figure 6.

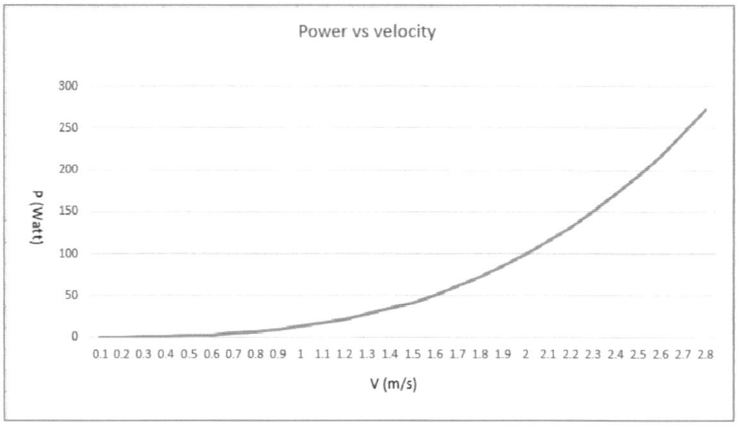

Figure 6. *Power vs Velocity Diagram*

Based on the graph shown on Figure 6, the prototype can have a speed of up to 2.8 m/s with a power of 271.24 W on both motors.

4.5. Selecting Electronic Components

After obtaining the required power in the previous discussion, a motor with a brushless motor type A2212/13T 1000KV is selected, which has a maximum power of up to 150 W. The selection of a 1000 KV motor aims to obtain greater torque; the lower the RPM, the greater the torque produced. The detail motor specifications is presented on Table 3 as follow:

Table 3. *Motor Specifications*

Cells	2-3 Li-Po
Kv	1000 RPM/V
Max efficiency	80%
Max efficiency current	4-10 A
No load Current	0.5A@10V
Resistance	0.090 ohms
Max current	13A for 60S
Max watts	150W
Weight	52.7 g / 1.86 oz

Based on the specifications shown in Table 3, the recommended battery size is between 2 and 3 cells, and then the lipo battery 3S 2200 mAh 25 C are selected. The calculation of battery power obtained through the results of the multiplication between current (I) and voltage (V) is described as follows:

$$
\begin{aligned}
2200\, mAh &= 2.2\, Ah \\
2.2\, Ah \; x \; 25\, C &= 55\, A \\
1S &= 3.7\, V \\
3S &= 11.1\, V \\
P &= I \; x \; V \qquad (3) \\
55\, A \; x \; 11.1 V &= 610.6\, W
\end{aligned}
$$

Based on the calculation of the battery power of 610.6 W, the value is higher than that required by the motor. The operational time of the vehicle is planned for 30 minutes. The calculation of the current strength is obtained as follows:

$$I = (2.2 / 30) \, x \, 60 = 4.4 \, A$$

Motor operation at 4.4A current can last up to 30 minutes of operating time. The selection of the ESC component must have a current value greater than that required by the motor. The maximum current of the motor is 13A so that the use of the 30A ESC can be selected.

testing is done using RPM of 1000 RPM to save battery usage. The motor will rotate at this speed at a voltage of 1 V, so the possible speeds are as follows.

$$P = I \, x \, V$$

$$P = 1 \, x \, 4.4 = 4.4 \, Watt$$

Based on the interpolation calculations, with a power of 4.4 watts, the motor can move at a speed of 0.708 m/s. Due to an error in production, at the rear of the vehicle there is an additional length so that the flow from the propeller is blocked. In addition, the characteristics of the propeller are not taken into account because it uses an existing design. This can reduce the estimated speed that has been calculated. Here's a picture of the electronic workflow.

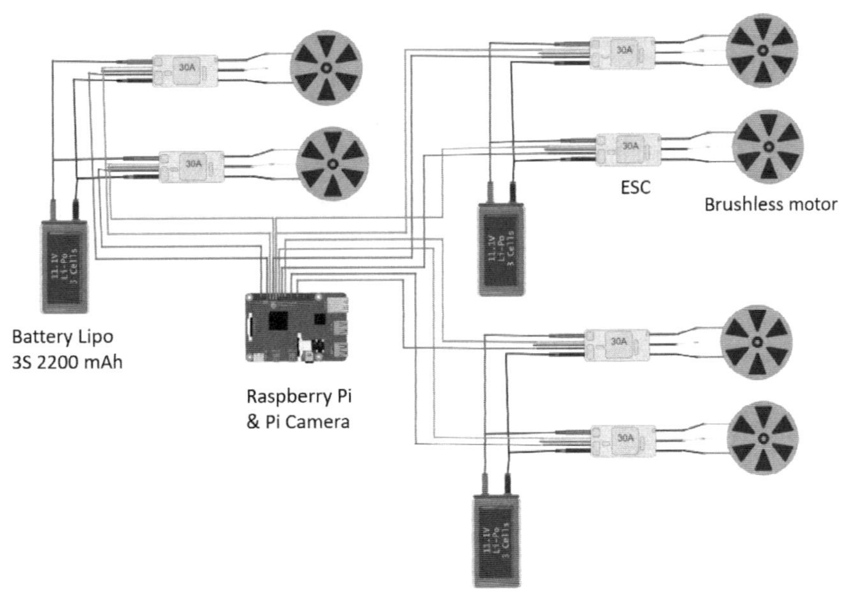

Figure 7. *Wiring Diagram*

4.6. Color Detection System and Application

Color detection is the process of identifying and detecting colors in digital images or videos. In this process, the image captured by the camera will be processed by the Raspberry Pi. Initially, the image captured by the camera in the form of BGR (blue-green-red) color space will be converted into HSV (Hue, Saturation, and Value) color space, which aims to facilitate the color segmentation process. Images in HSV form will go through a masking process that aims to distinguish the background color from the color of the object you want to detect. The detected color will be used as a reference for the navigation system to control the AUV. The object detection system flow diagram and object detection result are shown in Figure 8 and Figure 9 respectively.

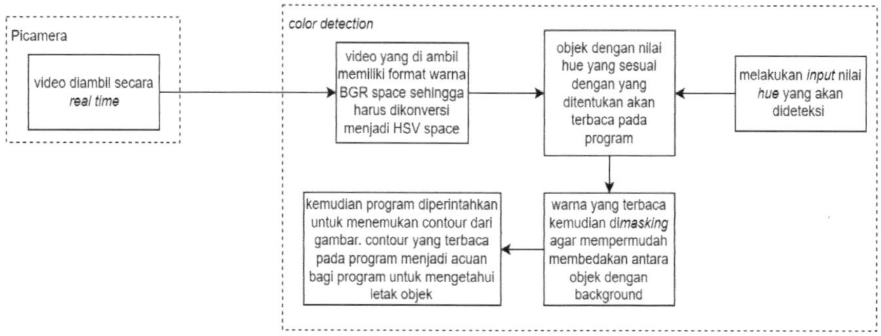

Figure 8. *Object Detection Process Flow Diagram*

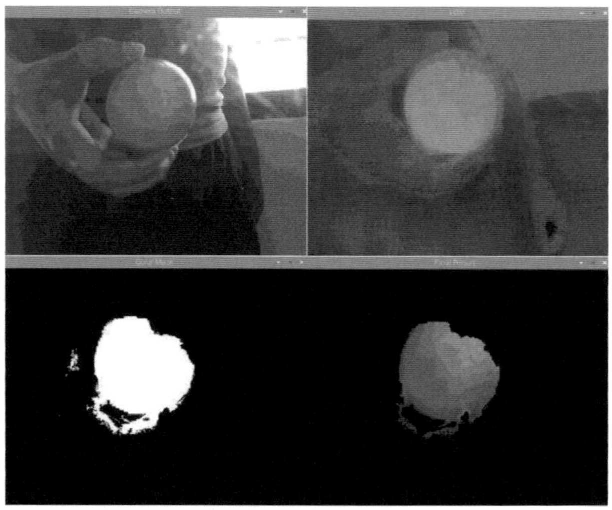

Figure 9. *Object Detection Processing Results*

4.7. Navigation System and Application by Using Object Tracking Syatem

In this process, the image results in masking, which is the output of the color detection process, is used as a reference for moving the AUV. If this program is visualized, then on the screen you will see two vertical lines spaced at 1/3 of the screen width measured from the center of the screen, which is made as a motion parameter of the AUV. Navigation and control system diagram for AUV used in this research is shown in Figure 10.

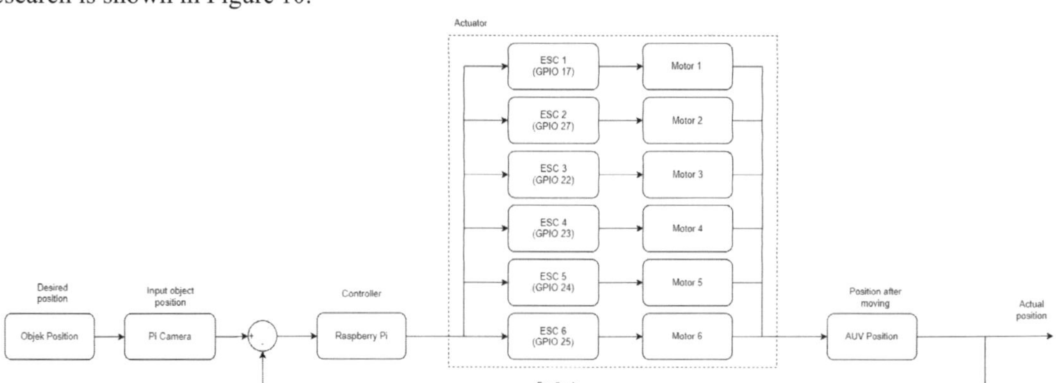

Figure 10. *Block Diagram System*

When the size of the object captured by the camera is less than the parameters set in the program, the AUV will move forward towards the object. The AUV can move backwards to avoid collisions with objects when the size of the object on the screen exceeds the parameters set in the program. The right and left movements are made with a parameter in the form of a vertical line one third of the width of the image, made on the right and left of the center of the image. The automatic object tracking process is shown in Figure 11 while the result of object tracking process can be seen in Figure 12.

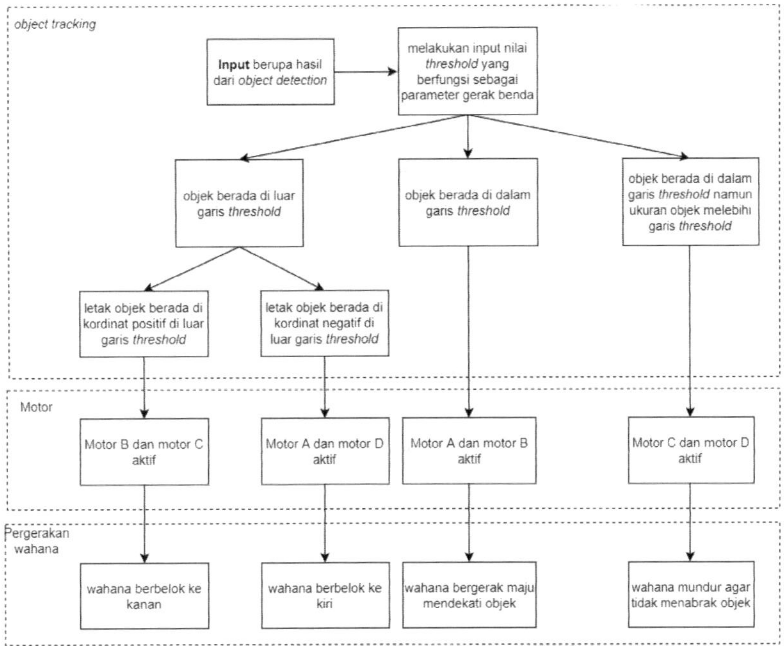

Figure 11. *Object Tracking Process*

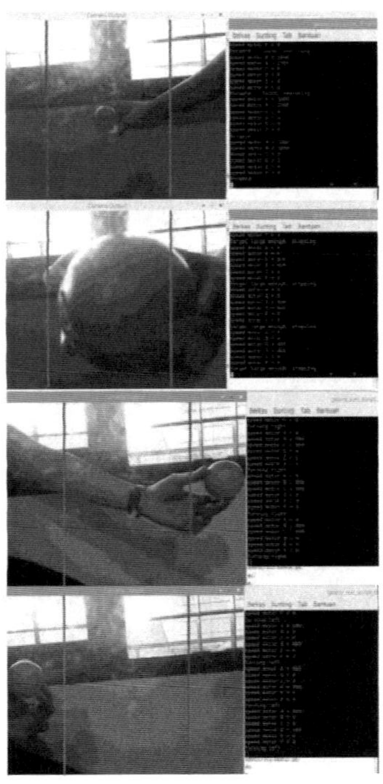

Figure 12. *Object Tracking Process Results*

IOP Conf. Series: Earth and Environmental Science 1166 (2023) 012016 doi:10.1088/1755-1315/1166/1/012016

5. Trial and Underwater Testing

5.1. Underwater Visibility Testing

Underwater visibility testing is carried out to ensure the developed system can work well under water. The underwater testing results is presented on Table 4.

Table 4. *Underwater Visibility Test Results*

Color	Hue value	Detection result	Detection status	Distance
Yellow	25	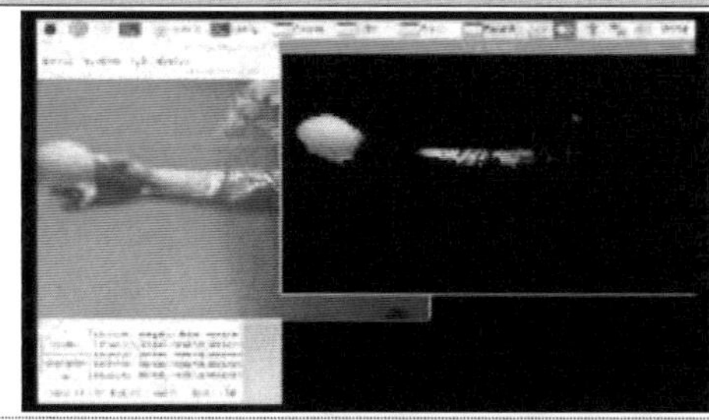	Detected	115 cm
Green	70	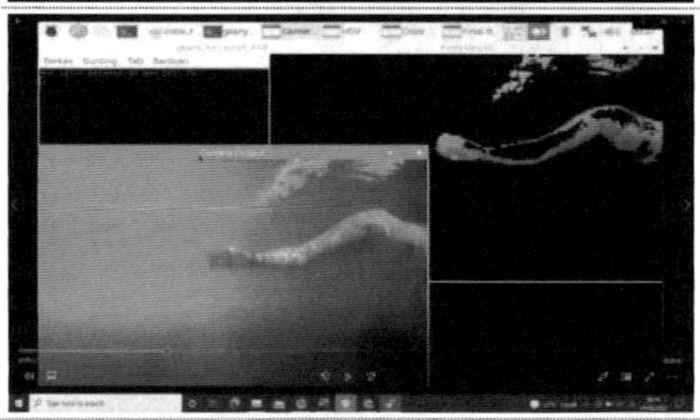	Detected	152 cm
Blue	100	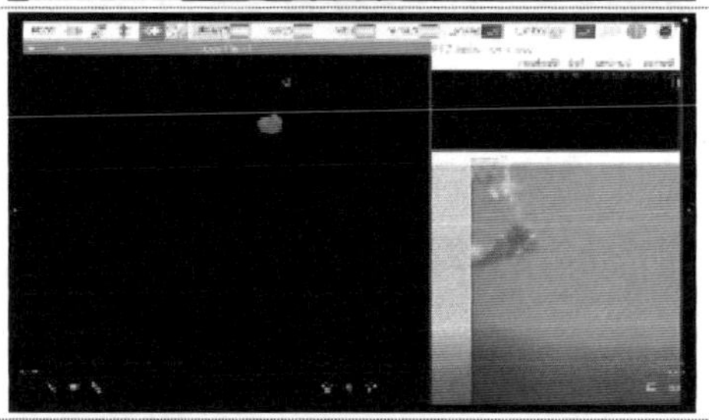	Detected	128 cm
Magenta	165	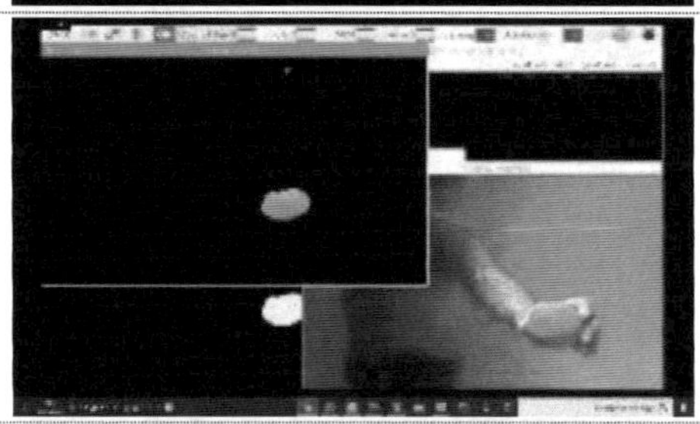	Detected	78 cm
Red	180	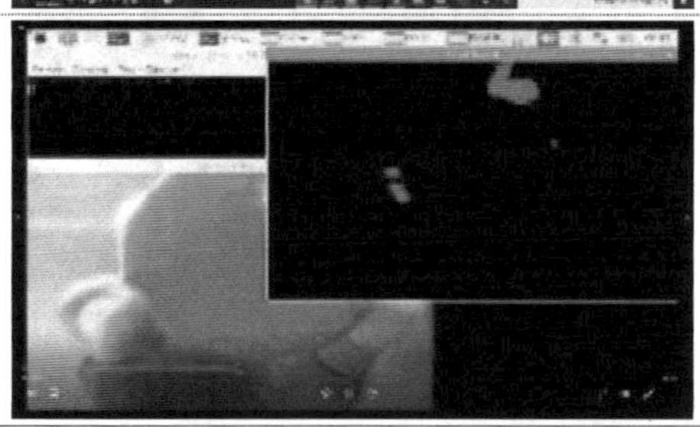	Not detected perfectly	0 cm

Based on the Table 4 above, the yellow color can be read well up to a distance of 115 cm from the vehicle, but the color of the hand emitted by light in the water is also read as a yellow object. So, if the AUV navigation test uses yellow, it is feared that other objects that emit light can divert the AUV. The green color can be read well up to a distance of 152 cm from the vehicle, but the color of the hand in the water is also read as a green object. So it is feared that if the AUV navigation test uses green, it could divert the AUV. The blue color can be read well up to a distance of 128 cm from the vehicle. The blue color also does not read other objects other than the ball, so it is suitable for tracking targets by the AUV. The magenta color can be read well up to a distance of 78 cm from the vehicle. The magenta color also does not read other objects other than the ball, but its range is still lower than the blue color. The magenta color can be read well up to a distance of 78 cm from the ride. The red color cannot be read perfectly in the water, even if it is brought closer to the dome.

5.2. Speed Testing by Following Objects on a Straight Track

Speed testing by following object on a straight tracck was conducted to determine the performance of the AUV in moving forward on a straight track as far as 2 m. The test was carried out five times based on the experimental results obtained from the data presented in Table 5 as follow:

Table 5. *Forward Motion Test*

No	Test	Time
1.	1st test	13.55 s
2.	2nd test	14.95 s
3.	3rd test	13.47 s
4.	4th test	14.45 s
5.	5th test	14.60 s

Based on the data obtained from the forward motion test, the average travel time obtained is 14.20 seconds, meaning that the average speed of the AUV in moving forward is 0.14 m/s.

Then this test is carried out to determine the performance of the AUV in moving backwards on a straight track as far as 2 m. The test was carried out five times based on the experimental results obtained from the data presented in Table 6 as follow:

Table 6. *Reverse Motion Test*

No	Test	Time
1.	1st test	19.07 s
2.	2nd test	18.97 s
3.	3rd test	21.98 s
4.	4th test	19.72 s
5.	5th test	19.34 s

Based on the data obtained from the backward motion test, the average travel time obtained is 19.82 seconds, meaning that the average speed of the AUV in moving backward is 0.1 m/s.

5.3. The 180-Degree Maneuver Test

Testing the maneuver 180 degrees to the right aims to determine the travel time of the vehicle when turning back to the right. The test was carried out five times, based on the experimental results obtained, the data presented in Table 7 as follow:

Table 7. *Right Rotation Test*

No	Test	Time
1.	1st test	14.60 s
2.	2nd test	15.80 s
3.	3rd test	14.28 s
4.	4th test	15.18 s
5.	5th test	16.97 s

Based on the data obtained from testing the maneuver 180 degrees to the right, the average travel time obtained is 15.36 seconds. With the fastest movement time of 14.28 seconds.

The 180-degree maneuver test to the left aims to determine the travel time of the vehicle when turning back to the left. The test was carried out five times, based on the experimental results obtained, the data presented in Table 8 as follow:

Table 8. *Left Rotation Test*

No	Test	Time
1.	1st test	13.55 s
2.	2nd test	12.63 s
3.	3rd test	13.78 s
4.	4th test	13.36 s
5.	5th test	14.60 s

Based on the data obtained from testing the maneuver 180 degrees to the left, the average travel time obtained is 13.58 seconds. With the fastest movement time of 12.63 seconds,

6. Conclusion

From all the discussions that have been described in the previous chapters, the following conclusions can be drawn:

1. The design of the penta tubular vehicle was chosen because this type of design is included in the stylish AUV, consisting of 1 main pipe and 4 pipes around it as ballast. These four ballast pipes not only function as fixed ballasts but also act as balancers for the main body.
2. When the size of the object captured by the camera is less than the parameters set in the program, the AUV will move forward towards the object. The AUV can move backwards to avoid collisions with objects when the size of the object on the screen exceeds the parameters set in the program. The right and left movements are made with a parameter in the form of a vertical line spaced one third of the width of the image, which is made on the right and left of the center of the image.
3. Based on the data obtained from the forward motion test, the average travel time obtained is 14.20 seconds and the average speed is 0.14 m/s. The data obtained from the backward motion test shows that the average travel time obtained is 19.82 seconds and the average speed is 0.1 m/s. Based on the data obtained from testing the maneuver 180 degrees to the right, the average travel time obtained is 15.36 seconds. With the fastest movement time of 14.28 seconds, while the data obtained from testing the maneuver 180 degrees to the left, the average travel time obtained is 13.58 seconds. With the fastest movement time of 12.63 seconds.

7. Acknowledgement

The works in this paper were supported by the Institut Teknologi Sepuluh Nopember (ITS) Research Fund with contract numbers: 1854/PKS/ITS/2022.

References

[1] Susanto T 2019 Rancang Bangun Metode Tracking Object Pada Underwater Remotely Operated Vehicle. Undergraduate Final Project, Institut Teknologi Sepuluh Nopember. Available on: https://Repository.Its.Ac.Id/Id/Eprint/60856.

[2] Sattar J., Dudek G., Chiu O., Rekleitis I, Giguere P, Mills A, Plamondon N, Prahacs C, Girdhar Y, Nahon M, and Lobos J P 2008 Enabling Autonomous Capabilities in Underwater Robotics *IEEE/RSJ International Conference On Intelligent Robots And Systems*, Nice, September 2008, pages. 3628–3634. doi: 10.1109/IROS.2008.4651158.

[3] Xia Y. and Sattar J. 2019 Visual Diver Recognition For Underwater Human-Robot Collaboration *2019 International Conference On Robotics And Automation (ICRA)*, Montreal, QC, Canada, Mei 2019, pages. 6839–6845. doi: 10.1109/ICRA.2019.8794290.

[4] Ortiz A. and Antich J. 2009 Bayesian Visual Tracking For Inspection of Undersea Power and Telecommunication Cables *J. Marit. Res.*, Vol. 6, No. 2, pages. 83–98.

[5] Christ R. D. and Wernli R. L. 2007 *The ROV Manual: A User Guide To Observation-Class Remotely Operated Vehicles*, 1st Ed. Amsterdam ; Boston: Butterworth-Heinemann.

[6] Roslan R., Said M., and Bakar A. 2015 Conceptual Design of Remotely Operated Underwater Vehicle.

[7] Rout R. 2013 Control Of Autonomous Underwater Vehicles," page. 91.

[8] Pulungan A.B., Nafis Z., Anwar M., Hastuti, Hamdani, and D. E. M. 2021 Object Detection with A Webcam Using the Python Programming Language *J. Appl. Eng. Technol. Sci. JAETS*, Vol. 2, No. 2, pages. 103–111, Mei 2021, doi: 10.37385/Jaets.V2i2.247.

[9] Supardi N.I. 2012 Penggunaan PLC Dalam Pengontrolan Temperatur, Simulasi Pada Prototype Ruangan Vol. 2, No. 2, page. 8.

[10] Ingole K and Khedkar S. R. Automatic Human Following Trolley Using Raspberry Pi Vol. 08, No. 07, page. 6.

[11] Chakravarthy K. V. K. and Krishna M. V. 2021 Color Objects Detection In Real-Time With Raspberry Pi And Image Processing page. 3.

[12] Hanumantharaju M. C., Vishalakshi G. R., Halvi S., and Satish S.B 2012 A Novel FPGA Based Reconfigurable Architecture For Image Color Space Conversion," Dalam *Global Trends In Information Systems And Software Applications*, Vol. 270.

Analysis of silica gel desiccant application in fuel storage tanks model to reduce palm oil-based biodiesel degradation

M F R Azhad[1], A Z M Fathallah[1], and N H Alami[2]

[1] Department of Marine Engineering, Institut Teknologi Sepuluh Nopember, Surabaya, 60117, Indonesia
[2] Department of Biology, Institut Teknologi Sepuluh Nopember, Surabaya, 60117, Indonesia

faizurazhad@gmail.com

Abstract. Biodiesel fuel is subject to degradation as a result of its prolonged storage period. The degradation of biodiesel fuel can have a negative impact on its qualities as well as the performance of the engine. These four stages of degradation are hydrolysis, oxidation, thermal decomposition, and bacterial contamination. A consequence of the hygroscopic qualities of biodiesel is the possibility of hydrolysis processes, which might result in the creation of free fatty acids, which can lead to engine corrosion. The presence of water encourages bacterial development in biodiesel, which might result in the formation of deposits. This study aims to investigate the application of silica gel desiccant in fuel tanks with varying biodiesel concentrations. The total acid number is used as the degradation rate parameter, while the number of bacterial colonies is used as the bacterial growth rate parameter. The research was carried out through experiments using tank models over a period of six weeks, with six tank conditioning. With the experiment results, reducing humidity in the fuel tank by applying desiccant can lead to the reduction of biodiesel degradation rate and micro-bacteria growth.

Keywords: Bacteria, Biodiesel, Degradation Rate, Silica Gel, Total Acid Number.

1. Introduction

Due to the nature of biodiesel, it is prone to quality loss during storage, particularly during marine transportation. Degradation of biodiesel can create insoluble deposits, clogging fuel filters [1,2] increase saponification rates and total glycerol levels,rop in iodine number ester content, and stored heating value [3]. Degraded fuel qualities affect the engine's performance and condition, as well as the state of its auxiliary systems. It is possible for the engine filter to become clogged and the fuel injectors to become damaged, resulting in total engine failure; it may also result in increased oil consumption during operation [2,4,5]. Biodiesel degradation occurs via four mechanisms: hydrolysis, which occurs when fuel comes into contact with water or humid air; oxidation, which occurs when biodiesel comes into contact with oxygen in the air; thermal decomposition due to overheating; and bacterial contamination brought about by external exposure to dust, air, or water [2,6,7]. This research focuses on hydrolysis and bacterial contamination oxidation.

The concentration of acidity increases in degraded biodiesel; this is caused by the breakdown of the fatty acid methyl ester chain to fatty acids [3]. This acid concentration can be expressed as a total acid number (TAN), which shows how much mg KOH is needed to neutralize the fat contained in 1 mg of the object (Biodiesel). According to previous research conducted, figure 1 shows that the total acid number in biodiesel increases directly with the storage time regardless of the research variables, which illustrates the rate of fuel degradation. Biodiesel degradation can also have an impact on fuel properties. Changes in chemical composition due to degradation will result in significant changes in engine performance and the resulting emissions. Fuel properties that will be affected include flashpoints, viscosity, and cetane number [7]. If the properties change on the fuel is significant, the fuel specification will not comply with regulation standards and emission certification. Sludge or sediment produced by degrading biodiesel will have an impact on the lubricating oil. Changes that occur to lubricating oil properties will also accelerate bearing, seal, and filter wear rate.

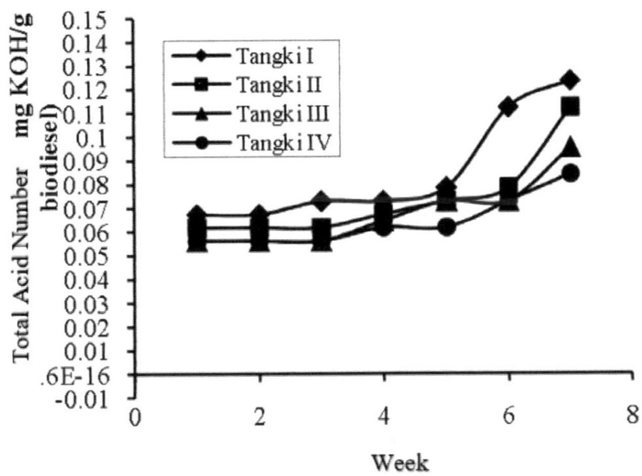

Figure 1. Increase in Acid Numbers in biodiesel storage [3].

Biodiesel has hygroscopic properties, which is the ability to absorb water molecules from the environment, especially the free humid air that enters the fuel tank through the vent pipe. Biodiesel has a 30 times greater capacity to absorb water than petroleum diesel fuels [8]. The presence of this water content can trigger a hydrolysis reaction in fatty acid methyl esters (FAME) to form free fatty acids that can cause corrosion to metals and degrade elastic polymeric materials (elastomers) such as rubber seals [5,9]. The metal heating coil in fuel tanks, tank coatings containing elastomers, as well as gaskets, and seals will suffer damage within a certain period due to the presence of high levels of acid. The presence of water content also triggers the growth of bacteria on biodiesel. Bacterial activity can form deposits in tanks, filters, pipes, and purifier units in ship material systems.

Figure 2. Clogging at Purifier Unit [10]

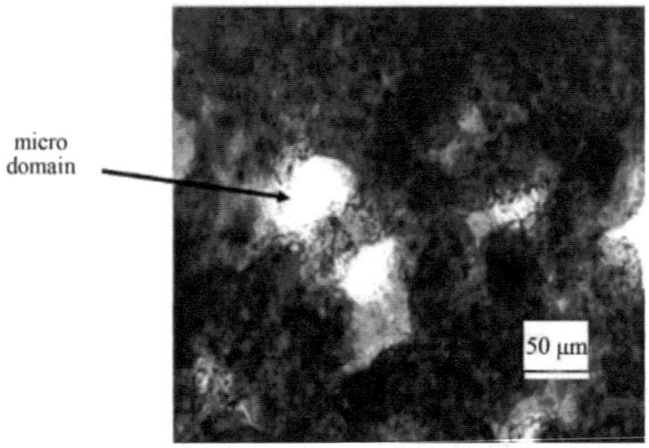

Figure 3. Magnification from Purifier Unit Deposits [10].

Examples of engine auxiliary system damage caused by bacteria are shown in detail in figure 2, figure 3, and figure 4. Deposits on the inner wall of the purifier shown in figure 2 occur on one year of use of B5 in a ferry vessel. Figure 3 shows a microscopic image of the deposited sample. Figure 4 shows bacterial activity at the magnification of the micro domain image area from figure 3. Micro bacteria tend to develop in areas with high water content, for example, in figure 4, which has H_2O levels of 17.2%. Bacteria are not found in fresh biodiesel blends but later found in the resulting sludge. These results indicate that biodiesel and fuel mixtures are not the sources of microbial contamination. The marine environment (air and water) and the ambient air's humidity strongly support the growth of bacteria. Aside from degradation, the deposition in biodiesel is also caused by the higher crystallization temperature compared to petro-diesel. At lower temperatures, biodiesel will be formed like jelly and cannot be pumped [4,10].

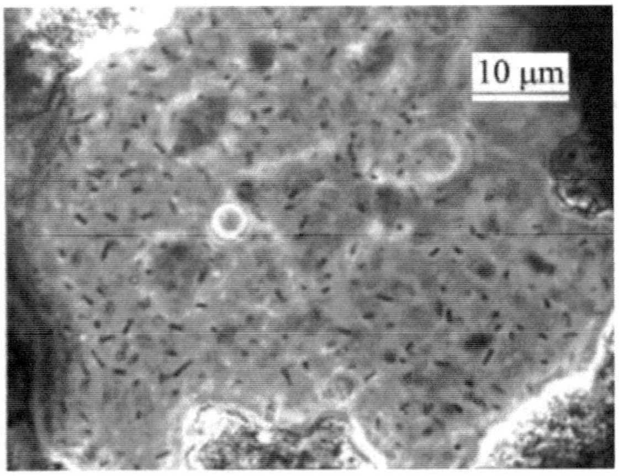

Figure 4. Magnification of Micro Domain, Revealing Bacterial Activities [10].

Methanol and other common adsorbates, silica-gel is widely used for dehumidification, gas separation, desalination, and adsorption in refrigeration due to its great capacity for adsorbing water [11]. Decreasing the atmosphere's relative humidity by utilizing desiccant is prevalent on both industry and individual scales, as the level of environmental humidity influences the development of bacterial activity. Silica-gel is mesoporous, amorphous silica that is stiff due to a constant mesh of colloidal silica composed of tiny hydrated SiO_4 granules. This desiccant is manufactured synthetically and classified as RD, Chromatorex, A, Band home silica gel [12]. Home silica gel is a material in the form of transparent glass granules colored blue and pink, where the colors will serve as indicators of the amount of water absorbed or the saturation state (see figure 5). Since the hydroxyl group in the composition is polar and capable of making hydrogen bonds with polar oxides (such as water), it contributes greatly to adsorption. The capacity of silica-adsorption gel is directly proportional to its polarity [13]. Silica-gel has a larger water adsorption capacity, but its low packing density and thermal conductivity (0.06–0.26 W/m K) negatively impact the adsorption system's overall efficiency[14]. Silica gel use is often found to maintain storage conditions of moist air in electronics, leather clothing, and food ingredients. Silica gel is also non-reactive with Carbon Dioxide (CO_2) gas which causes crust when reacting with KOH, NaOH, and BaO [15].

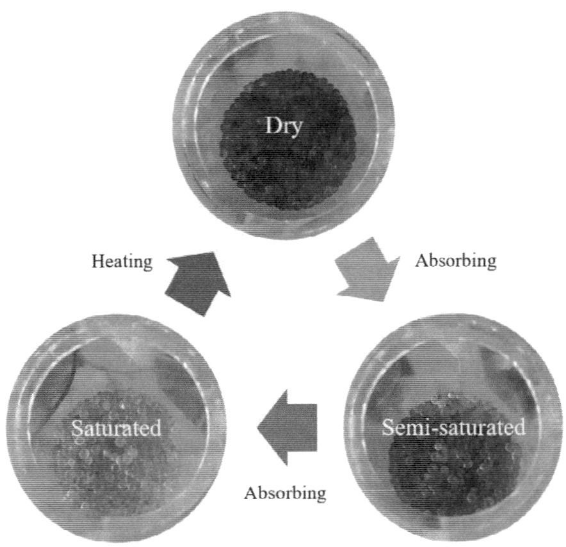

Figure 5. Silica Gel Colour Based on Saturation Rate.

2. Method

Aware of the problems regarding biodiesel storage, steps are taken to minimize the rate of biodiesel degradation in the fuel tank. This research was conducted to analyze silica gel treatment's effect to reduce degradation rate and bacterial abundance on biodiesel blend. This research employs an experimental methodology, with experiments performed using a fuel storage simulation using a tank model. The research object was divided into the treatment group and the control group. Data was obtained from Total Acid Number (TAN) and Bacteria Density observations. Additionally, the data collected is processed to produce research findings and conclusions. The research was conducted in Sepuluh Nopember Institute of Technology, Indonesia. Bacteria Population Test was conducted at the Department of Biology's Microbiology Laboratory. The total acid number test was conducted at the Department of Chemical Engineering Laboratory. The duration of the storage simulation is six weeks in the Marine Engineering Workshop.

Figure 6. Photograph of Fuel Tank Models.

Biodiesel blends are denoted as "BXX" with "XX" is representing the percentage content of biodiesel in the blend (ie: B40 is 40% biodiesel, and 60% petroleum diesel). The tank models depicted in figure 6 is categorized into 6 variables, as follows:

- "A" Tank : Contains B30 mixture with silica gel treatment.
- "B" Tank : Contains B40 mixture with silica gel treatment.
- "C" Tank : Contains B50 mixture with silica gel treatment.
- "D" Tank : Contains B30 mixture, acting as a control.
- "E" Tank : Contains B40 mixture, acting as a control.
- "F" Tank : Contains B50 mixture, acting as a control.

The biodiesel used is made from palm oil. Pure palm oil biodiesel was made using methods taken from previous research in the marine engineering department, which comply with international journals [16]. Palm oil processing to biodiesel starts from methoxide solution, transesterification process, and biodiesel purification. The pure palm biodiesel made comply with the national biodiesel quality standard (SNI) 7182:2015.

Bacteria density is expressed in Colony Formed Unit (CFU) per millilitres. The test was conducted in the Microbiology laboratory using the Total Plate Count (TPC) – Pour Plate method. The initial stage involved estimating the number of microorganisms present in a specific sample using a microscope and then determining the dilution factor required to transfer a single organism into a suitable growth medium [2]. Bacteria colonies were developed in the petri dish by giving a specific amount of biodiesel blend sample and nutrient agar. Bacteria colony formed in petri dish then calculated to depict original cell density with Equation 1.

$$OCD = CFU / (D \times V) \qquad (1)$$

Where OCD is original cell density in CFU/ml, CFU is colony formed in a petri dish in CFU unit, D is dilution (expressed in 10-n), and V is the volume of sample poured in ml (this will be 1 ml) [17]. An example of a colony formed in a petri dish is shown in figure 7.

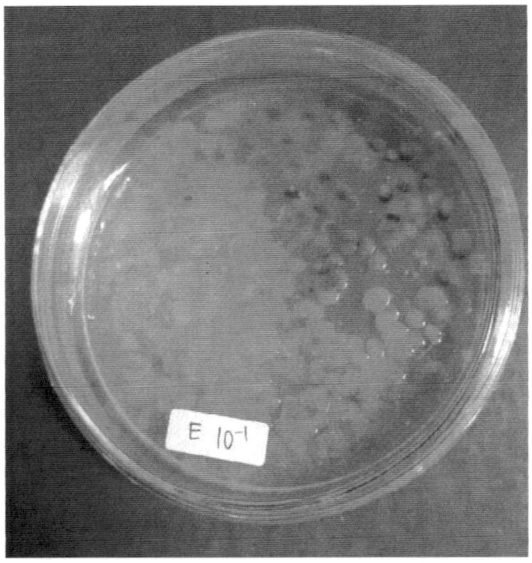

Figure 7. Photograph Example of Bacteria Colony Formed in Petri Dish.

3. Results and Discussion

3.1. Tank Model Design

The tank model designed has a volume of 4 litres, as illustrated in figure 8. Each tank was filled with biodiesel blends (B30, B40, and B50) at a volume of 2 litres per tank. The fuel tank model is constructed entirely of non-metallic materials, including glass and acrylic, to prevent chemical interactions between the fuel tank and the fuel blend.

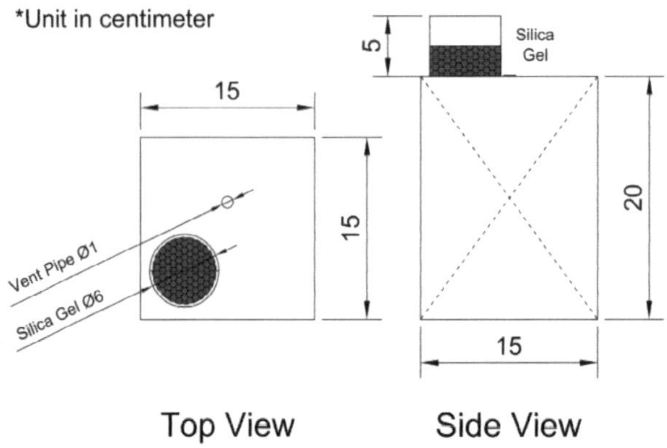

Figure 8. Fuel Tank Model Design.

On the roof of the tank model, ventilation holes represent the vent pipe in the actual fuel tank, and the tube serves as a container for silica gel as a desiccant that is separated from the fuel tank by an aluminium mesh. Fuel tank built as shown in figure 6 and figure 8 to allow silica gel to reduce tank humidity. If the silica gel becomes saturated, it turns pink and must be dried by heat to return to its original blue colour (unsaturated). 34.9 grams or 40 millilitres of dry silica gel are utilized. Silica gel can maintain humidity levels between 28% and 42% for four days with this amount and the fuel tank design. Bacterial life is drastically reduced when relative humidity is reduced from 90% to roughly 20% to 30%. The bacterial decay rate is slowed between 20% and 30% humidity [18]. The humidity and temperature of a silica gel-treated fuel tank model, a control fuel tank model, and the surrounding environment are detailed in table 1. Figure 8 illustrates the tank model concept in greater detail.

Table 1. Relative Humidity and Temperature Data.

Condition	Relative Humidity (%)		Temperature (°C)	
	min	max	min	max
Tank Model With Treatement (A,B,C)	28	42	30.5	34.5
Control Tank Model (D,E,F)	54	76	30.6	34.1
Environment	56	79	30.8	32.8

3.2. Degradation Rate Analysis

The degradation rate parameter is the increase of the total acid number in biodiesel, and the test result is shown in table 2. The initial TAN was obtained from the test result of B100 of 0.313 mgKOH / gr and the petroleum diesel fuel acid number of 0.1118 mgKOH / gr. The initial TAN of B30, B40, and B50 is calculated using Equation 2 from two test results.

$$TAN = (TAN1 \times C1) + (TAN2 \times C2) \quad (2)$$

Table 2. Total Acid Number Test Result.

Code	Detail	TAN (mgKOH/g)		
		Initial	3rd Week	6th Week

-	Petroleum diesel Fuel	0.1118	-	-
-	B100	0.3130	-	-
A	B30 with Silica Gel	0.1722	0.1679	0.1682
B	B40 with Silica Gel	0.1923	0.1902	0.3364
C	B50 with Silica Gel	0.2124	0.1904	0.2803
D	B30 control	0.1722	0.1120	0.2236
E	B40 control	0.1923	0.1232	0.2771
F	B50 control	0.2124	0.1675	0.3343

Where TAN1 is pure palm oil TAN value, TAN2 is petrol diesel TAN value, C1 is biodiesel mixed content, and C2 is petrol diesel mixed content. TAN value is taken from at the start of the experiment, third week, and sixth week is shown in table 2. Furthermore, the total acid number test data is converted into the change of TAN percentage between the initial test and the end of the experiment with Equation 3 as follows:

$$\%TAN = TAN1 / TAN0 \quad (3)$$

Where %TAN is the TAN increase percentage (%), TAN0 is the initial TAN value (mgKOH/g), and TAN1 is the TAN value at the end of the experiment (mgKOH/g). Then, these percentages are subtracted between control tanks (DEF) and treated tanks (ABC). This subtraction value is further defined as the silica gel performance, expressed in Equation 4 as follows:

$$X = \%TAN1 - \%TAN0 \quad (4)$$

Table 3. Changes of Total Acid Number and Silica Gel Performance.

Code	Detail	Change of Total Acid Number	Silica Gel Performance
A	B30 with silica gel treatment	97.7%	-32.2%
D	B30 Control	129.8%	
B	B40 with silica gel treatment	174.9%	30.8%
E	B40 Control	144.1%	
C	B50 with silica gel treatment	132.0%	-25.4%
F	B50 Control	157.4%	

Where X is silica gel performance (%), %TAN0 is TAN increase percentage on control tanks (%), and %TAN1 is TAN increase percentage on treated tanks (%). The result from Equations 3 and Equation 4 is shown in table 3. Figure 9 contains a graphical depiction of the percentage change in acid number data from table 3. Figure 9 shows that the silica gel treatment has succeeded in reducing the rate of degradation on 2 of 3 variable levels of biodiesel blends. The degradation rate with the change in acid number parameter was reduced by 32.2% on B30 and 25.4% on B50. The degradation rate is increasing on B40 by 30.8% with the application of silica gel. A decrease in the degradation rate is caused by lower tank relative humidity, resulting in a minimal hydrolysis reaction between fuel with moist air and the activity of oil-degrading bacteria (biodegradation) [7]. Water is required for the chemical or microbiological hydrolysis of fatty acid methyl esters [9].

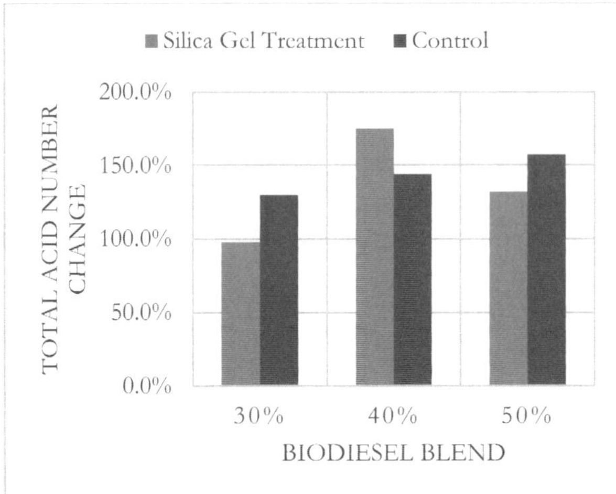

Figure 9. Percentage of TAN Value Change Diagram.

In the B40 blend, the phenomenon was purportedly caused by a measurement inaccuracy in sample tank B (B40 with treatment). The control variable data are consistent with the nature of biodiesel, which is more prone to deterioration than petroleum diesel; thus, the larger the biodiesel content, the faster the degradation rate, as indicated by an increase in the acid number [19]. The source of the measurement inaccuracy is almost certainly a non-homogeneous test sample. Each tank was sampled for up to 40 ml of 2 liters of fuel. The sample is representative of the condition of the biodiesel in the tank as a whole. The deterioration of biodiesel does not occur simultaneously, but in stages. Inhomogeneous tank conditions might result in erroneous test results, as the sample cannot accurately represent the fuel tank's total biodiesel state.

3.3. Bacterial Colony Growth Analysis

Total Plate Count test was conducted to detect bacteria colony presence in the sample. The test result is shown in table 4. The initial test was taken only from 3 biodiesel blend variables: B30, B40, and B50. Test sample for the 3rd and 6th week was taken from six tank models. Original cell density is calculated using Eq. (1), and the dilution used is 10-1.

Furthermore, the number of bacterial colonies amount data is converted into the number of bacterial colony differences by subtracting the amount of colonies in the 6th week and the initial number of colonies expressed as Eq. (5).

$$\Delta OCD = OCD1 - OCD \qquad (5)$$

ΔOCD is bacterial colony differences (CFU/ml), OCD0 is the initial bacterial colony amount (CFU/ml), and OCD1 is the bacterial colony amount at the sixth week. Then, these bacterial colony differences are compared between the control tank models and tank models with silica gel treatment, expressed as Eq. (6).

$$Z = 1 - (\Delta OCD1 / \Delta OCD0) \qquad (6)$$

Z is bacterial colony differences comparison (%), ΔOCD0 is bacterial colony differences in control tank models, and ΔOCD1 is bacterial colony differences in silica gel treated tank models. Results are shown in table 5.

Table 4. TPC Test Results

Code	Detail	Total Plate Count (cfu/ml)		
		Initial	3rd Week	6th Week
A	B30 with Silica Gel	1×10^1	2×10^1	5×10^1
B	B40 with Silica Gel	2×10^1	1×10^1	6.5×10^2
C	B50 with Silica Gel	0	0	4×10^1
D	B30 control	1×10^1	0	1.18×10^3
E	B40 control	2×10^1	1×10^1	1.06×10^3
F	B50 control	0	2×10^1	4×10^1

Table 5. Comparison and Increase of Bacterial Colony Number

Code	Detail	Bacterial Colony Differenc e (CFU/ml)	Treated Tank and Control Tank Comparison
A	B30 with silica gel treatment	40	96.6%
D	B30 Control	1170	
B	B40 with silica gel treatment	630	39.4%
E	B40 Control	1040	
C	B50 with silica gel treatment	40	0%
F	B50 Control	40	

figure 10 shows that silica gel treatment succeeded in decreasing the number of bacterial colonies in 2 of 3 variable levels of biodiesel mixture, with no change on one variable. Derived from table 5, silica gel succeeded in reducing bacterial colony changes by 96.6% on B30 and 39.4% on B40, but there was no change of bacterial colony on B50. At humidity of 20% to 30%, the is an increase in the bacterial decay rate, where the experiment relative humidity on the tank with treatment can reach 28% [18].

Low bacterial activity at 50% (B50) mixture both in the control tank (F) and with treatment (C) was predicted to be caused by the absence of bacterial colonies at the start of the experiment (see table 4). At the beginning of the experiment, Bacteria were allegedly petroleum diesel fuel congenital bacteria, where several types of bacteria can survive and decompose diesel fuel, such as *Bacillus sp., Pseudomonas Aeruginosa and Alkaligenes Feacalis* [20]. Palm oil biodiesel is made from methanol and has been through a drying process with temperatures up to 130 °C, which is not the sources of microbial contamination [10].

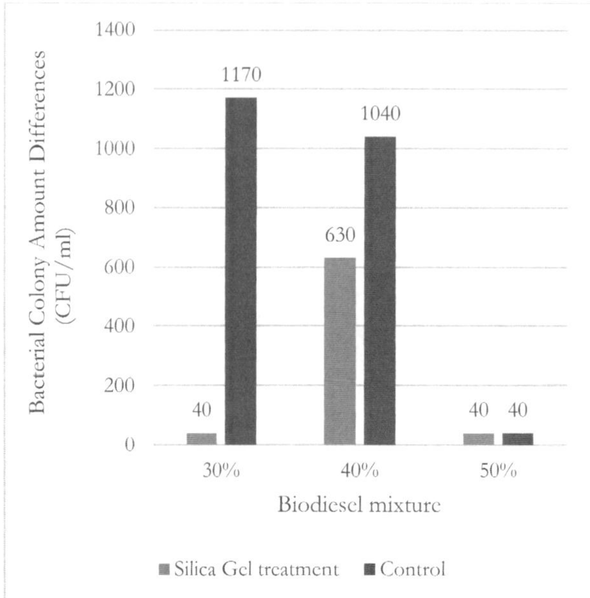

Figure 10. Bacterial Colony Amount Differences Diagram.

4. Conclusion and Recommendation

Analyzing silica gel desiccant in fuel tanks with different concentrations of biodiesel is the aim of the research considering biodiesel fuel degrades over time. The degradation rate is indicated by the change in total acid number, while the number of bacterial colonies determines the growth rate of bacteria abundance. The use of silica gel in fuel tanks can reduce biodiesel degradation rate by 32.2% on B30 and 25.4% on B50, but an increase in degradation rate by 30.8% occurred on B40. Silica gel treatment decreased the degradation rate in 2 of 3 biodiesel blend mixture variables. Furthermore, silica gel can reduce bacterial colonies by 96.6% at B30 and 39.4% at B40, but no reduction at B50. The performance of silica gel in reducing the amount of bacteria is not related to biodiesel mixture levels. With these results, reduction of humidity in the fuel tank can lead to the reduction of biodiesel degradation rate, especially microbial degradation.

The advancement of reducing biodiesel degradation rate method through desiccant still requires further study in the future. Future research can be conducted using a more advanced counting method to ensure better accuracy of bacterial colony count. A smaller fuel volume in the fuel tank is suggested to avoid inhomogeneity in bacteria spread.

Acknowledgement

We would like to express our gratitude to Mr. Anggi and Mrs. Rahma as laboratory assistants from the Microbiology Laboratory, Department of Biology, ITS for the assistance. We also would like to thank Mr. Nurafandi as an assistant in the Marine Power Plant laboratory, Department of Marine Engineering, ITS for preparing the equipment for this study. Best regards to all colleagues who support this research until finished and adequately documented.

References

[1] Monyem A and H. Van Gerpen J 2001 The effect of biodiesel oxidation on engine performance and emissions *Biomass Bioenergy* **20** 317–25

[2] Komariah L N, Hadiah F, Aprianjaya F and Nevriadi F 2018 Biodiesel effects on fuel filter; assessment of clogging characteristics *J Phys Conf Ser* **1095** 012017
[3] Silviana S and Buchori L 2015 Efek Penyimpanan Biodiesel Berdasarkan Studi Kajian Degradasi Biodiesel CPO *Reaktor* **15** 148
[4] Cardeño F, Lapuerta M, Rios L and Agudelo J R 2020 Reconsideration of regulated contamination limits to improve filterability of biodiesel and blends with diesel fuels *Renew Energy* **159** 1243–51
[5] Matbouei M, Weston D P, Liang X and Hainsworth S V 2021 An investigation of the effect of temperature on the oxidation processes of metallic diesel engine fuel system materials and B100 biodiesel from used cooking oil in exposure testing *Fuel* **285** 119063
[6] Fazal M A, Rubaiee S, Al-Zahrani A and Ghazali S 2022 Biodiesel degradation mechanism upon exposure of metal surfaces: A study on biodiesel sustainability *Fuel* **310** 122341
[7] Pullen J and Saeed K 2012 An overview of biodiesel oxidation stability *Renewable and Sustainable Energy Reviews* **16** 5924–50
[8] Maru M M, Lucchese M M, Legnani C, Quirino W G, Balbo A, Aranha I B, Costa L T, Vilani C, de Sena L Á, Damasceno J C, dos Santos Cruz T, Lidízio L R, Ferreira e Silva R, Jorio A and Achete C A 2009 Biodiesel compatibility with carbon steel and HDPE parts *Fuel Processing Technology* **90** 1175–82
[9] Aktas D F, Lee J S, Little B J, Ray R I, Davidova I A, Lyles C N and Suflita J M 2010 Anaerobic Metabolism of Biodiesel and Its Impact on Metal Corrosion *Energy & Fuels* **24** 2924–8
[10] Zhang T, Chao Y, Liu N, Thompson J, Garcia M, He B B, Gerpen J van and Chen S 2011 Case Study of Biodiesel-Diesel Blends as a Fuel in Marine Environment *Advances in Chemical Engineering and Science* **01** 65–71
[11] Askalany A A, Salem M, Ismael I M, Ali A H H, Morsy M G and Saha B B 2013 An overview on adsorption pairs for cooling *Renewable and Sustainable Energy Reviews* **19** 565–72
[12] Islam Md A, Pal A and Saha B B 2020 Experimental study on thermophysical and porous properties of silica gels *International Journal of Refrigeration* **110** 277–85
[13] Ruzhu Wang, Liwei Wang and Jingyi Wu 2014 *Adsorption Refrigeration Technology: Theory and Application* (Singapore: John Wiley & Sons)
[14] Rocky K A, Pal A, Rupam T H, Palash M L and Saha B B 2021 Recent advances of composite adsorbents for heat transformation applications *Thermal Science and Engineering Progress* **23** 100900
[15] Fathallah A Z M, Alami N H and Muzaffar R C 2020 Analysis of the Application of Silica Gel and Iron Powder in Fuel Tank Storage to Reduce Biodiesel Degradation Rate *IOP Conf Ser Earth Environ Sci* **557** 012066
[16] Lakshmi G, Rao N, Mamilla V R and Mallikarjun M v 2012 *Biodiesel production from palm oil by transesterification method*
[17] Leboffe M. J. and Pierce B. E. 2014 *Microbiology Laboratory Theory & Application* (San Diego: Morton Publishing)
[18] Bateman J B, McCaffrey P A, O'Connor R J and Monk G W 1961 Relative Humidity and the Killing of Bacteria *Appl Microbiol* **9** 567–71
[19] Tang H, Abunasser N, Wang A, Clark B, Wadumesthrige K, Zeng S, Kim M, Salley S, Hirschlieb G and Wilson J 2008 Quality survey of biodiesel blends sold at retail stations *Fuel* **87** 2951–5
[20] Natsir Djide M, Farid Samawi M, Biologi J, Sains dan Teknologi F, Alauddin Makassar Jl Sultan Alauddin No U, Gowa K, Selatan S, Perintis Kemerdekaan Km J, Ilmu Kelautan dan Perikanan F and Hasanuddin Jl Perintis Kemerdekaan Km U 2016 *Isolasi Bakteri Pendegradasi Minyak Solar Dari Perairan Teluk Pare-Pare* vol 4

STUDY OF SUPPLY CHAIN MANAGEMENT OF INDUSTRIAL PLAN MANUFACTURING DEVELOPMENT OF MARINE POWER TURBINE IN INDONESIA

Afian Kasharjanto[1], Erwandi[1], Cahyadi Sugeng Jati Mintarso [1], Eko Marta Suyanto[1] Daif Rahuna[1]

[1]The National Research and Innovation Agency, Republic of Indonesia

*Email correspondent: afian.kasharjanto@gmail.com

Abstract. Indonesia had announced its target to reduce carbon emission by achieving 0% Net Zero Emission (NZE) by 2060. So starting in 2022, the Government must update the regulations governing the energy sector to encourage the implementation of renewable energy. As an archipelagic country, one of the renewable energy potentials is Ocean Renewable Energy (ORE). The contribution of the ORE to Indonesia's National Energy Mix is almost none. Thus, it is a good chance for the government to give more incentives for implementing ORE, especially in eastern Indonesia where they are abundantly available. Since the infrastructure and supporting facilities there, including industries, are sparse and the supply chain management depends on Java Island, it needs special regulations to speed up the implementation of ORE. This paper studies the supply chain management of the implementation of marine current turbines base on secondary data and the experience of engineers from the National Research and Innovation Agency when they implemented the converter technology of the marine current turbine at Larantuka Strait, Nusa Tenggara Timur. Results In this study, the implementation of ocean current turbines must be integrated with government policies related to maritime connectivity and security because the energy produced is large enough and suitable for execution.

1. Introduction

One of the 3 (three) issues that must be followed up in the G20 Presidency is the transition to sustainable energy. Indonesia has committed to achieving the target of 0% gas emission (Net Zero Emission / NZE) in 2060. This target is very difficult to achieve so from now on we must work hard to carry out real efforts to achieve this target, while the target soon is to mix new renewable energy sources in 2025 of 23% is very difficult to achieve. The only solution is that we must have the courage to immediately switch to renewable energy sources. Another effort is the Government's commitment to gradually reduce the number of Diesel Powered Electric Companies (PLTD) which currently number around 1700 units spread throughout Indonesia to be replaced with power plants from renewable energy (low emission) sources.

The source of renewable energy in Indonesia is very large if it can be utilized optimally. Potential sources of renewable energy from solar, geothermal, water, wind, and sea are available in Indonesia. The table below shows the potential sources of renewable energy that can be utilized:

Table 1. Potential sources of new renewable energy [1].

Energy sources	Potential (GW)	Utilization (MW)
Solar	3,295	217
Bioenergy	95	6,637
Wind	155	154
Bioenergy	57	2,284
Geothermal	24	2,293
Marine	60	0
Total	3.686	11,585

According to [2], We suffer from the lack of discussion about our energy planning in the future Especiallcy as an archipelago country we have no plan of our potential of renewable energy source from sea,.So it is still not fully utilized (0 MW) as shown in Table 1. This phenomenon is very concerning and needs to be immediately found a solution. [2] Also stated that energy security' requires the availability of sufficient quantity of energy resource at affordable prices, implicating protection of the entire supply chain component and the infrastructure. In fact, our energy price are become higher from day to day like what we are experiencing right now.

Government needs to immediately evaluate the existing policies so that the government starts paying more attention to renewable energy from sea. The advantages of renewable energy from sea includes:

1. It has a large energy density (ρ) at low current speeds ($\rho_{water} = 830\ \rho_{air}$).
2. It can be estimated theoretically
3. No need for "Over-engineering" like other sources.
4. Smooth / not noisy.
5. Always Renewable and sustainable.

The energy potential of ocean currents is mostly found at the Eastern Indonesia [3], as shown in the figure below :

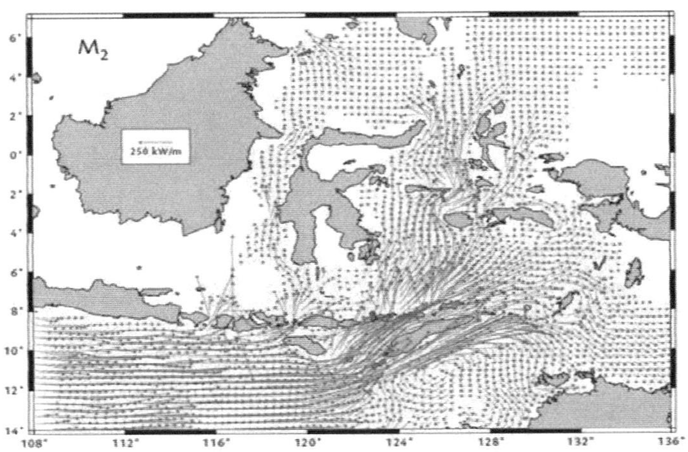

Figure 1. Potential ocean currents in eastern Indonesia

Based on this potential (Figure 1), most of the research was carried out in the Eastern Indonesia region. Ai Yuningsih et al from P3GL-ESDM [4] have studied the potential of ocean currents in Larantuka Strait, East Nusa Tenggara which explained that the potential source of ocean current energy in Larantuka Strait, NTT is very potential to be used as a source of electricity generation. This supports the research report carried out by Erwandi et al from BRIN [5], which has initiated several measurements in Larantuka Strait - NTT area about the energy potential of ocean currents in the same location as a potential location for the installation of Ocean Current Turbine.Meanwhile, Yulhendri from the Center for the Study and Engineering of Marine and Fisheries Technology [6] conducted studies at different locations at western part of Indonesia, namely the Riau Archipelago. The results of the study explain that Sugi Island is the most suitable area for the implementation of ocean current power plants. Other data regarding the study of the potential for ocean current power plant (PLTAL) following table 2:

Table 2. Potential ocean currents in a number of straits [7]

No	Area	Max tidal speed	Resource Density	Width of potential area (Width above 1,5 ms in sqm)	Theoritical resource (MW)	Tehnical resource (MW)	Practical Reource (MW)
1	Riau Strait	1,39	1,38	55,751,111	96,432	24,108	6,027
2	Sunda Strait	2,63	9,32	21,025,000	36,367	9,092	2,27
3	Toyopakeh-Bali	3	13,84	2,959,360	5,189	1,280	320
4	Lombok-NTB	3	13,84	2,959,360	33,049	8,262	2,066
5	Alas Strait-NTT	2,9	12,5	60,853,994	105,258	26,315	6,579
6	Larantuka-NTT	3	13,84	287,500	497	124	31
7	Pantar Strait-NTT	2,91	12,63	921,600	1,594	399	100
8	Boleng Strait-NTT	1,5	1,73	1,658,610	2,868	717	179
9	Molo Strait-NTT	1,85	3,24	216,400	374	93	23
10	Mansuar-West Papua	1,79	2,94	3,619,998	6,261	1,565	392
Total					287,889	71,955	17,990

Cooperation activities for the implementation of the study of potential for ocean currents have also been carried out as follows:

1. Cooperation Project for the construction of the Pancasila Bridge in Larantuka - NTT between Indonesia (PLN - the Netherlands), to plan the construction of a bridge connecting Larantuka - Adonara Island, NTT, where the lower side of the Bridge column is installed with a sea current turbine. The total capacity of the turbine is 40 MW, with an energy of 100 GWh/year.
2. The potential development project for tidal power plants (UK Cooperation and PT.Pertamina Indonesia Power) with a turbine capacity of 100 KW per Turbine.
3. Study of potential ocean currents in Lombok Strait and Sape Strait, as a research collaboration between Indonesia and France (2016-201),
4. MOU on the cooperation plan between P3GL ESDM and Andritz Hydro, in the development of ocean currents in Indonesia, (2014).
5. MOU between PT.SBS and PT.PLN (Persero) in the PLTAL development plan in the Maluku region (2015).
6. Development cooperation between PT. PAL (Indonesia) and PT. PLN (Persero) in the construction of marine energy turbines as a major industry commitment to the construction of marine turbines.

From the opportunities above, we should start opening up more opportunities in the turbine industry from the sea through a series of supply chains.

The purpose of the paper is to conduct an initial study of the extent of the supply chain of the marine renewable energy industry through the collection of secondary data on the state of industrial readiness, political support from the Government and other support from related industries (supporting component industries, banking and transportation industries

2. Supply Chain Management

Supply Chain Management is the process of planning, designing, and controlling the flow of information and materials along the supply chain in satisfying consumer demands now or in the future in an efficient and effective way [8]. Supply chain management (SCM) is an important business strategy, particularly in the manufacturing process. SCM assists the company to control its products and understand consumer needs.

According to [9] Supply Chain Management (SCM) is an optimization management of a supply chain operation. Implementation of Management is carried out by a (Manager) to optimize existing resources in a supply chain. A supply chain manager is tasked with identifying critical resource conditions in a supply chain, identifying potential strengths and influencing resource performance and can utilize a number of information to develop goals, objectives, and optimize supply chain flow. SCM managers can view the future as an opportunity.

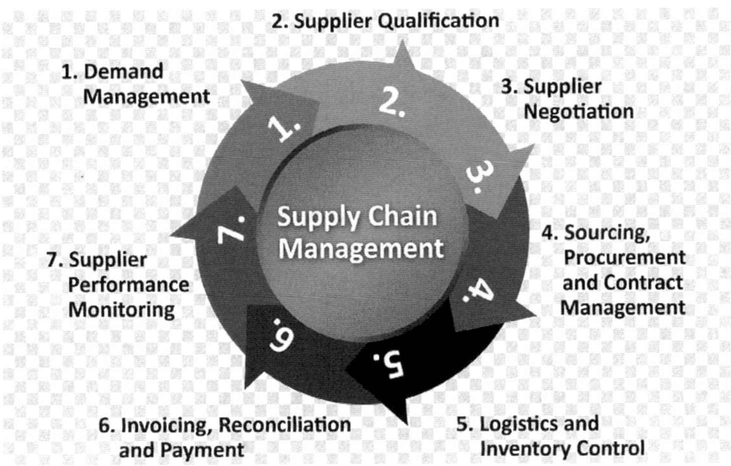

Figure 2. Supply Chain Management [10]

From Figure 2 above, SCM activities include all supporting components to form activities that can guarantee a product is easy to get, affordable prices and sustainable production through a series of collaborations all related components.

3. Methodology

The method of writing is in the form of secondary data collection on the potential capabilities of domestic industries as well as Government support in the marine new renewable energy sector in forming industrial supply chains to increase the utilization of local content (TKDN).

4. Literature review

As stated in introduction, another reason in Indonesia energy's problem is the Government's failure in energy policy due to the policy error of the National energy subsidy program. Energy subsidies lead to an increase in dependence on fossil energy consumption, it is not encouraging people to reduce fossil energy consumption and switch to environmentally friendly energy consumption. Energy subsidies lead to a reduction in budgetary subsidies for Health, Education and an increase in corruption, poverty, and income inequality.[11]. To solve these problems, the Government is determined to carry out energy reforms through a transitional policy of achieving the 0% Net Zero Emission (NZE) target of carbon gas emissions by 2060 as mandated in the Kyoto Protocol and Paris Agreement, calmly agreeing on a solution to the threat of climate change and maintaining an increase in climate change. global temperature below 15 degrees [11].

Renewable energy sources are expected to be a solution to be able to contribute to the reduction of air pollution and dependence on fossil energy sources. In this case, cooperation between the Government, researchers/universities, and the private sector (triple helix) is urgently needed to immediately find a solution in generating findings of renewable energy sources to reduce the threats mentioned above so that the target of increasing the contribution of the NRE mix from 19% in 2017 to 65% in 2050 can be achieved [12].

To save energy consumption one of its efforts is the use of renewable energy. However, there are still obstacles to its utilization, such as geographical constraints of the location of the source of renewable energy, distribution or delivery of renewable energy products, imbalances in the scale of economy in each region, and government policies that are not balanced in the application of subsidies and taxes on renewable energy products. The application of the renewable energy supply chain is one solution. The implementation of good supply chain management will result in more affordable and competitive renewable energy prices. Off course the obstacle is that in the early stages, namely the renewable energy industry's initial investment cost will be more expensive than the fossil energy industry. But in the long term, the renewable energy industry will produce more economical and profitable product prices. Production costs will decrease in the long run. This is due to the excess of renewable energy resources which is always renewable and will never run out. Long-term production efficiency will be used to cover investment costs and production cost savings. Therefore, renewable energy sources will be a solution to the problem of global energy consumption and overcome the effects of environmental impacts (air pollution) [13].

The application of supply chain management can also be applied to industrial concepts that convert waste into renewable energy sources (Waste to Energy/WTE). The WTE concept is also known as the circular economy concept 5R (Reuse, Recycling, Reduction, Repurpose and Recovery) which is proven to be able to create profit from something that was previously waste production. This shows that the implementation of good supply chain management in every industry will provide benefits [12].

According to [14], the supply chain is a network of companies that work together to create and deliver a product into hands of end users. Supply Chain Management is an integrated method between the parties involved either directly or indirectly in producing products or services. Where it starts in the process of getting raw materials from suppliers, goes to the production process, and ends in the delivery process to end users based on the spirit of collaboration to realize a common goal, namely customer satisfaction. The supply chain system can be guaranteed over [15]:

- Economical price guarantee
- Guarantee the continuity of material supply in the production process
- Guarantee the quality of production goods
- Guaranteed delivery time to consumers.

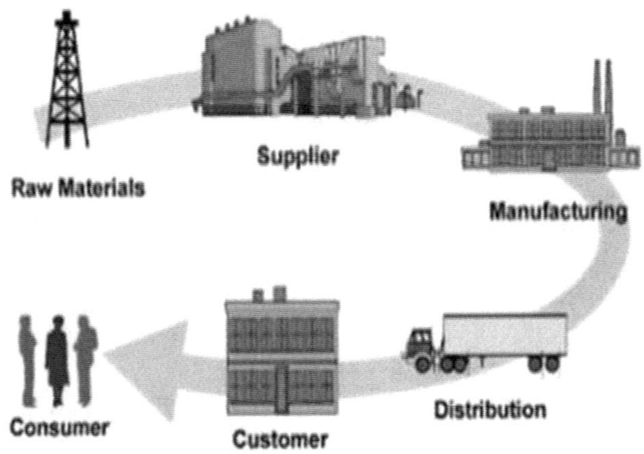

Figure 3. Supply chain process flow diagram.[16]

In supply chain management, there is a flow of material and information. Supply chain management must keep the flow of materials flowing from the source to the final consumer. In the flow of information, information technology allows demand and supply data to be obtained quickly and can increase the level of detail of a product.

Figure 3. above shows the flow of relationships between supporting components in the formation of industrial supply chain. Each component plays an important role in the success of the supply chain plan.

5. Experience in Testing and Potential Development of energy Resources from Sea

BRIN researchers have experience in designing, testing models, and testing prototypes of marine current turbines from a scale of 2 KW to a scale of 10 kW in Larantuka Strait-NTT and Madura Strait under the Suramadu Bridge, East Java. From field experience, the constraints of limited facilities and infrastructure outside Java in supporting the assembly and testing process at sea have become a separate obstacle starting from the loading and unloading process from ships at port, turbine assembly, and testing process at sea. Besides the limitations of repair shops, if there are components that are damaged so the component required to be repaired in Surabaya, it is also an experience in itself. The limited support for facilities and infrastructure is still a problem in the construction of ocean current turbines in the region. Figure 4. and Figure 5. show the experience during process of shipping goods from Surabaya to Larantuka,.where from Surabaya to port of larantuka transported by shipping via sea transportation, and then continued via trucking to assembling area.

Figure 4. The delivery process at the assembly site at Larantuka, Flores-NTT

Figure 5. The Process of unloading goods from port to assembly site at Larantuka, Flores-NTT

IOP Conf. Series: Earth and Environmental Science 1166 (2023) 012018 doi:10.1088/1755-1315/1166/1/012018

6. Support

6.1. Government support

The Central Government has tried to issue policies on the use of renewable energy resources, the problem is that supports is still not able to touch renewable energy from sea. Several incentives are needed to stimulate the industry to try the renewable energy industry from the sea. Local governments must also be more creative in seeking industrial opportunities in their respective regions that have potential marine energy sources to provide support for ease of business, funding, and other support in their respective regions. Synergies between the Central and Regional Governments also need to be maximized so that the potential of renewable energy sources in regions can be identified by central government for assistance in developing solutions.

Government through Bappenas has planned the development of renewable energy industry by optimizing the industrial supply chain sector as stated in the 2019-2024 (Indonesian Sharia economic master plan) as shown in It is mentioned about the plan to optimize the supply chain of the renewable energy industry is one of the specific strategies for developing renewable energy in Indonesia:

Tabel 3. Strategy for the formation of renewable energy industry supply chain [17]

	Strategy/Program/ Activity	Target indicators	year						Stakeholders
			2019	2020	2021	2022	2023	2024	
Strategy 1 : The occurrence of a sustainable relationship in the value chain of the renewable energy industry to produce sustainable energy									
2	Supply-Chain optimization of the renewable energy industry by strengthening energy industry partnerships	Mapping of the supply chain and demand for renewable energy as well as incentives for renewable energy industry partners							Ministry of Energy and Mineral Resources, Ministry of Finance, Energy BUMN, Renewable Energy NGO, Associations, Local Governments
	a. Renewable industry supply chain scale mapping		-	√	√	√	√	√	
	b. Mapping of renewable energy supply chain development priorities		-	√	√	√	√	√	
	c. Incentives for renewable energy industry partners to achieve sustainable energy targets		-	-	√	√	√	√	

Government Support which is through Government Regulations / Ministerial Regulations as below:

- Regulation of the Minister of Energy and Mineral Resources number 50/2017 in conjunction with Minister of Energy and Mineral Resources No. 04/2020 regarding the use of renewable energy sources for the provision of electricity
- Providing fiscal and non-fiscal incentives for renewable energy
- Ease of granting business permits in the renewable energy sector

6.2. Banking Support

In banking sector, although there have been some commitments to support funding from banks in the NRE industry. For example, BCA has disbursed a total of 4.7 trillion as of March 2021, while bank Mandiri has disbursed loans of 171 trillion as support to the government in its efforts to shift from fossil energy to renewable energy. , HSBC participates in environmental pollution programs (reducing gas emissions) through credit distribution, through the Bank's Net-zero program through credit, business actors in Indonesia are transitioning to the use of renewable energy such as geothermal, hydro, wind, and solar. but for renewable energy sector from sea , banking sector is still not too optimistic, because [18] :

- The financing of the renewable energy industry is a long-term project, so there is a mismatch in the banking program because banks want a quick fund turnover (short term).
- Most of the renewable energy industry is expensive, so banking sector are still worried about the speed of loan repayment from industry (as a customer).
- Considered to have a high level of risk. The industry must be assured of risk of failure

6.3. Industry support

Marine renewable energy is mostly located in Eastern Indonesia area, but the main industry and supporting industries are still concentrated in Java Island Region (especially East Java). Through the supply chain, all supporting components that are scattered in many places in a supply chain coordination will be coordinated.

1. Data on the number of metal industries in the NTB/NTT area : 0 (none)
2. Data on the number of shipbuilding industries in the NTB/NTT area : 1 (one)
3. Data on the number of metal industries in East Java : 39 (thirty-nine)
4. Data on the number of shipbuilding industries in East Java : 30 (Thirty)

7. Problems

From a number of opportunities above, there are a number of challenges faced in the marine energy development plan, including:

Table 4. Constraints in developing renewable energy from sea [1]

Aspect	Challenges
Technology	- Technology that is considered immature/not yet proven to operate commercial
Finance	- The initial capital cost is quite high - Difficulty in obtaining financing (high-risk projects) - The initial price of IDR/Kwh is quite high (this applies to almost the new renewable energy industry
Market	- Still not competitive compared to other new renewable energy sources - No Supply Chain yet
Regulations/Policies	- There are still no regulations that can be referred to, especially special support for the development of marine renewable energy in the Eastern part of Indonesia renewable energy - There is no clear framework for including Marine new renewable energy resources in the NDC (Nationally Determined Contribution) document
Infrastructure	- Poor network infrastructure

From table.4 it is stated that there are still a number of obstacles in the development of marine renewable energy. But there are still opportunities from a number of obstacles above, namely: In terms of technology, from the results of studies by researchers, it has been proven that Renewable energy technology from sea is feasible to be developed and realized considering the very open potential for

Indonesia as an archipelagic country. The central and regional governments have also begun to issue a number of regulations in order to provide as many opportunities as possible for interested industries to try out the Renewable energy industry. Likewise, the banking sector, as stated above, also supported and encouraged by government to provide soft loans support to the industry in renewable energy sector. The problem is only on market opportunities possibility that are still waiting for well facility and infrastructure support can be achieved to open up market opportunities through the availability / ease of use of local facilities and infrastructure.This support may be achieved by the readiness of the marine renewable turbine industry's supply chain.

8. Conclusion

From the article, it can be concluded that:

1. The supply chain of the marine renewable energy turbine industry is still not ready to be implemented due to:
   - The legal protection for government policies is still not optimal in favor of developing the renewable energy industry from the sea.
   - The existence of the marine renewable energy industry in the Eastern part of Indonesia renewable energy area still does not exist, (turbine manufacturing industry and its components.), Most of it still depends on the industry on the island of Java.
   - Even though there is already a signal of bank credit support for the new renewable energy industry, banks are still not too brave to invest in the new renewable energy industry from sea because it has not been proven concretely (There are no success story)
   - Support from the transportation industry is abundant and ready for shipping outside Java via shipping routes to small ports (sea Toll program), but support for facilities and infrastructure at ports and assembly sites outside Java needs to be more prepared.
2. Even so, as a very large archipelagic country, it is necessary to start a brave industry pioneer in the development of the marine new renewable energy industry (such as PT. PAL Indonesia), although it requires a commitment to banking support, especially for industries that are not as big as PT. PAL Indonesia.

9. Recommendation:

1. The government (central/regional government) needs to immediately prepare a legal protection for support of the new renewable energy industry from sea
2. In terms of the readiness of the shipbuilding industry and the metal industry (especially in Java), it is necessary to immediately prepare for the establishment of a supply chain for industry from sea.

References

[1] D. Kusdiana, "FGD roadmap pengembangan energi laut untuk mendukung transisi energi nasional serta peluncuran," *Balai Besar Surv. dan Pemetaan Geol. Kelaut.*, 2022, [Online]. Available: https://mgi.esdm.go.id/2022/08/01/fgd-roadmap-pengembangan-energi-laut-untuk-mendukung-transisi-energi-nasional-serta-peluncuran-peta-potensi-energi-laut-indonesia-dan-peta-sumber-daya-mineral-kelautan-indonesia/

[2] A. Rahman, P. Dargusch, and D. Wadley, "The political economy of oil supply in Indonesia and the implications for renewable energy development," *Renew. Sustain. Energy Rev.*, vol. 144, no. March, p. 111027, 2021, doi: 10.1016/j.rser.2021.111027.

[3] R. Ray and G. Egbert, "Tides in the Indonesian Seas," *Oceanography*, vol. 18, no. 4, pp. 74–79, 2005, [Online]. Available: http://scholar.google.com/scholar?hl=en&btnG=Search&q=intitle:Tides+in+the+Indonesian+Seas#6

[4] A. Yuningsih and A. Masduki, "Potential Energy of Ocean Current for Electric Power Generationin Coastal Areas of East Flores, Ntt," *J. Ilmu dan Teknol. Kelaut. Trop.*, vol. 3, no. 1, pp. 13–25, 2011, doi: 10.29244/jitkt.v3i1.7831.

[5] Erwandi, D. Rahuna, and C. S. J. Mintarso, "The development of the IHL wave-current turbine to convert the ocean energy from both wave and current," *IOP Conf. Ser. Mater. Sci. Eng.*, vol. 588, no. 1, 2019, doi: 10.1088/1757-899X/588/1/012008.

[6] Y. Suryansyah, "Potensi Energi Arus Laut untuk Pembangkit Tenaga Listrik di Pulau-Pulau Keci (Studi :Pulau Mantang di Bintan, Pulau Abang di Batam, dan Pulau Sugi di Karimun, Propinsi Kepulauan Riau)," *J. Kelaut. Nas.*, vol. 8, no. 1, p. 27, 2013, doi: 10.15578/jkn.v8i1.6218.

[7] PLN, "Rencana Pengembangan Pembangkit Listrik Tenaga Arus Laut (PLTAL) untuk Sistem Kelistrikan Nasional," *(www.pln.co.id)*, 2022.

[8] R. G. Schroeder, "Operations management: Contemporary concepts and cases," *FEB-UI*, 2007.

[9] G. Plenert, *What is supply chain management (SCM)?*, no. August 2005. 2007. doi: 10.1016/b978-012370517-4/50004-6.

[10] "Logistics & Supply Chain Management," *PGWING*. https://www.pngwing.com/id/free-png-tmwec

[11] T. B. Sumarno, P. Sihotang, and W. Prawiraatmadja, "Exploring Indonesia's energy policy failures through the JUST framework," *Energy Policy*, vol. 164, no. March, p. 112914, 2022, doi: 10.1016/j.enpol.2022.112914.

[12] Y. Fernando, M.-L. Tseng, N. Aziz, R. B. Ikhsan, and I. S. Wahyuni-TD, "Waste-to-energy supply chain management on circular economy capability: An empirical study," *Sustain. Prod. Consum.*, vol. 31, pp. 26–38, 2022.

[13] Y. Fernando, P. S. Bee, C. J. C. Jabbour, and A. M. T. Thomé, "Understanding the effects of energy management practices on renewable energy supply chains: Implications for energy policy in emerging economies," *Energy Policy*, vol. 118, no. August 2017, pp. 418–428, 2018, doi: 10.1016/j.enpol.2018.03.043.

[14] IN Pujawan and E. Mahendrawathi, "Supply Chain Management. Edisi Kedua. Surabaya: Guna Widya," 2010.

[15] "MODUL 4 - Manajemen rantai pasok pada industri konstruksi."

[16] V. D. V. D. Aropah, "Materi 3-Pelaku PBJ-Versi 4," 2021, [Online]. Available: https://virtual-library.lkpp.go.id/id/eprint/25/%0Ahttps://virtual-library.lkpp.go.id/id/eprint/25/1/Materi 3 - Pelaku PBJ- Versi 4.pdf

[17] Komite Nasional Keuangan Syariah, "Masterplan Ekonomi Syariah Indonesia 2019-2024," *Kementeri. Perenc. Pembang. Nasional/ Badan Perenc. Pembang. Nas.*, pp. 1–443, 2018, [Online]. Available: https://knks.go.id/storage/upload/1573459280-Masterplan Eksyar_Preview.pdf

[18] Herlina Kartika Dewi, "Bankir Menilai Kredit ke Sektor EBT Tetap Prospektif Tahun Ini," *Kontan co.id*. https://keuangan.kontan.co.id/news/bankir-menilai-kredit-ke-sektor-ebt-tetap-prospektif-tahun-ini

Effect of Positioning a Circular Flow Disturbance in Front of the Darrieus Turbine

D Satrio[1,*], Suntoyo[1], F Albatinusa[1], S Junianto[2], S Musabikha[3], Madi[4] and F O Setyawan[5]

[1] Institut Teknologi Sepuluh Nopember, Surabaya 60111, Indonesia
[2] Politeknik Elektronika Negeri Surabaya, Surabaya 60111, Indonesia
[3] National Research and Innovation Agency (BRIN), Jakarta 10340, Indonesia
[4] Institut Teknologi Sumatera, Lampung 35365, Indonesia
[5] Universitas Brawijaya, Malang 65145, Indonesia

*E-mail address: dendy.satrio@its.ac.id

Abstract. Darrieus turbine has a vertical-axis configuration that is suitable for current from any direction. Placing a circular flow disturbance in front of the turbine can influence the performance. This study wants to investigate the positioning of the flow disturbance to find the best configuration. The study method used numerical simulation with Computational Fluid Dynamics (CFD) software. There are three variations of the cylinder position in front of the turbine with the angle of 0°, 60°, and -60°. As a result, the configuration can improve or disrupt the performance depending on the position installation. In position 0°, the performance was a decrease of 18%. The best performance is at the flow disturbance position of -60°. The maximum coefficient of power (Cp) value of the turbine was 0.19, with increasing of 27%. Therefore, it is recommended to use the position of -60° in front of the turbine.

1. Introduction

The Indonesian government intends to encourage an increase in the supply of New and Renewable Energy (EBT) by ratifying the Regulation of the Minister of Energy and Mineral Resources of the Republic of Indonesia Number 39 of 2017. One of the biggest trends in the world is now using the ocean as a source of renewable energy or Ocean Renewable Energy (ORE). Using the ocean as renewable energy was triggered by the fact that exploitation of renewable energy resources on land often results in conflicts over land use [1]. Although at sea it can be threatened by corrosion, but efforts to overcome it already exist [2]. Apart from that, research on increasing the efficiency of conventional power plants is still needed [3–5].

According to Purba, et al. (2015), the characteristics of currents in Indonesia are bidirectional (two directions) and do not have large enough currents [6]. Thus, to use ocean currents in Indonesia as an energy source, it is necessary to adapt technology to the existing characteristic of ocean currents in Indonesia. Therefore, the Darrieus turbine is considered beneficial because it can rotate from any direction [7]. And this type is suitable for floating platform [8,9].

Fajar (2008) stated that a simulation result regarding the characteristics of cross-flow passing through a circular cylinder is that the flow for Re > 40 depends on time (unsteady) caused by the vortex released in the back of the cylinder (vortex shedding) [10]. This will affect the value of the drag coefficient, which is influenced by the wake area behind the cylinder. A flow disturber in front of the turbine can improve performance [11].

This can be applied as an innovation model by utilizing flow disturbance as performance optimization of the VAT (Vertical Axis Turbine). For example, in a research conducted by Sahim (2014) in conducting an experimental study on the effect of installing deflectors on the performance of the Darrieus-Savonius turbine, it was found that there was an increase in the coefficient of power (Cp) and the coefficient of the moment (Cm) are significant [12]. The Savonius turbine is a drag based type turbine which can be adjusted in a cross-flow axis position [13,14].

Bachtiar (2019) revealed that adding a triangular-shaped barrier structure in front of the Darrieus Turbine increased Cp with a peak value of 0.377 or an increase of about 7.08% [15]. Another study regarding the effect of the barrier cylinder diameter ratio in front of the Savonius Turbine can increase the Cp value by 28% compared to conventional turbines [16]. The results indicate a reduction in pressure and an increase in speed on the convex advancing blade.

Therefore, this research is interested in examining the effect of variations in the flow disturbance position on the Darrieus turbine performance. The circular flow disturbance is placed in front of the turbine, and then changed the angle configuration. This research uses Computational Fluid Dynamics (CFD) software to model the flow around the turbine and flow disturbance.

2. CFD Simulation

2.1. Turbine Specifications

The Darrieus turbine is a type of vertical axis turbine produces mechanical energy by rotating the turbine, which is driven by the kinetic energy of ocean currents. This type is a turbine driven by a lift force formed by the presence of foil blades so that the blades can rotate faster than the current speed. The Darrieus turbine is suitable for application in floating platform [8]. The turbine used in this study was a Darrieus turbine, as in reference [7]. This experiment was conducted in the Hydrodynamics Laboratory of Institut Teknologi Sepuluh Nopember. The airfoil shape of flow disturbance is changed to be cylindrical for more applicative such as a leg ocean platform [11]. The turbine and flow disturbance specifications are in Table 1.

Table 1. Turbine and flow disturbance specifications.

Parameter	Specifications
Number of blades (N)	3
Foil type	NACA 63(4)021
Chord length (C)	0.07 m
Turbine diameter (D)	0.50 m
Flow disturbance diameter (ds)	0.15 m
Flow disturbance ratio (ds/D)	0.3

2.2. Boundary Conditions

Computational Fluid Dynamics (CFD) is a numerical analysis method that is commonly used in performing fluid mechanics analysis. The working principle of CFD is to perform discretization, which is a form of breaking the domain into several small areas called grids, meshes, or cells. It is necessary to first set the value for the boundary condition.

There are several boundary conditions set in the simulation in this study that shown in Figure 1. Some of these conditions are the inlet velocity and the condition of the airfoil walls. According to the experimental conditions, the entered inlet velocity is the fluid flow velocity. As for the airfoil walls, it is necessary to arrange them to have wall motion moving walls and shear conditions of no slip.

The turbine is simulated in a rotating condition. Therefore, the computational domain is divided into two: the stationary domain and the rotary domain. In the rotary domain, settings need to be made to make the simulation using the sliding mesh technique by entering the mesh motion input at the turbine rotational speed according to the experimental data to be tested.

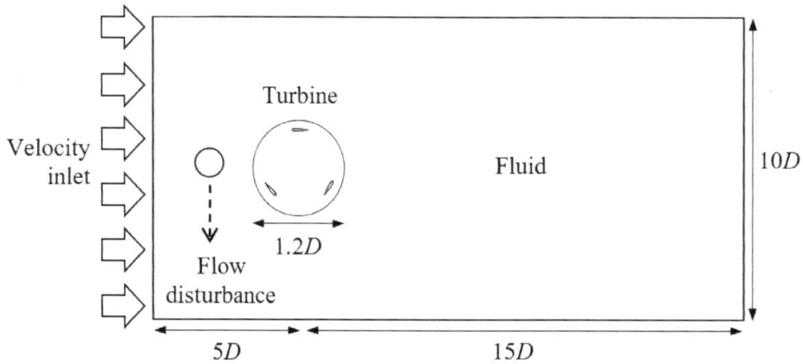

Figure 1. Numerical model of a turbine modified with a flow disturbance in position of 0°.

2.3. Meshing

In general, the number of meshes will affect the detailed accuracy of the analysis, so the more the number of meshes, the higher the accuracy of the analysis results. However, the higher the meshing accuracy, the longer the computation time. Thus, a mesh independence study was carried out to verify the most effective number of mesh elements with the fastest processing time. The simulation was performed at 0.5 m/s velocity flow without flow disturbance intake. The number of mesh elements has a good result at 257,000 as illustrated in Figure 2.

In this study, the meshing used is of the triangle mesh type because this type can represent and has a sufficient level of accuracy in CFD simulations which are usually carried out on vertical turbines. The mesh on the foil section is set to be denser and smoother than the fluid area. The foil area is also given an inflation layer which functions to get the value of $y+ < 1$, following the provisions of the turbulent model used in the simulation, namely k-ω SST [11,17].

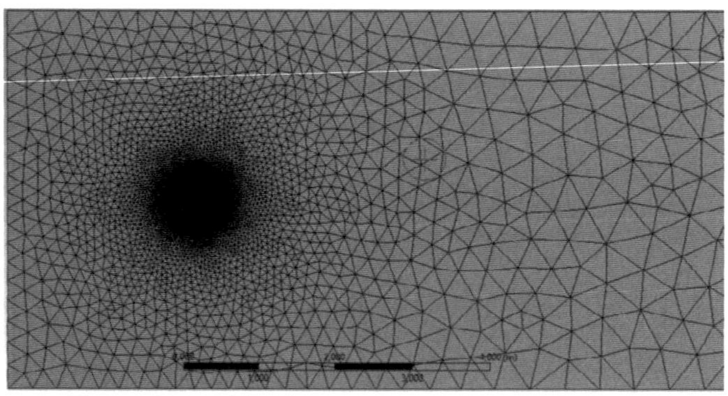

Figure 2. Meshing in overall the numerical domain.

3. Results and Discussion

3.1. Validations

CFD simulation is numerical modelling which, in this study, was carried out based on experimental data [7]. The result of the CFD simulation is the turbine moment; this data is multiplied by the turbine rotation speed value so that power can be generated. The power extracted from the numerical simulation is compared with the experimental data, as shown in Figure 3.

The results of CFD simulations were performed in several conditions and obtained results where the power at low current velocity conditions had the same value as the experimental results. The results of 2D CFD modelling are quite acceptable compared with the other research [13,17]. Meanwhile, in the turbine operating conditions at a higher current velocity (0.6 m/s), the power value is higher than the experimental data (which is in 3D condition) due to the weakness of 2D numerical models, which ignores the shaft, arm, and circular plate models at the ends of the foil.

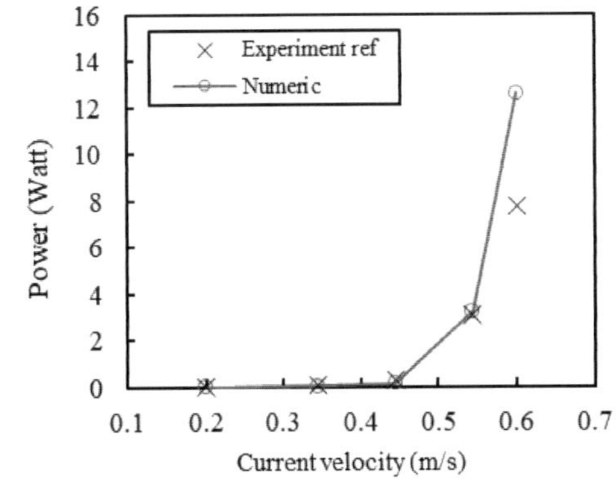

Figure 3. Numerical validation with experimental published data [7].

3.2. Variations

A cylindrical flow disturbance with a diameter ratio (ds/D) of 0.3 is placed in front of the turbine shaft. The effect of the diameter ratio will be analysed in the next section. The distance from the flow disturbance shaft to the turbine shaft is 1D (0.5 m). The distance used was chosen because it refers to the recommended distance for the use of a flow disturber [16].

Adding a flow disturbance will make the flow chaotic with changes in the current flow through the structure. The simulation to determine the effect of the addition of the flow disturbance was carried out at current conditions with a velocity of 0.5 m/s, according to experimental data, with TSR = 1.43 [7]. The analysis is carried out by calculating the change in the value of the power coefficient (Cp) produced by the turbine and comparing it between a turbine without a flow disturbance and a turbine with a flow disturbance that shown in Figure 4. Without the flow disturbance, the Darrieus type turbine has a Cp value of 0.156, while the turbine given a flow disturbance of 0° gets a Cp value decreased to 0.127.

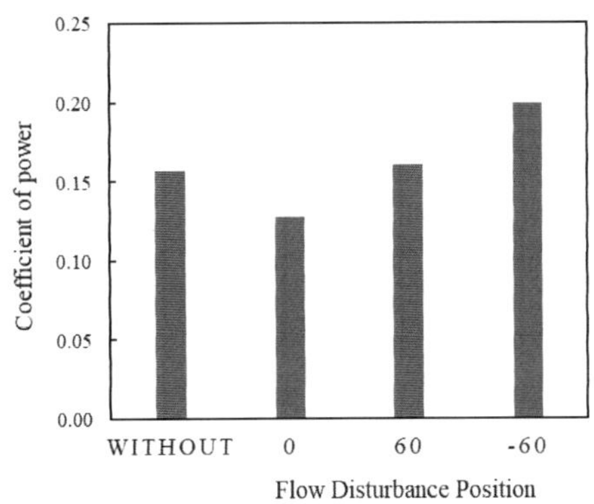

Figure 4. Cp of the original turbine (without a flow disturbance) compared with the modified turbine with a flow disturbance.

The use of the flow disturbance at 0° disturbs the flow and decreases the turbine performance. Therefore, the flow disturbance position is moved so that it is in a position that does not block the flow from entering the turbine. In accordance with the recommendations in Setiawan (2019), the best position obtained is the position of the flow disturbance, which forms an angle of 60° to the turbine shaft [16]. Simulations are carried out with the same distance and diameter as in the flow disturbance experiment 0°; however, it was modified in -60° and 60° as shown in Figure 5.

The positioning of the flow disturbance 60° positively impacts the addition of torque. This resulted in an increase in the value of Cp, which increased by 3% compared to conventional Darrieus turbines (without flow disturbance). The turbine's performance, which was given a flow disturbance at the -60° position, also increased. The increasing Cp was more significant with other position variations, which was 27%. The results of this simulation show that the best turbine performance increase is obtained with the addition of the flow disturbance at the -60° position.

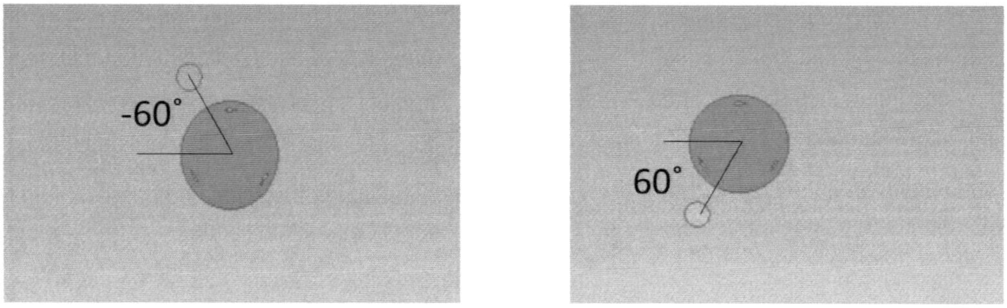

Figure 5. Changing the flow disturbance position with angle variation of 60° and -60°.

The placement of the cylinder right in the middle of the turbine shaft will cause the flow that enters the turbine to be a wake formed from the flow that passes through the cylinder (Figure 7). At some points of turbine rotation, this vortex phenomenon can produce greater torque, as in 1140° (Figure 6), but the wake flow will cause a decrease in torque compared to conventional turbines. The negative torque is also higher than the turbine without flow disturbance.

Meanwhile, if the flow disturbance is placed in the 60° and -60° positions, the wake flow, which has a low speed, does not become the flow that drives the turbine (Figure 7). The vortex phenomenon, which has a higher speed than the freestream velocity, will generate greater torque than conventional turbines. This result is supported by previous research [11]. The flow disturbance position -60° has a significantly increased turbine performance because the resulting velocity acceleration provides a stronger thrust that hits the foil's leading edge. The resulting boost from the flow can be seen in the generation of significant torque values compared to conventional turbines. This is different from the 60° position, which provides an impetus to the trailing edge in this study. Although it provides a positive impulse, it will not have much impact because it does not generate a significant force.

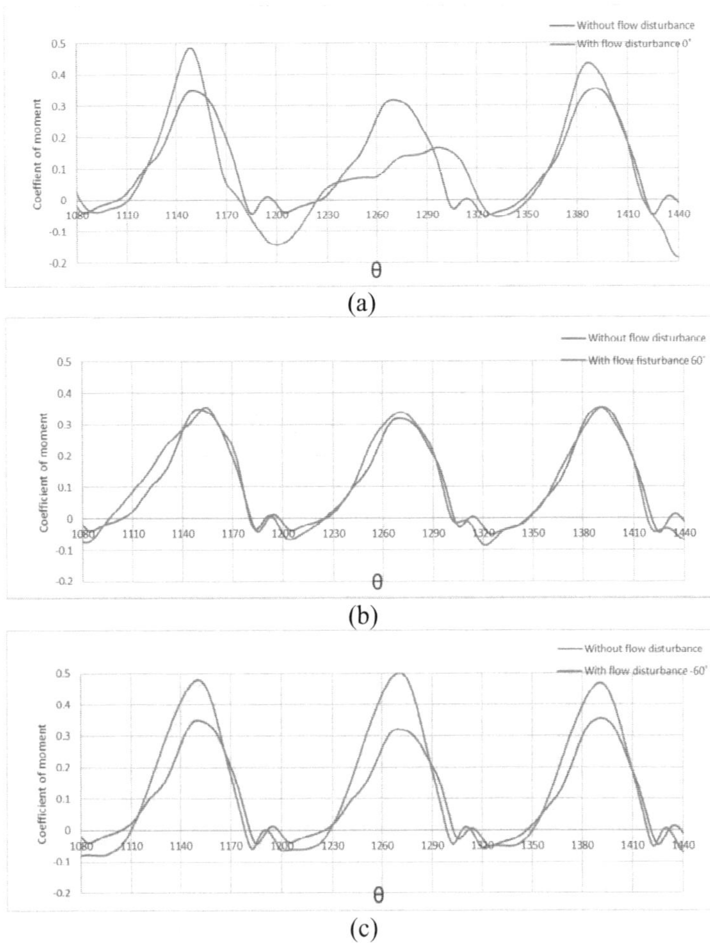

Figure 6. Cm of the original turbine (without a flow disturbance) compared with the modified turbine with a flow disturbance of (a) 0°; (b) 60°; (c) -60°.

3.3. Flow Contours

The flow contours in the simulation results can be compared to show the reasoning analysis. The velocity contour and streamline particle are showed in Figure 7. This was done to capture the flow phenomena that occurred in different configuration angle of the flow disturbance.

This decrease in turbine performance occurs because the flow behind the flow disturbance placed on the turbine shaft (0°) becomes a wake that experiences speed deceleration. This wake flow, shown in Figure 7, causes a disturbance in the flow that should be the power to drive the turbine. This has also been investigated by Setiawan (2019), who stated that using a flow disturbance right at the turbine shaft would disrupt the turbine's performance [16]. This phenomenon is defined as a blockage effect that blocks the upstream flow to the turbine.

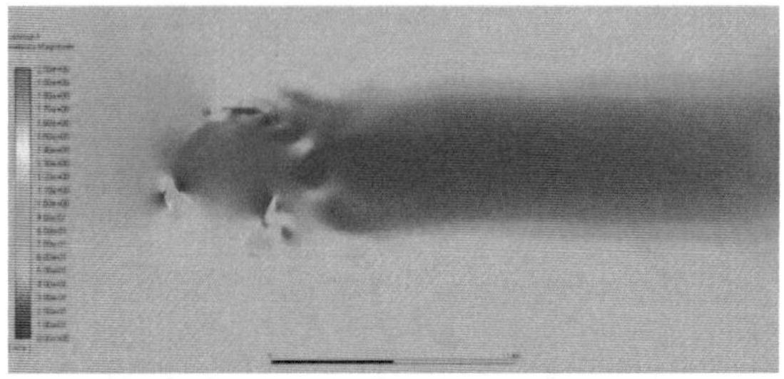

(a) Turbine without a flow disturbance

(b) Flow disturbance with dimension of ds/D 0.3 and angle of 0°

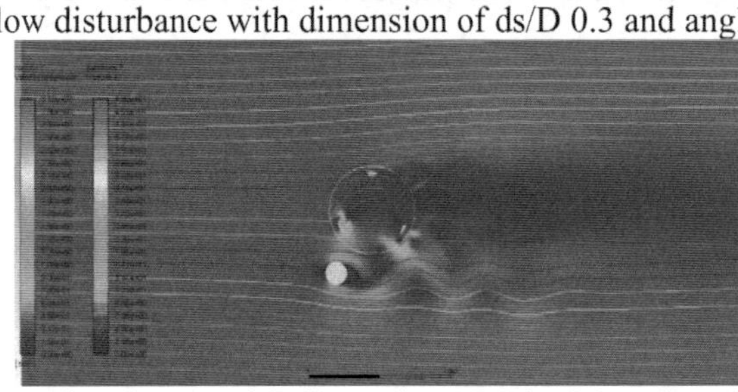

(c) Flow disturbance with dimension of ds/D 0.3 and angle of 60°

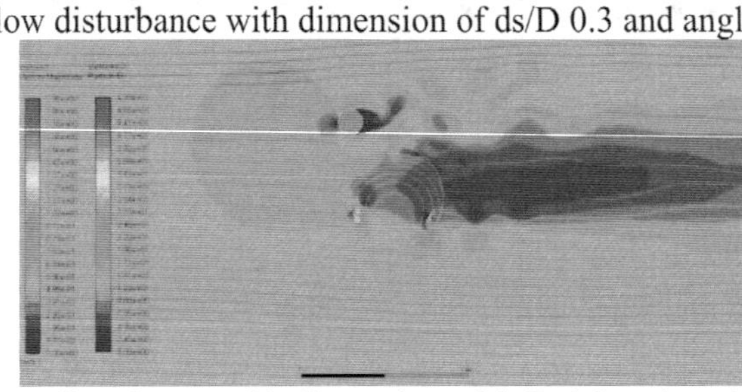

(d) Flow disturbance with dimension of ds/D 0.3 and angle of -60°

Figure 7. Flow contours around the turbine that are influenced by a flow disturbance ds/D 0.3 with variations angle of (b) 0°; (c) 60°; (d) -60° at TSR 1.43.

4. Conclusions

CFD simulation has been done for the Darrieus turbine without and with a flow disturbance. The 2D numerical model of the original turbine (without flow disturbance) has been validated with published experimental data. There are three positions of the flow disturbance in front of the turbine with an angle of 0°, 60°, and -60°. Based on those variations, the best performance results from the flow disturbance with the angle of -60°. The maximum coefficient of power value of the turbine was 0.19, with increasing of 27% compared with the original turbine. However, in the position of 0°, the flow disturbance has an impact on decreasing the performance. Using a flow disturbance in front of the turbine could enhance the turbine's performance with the appropriate configuration.

Acknowledgements
The authors would like to convey their appreciation to the Directorate of Research, Technology and Community Service; Ministry of Higher Education, Research, and Technology, the Republic of Indonesia, which funded the current project under a scheme called Basic Research under contract number 1514/PKS/ITS/2022. The authors also thank Institut Teknologi Sepuluh Nopember, for supporting the relevant project under a scheme called Scientific Research under contract number 980/PKS/ITS/2022.

References

[1] Yuningsih A and Masduki D A 2011 Potential Energy of Ocean Current for Electric Power Generationin Coastal Areas of East Flores, NTT *J. Ilmu dan Teknol. Kelaut. Trop.* **3** 13–25
[2] Musabikha S, Utama I K A P and Mukhtasor 2018 State of the art in protection of erosion-corrosion on vertical axis tidal current turbine *AIP Conference Proceedings* vol 1964
[3] Kurniawati I, Pratilastiarso J and Satrio D 2020 Analysis of the Effect of Cooling Water Condenser to Power Plant Cycle Using Cycle-Tempo Software *IES 2020 - International Electronics Symposium* (Surabaya: IEEE) pp 37–42
[4] Satrio D, Trisnayanthi N N A I and Pratilastiarso J 2021 Analysis of the Effects of Fuel Type Selection on the Performance and Fuel Consumption of a Steam Power Plant *Int. J. Adv. Sci. Eng. Inf. Technol.* **11** 2046–54
[5] Nyoman Ayu Indah Trisnayanthi N, Pratilastiarso J and Satrio D 2020 Effects of Fuel Replacement on the Performance and Fuel Consumption of Steam Power Plant *IES 2020 - International Electronics Symposium* (Surabaya: IEEE) pp 67–73
[6] Purba N P, Kelvin J, Sandro R, Gibran S, Permata R A I, Maulida F and Martasuganda M K 2015 Suitable Locations of Ocean Renewable Energy (ORE) in Indonesia Region-GIS Approached *Energy Procedia* **65** 230–8
[7] Satrio D and Utama I K A P 2021 Experimental investigation into the improvement of self-starting capability of vertical-axis tidal current turbine *Energy Reports* **7** 4587–94
[8] Junianto S, Fadilah W N, Adila A F, Tuswan T, Satrio D and Musabikha S 2022 State of the Art in Floating Tidal Current Power Plant Using Multi-Vertical-Axis-Turbines *IES 2022* (Surabaya: IEEE) pp 1–7
[9] Sony J, Mukhtasor, Prastianto R W and Jo C H 2020 Effects of demi-hull separation ratios on motion responses of tidal current turbines-loaded catamaran *Ocean Syst. Eng.* **10** 87–110
[10] Fajar B 2008 Numerical Study of Cross-Flow Characteristics in Single Cylinders (in Bahasa) *Mesin* **23**
[11] Satrio D, Suntoyo and Ramadhan L I 2022 The advantage of flow disturbance for vertical-axis turbine in low current velocity *Sustain. Energy Technol. Assessments* **49** 101692 1-9
[12] Sahim K, Ihtisan K, Santoso D and Sipahutar R 2014 Experimental study of darrieus-savonius water turbine with deflector: Effect of deflector on the performance *Int. J. Rotating Mach.* **2014**
[13] Satrio D, Firdaus Yusri M, Mukhtasor, Rahmawati S, Junianto S and Musabikha S 2022 Numerical simulation of cross-flow Savonius turbine for locations with low current velocity in Indonesia *J. Brazilian Soc. Mech. Sci. Eng.* **44** 1–11
[14] Satrio D, Wiyanto A A and Mukhtasor M 2022 In-Situ Experiment of Cross-Flow Savonius Hydrokinetic Turbine With a Deflector *J. Mar. Sci. Technol.* **3** 1–4
[15] Bachtiar B and Ariwiyono N 2019 Numerical Study of the Effect of Changes in the Front-mounted Triangle Obstacle of the Returning Blade on the Performance of the Darrieus Wind Turbine Type Naca 0015 (in Bahasa) *Inovtek Polbeng* **9** 79
[16] Setiawan P A, Yuwono T, Widodo W A, Julianto E and Santoso M 2019 Numerical study of a circular cylinder effect on the vertical axis savonius water turbine performance at the side of the advancing blade with horizontal distance variations *Int. J. Renew. Energy Res.* **9**
[17] Marsh P, Ranmuthugala D, Penesis I and Thomas G 2017 The influence of turbulence model and two and three-dimensional domain selection on the simulated performance characteristics of vertical axis tidal turbines *Renew. Energy* **105** 106–16

Experimental Study of the Influence Wave Period on the Performance Darrieus Turbine with Modification of the Blade Angle

Eko Marta Suyanto[1,2], Sutopo Purwono Fitri[1,*], Erwandi[2], Daif Rahuna[2], and Afian Kasharjanto[2]

[1]Marine Engineering, Institut Teknologi Sepuluh Nopember, Surabaya, Indonesia
[2]National Research and Innovation Agency, Surabaya, Indonesia

*Email correspondent: sutopopf@its.ac.id

Abstract. Indonesia has a long coastline that has the potential for renewable energy sources from waves to be used as an energy source for power generation. To take advantage of this potential, research has been carried out by testing the Darrieus-type vertical axis turbine model at the Indonesian Hydrodynamics Laboratory. The weakness of the wave turbine is that it cannot rotate due to the wave force due to the position of the turbine rotor which is perpendicular to the turbine axis. By changing the angle of the turbine blade in an inclined position, it is expected to produce optimal efficiency to produce rotor rotation. This paper describes the testing of a turbine model with an angled turbine blade that produces the best efficiency and rotation due to wave forces and ocean currents. The test is carried out by varying the angle of the turbine blade tilt (45°, 60°, and 75°). Each is given a variation in height and wave period but still, maintains the turbine rotor area. Furthermore, the torque, efficiency, rotation, and resulting power are measured. wave height of 0.15 meters with variations in wave periods ranging from 1.2 - 2.5 seconds. The test results show that a turbine with a rotor angle of 60° produces the best performance.

1. Introduction

The Studies on converting ocean renewable energy are developing rapidly. One of them is the energy conversion technology of seawater circulation/movement. There are two types of seawater movement. The first is the motion of ocean currents caused by ocean tides due to the interaction of the gravitational force between the earth, moon, and sun. The second is the wave where the water surface undergoes the up and down motion of ocean waves caused by wind blowing on the sea surface due to changes in weather. The development of technology to capture the two types of movement of seawater goes independently, where the technology is different from each other [1]. In contrast to ocean current turbines, which can convert kinetic energy using horizontal axis turbines (such as windmills), as well as vertical axis turbines (Darrieus, Savonius, Gorlov) [1] [2], Wave turbines are used to convert wave motion which always moves back and forth in two directions in one wave period that lasts in seconds. The turbines used are usually Wells turbines of the oscillating water column type, both those that stay on the beach and those that float in the middle of the sea, floating turbines of the heave rider or wave rider type, or the pendulum type turbines [3]. This paper discusses the vertical axis Darrieus type turbine which is designed to increase the ability to convert the kinetic energy of ocean currents and at the same time convert the potential energy of waves. With the use of a vertical Darrieus turbine, absorption of both energies can be carried out, but in this paper, the author only presents an experimental wave energy absorption by a vertical Darrieus turbine.

One of the world's cleanest, most plentiful, sustainable, and renewable energy sources is ocean wave energy, and it is anticipated to have greater potential to compete with the usage of non-renewable energy sources now being used to gradually lessen reliance on energy derived from fossil fuels [4]. To solve issues with in part turbin as a result of corrosion, the equipment conditions quickly destroyed owing to water conditions have been researched [5], then the part is temporarily protected by paint.

A vertical axis turbine is one of the many technologies that may transform wave energy sources into electrical power. Studies by the Indonesian Hydrodynamics Laboratory on the design of a wave current rotor installed under the Suramadu Bridge in the Madura Strait indicate that the wave period affects the number of turbine revolutions [6]. Straits with weak ocean currents, usually the sea is wavy. Furthermore, it turns out that the ocean current power plant from the Indonesian Hydrodynamic Laboratory can also rotate when exposed to waves which prove that the wave period affects the number of turbine rotations [1][6].

Patent [1]. describes the present invention as relating to an ocean current and wave power plant, i.e. a power plant that converts, either jointly or individually, the kinetic energy of the ocean currents velocity as well as the potential energy of the difference in water level due to ocean waves into electric power, wherein the plant there are six horizontal Wells turbine blades and three Darrieus turbine blades to rotate the main shaft vertically and horizontally and the shaft rotation is connected through a gearbox to the generator so that electricity can be generated from the generator rotation. The results of a patent search similar to the present invention also show that the patent [7] is only for converting wave energy, the turbine is held by three upper radial arms which are placed in the wells rotor, and the vertical turbine blades are connected to the radial arms so that the blades cannot rotate.

While the patent [8]. Darrieus turbine blades are tilted to reduce the projection area of the turbine which results in reduced electrical power generated. In some cases, studies have shown that using a straight blade Darrieus Turbine (90°) is very good at absorbing the energy of ocean currents but cannot absorb pure ocean wave energy so the turbine cannot rotate if there is only wave energy [9][10].

When the force taken from the water wave is equal to or higher than the mass load of the turbine, the turbine can start rotating and generating energy. In contrast to regular operating turbines as a power generation system, which includes a connected generator that also serves as an additional load, this situation does not exist. As a result, it is simpler to establish the initial conditions in our experimental setup than it is in turbines that are working normally. The shortest wave time necessary to drive the turbine is typically a good indicator of rotation ability [11].

Experimental research is more trustworthy because it measures actual turbine performance rather than making assumptions. There are three experimental studies on improving the ability of vertical axis turbines to start on their own by combining wheels, Darrieus, or Savonius turbines [2][9][12]. However, in this study, only using a vertical axis Darrieus turbine [12] by providing several variations of waves to be able to rotate the turbine, because the Darrieus turbine has enough good performance coefficients to be able to rotate with some modification of the turbine blade angle.

The concept put out in this study was based on the helical blade turbine's basic operating principle but was simplified to use a tilted blade turbine to capture wave energy. The turbine blades are immediately at an angle from the upper arm position to the forearm position and have a specific angle. The Darrieus turbine, in theory, seeks to evenly disperse the hydrodynamic load at each point. Beveled blades, on the other hand, have the benefit of a more simple shape, making them simpler to produce.

From some of these experiences, it is necessary to develop research for the Darrieus turbine by changing the blade angle which was initially 90° or perpendicular [9] to be tilted (angled) with the aim of being able to convert ocean wave energy optimally in addition to ocean current energy. Therefore, in this study, a number of changes in turbine blade angle changes with various wave variations will be tested to optimize the effectiveness of turbine performance.

IOP Conf. Series: Earth and Environmental Science 1166 (2023) 012020 doi:10.1088/1755-1315/1166/1/012020

2. Model Test and method

Section 2 describes turbine details and experiment methods including turbine model, Testing setup, Calibration, and validation of measuring equipment

2.1 Turbine Model

Turbine model was chosen based on reference [9] which uses a turbine blade type NACA 0021 which was initially installed vertically symmetrically because it could not rotate if given a wave of water. And according to [13] Tests to determine how much influence the waves and tilted flow to the horizontal axis turbine Torque and turbine thrust are almost not affected by waves compared to current flow.

So in this study, the model was made according to Figure 1, Figure 2, and Figure 3. with the change of the Darrieus turbine blade from a straight blade to an angled blade to absorb wave energy. Table 1 uses to explain the blade angle due to the difference between upper and lower diameter when installed with three different variations of the turbine angle but the area is fixed. For blade installation with an angle of 45°, there will be a difference in the top diameter of 1202.02mm with a bottom diameter of 952.02 mm, for the installation of a blade with an angle of 60° there will be a difference in the top diameter of 981.15mm with a bottom diameter of 731.15 mm, the installation of a blade with an angle of 75° will there is a difference between the top diameter of 879.62 mm and the bottom diameter of 629.62 mm.

Figure 1. Vertical-axis darrieus turbine experimental model with Position blade 45°

Figure 2. Vertical-axis darrieus turbine experimental model with Position blade 60°

Figure 3. Vertical-axis darrieus turbine experimental model with Position blade 75°.

Table 1. Specification of the model turbine

Parameters	Specifications		
Angle Blade	45°	60°	75°
Number of blades (N1)	3	3	3
Chord length (C)	83 mm	83 mm	83 mm
Blade span (S)	980 mm	980 mm	980 mm
Foil type NACA	0021	0021	0021
Area (A)	416500 mm²	416500 mm²	416500 mm²
Turbine diameter (D1) Up	1202 mm	981 mm	879 mm
Turbine diameter (D2) Down	952 mm	731 mm	629 mm
Number of arms (N2)	6	6	6
Shaft diameter (d)	50 mm	50 mm	50 mm

2.2 Testing setup

Testing setup on the Maneuvering Ocean and engineering Basin (MOB) wave test tank, Indonesian Hydrodynamic Laboratory (IHL). Figure 4. is a photo of the MOB facility which consists of a wave generator, carriage, and sub-carriage with specifications as shown in table 2. Figure 5. is the installation of a Darrieus Turbine which is placed perpendicularly under the sub-carriage and is facing straight with the wave generator at a distance of 5 m to obtain a constant wave height and period.

Table 2. Specification of the MOB wave tank

Parameters	Specifications
Deep part	60 x 35 m, maximum water depth 2.5 m
Shallow part	45 x 35 m, maximum depth 1.25 m
Main Carriage	Maximum Accelaration 0.8 m/sec Maximum Speed 2.00 m/sec
Sub Carriage	Maximum Accelaration 0.8 m/sec Maximum Speed 2.00 m/sec
Wave Generator Capability	Regular and irregular waves Wave period 0.5 – 3 sec Wave height up to 0.3 m (significant)
Wave Maker	Single wave board typically consists of 72 pieces Electric motor driven Simultaneously control of electric motor and frequency by computer

Figure 4. MOB facilities: water tank, wave generator, carriage, and sub-carriage.

Figure 5. Setting 3 variation darrieus Turbin in MOB facilities: water tank, wave generator, carriage, and sub carriage.

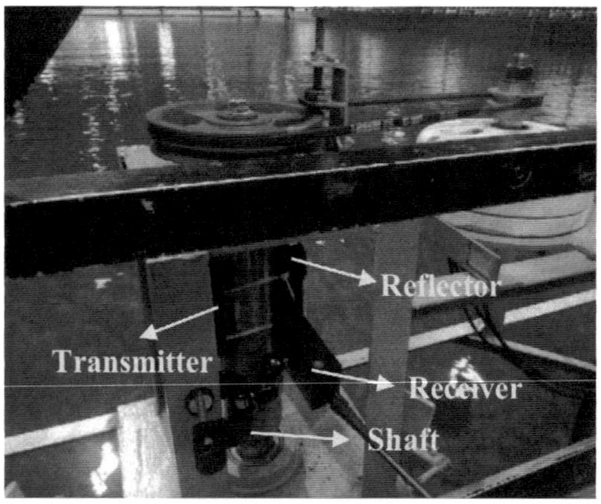

Figure 6. Setting of measuring equipment on shaft turbine in MOB facilities.

Figure 7. Setting of measuring equipment power shaft datum series 430-compact on computer

figure 6. shows the installation of a rotation reflector (rpm) and the installation of a strain gauge for measuring the torque and output power of the Darrieus turbine shaft. Figure 7. is a power shaft equipment designed to measure rotation, power, and torque applied to a rotating Darrieus turbine shaft. This equipment system uses a strain gauge sensor element to measure the torsional torque of the shaft. The output voltage of the strain gauges is sent through the transmitter (which also functions as a strain gauge amplifier) to the receiver. The receiver also functions as a shaft rotational speed sensor. From the receiver, torque and rotational speed data are processed to get the power value using the Data User Interface (DUI) tool. The DUI can be used to display and log torque, rotational speed, and power values or pass these three parameters to a computer for display and log. For the specifications for the power shaft equipment, the datum series 430 is used as shown in table 3.

Table 3. Specification of the power shaft equipment

Name Part	Standard
Shaft Fit Dimensions	90-11000mm
Electronics Accuracy	0.1%
IP Rating	IP67
Operating Temperature	0-70°C
Data Sampling Rate	10 SPS as Standard
Data Output	RS485 & USB for Torque and speed. Analog available as an extra
Battery Life	30 Days as Standard.
Battery Type	Rechargeable Hot Swap. N.B Faster sample rate reduces battery life
Cable Length	4 m as Standard

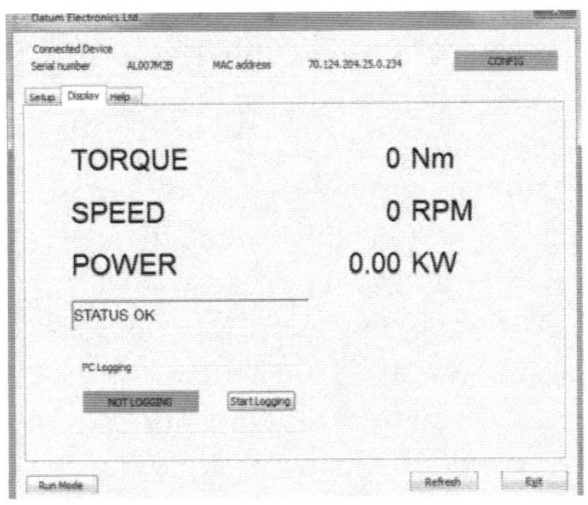

Figure 8. Package includes Software DUI Config power shaft datum series 430-compact on turbine

In figure 8. the DUI Config Software is used on the computer to enter the calibration factor of the voltage generated by the strain gage into the torque value of the Instrument DUI and to display and save the torque, rotational speed, and power data sent from the DUI Instrument.

Shaft OD	88	mm
Shaft ID	42	mm
Max Expected Torque	1.000	Nm
Tool Steel	**ST37**	
Youngs Modulus	2,00E+11	
Poissons Ratio	0,285	
Gauge Factor	2,09	
OD	*0,088*	*M*
ID	*0,042*	*M*
*T*16000*GF*(1+PR)*OD*	*3,78E+06*	
*T*16000*GF*(1+PR)*ID*	*1,80E+06*	
*((OD^4)-(ID^4))*π*YM*	*3,57E+07*	
Calibration Value	**0,106**	**mV/V**

Figure 9. calculation of the calibration factor for torque value in shaft turbine

In figure 9. the turbine shaft calibration value from the voltage generated by the strain gauge becomes the torque value. Cell calibration values are filled with values according to the diameter of the turbine shaft whose torque is measured, the estimated maximum torque, and the value of the measuring factor of the strain gauge used. The theoretical basis for calculating the value of mechanical (shaft) power (P_{mech}) in Watts from the value of torque (T) in Nm and Shaft speed (ω_{mech}) in radians/sec according to equations (1), (2), and (3) and rotation shaft turbine (n) in :

$$P_{mech} = T * \omega_{mech} \quad (1)$$

$$\omega mech = n * 2 * \frac{\pi}{60} \quad (2)$$

2.3 Calibration and validation of measuring equipment

The instrumentation in the MOB facility to support this experiment is a wave generator which is measured using a wave high sensor. Another measurement instrumentation is using a power shaft meter to measure rotation, torque, and turbine power output used in this research. To ensure that each sensor and its components work properly, calibration data is discussed [11]. Data measured are turbine rotation and wave generation in the wave test tank because this research is focused on analyzing torque and power output and turbine shaft rotation. For performance effect, other researchers additionally measure torque or power output [2][14].

The three data (turbine speed, torque, and power output) are displayed in one computer interface. The captured data can be monitored at one time at the same time. Next, the data is extracted at a steady state and then the values are averaged. Calibration is provided for the strain gauge sensor to measure turbine power output and torque. The tools used as calibrators are the tachometer and strain gauge simulator. Figure 7. shows calibration data of the strain gage on the Darrieus turbine shaft with Full-Scale Torque 1000 Nm and Full-Scale Strain 0.105847802 mV/V, which is seen as increasing the torque (Nm) will increase the output signal output in the strain gage mV/V significantly. linear. For the measurement of turbine rotation, the rpm meter is used which is integrated with the shaft meter sensor.

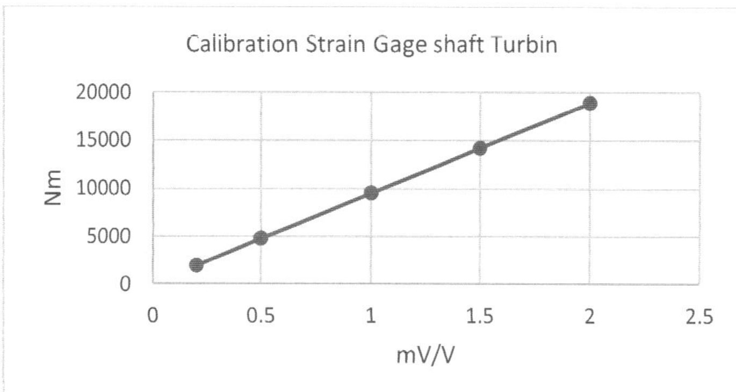

Figure 10. Calibrate the value of strain gage (torque) with input from strain gauge simulator

3. Results and discussion

3.1 Experiment 3 variation inclined blade capability

This experimental study of the tilted blade Darrieus turbine with regular wave generation was carried out in the MOB test pool which is one of the Indonesian Hidrodinamic Laboratory, Surabaya. This study aims to determine changes in turbine blade angle on rotation, torque, and the output power of the Darrieus turbine shaft by providing some regular wave variation. The process of testing three variations of the 45°, 60°, and 75° Turbine models, they are carried out one by one, the overall test setting system is the same, namely, the turbine model is inserted into the pool water up to a limit of 15 cm at the upper turbine blade limit which cannot be immersed in the water.

In Figure 11. This is a test of a Darrieus turbine with a blade angle of 45°, and Figure 12. This is a test of a Darrieus turbine with a blade angle of 60°, and Figure 13. This is a test of a Darrieus turbine with a blade angle of 75°. The video was recorded using an underwater camera with a variation of 0.15 m wave height generation and a period variation of 1.2 seconds, 1.5 seconds, 1.8; 2 seconds, 2.3 seconds, and 2.5 seconds. In this experiment, changing the conditions of the three variations of the turbine blade angle from 45° to 60° and from 60° to 75° was only done by replacing the three upper turbine radial arms and the three lower turbine radial arms. The angle change will be adjusted after each test completion of one variation of the tilted turbine blade model test has been completed.

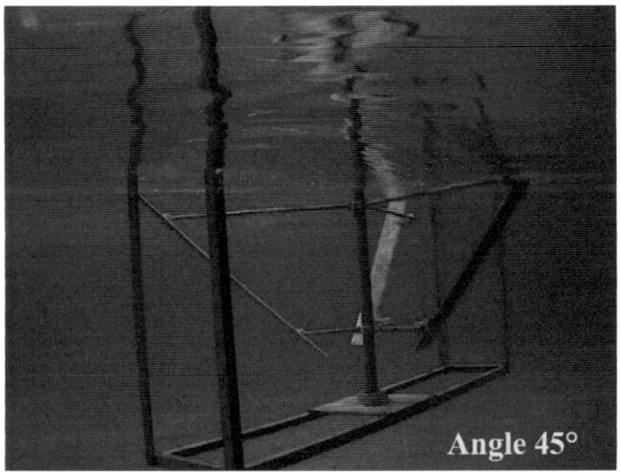

Figure 11. Vertical-axis darrieus turbine experimental model in action on MOB IHL with Position blade 45°.

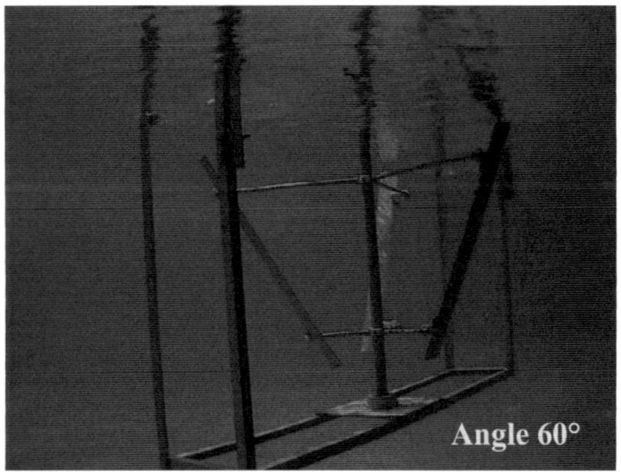

Figure 12. Vertical-axis darrieus turbine experimental model in action on MOB IHL with Position blade 60°.

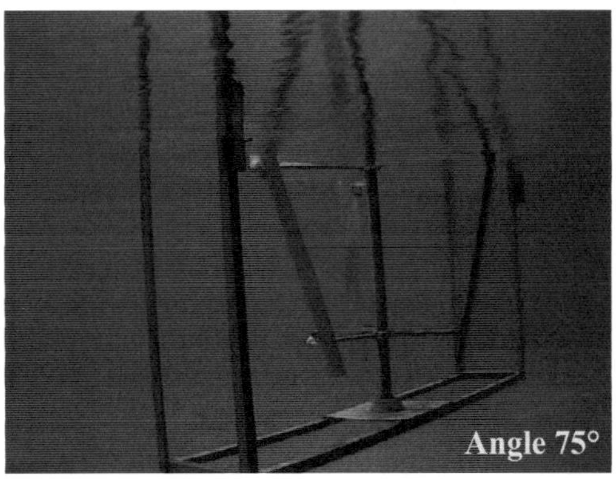

Figure 13. Vertical-axis darrieus turbine experimental model in action on MOB IHL with Position blade 75°.

In the test result with a turbine blade tilted 45° in table 4. the wave generators generate a wave period variations from 1.2 seconds, 1.5 seconds, 1.8 seconds, 2 seconds, 2.3 seconds, and 2.5 seconds producing rotation, torque, and output power of the turbine shaft. Where the increasing wave period from 1.2 seconds to 1.5 seconds will increase the power and rotation of the turbine rotor and after that from the wave period from 1.5 seconds to 1.8 seconds and above the power and rotation of the turbine rotor will decrease. But the torque value will increase after the wave period increases.

Table 4. Test results for a 45° blade turbine with variations in wave period

Wave Period with Wave High 0.15 m (Seconds)	Torque (Nm)	Rotation speed (rpm)	Power (Watt)	Tip speed ratio (TSR)
1.2	6.5525	8.3267	5.7136	0.917384
1.5	6.4789	10.3283	6.9555	0.587126
1.8	6.4745	7.1443	4.8196	0.407726
2	6.4903	4.7271	3.2278	0.330258
2.3	6.6525	8.3267	5.4391	0.249723
2.5	6.8172	7.1924	4.8003	0.211365

Table 4. shows that the 45° angle turbine test with 0.15 meters wave generator conditions in period 1.5 seconds got the highest turbine rotation and power among wave generation with 5 other wave period variations.

In the test result with a turbine blade angle of 60° according to table 5. the wave generators generate a wave period variations from 1.2 seconds, 1.5 seconds, 1.8 seconds, 2 seconds, 2.3 seconds, and 2.5 seconds producing rotation, torque, and output power of the turbine shaft. The increasing wave period from 1.2s to 1.5s will increase the power and rotation of the turbine rotor. After a change in the wave period from 1.5s to 1.8s and above, the turbine rotor power and rotation will fluctuate and decrease. Torque will slowly decrease fluctuating with increasing wave periods.

Table 5. Test results for a 60° blade turbine with variations in wave period

Wave Period with Wave High 0.15 m (Seconds)	Torque (Nm)	Rotation speed (rpm)	Power (Watt)	Tip speed ratio (TSR)
1.2	5.9500	12.2556	7.2021	0.729255
1.5	5.2500	22.1429	12.1737	0.466723
1.8	5.2500	13.9365	7.6970	0.324113
2	5.4820	5.8343	3.2805	0.262532
2.3	5.5101	7.9909	4.5892	0.198512
2.5	5.3800	7.7300	4.3550	0.16802

In table 5. It can be seen that the 60° angle turbine test with 0.15-meter wave generation conditions in a 1.5 seconds period gets the highest turbine rotation and power among wave generation with 5 other wave period variations.

The test with a turbine blade slope of 75° according to table 6. with the provisions of the variation of the wave period from 1.2 seconds to 2.5 seconds produces rotation, torque, and output power of the turbine shaft. Where the increasing wave period from 1.2 seconds to 1.5 seconds will increase the torque, power, and rotation of the turbine rotor, and after that the generation of the wave period from 1.5 seconds to 1.8 seconds and above will decrease the torque, power, and turbine rotation fluctuating.

Table 6. Test results for a 75° blade turbine with variations in wave period

Wave Period with Wave High 0.15 m (Seconds)	Torque (Nm)	Rotation speed (rpm)	Power (Watt)	Tip speed ratio (TSR)
1.2	6.4586	5.7842	3.9158	0.64277
1.5	6.9713	6.3402	4.3599	0.411373
1.8	6.5800	5.7037	3.8191	0.285675
2	6.2958	4.6026	2.8456	0.231397
2.3	6.2620	4.7289	3.0104	0.174969
2.5	5.7622	4.7195	2.8362	0.148094

In table 6. It can be seen that the 75° angle blade turbine test with 0.15m wave generation conditions in a 1.5 s period gets the highest turbine rotation and power among wave generators with 5 other wave period variations.

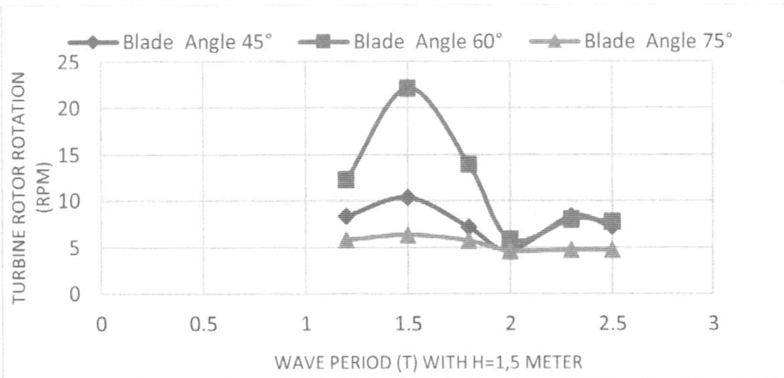

Figure 14. Rotation turbine on 3 variations of blade angle

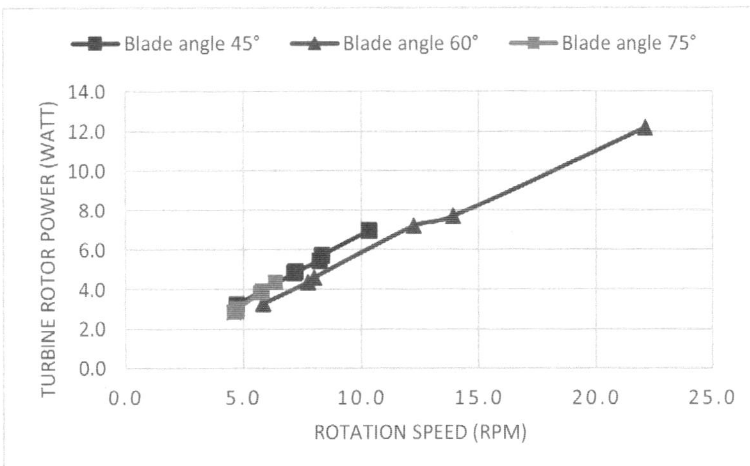

Figure 15. Output power turbine on 3 variations of blade angle

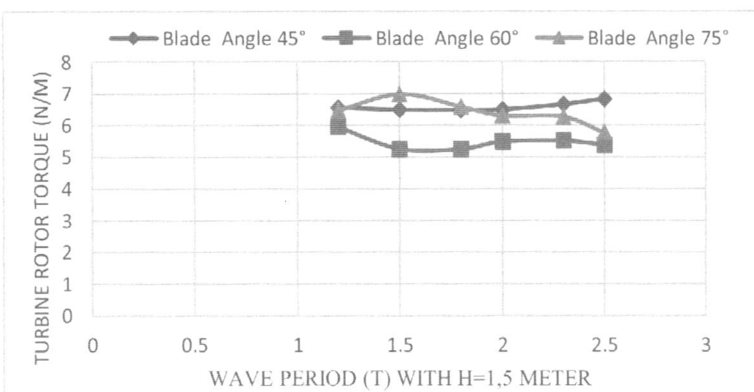

Figure 16. Torque on 3 variations of blade angle

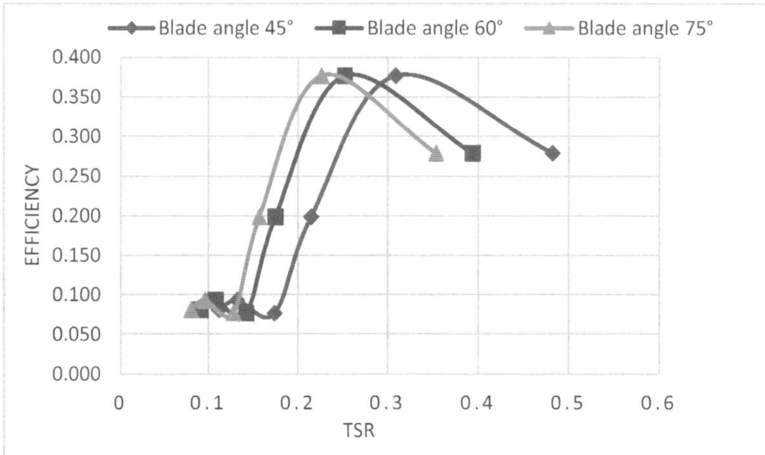

Figure 17. TSR and Efficiency on 3 variations of blade angle

Rotation, torque, and turbine power are the results of this model test. The resulting value is the average value obtained from several tests for three types of turbine models. In Figure 14. it can be seen that with the wave height of 0.15 meters with a wave period of 1.5 seconds for three turbines test with a blade angle of 45°, 60°, and 75° will have the highest rotation but turbine rotation will decrease after a given wave period below and above 1.5 seconds. But the blade angle of 60° has the highest rotation compared to others (45° and 75°).

In Figure 15. diagram of power for the three turbines with angles of 45°, 60°, and 75°, it is also seen that with a wave height of 0.15 meters and a wave period of 1.5 seconds will have a higher output power than given the wave period below and above 1.5 seconds. But the turbine with an angle of 60° has the highest power output rather than the angle of 45° and 75°. So it can be concluded that an increase in turbine rotation will increase the turbine power as well, as in Figure 14. and Figure 15.

In Figure 16. performance diagram of torque for the blade angle of 45°, 60°, and 75°, it is also seen that the arrival of a wave with a height of 0.15 m with 6 type wave period to 3 turbines will create varying torque. But the blade angle of 75° has the highest torque compared to others (45° and 60°).

Figure 17. The efficiency of the three turbine models increases with the addition of the TSR value up to 0.3, but the value decreases with the addition of wave speed above 7.68 m/s (TSR above 0.3). The maximum efficiency (37.7%) of the three turbines is at a wave speed of 7.68 m/s (TSR 0.22 to 0.3). Low turbine efficiency due to high wave speed (above 7.68 m/s), meaning that the turbine power efficiency at low wave speed is 7.68 m/s with a period of 1.5 seconds more optimum than at high wave speed above 7.68 m/s. So it can be seen that the Tip speed ratio (TSR) and efficiency three values of the turbine variation decrease with the increase in the wave period. There is a study that is almost the same as this study to increase the ability of vertical axis turbines by using angled blades to absorb wave energy other than marine energy in the Madura strait although with different blade types (Blade type-LHI 166) [6]. And it turns out that by using the tilt angle, the turbine can rotate if it is exposed to ocean waves even though the efficiency of the turbine test is lower than the theoretical efficiency (0.593). TSR is the ratio of the speed between the rotation of the Rotor Angle to the speed of the wave. (ω) is the angular speed of the rotor, (V) is the wave speed and (r) is the Turbine Radius. To get TSR, use the formula in the equation below:

$$\mathrm{TSR} = \frac{\omega r}{\nu} \tag{3}$$

3.2 Rotation and torque turbine characteristics

Figure 14. is a comparison of rotation conditions and wave period, where the highest rotation of three turbine blade angle variations in the wave period is 1.5 seconds. Although the 3 variations of the turbine blade angle are given the water waves continue to rotate, angle blade 60° rotates higher than angle blades 45° and 75°. So that by testing the model 3 turbine above with three different blade angles capable of absorbing wave energy, it has also been proven by other studies in the absorption of wave energy even with tilted turbines with different blade types and installation positions [15]. The Darrieus turbine will continue to rotate by tilting the blade under specific circumstances in conjunction with the arrival of the water wave. However, each turbine model will have a varied turbine rotation capability due to the effect of the blade angle position and the speed of the water waves striking it [11]. Turbine blades tilted 75° in figure 3., Figure 13., and Figure 14. has a low yield in turbine shaft rotation given 6 variations of wave period. But the turbine blades are inclined at 45° in figure 1., Figure 11., and Figure 14. has a higher rotational yield than the turbine with a blade tilt of 75° and less than the turbine with a blade tilt of 60°.

So it can be concluded that the characteristics of the three tilted turbines are the same if they are struck by waves with a period of up to 1.5 seconds, they will experience a good increase in rotation, and will experience a decrease in a rotation that fluctuates if they are struck by water waves a period over of 1.5 seconds. In this study, the adjustment of the turbine blade tilt position will be used to provide a vertical-type Darrieus turbine installation solution with a 60° tilted blade which has a high operational rotation in the absorption of wave energy which will be recommended for installation on the coast. Because the angle between the blades and the incoming wave makes the turbine produce a larger rotation than other turbine angle models (45° and 75°). The results of this test are also almost the same as the observations of other studies that use a blade angle of 60 to absorb ocean wave energy [6] [15].

4. Conclusions

Experiments of three variations of the vertical axis turbine angle at 45°, 60°, and 75° angles have shown the effect of turbine blade tilt and wave period to increase turbine rotation. The maximum efficiency (37.7%) of the three turbines is at a wave speed of 7.68 m/s (TSR 0.22 to 0.3). The power shaft meter instrumentation used to support this research worked well, with a calibration error of 0.1% in the torque measurement process. Darrieus turbine blades tilted at an angle of 60° have better rotation capability on the effect of water waves period than turbine blades tilted at 45° and 75°. The tilted-blade turbine in this experiment has its efficiency characteristics, where the three tilt-blade turbine models will rotate maximally at a wave period of 1.5 seconds but there is a fluctuating decrease in rotation after the generation of the wave period exceeds 1.5 seconds. Therefore, the turbine blade angle of 60° is suitable for applications in bumpy area locations.

References

[1] Erwandi *et al.*, 2018 "Power plant for ocean currents and waves to convert the kinetic energy of ocean currents and or potential energy of ocean waves into Electrical Energy (Indonesian Language)," IDP000054295.
[2] M. J. Alam and M.T. Iqbal., no. December, 2020 "A low cut-in speed marine current turbine,", [Online]. Available: https://www.researchgate.net/publication/288704099_A_low_cut-in_speed_marine_current_turbine
[3] A. Falcao and J. C. C. Henriques., no. January, 2016 "Oscillating-water-column wave

energy converters and air turbines : A review Oscillating-water-column wave energy converters and air turbines : A review,", doi: 10.1016/j.renene.2015.07.086.

[4] M. A. Mustapa, O. B. Yaakob, Y. M. Ahmed, C. Rheem, and K. K. Koh., 2017, "Wave energy device and breakwater integration : A review," *Renew. Sustain. Energy Rev.*, vol. 77, no. March, pp. 43–58, doi: 10.1016/j.rser.2017.03.110.

[5] S. Musabikha, I. K. Aria, and P. Utama., 2018, "State of the Art in Protection of Erosion-Corrosion on Vertical Axis Tidal Current Turbine," *AIP Conf. Proc.*, vol. 020047, doi: 10.1063/1.5038329.

[6] Erwandi, D. Rahuna, and C. S. J. Mintarso., 2008 "The development of the IHL wave-current turbine to convert the ocean energy from both wave and current," *IOP Conf. Ser. Mater. Sci. Eng.*, vol. 588, no. 1, pp. 5–13, 2019, doi: 10.1088/1757-899X/588/1/012008.

[7] Scheijgrond, Peter, and Cornelis., 2010 "a Device for the Utilisation of Wave Energy and a Method," WO 2010/011133 Al.

[8] Scheijgrond *et al.*, 2010 "an Apparatus for Harvesting Energy from a Body of Water and a Method," WO 2010/062170 A1.

[9] D. Rahuna., 2016, "Experimental Study of Wave Energy Conversion Using a Combination of Wells Rotor and Darrieus Turbine (Indonesian Language)," *Wave J. Ilm. Teknol. Marit.*, vol. 10, no. 1, pp. 31–38, doi: 10.29122/jurnalwave.v10i1.2631.

[10] E. J. Soons., 2007 "Improving the Hydrodynamic Performance of a Combined Wells-Darrieus Rotor,".

[11] D. Satrio, I. K. Aria, and P. Utama., 2021, "Experimental investigation into the improvement of self-starting capability of vertical-axis tidal current turbine," *Energy Reports*, vol. 7, pp. 4587–4594, doi: 10.1016/j.egyr.2021.07.027.

[12] G. Saini and R. P. Saini., 2020, "Comparative investigations for performance and self-starting characteristics of hybrid and single Darrieus hydrokinetic turbine," *Energy Reports*, vol. 6, pp. 96–100, doi: 10.1016/j.egyr.2019.11.047.

[13] R. Martinez, G. S. Payne, and T. Bruce., no. December 2017, "The e ff ects of oblique waves and currents on the loadings and performance of tidal turbines," *Ocean Eng.*, vol. 164, pp. 55–64, 2018, doi: 10.1016/j.oceaneng.2018.05.057.

[14] D. H. Zeiner-gundersen., 2015, "A novel flexible foil vertical axis turbine for river , ocean , and tidal applications," *Appl. Energy*, vol. 151, pp. 60–66, doi: 10.1016/j.apenergy.2015.04.005.

[15] R. Dominy, P. Lunt, A. Bickerdyke, and J. Dominy., Feb. 2007, "Self-starting capability of a Darrieus turbine," *Proc. Inst. Mech. Eng. Part A J. Power Energy*, vol. 221, no. 1, pp. 111–120, doi: 10.1243/09576509JPE340.

Experimental Study on Hydrodynamic Aspects of Turbine which Convert Hydrokinetic and Potential Coastal Wave Energy

Daif Rahuna[1], Erwandi[1], Afian Kasharjanto[1], Eko Marta Suyanto[1,2], Cahyadi Sugeng Jati Mintarso[1,2]

[1] The National Research and Innovation Agency, Republic of Indonesia
[2] Departement of Marine Engineering, Institut Teknologi Sepuluh Nopember, Surabaya Indonesia

*Email correspondent: rahuna.daif@gmail.com

Abstract. Currently, the exploration of ocean renewable energy sources was mostly carried out to obtain optimal results and low cost. The waves that arrive on the beach consist of both potential energy where the water surface moves up and down and hydrokinetic energy where the volume of water comes and goes into the beach sand. They had the potential to be converted to electricity. This paper explained the study of the hydrodynamic aspects of turbines that convert hydrokinetic and potential coastal wave energy. The vertical axis darrieus turbine was modified to catch both energies. It could convert energies from hydrokinetics and the potential of waves simultaneously, whereas a vertical axis turbine with 6 horizontal blades and 3 vertical blades in a shaft. Testing was done at the testing Tank, Hydrodynamic Technology Research Center, National Research and Innovation Agency. The hydrodynamic tests were with 3 turbine variations, wave variations, and current velocity. The test results, vertical axis turbines with horizontal blades could receive wave energy, due to the orbital motion of water particles and vertical blades were very effective in receiving current energy so that turbines with 2 types of vertical and horizontal blades could convert wave and current energy.

1 Introduction

The danger of using energy sourced from fossil fuels will gradually run out, and the use of alternative renewable and environmentally friendly energy sources continues to increase [1]. Hydro energy is the oldest source of energy used by humans. dam technology to utilize water as an energy source to drive turbines. However, hydrokinetic turbines utilize the kinetic energy of the current at a much lower power density [2]. The energy produced from wave energy is very large, where the energy of the ocean waves produced is compared to the amplitude and period of movement of ocean waves, where the length of the period (10s) and the magnitude of the amplitude (2m). The energy produced is generally 50 to 70 kW m^{-1} of wavelength [3]. While ocean currents have considerable potential for stored energy, as seen in [4], the interaction that occurs between the earth and the moon is estimated to produce current and tidal energy of 3.17 TW (Tetawatt) every day throughout the world slightly larger than all plants that existed throughout the world in 1995, which was 2.92 TW.

Theoretically, the total national marine energy resources, including ocean heat, ocean waves, and ocean currents, are huge, but the potential of ocean energy that can be utilized is small. The most prepared marine energy industry is an industry based on wave technology and tidal current technology, with a practical potential of 6,000 MW. Research has been conducted on the condition of the Larantuka strait, East Nusa Tenggara which includes tidal conditions, morphological conditions and current conditions, and the placement of a turbine to get the best concentration at the moment [5]. Ocean current energy as renewable energy is fairly potential energy in coastal areas, especially small islands in the eastern region [6]. The most potential areas for the development of marine current power plants suggested by Marine Current Turbine Ltd. are those with a minimum speed value of 2m/s - 2.5 m/s [7].

Nearshore phenomena that occur due to the current circulation around the coast, can be classified into three types, namely: longshore currents, rip currents, and backflow/cross-shore flows. The current circulation system is often not uniform between the three depending on the direction/angle of the wave affected by the wind [8]. Ocean waves produce potential energy and kinetic energy, the potential energy of ocean waves is affected by wave height while kinetic energy depends on the movement of the number of seawater particles [9]. Waves that come towards the coast carry the water mass and momentum, in the wave direction propagation. This causes currents to occur around the coastal area. Wave propagation towards the coast will cross the offshore zone, surf zone, and swash zone. The motion of water particles stores a very large potential energy, making it possible to use it as a new energy source, where waves store potential energy and kinetic energy while currents store kinetic energy [10]. The total energy of the resulting ocean waves is as follows:

$$E_T = E_p - E_k \tag{1}$$

where, E_T = total energy [J/m]
E_p = potential energy [J/m]
E_k = kinetic energy [J/m]

A type of turbine with a horizontal blade and a vertical shaft or vice versa was invented by Dr. A. A. Wells [11], commonly known as the Wells turbine. Various technologies proposed in the last decade to convert wave energy into electrical energy are based on the principle of Oscillating Water Columns (OWC) located near or on the coast by harnessing the power of coastal waves [12]. The principle of the Open Water Column consists of two distinct parts: an air chamber that converts wave energy into pneumatic energy and converts pneumatic energy into mechanical energy, in which air flows due to changes in pressure generated by waves in the chamber, and air moves to rotate the turbine shaft. Many studies have been conducted in the development of turbine well designs based on experimental studies and numerical analysis to determine the influence and determine design parameters [13]; [14]; [15]; [16]; [17].

The wells rotor converts the wave energy by an oscillating water coulomb (OWC) system, where the turbine is in a compartment. Turbines run depending on fluid motion, some directly utilize the wave motion of water due to waves, and some use a pneumatic system. The rotor has two main parts, shaft, and blades. It does not have a radial arm because the blade is perpendicular to the shaft and is connected to the hub and shaft in the center. It can convert wave energy into mechanical energy due to the orbital motion of water particles.

On vertical axis turbines with vertical blades were invented by George Jean Marie Darrieus [18], known as the darrieus turbine. Darrieus turbine character is affected by the parameters such as blade dimension, blade shape, angle attack, etc. [19]. This darrieus turbine has advantages including not taking into account the direction of flow due to its symmetrical shape, gravitational pressure is not able to reverse the shape of its blades, is able to operate at a head and low speed [20], while the disadvantages are the inability to self-start, and high vibrations.

Experimental tests on the hydro-kinetic turbine are widely carried out to improve the efficiency of the turbine in the optimization of energy conversion, among others, an Experimental test of a prototype of a 2KW ocean current turbine in the Larantuka strait in the utilization of ocean current energy into electrical energy [6], comparing straight-blades and inclined blade vertical axis [21], the shear flow effect on tidal turbine blade [22], performance test twin hull turbine 10KW in Suramadu bridge, energy source of ocean currents based on the moon phases within 30 days [23].

Research by combining two types of turbines, namely darrieus and wells rotors to determine the effectiveness of turbines in the absorption of wave and current energy. the efficiency of darrieus is still low compared to wells rotors in wave energy conversion, the results of the study explain that the efficiency will increase if darrieus turbine is combined with wells rotors in a shaft [24].

Wells rotor converts the wave energy by oscillating water coulomb (OWC) system, where the turbine is in a compartment. Turbines run depending on fluid motion, some directly utilize wave motion of water due to waves and some use a pneumatic system. The rotor well has two main parts, shaft, and blades. The wells rotor does not have a radial arm because the blade is perpendicular to the shaft and is connected to the hub and shaft in the center. The wells rotor is capable of converting wave energy into mechanical energy caused by the orbital motion of water particles.

From the concept of performance in each type of turbine, where the wells rotor has the ability to absorb energy from the vertical direction and the darrieus turbine is able to absorb energy from the horizontal direction, the combination of the two is carried out in a shaft. The purpose of this experimental study: optimum waves, characteristic turbine, rotary and torque.

2 Methodology

2.1 Model geometry

Wells rotor converts potential energy and turbine darrieus converts kinetic energy, to determine the performance of two turbines when combined in absorbing hydro-kinetic and wave potential simultaneously, thus simulating turbine testing on a laboratory scale.

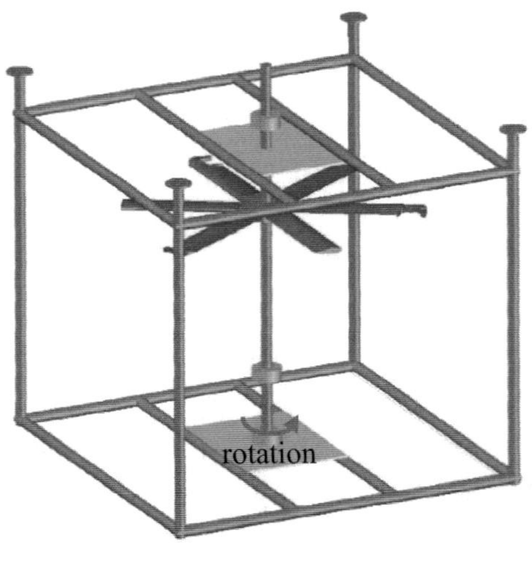

Figure 1. Horizontal Blade

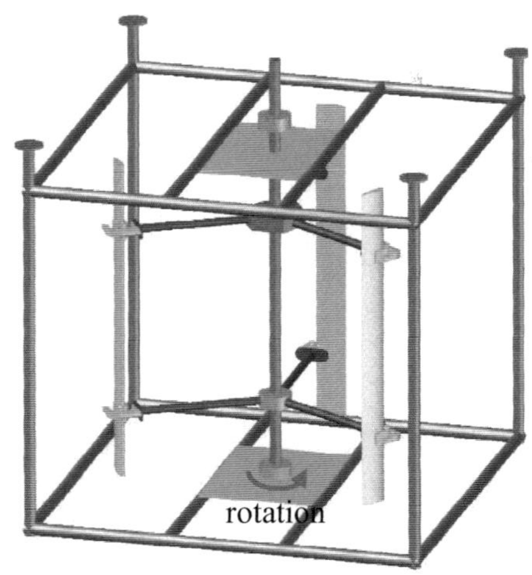

Figure 2. Vertical Blade

The experimental model of the vertical and horizontal blade turbines in a shaft has specifications, horizontal blades as shown in figure 1, NACA 0021 airfoil type, a number of blades 6, diameters 0.910 m, chord length 0.180 m, solidity 0.62 with a pitch angle of 0°; vertical blade as shown in figure 2. NACA 0021 foil type, diameter 1.04 m, chord length 0.083 m, length 0.980 m, number of blades 3, angle attack 15°, and solidity 0.41.

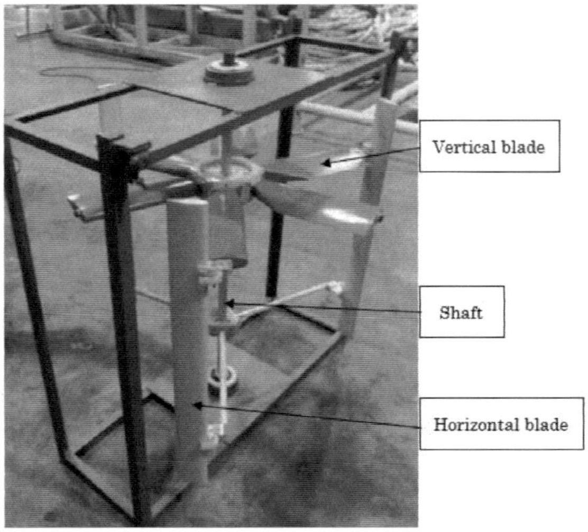

Figure 3. Experiment model

Figure 3. shows an experiment model of a turbine consisting of vertical blades (wells rotor), horizontal blades (darrieus turbine), and a radial arm as a connection to a shaft. In the experiment carried out separately, the vertical blades, horizontal blades, and vertical-horizontal blades (wells-darrieus). each part is bolted together, so experimental model assembly to easy by changing the experimental configurations.

2.2 *Testing setup*

Tests are carried out at the Maneuvering Ocean Basin (MOB), Hydrodynamic Technology Research Center, National Research and Development Agency, as shown in figure 4, where it has a wave generator and can simulate actual ocean waves on a laboratory scale, test tank facility filled with fresh water.

Figure 4. Maneuvering Ocean Basin facility

In the planning of laboratory-scale experimental models, what needs to be considered are the specifics of the waves that can be generated and the conditions in the test tank. Secondary data collection is used in determining the initial design and determining the dimensions of the experimental model to adjust the conditions of the test tank. Secondary data obtained from the condition of the test tank and the resulting waves are *D*, depth 1.2m; *H*, wave high 0.08m-0.18m; *T*, wave period 1.0s-2.0s.

Figure 5. horizontal-vertical blade test with a wave

Figure 6. Vertical blades with a wave

As shown in figure 5 and figure 6, an experimental horizontal-vertical blade turbine and horizontal blade in a shaft with a wave, where turbine models are immersed in water positioned at the bottom of the pool, with a depth of 1.2m. During the experiment, the wave by the wave generator, and there is an interaction between the wave and the turbine.

Table 1. Experimental wave data

Wave	Specifications
H, wave high	0.08m; 0.10m; 0.12m; 0.14m; 0.16m; 0.18m
T, wave period	1.0s; 1.5s; 2.0s

Experiment with 3 test model configurations, including horizontal blade turbine, vertical blade turbine, and horizontal-vertical blade turbine in a shaft. This test is to determine the amount of the turbine rotation and torque produced by each wave variation, presented in Table 2, the data is recorded from the turbine not running until runs.

Figure 7. Towing Tank facility

The turbine test model with a current is carried out at the Towing Tank (TT), Hydrodynamic Technology Research Center, National Research and Development Agency, as shown in figure 7, where test tank facilities contain fresh water with a towing train model test. The test model will be placed on the carriage test frame and carriage speed of 0.2m/s – 1.4m/s. As shown in figure 8, The turbine model is immersed in water, distancing between the turbine and the water surface is 0.10 m. During the experiment, no waves were formed due to the low towing speed so there was no significant interaction. The distance between the water surface and bottom tank is 5.4m. the turbine shaft is held with two bearings above the water of 0.12m and a small bearing at the end of the shaft in the water. This is done because it does not have facilities for the rated current, so carriage velocity alternative current velocity.

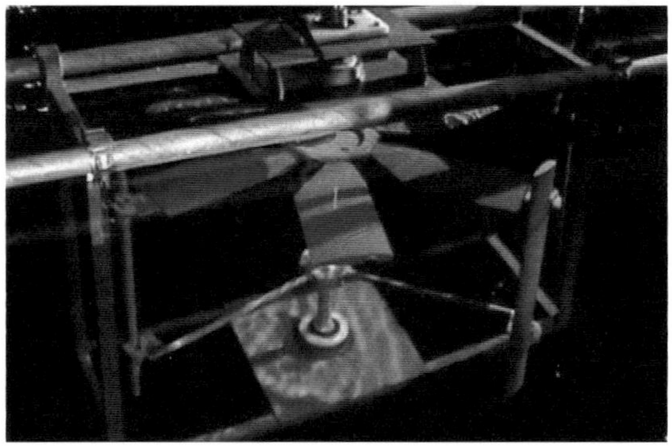

Figure 8. Turbine experiment with current

Experimental needs several measuring instruments to collect test data, such as wave high spectrum (WHS) is a wave measuring instrument, which is in the form of a sensor connected to a monitor to find out the spectrum of waves, a tachometer is a rotational measuring instrument, which will be used to find out the rotation that occurs in the turbine during the test, a torque meter is used to measuring turbine rotation torque, a current meter is a horizontal measure of water current, which will be installed in the carriage and part of it dyed into the water and video camera to visually record conduct phenomenon of the test model during testing.

According to [25] [26], calibration is an activity that relates the values indicated by the measuring instrument or values represented by the measuring material to values that are already known for their degree of correctness (relating to the measured quantity) under certain conditions. All Instruments must be well-calibrated to minimize measurement errors.

2.3 horizontal and vertical velocities of water particles

The horizontal velocity of water particles moving orbitally will always decrease to a certain depth at each of the same wave heights. The horizontal and vertical velocity components of the water particle (you and w) can be determined based on the following equation

Kecepatan horisontal partikel air yang bergerak secara orbital akan selalu menurun sampai dengan kedalaman tertentu pada setiap tinggi gelombang yang sama. Komponen dari kecepatan horisontal dan vertikal pertikel air (*u* dan *w*) dapat ditentukan berdasarkan persamaan sebagai berikut:

- Kecepatan horisontal partikel air

$$u = \frac{\delta\Phi}{\delta x}$$

$$u = \left(\frac{\pi H}{T}\right)\frac{\cosh k(d+y)}{\sinh kd}\sin(kx - \sigma t)$$

- Kecepatan vertikal partikel air

$$w = \frac{\delta \Phi}{\delta z}$$
$$w = \left(\frac{\pi H}{T}\right) \frac{\sinh k(d+y)}{\sinh kd} \sin(kx - \sigma t)$$

2.4 Power turbine of capacity

The magnitude of the energy potential generated from these ocean currents can use in the equation below. because in the actual situation there are various kinds of losses, friction losses, trailing edge losses of the blade, or others.

$$P = {}^{1}\!/_{2}\, \rho A V^3 C_p \tag{2}$$

Where:
P = power [watt]
ρ = density [Kg/m^3]
A = area [m^2]
V = velocity [m/s]
C_p = coefficient power

Which is the value of the thrust force if multiplied by the radius of the turbine will produce the torque value of the Darrieus turbine which is expressed by:

$$\tau = Fr \tag{3}$$

Where:
τ = torque [Nm]
F = force [N]
r = radius [m]

So that the turbine power can be stated as follows:

$$P_T = \tau\omega \tag{4}$$

Where:
P_T = turbine power [watt]
τ = torque [Kg/m^3]
ω = angular velocity [m^2]

Turbine efficiency is the ratio between the turbine power generated and the power from the water flow. This relationship is expressed in the equation:

$$\eta_T = \frac{P_T}{P_N} \tag{5}$$

Where:
η_T = efficiency
P_T = turbine power [watt]
P_N = machine power [watt]

Tip speed ratio (λ) is one of the variables in the calculation of the chord's length that can affect the width of the blade and gives impact the power generated by the turbine.

$$\lambda = \frac{v}{V} = \frac{\omega r}{V} = \frac{2\pi f r}{V} \tag{6}$$

where, λ = Tip Speed Ratio [J/m]
v = Rotor velocity [rpm]
V = Current velocity [m/s]
ω = Angular velocity [rad/s]
r = radius of the turbine [m]
f = wave frequency [/s]

In theory, Betz limit is the maximum efficiency, while the constant number is 16/27 (59.3%). The Betz value is achieved by the turbine rotor in absorbing energy.

3 Result and Discussion

3.1 Selection of the optimal wave

Waves are the main factor in wave energy conversion because waves are energy inputs in the turbine experiment in wave energy conversion. In this test, waves are generated by a wave generator to produce the appropriate experimental wave for the drive turbine.

Data input in wave generation does not necessarily get the output of the wave spectrum according to the data input. Therefore, it is necessary to validate the wave using the WHS (wave high spectrum) wave measuring instrument. The test wave input data includes *H* wave height, *T* wave period, and wave factors.

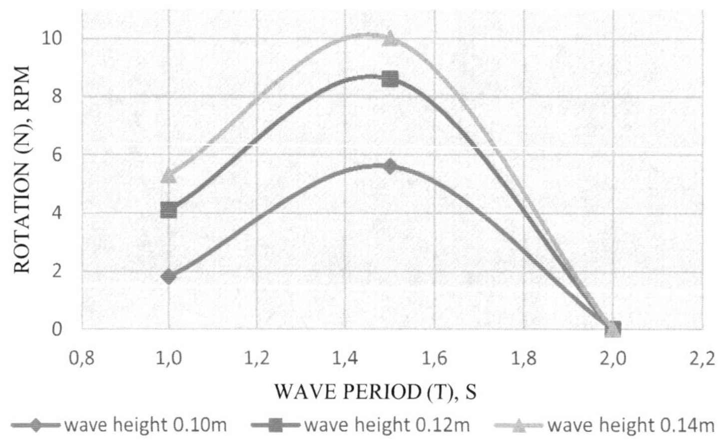

Figure 9. Correlation between wave periods and turbine rotations

The purpose of the wave period determination is to obtain the optimum wave period in producing maximum turbine rotation. By varying several wave heights and periods, some rotational data from the turbine was obtained, resulting rotation data were as shown in Figure 9, wave height, *H*=0.10m, wave period, *T*=1.0s, turbine rotation produced 1.8rpm, wave period, *T*=1.5s and turbine rotation produced 5.6rpm, wave period, *T*=2.0s and turbine do not rotate, then wave height, *H*=0.120m, wave period, *T*=1.0s and turbine rotation produced 4.1rpm, period *T*=1.5s and turbine revolutions generated 8.6rpm, wave period, *T*=2.0s and turbine do not rotate, and then wave height, *H*=0.140m, wave period, *T*=1.0s and turbine rotation generated 5.3rpm, wave period, *T*=1.5s and turbine rotation produced 10.0rpm, wave period, *T*=2.0s, and turbine do not rotate. Tests 1, 2, and 3 experienced the same trend where there are an increase in turbine rotation at T= 1.0s-1.5s and a decrease in T= 2.0s. Then the result of determining the optimum wave period in producing maximum turbine rotation is the specified wave period of 1.5s.

3.2 *Turbines experiment with waves*

The optimum wave period is used in turbine testing by varying the wave height, then obtaining rotation and torque data. With a wave period of 1.5 seconds and varying wave height, the resulting turbine rotation and torque data are obtained from the testing of each turbine with horizontal - vertical blades and horizontal blade of the turbine, as shown in figure 10, where horizontal - vertical blades of the turbine at wave height, H=0.08m produces n=3.6rpm, at wave height, H=0.10m produces n=5.6rpm, at wave height, H=0.12m produces n=8.6rpm, at wave height, H=0.14m produces n=10rpm, at wave height, H=0.16m produces n=11.2rpm, and at wave height, H=0.18m for turbines produces n=12rpm. The horizontal blade of the turbine obtain at wave height, H= 0.081m produces n=4.4rpm, at wave height, H=0.104m produces n=8.3rpm, at wave height, H=0.123m produces n=10.7rpm, at wave height, H=0.141m produces n= 11.4rpm, at wave height, H=0.163m it produces n=12.3rpm, and at wave height, H=0.182m wave heights for turbines it produces n=13rpm.

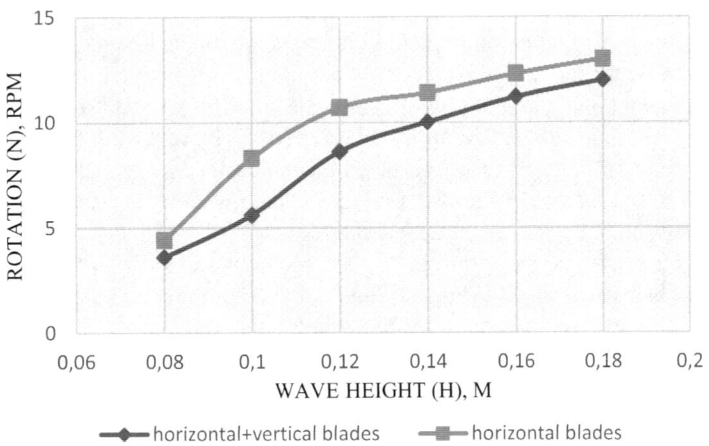

Figure 10. Correlation between wave height and turbine rotations

As shown in figure 11, the Correlation between wave height and torque horizontal - vertical blades of the turbine, at wave height, H=0.08m produces τ=0.5Nm, at wave height, H=0.10m produces τ=0.93Nm, at wave height, H=0.12m produces τ=1.28Nm, at wave height, H=0.14m produces τ=1.48Nm, at wave height, H=0.16m produces τ=1.64Nm, and at wave height, H=0.18m produces τ=1.76Nm. The horizontal blades are at wave height, H=0.081m produces τ=0.4Nm, at wave height, H=0.104m produces τ=0.77Nm, at wave height, H=0.123m produces τ=1.07Nm, at wave height, H=0.141m produces τ=1.32Nm, at wave height, H=0.163m produces τ=1.45Nm, and at wave height, H=0.182m produce τ=1.61Nm.

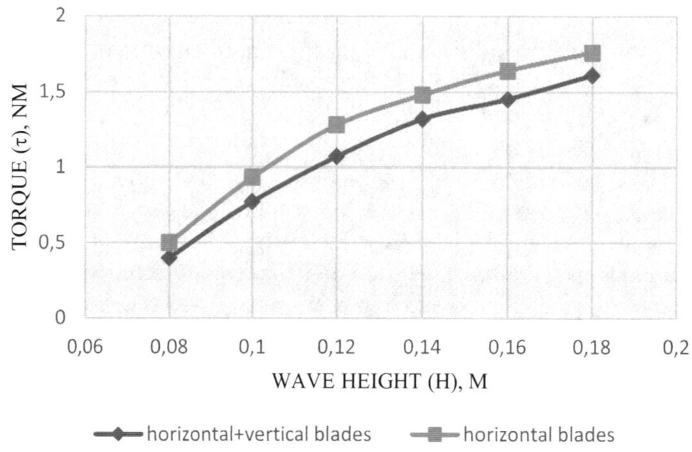

Figure 11. Correlation between wave height and turbine torque

As shown in figure10 and figure 11, the correlation between the high wave, torque, and rotation of the turbine, and wave period is the same, that with increasing wave height the greater will always be followed by the increased torque and rotation of the turbine. It is the orbital velocity of water particles affected by the wave height and wave period. Because the results obtained are the horizontal speed and vertical speed of the water particles, calculations are carried out by resultant determining the of the two speeds, and the resulting velocity of water particles on the surface is greater than the speed of water particles that are below the surface of the water so that in determining the speed of water particles, the speed of water particles on the surface is used [24].

Horizontal blade turbines can produce higher rotation and torque compared to the rotation produced by horizontal-vertical blade turbines. This is due to the fact that the orbital motion of water particles can only be captured by horizontal blade turbines, and vertical blades become resistant.

Experiments of the turbines with vertical blades with wave variations do not produce rotation. This is evidence that the sinusoidal wave motion of circular water particles has amplitude for deep water waves. The vertical blade turbine is aerofoil-shaped so that the pressure and speed of water particles crossing the foil surface on both sides of the blade are equally large.

3.3 Turbine experiment with current

The purpose of the turbine experimental with a current speed is to determine the rotation of the current against the variation in the speed of the current at which the speed of the current is simulated by the carriage speed in the towing tank. In this experiment, hence used 3 test configurations, including horizontal blades, vertical blades, and horizontal-vertical blades with a carriage speed of 0m/s-1.4m/s, rotation, and torque as output data.

As shown in figure 12, test results of horizontal-vertical blades turbine are as follows, the velocity v=0.2m/s produces turbine rotation n=1.7rpm, the velocity v=0.4m/s produces turbine rotation n=6.4rpm, the velocity v=0.6m/s produces turbine rotation n=20.8rpm, the velocity v=0.8m/s produces turbine rotation n=31.8rpm, the velocity v=1.0m/s produces turbine rotation n=41.5rpm, the velocity v=1.2m/s produces turbine rotation n=49.1rpm, and current speed v=1.4m/s produces turbine rotation n=58.7rpm. Test result Test results of vertical blades turbine are as follows, the velocity v=0.2m/s produces the rotation of the turbine n=2.3rpm, the velocity v=0.4m/s produces the rotation of the turbine n=14.5rpm, the velocity v=0.6m/s produces the rotation of the turbine n=27.3rpm, the velocity v=0.8m/s produces the rotation of the turbine n=37.8rpm, the velocity v=1.0m/s produces turbine rotation n=47.2rpm, the velocity v=1.2m/s produces turbine rotation n=56.1rpm, and at the velocity, v=1.4m/s produces turbine rotation n=64.3rpm.

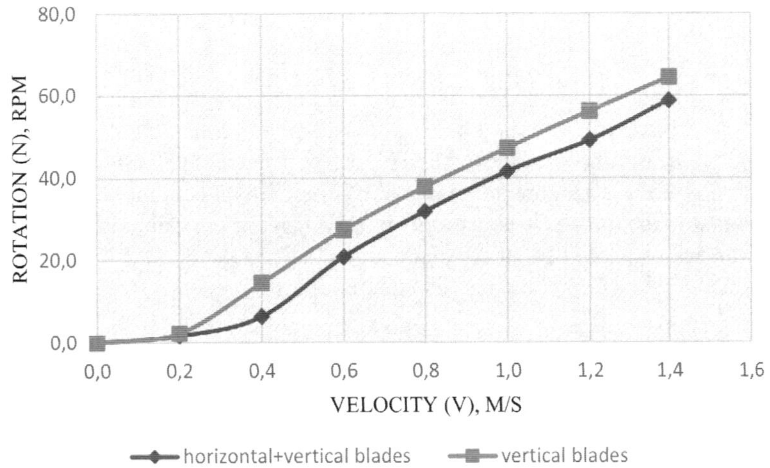

Figure 12. Correlation between velocity and rotation

As shown in figure 13, the test results of horizontal-vertical blades turbine are as follows, the velocity v=0.2m/s produces torque τ=1.7Nm, the velocity v=0.4m/s produces torque τ=1.29Nm, the velocity v=0.6m/s produces torque τ=1.76Nm, the velocity v=0.8m/s produces torque τ=3.25Nm, the velocity v=1.0m/s produces torque τ=5.52Nm, the velocity v=1.2m/s produces torque τ=8.76Nm and current speed v=1.4m/s produces torque τ=12.62Nm. Test result Test results of vertical blades turbine are as follows, the velocity v=0.2m/s produces torque τ=1.77Nm, the velocity v=0.4m/s produces torque τ=3.84Nm, the velocity v=0.6m/s produces torque τ=4.38Nm, the velocity v=0.8m/s produces torque τ=6.94Nm, the velocity v=1.0m/s produces torque τ=9.85Nm, the velocity v=1.2m/s produces torque τ=14.31Nm and current speed v=1.4m/s produces torque τ=18.32Nm.

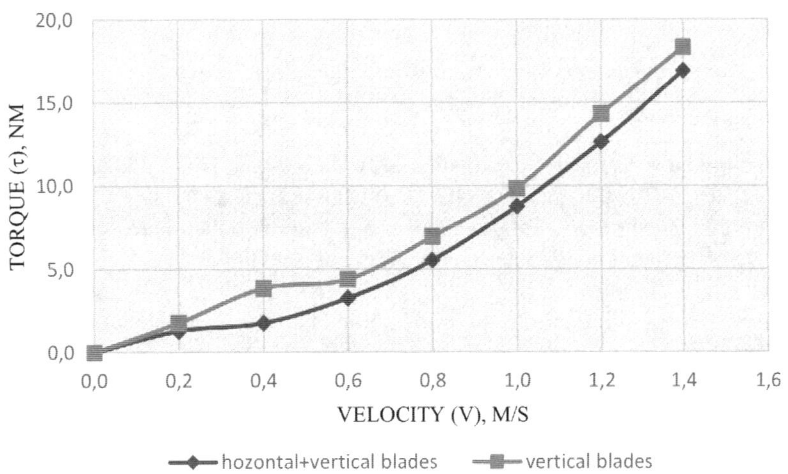

Figure 13. Correlation between velocity and torque

Hydrokinetics is influenced by the mass and speed of water particles, the test of turbines with current is known that horizontal-vertical blades turbine and vertical blades turbine can produce rotation and torque, where the test results as shown in figure 12 and figure 13 with an increase in the speed of current, rotation and torque increase.

Vertical blade turbines produce higher rotation and torque compared to the revolutions produced by horizontal-vertical blade turbines. This is due to the fact that the horizontal movement of water particles can only be captured by vertical blade turbines, and horizontal blades become resistant.

Experiments of the turbines with horizontal blades with current velocity variations do not produce rotation. Prove that hydrokinetic force is generated by the current without being affected by potential forces. The horizontal blade turbine is aerofoil-shaped so that the pressure and speed of water particles crossing the foil surface on both sides of the blade are equally large.

4 Conclusion

The horizontal blade turbine can only capture the potential energy of the wave due to orbital motion and cannot capture the hydrokinetic energy, causing the force direction is the same as a turbine blade direction, as well as for the vertical blade turbine can capture the hydrokinetic energy of the current, where the current force direction is perpendicular to the turbine blades and cannot capture the energy of the wave potential in the force direction is the same as a turbine direction. Both experience the same phenomenon where the pressure P_1 and the speed of V_1 are the same as the pressure of P_2 and the speed of V_2 against the blade surface.

Horizontal-vertical blades turbine produce lower rotation and torque than horizontal blades and vertical blade turbines. However, these turbines can capture the wave potential energy and hydrokinetic energy of the current, where the total energy of the coastal wave is the summation of the wave's potential energy and hydrokinetic energy. Self-starts of the turbine with horizontal-vertical blades in a shaft have a low where at a current speed of 0.2m/s and a wave height of 0.08 m. The turbine can produce rotation, n=1.7rpm, and torque, τ=1.7Nm.

References

[1] O. Ellabban, H. Abu-Rub, and F. Blaabjerg, "Renewable energy resources: Current status, future prospects and their enabling technology," *Renew. Sustain. Energy Rev.*, vol. 39, pp. 748–764, Nov. 2014, doi: 10.1016/J.RSER.2014.07.113.

[2] H. Alizadeh, M. H. Jahangir, and R. Ghasempour, "CFD-based improvement of Savonius type hydrokinetic turbine using optimized barrier at the low-speed flows," *Ocean Eng.*, vol. 202, p. 107178, Apr. 2020, doi: 10.1016/J.OCEANENG.2020.107178.

[3] J. Twidell, *Renewable energy resources*. Routledge, 2021.

[4] L. H. Kantha and C. A. Clayson, *Numerical models of oceans and oceanic processes*. Elsevier, 2000.

[5] A. Yuningsih, "DI KAWASAN PESISIR FLORES TIMUR , NTT POTENTIAL ENERGY OF OCEAN CURRENT FOR ELECTRIC POWER GENERATIONIN COASTAL AREAS OF EAST FLORES , NTT I . PENDAHULUAN Permintaan energi di Indonesia cenderung meningkat pesat sejalan dengan pertumbuhan ekonomi dan pert," vol. 3, no. 1, pp. 13–25, 2011.

[6] A. K. Erwandi, P. Sasoko, W. B. Rina, E. Marta, and D. Rahuna, "Vertical axis marine current turbine development in indonesian hydrodynamic laboratory-surabaya for tidal power plant," in *International conference and exhibition on sustainable energy and advanced materials (ICE SEAM 2011). Solo, Indonesia*, 2011, pp. 15–23.

[7] P. L. Fraenkel, "Power from marine currents," *Proc. Inst. Mech. Eng. Part A J. Power Energy*, vol. 216, no. 1, pp. 1–14, 2002.

[8] R. G. Dean and R. A. Dalrymple, *Coastal processes with engineering applications*. Cambridge University Press, 2004.

[9] B. Le Méhauté, *An introduction to hydrodynamics and water waves*. Springer Science & Business Media, 2013.

[10] B. Triatmodjo, "Teknik pantai," 1999.
[11] A. A. Wells, "Fluid driven rotary transducer," *Br. Pat. spec*, vol. 1, pp. 595–700, 1976.
[12] A. F. d. O. Falcão, "Wave energy utilization: A review of the technologies," *Renew. Sustain. Energy Rev.*, vol. 14, no. 3, pp. 899–918, Apr. 2010, doi: 10.1016/J.RSER.2009.11.003.
[13] A. Brito-Melo, L. M. C. Gato, and A. J. N. A. Sarmento, "Analysis of Wells turbine design parameters by numerical simulation of the OWC performance," *Ocean Eng.*, vol. 29, no. 12, pp. 1463–1477, Sep. 2002, doi: 10.1016/S0029-8018(01)00099-3.
[14] S. Raghunathan, "The wells air turbine for wave energy conversion," *Prog. Aerosp. Sci.*, vol. 31, no. 4, pp. 335–386, Jan. 1995, doi: 10.1016/0376-0421(95)00001-F.
[15] Z. Taha, Sugiyono, and T. Sawada, "A comparison of computational and experimental results of Wells turbine performance for wave energy conversion," *Appl. Ocean Res.*, vol. 32, no. 1, pp. 83–90, Feb. 2010, doi: 10.1016/J.APOR.2010.04.002.
[16] M. Torresi, S. M. Camporeale, P. D. Strippoli, and G. Pascazio, "Accurate numerical simulation of a high solidity Wells turbine," *Renew. Energy*, vol. 33, no. 4, pp. 735–747, Apr. 2008, doi: 10.1016/J.RENENE.2007.04.006.
[17] P. Halder, A. Samad, J. H. Kim, and Y. S. Choi, "High performance ocean energy harvesting turbine design–A new casing treatment scheme," *Energy*, vol. 86, pp. 219–231, Jun. 2015, doi: 10.1016/J.ENERGY.2015.03.131.
[18] D. G. J. Marie, "Turbine having its rotating shaft transverse to the flow of the current." Google Patents, Dec. 08, 1931.
[19] N. Kolekar, S. S. Mukherji, and A. Banerjee, "Numerical modeling and optimization of hydrokinetic turbine," in *Energy Sustainability*, 2011, vol. 54686, pp. 1211–1218.
[20] A. Abdel Azim El-Sayed, C. Hirsch, and R. Derdelinckx, "Dynamics of vertical axis wind turbines (darrieus type)," *J. Rotating Mach.*, vol. 31, no. 1, pp. 33–41, 1995.
[21] D. Satrio and I. K. A. P. Utama, "Experimental investigation into the improvement of self-starting capability of vertical-axis tidal current turbine," *Energy Reports*, vol. 7, pp. 4587–4594, Nov. 2021, doi: 10.1016/J.EGYR.2021.07.027.
[22] M. Magnier, N. Delette, P. Druault, B. Gaurier, and G. Germain, "Experimental study of the shear flow effect on tidal turbine blade loading variation," *Renew. Energy*, vol. 193, pp. 744–757, Jun. 2022, doi: 10.1016/J.RENENE.2022.05.042.
[23] A. Kasharjanto, D. Rahuna, and R. B. Aditya, "Performance test of the 10 Kilowatt Twin Hull Turbine on the Suramadu Bridge (in Bahasa)," *Kapal J. Ilmu Pengetah. dan Teknol. Kelaut.*, vol. 14, no. 3, pp. 79–86.
[24] P. Scheijgrond, "A device for the utilisation of wave energy and a method," *Int. Pat. WO/2010/011133 (28/01/2010), accessed*, 2009.
[25] A. Hadi, *Pemahmn & Penerapan ISO/IEC 17025: 2005*. Gramedia Pustaka Utama, 2007.
[26] P. De Bièvre, "The 2012 International Vocabulary of Metrology:"'VIM,"'" *Chem. Int. IUPAC*, vol. 34, no. 3, pp. 26–27, 2012.

IOP Conf. Series: Earth and Environmental Science 1166 (2023) 012022 doi:10.1088/1755-1315/1166/1/012022

Geology and Seabed Topography Research to Support the Implementation of Indonesia's Submarine Cable Corridor Buton Island to EEZ (Hub Indonesia - USA)

Djunaedi Muljawan[1], Buddin Al Hakim[2,*)], Nugraheni Setyaningrum[3], Wahyu Widodo Pandoe[2], Alfi Rusdiansyah[4], Joko Widodo[3],Trinugroho [4]

[1] Research Center for Geospatial, BRIN, West Java, Indonesia
[2] Research Center for Hydrodynamic Technology, BRIN, East Java, Indonesia
[3] Research Center for Remote Sensing, BRIN, West Java, Indonesia
[4] Research Center for Geological Disaster, BRIN, West Java, Indonesia

* corresponding authors: budd001@brin.go.id

Abstract. The Indonesian Government has established pipelines and cable corridors through the Decree of the Minister of Marine Affairs and Fisheries Number 14 of 2021 concerning Submarine Pipelines and Cables. Indonesia's strategic position in the world has become a favorite route for laying submarine cables, connecting Australia and Indonesia with America and other countries. The purpose of this study is to conduct a technical study of the existing ratified route corridor on the geological conditions of the Indonesian waters. The method used is path analysis of geological and geomorphology conditions by compiling class criteria related to and increasing risks to geological conditions. The results of the geological analysis show that several submarine cable lines pass through volcanic areas and faults; also, there are areas with a seabed side slope of >20%. This condition can affect the cable laying process, operations, and vulnerability. In addition, it is necessary to rearrange the paths that have a high hazard vulnerability to disasters and other activities. The results of this study are in the form of recommendations for proposed corridor revisions that reduce constraints related to geology and geomorphology.

1. Introduction

The national medium-term development plan (RPJMN) 2020–2024 states that Indonesia will implement 41 national priority programs (PSN) to create a strong economic foundation based on competitive advantage. One of these PSNs is concerned with the digital transition's Information and Communication Technology (ICT) infrastructure. BPS figures for 2021 [1] show that the information and communication technology sector has significantly risen. 78.18% of Indonesia's total number of households has internet connectivity as of 2020. Compared to 2016, when the data peaked at 47.22 %, this data is higher.

Moreover, there was a significant increase in the number of internet users. 53.73% of Indonesia's population used the internet in 2020 using their mobile devices. Compared to 2016, this condition worsened when it was 25.37%. Over the past five years, Indonesia's access improvements have multiplied.

The Indonesian Submarine Cable Communication System (SKKL) currently has a length of 115,000 km, including 55,000 km, in the Exclusive Economic Zone (EEZ) [2]. As an archipelagic country, Indonesia is closely related to developing technology and operating submarine cables. Marine cables are essential to connect an island's communications and other needs. Indonesia's strategic position in the world has become a favourite route for laying submarine cables, connecting Australia and Indonesia with America and other countries.

The Government of Indonesia issued a regulation related to installing submarine pipelines and/or cables through the Decree of the Minister of Marine Affairs and Fisheries Number 14 of 2021 concerning Pipelines and/or Underwater Cables. This decree has 217 national submarine cable lines, 43 national subsea pipelines, four landing stations, and 209 Beach Man Hole (BMH) points throughout Indonesia (Figure 1).

The enactment of this Ministerial Decree for deploying submarine cables crossing Indonesian territory must follow a predetermined path. In general, this regulation has considered the existence of existing submarine cables. Only a few corridors or segments are counted as new corridors. One of the new corridors of this regulation is corridor 72, which connects Buton Island in Southeast Sulawesi with the EEZ. This EEZ point will be connected to a corridor outside the territory of Indonesia; from several international cable plans, the corridor line will be connected to Los Angeles, USA.

The best possible cable route can be suggested to avoid submarine obstacles and hazards, such as Existing cables and pipelines, Fishing and trawling, Shipping lanes, harbour entries, anchorages, Dumping and dredging, Seabed contamination, Oil and gas fields, Military activities, Ammunition dumping grounds, Shipwrecks, and other submarine junk, abandoned cables, and Areas of sand or gravel extraction. In addition, the contribution of anchoring activities to cable faults increased to 48%, becoming the most significant cause of submarine cable faults in recent years. Once the cable is impacted to some extent, the electric transmission and/or signal transmission will be cut off, seriously affecting the crowd is production and life [3].

Based on several studies, fishing activity is the leading cause of damage to submarine cables. However, damage due to natural hazards also needs to be considered and taken into account by looking at geological and geomorphological conditions before deploying submarine cables. Implementing cables at sea is often problematic because the seabed conditions are unknown, causing various risks when laying cables [4]. Almost every cable failure carries a high service recovery and repair cost, which must be passed on to cable service users [5]. Potential risks caused by geological conditions can be due to faults, seismic activity, or volcanoes. Meanwhile, geomorphological conditions are caused by seabed slopes, sand waves, pockmarks, shoals, reefs, and others.

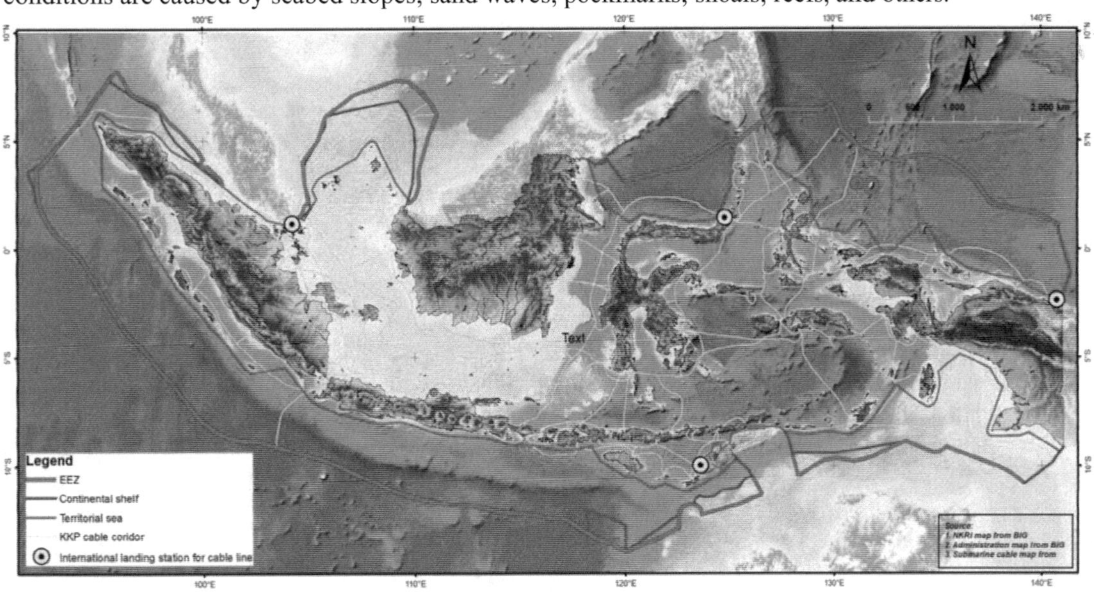

Figure 1. Submarine Cable Map [Decree of the Minister of Marine and Fisheries of the Republic of Indonesia No.14, Feb. 2021]

Based on records spanning several decades, it may be estimated that c.100– 150 cable faults occur annually worldwide, the distribution of faults caused by external forces (external aggression), including seabed movement and abrasion. These patterns were taken from a global database of 2,162 cable faults back to 1959. Most faults occur on the continental shelf, near land in water depths of less than 100 m [5]. Some geological and seismic influences on the cable break include: on August 2, 2022, Telkom's marine segment fiber optic cable broke at 23.8 km off the coast of Bakongan, allegedly due to a natural disaster, an earthquake.

Information and communication technology (ICT) in 2020 showed the second-highest growth rate in the gross domestic product after the health sector. The addition and development of new ICT infrastructure, such as the construction of submarine cable infrastructure, especially the Buton Island corridor, Southeast Sulawesi to the EEZ to America (based on spatial data from the website https://www.submarinecablemap.com/), will have a positive impact on GDP and as well as the ease of access to communication for the peoples, especially in Eastern Indonesia.

For this reason, in deploying submarine cables, it is necessary to pay attention to several aspects, including engineering, shipping safety, defense and security, protection of the marine environment, and geopolitical aspects. The research is part of the technical aspect, especially related to geological and geomorphological conditions. This study aims to obtain an overview of the geological, bathymetric, and morphological conditions along corridor-72 (Buton Island – To EEZ/Los Angeles USA), analyze its risk and provide input and mitigation of the corridor with the least potential risk.

2. Methodology

The method used in reducing potential risks is to analyze risk and make path mitigation from the recommended path. In this study, only risks due to geological and geomorphological conditions are carried out in corridor-72, Ministry of Marine Affairs and Fisheries Decree number 14 of 2021. Geological and geomorphological studies include faults, seismic activity, volcanoes, and slopes in Buton Island's cable segment/corridor to EEZ.

2.1. Study Area

This study's location is one of the segments/corridors stipulated in the Minister of Marine Affairs and Fisheries Decree No. 14 of 2021, namely Buton Island to the EEZ. This segment passes through the waters of the Banda Sea, Seram Sea, and the Halmahera Sea. This line corridor segment will facilitate underwater infrastructure (cables, pipelines, Etc.) from Indonesia to Los Angeles (USA) via the EEZ. The study on this corridor is limited to the Buton Island to EEZ segment.

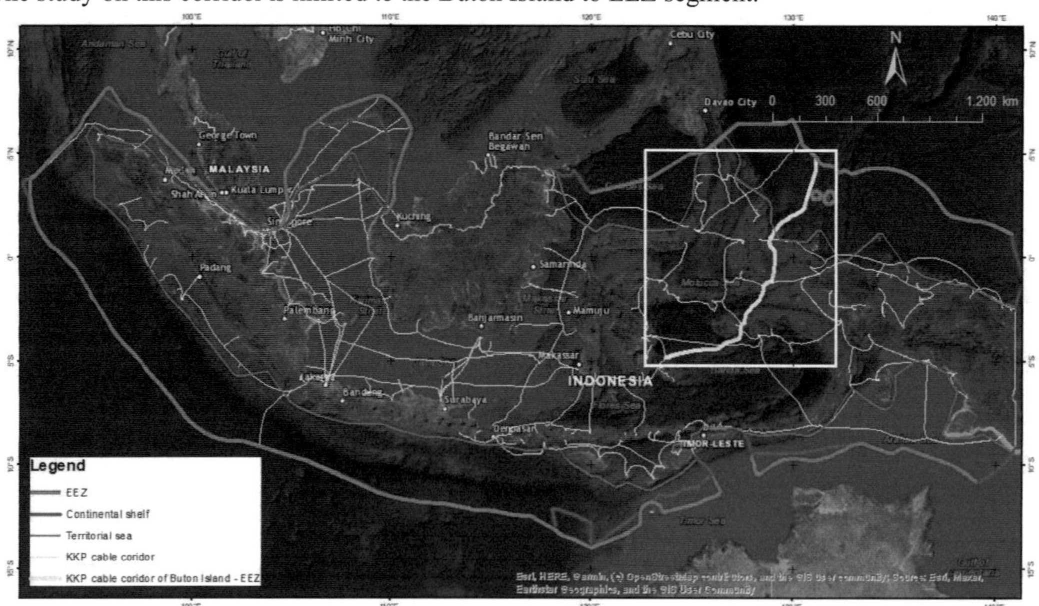

Figure 2. Cable corridor of the Buton Island - EEZ in the KKP corridors (Minister of Marine Affairs and Fisheries Decree No. 14 of 2021)

Location route positioning list (RPL) Buton Island - ZEE, according to the Minister of Marine Affairs and Fisheries Decree No. 14 of 2021, can be seen in Tables 1 and 2.

Table 1. Position of start and end point corridor 72, Buton Island – ZEE

Corridor Number	Start Corridor		End Corridor	
	Buton Island, Southeast Sulawesi		EEZ	
	Lat	Long	Lat	Long
72	4° 50' 36.4" S	123° 39' 03.2" E	4° 48' 17.7" N	131° 07' 41.5" E

(Sources: Decree of KKP No. 14, Feb 2021)

Table 2. Location Cable Route 72, Buton Island – ZEE

No	Waypoint Coordinate		No	Waypoint Coordinate	
	Latitude	Longitude		Latitude	Longitude
1	4° 48' 45.7" S	123° 47' 25.4" E	23	1° 59' 50.2" S	128° 20' 12.7" E
2	4° 43' 44.1" S	123° 56' 51.9" E	24	1° 46' 39.6" S	128° 27' 34.7" E
3	4° 29' 37.9" S	125° 07' 01.9" E	25	1° 31' 36.5" S	128° 31' 37.6" E
4	4° 25' 34.4" S	126° 02' 56.5" E	26	1° 27' 11.2" S	128° 30' 46.0" E
5	4° 20' 39.6" S	126° 20' 58.6" E	27	1° 22' 11.9" S	128° 32' 16.0" E
6	4° 14' 08.2" S	126° 30' 06.9" E	28	1° 11' 52.3" S	128° 41' 05.0" E
7	4° 09' 00.0" S	126° 32' 47.7" E	29	1° 09' 48.0" S	128° 45' 11.6" E
8	4° 05' 55.7" S	126° 37' 51.6" E	30	1° 09' 05.4" S	128° 51' 05.7" E
9	3° 58' 59.0" S	127° 06' 28.7" E	31	1° 07' 23.0" S	128° 54' 09.5" E
10	3° 56' 06.1" S	127° 15' 46.8" E	32	1° 02' 56.0" S	128° 57' 20.3" E
11	3° 51' 10.9" S	127° 20' 47.1" E	33	0° 03' 59.0" N	129° 07' 18.8" E
12	3° 44' 50.2" S	127° 23' 20.8" E	34	0° 14' 36.8" N	129° 06' 37.3" E
13	3° 34' 33.9" S	127° 24' 05.1" E	35	0° 41' 51.3" N	128° 48' 32.4" E
14	3° 26' 06.1" S	127° 23' 25.3" E	36	0° 46' 15.2" N	128° 47' 44.8" E
15	3° 19' 04.7" S	127° 24' 19.7" E	37	0° 52' 49.9" N	128° 50' 10.4" E
16	3° 16' 23.0" S	127° 25' 22.8" E	38	1° 10' 16.4" N	128° 59' 32.7" E
17	3° 12' 03.2" S	127° 29' 11.4" E	39	1° 43' 14.5" N	129° 23' 39.8" E
18	3° 10' 00.4" S	127° 32' 24.0" E	40	1° 57' 40.8" N	129° 51' 47.1" E
19	3° 09' 25.1" S	127° 35' 07.7" E	41	2° 10' 50.8" N	130° 19' 52.0" E
20	2° 55' 02.9" S	127° 34' 54.8" E	42	2° 20' 40.9" N	130° 29' 57.3" E
21	2° 43' 13.1" S	127° 38' 31.0" E	43	4° 09' 06.1" N	131° 05' 36.6" E
22	2° 24' 52.6" S	127° 52' 25.7" E			

(Sources: Decree of KKP No. 14, Feb 2021)

2.2. Research Data

The data used in this study is data in raster, vector, and tabular formats from various sources, including:

1. National bathymetry data was obtained from the Geospatial Information Agency, which was formed from the inversion of gravity anomaly data with six arc-second spatial resolution using MSL datum, which was downloaded via https://tanahair.indonesia.go.id.
2. Fault data was obtained from the Geoportal data of the Ministry of Energy and Mineral Resources of the Republic of Indonesia. This data is openly available and can be accessed via https://geoportal.esdm.go.id.
3. Seismic data from 2007 to 2022 were obtained from GeoForschungsZentrum (GFZ), Germany https://geofon.gfz-potsdam.de/waveform/archive/.
4. Administrative data and distribution of underwater volcanoes are obtained from the Geospatial Information Agency via https://tanahair.indonesia.go.id.
5. Cable corridor data is obtained based on Decree of Minister of Marine Affairs and Fisheries of the Republic of Indonesia Number 14/2021 on Undersea Pipes and/or Cable Lines.

2.3. Data processing and analysis of the result

Spatial data processing in this study uses the QGIS software version 3.22. QGIS is a user-friendly Open-Source Geographic Information System (GIS) licensed under the GNU General Public License. QGIS is an official project of the Open-Source Geospatial Foundation (OSGeo). It runs on Linux, Unix, Mac OSX, Windows, and Android and supports numerous vector, raster, and database formats and functionalities. Spatial analysis was used to calculate the slope and overlap of various data used in this study. In addition, it also uses the nearest distance analysis approach to analyze segments/corridors of Buton Island to the EEZ on fault lines and seismic points. The stages of data processing and analysis are as follows:

1. Fault and seismicity
2. The slope of the seabed
3. Seamount

2.3.1. Fault and seismicity. Geological fault data related to faults uses data from The Ministry of Energy and Mineral Resources [6], which is superimposed on the study corridor and then interprets the hazards for corridor mitigation efforts. Meanwhile, seismic information based on GFZ data is classified according to earthquake strength into five classes, namely:

- magnitude 3 - 4 with a small classification
- magnitude 4 - 5 with light classification
- magnitude 5 - 6 with medium classification
- magnitude 6 - 7 with strong classification
- magnitude > 7 with large classification

2.3.2. The slope of the seabed. The seabed slope is processed from national bathymetric data with the classification of slope values based on the modified results divided into five classes, namely:

- slope class < 5 classified as flat
- slope class < 10 is classified as slightly tilted
- slope class < 15 classified as oblique
- slope class < 20 classified as a rather steep
- slope class > 20 classified as steep

With the complex geological conditions traversed by the MPA corridor, steep geomorphological paths have the potential for landslides or sagging of marine infrastructures such as cables or pipes that can damage cables or pipes. Cable and pipe corridor paths are managed through slopes of the location that are not too steep (less than 20 degrees) to avoid landslides and sagging of cables or pipes. If conditions due to slopes cannot be avoided, it is necessary to anticipate technological interventions such as the use of anchors on cables or pipes.

2.3.3. Seamount. The potential risk of the presence of an underwater volcano that the route passes through must be carefully considered. Active underwater volcanoes can generate heat around the volcano's location and potential eruptions. Volcanoes need to be avoided to reduce the risk of damage to underwater infrastructure, especially submarine cables, so it is necessary to know the distance between the cable lines and the position of the underwater volcano.

2.4. Flowchart Research

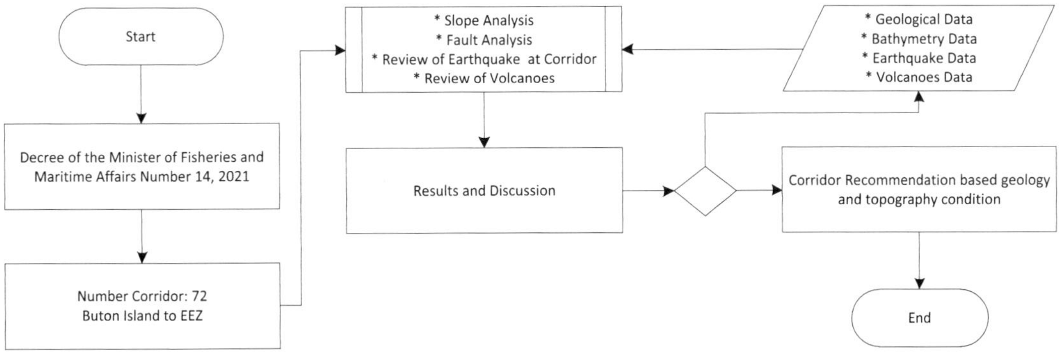

Figure 3. Flowchart of Research Geology and seabed topography research to support the implementation of Indonesia's submarine cable

3. Result and Discussion

3.1. Review of geology along the corridor

Eastern Indonesia is a tectonically active region resulting from the convergence of the Eurasia, Australia, and Pacific plates [7][8][9][10]. The KKP cable and pipeline corridor from Buton Island to EEZ is located in eastern Indonesia through the Banda Sea, Seram Sea, and the Halmahera Sea. This area is part of the Eurasian Plate, which moves relative to the Southeast and interacts with the Indies-Australian Plate, which is moving relative to the North-West, and the Pacific Plate, which is moving relative to the west. The meeting of the three plates produces a complex geological system: a subduction zone that forms a trough around the area with a depth varying between 4500 meters to 7000 meters. The Australian plate moves to the north-northeast relative to Eurasia at about 70 mm/yr.

In contrast, the Pacific motion relative to Australia is approximately 120 mm/yr in the west direction [11]. This complex geological system produces faults and folds in various directions with the dominance of North-south, West-east, and Northwest-Southeast, as well as steep geomorphology. (Figure 4 and Figure 5).

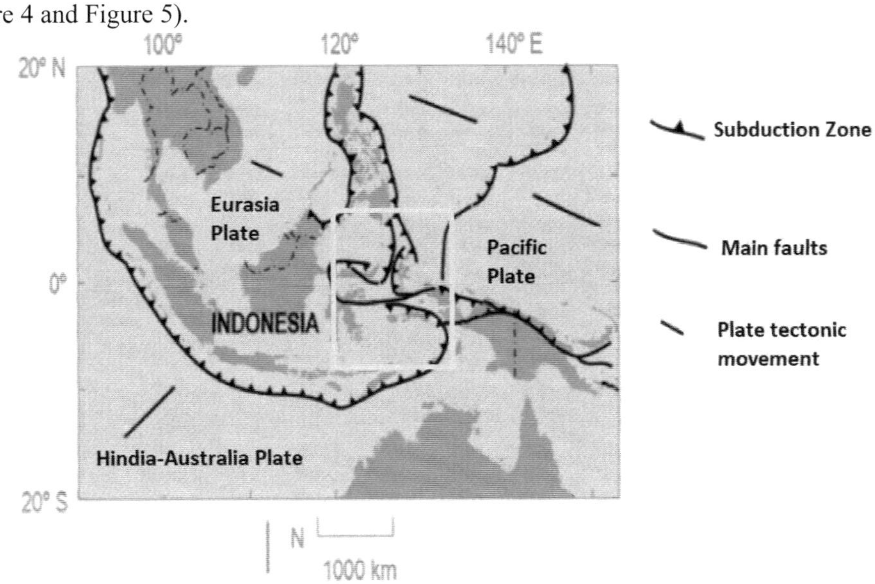

Figure 4. The Buton Island- EEZ route through the region with the geological system as a result of the collision of three plate tectonics (Yellow Box) [12]

3.1.1. Fault. The Buton Island - EEZ Corridor line passes through several faults, including the South horizontal Fault of Buru Island, which crosses the Southeast-Northwest direction. The rising fault is the North Seram Island Subduction, which runs from west to east, and the Sorong Horizontal Fault of the Obi – Bacan Islands (North), which runs from west to east. (Figure 5).

3.1.2. Seismicity. Complex geological conditions are causing high levels of seismic activity around the Seram Sea, Banda Sea, and the Halmahera Sea. The MPA corridor path from Buton Island -EEZ seems to avoid seismic activity points with seismic intensity below 5 on the Richter Scale. However, this corridor still passes through a seismic intensity point of more than 7 to the south of Buru Island. (Figure 6).

3.1.3. Seamount. The identification results from PUSHIDROSAL show that several seamounts are located around the Banda Sea through the Jalacitra II Banda expedition in July 2022, as shown in the image below. Several seamounts are close (approximately 15 - 20 km) from the trail, namely Seamount Moro Sagoro (east of Buru Island) and Seamount Gapuro Sagoro (northwest of Halmahera Island). Although the seamount is quite far, approximately 20 km from the route, it still has the potential risk of damage to underwater infrastructure (cables, pipes, etc.), although the risk that will be caused is minimal. (Figure 7).

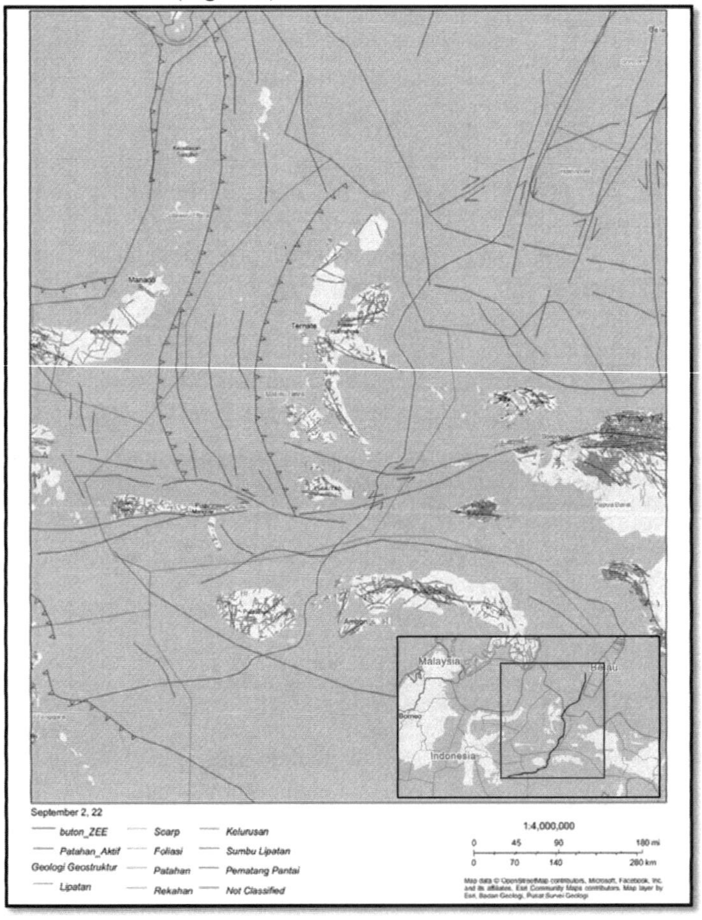

Figure 5. The Buton Island -EEZ corridor route (Blue line) cut several faults (red line). The faults as a result of the collision of three tectonic plates [6]

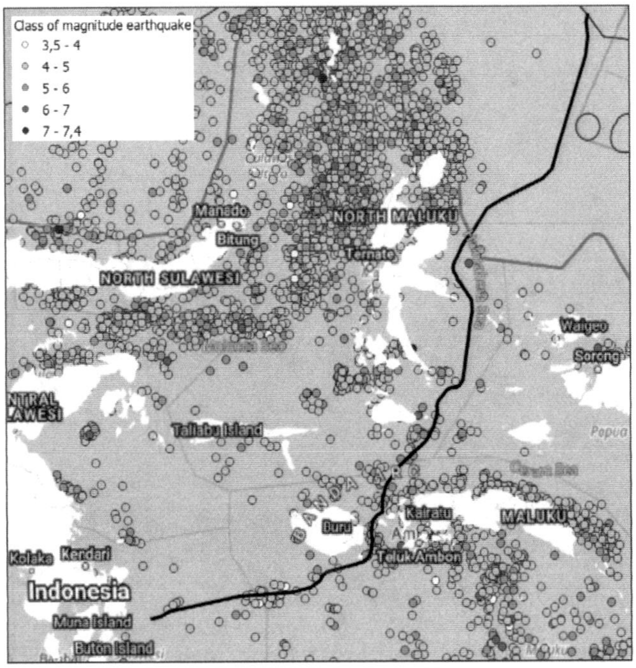

Figure 6. seismic activity on the Button Island -EEZ route [13]

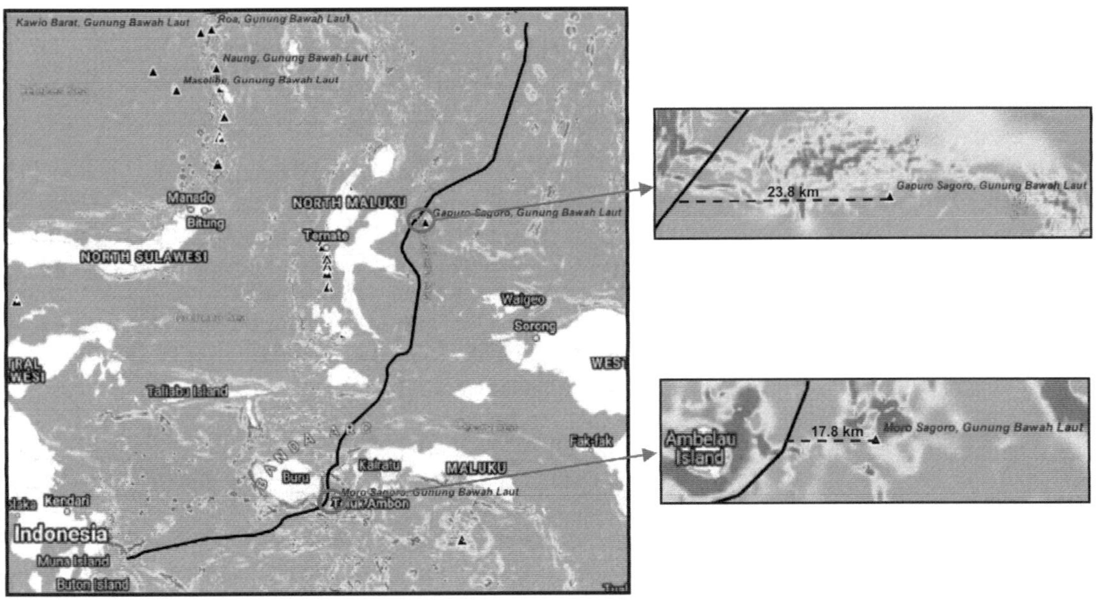

Figure 7. Position seamounts along the corridor, (a) Seamount Gapuro Sagoro, and (b) Seamount Moro Sagoro.

3.2. Review of geomorphology and Bathymetry Area

The slope processing results on the Buton Island-EEZ corridor show that the path passes through several locations that are quite steep (> 20 degrees). This condition has the potential for landslides or the deterioration of underwater infrastructure, as shown by cables or pipes to the east of the Buru and Halmahera Islands (Figure 8).

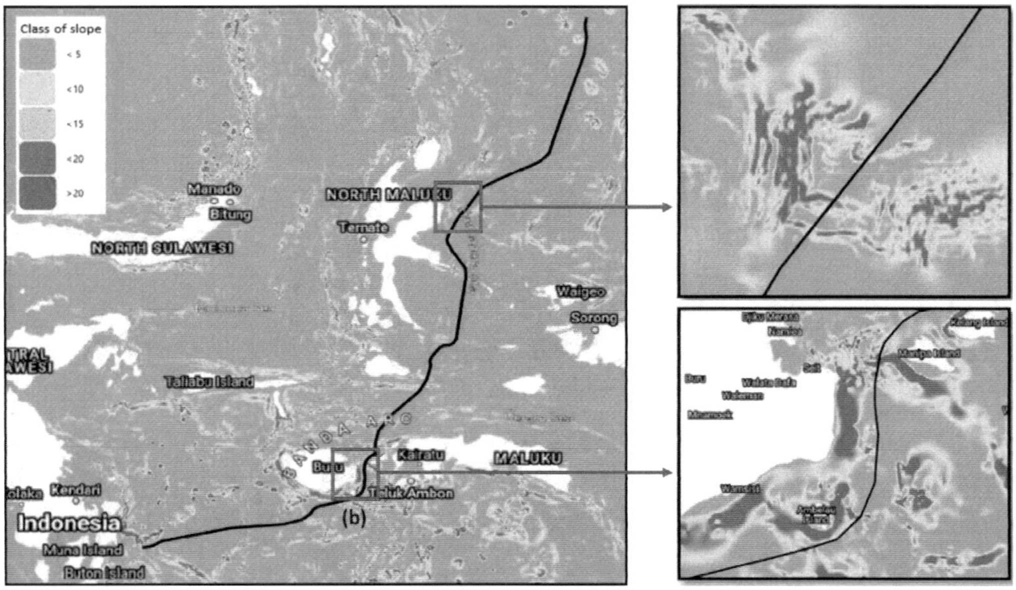

Figure 8. Buton Island - EEZ corridor (Black line) through several locations that have a steep morphology (red color) - From Batnas 2019 bathymetric data

Eastern Indonesia is a tectonically active region resulting from the convergence of the Eurasia, Australia, and Pacific plates [7][8][9][10]. The KKP cable and pipeline corridor from Buton island to EEZ is located in eastern Indonesia through the Banda Sea, Seram Sea, and the Halmahera Sea. This area is part of the Eurasian Plate, which moves relative to the Southeast and interacts with the Indies-Australian Plate, which is moving relative to the North-West, and the Pacific Plate, which is moving.

3.3. Discussion and Recommendation

From the study results on the path through the fault, it is impossible to avoid because the area is a fault zone which is quite complicated, so changing the path to avoid faults will result in a longer path. In addition, faults generally occur far below the surface of the seabed, so it is unlikely to cause direct damage to underwater infrastructures such as cables and pipes.

In addition, viewed from the slope, it appears that there is a steepness to the east of Buru Island, where the MPA corridor route passes through a location with a steepness morphology of more than 20 degrees. It is recommended that there be a slight change of course further east to avoid a slope of more than 20 degrees. To reduce the risk due to high seismic activity, it is recommended that this corridor route changes northward through the Banda basin and continues to the north of Buru Island (Figure 9).

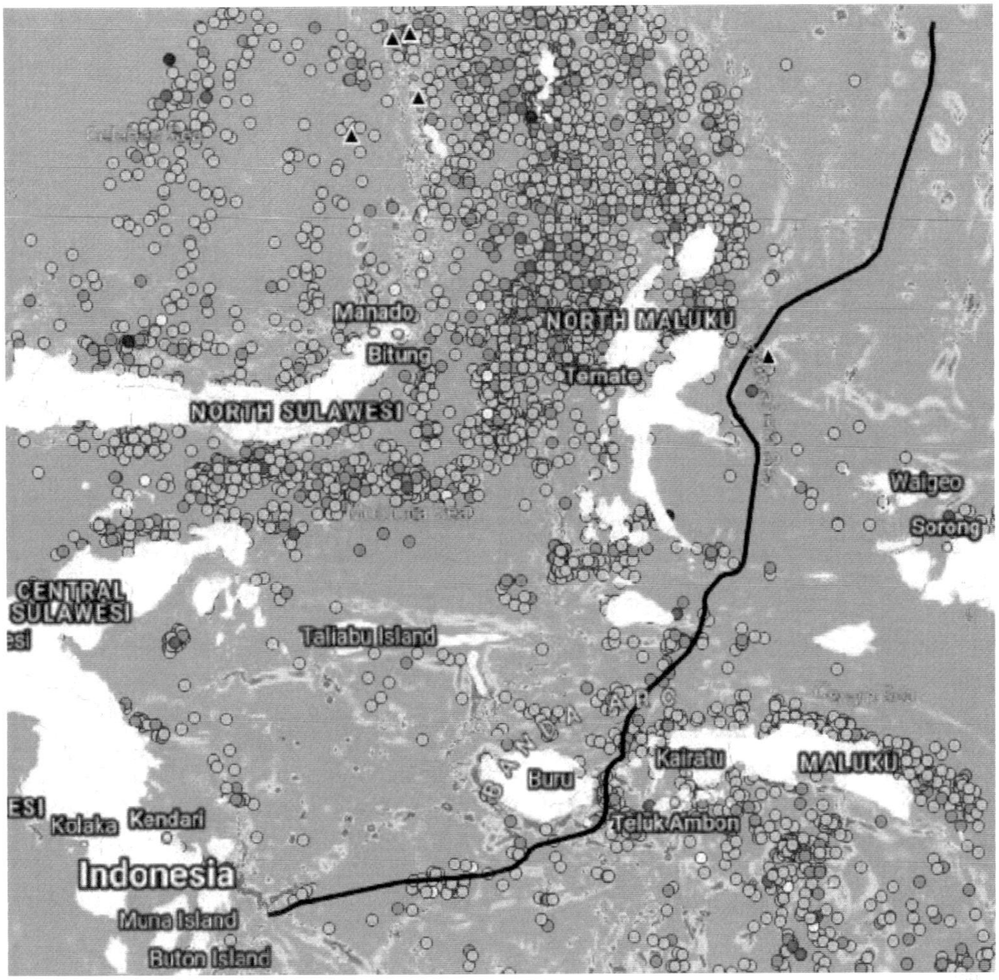

Figure 9. Recommendation changes to the route of the Button Island -EEZ (blue line)

In general, in terms of fault distribution data, the presence of volcanoes, slopes, and seismic activity of corridor 72 connecting Buton Island-EEZ is relatively safe. It is only necessary to pay attention to seamounts, namely Seamount Moro Sagoro (east side), Buru Island, and Seamount Gapuro Sagoro (northwest of Halmahera Island). For a detailed geomorphological review, it is necessary to carry out a marine survey. The existence of natural objects that make up the geomorphology of the sea was identified through marine surveys to determine the route of the cable or seabed pipeline, including sand waves, pockmarks, and shoal/reef. Sand waves are the shape of the seabed perpendicular to the direction of ocean currents with a length of hundreds of meters and a height generally between 1 and 15 m, with a gentle slope of mostly 2-5 degrees [14]. Sand waves are generally easy to change depending on the conditions of ocean currents. Thus, the cable path as much as possible to avoid sand waves. Pockmark or eruption crater is among the most common seabed geomorphologies in hydrographic survey activities. Eruption craters have various shapes and sizes. [15] divide eruption craters into six relatively common morphological classes: normal pockmarks, unit pockmarks, elongated pockmarks, eyed pockmarks, strings of pockmarks, and complex pockmarks. The typical eruption crater (normal pockmark) is a circular basin with a diameter of 10-700 m with a depth of 1-45 m. Pockmarks are unstable parts of the seabed; therefore, cable lines must avoid these objects. Shoal is a shallow ridge that appears on the surface of the water, which is covered by unconsolidated sand. Shoals need to be kept out of the way like shipwrecks.

The existence of sand waves, pockmarks, shoals, reefs, and others can cause damage to infrastructure because these objects are unstable. In addition, man-made objects such as shipwrecks, existing pipes or cables, abandon rigs, etc., can cause damage or damage to existing infrastructure, so it can cause losses. It should be avoided, but because the size of the existence of these objects is relatively small (1 meter to 1000 meters), they can only be identified through marine surveys. Therefore, in this paper, no review is carried out. By conducting a sea survey with a corridor width of 500 meters along the route, it is expected to be able to identify objects and make lane changes in the 500-meter corridor. Geological engineering can be done if it is impossible to avoid these objects, but it will result in high costs. Man-made objects can be identified through marine surveys. The types of sensors in marine surveys to identify these objects consist of Multibeam Echosounder, Side Scan Sonar, Sub Bottom Profiler, Magnetometer, and Coring.

3.4. Conclusion

Several matters related to the geological and bathymetric review of the corridor path for the installation of underwater infrastructure, both cables and pipes, have been determined through the Decree of the Minister of Maritime Affairs and Fisheries (KepMen KKP no 14 of 2021) as follows:

1. The geological condition of the corridor from Buton Island- EEZ to Los Angeles (USA) is quite complicated because the area collides with three tectonic plates, namely Eurasia, Indo-Australia, and the Pacific. This condition causes the emergence of several faults, and seismic activity in the area is very much.
2. The collision of three oceanic plates formed a trough resulting in several locations with steep geomorphology that the Buton Island -EEZ corridor traversed.
3. Although the Buton Island -EEZ corridor route was determined through a study by the national team, there are still potential risks on this route. A deeper study needs to be done to determine the more severe risks to infrastructure on this corridor to minimize potential risks or mitigate risks. Risk reduction can be in the form of changing the proposed path through changes to the attachments to the Ministerial Decree or anticipation through technological intervention if it is impossible to change the path.
4. The existence of morphology in the form of pockmarks, sand waves, shoals, and reefs cannot be identified through secondary data, and it can only be identified through marine surveys
5. Path evaluation that gives optimal results does not only consider geological and geomorphological factors.

Author contribution statement

Djunaedi Muljawan: conceptualization, analysis, writing and supervision, **Buddin Al Hakim**: analysis and writing, conceptualization, original draft preparation, review, **Nugraheni Setyaningrum**: analysis, writing and editing, review, conceptualization, **Wahyu Widodo Pandoe**: funding acquisition, supervision, **Alfi Rusdiansyah**:data collection, **Joko Widodo**: conceptualization and curation, **Trinugroho**: editing and data collection.

Acknowledgment

This research was conducted under the resources available at the National Research and Innovation Agency (BRIN). We also thank BIG, KKP, ESDM, and Pushidrosal for providing the data. Our appreciation goes to the reviewers who have provided much input for this paper and the entire BRIN's submarine cable research group, who have provided data support and discussions.

References

[1] Badan Pusat Statistika (BPS), *Telecommunication Statistics in Indonesia 2020*. 2021.
[2] Kementerian Komunikasi dan Informatika, "Siaran Pers No. 36/HM/KOMINFO/02/2022 tentang Potensial Jadi Titik Penghubung Dunia, Menteri Johnny: Optimalkan Trak dan Utilisasi Fibre Optik Indonesia," 2022.

[3] H. S. Yoon and W. B. Na, "Anchor drop tests for a submarine power-cable protector," *Mar. Technol. Soc. J.*, vol. 47, no. 3, pp. 72–80, 2013, doi: 10.4031/MTSJ.47.3.6.

[4] D. Muljawan *et al.*, "Marine Survey To Minimize the Risks of the Telecomunication Marine Calbe," *J. Rekayasa Lingkung.*, vol. 12, no. 1, pp. 71–81, 2019.

[5] L. Carter, Burnett;, Douglas, S. Drew;, and G. Marle, *Submarine cables and the oceans: connecting the world.* 2009. doi: 10.7312/clar91608-007.

[6] Kementerian Energi dan Sumber Daya Mineral, "One Map ESDM," *TIM GIS KESDM*, 2022. https://onemap.esdm.go.id/

[7] J. A. Katili, "Tectonic evolution of eastern Indonesia and its bearing on the occurrence of hydrocarbons," *Mar. Pet. Geol.*, vol. 8, no. 1, pp. 70–83, 1991, doi: 10.1016/0264-8172(91)90046-4.

[8] F. Hinschberger, J. A. Malod, J. P. Réhault, M. Villeneuve, J. Y. Royer, and S. Burhanuddin, "Late Cenozoic geodynamic evolution of eastern Indonesia," *Tectonophysics*, vol. 404, no. 1–2, pp. 91–118, 2005, doi: 10.1016/j.tecto.2005.05.005.

[9] R. Hall, "Late Jurassic-Cenozoic reconstructions of the Indonesian region and the Indian Ocean," *Tectonophysics*, vol. 570–571, pp. 1–41, 2012, doi: 10.1016/j.tecto.2012.04.021.

[10] A. Patria, H. Tsutsumi, and D. H. Natawidjaja, "Active fault mapping in the onshore northern Banda Arc, Indonesia: Implications for active tectonics and seismic potential," *J. Asian Earth Sci.*, vol. 218, no. April, p. 104881, 2021, doi: 10.1016/j.jseaes.2021.104881.

[11] C. DeMets, R. G. Gordon, and D. F. Argus, "Geologically current plate motions," *Geophys. J. Int.*, vol. 181, no. 1, pp. 1–80, 2010, doi: 10.1111/j.1365-246X.2009.04491.x.

[12] H. Latief *et al.*, *Air Turun Naik Di Tiga Negeri: Mengingat Tsunami Ambon 1950*, no. January. Indian Ocean Tsunami Information Center (IOTIC), UNESCO/IOC, UNESCO, 2016.

[13] Geofon GFZ, "GFZ Seismological Data Archive," *GFZ German Research Centre for Geosciences*, 2022. https://geofon.gfz-potsdam.de/waveform/archive/

[14]. .. Andreas C.W Thaiënne A.G.P. van Dijk, "Sand Wave: Small sand waves and sand ripples are common in sandy areas of the banks." 2012.

[15] M. Hovland, J. V. Gardner, and A. G. Judd, "The significance of pockmarks to understanding fluid flow processes and geohazards," *Geofluids*, vol. 2, no. 2, pp. 127–136, 2002, doi: 10.1046/j.1468-8123.2002.00028.x.

IOP Conf. Series: Earth and Environmental Science 1166 (2023) 012023 doi:10.1088/1755-1315/1166/1/012023

Performance Evaluation Of Permanent Magnet Synchronous Generator (PMSG) On Tidal Power Generation Optimization

C R Handoko[1] , Mukhtasor[1] and E S Koenhardono[2]

[1] Dept. of Ocean Engineering, Sepuluh Nopember Institute of Technology, Indonesia
[2] Dept. of Marine Engineering, Sepuluh Nopember Institute of Technology, Indonesia

E-mail: catur.handoko@ppns.ac.id

Abstract
In a tidal current energy generation system, optimization of power generation can be done through turbine design, system design, and control of mechanical transmission from the turbine to the generator, as well as from the aspect of electronic generator output control. This research aims to analyze the performance of the PMSG generator from a prototype of the tidal current energy power plant. Turbine rotation is simulated using an ac motor equipped with an ac drive that supports separate speed and torque control. The hydraulic transmission system consists of a pump and a hydraulic motor that transfers the rotation of the turbine to the generator, where the power output is observed with variations in turbine speed and torque. The results indicate that the transmission system has worked well in transmitting the turbine's mechanical power and increasing the rotational speed. The decrease in speed with increasing load and low average efficiency (less than 20%) occurs, mainly due to the characteristics of the generator being tested. The generator has a large rated torque requirement of 32 Nm, which is much higher than the main drive capacity of the turbine simulator. The design of a tidal energy conversion system, in which the tidal current speed fluctuates, requires careful selection of the generator, not only from the aspect of power capacity and rated rotational speed, but also from the initial torque value and required rated torque.

1. Introduction

Tidal current energy has attracted a considerable amount of attention as one of the maritime energy sources around the world. As one of the world's largest archipelagic nations, Indonesia has several of straits with a total estimated tidal current energy potential of 17.9 GW [1]. Several studies have researched and modeled resource potential in several locations, such as the Alas Strait [2], Patinti Strait [3] and the Larantuka Strait [4]. In the field implementation, the Indonesian Hydrodynamics Laboratory team has field tested a prototype 2 kW ocean current turbine with a turbine diameter of 2 m in the

Larantuka Strait [5]. One of the problems inhibiting the competitiveness of renewable energy is the intermittent nature due to the nature of the energy source. In the context of ocean current energy, tidal fluctuations cause the average power production to be far below the design rated power. To increase this power, optimization is the key so that ocean current energy can compete in the market.

The technology used in the tidal power generation system is similar to the technology used in the wind power generation system. Included in this is the type of generator used[6].

First, the technology used in a renewable energy conversion system is a permanent magnet synchronous generator (PMSG). In this system, the excitation of the electric field generated uses a permanent magnet so that no access to the rotor is required like a synchronous generator in general. With the structure of using a larger number of field poles, the need for rotating speed is much reduced and even the gearbox can be reduced or eliminated. This type of generator also offers full-range speed control. This machine has the characteristics of low maintenance and high power density.[7].

Second, generator that is also widely used for wind turbines is the dual induction generator (DFIG) because it offers the great advantage that the power generated by the engine can be connected directly to the grid. The power electronics converter used has a smaller scale which is connected to the rotor circuit for control purposes. However, the speed range of this generator is limited to ±30% of the nominal speed, so a gearbox must be used to connect the turbine to the generator[6]. In this system, excitation is used or the provision of a magnetic field current to the engine rotor using slip rings. All these mechanical parts decrease system reliability and increase maintenance requirements, which are considered key features for marine turbine development, which makes using these generators in tidal energy systems in the long term expensive.[8].

Most of the work on tidal energy conversion focuses on the use of PMSG. [9] design and build a tested PMSG prototype with three different fixed pitch horizontal axis turbines, and [10] designed and tested a 600 kW horizontal axis turbine connected to the PMSG via a low transmission ratio gearbox. Another study, [11], conducted a comparative study of tidal current turbines driving DFIG and PMSG. Stability testing shows the advantages of using PMSG over DFIG.

In a direct-drive transmission system, where the turbine and generator are connected directly without a gearbox, a generator with a very low speed is required. This generator must be specially designed by multiplying the poles and a strong permanent magnet. OpenHydro tidal turbine is an example of the development of a special low-speed generator for direct-drive tidal turbines.

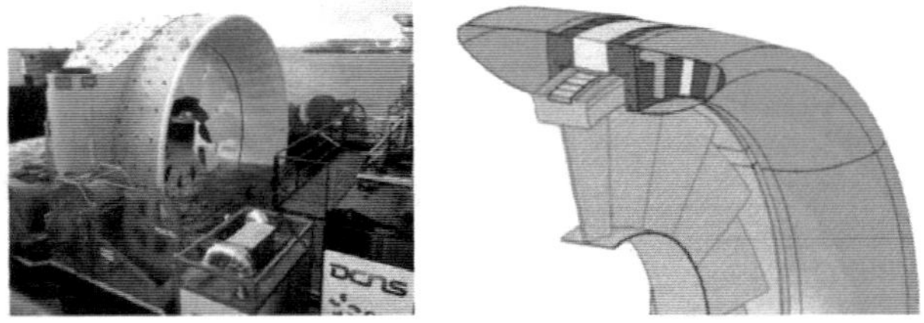

(a) (b)

Figure 1 (a) OpenHydro Tidal Turbine System[12]
(b) Arrangement of Permanent Magnets on the rotor[13]

2. Prototype Overview

The prototype of this ocean current power generation system consists of several parts, namely a turbine simulator, a hydraulic transmission system and a generator.

IOP Conf. Series: Earth and Environmental Science 1166 (2023) 012023 doi:10.1088/1755-1315/1166/1/012023

2.1 Turbine Simulator

In this test, an electric motor is used as a substitute for a turbine or turbine equipped with a gearbox (geared turbine). The driving motor used must have the ability to simulate turbine rotation, which is rotating with a certain rotational speed and torque. For this reason, the options available for electric motors are dc motors with separate amplifiers, hydraulic motors, and 3-phase ac electric motors with variable drives that support separate speed and torque settings.

Separate amplifier DC electric motors are now rarely found on the market. While this type of electric motor in the laboratory is generally quite old and has a low capacity (only 0.75 kW maximum or less). This motor must also be supported by two DC power supplies each for armature current and magnetic field current. Meanwhile, a variable displacement hydraulic motor can be an alternative, even at the most expensive price.

The next option is to use an AC motor with VSD (Variable Speed Drive) that supports torque control. So finally here, the test uses the Mitsubishi FR-A700 VSD with specifications of 1.5kW (2HP) 400V 3 phase, Speed Control Range: 0.2Hz - 400Hz with the support of a technology called vector control or Field Oriented Control (FOC). Table 3 shows the specifications of the 3-phase motor used in this test.

Table 1 Parameters of 3-Phase Induction Motor

Description	Symbol	Value	Units
Power	P	1.5	kW
Speed	n	1480	rpm
Class	C	C	-
Voltage	V	220/380	Volt
Cos Phi	φ	0.78	-
Efficiency	η	78.5%	-
Current	I	6.4/3.7	A
IP	IP	55	-

2.2 Hydraulic transmission system

This transmission system, also known as Hydrostatic Transmission (HST), transmits the rotation of the turbine to the generator. This system consists of a hydraulic pump and hydraulic motor and is equipped with a relief valve and a check valve. It is planned that the rotation of the hydraulic motor is greater than about three times the speed of the turbine or pump shaft. So here, however, this is adjusted to the availability of hydraulic system components available in the market.

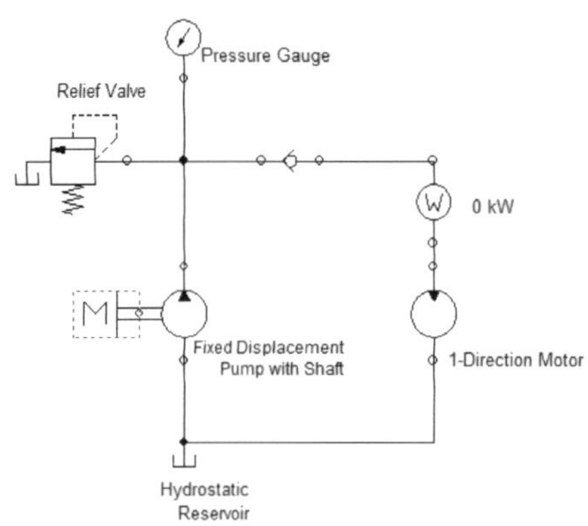

Figure 2
Hydraulic transmission system schematic

Table 2 Hydraulic Transmission System Parameters

Description	Symbol	Value	Units
Displacement of Pump	Dp	50	cm3
Displacement of Motor	Dm	20	cm3
Oil Tank Capacity	Vt	5	litre

2.3 Permanent Magnet Synchronous Generator (PMSG)

In this study, a permanent magnet synchronous generator (PMSG) which is available on the market is used. The following figure shows a selection of available generators with their technical specification data. This prototype takes a generator with a rated power capacity of 600 Watt.

Table 3 Specifications of PMSG

Model	M-600
Rated Power	600W
Max Power	650 W
Rated Speed	750 rpm
Rated Voltage	24 V
Start Torque	3.7
Rated Torque	32

2.4 Torque sensor

In this research, a torsimeter is also used which is installed between the turbine simulator and the hydraulic pump. Torsimeter is needed to see the amount of mechanical power, torque, and speed generated by the turbine simulator to drive the generator through the hydraulic transmission system.

And the whole prototype of the stranded generator system is shown in the following figure

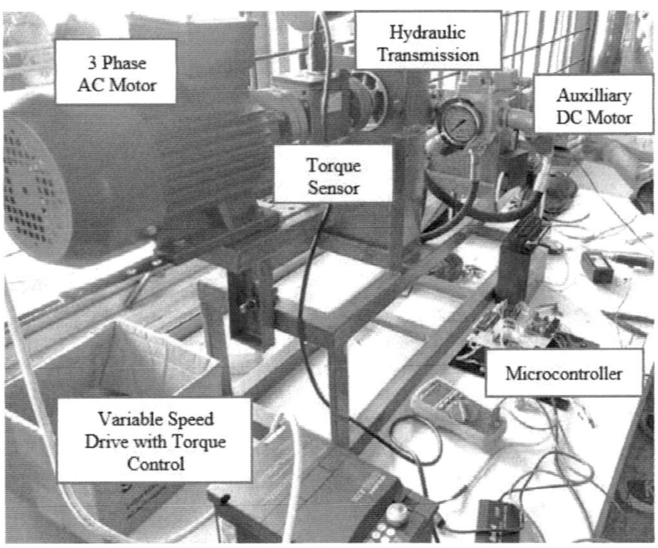

Figure 3 Generating System Prototype

Figure 4 Torque Sensor

3. Tests and Results

The following is a description of the process and the results of the tests carried out on the PLTAL system prototype. Before testing, so that the equipment avoids problems such as overcurrent and over-voltage which can cause overheating and damage, an initial calculation is carried out on the specifications of each component in this prototype system. After this stage, it is continued with the first test, namely the no-load test. From this test, it will be seen whether the turbine rotation can be transmitted properly to the generator shaft. Also to see if the generator rotation is greater than the turbine shaft rotation as planned. In the second test, the generator will be loaded with incandescent lamps, gradually to see changes in voltage, power, and also the rotational speed of the generator.

3.1 Torque Setting

Calculation of the power of a 3-phase induction motor and a variable speed drive (turbine simulator) is needed to determine the amount of torque that can be generated according to the motor's power capacity. With a capacity of 1.5 kW and a nominal speed of 1400 rpm. Then the magnitude of the full load torque can be calculated as follows:

$$P = T.\omega \quad (1)$$

$$P = T.\frac{2\pi}{60}.n \quad (2)$$

Where:
P = Power (Watts)
T = Torque (Nm)
ω= angular velocity (rad/s)
N = Rotating speed (rpm)

With the above formula, the full load motor torque can be calculated:

$$T = \frac{1500x60}{2x\ 3.14x1500} = 9.55\ N.m$$

And if the motor efficiency factor is taken into account, which is stated on the motor nameplate at 0.785, the full load torque is obtained at 7.5 Nm. In the torque setting using VSD, the setting input is % of full load torque, then it has been calculated and measured through a torque meter, the results of which can be seen in Fig. the following table.

Table 4 Torque Setting Table for Testing

No.	% Full Load	Calculation Result Torque (Nm)	Torque Result Setting VSD and Measurement (Nm)
1	50%	3.75	3.9
2	75%	5.6	5.9
3	100%	7.50	7.3
4	125%	9.37	9.6

From the table above, it can be seen that the results of calculations and measurements are relatively not too far apart. The VSD setting can be done up to a maximum of 400%, but in testing the torque boosting at around 170% has approached the maximum motor current limit and has caused overheating of the motor body. For this reason, in this test, the addition of torque is limited to 125% only.

The next most important part that must be considered in the generator specifications, in addition to power capacity, rotational speed, and the output voltage is the nominal torque (rated torque). This torque will determine how much power (in Watts) is needed to drive the generator to reach the rated speed and rated power.

In this generator, the rated torque is 32 Nm. So when calculated by equation (2), the required power will be:

$$P = \frac{2\pi}{60} x32x750 = 2512\ Watt$$

$$P = 3.37\ HP$$

So with an induction motor and VSD with a capacity of only 2 HP, the nominal power (power rated) cannot be achieved. For this reason, the maximum power measured in this test will be limited by considering the limitations calculated above.

3.2 No-load Test

The no-load test is a test on the condition of the generator being unloaded. In this test, the driving motor (turbine simulator) is rotated with a frequency of 10 Hz, which means the motor rotates at approximately 300 rpm when it is not connected to the mechanical load. In a real tidal energy conversion system, this can mean the speed of the turbine which has been accelerated by the gearbox (geared turbine). This speed, which is the minimum rotational speed, is estimated that the hydraulic transmission system has sufficient pressure to produce flow, which then flows to produce a strong enough torque on the hydraulic motor. The results are shown in the following table.

Table 5 No-load Test

No.	Torque (Nm)	Turbine Speed (RPM)	Generator Speed (RPM)	Voltage (V)
1	3.9	158	461	19.2
2	5.9	230	654	21.26
3	7.3	272	768	23.7
4	9.6	268	758	38.46

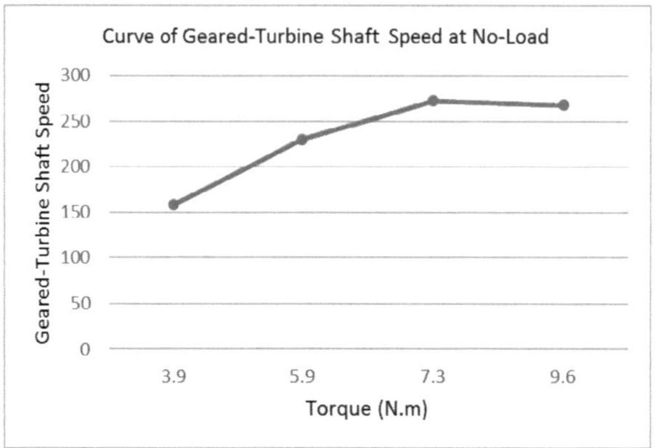

Figure 5 Turbine Shaft Speed at No-Load Test

From the above results, it appears that the hydraulic transmission system and the generator without load, produce a turbine shaft rotational speed that is proportional to the applied torque. And the generator shaft speed is 2.85 times the average turbine shaft speed. And the generator output voltage level increases with increasing rotation. Figure 5 also shows that the base-speed of the turbine, has no effect on the final speed of the turbine due to the load. The speed for each torque is almost the same. This means that turbine torque has more influence on other quantities.

3.3 Test with load

In testing with loads, the generator is loaded with a purely resistive load, in the form of a row of incandescent lamps. In this test, the turbine simulator is rotated with 3 different rotational base-speeds, to accommodate the relatively low, medium, and high speeds of the ocean current turbine. For this reason, the driving motor is rotated with a frequency of 10 Hz, 20 Hz, and 30 Hz, with the torque being increased gradually. The results are shown in the following tables.

Table 6 Tests with a load at a frequency of 10 Hz

No.	Torque (Nm)	Mechanical Power (W)	Turbine Shaft Speed (RPM)	Generator Shaft Speed (RPM)	Load Current (A)	Load Voltage (V)	Electric Power (W)
1	3.9	46	117	391	0.35	12.7	4.4
2	5.9	94	156	408	0.9	11.5	10.7
3	7.3	175	190	474	1.67	13.13	21.9
4	9.6	213	212	527	2.07	14.16	29.3

Table 7 Tests with a load at a frequency of 20 Hz

No.	Torque (Nm)	Mechanical Power (W)	Turbine Shaft Speed (RPM)	Generator Shaft Speed (RPM)	Load Current (A)	Load Voltage (V)	Electric Power (W)
1	3.9	47	120	323	0.37	8.99	3.3
2	5.9	97	161	420	0.96	11.93	11.4
3	7.3	153	188	482	1.55	15.2	23.5
4	9.6	216	215	530	20.57	20.57	41.7

Table 8 Tests with a load at a frequency of 30 Hz

No.	Torque (Nm)	Mechanical Power (W)	Turbine Shaft Speed (RPM)	Generator Shaft Speed (RPM)	Load Current (A)	Load Voltage (V)	Electric Power (W)
1	3.9	47	119	322	0.35	8.6	3
2	5.9	79	154	331	1.1	9.76	10.8
3	7.3	113	176	399	1.77	12	21.2
4	9.6	194	205	528	2.18	18.03	39.3

And for each base-speed, figure 6 plots the turbine speed as the function of the torque values. It also shows that the base-speed of the turbine, has no effect on the final speed of the turbine due to the load. The speed for each torque is almost the same. This means that turbine torque has more influence on other quantities.

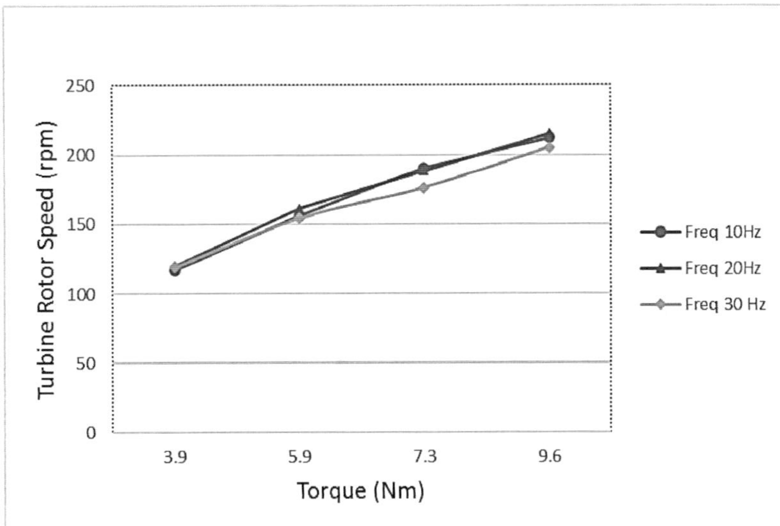

Figure 6 Turbine rotor speed for each base speed

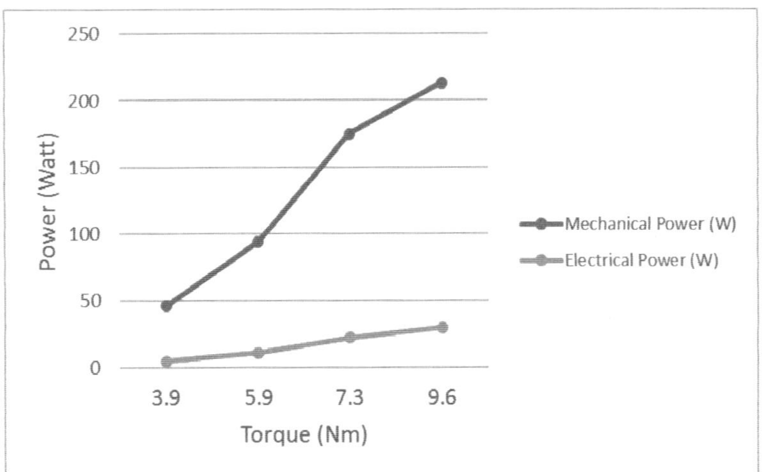

Figure 7 Mechanical Power and Electric Power

The tables above also show that the speed of the generator shaft immediately drops as the load increases. The power generated is relatively the same. This shows that the torque generated by the turbine simulator determines the power output generated by the generator. While the rotational speed of the hydraulic motor is relatively the same as the turbine shaft speed changes. By recalling the turbine specifications where the rated torque required to produce the rated power is up to 32 Nm, it is reasonable to suspect that the cause of the drop in the rotational speed of the generator shaft and also the turbine shaft is the inability of the system to provide the required torque for the generator.

Figure 7 is also indicated by the average efficiency of the electrical power generated by the generator, compared to the mechanical power generated by the turbine. The average efficiency value of the generator is measured at only 14%.

4. Discussion

The test results of the tidal energy conversion prototype above show several things related to the performance of the system and in particular the permanent magnet synchronous generator it drives. First, in the no-load test, the hydraulic transmission system worked well, transmitting power from the turbine to the generator. The rotational speed of the generator shaft increased by an average of 2.8 times the turbine shaft speed, and there was no significant speed drop.

A turbine simulator, in this case, a 3-phase induction motor with torque control VSD can simulate not only turbine rotation speed, but also torque. Through the torque control feature, constant torque can be generated to simulate turbine torque.

The generator used in this experiment lacks efficient loading characteristics. To produce nominal power, it takes a very large torque, which is 32 Nm (from the generator datasheet). From the calculation of power, torque, and rotational speed, it takes a large amount of power compared to the generator output power. For further research, it is necessary to select a generator with better loading characteristics.

The torque generated by the turbine simulator determines the power output generated by the generator. The initial speed of the different turbines immediately drops and produces relatively the same rotational speed and relatively the same power.

5. Conclusion

From the testing and analysis of the data above, the following conclusions can be drawn:

- An electrical system with a choice of hydraulic transmission configurations and a low-speed PMSG generator has been built and tested to simulate the generation of ocean currents, especially at low speeds.
- The torque produced by the turbine will determine the power output of the generator. However, a minimum speed of 300 rpm is required to allow the hydraulic system to operate to produce sufficient pressure and flow to produce a torque on the hydraulic motor and generator side. For low-speed ocean currents, this minimum speed requirement can be met by adding a gearbox or using a marine wear turbine that is equipped with a gearbox (geared turbine).
- For the study of power optimization, the capacity of tidal turbine simulator should be increase to cover the range of the generator full load capacity or using the generator with the better rated torque characteristic.

Reference

[1] Mukhtasor, 2014 *Mengenal Energi Laut* 1st ed. Surabaya.
[2] Blunden L S Bahaj A S and Aziz N S, Jan. 2013 Tidal current power for Indonesia? An initial resource estimation for the Alas Strait *Renew. Energy* **49** p. 137–142.
[3] Novico F Sudjono E H Egon A Menier D Methew M and Pratama M B, 2021 Tidal current energy resources assessment in the patinti strait, indonesia *Int. J. Renew. Energy Dev.* **10**, 3 p.

517–525.
[4] Firdaus A M Houlsby G T and Adcock T A A, 2020 Tidal energy resource in Larantuka Strait, Indonesia *Proc. Inst. Civ. Eng. - Energy* **173**, 2 p. 81–92.
[5] Erwandi A K Sasoko P Rina W B Marta E and Rahuna D, 2011 Vertical axis marine current turbine development in indonesian hydrodynamic laboratory-surabaya for tidal power plant in *International conference and exhibition on sustainable energy and advanced materials (ICE SEAM 2011). Solo, Indonesia* p. 15–23.
[6] Amirouche E Ghedamsi K and Aouzellag D, 2019 *Tidal power systems: A review* .
[7] Jain A Shankar S and Vanitha V, 2018 Power generation using Permanent Magnet Synchronous Generator (PMSG) based variable speed wind energy conversion system (WECS): An overview *J. Green Eng.* **7**, 4 p. 477–504.
[8] Wani F Dong J and Polinder H, 2020 Tidal Turbine Generators *Adv. Model. Control Wind Hydrog.* p. 1–15.
[9] Yuen K *et al.*, Feb. 2009 Matching a Permanent Magnet Synchronous Generator to a Fixed Pitch Vertical Axis Turbine for Marine Current Energy Conversion *Ocean. Eng. IEEE J.* **34** p. 24–31.
[10] Li Y Liu H Lin Y Li W and Gu Y, 2019 Design and test of a 600-kW horizontal-axis tidal current turbine *Energy* **182** p. 177–186.
[11] Aly H, Dec. 2016 Dynamic modeling and control of the tidal current turbine using DFIG and DDPMSG for power system stability analysis *Int. J. Electr. Power Energy Syst.* **83** p. 525–540.
[12] Patel S, First Turbine Deployed at French Tidal Power Farm. [Online]. Available: https://www.powermag.com/first-turbine-deployed-french-tidal-power-farm/. [Accessed: 27-Oct-2021].
[13] Baker N Hodge E Cawthorne S and Spooner E, 2014 *Modelling of the generator for OpenHydro's tidal energy system* **2014** .

Spiral analysis of vertical currents in Lamong bay

Mahmud Mustain[1*] and Sujantoko[1*]

[1]Department of Ocean Engineering, Institut Teknologi Sepuluh Nopember, Indonesia
*Corresponding author: mmustain@oe.its.ac.id, sujantoko@oe.its.ac.id

Abstract. Coastal management is interested in restoring the natural conditions of increasing the comfort of the community after being degraded due to reclamation, as in Lamong Bay. Therefore, it is necessary to know the character of the current pattern vertically in the study area. This paper aims to find out the current vertical pattern by referring to the shape of the Ekman Spiral. This concept is based on the implementation of surveys and the presentation of currents with hydrodynamic analysis in Lamong Bay. The survey has been carried out at five points, presented with measurement results, and analyzed to the character of the current spiral pattern to depth. Tidal measurement and modeling are also carried out to determine the validation of current measurements synchronized with their position and time. The results obtained in the form of a spiral are not exactly according to the shape of the Ekman Spiral theory, namely on the Southern Hemisphere Earth rotating towards the left. The speed is getting lower and smaller i.e. getting smaller and smaller to the depths (average of CM-1: 0.24 m/s for 0.2D, 0.21 m/s for 0.6D, and 0.18 m/s for 0.8D). Validation of tidal and current measurements from measurements and models qualitatively obtained minimal errors.

Keywords: Current, Spiral-Ekman, Lamong bay, Tide

1. Introduction

Lamong bay is located on the border between the city of Surabaya and Gresik (Figure 1). It is a multipurpose terminal area flanked by two ports belonging to PT. Pelabuhan Indonesia III (Persero), namely the Gresik port in the west and the main port of Tanjung Perak in the east (Lamong bay terminal). Over time, the activity at Lamong terminal has become quite congested and very busy, making it one of the busiest ports in Indonesia. For supporting activities in Lamong bay terminal, PT. Pelabuhan Indonesia III carried out the construction and development of the Teluk Lamong terminal in Surabaya. However, the development at Teluk Lamong Terminal has a negative impact resulting in changes in current patterns and the potential for sediment distribution in the Lamong Bay area. It will hamper activities in the Lamong bay area.

Sediment generally comes from the earth's crust, which is transported through hydrological processes from one place to another vertically and horizontally. Sediment is a mixture of various substrates with different fractions, such as gravel, sand, silt, and clay. Most watersheds carry mud caused by natural erosion from rivers, and almost all sediment content will continue to increase with erosion from agricultural, forestry, construction, and mining soils. The movement of sediment in the waters is influenced by the current's speed and the sediment's grain size. The larger the grain size of the sediment requires a more incredible current velocity to transport the sediment. Thus, the direction and speed of the currents in water will affect the distribution pattern of sediment transport in these waters. In general, the coarse-sized sediment material (bedload) will settle not far from the source, while the fine-sized sediment material (TSS) will be deposited further from the source [1]. Some of the negative impacts resulting from the presence of sedimentation in the coastal area of Lamong Bay [2] are that it can interfere with the shipping activities of the Lamong bay terminal, and it can also hamper shipping lanes on the Lamong River itself. In addition, activities in Lamong bay will have an impact on the development of tourism around it, namely Kenjeran beach tourism [3,4], mangrove forest tourism [5,6], tourism on Bawean Island [7], and others.

This paper aims to determine the pattern of horizontal and vertical currents at the current sampling measurement point in the Lamong bay area. The results of current measurements at each predetermined depth are plotted at each measurement of the sampling position.

Figure 1. Lamong bay

2. Literature and Method Review

In oceanographic theory, it is stated that the Ekman spiral on the surface due to wind shear force is a balance of three components of force, namely the shear stress force on the surface, the Coriolis force, and the Drag force. This balance of forces causes differences in ocean currents. In the ocean, there are two places where the Ekman spiral can be observed. At sea level, where the required shear stress is equal to the wind stress force, the shear stress force is created by friction with the seabed. Norwegian oceanographer Nansen first observed this phenomenon on the surface during his expedition. He noticed that the iceberg was not drifting in the same direction as the wind. Later his student, the Swedish oceanographer Ekman, described this process physically [8].

The solution to this spiral Ekman results from the equation of shear stress in contact with the base known as flow by the base (bottom). So this spiral Ekman formula is a particular case for air-water interaction. The equations of motion for a geostrophic flow, which are the same as stated in the bottom spiral section, can be reduced to [9]:

$$-f(v - \bar{v}) = v_E \frac{\partial^2 v}{\partial z^2} \quad (1)$$

$$f(u - \bar{u}) = v_E \frac{\partial^2 v}{\partial z^2} \quad (2)$$

IOP Conf. Series: Earth and Environmental Science 1166 (2023) 012024 doi:10.1088/1755-1315/1166/1/012024

Boundary condition in the case:

Surface (z=0): $\rho_0 v_E \frac{\partial u}{\partial z} = \tau_x$ and $\rho_0 v_E \frac{\partial v}{\partial z} = \tau_y$

Towards interior ($z \rightarrow -\infty$): $u = \bar{u}$, $v = \bar{v}$

The Solution can be determined:

$$u = \bar{u} + \frac{\sqrt{2}}{\rho_0 f d} e^{z/d} \left[\tau_x \cos\left(\frac{z}{d} - \frac{\pi}{4}\right) - \tau_y \sin\left(\frac{z}{d} - \frac{\pi}{4}\right) \right] \quad (3)$$

$$v = \bar{v} + \frac{\sqrt{2}}{\rho_0 f d} e^{z/d} \left[\tau_x \cos\left(\frac{z}{d} - \frac{\pi}{4}\right) + \tau_y \cos\left(\frac{z}{d} - \frac{\pi}{4}\right) \right] \quad (4)$$

The wind-driven flow component is inversely proportional to the thickness of the Ekman layer. Accordingly, if the thickness of the coating is small due to the small viscosity of the liquid, for example, this component can be substantial. On the surface, flow occurs 45 degrees to the right in the Northern Hemisphere and 45 degrees to the left in the Southern Hemisphere in the wind direction. In the case of the lower Ekman spiral, this is the opposite. Figure 2 gives the illustration of the Ekman Spiral.

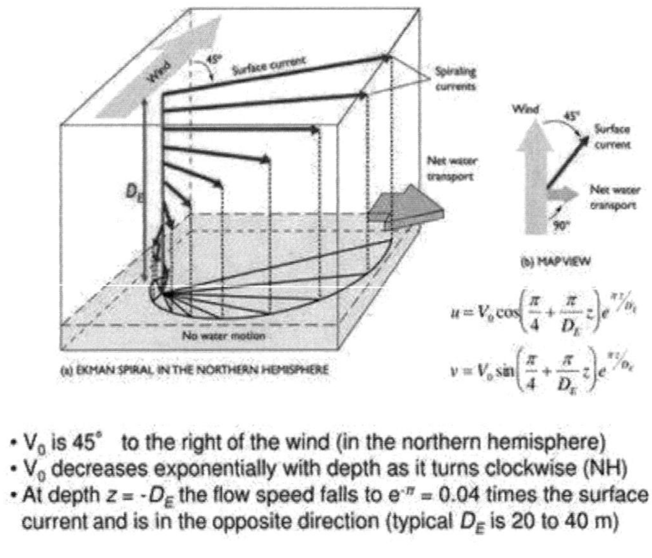

Figure 2. Illustration of Ekman spiral [10]

3. Current Measurement Result

This paper is the second year of research, continuing the first research in 2021. In the first year of research, bathymetry and tidal surveys have been successfully carried out, so this year it is continued with the creation of a current model in Lamong bay. So, bathymetric data of tidal data were used in this study. Current measurements are carried out as many as 5 points with an interval of every 1 hour at three depths. The measurements of the three depths are; 0.2D, 0.6D, and 0.8D. The total length of this measurement is carried out for 24 hours. The equipment used in the current measurement is; Current flowatch FL-03, handheld GPS, and Boat.

A. Implementation and Results of The Flow Survey

IOP Conf. Series: Earth and Environmental Science 1166 (2023) 012024 doi:10.1088/1755-1315/1166/1/012024

Five current measurement positions are recorded with adjacent coordinates, namely CM-1 – CM-5 (Table 1). Figure 3 provides a map of the location of the current measurement. Measurements are carried out by measuring the point's depth, then inserting the current meter into the waters gradually, at depths of 0.2D, 0.6D, and 0.8D. Figure 3 provides measurement conditions for 24 hours at intervals of 1 hour. The speed and direction are recorded on the current meter at the same time. This current speed recording is carried out at the location point that has been plotted in the depth of the handheld GPS and is carried out every 1-hour interval for 24 consecutive hours. The results of the recapitulation of overall current measurements are given in table 2. This current measurement was carried out on June 18, 2021, at 08:00 WIB until June 19, 2021, at 07:00 WIB. Figure 4 provides the conditions when measuring the 24-hour current (Figure 4). Table 2 shows current measurements' results that include direction and speed at each depth point.

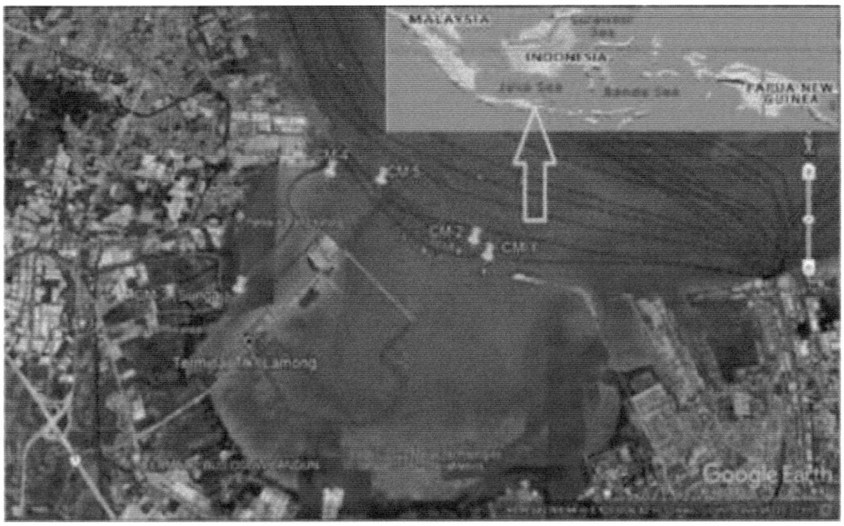

Figure 3. Current measurement location

Figure 4. Current Measurement over 24 hours

Table 1. Coordinates of the location of the measurement of current

Point Measurement	Geographic Location		Depth LWS (m)
	x	y	
CM-1	687479.00 mE	9204385.00 mS	-6
CM-2	687283.00 mE	9204610.00 mS	-5
CM-3	683969.00 mE	9203955.00 mS	-6
CM-4	685223.00 mE	9205555.00 mS	-4
CM-5	685963.00 mE	9205442.00 mS	-4

Table 2. Results of current measurement

Location	Parameter	0.2 D		0.6 D		0.8 D	
		Speed	Direction	Speed	Direction	Speed	Direction
CM-1	Max	0.42	248	0.40	248	0.40	248
	Min	0.08	31	0.05	32	0.04	36
	Avg	0.24	-	0.21	-	0.18	-
CM-2	Max	0.34	45	0.40	246	0.40	246
	Min	0.09	58	0.05	50	0.06	52
	Avg	0.23	-	0.22	-	0.22	-
CM-3	Max	1.12	55	1.15	56	0.92	44.3
	Min	0.09	53	0.10	60	0.10	57.7
	Avg	0.57	-	0.57	-	0.51	-
CM-4	Max	0.56	74	0.48	76	0.44	710
	Min	0.09	197.5	0.09	202	0.09	208
	Avg	0.28	-	0.26	-	0.24	-
CM-5	Max	0.40	122	0.38	124	0.37	126
	Min	0.09	117	0.10	126	0.08	123
	Avg	0.24	-	0.24	-	0.21	-

B. Current analysis

Analysis of the results of current measurements is carried out by describing the movement of the speed and direction of the current at each depth measured every hour. The current analysis is also carried out by showing the relationship between currents and tides. Figure 5 gives the current measurement position when the tidal position is in the neap condition. When we analyze that the neap tide is the condition of the lowest tidal range, which is only about one meter, as seen in figure 5. So it is different compared to spring tide conditions, whose tidal range in figure 5 is almost 2.5 meters. This fact is undoubtedly in spring tide conditions more freely to see the current vertical profile at a certain point of measurement. Because the position of this measurement is between 1 to 2 meters, while the form of measurement at each point is at a depth of 0.2D, 0.6D, and 0.8D, logically, we can still see the Ekman spiral profile in the current measurement results. However, it is not as easy as if it is on a spring tide consignment with a more extensive tidal range. The main thing in Figure 5 is that there is a match between the current measurement time and the tidal data recording time. It means that under certain conditions, the water level at the tide can be seen at the same time as the results of current measurements at the same time in direction and speed. Then we hope to be able to compare each point for variation against its depth. Ultimately, we can see the current vertical profile as illustrated in Ekman's spiral theory.

Here current events are presented in vector diagrams, and the distribution of current velocity and direction is according to tidal conditions. Figures 6 to 15 show the plotting results of current measurements at points CM-1 to CM-5, including speed-vector and current distribution. The speed vector is displayed in this form to show how the Ekman spiral current profile can easily be seen on each

CM-1 to CM-5. Meanwhile, the current distribution reflects the direction and size distribution of speed.

Figure 6 provides the results of current measurements at point CM-1, which shows the highest current at low tide, with the highest current speed of 0.42 m/s at a depth of 0.2D with a direction of 248° (southwest). Generally, the highest current speed occurs when the current moves to the southwest from the measurement of 18/06/21 at 19.00. It corresponds to the tidal condition in the most significant speed change slope seen in the tidal record at a water level of 1.5 meters. At high tide, the current moves to the southwest, and when it comes to low tide, it moves to the northeast. At the time of the highest and the lowest low tide, the current velocity is relatively lower than when the water moves at the lowest low tide or in the direction of the tide. It is very reasonable that there must be a minimum speed or even zero in peak conditions and tidal valleys due to the moment of occurrence of a change in the opposite direction. The highest current velocity in the vertical position level is on the surface, represented in the 0.2D measurement. At the same time, the middle current and bottom base current are represented in measurements of 0.6D and 0.8D. Figure 7 shows the current distribution below, which shows the velocity distribution in the southwest and northeast directions and the relatively high current velocity at 0.2D.

Figure 8 shows that the current velocity at point cm-2 has a high value towards high and low tides. When it comes to high tide, the current moves to the southwest, and when it comes to low tide, the current moves to the northeast. The maximum current velocity is 0.4 m/s which occurs in the middle (0.2D) and bottom (0.8D), which moves in a southwest direction (246°). The lowest current speed is 0.05 m/s with a direction to the northeast (50°) located in the middle (0.6D). The average current velocity in 0.2D is 0.23 m/s, 0.6D is 0.22 m/s and at the bottom (0.8D) is 0.22 m/s. Figure 7 provides a current vector curve of CM-2 at each depth under tidal conditions. The current distribution at point CM-2 shows the speed of distribution and direction of currents almost the same as point CM-1, namely in the southwest and northeast directions—Figure 9 distribution of current movement of CM-2.

Figure 10 shows the high current velocity at low tide in CM-3. The current velocity is influenced by the position of the measurement point located at the mouth of the river. So that at low tide, the water impulse from the river flow and the water movement towards the sea reinforce each other, the current velocity is higher when the water recedes, and towards low tide, the current moves to the northeast. When the sea water conditions are high tide, the current moves towards upstream of the river, namely to the west-northwest at low speed. Figure 11 gives the distribution of the movement of current CM-3. This is influenced by the encouragement of the flow of the river that meets the tide moving towards the land. The maximum speed occurring at low tide is 1.15 m/s which occurs in the middle (0.6D) the minimum current speed is 0.09 m/s occurring on the surface (0.2D). The average current velocity is 0.57 m/s at the surface and center (0.2D and 0.6D) and 0.51 m/s (0.8D).

Figure 12 shows that the CM-4 point has the same current movement as the CM-3 point because it is still close to the river mouth. High current velocity occurs at low tide. The direction of the current movement is still moving towards the southwest, south, east to northeast. The maximum speed at low tide is 0.56 m/s which occurs on the surface (0.2D) with a movement direction of 74° (east). The minimum current velocity is 0.09 m/s which occurs both at the surface (0.2D), in the middle (0.6D), and below (0.8D). the average current velocity is 0.28 m/s at the surface, in the middle (0.2D) is 0.26 m/s and at the bottom 0.24 m/s (0.8D). Figure 13 shows a graph of the current distribution showing the flow velocity, which is distributed to the southwest, south then, east, and northeast, and the relatively high current velocity at the surface (0.6D).

Figure 14 shows the CM-5 point moved to the southeast, north, and northwest. High current velocity occurs at low tide with a maximum speed of 0.4 m/s in the 122° (southeast) direction. The high current velocity occurs when heading in the receding direction. Figure 13 provides the direction of the current movement, which is still moving to the southeast, east, and north. The maximum speed that occurs at low tide is 0.4 m/s which occurs at the surface (0.2D), the minimum current speed is 0.09 m/s which occurs at the surface (0.2D), 0.38 m/s occurs at the middle (0.6 D), and 0.37 m/s occur at the bottom (0.8D). the average current velocities are 0.24 m/s at the surface and center (0.2D and 0.6D) and 0.21 m/s (0.8D). Figure 15 provides a current velocity distribution showing the movement of the current to the southeast, north to the northwest, and the velocity of the current at the higher surface and center.

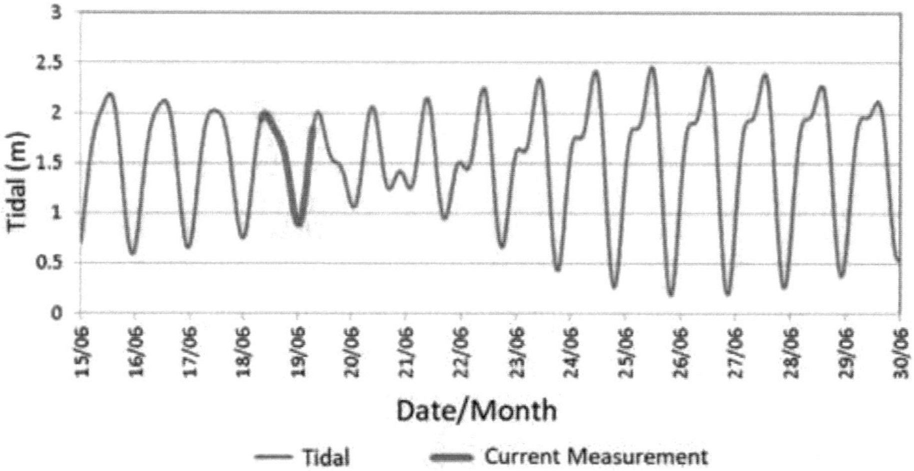

Figure 5. Measurement results at the position of measuring current at neap tide

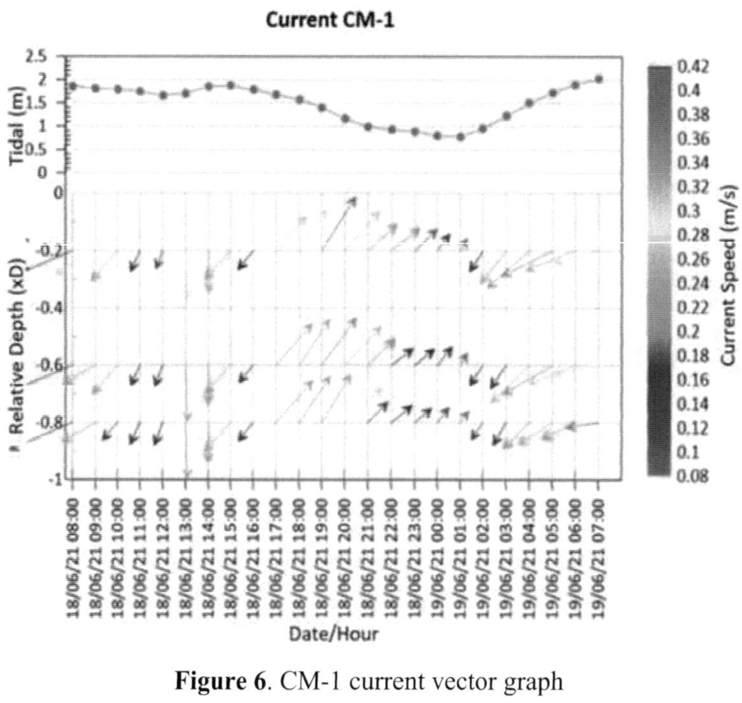

Figure 6. CM-1 current vector graph

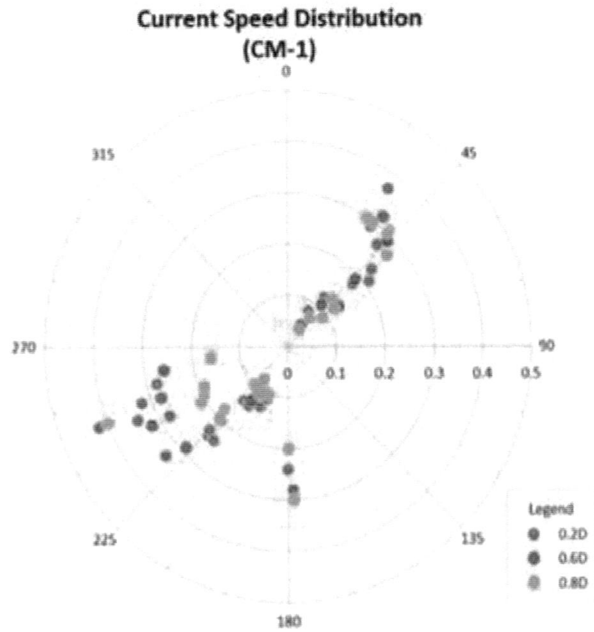

Figure 7. The distribution of current movement CM-1

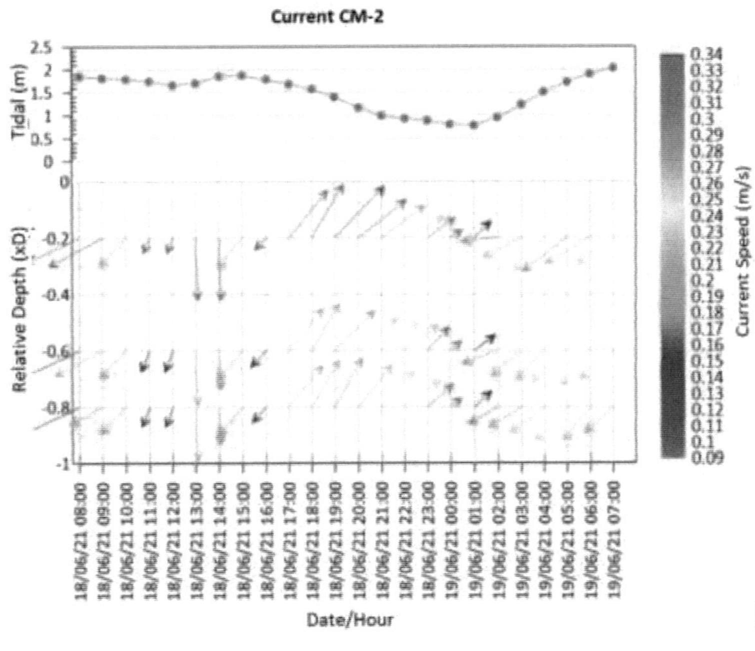

Figure 8. CM-2 current vector graph

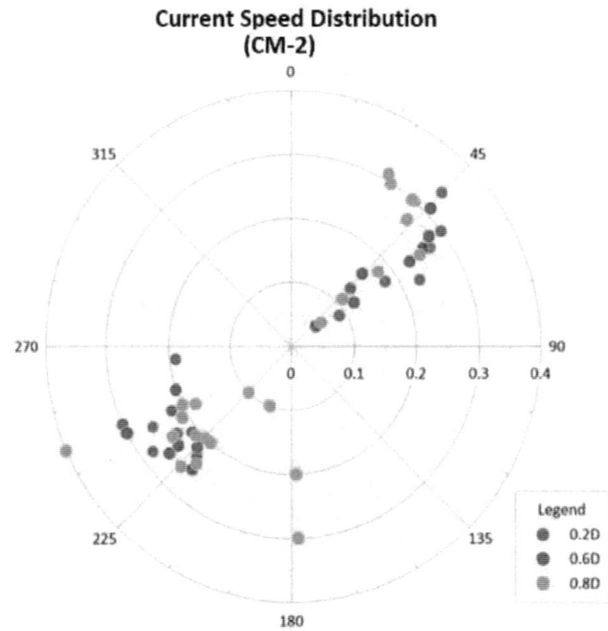

Figure 9. The distribution of current movement CM-2

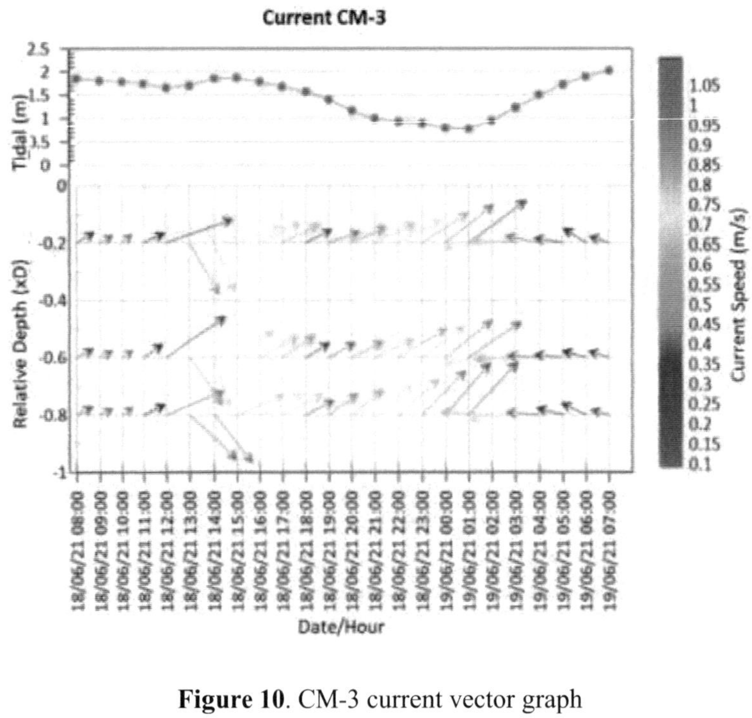

Figure 10. CM-3 current vector graph

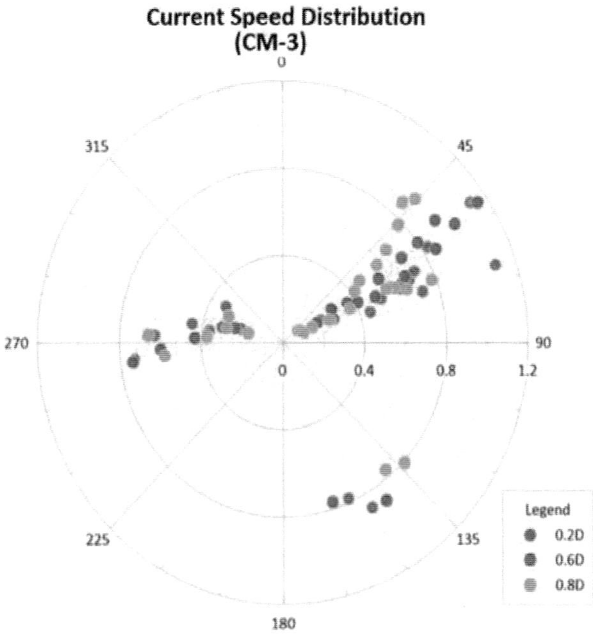

Figure 11. The distribution of current movement CM-3

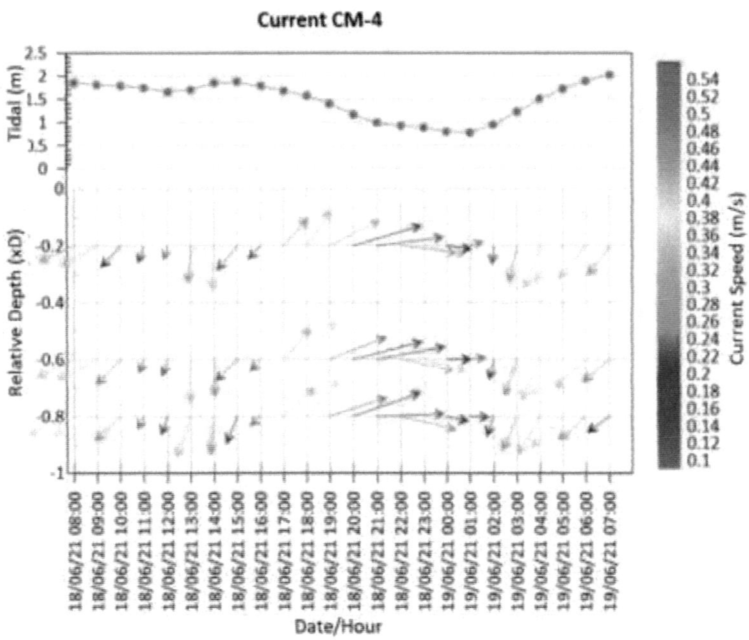

Figure 12. CM-4 current vector graph

IOP Conf. Series: Earth and Environmental Science 1166 (2023) 012024 doi:10.1088/1755-1315/1166/1/012024

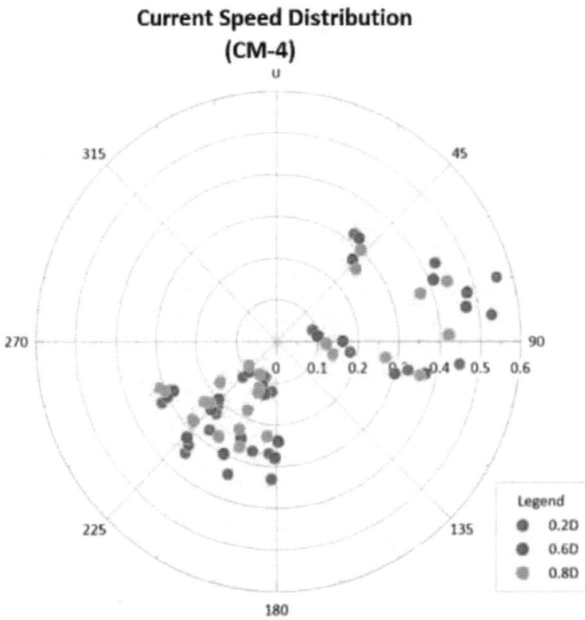

Figure 13. The distribution of current movement CM-4

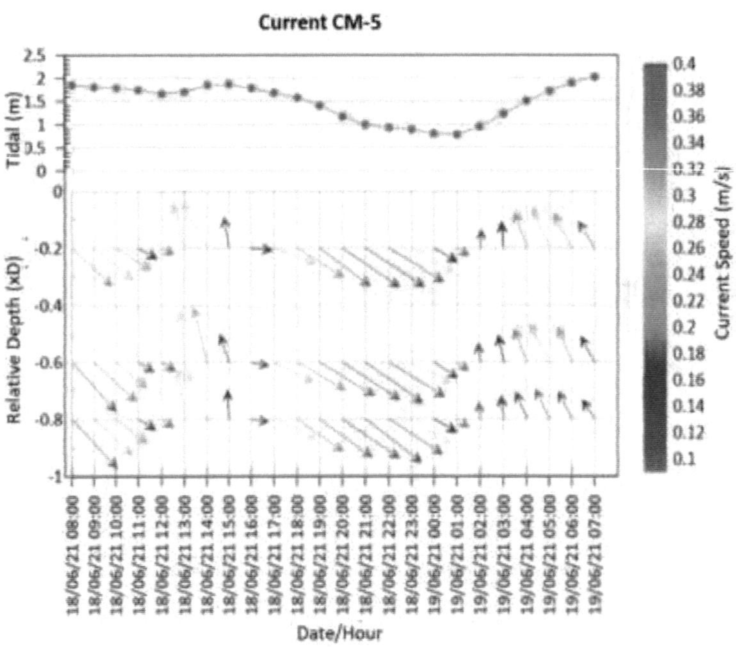

Figure 14. CM-5 current vector graph

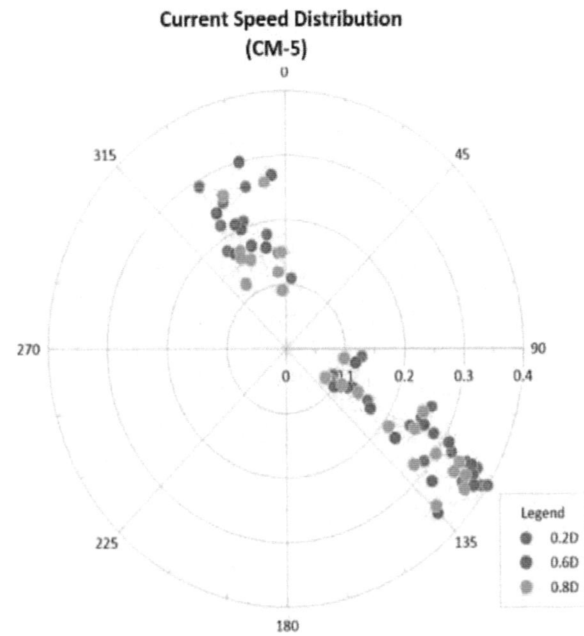

Figure 15. The distribution of current movement CM-5

C. Validate Measurements with Tidal Data

Validation with tide data is carried out to determine whether the results of running simulations follow the tidal data from observations in the field. Validation of the appropriate point is carried out inside the exact coordinates. In validating the tides, the results of this run are to be compared with the initial tidal data. To find out how much the data difference is, the RMSE (Root Mean Square Error) method is used. On the RMSE error test, it is said to be good if the value is close to ZERO. The results of this validation give an RMSE value of 0.301611, which is included in the RMSE interpretation category in the medium-to-small group. Figure 15 provides a graph of validation (calibration) of the tide model against the observation tide. While Figure 16 provides a current velocity validation graph.

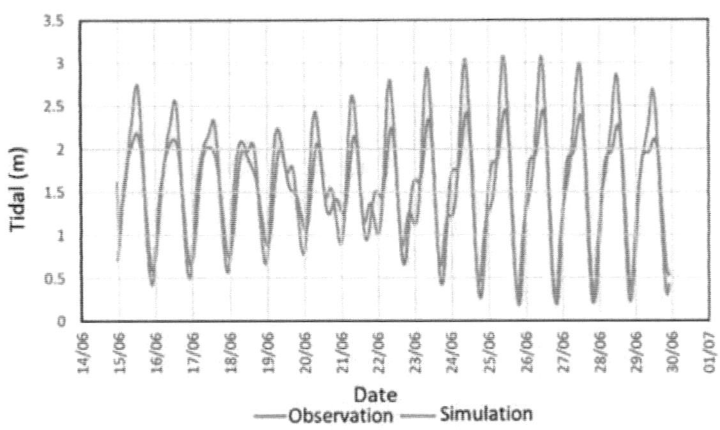

Figure 16. Calibration graph of tidal model and tidal observation

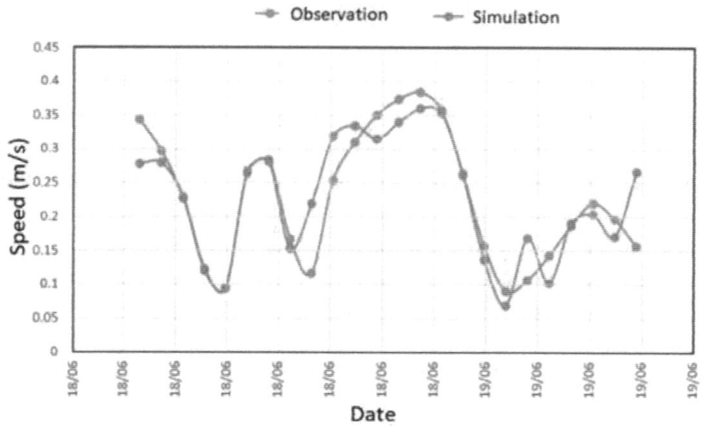

Figure 17. Current velocity validation graph

4. Discussion

Following the general pattern of tidal currents in the Madura Strait, the surface current goes westward at high tide, while at low tide, the surface current goes east [5 - 12]. Based on the results of current measurements on CM-1 and CM-2, both have almost the same direction; namely, in the direction of inclination to the maximum tide the direction of the current is towards the southwest while it is in the direction of the slope of the low tide the current goes to the northeast. The measurement results indicate that the maximum current velocity occurs at the maximum slope to the low tide, which is greater than at the maximum slope to the high tide. This phenomenon is interesting to be the subject of discussion.

The measurement results on CM-5 and CM-4 experience changes in the direction of the current, which is generally at high tide; there is a deflection to the north. This change in current direction is caused by the shape of the beach profile, namely the current measurement position at the end of the mouth of the bay, which causes the current to turn north. Current will pass through CM-5 first and then CM-4. The results of current measurements on CM-3 give a different shape; this position is in the form of a river estuary. The current pattern is the interaction between river currents and tidal water. Logically it is easy to understand that the maximum speed occurs at low tide due to the superposition of the strengthening of direction and speed that leads to the high seas. While the slightest minimum current is at high tide due to superposition in the opposite direction, the tide goes towards the river mouth, and the river water goes offshore. This condition has been proven in the measurement results.

The subsequent discussion is related to the direction of the vertical current at each measurement point. It is not very easy to see from the plotting results to distinguish the change in the direction of a measurement at a certain point at a particular hour. According to Ekman's Spiral theory, the current direction will turn right in the Northern Hemisphere and turn left for measurements of the Southern Hemisphere to the inner direction [1, 4]. In this study, the available data are located in the southern hemisphere; table 1 is at a latitude of 10 degrees south, so the current direction is deflected to the left for depth.

In the measurement results, the available data are plotting images and tables, the maximum average value, and the minimum average value. Based on the plotted images, it is not easy to distinguish the direction-to-depth changes that occur almost equally from 0.2D, 0.6D, and 0.8D depths. This condition occurs because there is a shortage of various depths. However, a record clearly shows a different direction, namely from measurements at the CM-4 point at 21.00, more precisely at a depth of 0.2D and 0.6D to 0.8D. In the southern hemisphere, the current is deflected to the left to depth. At depths of 0.2D and 0.6D, the current direction is about 105 degrees (not 75 degrees because two-dimensional plots can be projected from both directions). In contrast, at depths of 0.8D, the current direction is 90 degrees.

According to the researcher's analysis, the cause of this is the shape of the shoreline profile at the Lamong Bay estuary. At high tide, the current becomes shallow until it enters Lamong bay, which is primary friction commonly called siltation. In contrast, the water will flow freely at low tide without any obstacles back toward the sea. It is to the Ekman spiral current, predominantly based on currents due to wind friction rather than tidal currents [18 - 20]. Thus it makes perfect sense that the record of the distribution of current looks predominantly patterned in the direction in which the current comes when it goes towards the tide and its reverse direction towards receding. However, analysis based on Ekman spiral theory can still be done because measurements are not made on the surface at all. Ekman's spiral theory states that surface currents due to wind will turn 45 degrees to the right for the earth's northern hemisphere from the wind direction on the surface [20]. The location of this study, namely Lamong Bay, is at a position of about 7 degrees in a southern latitude, so the Ekman spiral current on the surface should be turned left 45 degrees from the wind direction on the surface. This study does not use wind data, and current measurements are not taken on the surface. However, in 0.2D, thus Ekman's spiral theory-based analysis and interpretation become less easy to explain. However, the results of current measurements can still provide a spiral shape from measurements at the levels of 0.2D, 0.6D, and 0.8D. It can be seen in the figure that the current velocity curve decreases from top to bottom and is accompanied by a change in direction. However, it is relatively difficult to determine movement patterns and shrink speed.

As a result, this study can show that the measurement results can provide a current vertical profile even though it is less relevant when using Ekman spiral theory. The condition of the bay profile or coastline around the measurement site is also thought to affect the current measurement results. There is one more thing that is interesting to analyze in more detail. Ekman's spiral theory uses the only rule that the force that deflects the direction of the current is the Coriolis force. The earth's rotation generates this force, and its magnitude depends on the latitude of a place, namely at the equator. It is zero and increases towards the poles [19, 20]. The exciting thing here is that the existence of the research location is at a latitude close to zero (equator), which is about 7 degrees in the southern latitude. Because the influence of the Coriolis force is minimal, the measurement results according to the plotting speed appear to be very vertical. There is a change in speed and direction, namely shrinking towards the bottom and the direction turning to the left (it still needs careful and additional measurements to add more gradations towards the vertical).

5. Conclusions

This study produced recordings from current measurements at five points CM-1 to CM-5, with each getting a change in the speed value shrinking to depth. Meanwhile, the change in direction to depth is still not so convincing to be seen by Ekman's spiral theory. This theory is based on the source of motion from the wind friction force, while this study uses the force and tidal currents. Although unclear, the current recording results can still show a spiral pattern against depth.

In general, the current measured at points CM-1 and CM-2 gives a horizontal result to the southwest, namely from the Madura strait to Lamong bay at the time of the tide. While at low tide, the direction is opposite. The current velocity at the maximum slope to low tide is more significant than at the maximum slope to high tide. This condition is due to siltation at the mouth of Lamong Bay. The measurements of ocean currents at points CM-4 and CM-5 indicate a change in sea currents towards the north during high tide conditions because the measurement point is at the end of the mouth of Lamong bay. While the measurement results at the CM-3 position give different results, the receding speed has a much greater value when compared to measurements at the other 4 points. This phenomenon is because the superposition of two currents is mutually reinforcing when it comes to low tide, namely the tidal current and the river's current.The results of the backfill flow in the vertical direction of each measurement point are not very easy to see in plotting. Namely, there is a left-leaning deflection to depth. However, there is still evidence that the Ekman Spiral is shown in one position, namely at the CM-4 point at 21:00. At depths of 0.2D and 0.6D, the Ekman Spiral is at about 105 degrees, and 0.8D is at 90 degrees.

Acknowledgment

This research was supported fully by Department of Ocean Engineering, DRPM-ITS, and the Minister of Research and Technology (Indonesia). Therefore, the authors thank all institutions for the fund and the facilities.

References

[1] Stewart, R. H., 2000. Introduction to physical oceanography. Texas: Department of Oceanography.

[2] Sujantoko, Pratikto,W.A, Prastianto, R.W, Maulana, M.I, Vibriyanti, A. 2022. Study of Changes in Coastal Morphology Due to Utilization of the Surabaya City Coastal Area. *International Journal of Marine Engineering Inovation and Research*. Vol.7, No.1 pp. 26-32.

[3] Mahmud, M., Sujantoko, Prasetyo, M. A., 2022, Empowerment and optimalisation of Kenjeran beach tourism potential, Surabaya-East Java. *International Journal on Engineering Applications (IREA)*, vol 13, no.4.

[4] Mustain, M., Mukhtasor, Sujantoko, Pustika, N., 2014, Community based management analysis of Kenjeran coastal area, *Proceeding Marine technology*.

[5] Mahmud, M., Garang, I. J., Ikhwani, H. Sujantoko, 2022. Analysis and maximize the ecotourism potential of the Wonorejo mangrove in East-Surabaya. *International Journal on Engineering Applications (IREA)*, Vol 10, No.2, pp. 158-166. https://doi.org/10.15866/irea.v10i2.21158.

[6] Mustain M., R. Leonard, 2019. Coastal zone management: the field approach to Wonorejo-East Surabaya, *International Journal on Engineering Applications (IREA)*, Vol 7, No 4, pp. 145-151. https://doi.org/10.15866/irea.v7i4.17552

[7] Armono, H.D., Sujantoko, Hidayah, Z., Nuzula, N. I., 2021. Hydro-oceanographic mapping to support coastal eco-tourism activities in Bawean Island, East Java. *Sustainable Islands Development Initiatives – International Conference*, September 2019, pp.A39 - A52.

[8] Ekman, V. W. 1905. On the influence of the Earth's rotation on ocean currents. *Arch. Math. Astron. Phys*., 2, pp. 1-52.

[9] Cushman-Roisin, Benoit; Beckers, Jean-Marie, 2009. Introduction to Geophysical Fluid Dynamics,

[10] Ekman spiral formula - Bing images, Acces 4 September 2022. https://www.bing.com/images/search?view=detailV2&ccid%20=qRLK%2F4vR&id=225AE567B7FAF3E26DACED04D5A8C60B67790F6B&thid=OIP.qRLK_4vR7ERiYF2EJZ0UegHaFj&mediaurl=https%3A%2F%2Fimage.slideserve.com%2F663321%2Fslide14-l.jpg&exph=768&expw=1024&q=ekman+spiral+formula&form=IRPRST&ck=C71D4A492FE177C1D8F8EF5AB755C700&selectedindex=2&ajaxhist=0&ajaxserp=0&vt=2&pivotparams=insightsToken%3Dccid_92kgLeA2*cp_AAAC5A5C07AC78C4C4297FBFCA6851F9*mid_98FEBCCC7EB56BF4901432A7B92678A7F00C95FA*simid_608032997223199144*thid_OIP.92kgLeA2MTVgJcpZjoN5tgFEDA&sim=1&iss=VSI&ajaxhist=0&ajaxserp=0

[11] Mustain, M., Sujantoko, 2022, Simple analysis of determination of tidal types of sea water: a case study in Lamong bay waters, East Java, *2021 IEEE Ocean Engineering Technology and Innovation Conference: Ocean Observation, Technology, and Innovation in Support of Ocean Decade of Science* (OETIC), pp. 13-17. https://doi.org/10.1109/OETIC53770.2021.9733727.

[12] Mustain, M., 2008. Analisa pola arus pasang surut pada alur pelayaran Tanjung Perak di Selat Madura, *Jurnal Teknologi Kelautan ITS*, vol.12 (2), 121-127

[13] Mustain, M., 2018. The Model of current pattern at the mouth of Porong river Sidoarjo, International Journal of Civil Engineering and Technology (IJCIET), vol. 9, Issue 1, 688–701.

[14] Mustain M., A. Sufyan, R Akhwady, 2018. Modeling Distribution of BOD, DO, Phosphate, and Nitrate of Bengawan Solo River in Ujung Pangkah, Gresik East Java, *International Journal of Civil Engineering and Technology (IJCIET)*, vol. 9, Issue 2, pp.565–576.

[15] Mustain M., 2019, Model of sedimentation and bed-load concentration patterns in mouth of Porong River of Sidoarjo Post Mudflow, *International Journal on Engineering Applications (IREA)*, vol 7, no 5, pp. 169-178.

[16] Mustain, M., 2017. The Model of Shoreline Change in Estuary of Porong River after Mud Volcano, *International Journal of Engineering and Applied Sciences (IJEAS)*, vol.4, issue-11.
[17] Mustain, M., 2011, Sidoarjo mud phenomenon as a permanent mul-vocano for preliminary hypothesis, *Jurnal Lingkungan Hidup Bumi Lestari Universitas*, Vol. 11, No.2, pp. 290-296.
[18] van den Bremer TS, BreivikØ. 2017 Stokes drift Phil. Trans. R. Soc. A376:20170104, http://dx.doi.org/10.1098/rsta.2017.0104
[19] Cushman-Roisin, Benoit; Beckers, Jean-Marie (2009). Introduction to Geophysical Fluid Dynamics, ACADEMIC PRESS, http://www.ccpo.odu.edu/~klinck/Reprints/PDF/roisinGFD2010.pdf.
[20] Ekman, V. W. 1905. On the influence of the Earth's rotation on ocean currents. Arch. Math. Astron. Phys., 2, 1-52. [1], https://empslocal.ex.ac.uk/people/staff/gv219/classics.d/Ekman05.pdf

IOP Conf. Series: Earth and Environmental Science 1166 (2023) 012025 doi:10.1088/1755-1315/1166/1/012025

Study of Diameter-to-Draft Ratio of Pitch Point Absorber as Wave Energy Converter in Indonesia Waters Based on Boundary Element Method

A Aulia[1], A G Athallah[1], M A Albasyir[1], S P Selvia[1], H A Darisman[1] and D Satrio[1,*]

[1] Department of Ocean Engineering, Institut Teknologi Sepuluh Nopember, Surabaya, Indonesia

*E-mail address: dendy.satrio@its.ac.id

Abstract. Recently, harnessing the energy from low wave energy density areas is concerned to deal with renewable energy targets. The previous study proposed a pitch point absorber with a submerged sectional equivalent area as the design parameter. However, that parameter did not suggest the best sectional area and did not directly correlate with the theory of structure hydrodynamic. Thus, this research proposes the diameter-to-draft ratio as a design parameter for the pitch point absorber. The study was conducted numerically using Boundary Element Method software to investigate diffraction characteristics of the device and analyze structure response in irregular waves. The model was modified with five different diameters and ratios. JONSWAP Spectrum was used to generate wave elevation with a 2-m significant wave height and 10-second peak period. The time domain simulation was set at 10.800 seconds. The result of this study showed that the highest responses occurred when the diameter-to-draft ratio was 5 because it has the closest structure natural frequency to assumed wave frequency, which makes it easier to resonate. In all diameters, the higher ratio affects the range of the structure natural frequency getting farther from the assumed wave, so the responses become smaller.

1. Introduction

Fossil fuels mainly cause the increase of climate change for about 86% of total global emissions over the last decade [1]. Almost all developing countries worldwide declared to reach net-zero emissions before entering 2050 [2]. Besides, it is also based on adopted Sustainable Development Goals (SDGs) number 7 and 13, which states that having affordable and clean energy and taking action on climate change [3]. Responding to this background, [4] suggested that fossil fuels must be replaced by renewable energy sources. Although on the other hand, research on conventional power plants is still needed [5,6]. Concerning the solution, [7] implied that ocean energy is one of the most potential renewable energies with large quantities, predictable, and environmentally friendly to fulfil the global energy demand. According to research by [8], the estimated theoretical potential of ocean energy in Indonesia reaches 18—72 GW. Ocean current energy has a large enough potential that can be converted with turbine technology [9–11]. However, [12] suggest that wave energy is one of the Ocean Renewable Energy which potentially exploited because it offers high energy density. However, the higher energy harnessed, the larger the WEC is, which increases the cost of a complex structure. To overcome these aspects, WEC must be designed to be more versatile in a small size. [12] recommend low wave energy density sea as an ideal area to install the small size WEC. This field has recently been concerned about research to find the effectiveness of harnessing energy from low wave energy density areas.

However, the utilization of a wave energy converter depends on its wave height and wave period at the projected area. A study from [13] concluded that a wave energy converter might result in different power and efficiency in a different area. Hence, a wave energy converter needs to be studied in specific sea waters based on its wave height, wave period, and sea depth. The installation area of a wave energy converter is described by a wave power density as a function of wave height and period. The low wave energy density area is considered around 15 – 25 kW/m.

A previous study proposed a wave energy converter type of point absorber by utilizing the pitch motion of the device to rotate the spin electric generator [14]. The study found that the design of the pitch point absorber depends on its horizontal dimension to generate good motion in short waves and high frequency. Another study concluded that the pitch point absorber is sensitive to short wave periods due to its small size or diameter compared to wavelength [15]. Therefore, a pitch point absorber is considered a good option for low wave energy density areas.

The initial research on the pitch point absorber was about the mechanism of sliding mass or generator of the device [14,16]. However, the hydrodynamic study of pitch point absorbers was mainly started around 2018. The study by [17] analyzed the effect of buoy diameter on the pitch response amplitude operator (RAO) of the device. That study found that the maximum value of pitch RAO decreases when the diameter is larger. Then, the parametric design of pitch point absorber was proposed with an underwater equivalent area by [18]. Besides, that research also studied the effect of the shape and diameter of the device on its pitch response amplitude operator (RAO). It was shown that the higher equivalent area results in a higher maximum value of pitch RAO under the same buoy diameter, yet the study did not suggest the best sectional area. In addition, cylindrical-shaped buoy has a higher pitch RAO than hemisphere-shaped buoy, so the cylindrical-shaped buoy was considered the best shape for pitch point absorber [18]. The study of pitch point absorber motion in irregular waves has been done by [15] based on JONSWAP spectrum. The study analyzed the effect of buoy diameter on pitch RAO and time-domain pitch motion. It was shown that the higher diameter affects the pitch RAO, and pitch motion becomes smaller.

Based on the above explanation about the research of pitch point absorber, the previous research has not concluded on the best parametric design for pitch point absorber. A comprehensive study about the effect of pitch motion characteristics and its pitch motion response in irregular waves also has not been done. Thus, this study proposes the diameter-to-draft ratio as a design parameter for the cylindrical pitch point absorber. That parameter is based on the theory of structure hydrodynamic that the underwater volume is the primary consideration for the hydrodynamic characteristics of the device. The comprehensive diffraction analysis of the buoy is carried out in this study. The pitch RAO, damping value, added mass, and wave excitation force are analyzed to study its effect on pitch motion in irregular waves.

2. Methodology

This research is conducted numerically using Boundary Element Method software. That method is taken as the basis numerical model due to its ability to represent the surface of the device with many elements so that wave velocity potential could be modelled better [19]. This study modelled 25 configurations which were modified from five different diameters and ratios. The diffraction analysis of all models is carried out by considering its pitch RAO, damping value, added mass, and wave excitation. After that, all models are simulated in irregular waves for a 10.800-seconds time domain. The time domain simulation is to analyze its pitch motion through simulation time. Finally, the effect of diameter-to-draft ratio on pitch RAO, damping value, added mass, wave excitation force, and pitch time-domain response are studied. These stages will explain the detail of the research methodology in the form of data collection, validation of numerical modelling, and variation of the study.

2.1. Data Collection

The data needed for this study are the common size of the point absorber wave energy converter and environment data. The size of the point absorber in this study was taken based on research review from [20]. The diameter of point absorber usually ranges between 5 – 15 meters, and other considerations such as structure simplicity also becomes the reason to select the size of point absorber. Therefore, this research takes 3 – 11 meters diameters for variation of the study. On the other side, ratio of diameter-to-draft ratio was taken based on a few configurations of the previous research about point absorbers.

Usually, the heave point absorber has a ratio of diameter-to-draft in the range of 1 to 3, but the pitch point absorber has ratio values of more than 5. Then, this research takes 5 – 25 ratio values of diameter-to-draft for the variation of this study.

The present study used the environment data in Sumba, Indonesia, which has a wave power density around 20 kW/m [21]. The average significant wave height of the selected area is 2-meters with a peak period of 10 seconds. These data had been studied for over 16 years from 1991 – 2015. The water depth is 30 meters. The environmental data will be applied in the motion response simulation to generate irregular waves with JONSWAP spectrum. The wind and current speed were not modelled in this research.

2.2. Validation of Numerical Model

The validation of numerical model was conducted by performing numerical results from the previous study. A study from [15] was taken as the reference to ensure the accuracy of the present numerical model. Those studies have validated their numerical model to the experimental model; hence the previous numerical model succeeded in describing the physical result.

The validation was done by re-modelling the configuration from [15], which can be seen in table 1. The output of pitch RAO from the previous study will be compared to the RAO of this present study. The average error and relation between the curve were taken for the validation parameter, where the feasible validation must have an average error of less than 5% and relation of more than 99%.

Table 1. Model configuration for validation of numerical model.

Diameter	Draft	Height	COG (from base)	Mesh Element
5 m	0.27 m	3 m	2.25 m	14000
10 m	0.12 m	3 m	2.25 m	14000
15 m	0.094 m	3 m	2.25 m	14000
20 m	0.081 m	3 m	2.25 m	14000

2.3. Variation of the Study

After the validation of numerical model meets the criteria, the research will be continued to simulate and analyze the variation of numerical model. These stages will explain the buoy geometry, diffraction setup, and motion simulation setup.

2.3.1. Model Geometry

The present study modified the pitch point absorber with five diameters and five diameter-to-draft ratios, which can be seen in table 2 and figure 1. The height of the buoy was 3-meters for all models. The stability consideration has been carried out in this study, in which all the models have righting lever of more than 0.15-meters. This study focuses on the effects of motion characteristics of the buoy's overall structure; therefore, the sliding mass inside the buoy was not modelled due to its small effect of gravity center change and damping effect compared to buoy large diameter and weight.

Table 2. Model configuration for variation of numerical model.

Diameter (m)	Ratio D/d
3	5, 10, 15, 20, 25
5	5, 10, 15, 20, 25
7	5, 10, 15, 20, 25
9	5, 10, 15, 20, 25
11	5, 10, 15, 20, 25

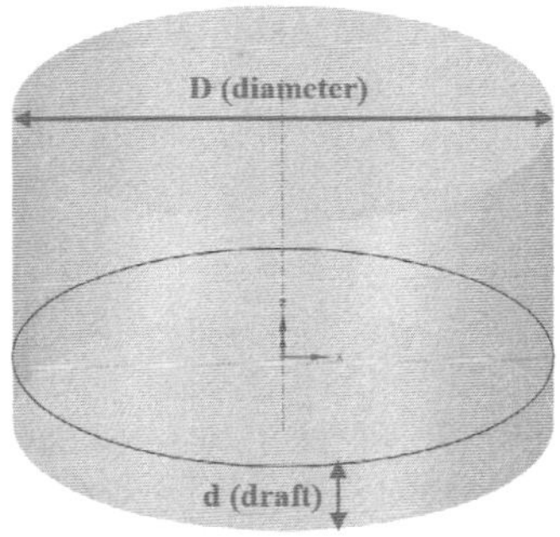

Figure 1. Drawing of pitch point absorber.

2.3.2. Diffraction Setup

The diffraction setup consists of buoy inertia or radius of gyration, number of mesh elements, and wave heading angle. The formulas of buoy inertia are presented in equation (1) and equation (2).

$$\mathrm{I_x} = \mathrm{I_y} = \frac{1}{12}\mathrm{m}(3\mathrm{r}^2 + \mathrm{h}^2) \tag{1}$$

$$\mathrm{I_z} = \frac{1}{2}\mathrm{mr}^2 \tag{2}$$

Where m is the mass of buoy, r is the radius of buoy, and h is the height of buoy.

The number of mesh elements was set to 14.000 elements. The wave heading angle was – 180 degrees since the cylindrical buoy is symmetric and has the same pitch motion characteristics for all heading angles. That diffraction setup was used to solve the response amplitude operator (RAO) in Equation 3, which is based on regular waves. $F_{exj}(\omega)$ is wave pitch excitation moment, $M_a(\omega)$ is pitch added mass, $R_r(\omega)$ is pitch damping value, K is stiffness, and M is structure mass. The excitation moment, added mass, and pitch damping is in frequency domain function, yet stiffness and mass are fixed. Therefore, the result of pitch RAO, pitch damping value, pitch added mass, and wave excitation was taken to be analyzed.

$$RAO\ (\omega) = \frac{F_{exj}(\omega)}{K - \omega^2\big(M + M_a(\omega)\big) + j\omega R_r(\omega)} \tag{3}$$

2.3.3. Motion Simulation Setup

The motion simulation needs the diffraction result in frequency domain as an input, such as motion RAO, damping value, added mass, wave excitation force, wave drift, and so on. The frequency domain will be the input to solve the motion simulation since the irregular wave is superposition of infinity of regular waves. All 25 models will be simulated in irregular waves to analyze its motion in time domain. The irregular wave was generated from JONSWAP spectrum based on environment data in section 2.1. JONSWAP spectrum equation is shown in equation (4). The simulation time domain was set to 10.800 seconds based on DNV OS E301.

$$\mathrm{S_{JON}}(\omega) = \frac{5}{16}(1 - 0.287\ln{(r)})\mathrm{H_S^2}\omega_\mathrm{p}^4\omega^{-5}\exp\left(-\frac{5}{4}\left(\frac{\omega}{\omega_\mathrm{p}}\right)^{-4}\right)\mathrm{r}^{\exp\left(-\frac{1}{2}\left(\frac{\omega-\omega_\mathrm{p}}{\psi\omega_\mathrm{p}}\right)^2\right)} \tag{4}$$

where, Hs is significant wave height (m), ω_p is peak wave frequency (rad/s), and r is peak elevation factor (r = 1, $5 \leq T_p/\sqrt{H_s}$). Ψ is 0.09 (when $\omega > \omega_p$) and 0.07 (when $\omega < \omega_p$).

The mooring line was set perpendicular to the seabed 31-meters in length, which connected to bottom of the buoy and fixed in the seabed. That configuration allows the buoy to move freely in pitch direction and limited in translation direction, also makes the mooring line relatively taut every time. The mooring properties is shown in table 3. The consideration of mooring line strength was also carried out, which resulted in the selected properties already meeting the strength requirement from DNV OS E301. The overall illustration of the motion simulation setup is shown in figure 2. The expected result from motion simulation is pitch motion of the buoy in time domain.

Table 3. Mooring line properties.

Parameters	Value / Description
Mooring Material	Wire Rope
Length (m)	31
Nominal Diameter (mm)	38
Equivalent Diameter (mm)	30
Max. Breaking Load (kN)	1010

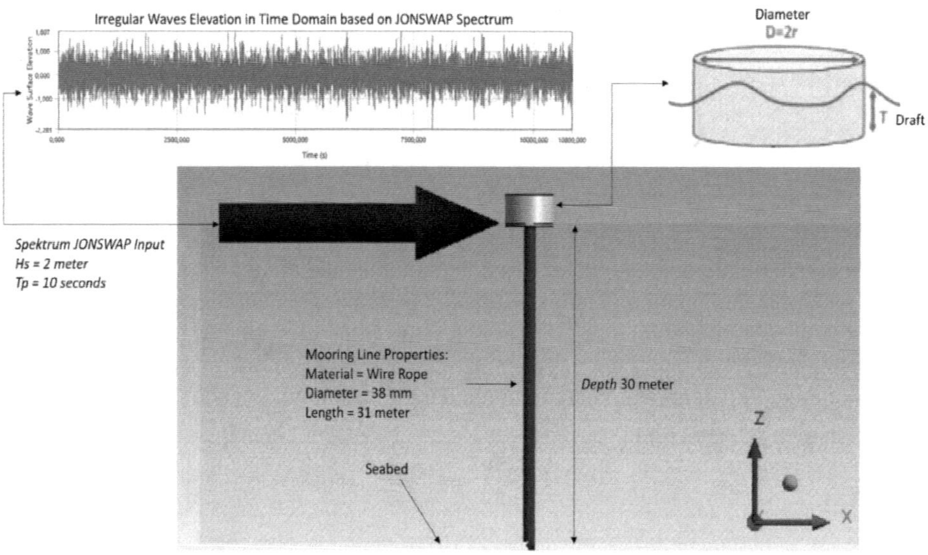

Figure 2. Illustration of motion simulation setup.

3. Results

3.1. Validation

The previous numerical model from [15] had been validated with experimental results, so the methodology of the previous numerical model could be the validation for future numerical model. This study re-modelled all four configurations which can be seen in Table 1. Based on those dimension characteristics, the 5-meters to 20-meters diameter model has diameter-to-draft ratios of 18.5, 83, 159, and 246, respectively.

The comparison results of present numerical model with the previous numerical model for pitch RAO can be seen in figure 3. In general, the result of present numerical model has a lower result than the previous numerical model. The error value of diameter 5-meters to 20-meters model were 3.62%, 4.08%, 2.2%, and 0.24%, respectively. On the other hand, the relation of the present numerical model is very similar compared to the previous model. The relation value between the curve of diameters 5-meters to 20-meters model were 99.73%, 99.54%, 99.84%, and 99.49%, respectively. The small disparity between the curves may come from different meshing method, even with the same number of meshing elements that have been implemented. Based on the validation result, all the models have met the criteria which have errors value of less than 5% and relation value of more than 99%. Therefore, the current numerical model is feasible for further analysis.

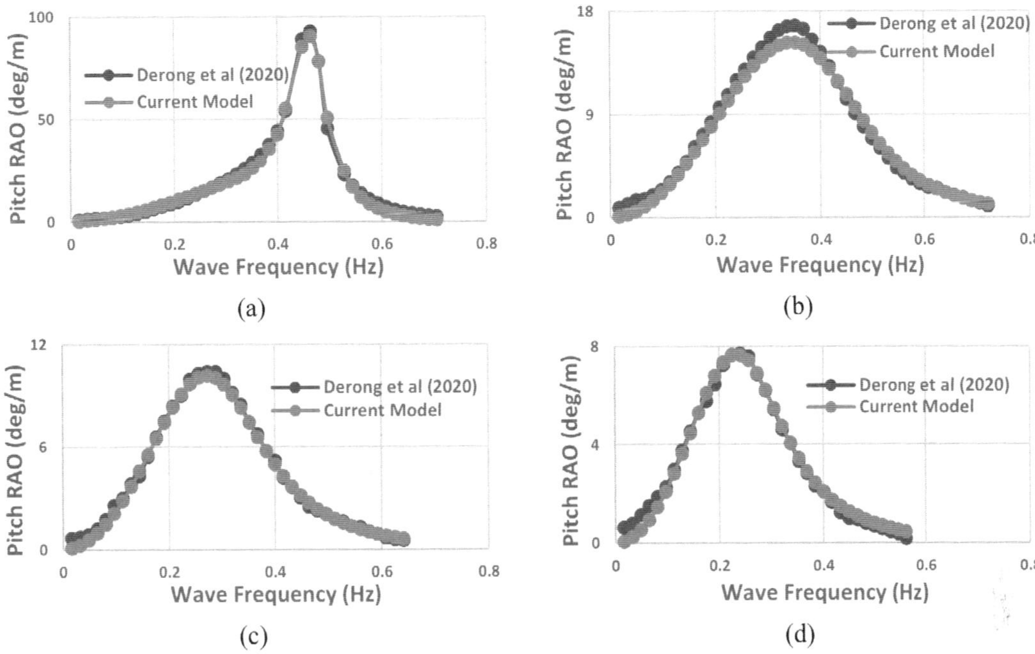

Figure 3. Pitch RAO for numerical validation at (a) D=5 m, (b) D=10 m, (c) D=15 m, (d) D=20 m.

3.2. Pitch Response Amplitude Operator

Pitch RAO as the main motion characteristic of pitch point absorber will be discussed in this section. The RAO curve is a function of wave frequency and wave amplitude. Figure 4 shows the pitch RAO for all configurations. All the configurations have the same curve pattern. The result indicates that all configurations have similar motion characteristics in pitch motion, yet differences in its natural frequencies and maximum pitch RAO values. Hence, the main parameters that need to be considered from pitch RAO are the natural frequency and its maximum value.

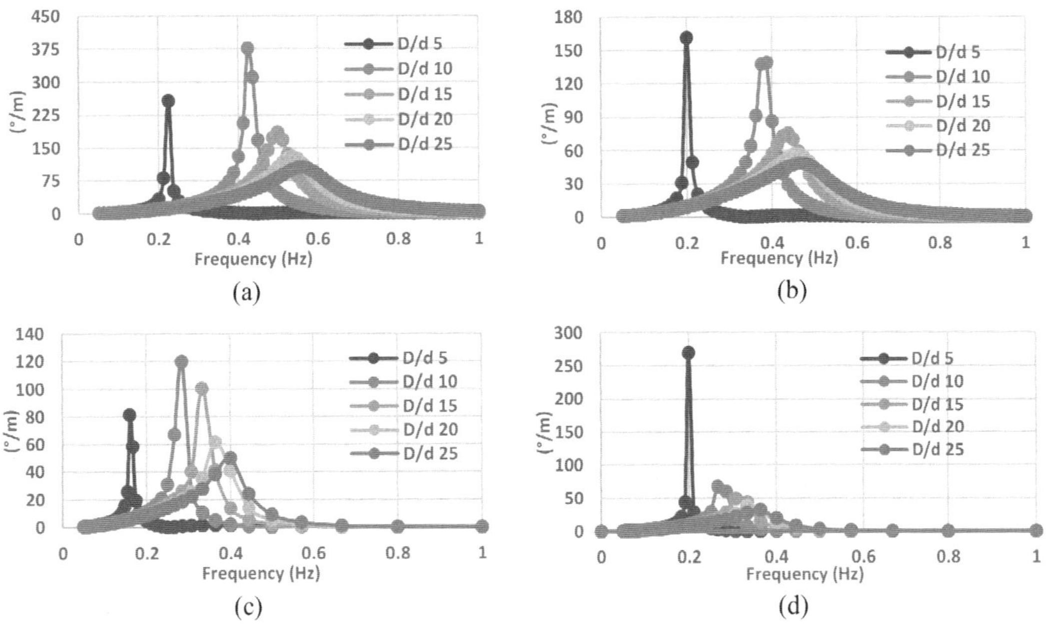

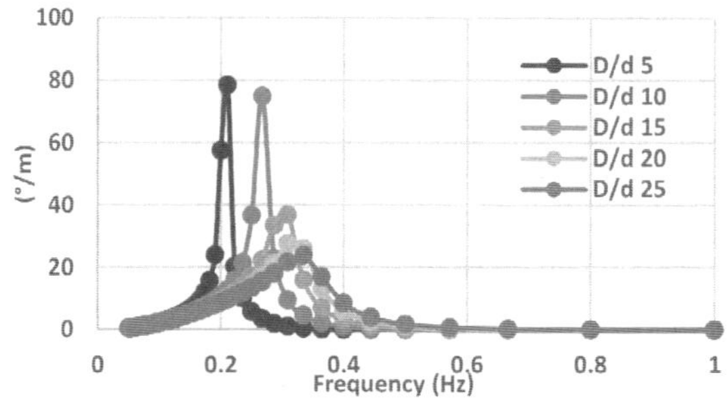

(e)

Figure 4. Pitch RAO on (a) D=3 m, (b) D=5 m, (c) D=7 m, (d) D=9 m, (e) D=11 m.

The relation of buoy natural frequency and its ratio is shown in Figure 5. The natural frequency becomes smaller as the ratio decreases so that the ratio of 5 has the smallest natural frequency compared to other ratios in all diameters. On the other hand, the wave frequency that potentially occurred at 2-meter wave height is more than 4 seconds or less than 0.25 Hz [22]. Therefore, the natural frequency in the ratio of 5 is within the range of possible wave frequency. That condition makes the ratio of 5 easier to resonate compared to other ratios. It is also noted that the natural frequency in the ratio of 5 significantly decreases from the ratio of 10, meanwhile the natural frequency from the ratio of 25 to 10 tends to be closer. That phenomenon is in line with the draft change of each ratio. For example, in 3-meters diameter, the draft of ratio of 5 to 25 are 0.6, 0.3, 0.2, 0.15, 0.12, respectively. From this information, the draft change between the ratio of 5 to 10 is greater than the other ratio. Therefore, the larger ratio could make the change of natural frequency becomes smaller.

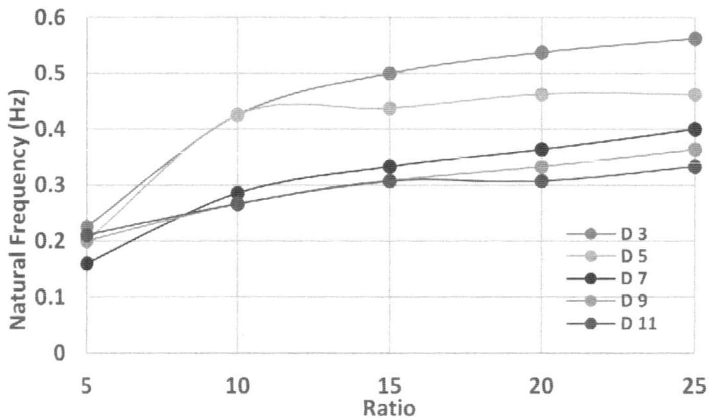

Figure 5. Natural frequency of pitch motion on all configurations.

Figure 6 shows the summary of maximum pitch RAO. Based on the overall curve pattern, the larger ratio results in a smaller maximum pitch RAO, yet there are some configurations that are out of the pattern, especially in the ratio of 10. However, the maximum pitch RAO does not affect the motion response directly since the maximum RAO only occurs if the buoy resonates with the wave. Therefore, the natural frequency of the buoy has more influence than maximum pitch RAO in generating high motion response. It also noted that the other hydrodynamic components, such as damping and added mass, also affect the motion response in irregular waves. Based on the explanation in this section, it is expected that the ratio of 5 is more favorable than the other ratios in resulting good motion response.

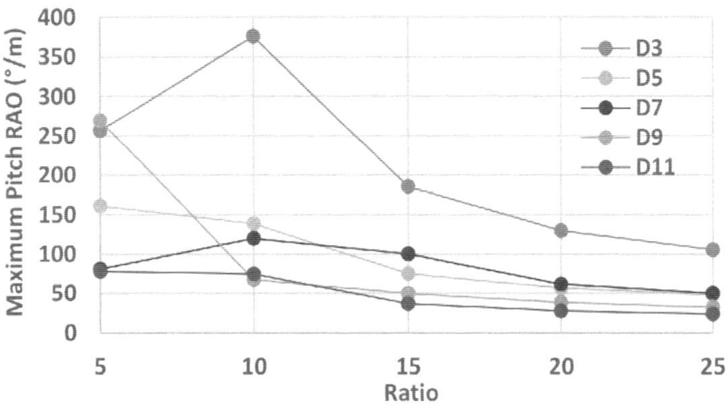

Figure 6. Maximum pitch RAO on all configurations.

3.3. Pitch Damping Value

Radiation damping is a damping force acting on the float which influences a structure's response to waves. Figure 7 shows the maximum damping value of all the configurations for pitch motion. In general, higher ratios result in higher radiation damping, yet there are 7-meters and 11-meters that have a different pattern from the others. For all diameters, the highest radiation damping occurs at a diameter-to-draft ratio of 25 except for a 7-meters diameter. A significant increase in maximum damping occurs on the ratio of 15-20 at 9-meters and 11-meters diameters. Moreover, the ratio of 5 and 10 at 7-meters diameter is likely to have higher radiation damping than 9-meters diameter. It is observed that there are fluctuations in maximum damping value at high ratio dan diameter, such as the ratio of 10 to 20 and diameter 7 - 11 meters. Hence, it needs special attention in designing large diameters at high diameter-to-draft ratios. It is known that radiation damping tends to reduce the structure's performance. Regardless of the maximum damping value fluctuation, the ratio of 5 has the lowest damping compared to other ratios. Therefore, a diameter-to-draft ratio of 5 is likely to give a better structure pitch performance.

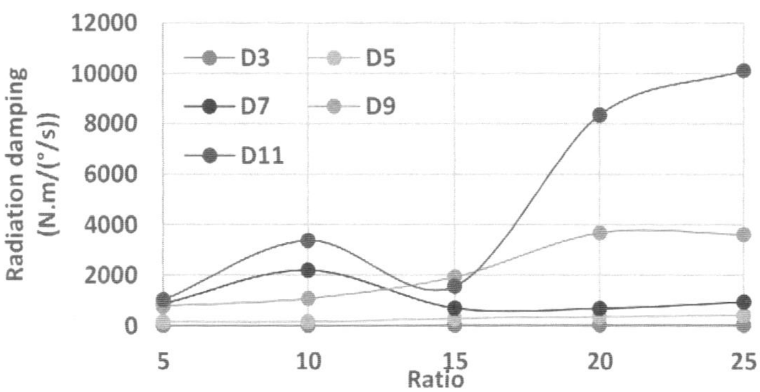

Figure 7. Maximum radiation damping of all configurations.

3.4. Pitch Added Mass

Added mass contributes to the inertial force by the buoy's pitching motion. Figure 8 shows the pitch added mass as a function of wave frequency for all diameters. A larger diameter-to-draft ratio tends to increase the maximum added mass. Diameter-to-draft ratio of 25 generates added mass 18-29% larger than the diameter-to-draft ratio of 5 for each diameter. The maximum added mass occurs around 0.18-0.4 Hz and slightly reduces on a higher frequency. In addition, added mass increases dramatically in diameter of 7-11 meters. It is observed that added mass increases 99.8% from 3-meters diameter to 11-meters. It indicates that the wider buoy diameter and diameter-to-draft ratio generate greater added inertia force.

Added mass is related to the deflection of surrounding fluid due to acceleration or deceleration of the buoy. It is practically easier to understand this parameter as a curve in the frequency domain shown in figure 8. The added mass at the ratio of 25 varies greatly. It indicates that the flow patterns around the buoy can change dramatically with wave frequency, causing a reduction in pitching response performance so that less-fluctuated added mass is preferable to generate good pitching motion. Therefore, the diameter-to-draft ratio of 5 is more favourable than the other ratio due to its stable added mass value in the frequency domain.

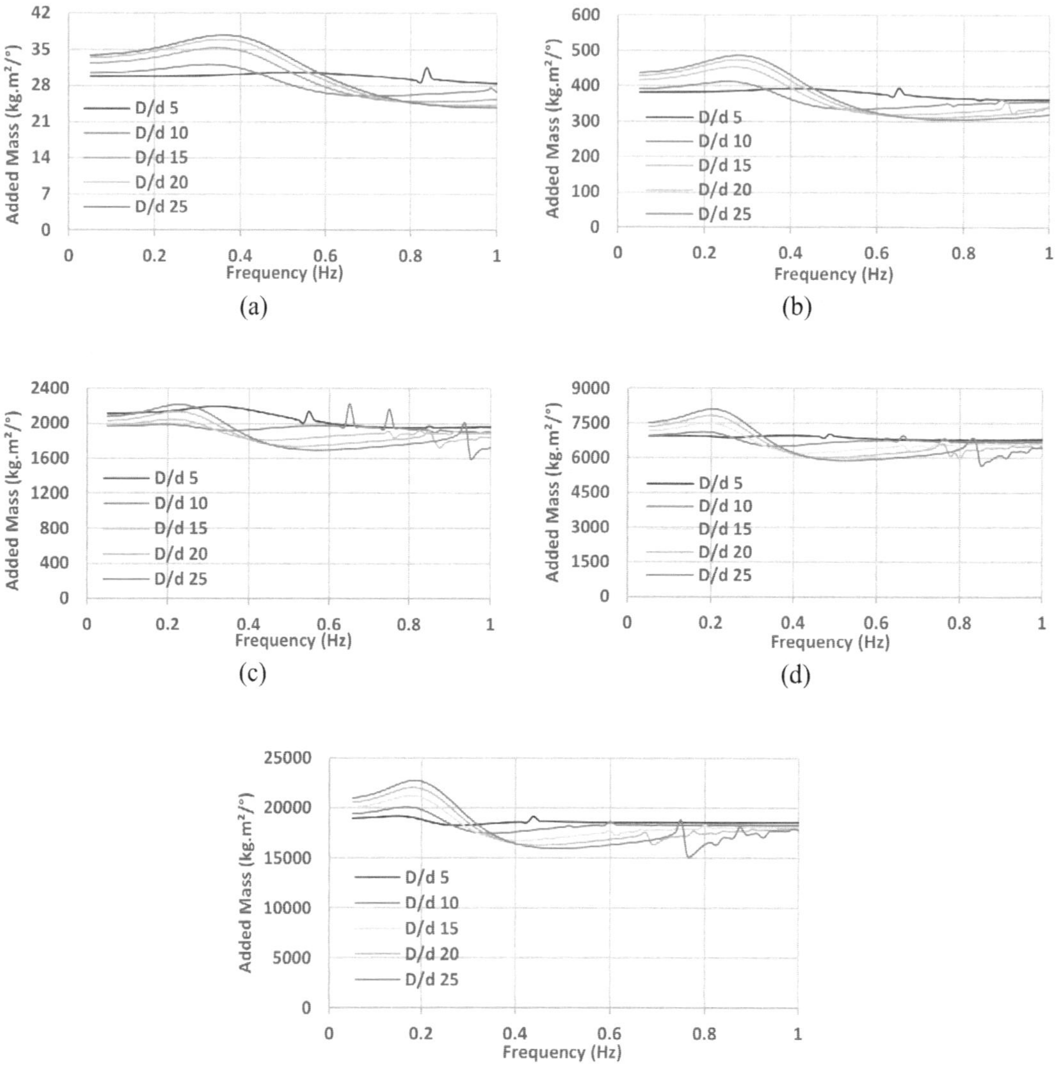

Figure 8. Frequency-domain pitch added mass on (a) D=3 m, (b) D=5 m, (c) D=7 m, (d) D=9 m, (e) D=11 m.

3.5. Wave Excitation for Pitch Motion

When a float or wave energy converter was working, there was a force due to the waves that affects the float. The force is called wave excitation, a combination of Froude-Krylov and diffraction forces. The potential theory is used to calculate the force for each degree of freedom. However, this study only focuses on the pitch motion in which the pitch moment of excitation has the greatest effect on energy generation. Figure 9 shows the maximum excitation moment for each diameter and diameter-to-draft ratio. It showed that higher ratio results in a higher excitation moment. The maximum excitation moment reaches approximately 658.8 kN/m in 11-meters diameter with a ratio of 25. The pitch excitation moment does not directly correlate with pitch response since there are other linear parameters, such as RAO, damping, and added mass. Therefore, time domain simulation is used to analyse the motion response of the pitch point absorber and select the best diameter-to-draft ratios.

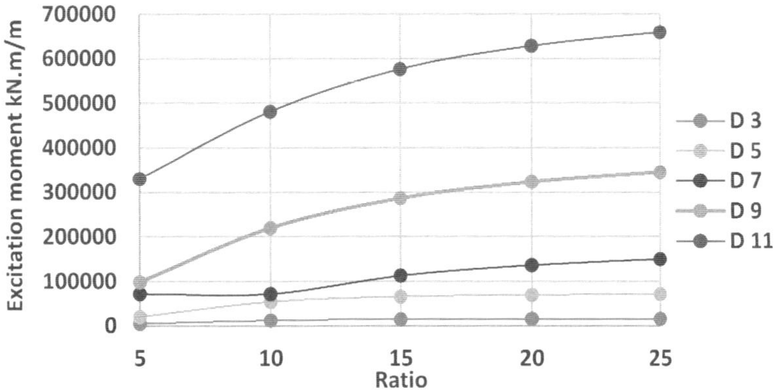

Figure 9. Excitation moment on all configurations.

3.6. Motion Response in Irregular Waves

The motion response of pitch point absorber is a very important analysis since it will affect the generator to generate electricity. The simulation, as previously described in section 2.3.3, will be analyzed in this section. Figure 10 shows the irregular wave elevation for a 2-meters significant wave height, 10-seconds peak period, and 10800-seconds time domain. Based on irregular wave elevation, the waves may have different heights, steepness, and period, so the motion result will be non-linear parameter. Since the irregular waves are superposition of infinity of regular waves, which is linear parameter, the analysis of all the linear parameter inputs will affect the motion response of each configuration. Therefore, this section also analyzed the effect of diffraction characteristics on motion response in irregular waves.

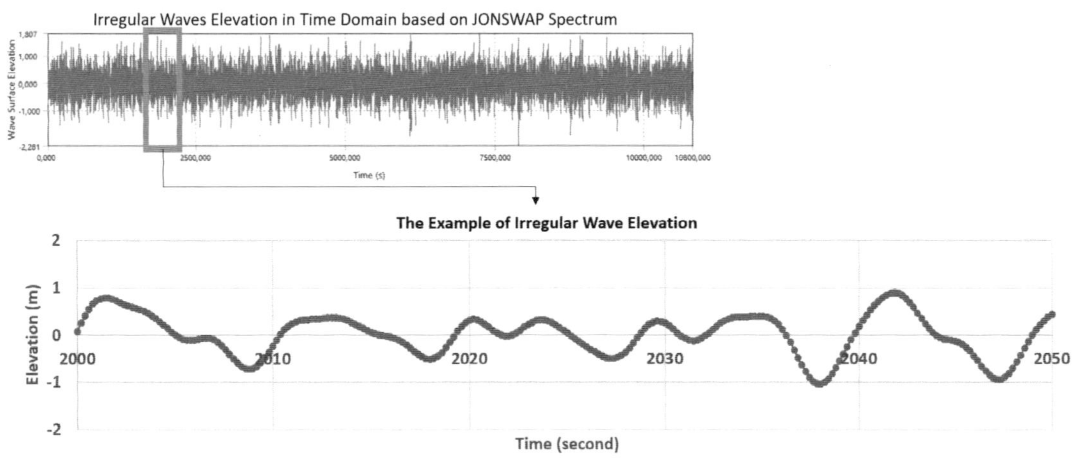

Figure 10. Irregular wave elevation.

The pitch rotation result of pitch point absorber is disordered throughout the simulation time. The example of pitch rotation results in time domain can be seen in figure 11. Therefore, to simplify the analysis, this paper analyzes the graph by making the new parameter called "Amplitude". That parameter is used to know which buoy has the higher or lower response. The parameter describes the 75% of data between a certain value. For example, if a curve has an amplitude value of 20 degrees, it means that 75% of buoy motions are in the range of 20 to -20 degrees.

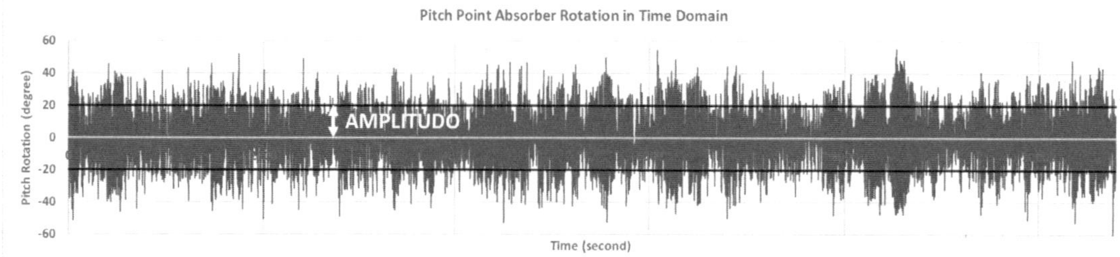

Figure 11. Example of pitch rotation result in time domain.

Figure 12 illustrates the result of the amplitude parameter for every model variation. It can be seen from the graph that the higher the ratio, the motion response will be smaller. Hence, the diameter-to-draft ratio of 5 is suggested as the best ratio to generate a large pitch motion. The suggested ratio is linear with the previous explanation about diffraction analysis. The ratio of 5 has the lowest natural frequency and the closest to wave period, so the buoy with a ratio of 5 is easier to resonate than the other ratios. The ratio of 5 also has a larger maximum pitch RAO than the other ratios. Therefore, the ratio of 5 has the potential to resonate in high pitch RAO which leads to high response of the buoy. In addition, based on the linear hydrodynamic analysis, the ratio of 5 has the lowest radiation damping and generates non-fluctuating added mass. It means that ratio of 5 has good pitch motion performance due to its capability to resonate in high value of RAO with reasonable damping value and added mass.

Another interesting phenomenon is that the higher ratio makes the disparity of amplitude parameters between the diameter become smaller. For example, in the ratio of 5, the maximum amplitude is 20.21 degrees at 11-meters diameters and the minimum amplitude is around 13.26 degrees at 9-meters diameters. However, in the ratio of 25, the value of the amplitude is very similar around 2.5 degrees for all diameters. The phenomenon is linear with the explanation of added mass analysis. The previous section has shown that ratio of 25 could produce 29% of added mass than the ratio of 5, hence the response disparity in the ratio of 25 becomes smaller.

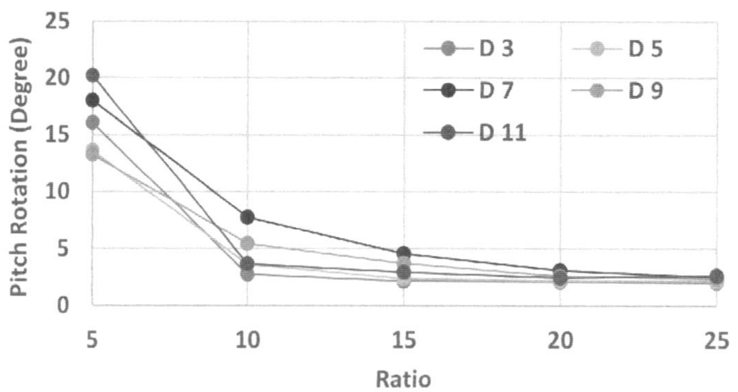

Figure 12. Comparison of amplitude parameter to D/d ratio.

The number of pitch motions in a wave period is also carried out in this study. Based on [23], the number of pitch motions is also important information to calculate the potential generated electricity. This study examines the number of pitch motions in a few waves period. Figure 13 shows three irregular wave elevations and their pitch motion elevation. This study takes the configuration of diameter-to-draft ratio of 5 as an example. In the first wave period, buoy with a 3-, 5-, 9-, and 11-meters diameter has 1.5 times pitch rotation, meanwhile 7-meter diameter has 1.35 times pitch rotation. During the second wave period, all diameters have 2 times pitch rotation. As of the third wave period, all diameters have 1.5 times pitch rotation. Based on the number of pitch motions in three waves elevation, a buoy may have different numbers of pitch rotation in different waves. A buoy also has a different number of pitch rotations compared to other configurations. Therefore, the generated electricity of a buoy may not be the same as the time changes. The number of pitch rotations is not explained further in this research. However, based on the examination above, 1.5 times number of pitch rotations may be the assumption in the preliminary calculation of generated electricity in further research.

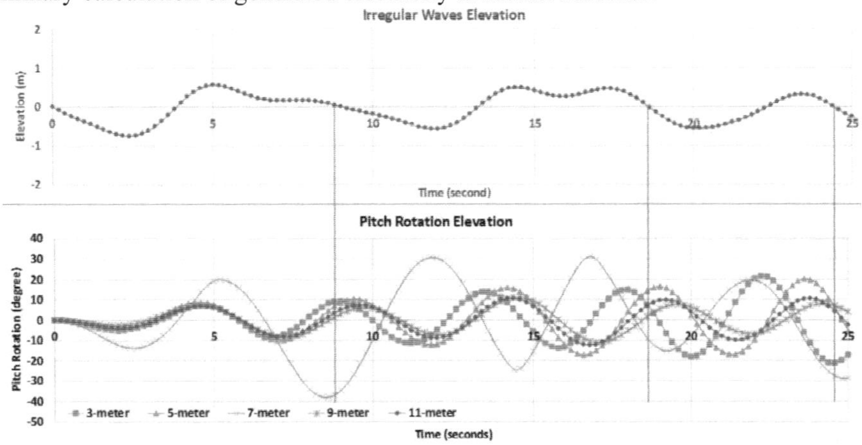

Figure 13. Pitch motion elevation in certain wave period.

4. Conclusion

In this study, diffraction simulations were analyzed to get the pitch motion characteristic of pitch-point absorber. Furthermore, the dynamic performance for each variation was studied using time-domain simulation under the condition of irregular waves. The conclusions of this study are as follows:

- From each variation of diameter-to-draft ratio, the maximum pitch RAO tends to decrease as the ratio increases. It was found that ratio of 5 has the smallest natural frequency. Therefore, the ratio of 5 is closer to the selected wave frequency range making it easier to resonate.
- The radiation damping, added mass, and wave excitation of pitch motion have a correlation with the diameter-to-draft ratio. A lower diameter-to-draft ratio has smaller radiation damping which is likely to give better pitch performance in motion simulation. The fluctuation of added mass correlates to the change of flow pattern around the device. Diameter-to-draft ratio of 5 has less-fluctuated added mass which indicates more stable performance along wave frequency than the other ratios. Diameter-to-draft ratio of 25 has the highest excitation moment for each diameter.
- The pitch response under selected irregular conditions gives an actual response in low-density energy areas. Diameter-to-draft ratio of 5 produces a higher amplitude pitch response than the other ratios. A higher diameter to draft ratio tends to produce lower pitch response amplitude.

Among all diameter-to-draft ratios, the ratio of 5 in linear hydrodynamic shows reasonable results and is likely to give better pitch performance. It is validated in time-domain simulation under irregular waves that ratio of 5 results in higher pitching motion than the other ratio. Therefore, the diameter-to-draft ratio of 5 is recommended for the configuration of pitch point absorber in further research.

Acknowledgements

The authors gratefully acknowledge the Ministry of Education, Culture, Research, and Technology for funding this research under a scheme called PKM (Program Kreativitas Mahasiswa). The support from Institut Teknologi Sepuluh Nopember for guidance this program is also acknowledged.

References

[1] McKenzie J and Carter A V. 2021 Stepping stones to keep fossil fuels in the ground: Insights for a global wind down from Ireland *Extr. Ind. Soc.* **8** 101002
[2] Zhang Q, Wu X and Chen Y 2022 Is economic crisis an opportunity for realizing the low-carbon transition ? A simulation study on the interaction between economic cycle and energy regulation policy *Energy Policy* **168** 113114
[3] Poudyal R, Loskot P, Nepal R, Parajuli R and Khadka S K 2019 Mitigating the current energy

crisis in Nepal with renewable energy sources *Renew. Sustain. Energy Rev.* **116** 109388

[4] Quirapas M A J R and Taeihagh A 2021 Ocean renewable energy development in Southeast Asia: Opportunities, risks and unintended consequences *Renew. Sustain. Energy Rev.* **137** 110403

[5] Satrio D, Trisnayanthi N N A I and Pratilastiarso J 2021 Analysis of the Effects of Fuel Type Selection on the Performance and Fuel Consumption of a Steam Power Plant *Int. J. Adv. Sci. Eng. Inf. Technol.* **11** 2046–54

[6] Kurniawati I, Pratilastiarso J and Satrio D 2020 Analysis of the Effect of Cooling Water Condenser to Power Plant Cycle Using Cycle-Tempo Software *IES 2020 - International Electronics Symposium: The Role of Autonomous and Intelligent Systems for Human Life and Comfort* (Surabaya: IEEE) pp 37–42

[7] Azam A, Ahmed A, Li H, Tairab A M, Jia C, Li N and Zhang Z 2022 Design and analysis of the optimal spinning top-shaped buoy for wave energy harvesting in low energy density seas for sustainable marine aquaculture *Ocean Eng.* **255** 111434

[8] Langer J, Quist J and Blok K 2021 Review of renewable energy potentials in indonesia and their contribution to a 100% renewable electricity system *Energies* **14**

[9] Satrio D, Suntoyo and Ramadhan L I 2022 The advantage of flow disturbance for vertical-axis turbine in low current velocity *Sustain. Energy Technol. Assessments* **49** 101692 1-9

[10] Satrio D, Firdaus Yusri M, Mukhtasor, Rahmawati S, Junianto S and Musabikha S 2022 Numerical simulation of cross-flow Savonius turbine for locations with low current velocity in Indonesia *J. Brazilian Soc. Mech. Sci. Eng.* **44** 1–11

[11] Satrio D and Utama I K A P 2021 Experimental investigation into the improvement of self-starting capability of vertical-axis tidal current turbine *Energy Reports* **7** 4587–94

[12] Foteinis S 2022 Wave energy converters in low energy seas: Current state and opportunities *Renew. Sustain. Energy Rev.* **162** 112448

[13] Zhang Y, Zhao Y, Sun W and Li J 2021 Ocean wave energy converters: Technical principle, device realization, and performance evaluation *Renew. Sustain. Energy Rev.* **141** 110764

[14] Chen H M and Delbalzo D R 2013 Circular-slide wave energy converter in random waves *Ocean. 2013 MTS/IEEE - San Diego An Ocean Common* 4–8

[15] Derong D, Fei C, Hui Z, Xuefeng Y and Fang Z 2020 Study on capture power of the sealed-buoy wave energy converter in low energy flow density area *Renew. Energy* **152** 1024–34

[16] Liu S, Li D and Chen H M 2015 Circular-Slide Wave Energy Converter *IPEMEC 2015* pp 614–9

[17] Chen F, Duan D R, Zhang H, Yang X F and Zhao F 2018 Study on hydrodynamic performance of the sealed-buoy wave energy converter *IOP Conference Series: Earth and Environmental Science* vol 199 p 032067

[18] Chen F, Duan D, Han Q, Yang X and Zhao F 2019 Study on force and wave energy conversion efficiency of buoys in low wave energy density seas *Energy Convers. Manag.* **182** 191–200

[19] Nguyen H P and Wang C M 2022 Finite element - Multi-domain boundary element method for hydroelastic analysis of large floating pontoons with perforated plates *Ocean Eng.* **246** 110659

[20] Babarit A 2015 A database of capture width ratio of wave energy converters *Renew. Energy* **80** 610–28

[21] Rizal A M, Ningsih N S, Sofian I, Hanifah F and Hilmi I 2019 Preliminary study of wave energy resource assessment and its seasonal variation along the southern coasts of Java, Bali, and Nusa Tenggara waters *J. Renew. Sustain. Energy* **11** 014502

[22] Chun H and Suh K D 2018 Estimation of significant wave period from wave spectrum *Ocean Eng.* **163** 609–16

[23] Zhu L, Wang Y, Yang Z and Qu Y 2013 Buoy-rope-drum wave power system *Adv. Mech. Eng.* **2013** 2–6

Sail technology on fishing vessel for blue swimming crab catching application – *A review*

N Setiyobudi[1,3], A Santoso[2], D P Sari[3], T Muttaqie[3], Muryadin[3], F M Noor[3], Zarochman[4], S Sasmita[4], Z Wassahua[4], A B Setiawan[3], S S Sukoraharjo[4] and A Kuncoro[4]

[1] Master Program in Marine Engineering Department, Faculty of Marine Technology ITS, Surabaya, Indonesia
[2] Marine Engineering Department, Faculty of Marine Technology ITS, Surabaya, Indonesia
[3] Research Center for Hydrodynamics Technology, National Research and Innovation Agency (BRIN), Surabaya, Indonesia
[4] Research Center for Fishery, National Research and Innovation Agency (BRIN), Cibinong, Indonesia

Corresponding author's e-mail: 6019221001@mhs.its.ac.id

Abstract. The increase in fuel prices will gradually have an impact on the capture fisheries sub-sector, especially the fishermen. Sail technology is an alternative that can be used to save fuel oil, although it is realized that this technology cannot completely replace the diesel motor that is commonly used in fishing operations. This study will explain the application opportunities of sail technology on fishing vessels for blue swimming crab *(Portunus pelagicus)* in the deep waters of the Java Sea. This fishing ground is more than 12 nautical miles from the coast. Using sail technology on fishing vessels is very dependent on the characteristics of wind speed and direction. This discussion explores the relationship between the west and east monsoons in the Java Sea with the direction of fishing trip. Technological difficulties and current challenge in sail technology on fishing vessel for blue swimming crab are identified. The possible ideas for future research in this field are also discussed.

Keywords: Blue Swimming Crab, Fishing Vessel, Java Sea, Sail Technology

1. Introduction

The increase in fuel prices will gradually have an impact on the capture fisheries sub-sector, especially the fishermen. It is estimated that there are 2.2 million fishermen who lose money because they have to pay more for operational costs [1]. The impact of rising fuel prices on the capture fisheries industry is greater than on aquaculture since fuel expenses account for 45% of the overall costs in the capture fisheries industry. This is due to the fact that demand for fish is inelastic (price increases do not necessarily result in a reduction in demand), and the price increase is also influenced by seasonal factors [2]. In addition, the scarcity or crisis of fuel oil in various regions that has occurred recently has often forced fishermen not to go to sea, including the sudden increase in the price of basic necessities which also causes fishermen to find it difficult to go fishing [3]. The scarcity of diesel oil is due to the reduction of the diesel quota where in 2021 it will be around 15.8 million kilo liters while this year it is 15.1 million kilo liters [4]. The rise in diesel/gasoline prices will have a cumulative effect on rising production costs across all economic sectors, leading to price increases across the board and collectively driving up inflation nationally. With the increase in fuel prices, inflation will also increase. [5].

Sail technology is an alternative that can be used to save fuel oil, although it is realized that this technology cannot completely replace the diesel motor that is commonly used in fishing operations. Savings fuel on the use of sail is very dependent on wind conditions. In addition, sail usage takes almost 2 times if compared to machine use. But the sail can be used as an alternative for fishing vessel propulsion [6].

In this paper, the application opportunities of sail technology on fishing vessels for blue swimming crab *(Portunus pelagicus)* in deep waters of the Java Sea are explained. This fishing ground is more than 12 nautical miles from the coast. By discussing the relationship between the east and west monsoons in the Java Sea with the direction of the trip from fishing port to fishing ground and vice versa, where the angle of the wind to the sail affects the speed of the vessel. In addition, the technological difficulties and current challenges in sail technology on a fishing vessel for blue swimming crab are identified and the possible ideas for future research in this field are also discussed.

2. Sail technology for fishing vessel

2.1. A possibility for fishing boats with sails from previous research

Several types of sails commonly used on ships are soft sails (canvas), rigid sails, and mechanical sails. Most of the fishing boats use soft sails (canvas), where Japan has developed the "Enoshima-Maru" fishing boat with a GT 160 and a sail area of 73 m^2. In Japan, fishing boats use such basic sails. The sail technology was created by Aburatsubo Port Service Co., and a lot of sail-powered fishing vessel were built about 1985 [8]. Fishing Technology Center-Ministry of Marine Affairs and Fisheries of Indonesia has tested a fishing boat using a sail on KM. Maju with GT 7 and sail area of 44 m^2 operated in Probolinggo, East Java in 2010 [6]. Another study revealed that the analysis of the sail's thrust was done using varying wind speeds, specifically 4–12 knot in the sail's direction of attack (Angle of Attack) of 0^0, 30^0, 60^0, and 90^0. According to the findings of the CFD research, the sail's highest contribution to the ship's speed is 3.79 knots for rectangles, 4.3 knots for trapezoidal sails, and 4.4 knots for triangular sails. The sails 90^0 location provides the wind direction's contribution (Angle of attack) [3]. However, due to the high costs of maintenance and investment, there have been numerous attempts to use sails as a propulsion alternative for fishing boats. However, these prior sail-assisted ship experiences and methods propose one of the answers to combat global warming [8].

2.2. The physics of sailing

Just like other sailing boats, fishing vessel have the potential to use sails as propulsion. The case is similar for sailboats. Sailboats can move forward of course by capturing the wind speed at a certain angle. And the best wind-in angle is from the side of the boat or 90° from the direction of the boat or what is commonly called "abeam" (from the side). In extreme terms, the speed of the sailing ship can be 3 times the wind speed with a record of the ship's resistance being as small as possible, but some racing sailing boats are designed to be more than 2 times the wind speed [7].

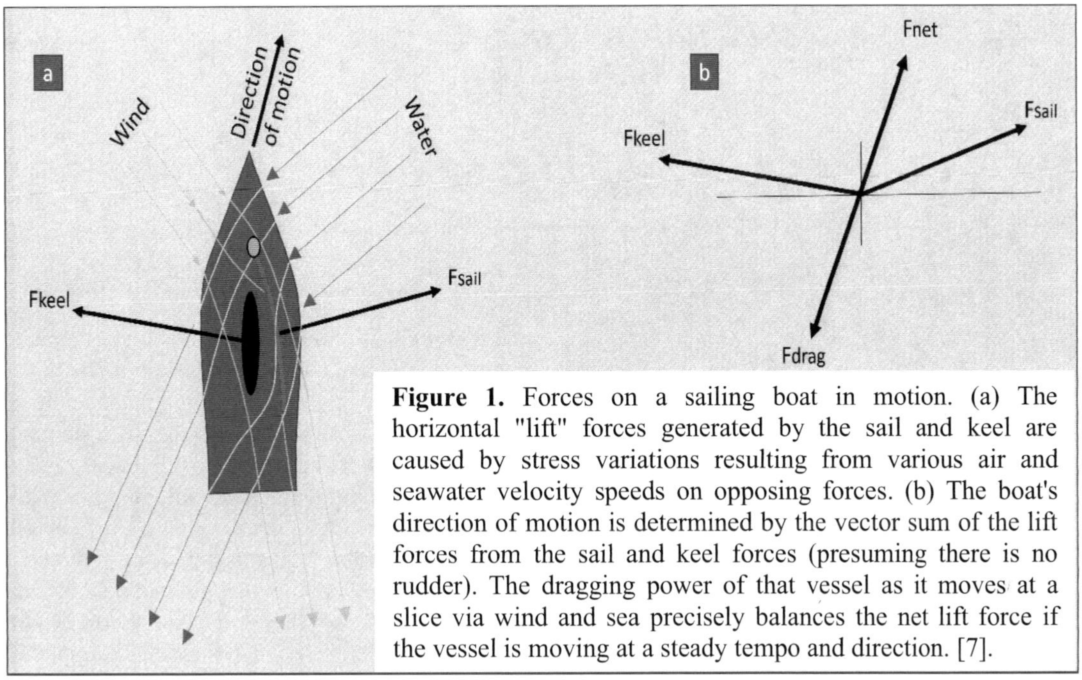

Figure 1. Forces on a sailing boat in motion. (a) The horizontal "lift" forces generated by the sail and keel are caused by stress variations resulting from various air and seawater velocity speeds on opposing forces. (b) The boat's direction of motion is determined by the vector sum of the lift forces from the sail and keel forces (presuming there is no rudder). The dragging power of that vessel as it moves at a slice via wind and sea precisely balances the net lift force if the vessel is moving at a steady tempo and direction. [7].

2.3. Predicting speed

Winds energy on sail rigging is assumed being proportionate squared perceived airflow rate by the entire sails surface, according to Bernoulli's principle. Then, using empirical factors for lifting and dragging that are equations of sails configuration and angles of entry, the real forces are determined. The wetted surface area of the hull determines frictional resistance, which rises with the square of boat speed. Empirical coefficients are used in each and every one of the numerous resistance contributions. The "prismatic coefficient" of the hull, which is an inverse measurement of the taper of its ends, is used to express wave and form resistance.

Computer programs for predicting speed range in complexity. Some that have been developed over many years for use in racing are kept secret and under careful guard. Figure 2 displays the outcomes of calculations using a freely accessible program for sailing yacht at 30 feet (10 meters). The calculated boat speed is depicted in the figure based on the angle of sails and the air strength. When cruising at around 90 degrees against that wind, the projected sailboat rates are the highest. Beam reaching is how sailors refer to it. It produces a boat speed that is roughly half that of the wind. [7].

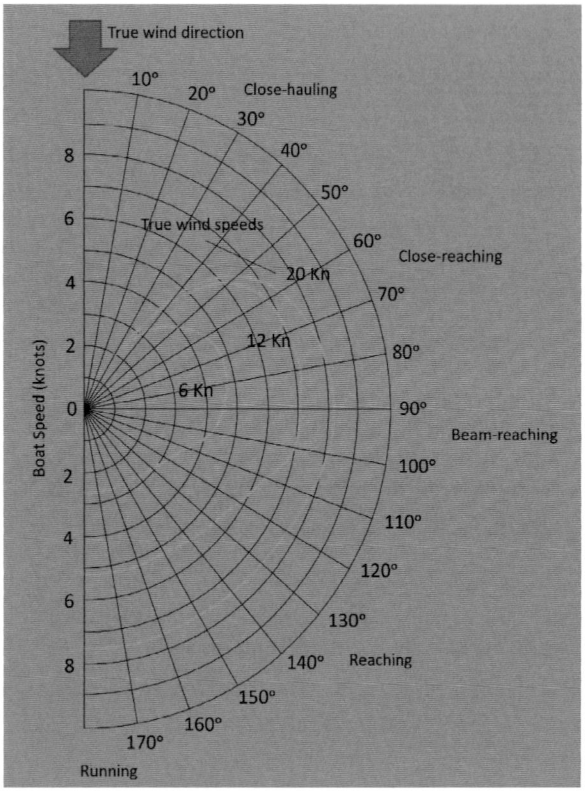

Figure 2. Plotted according to its boat's movement position in relation to the blow vector are the speeds anticipated using a computational simulation for three distinct air rates between 6 to 20 mph with a 10-meter sailing boat. (18.5 km/h = 10 mph). When its angle 180 degrees, this vessel was "running" against the flow directly behind it. While the vessel is "beam reaching," or traveling on an angle of around 90° to the wind, the fastest speeds are projected. Even when "close hauling" almost directly into the wind, the boat makes some progress [7]

3. Blue swimming crabs

The blue swimming crab (BSC) *(Portunus pelagicus)*, a key economic fishing product with great demand and a high price, is a source for maritime biology [9-10]. BSC are found in estuary, lagoon, and sea grass environments on Eastern Asian and Southeastern Asian coastline area with mud or sand bottoms at a greatest depth of 40 m [11-12]. Furthermore, it is spread throughout Indonesia, including the south and east of Kalimantan, the southeast of Sulawesi, and the south-west of Papua [13]. The research found that BSC has a short lifespan (less than three years) and a high growth rate (100 mm in months). In the Java Sea, September, October, and November are the three months with the highest concentration of ovigerous females. This presupposes that the ovigerous females will hatch their eggs after a few days. And suggested that the off-season falls between March - May and August – September in order to ensure the long-term viability of this species [14]. On the east coast of Lampung, the optimal average crab-size carapace width is 145.54 mm caught in an area >12 miles from the shoreline at a depth of >10 meters [15].

IOP Conf. Series: Earth and Environmental Science 1166 (2023) 012026 doi:10.1088/1755-1315/1166/1/012026

Figure 3. Blue Swimming Crab (BSC) *(Portunus Pelagicus)* [19]

4. Wind characteristics in the Java Sea

There are 2 (two) wind seasons in the Java Sea: the west monsoon and the east monsoon. During the east monsoon, the wind blows from the east, while in the west season, the wind blows from the west. Typically, the east season runs from May to October, whereas the west season lasts from November to April. According to one study, the Java Sea region's surface wind speed (10 m) in the east monsoon averaged 4-6 m/s, whereas the west monsoon averaged 3 m/s during the course of 20 years (2000-2019) [16] (see figure 4.).

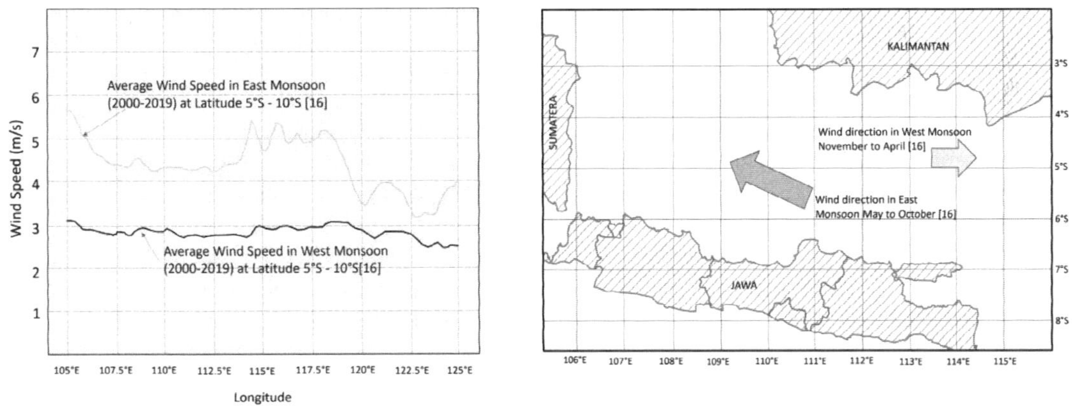

Figure 4. The average wind speed in the south latitude is 5° S to 10° S (left) and the general wind direction in the Java Sea (right) [16].

In the area surrounding the Seribu Islands of the Java Sea, according to another study, the maximum speed was 8.61 m/s and the average was 6.65 m/s over the course of 30 years (1984–2013) [17] (figure 5.). Additionally, according to [18], implementation in the field requires careful calculations because wind speed and wave height are closely related.

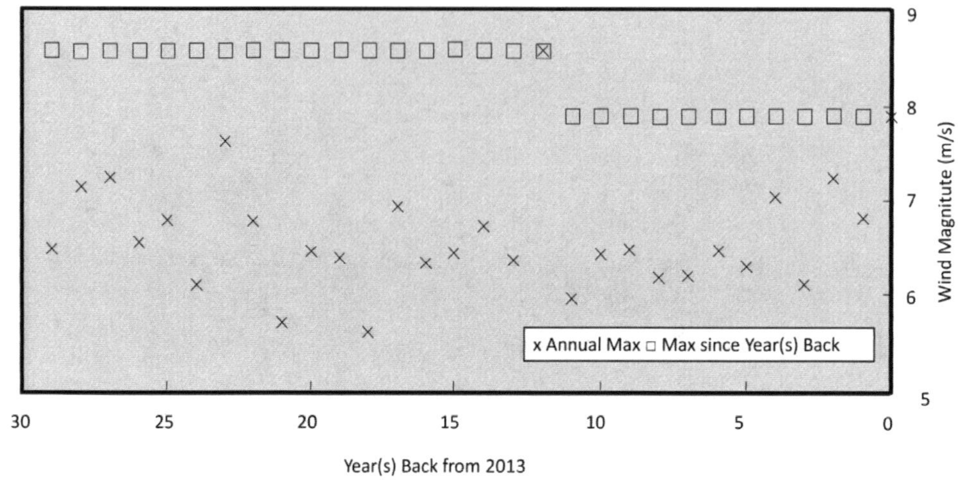

Figure 5. The annual maximum unidirectional speed for the 30-year wind data is displayed on the year(s) back series [17].

5. Result and discussion

A fishing vessel that satisfies the criteria of the three pillars of sustainability with regard to the social, economic, and environmental elements over the course of its life cycle is simply said to be sustainable [20]. Sail technology, used as a fishing vessel propulsion method, is one simple and established technology. With the advancement of technology in the capture fisheries sub-sector, including boats, fishing gear, and deck machinery, a fisherman will most likely consider employing technology that is more ecologically friendly from an economic standpoint. Like the puring net fishermen in Pulolampes, the 'ciker net' in Teluk Penyu, then the 'cantrang' on the north coast of Java [21,22,23], and the BSC fisherman in Rembang, they are basically aware that the use of eco-friendly and economical technology will be very helpful them in carrying out fishing activities, but again problems will always arise, both technical and non-technical which makes them return to old habits that tend to be inefficient and not environmentally friendly. Especially lately, the increase in fuel oil is felt by fishermen so they choose not to go to sea because the results obtained are not proportional to the capital spent. Additionally, they will suffer because the cost of essential goods will undoubtedly increase.

These issues have led to the development of technology that can help fishermen, one of which is the use of sails as a kind of boat propulsion for catching fish. Using a sail to move the boat when leaving for the fishing ground and when returning to the fishing base is considered very likely for fishermen to be able to return to the sea at a relatively low cost, especially for fishermen who are on the north coast of Java. It is understood that sails cannot completely replace the diesel engine. In addition, the fisherman is accustomed to using sails, but with a little instruction, they will undoubtedly be able to control the sails on their boats.

Geographically speaking, a fisherman from Java's north coast has an edge when using sails because boats travel south to north when leaving and north to south while returning. This is due to the fact that the Java Sea has two winds, the west monsoon and the east monsoon, which causes the wind to catch on the sail at an angle between 70 and 110 degrees, making it particularly effective for moving boats. The following is an illustration in general in the Java Sea:

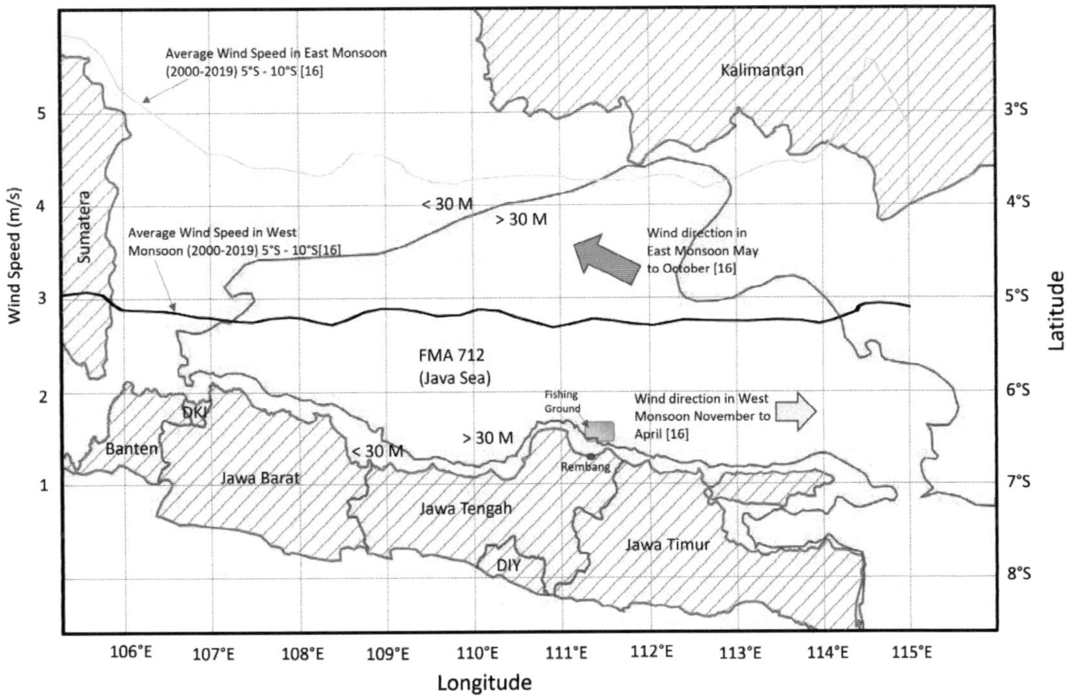

Figure 6. Depicts the overall location of the Java Sea in relation to the Java Sea's potential use of sails in terms of wind direction and speed.

Based on figure 6 above, it shows that with a potential wind speed of 4-6 m/s and a relative wind direction from east to west in the east season, from May to October, the boat will easily take advantage of the wind angle to move north to the fishing ground and vice versa. Meanwhile, during the west season (November – April), the wind speed is around 3 m/s relatively smaller than the east season. However, for BSC-catching vessels, it is recommended not to catch BSCs in March-May and August-September [14]. So that the BSC fisherman can rest, fix the boat, and prepare their fishing equipment in accordance with the light wind speed.

The following is an example of the scenario for the BSC capturing operation and goes into greater detail about the prospects for utilizing sails on crab catching vessels, notably in Rembang as one of the centers for BSC on Java Island:

IOP Conf. Series: Earth and Environmental Science 1166 (2023) 012026 doi:10.1088/1755-1315/1166/1/012026

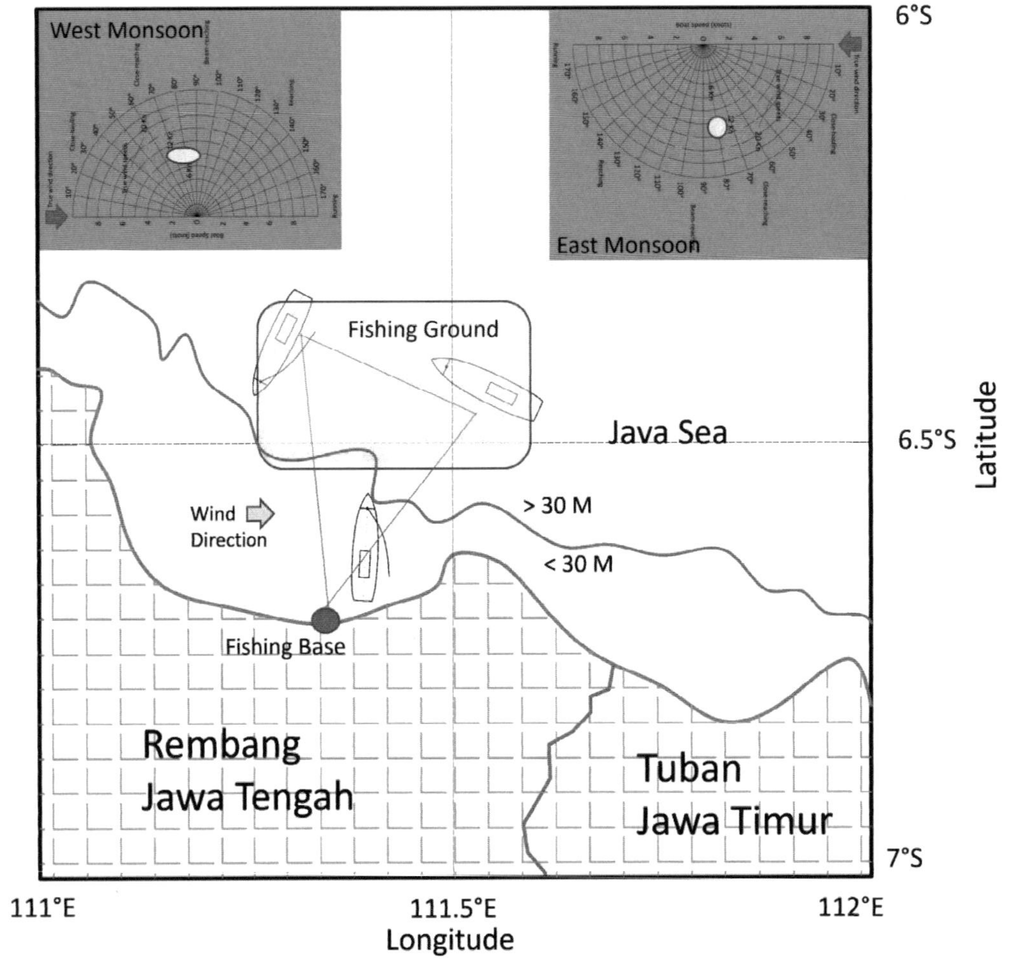

Figure 7. Is an example of a sail-powered BSC fishing operation in Rembang.

Based on figure 7, it is clear that BSC capturing vessels can readily use a sail to travel to the fishing site given the potential breeze velocity and direction in the Java Sea. The boat's speed can be expected to be around 5-7 knots with a wind speed of 4-6 m/s or 8-12 knots, and around 4-5 knots with a wind speed of 3 m/s or 6 knots. And for setting and hauling operations of fishing gear, the boat is driven by a diesel engine. Therefore, the following scenario best describes how the BSC's catching operation will proceed:

1. Using a diesel engine to leave the port
2. Using a sail to travel to a fishing area
3. Using a diesel engine to set up and hauling fishing gear
4. Using a sail, return to the fishing base
5. Start a diesel engine and enter the port area.

It is anticipated that if the distance between the fishing base and the fishing ground is 20 nautical miles, it will take 3–4 hours to travel there during the east season and 4-5 hours during the west season.

In addition to the benefits that fishermen can gain from employing sail technology, there are a number of difficulties associated with using it, such as the need for careful calculations when choosing the sail

and mast materials for people who are not accustomed to using sails on boats. Adding sails to boats is a problem in and of itself that calls for both accuracy and caution. One of the best ways to maximize the effectiveness of fishing operations is to design sailing boats from scratch.
BSC fishing vessels are extremely likely to utilize environmentally friendly equipment, particularly other new renewable energy technologies. For instance, solar cooling units can be used to store crabs so that the quality is maintained and fishermen can spend up to seven days at sea. The usage of electric motor drives in conjunction with a sail is also something that will be very interesting to study and implement in the future.

6. Conclusion

Based on the findings of the discussion above, it can be said that sail technology has the potential to be used by fishermen along Java's north coast, particularly those who capture BSCs in the Java Sea. In order to save operating expenses, research is also required to determine the best design for a sail-powered BSC fishing vessel.

7. Acknowledgement

This article is a part of our research project that was funded by the National Research and Innovation Agency (BRIN) in Degree by Research (DBR) Program. The authors appreciate all of the cooperation from the Head of Research Center for Hydrodynamics Technology and the Head of Marine Engineering Department, Faculty of Marine Technology, ITS.

8. References

[1] Jelita I N 2022 Harga BBM Naik, KKP Ungkap 2,2 Juta Nelayan Merugi *Media Indonesia*, https://mediaindonesia.com/ekonomi/510394/harga-bbm-naik-kkp-ungkap-22-juta-nelayan-merugi (accessed August 30, 2022).
[2] Mira Rahadian R Zulham A 2014 Impact of Fuel Prices Increased to The Performance of Marine and Fisheries Sector *J. Sosek KP* Vol. **9** N0. 2
[3] Anwar M U Trimulyono A Iqbal M 2016 Studi Perancangan Desain layar Berbasis Cfd Pada Perahu Motor Tempel Tipe Katir Di Pelabuhan Sadeng Guna Penghematan Bbm *F. Teknik and U. D. Semarang* vol. **4** no. 1 pp 306–313
[4] Lukmansyah O 2022 Solar Langka, Nelayan Ngeluh Susah Melaut *CNN Indonesia,* https://cnnindonesia.com/ekonomi/20220408130655-85-782149/solar-langka-nelayan-ngeluh-susah-melaut (accessed Apr. 08, 2022)
[5] Setyawan D 2014 The Impacts of the Domestic Fuel Increases on Prices of the Indonesian Economic Sectors *Energy Procedia* Volume **47** Pages 47-55 ISSN 1876-6102 https://doi.org/10.1016/j.egypro.2014.01.195. https://www.sciencedirect.com/science/article/pii/S1876610214002112
[6] Rianto Y Nasirudin A 2011 Analisa Ukuran dan Bentuk Layar Kapal Ikan Jenis Purse Seine; Studi Kasus: KM. Maju *Final Project* Jurusan Teknik Perkapalan Institut Teknologi Sepuluh Nopember Surabaya
[7] Bryon D. Anderson 2008 *Physics Today* 61, 2, 38; doi: 10.1063/1.2883908
[8] Yoshimura Yasuo 2001 A Prospect of Sail-Assisted Fishing Boats *Fisheries Science* **68** (Supplement 2) 1815-1818 https://doi.org/10.2331/fishsci.68.sup2_1815, 2002-06, Proceedings of International Commemorative Symposium - 70th Anniversary of the Japanese Society of Fisheries Science, October 1-5, 2001, Yokohama.
[9] Zairion Wardiatno Y and Fahrudin A 2015 Sexual maturity, reproductive pattern and spawning female population of the blue swimming crab, Portunus pelagicus (Brachyura: Portunidae) in East Lampung Coastal Waters *Indonesia Indian Journal of Science and Technology* **8**(7) 596-607
[10] Santoso D Karnan Japa L and Raksun 2016 Bio-ecological characteristic of blue swimming crab (Portunus pelagicus) at Dusun Ujung water, East Lombok *Jurnal Biologi Tropis* **16**(2) 94-105

[11] Lai J C Y Ng P K L and Davie P J F 2010 A revision of the Portunus pelagicus (Linnaeus, 1758) Species complex (Crustacea: Brachyura: Portunidae), with the recognition of four species *The Raffles Bulletin of Zoology* **58**(2) 199-237.

[12] Ng P K L 1998 Crabs. In: The Living Marine Resources of the Western Central Pacific, Volume 2: Cephalopods, Crustaceans, Holothurians and Sharks. In: Carpenter K E and Niem V A (Editors) *FAO Species Identification Guide for Fishery Purposes* (Rome: FAO of the United Nations)

[13] Sumiono B 1997 Fishing activities in relation to commercial and small-scale fisheries in Indonesia *Proc. of the Regional Workshop on Responsible Fishing, Bangkok, Thailand* pp 41-70 Samut Prakarn, Thailand: Training Department, Southeast Asian Fisheries Development Center.

[14] Ernawati T Sumiono B Madduppa H 2017 Reproductive ecology, spawning potential, and breeding season of blue swimming crab (Portunidae: Portunus pelagicus) in Java Sea, Indonesia *Biodiversitas Journal of Biological Diversity* **18**(4):1705-1713. DOI: 10.13057/biodiv/d180451

[15] Zairion Boer M Wardiatno Y Fahrudin A 2014 Composition and Size of The Blue Swimming Crab *(Portunus Pelagicus)* Caught at Several Bathymetric Stratifications in East Lampung Water *J. Lit. Perikan. Ind.* Vol.**20** No. 4 December 2014: 199-206.

[16] Sofiati I Yulihastin E Suaydhi Putranto MF 2020 Meridional variations of sea surface temperature and wind over southern sea of Java and its surroundi *Journal of Physics and Its Applications*, **3**(1) Pages: 129-135, http://ejournal2.undip.ac.id/index.php/jpa/index

[17] Poerbandono 2016 Wind characteristics and the associated risk of erosion in Seribu Islands patch reef complexes, Java Sea, Indonesia *AIP Conference Proceedings* 1730, 080001 (2016); doi: 10.1063/1.4947416

[18] Sugianto D N Zainuri M Darari A Suripin Darsono S and Yuwono N 2017 Wave Height Forecasting Using Measurement Wind Speed Distribution Equation In Java Sea, Indonesia *International Journal of Civil Engineering and Technology* **8**(5) pp 604-619 https://iaeme.com/Home/article_id/IJCIET_08_05_068

[19] Oleh self - Karya sendiri, CC BY-SA 3.0 https://commons.wikimedia.org/w/index.php?curid=4963391

[20] Wibawa I P A 2016 Sustainable Fishing Vessel Development by Prioritising Stakeholders Engagement in Indonesian Small-Scale Fisheries *A Thesis* School of Marine Science and Technology Newcastle University United Kingdom December 2016

[21] Sasmita S et al 2018 Distribusi Ukuran Ikan Teri Hasil Tangkapan Jaring Puring di Perairan Pulolampes, Kabupaten Brebes Jawa Tengah *Journal of Fisheries and Marine Science* Vol. **2** No. 2

[22] Wassahua Z Martasuganda S Baskoro M S Sondita M F A 2021 The comparison of sweeping trammel net area operated with different towing line *AACL Bioflux* **14**(1):91-102.

[23] Sasmita S Martasuganda S Purbayanto A 2012 Keragaan Desain Cantrang pada Kapal Ukuran <30 GT di Pantai Utara Jawa Tengah *Jurnal Perikanan dan Kelautan* Vol. **II** No. 2: 79-86

IOP Conf. Series: Earth and Environmental Science 1166 (2023) 012027 doi:10.1088/1755-1315/1166/1/012027

Oceanographic survey on current due to breaking wave for energy generations: a case study of Pacitan bay, East Java, Indonesia

S Lestari[1,*], Mukhtasor[1,*], D Satrio[1,*] and S Rahmawati[1]

[1] Department of Ocean Engineering, Institut Teknologi Sepuluh Nopember, Surabaya, Indonesia

*E-mail address: sri.lestari15@mhs.oe.its.ac.id; mukhtasor@oe.its.ac.id; dendy.satrio@its.ac.id

Abstract. The southern sea of Indonesia faces the open ocean i.e., the Indian Ocean. This provides an initial hypothesis that the location has a ti potential for wave energy throughout the year. The current due to the breaking wave is estimated to have a high velocity. This study focuses on analyzing the velocity of current oceanographic survey due to breaking waves with a case study in Pacitan bay, East Java, Indonesia. The oceanographic survey method uses ADCP and a Current Meter to capture wave and current data sampling with the purpose of the research was for the electrical energy generation near the coast. ADCP is used to record the occurrence of waves and ocean currents before the wave break. The Current Meter is used to record current occurrences after the wave breaks. The instantaneous absolute current velocity, ca, and return current velocity, U, reach a value of 2-3 m/s.

1. Introduction

Marine energy was chosen as an alternative power source because it has three advantages: renewable, sustainable, and carbon-free [1]. Indonesia as an archipelagic country has many straits with a tidal current energy potential of around 17,9GW [2]. Until nowadays, this potential has not met electricity demand. And this may be our time.

Research related to the tidal current that has been carried out includes a numerical evaluation to determine the characteristics of fish aggregating devices (FADs) as well as a tidal current energy conversion system using a vertical axis turbine and a horizontal axis turbine [3]. Then, laboratory experiments were carried out to determine the self-start capability of the vertical axis turbine [4]. The next research is the numerical simulation of the cross-flow Savonius turbine for locations with low current velocity in Indonesia [5] and the advantage of flow disturbance for vertical-axis turbines with low current velocity [6]. The next research is a floating tidal current power plant using multi-vertical-axis turbines [7]. With relatively less tidal current velocities, typically in the order of about 3,9 m/s at maximum, a vertical axis turbine is one of the options available for Indonesia.

On the other hand, the installation of a marine energy conversion system can produce corrosion, so several attempts have been made to select materials that support the system's performance [8]. Tidal current energy sources provide extensive research and development advantages for marine energy to be converted into electrical power [9], [10]. Especially with the density of seawater reaching 804 times denser than the density of air, it provides a great advantage for energy generation from the ocean [11].

Same to tidal current energy, Indonesian ocean wave energy has similar advantages. It is dominant in the southern sea of Indonesia and it faces the open ocean i.e the Indian Ocean. Ocean wave energy is generated from the transfer of wind energy that touches the sea surface and then moves it to generate the waves [12]. However, waves do not travel on motionless seawater. Waves in the ocean stroll with ocean currents (tidal currents) [13]. Where the waves are generated by wind energy, while the tidal current is caused by the earth's gravity. When both of them propagate closer to shallow water, a shoaling phenomenon occurs. The sea surface profile turns steeper along with the shallower seabed profile so that the wave height continues to increase, then becomes unstable and forms waves break [14]. The phenomenon of breaking waves is often explained as a separate event by not mentioning the role of the tidal current.

After the breaking waves, mixing occurs where the energy of currents and ocean waves interact in complex situations. Then the current due to breaking waves are generated. The current energy due to breaking waves is continuous and has a high energy density allowing the device to generate more power at a lower cost and a location close to the shoreline makes installation easier [15]. The current energy due to breaking waves is predicted to have a high velocity at certain locations and a certain type of breaker. By developing this energy, ocean current energy exploration will not only focus on tidal currents. The current energy on the move: current due to breaking wave.

In the last six decades, the development of science-based nearshore circulation induced by waves has continued. Early research places a great focus on being able to explain the physical phenomena that occur. This research attempts to explain the complex physical phenomenon of breaking waves. Then the wave character at the breaking point can be explained by two approaches i.e. breaking positions and breaker index [16]. Research then attempts to explain the stress that occurs in current and wave association after the wave breaks [17]. And, three types of breaking waves on the coast then be classified based on the relative bottom slope. These three types of breaking waves are associated with energy dissipation which affects the current velocity due to breaking waves and can be identified by qualitative and quantitative assessments [18]. All of these results are the basis for current research on energy generation.

Energy generation from currents due to breaking waves began to be developed in this decade. One of the current energy converter technology innovations due to breaking waves is Ducted WEC [19]. Ducted WEC utilizes current energy due to breaking waves to be converted into useful electrical power. In a significant wave height of 1 meter, the turbine can reach Cp 0,38 at maximum. The ducting structure in this system is in charge of directing the flow and increasing the flow velocity up to six times [11], [20]. This research was carried out until the WEC prototype was created which was installed on a relatively mild ocean at Kandooma Island, Maldives.

Current due to breaking waves energy is the source of energy for the future. It is a silent, invisible, predictable, clean, and constant source of energy at a certain location. The description leads to the current research that describes the Oceanographic Survey on Currents due to Breaking Waves for Energy Generation with case studies in the Indonesian sea. The topic of research on energy generation from currents due to breaking waves is a novelty, so research and development are very open to being carried out on the principle of mutual benefits.

2. Methodology

2.1 Site selections

This phase starts with the identification of problems in the research. The literature study obtains all relevant facts and is able to assist in selecting the right decisions for the work of this research. From this phase, three considerations were found for the site selection of research on Pacitan bay, East Java, Indonesia i.e.:

1. The location was chosen because it is on the southern sea of Indonesia. This provides an initial hypothesis that the location has a high potential for wave energy throughout the year.
2. The selected beaches have different types of breaking waves (based on the qualitative assessment at the beginning of the research phase). This is expected to provide information on the character of the current velocity due to breaking waves for the purpose of energy generation.
3. Pacitan Regency has two location points that have not been electrified by PLN (the national power company of Indonesia) from a total of 146 points throughout East Java [21].

2.2 Data Collections

The data needed are primary data. Primary data includes: wave parameters (Hs, Tp, v) and current parameters due to breaking waves (ca, U) obtained from the oceanographic survey. Figure 1. below is the primary data collection scenario.

IOP Conf. Series: Earth and Environmental Science 1166 (2023) 012027 doi:10.1088/1755-1315/1166/1/012027

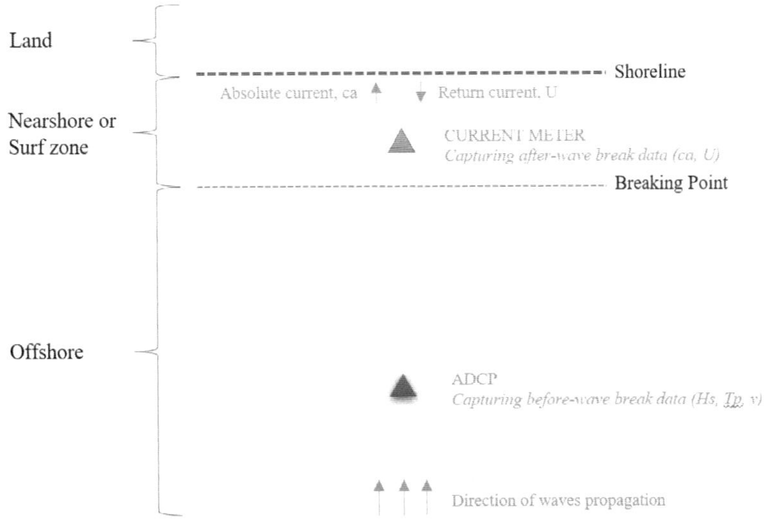

Figure 1. Oceanographic data collection scenario

The picture above is a scenario of the oceanographic survey in the Pacitan sea. The oceanographic data collection process utilizes two main oceanographic instruments, namely Current Meter and ADCP (Acoustic Doppler Current Profiler). The Valeport MIDAS ECM Electromagnetic Current Meter is planned to measure the current due to breaking waves at a sensor depth of 0.5 meters on a Mean Water Surface (MWS) as high as 0.8-1 meters. Current Meter records current velocity data per second. Current Meter data collection will be carried out simultaneously with the offshore ADCP data collection process. This is intended to analyze the relationship between ocean waves before the waves break with currents after the breaking wave. The team will simultaneously operate the GPS (Global Positioning System) every time data is taken with a Current Meter to mark the measurement location point.

Sontek Argonaut-XR 750kHz ADCP is placed at a depth of fewer than 30 meters with a distance of 800 meters from the Current Meter measurement point. ADCP is set to be able to record one data in every 10 minutes duration. So that 6 data will be obtained in the measurement duration of 1 hour. When in the deployment stage, the team was assisted by an Echosounder to obtain underwater bathymetry depth information in accordance with oceanographic data collection scenarios. Next, the team will simultaneously operate a GPS at the same time to mark the measurement location point. The following is a table specification of the oceanographic instruments:

Table 1. Detailed specifications of the oceanographic instruments

No	Subjects	Specifications	Detail Specifications
1	ADCP	Sontek Argonaut-XR 750kHz	Integrated piezoresistive pressure sensor 0,1% accuracy [22]
2	Current Meter	Valeport MIDAS ECM Electromagnetic Current Meter	Valeport 11cm discus EM sensor range 0-5 m/s; accuracy +- (1% of reading + 0,005 m/s) with resolution 0,001 m/s [23]
3	Echosounder	Garmin GPSmap 585	GPS 10 meters (33 feet) RMS 95% typical; DGPS (WAAS) < 3 meters (10 ft) 95% typical with DGPS corrections [24]
4	GPS	Garmin GPS 73	-

3. Results and Discussion

3.1 Hydrodynamic Analysis Before Wave Break

As stated in Sub-chapter 2.2, the oceanographic survey scenario in this study utilizes two main oceanographic instruments i.e., Current Meter and ADCP, respectively. The oceanographic survey was carried out on July 1, 2022. The dates are the 2nd of the lunar calendar, chosen because the ocean currents were at a lower velocity and the weather was sunny so it was relatively safe for the ADCP deployment process. ADCP is placed on a relatively flat sandy seabed at a sea depth of 27.7 meters (based on Echosounder data measurements) at a distance of 800 meters from the point where Current Meter data is collected on Ngiroboyo Beach. When the ADCP has been installed and has started recording current and wave occurrences, the Current Meter is operated to record the current due to breaking waves on the beach.

Oceanographic data collection activities are pursued in accordance with the planned scenario. The following Figure 2. is the implementation of the oceanographic survey scenario for this research. Then below is an illustration of oceanographic data that was successfully recorded as presented in Figure 3.

Figure 2. Implementation of the full oceanographic survey scenario for the research of breaking waves in Pacitan, Indonesia (Source: Google Earth, 2022)

IOP Conf. Series: Earth and Environmental Science 1166 (2023) 012027 doi:10.1088/1755-1315/1166/1/012027

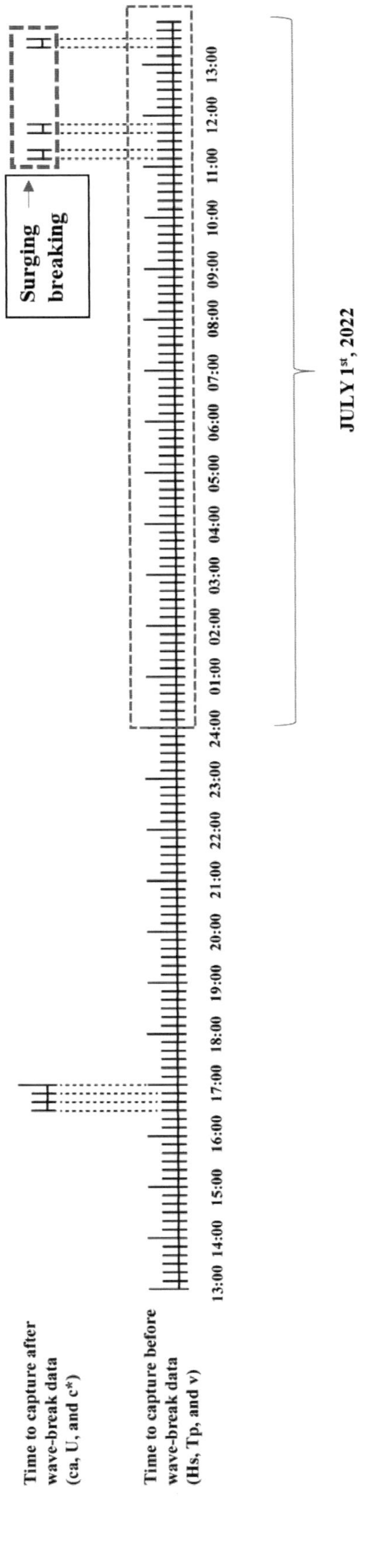

Figure 3. Data illustration from the oceanographic survey

The picture above attempts to show the part of measurement data by ADCP and Current Meter so as to predict the correlations. Wave measurements (Hs, Tp) by ADCP run for 14 hours. Along with wave measurements by ADCP, current measurements due to breaking waves have been collected from Ngiroboyo Beach (when the low tide). Data from Ngiroboyo Beach has been paused several times due to the water surface and high current velocity so it does not fully record 10 minutes of current events due to breaking waves.

The following is the distribution of the recorded oceanographic data as shown in Figure 4 and Figure 5.

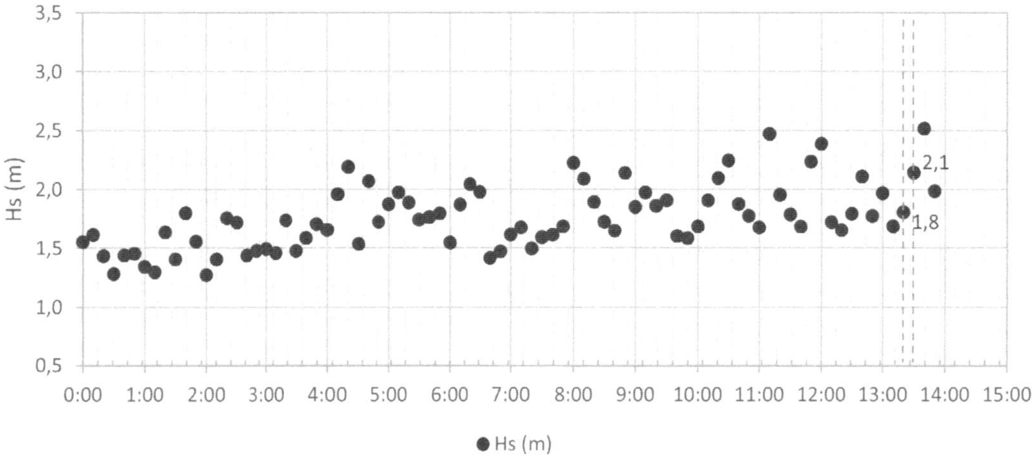

(a)

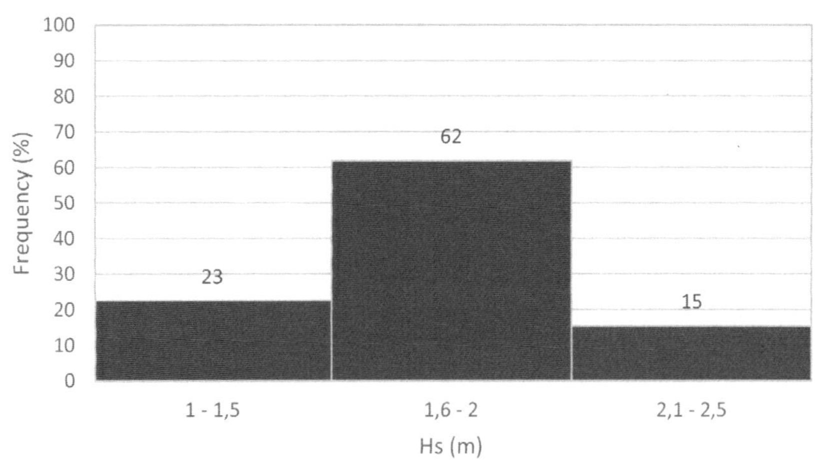

(b)

Figure 4. Significant wave height data distribution: (a). time series data and (b). histogram

This paper attempts to present data with good resolution and local occurrence to present clear correlations. Figure 4 shows that on July 1, 2022, at Pacitan sea, an oceanographic survey was carried out to obtain data on significant wave heights (Hs) per 10 minutes for 14 hours. As shown in Figure 4b, the distribution of Hs is dominated by a value of 1,6-2 meters with a frequency of 62%. While the Hs values recorded at 13.20 WIB and 13.30 WIB were worth 1,8 meters and 2,1 meters, respectively, as shown in Figure 4a.

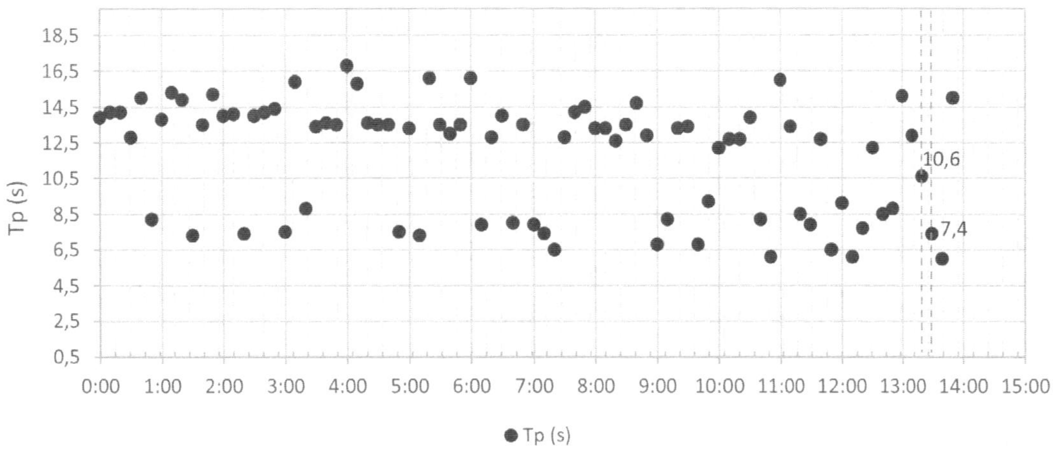

(a)

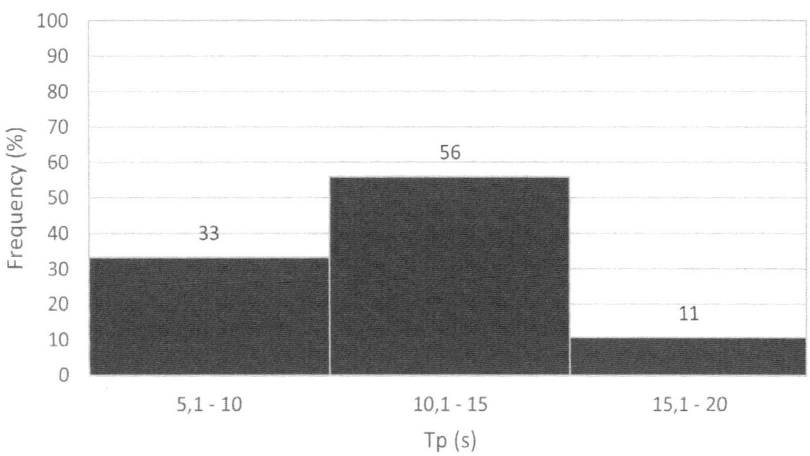

(b)

Figure 5. Peak period data distribution: (a). time series data and (b). histogram

Next, Figure 5 shows that on July 1, 2022, at Pacitan sea, an oceanographic survey was carried out to obtain data on peak period (Tp) per 10 minutes for 14 hours. As shown in Figure 5b, the distribution of Tp is dominated by a value of 10,1-15 seconds with a frequency of 56%. While the Tp values recorded at 13.20 WIB and 13.30 WIB were worth 10,6 seconds and 7,4 seconds, respectively, as shown in Figure 5a. This value is relatively in accordance with the hypothesis on wave energy potential in the southern sea of Indonesia as mentioned in Sub-chapter 2.1 Site Selections (point 1, pg. 3).

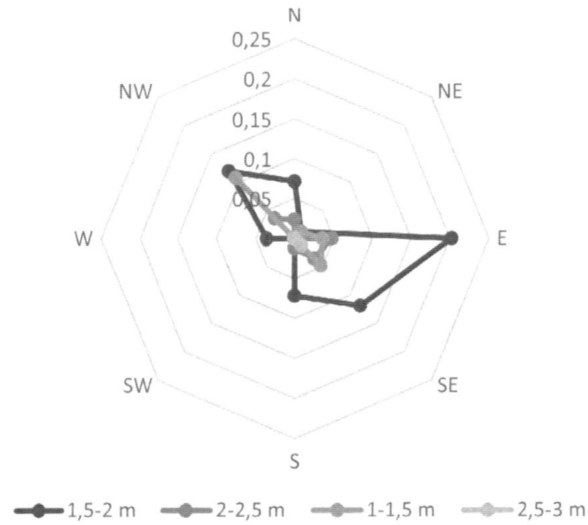

Figure 6. Wave rose

Table 2. Wave direction data distribution from the oceanographic survey

Year	Month	Day	Hour	Minutes	Hs (m)	Wave Dir (◦)	WD_
2022	7	1	13	20	1,8	273,8	W
2022	7	1	13	30	2,1	321,3	NW

Next, a wave rose analysis. A wave rose is applied for knowing the Hs value and the angle of the dominant wave in a time series data set. From Figure 6, it is stated that on July 1, 2022, at Pacitan sea, an oceanographic survey was carried out to obtain data on Hs and the magnitude of the angle of incidence of waves per 10 minutes for 14 hours. The result is the wave rose was dominant in the North West (NW), South East (SE) and East (E) direction as shown by the 'blue line' in Figure 6. The wave came from the direction with a range of values as high as 1,5-2 meters. While for local events, the direction of the incoming waves recorded in the oceanographic survey at 13.20 WIB and 13.30 WIB shows the direction of West (W) and North West (NW) respectively, as shown in Table 2.

Table 3. The wind direction data distribution from Yogyakarta Meteorological Station [25]

Date	V_{max} (m/s)	Max Wind Dir (◦)	V_{avg} (m/s)	Dominant Wind Dir (◦)
01-07-2022	12	130	6	SE

The results shown by the oceanographic survey data are then compared with the data obtained from the nearest wind measurement station i.e Yogyakarta Meteorological Station on the same date as shown in Table 3. The measurement results indicate the dominant wind direction is in the South East (SE) direction. This adds to our confidence that Hs and Tp on July 1, 2022, as shown in Figures 4 and Figure 5, are generated by the wind from the SE direction.

3.2 Hydrodynamic Analysis from Current due to Breaking Wave
When a wave first breaks at a certain depth, it will often break again until it reaches the shoreline (closer to zero depth). This area is then referred to as the surf zone. There is also a reference that calls it nearshore. Then in this area, an investigation will be carried out on the oceanographic survey on the current due to breaking waves for energy generation.

As we observed when standing on the beach, when the waves first broke, the propagation of waves experienced a rapid transition of the wave shape. The waves are no longer in the form of sine waves that propagate, but mixing events occur for some time until they eventually reach the shoreline. Currents due to breaking waves appear with a particular character that is formed from the association between current energy and ocean wave energy in a complex situation.

On beaches with Spilling Breaking and Plunging Breaking types, high turbulence and energy dissipation occur until the current flows reach the shoreline. However, it is different from the Surging breaking type where this type of breaking wave causes limited turbulence [18]. Because of this, there is a limited energy dissipation of the energy carried by the current due to the breaking wave (absolute current velocity, ca). This energy is eventually returned to the ocean as a return current velocity, U. If it is associated with the research objective i.e as a location for energy generation, information about the type of breaking wave becomes an important variable.

Ngiroboyo has a Surging breaking type [18]. Taken at low tide the following Figure 7 are current oceanographic measurements of the Ngiroboyo beach:

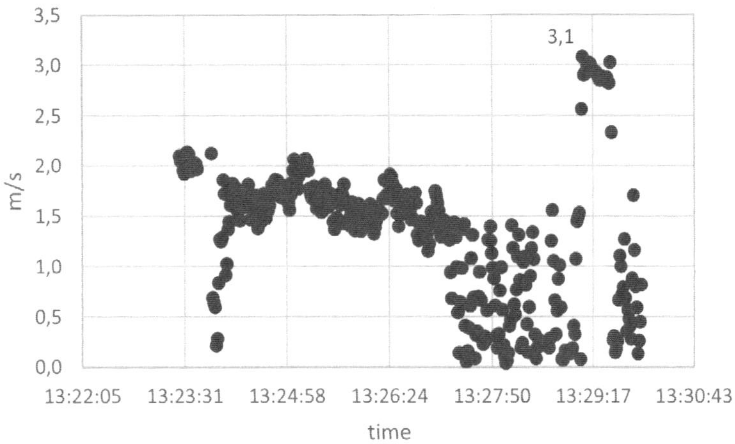

(a)

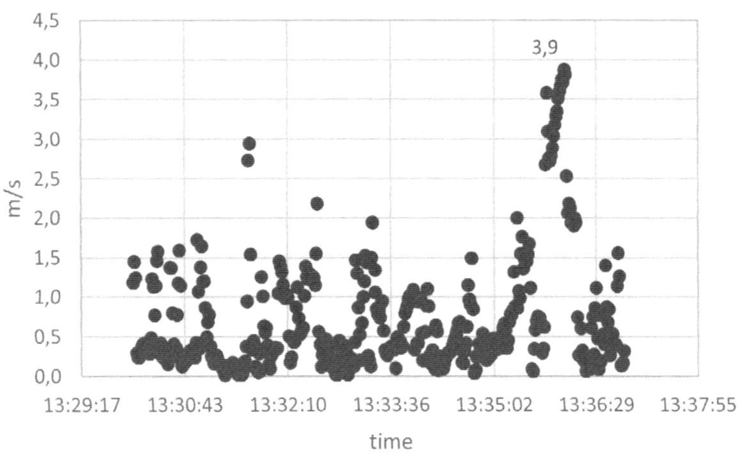

(b)

Figure 7. Ngiroboyo current due to breaking wave data: (a) time series at 13:23:27 – 13:29:59 WIB (as mentioned 13.20 WIB at Figures 4 and 5); (b) time series at 13:30:00 – 13:36:53 WIB (as mentioned 13.30 WIB at Figures 4 and 5)

Figure 7a-7b shows the current velocity values due to breaking waves at 13.20 WIB and 13.30 WIB, respectively, at Ngiroboyo beach. The ideal data is the current velocity data per second for 10 full minutes without a break, but due to safety considerations, this measurement is forced to be paused several times as shown in Figure 7a-7b. From the oceanographic survey of currents due to breaking waves, it is known that the velocity values at 13.20 WIB and 13.30 WIB are 3.1 and 3.9 m/s at maximum, respectively.

Next, from the current velocity value in figure 7a, the average values of ca 2.16 m/s and U 1.2 m/s are obtained. From the current velocity value in figure 7b, the average value of ca 0.72 m/s and U 1.69 m/s are obtained. The instantaneous absolute current velocity, ca, and return current velocity, U, reach values of 2-3 m/s.

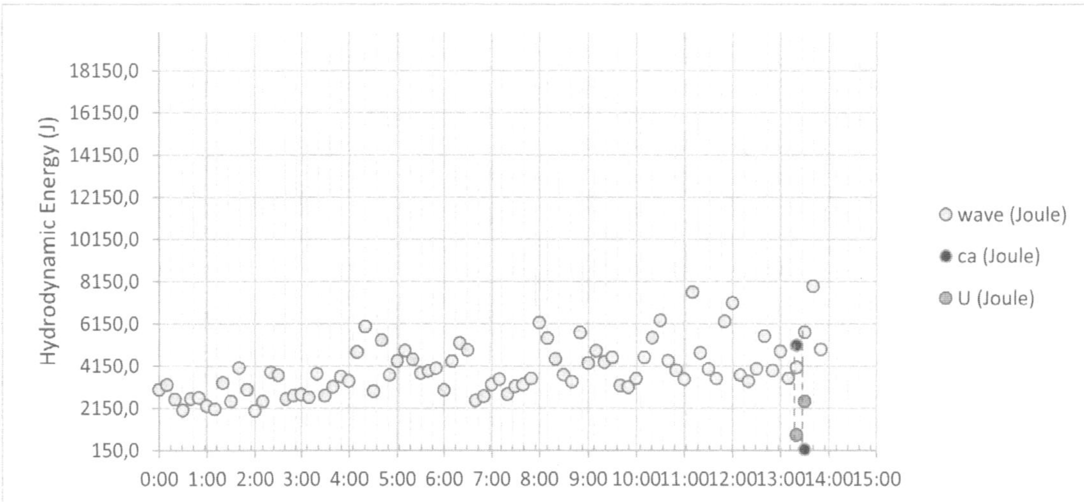

Figure 8. Hydrodynamic energy time series data distribution

Figure 8 shows the relationship between wave (Hs, Tp), ca, and U on a single graph. The hydrodynamic energy per unit area of wave (before the wave broke) at 13.20 WIB and 13.30 WIB was 4086,3 Joules and 5750.3 Joules respectively. While the hydrodynamic energy per unit area of ca is 5136.3 Joules and 190.6 Joules; and the hydrodynamic energy per unit area of U is 879.9 Joules and 2468.4 Joules, respectively. ca and U are the current variables after the wave breaks.

Table 4. Summary of the research on oceanography survey on current due to breaking wave for energy generation

Year	Month	Day	Hour	Minute	Hs (m)	Tp (s)	Ew (J)[26]	ca (m/s)	Ea (J)[27]	U (m/s)	Ea (J)[27]
2022	7	1	13	20	1,8	10,6	4086,3	2,2	5136,3	1,2	879,9
2022	7	1	13	30	2,1	7,4	5750,3	0,7	190,6	1,7	2468,4

A summary of all the results of the oceanographic survey on July 1, 2022, at Pacitan sea and Ngiroboyo beach, is presented in Table 4. From this study, it can be concluded that at a suitable location i.e Surging breaking, for example, Ngiroboyo beach, current energy due to breaking waves per unit area has the potential energy to generate electricity. Current energy due to breaking waves per unit area exceeds the wave energy per unit area before the wave breaks. Especially when both, ca and U, can be converted into electricity.

The results of the research make sense by observing that the current energy per unit area after the breaking wave is affected by the breaking wave phenomenon itself. Where the "shoaling" causes the wave steepness to reach a climax condition so that energy accumulation occurs just before the wave breaks. Coupled with the limited turbulence and energy dissipation that occurs on the beach with the type of Surging breaking as described in Sub-chapter 3.2.

4. Conclusions

From the results of the oceanographic survey, the following conclusions are obtained:

1. Oceanography survey on July 1, 2022, has been carried out for 14 hours in duration.
2. Based on this research, the Pacitan sea has the Hs 1,2-2,5 m and Tp 5,9-19,7 s propagates. Hs is dominated by a value of 1,6-2 meters with a frequency of 62%. Tp is dominated by a value of 10,1-15 seconds with a frequency of 56% based on this research.
3. The value of ocean current velocity due to breaking wave reaches 2-3 m/s at Surging breaking beach (Ngiroboyo beach) is potential for energy generation.
4. This potential was only taken and was in one location on the southern sea of Indonesia i.e., Ngiroboyo beach in Pacitan bay. While the commensurate amount was a lot and further exploration was needed.

The current energy potential due to the breaking wave is site-specific. Potential at location points with certain criteria. And as with other energy sources, this huge potential is directly proportional to some of the challenges encountered. Current energy due to breaking waves fluctuates daily or yearly due to irregular waves in the ocean. However, research and development are needed to obtain more alternative energy sources to meet mutual benefits to our better lives.

Acknowledgments

The authors would like to convey their appreciation to the Ministry of Education, Culture, Research and Technology of the Republic of Indonesia (Kemendikbudristek RI) which is funding the current project under a scheme called Master Thesis Research under main contract number: 084/E5/PG.02.00.PT/2022 and derivative contract number: 1452/PKS/ITS/2022.

References

[1] Paul A. Lynn, *Electricity from wave and tide*, 1st ed., vol. 1. London: Wiley, 2014.
[2] Mukhtasor *et al.*, "Indonesia ocean energy potential resources (in the Indonesian language)," Jakarta, 2014.
[3] S. Rahmawati, H. Mutsuda, Y. Doi, and Y. Moriyama, "Characteristics of a fish aggregating device with ocean energy harvester," *Journal of Marine Science and Technology (Japan)*, vol. 23, no. 3, pp. 435–452, Sep. 2018, doi: 10.1007/S00773-017-0482-6/FIGURES/33.
[4] D. Satrio and I. K. A. P. Utama, "Experimental investigation into the improvement of self-starting capability of vertical-axis tidal current turbine," *Energy Reports*, vol. 7, pp. 4587–4594, Nov. 2021, doi: 10.1016/J.EGYR.2021.07.027.
[5] D. Satrio, F. Muhammad, M. Mukhtasor, S. Rahmawati, S. Junianto, and S. Musabikha, "Numerical simulation of cross-flow Savonius turbine for locations with low current velocity in Indonesia," *Journal of the Brazilian Society of Mechanical Sciences and Engineering*, vol. 44, p. 315, Sep. 2022, doi: 10.1007/s40430-022-03620-w.
[6] D. Satrio, S. Suntoyo, and L. Ramadhan, "The advantage of flow disturbance for vertical-axis turbine in low current velocity," *Sustainable Energy Technologies and Assessments*, vol. 49, p. 101692, Sep. 2022, doi: 10.1016/j.seta.2021.101692.
[7] S. Junianto, W. N. Fadilah, A. Firyal Adila, Tuswan, D. Satrio, and S. Musabikha, "State of the art in floating tidal current power plant using multi-vertical-axis-turbines," *2022 International Electronics Symposium (IES)*, pp. 97–103, Aug. 2022, doi: 10.1109/IES55876.2022.9888749.
[8] S. Musabikha *et al.*, "Material selection for vertical axis tidal current turbine using Multiple Attribute Decision Making (MADM)," *MS&E*, vol. 1158, no. 1, p. 012001, Jun. 2021, doi: 10.1088/1757-899X/1158/1/012001.

[9] S. Junianto, M. Mukhtasor, R. Prastianto, and C.-H. Jo, "Effects of demi-hull separation ratios on motion responses of tidal current turbines-loaded catamaran," *Ocean Systems Engineering*, vol. 10, pp. 87–110, Mar. 2020, doi: 10.12989/ose.2020.10.1.087.

[10] D. Satrio, I. K. A. P. Utama, and Mukhtasor, "Numerical investigation of contra rotating vertical-axis tidal-current turbine," *Journal of Marine Science and Application*, vol. 17, no. 2, pp. 208–215, 2018, doi: 10.1007/s11804-018-0017-5.

[11] P. Halder, H. Takebe, K. Pawitan, J. Fujita, S. Misumi, and T. Shintake, "Turbine characteristics of wave energy conversion device for extraction power using breaking waves," *Energies 2020, Vol. 13, Page 966*, vol. 13, no. 4, p. 966, Feb. 2020, doi: 10.3390/EN13040966.

[12] S. Saincher and J. Banerjee, "Influence of wave breaking on the hydrodynamics of wave energy converters: a review," *Renewable and Sustainable Energy Reviews*, vol. 58, pp. 704–717, May 2016, doi: 10.1016/J.RSER.2015.12.301.

[13] T. S. Hedges, "Combinations of waves and currents: an introduction," *Proc. Instn Civ. Engrs*, vol. 82, Jun. 1987, pp. 567–585, 1987, doi: 10.1680/IICEP.1987.319.

[14] W. Mo, A. Jensen, and P. L. F. Liu, "Plunging solitary wave and its interaction with a slender cylinder on a sloping beach," *Ocean Engineering*, vol. 74, pp. 48–60, Dec. 2013, doi: 10.1016/J.OCEANENG.2013.09.011.

[15] F. M. Assidiq, R. W. Prastianto, and Mukhtasor, "A preliminary study of breaking waves phenomenon on a numerical wave tank," *IOP Conf Ser Earth Environ Sci*, vol. 557, no. 1, p. 012057, Aug. 2020, doi: 10.1088/1755-1315/557/1/012057.

[16] I. A. Svendsen and J. B. Hansen, "Deformation up to breaking of periodic waves on a beach," *Coastal Engineering Proceedings*, 1976, doi: 10.9753/icce.v15.26.

[17] U. K. Navera, "Development of a model for predicting wave-current interactions and sediment transport processes in nearshore coastal waters," Cardiff University, United State, 2004. Accessed: Oct. 09, 2022. [Online]. Available: https://www.semanticscholar.org/paper/Development-of-a-model-for-predicting-wave-current-Navera/dc123a341e7ab6a20f0d34b3852a12e5937c2e65

[18] I. A. Svendsen, *Introduction to nearshore hydrodynamics*, vol. 24. World Scientific Publishing Co., 2005. doi: 10.1142/5740.

[19] T. Shintake, "Technical R&D on a surf zone WEC," Aug. 2017. [Online]. Available: https://www.researchgate.net/publication/320072316

[20] K. A. Pawitan, H. Takebe, H. Andrean, S. Misumi, J. Fujita, and T. Shintake, "Duct attachment on improving breaking wave zone energy extractor device performance," *Energies 2021, Vol. 14, Page 6428*, vol. 14, no. 19, p. 6428, Oct. 2021, doi: 10.3390/EN14196428.

[21] ESDM Office of East Java Province, "East Java electrification data by PLN (in the Indonesian language)," Surabaya, May 2022.

[22] SonTek, Ysi, and Inc, "The practical solution to current and wave measurement," California, Sep. 2016. Accessed: Oct. 09, 2022. [Online]. Available: www.sontek.com

[23] Valeport Ltd., "MIDAS ECM operating manual," United Kingdom, Feb. 2019. [Online]. Available: www.valeport.co.uk

[24] Inc. Garmin International, "GPSMAP ® 580/585 chartplotter combo fishfinder owner's manual," USA, Jan. 2011. Accessed: Oct. 10, 2022. [Online]. Available: https://www.garmin.co.id/products/discontinued/gpsmap585/

[25] BMKG, "BMKG database center online data (in the Indonesian language)," *Meteorology Climatology and Geophysics Council*, 2015. http://dataonline.bmkg.go.id/home (accessed Oct. 02, 2022).

[26] E. B. Djatmiko, *Perilaku dan operabilitas bangunan laut di atas gelombang acak*, First., vol. 1. Surabaya: ITS Press, 2012.

[27] M. Burić, S. Grgurić, H. Mikulčić, and X. Wang, "A numerical investigation of tidal current energy resource potential in a sea strait," *Energy*, vol. 234, pp. 5–8, 2021, doi: https://doi.org/10.1016/j.energy.2021.121241.

IOP Conf. Series: Earth and Environmental Science 1166 (2023) 012028 doi:10.1088/1755-1315/1166/1/012028

Study of Vortex Effect on Vertical Axis Marine Current Turbine Arm

Amiadji[1], I S Arief, A Baidowi[1*]
[1]Department of Marine Engineering, Faculty of Marine Technology, Institut Teknologi Sepuluh Nopember (ITS), Surabaya, East Java 60111, Indonesia
*Corresponding author: ahmadbai@gmail.com

Abstract. The design of a marine current turbine generator specifically in deep water with unequal seabed which requires the designer to develop a floating platform for the turbine. It creates a massive challenge in improving the mechanical system, such as the turbine arm. In order to make the monitoring and maintenance accessible, the generator must be placed on the platform's deck. A long shaft is required to make the connection between the generator and turbine. The problems occur due to the vibration generated by the vortex around the shaft or the turbine arm. This paper aim is to investigate the vortex effect on the turbine arm and propose an advanced design to reduce its impact. The computational fluid dynamic method is approached to analyse the vortex generated by the turbine arm. The analysis compared the conventional cylindrical and the foil-shaped turbine arm. Moreover, the investigation includes force along the X-axis and Y-axis, flow, drag, lift coefficient, also fatigue lift near the turbine arm. The numerical output illustrates that the foil shaped on the turbine arm provided limited vortex and improved the performance compared to the conventional cylindrical turbine arm.

1. Introduction

Nowadays, the surge in oil prices occurred due to the Ukraine war. The oil price leveraged above 100 USD per barrel. European People struggle to face the winter due to the high cost of electricity [1]. In another place, the government of Indonesia had to cut the oil subsidised, and inflation happened. The government switched the policy from fossil fuel to renewable energy to find a breakthrough solution. The policy is called *bauran energi* (energy mixture), twenty-three percent of total energy must be generated from renewable sources [2]. With the immense Indonesia sea area, marine renewable energy is one of the vast potential electricity energy sources to be harnessed from the ocean.

Marine renewable energy has many types of energy conversion tools. One of the types is Wave Energy Converter (WEC), the research and development already started in Indonesia. The WEC designed with pendulum type has been developed by Arief et al. [3]. The tool platform is built with an outrigger and horizontal pendulum [4]. Another form of marine renewable energy is the marine current turbine, it uses ocean current to harness the electricity. In 2016, The Ministry of Mineral and Resources of Republic Indonesia published the ocean current potential map [5]. The study by Firdaus et al. stated that the highest velocity stream of the tidal current is located in Larantuka Strait, Flores [6].

Several times research and development held on the marine current turbine within Indonesia, but still has no approach to the commercial stage. A previous study by Baidowi et al. observed the motion response of quadrimaran platform marine current turbine as shown in figure 1, compared the various type of platforms and formulated an innovative design [7]. The different types of marine current turbines have already been analysed for the maintenance process assessment in the Larantuka strait [8]. It still needs extra research and development to reach the commercial product.

In addition, the Vortex problem in marine current turbine could occur during the operation stage. The importance of efficient design is vital due to the limitation of the vortex in each turbine part. In the past study of the Vortex effect on hydrofoil of marine current turbine held by Han et al. [9], the study successfully proved that the groove method reduces the vortex on the hydrofoil. Further investigation by Wang et al. on the design of a foil on the vertical marine current turbine with free vortex [10].

The scope of this study is the analysis of the vortex effect that occurred within the turbine arm of the vertical axis marine current turbine. The turbine arm's function is to cover (sleeve) the shaft that connects the generator and turbine. The analysis of vortex among the turbine arm has some limitations in both forces along the X-axis and Y-axis, flow characteristics, drag, also coefficient of lift. Afterwards, the quadrimaran platform will be installed on Nusa Penida Island.

Furthermore, the complete analysis of the vortex along the turbine arm is using the Computational Fluid Dynamic (CFD) method. The early study by Baidowi [11] compared the stress between the foil and cylindrical form of the turbine arm. Figures 3 and 4 show the design comparison of both turbine arms types. This research compares the vortex analysis between the cylindrical and foil type on the marine current turbine platform.

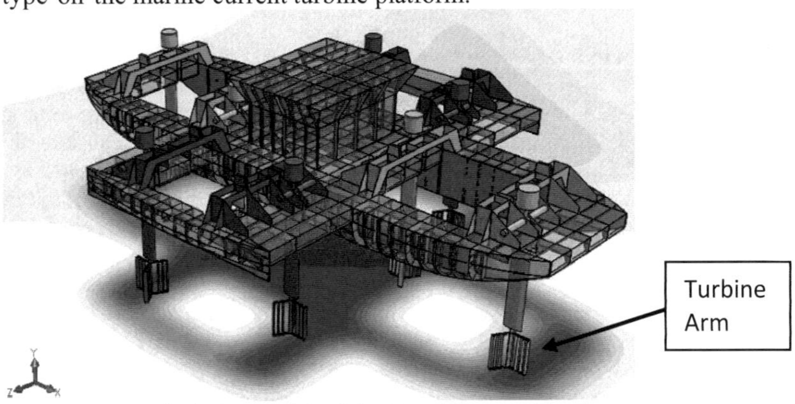

Figure 1. Detailed 3D Model of the Quadrimaran Platform

The detailed dimension of the platform is shown in Table 1.

Table 1. Platform Dimension

	Main Hull	Side Hull
Lpp	20 m	8 m
Breadth	6.6 m	4.17 m
Height	2 m	
Total Deck width	14.8 m	2 m
Displacement	44.93 ton	

The turbine will be held by a slewing bearing and the upper side of the bearing will be connected to the shaft so the weight of the turbine is held by the turbine arm.

Figure 1 shows the design of the platform of Ocean current energy generator with the turbine arm at its operational condition. The turbine arm holds the turbine weight and connect to the platform so the shaft is inside the casing or the arm as shown by Figure 2.

IOP Conf. Series: Earth and Environmental Science 1166 (2023) 012028 doi:10.1088/1755-1315/1166/1/012028

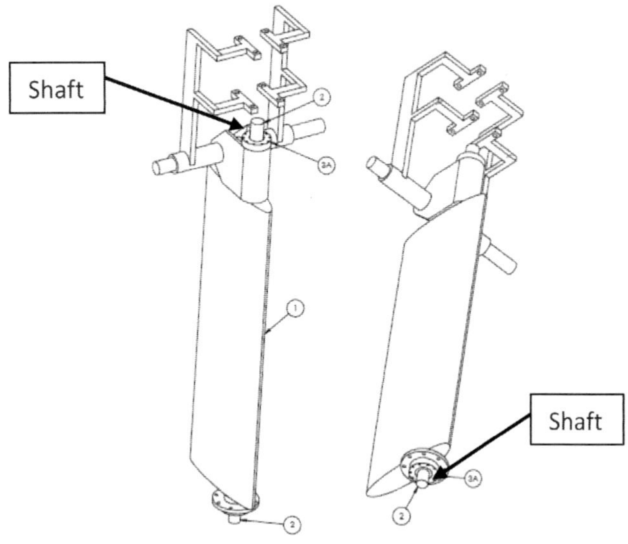

Figure 2. Turbine arm and shaft inside the turbine arm

2. Method and Theory

2.1. Computational Modelling

The modelling is finished by making two main areas. First is the turbine arm field, the turbine arm will become the shaft and then analyse a vortex that occurred among the turbine arm. The second is the domain field, which is used as a fluid medium or flow. The boundary condition defines the two areas which determine the fluid input, fluid output, turbine arm geometry, and external field. If it is complete, it can be continued with the meshing process. The determination of the meshing density does not require a high value. If the meshing is conducted at a high density, the process of running or simulation will need a longer time, and the results are not much different with a low-density level. The current speed tested in the simulation is 2.4 m/s. This numerical investigation uses Numeca Fine Marine to conduct the vortex analysis on the turbine arm.

2.1.1. Geometry

Based in Figure 1 below the turbine arm is located on the quadrimaran platform of marine current turbine. The platform consists of 8 turbines that work as parallel moves, each with a coupled generator. There is a shaft inside the turbine arm, as shown in Figure 2. The turbine arm's efficient design is vital to run adequate operation.

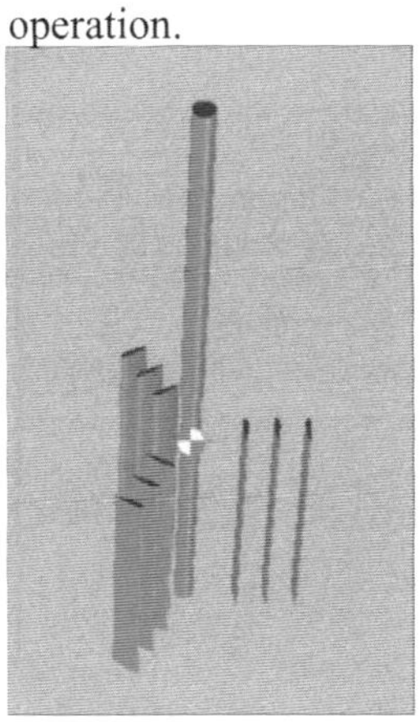

Figure 3. Turbine Arm -Cylindrical

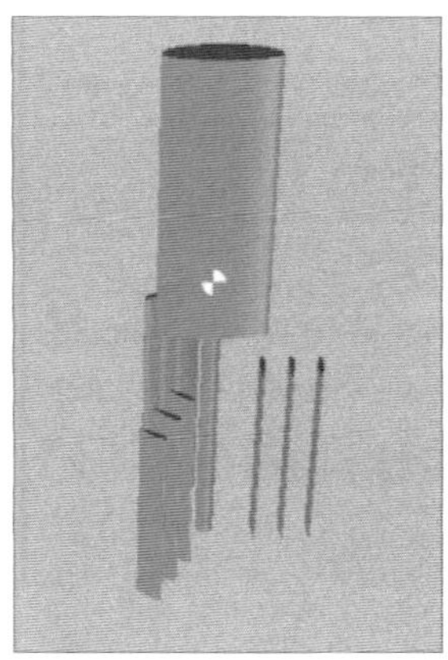

Figure 4. Turbine Arm-Foil (right)

IOP Conf. Series: Earth and Environmental Science 1166 (2023) 012028 doi:10.1088/1755-1315/1166/1/012028

2.1.2. Simulation Process

1) Pre-Processor

It consists of the input from a flow problem that will be simulated. This step includes the defining domain, fluid properties, and boundary condition. The object to be analysed also goes through a meshing process in this step. Besides meshing, there is also inputting the simulation parameter before the simulation can be run and solved. The method of making the domain was helped by the software called C-Wizard. The author input the geometry, and the C-Wizard tool automatically created the domain [12].

2) Processor

In the solver step, the calculation stage for every time step (iteration) is associated with the mesh configuration and the input parameters or method. The processing or running can be monitored from the monitor menu.

3) Post-Processor

Post-processing is an end-step process which consists of results from the calculation step. The visualisation from a formulation can also be seen in this step with a variation of 2 dimension or 3-dimensional view.

2.2. Vortex Theorem

Vorticity, a characteristic that is widely employed in hydrodynamics, is a vector that is twice as strong as the rate-of-rotation vector. In flows where viscous effects predominate, such as boundary layer, separated, and wake flows, the size of the vorticity shows the rotationality of a flow and is crucial [13]. As shown in Figure 5, the vortex occurred around the turbine arm.

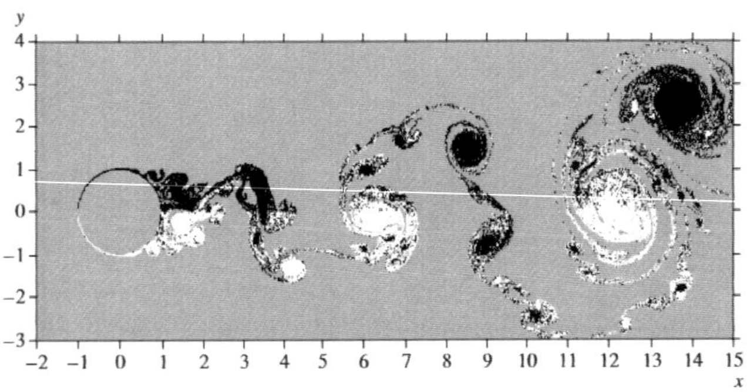

Figure 5. Vortex flow behind cylinder

2.3. Lift and Drag Theory

The lift is a force that is perpendicular to the oncoming flow direction thus this use Fy value. It is created by an object rotating through a moving fluid, resulting in a pressure difference among the body and creating a lift force. While drag is a force that acts opposite to the relative motion of an object moving concerning a surrounding fluid thus assume this use Fx value. Sometimes it is called air resistance or type of friction. These forces will be used as a reference for the force that the shaft will produce [14,15]. Below both equations (1), and (2) the coefficients lift and drag.

$$CL = \frac{2\,x\,L\,(N)}{\rho\left(\frac{kg}{m^3}\right)x\,u^2\left(\frac{m}{s}\right)x\,S\,(m^2)} \qquad (1)$$

$$CD = \frac{2\,x\,Fd\,(N)}{\rho\left(\frac{kg}{m^3}\right)x\,u^2\left(\frac{m}{s}\right)x\,S\,(m^2)} \qquad (2)$$

2.4. Reynold Number

The Reynolds number describes the relationship between inertial and viscous forces within a fluid experiencing relative internal motion as a result of varying fluid velocities used as a reference for the force that the shaft will produce [16].

$$Re = \frac{u\left(\frac{m}{s}\right) x\ \mathrm{L}\ (m)}{V\ (m^2/s)} = \frac{\rho\ (kg/m^3)\ x\ u\left(\frac{m}{s}\right) x\ \mathrm{L}\ (m)}{\mu\ (\mathrm{Pa.s})} \quad (3)$$

2.5. Fatigue Life

The fatigue life determines the life stage of the tool operation. The fatigue life can be calculated by determining Accumulative Fatigue Damage (DM), Palmgren-Miner law [17]. The DM.

$$DM = \sum_{i=1}^{k} \frac{n_i}{N_i} \quad (4)$$

And finally, the fatigue life can be calculated as,

$$\text{Fatigue life} = \frac{\text{Design life}}{\text{DM}} \quad (5)$$

The design life stated in 20 years of operation.

3. Result and Discussion

3.1. Cylinder Turbine Arm

As a result of this investigation, the vortex effect occurred along the turbine arm on both foil and cylindrical types. As shown on Figure 6 and 7, the flow around cylinder creates a vortex and the resultant forceis 2.22 kN.

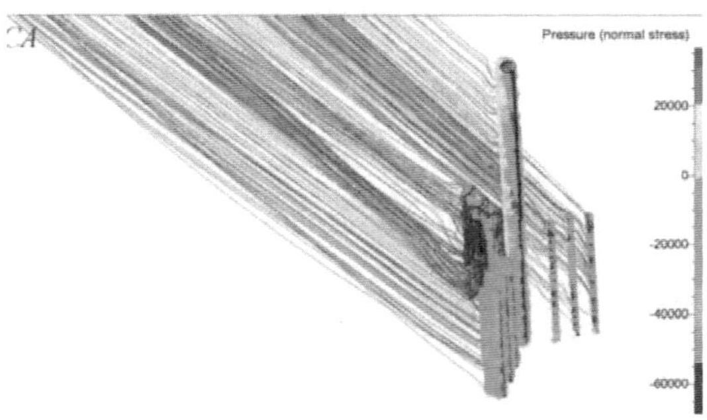

Figure 6. Vortex Effect of Foil Turbine Arm

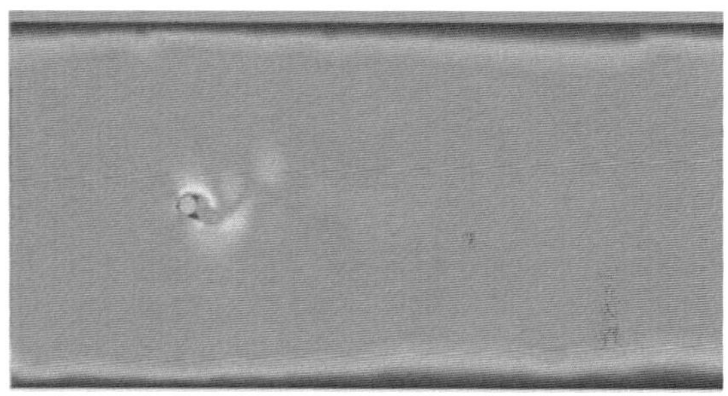

Figure 7. Vortex Effect of Cylindrical Turbine Arm

This force then apaplied to the structural strength of the turbine arm to obtain the maximum stress of the turbine arm. The maximum stress of the turbine arm is 211.4 Mpa as shown by Figure 8. While the maximum displacement of the structure is 7.675 10^{-2} mm (Figure 9). Since the material used in this study is AISI 4130 with yield stress 435 Mpa, it can be concluded that the design still far from the yield limit of the material.

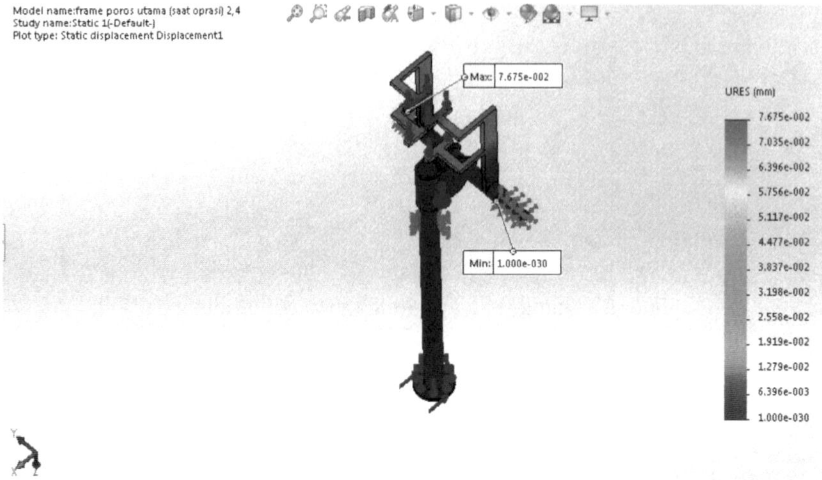

Figure 8. Stress distribution along the cylinder turbine arm

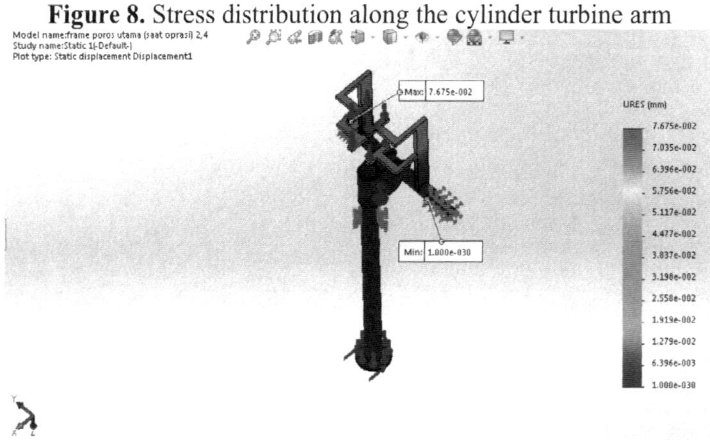

Figure 9. Dispacement distribution along the cylinder turbine arm

3.2. Foil Turbine Arm

In order to reduce the vortex formation around the turbine arm, a foil shape turbine arm also developed and analysed using CFD. The resultant force based on the simulation is 1918N or 1.918 kN. Based on the Figure 10 and Figure 11, it can be concluded that the vortex flow behind the turbine arm is dissipated and may reduce the resultant force actong on the turbine arm.

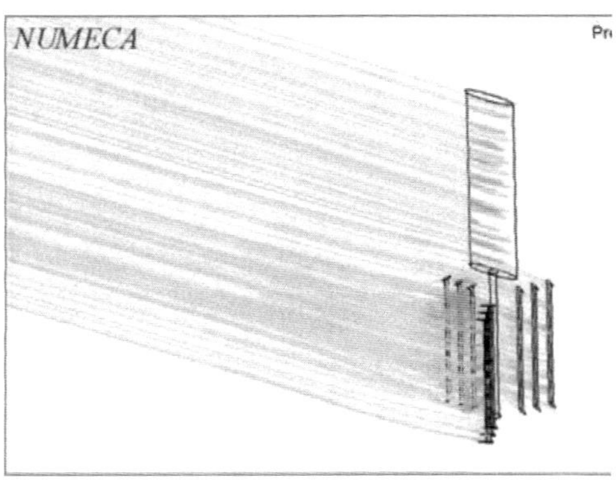

Figure 10. Vortex Effect of Foil Turbine Arm

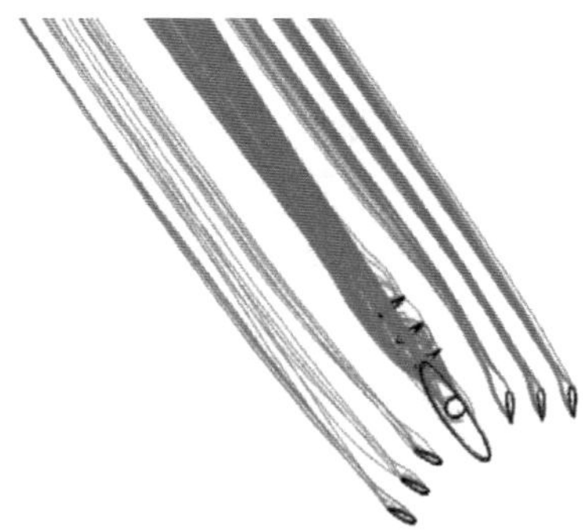

Figure 11. Vortex Effect of Cylindrical Turbine Arm

The maximum stress occurred on the turbine arm is 26.85 Mpa while the maimum displacement is 1.346 10^{-2}mm as shown in Figure 12 and Figure 13

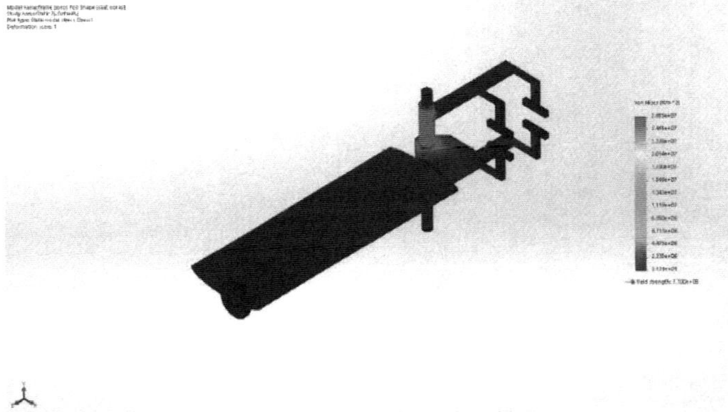

Figure 12. Stress Distribution Foil Turbine Arm

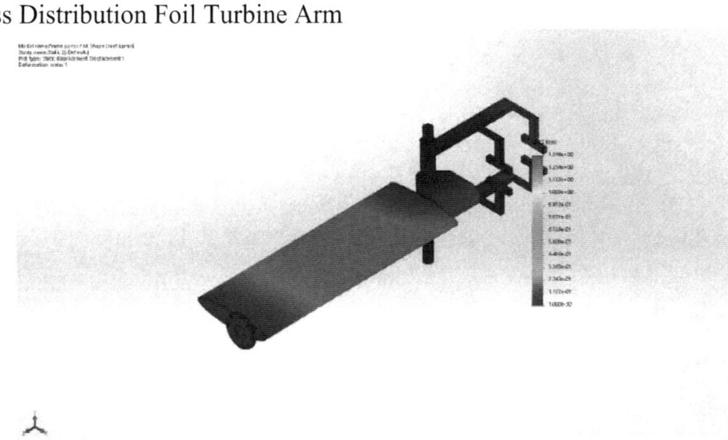

Figure 13. Displacement distribution Foil Turbine Arm

3.3. Discussion

Based on the analysis using finite element method software, the maximum stress for cylinder shaped is higher due to vortex that generates Fy Forces. The comparison is shown in Table 2, the table shows that the maximum stress is still under the allowable stress.

Table 2 Maximum Stress and Cycle Comparison Cylinder vs Foil

	Operasi	Allow stress	status	Cycle
	Mpa	Mpa		
Cylinder	211.4	347.9	Comply	5.0×10^6
Foil Shaped	26.85	347.9	Comply	1.7×10^7

While the analysis of fatigue life cycle shows that the foil shaped turbine arm resulted in higher number of cycle compared to cylinder shaped turbine arm as shown by Table 3, the fatigue life cycle of foil shape is 1.7×10^7 while the cylinder shape is 5.0×10^6.

4. Conclusion

In the current work, the vortex effect on the foil turbine arm has a lower value than cylindrical. The foil shaped turbine provides higher fatigue life and lower stress compared to cylinder shaped turbine arm, but it is necessary to analyze the turbine arm in an array that can be used for the next research.

Acknowledgement
The authors are grateful to the department of Marine Engineering, Institut Teknologi Sepuluh Nopember-Surabaya. The energy research group from Marine Manufacturing and Design (MMD) Laboratory.

References

[1] Belaïd F 2022 Implications of poorly designed climate policy on energy poverty: Global reflections on the current surge in energy prices *Energy Res Soc Sci* **92** 102790
[2] National Energy Council of Indonesia 2019 *Indonesia Energy Outlook 2019* (Jakarta)
[3] Arief I S, Utama I K A P, Hantoro R, Prananda J, Arvisa T R and Kusuma R F 2019 Mooring Experimental Study of Motion Response for Pendulum Wave Energy Converters *IOP Conf Ser Mater Sci Eng* **462** 012010
[4] Arief I S, Utama I K A P, Hantoro R, Prananda J, Safitri Y, Rachmattra T A and Rindu F K 2018 Response to Pontoon and Pendulum Motion at Wave Energy Converter Based on Pendulum System ed Ir Sunarno, A Pulung Sasmito and L Peng Hong *E3S Web of Conferences* **43** 01022
[5] Ministry of Energy and Mineral Resources Republic of Indonesia 2016 *The Ocean Current Energy Potential Locations Map of Indonesia* (Bandung)
[6] Firdaus A M, Houlsby G T and Adcock T A A 2020 Tidal energy resource in Larantuka Strait, Indonesia *Proceedings of the Institution of Civil Engineers - Energy* **173** 81–92
[7] Baidowi A, Semin S, Koenhardono E S, Amiadji A and Ainin F 2019 Motion Response of a Novel Platform of Vertical Axis Marine Current Turbine *International Review of Mechanical Engineering (IREME)* **13** 710
[8] Baidowi A, Semin and Koenhardono E S 2021 Harnesssing Technology Maintenance Process of Tidal Energy Converter *J Phys Conf Ser* **1908** 012034
[9] Han Y, Liu Y and Tan L 2022 Method of variable-depth groove on vortex and cavitation suppression for a NACA0009 hydrofoil with tip clearance in tidal energy *Renew Energy* **199** 546–59
[10] Wang L B, Zhang L and Zeng N D 2007 A potential flow 2-D vortex panel model: Applications to vertical axis straight blade tidal turbine *Energy Convers Manag* **48** 454–61
[11] Baidowi A 2020 *The Design of Onboard Mechanical System for Operational Flexibility of Marine Current Energy Converter* (Surabaya: Institut Teknologi Sepuluh Nopember)
[12] NUMECA 2020 *User Manual FINEMarine v.9.2*
[13] Crowe C T, Elger D F, Williams B C and Roberson J A 2009 *Engineering Fluid Mechanics* (New Jersey: John Wiley & Sons, Inc)
[14] Abbott I H and Doenhoff A E 1959 *Theory of Wing Sections Including a Summary of Airfoil Data* (New York: Dover Publications, Inc.)
[15] McCormick B W 1995 *Aerodynamics, Aeronautics, and Flight Mechanics* (Toronto: John Wiley & Sons, Inc.)
[16] Sommerfeld A 1909 *Ein beitrag zur hydrodynamischen erklaerung der turbulenten fluessigkeitsbewegungen* (Rome: Accademia dei Lincei)
[17] Zeiner-Gundersen D H 2015 Turbine design and field development concepts for tidal, ocean, and river applications *Energy Sci Eng* **3** 27–42

Fatigue Life Analysis on The Local Construction Near the Chain Stopper Area of Papa Terminal Floating Storage Offloading (FSO) With 4X2 Chain Line Spread Mooring Configuration

A Louis[1], Murdjito[1*], N Syahroni[1]

[1] Department of Ocean Engineering, Sepuluh Nopember Institute of Technology, Surabaya, Indonesia

[*] Corresponding author: murdjito@oe.its.ac.id

Abstract. This research aims to determine the fatigue life of the structure near the chain stopper of a storage tanker using the spectral fatigue analysis method. The wave spectrum is developed based on six significant wave height and period variations from the site wave scatter data using a modified spectrum of JONSWAP. The stress analysis is carried out using finite element analysis software. It can then be processed into Stress RAO from each loading direction. The response spectra are generated from the wave spectrum, multiplying with the square of the Stress RAO. These response spectra will determine the stochastic values of each stress range variation in obtaining the number of cycles that occur in each variation (nL). The accumulation of each ratio obtained from all existing variations will produce a fatigue index value (D). The American Bureau of Shipping's S-N curve D-Class is used in this fatigue life analysis. The Gamkonora Aframax tanker operated as a temporary storage tanker at Papa's field area is used as the case of this study.

Keywords: ***Fatigue, Chain Stopper, Stress, Wave Spectrum, Response Spectrum.***

1. Introduction

PAPA Field is an oil and gas production area off the coast of Karawang, West Java. To accommodate the needs for oil transfer in that area, immediate countermeasures need to be carried out by connecting oil access to a Temporary Storage Tanker which would be moored with a spread mooring scheme. Then the offloading scheme uses a side-by-side system with Export Tanker. Long-term planning is also carried out by installing a permanent spread mooring system which requires an analysis of the fatigue life of several components constructions on the FSO that will be used. The FSO to be used is a conversion from Tanker MT. Gamkonora.

Fatigue analysis is closely related to safety issues in the design and operation of offshore marine structures that are planned to operate for many years. Damage to marine structures is generally caused by fatigue in the structure, both in the main structural components and secondary structures [1]. Therefore, this is a benchmark for the quality of marine buildings in the aspect of their structural design. The cyclical and repetitive wave loads are mainly the main cause of marine structures experiencing fatigue. The structure also responds dynamically [2].

Many rules and procedures for fatigue analysis of floating marine structures have been published to calculate the fatigue life of the construction. For example, the latest guideline from the Indonesian Classification Bureau was published in 2021. BKI adopted the American Bureau Shipping rules titled 'Guidelines for Floating Productions Installations'. With the application of the Finite Element Method in analyzing the strength of the ship's construction, the calculation of the age of the building is designed according to the design factors that have been determined in the class rules used.

Many studies of fatigue analysis have been carried out. In addition to fatigue analysis on the anchor chain of a moored floating structure, fatigue analysis is also needed for local structures near the chain stopper area. Therefore, following the issue that has been explained previously, the author chose 'Fatigue Life Analysis on The Local Construction Near the Chain Stopper Area of Papa Terminal Floating Storage Offloading (FSO) with 4X2 Chain Line Spread Mooring Configuration' as the title of this paper. This research will provide the procedure for calculating the fatigue life of the construction around the FSO chain stopper concerning the rules and procedures set by BKI – FPI.

2. Research Methodology

2.1. Flow chart

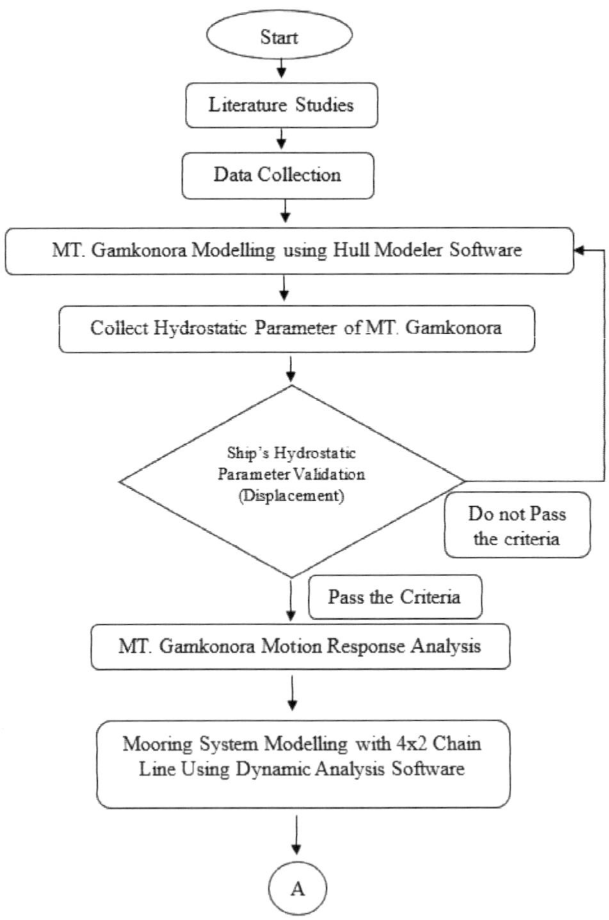

Figure 1. Flow chart

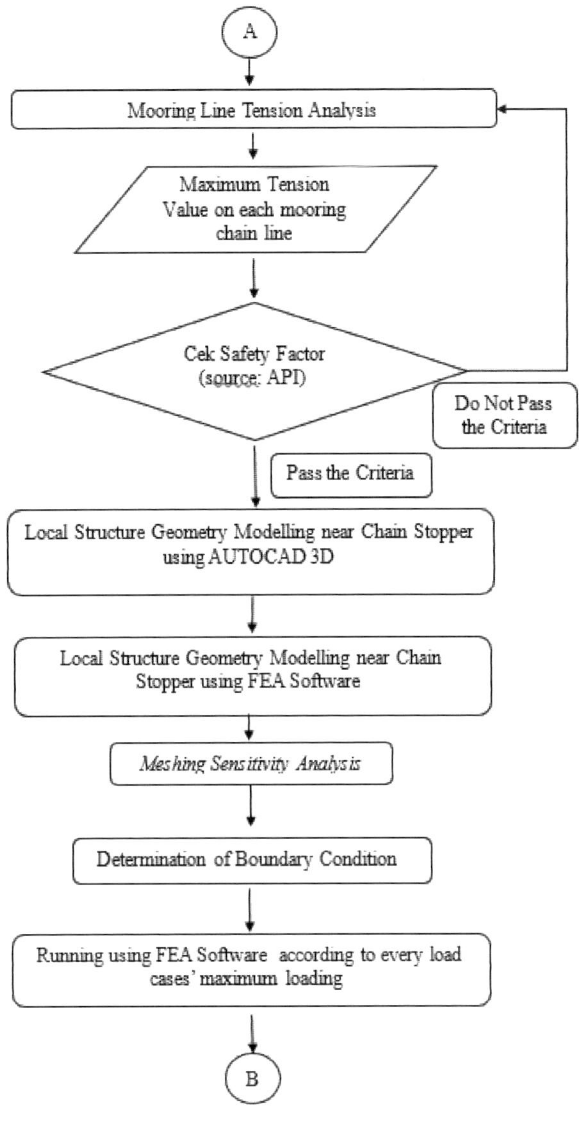

Figure 2. Flow chart Continuation 1

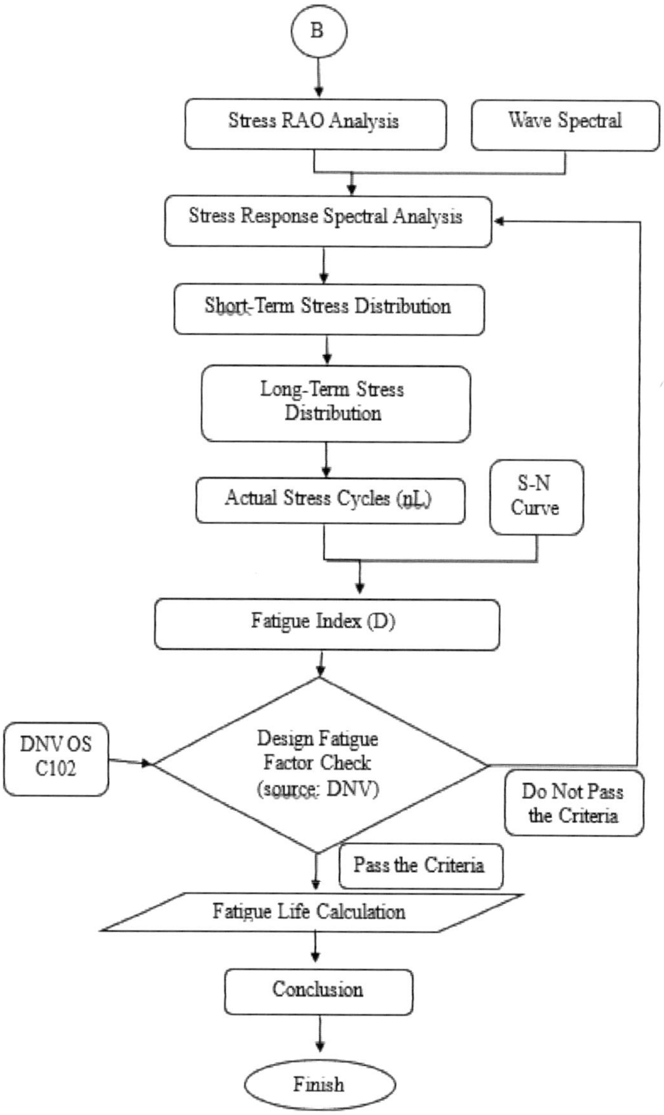

Figure 3. Flow chart Continuation 2

2.2. Data collection
In conducting this research, a set of supporting data is needed to perform theoretical calculations and software modeling. In this study, the authors collect data consisting of the principal particular of the ship used, the dimensions of the anchor line used, environmental data in the form of metocean data, and wave scatter data in the waters around Karawang, West Java, as well as the general arrangement of ship construction and chain stopper.

2.3. Ship modeling using hull modeler software

2.3.1. Obtaining MT Gamkonora Hydrostatic Parameters. Hull modeler software is used to design the ship redesign and obtain hydrostatic data in the process of modeling and analyzing FSO

GAMKONORA which will be used in this research. The structure modeling refers to the ship's Principal Particular and General Arrangement data, as well as several approach formulas for unknown data. In addition, the output obtained from modeling in the hull modeler software is used as the input to obtain the Response Amplitude Operator (RAO) of 6 types of ship movement on free-floating conditions according to a predetermined load variation.

2.3.2. Obtaining MT Gamkonora Hydrostatic Parameters. After the modeling phase on the hull modeler software is complete, validation is carried out first to determine the feasibility of the model that has been made to be used in the analysis inside the software.

2.3.3. Obtaining MT Gamkonora Hydrostatic Parameters. This study's structural motion response analysis will be done using hull modeler software with 2 variations of loading conditions, Full Load (98%) and Light Load (20%). Each condition will provide different hydrostatic data and produce different values for the motion response of the structure.

The response of the structure motion in free-floating conditions will produce 6 degrees of freedom (6 Degrees of Freedom) consisting of 3 translational motions (surge, sway, and, heave) and 3 rotational motions (roll, pitch, and, yaw).

2.4. Mooring system modeling using dynamic analysis software

The GAMKONORA FSO model which will be analyzed using dynamic analysis software will refer to the principal particular data (Table 1), the general arrangement, as well as outputs such as RAO, wave drift, and hydrostatic stiffness and damping data that have been previously earned in the hull modeler software. Meanwhile, mooring line data and environmental data can be seen in Table 2. and Table 3. Below.

Table 1. Principal Particular
MT. GAMKOMORA

Type	Tanker
LOA (m)	244
B (m)	44
H (m)	21.5
T actual (m)	12.7
LPP (m)	233
DWT (ton)	88258
Displacement (ton)	109431

Table 2. Anchor Line Dimension

Type	Chain, Studlink
Grade	R4
Length (m)	350
Diameter (mm)	87
Minimum Breaking Load (kN)	7682

Table 3. Metocean Data on MGRID 2-3
(5° 40' S - 5° 50'S / 106° 56' E - 107° 22' E)

	Return Period 1 year	Return Period 100 year
Hs (m)	2.00	3.20
Tp (s)	7.10	8.60
Wind Speed (m/s)	9.90	20.60
Current Speed (m/s)	0.70	1.13
Water Depth (m)	34.00	34.00

Mooring configuration modeling in dynamic analysis software is used to get the maximum tension value in each existing mooring chain to determine which chain stopper will then be analyzed later.

2.5. Modeling of ship's bow geometry for local structural analysis using finite element analysis (FEA) software

After the tension on the existing mooring lines has been obtained, the focus of the work will be continued on the local analysis of the chain stopper itself. As for before entering the FEA software, the modeling of the chain stopper will be made first using the help of Autocad 3D software. The model includes a chain stopper and other construction inside the ship's bow. After that, the model will be exported directly from AutoCAD 3D so that it can be used as input for FEA software.

Inside the FEA software, there are several things to do. The first is to determine the boundary conditions on the ship's bow structure. Second, the converted dynamic analysis software running results are to be inputted into the chain stopper. Third, the meshing process must be carried out by first conducting a meshing sensitivity analysis, so that accurate results can be obtained with efficient running time. Finally, the running process will be conducted on the FEA software to obtain a solution in the form of maximum principal stress on the 30 existing load cases to be converted into a stress hotspot.

2.6. Wave energy spectrum analysis

The wave spectrum for each of the 6 variations of the previously defined Hs-Tp class can be obtained by utilizing the JONSWAP spectrum graph feature which can be obtained using dynamic analysis software. From the wave spectrum graphs that have been obtained from the software, 11 frequency values equally distributed between 0.59 rad/s to 1.62 rad/s will be used to calculate the area under the curve using the Simpson 1 method.

2.7. Stress RAO analysis

RAO can be used as a transfer function to determine the magnitude of the structure's response. In this case, the main goal is to change from a regular wave approach to a random wave approach which is considered to be more representative of the real wave state in nature.

2.8. Stress response spectral analysis

The Stress Response Spectra is made by multiplying each value per frequency of the wave spectrum in each class (operating mode) that exists by the square of each Stress RAO value owned by the 5 variations of different environmental load directions. Briefly, the process is as can be seen in Figure 4.

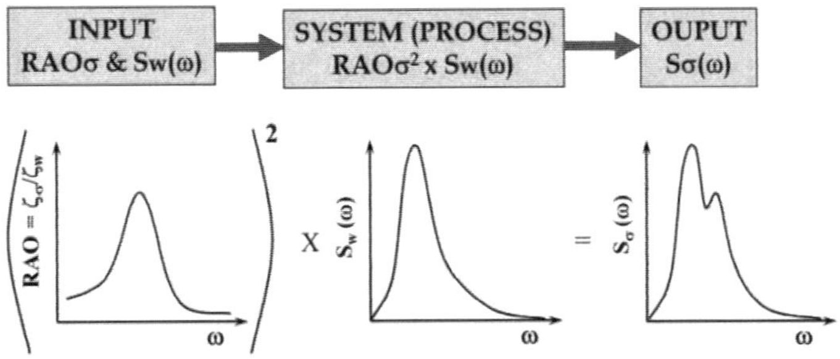

Figure 4. Stress response spectrum making process

2.9. S-N curve used

According to the rules set by DNV in determining the S-N used for fatigue index calculation, DNV RP C 203 states that the D-class S-N curve must be used for the chain stopper constructions [3]. There are at least 2 parameters that must be used in determining the value of N in the S-N curve as seen in formula (1), namely 'Log A, and 'm'. The selected D-class S-N curve gives a Log A value of 12.6007 and an m value of 3 [4].

$$N = AS^{-m} \tag{1}$$

2.10. Calculation of fatigue life using the spectral method

For each stress response spectrum, the following calculations are carried out:

- The area under the stress response spectral curve (m0) and the moment of its area (m2). [5]

$$m_0 = \int_0^{\infty} S\sigma\sigma(f)\,df \tag{2}$$

$$m_2 = \int_0^{\infty} S\sigma\sigma(f)f^2\,df \tag{3}$$

- Number of cycles of stress range in 1 second (n) [5]

$$n = \frac{1}{Tz} = \frac{1}{2x}\sqrt{\frac{m_2}{m_0}} \tag{4}$$

- Probability of occurrence of stress range (S) in the short-term using the probability density function formula from the Raileigh distribution.

$$p_{s(s)} = \frac{s}{m0} e^{\frac{-s^2}{2mo}} \tag{5}$$

- The number of cycles that occur in the specified operating mode during (T = 25 years of design life (nL). [4]

$$n_L = (\Sigma_a \Sigma_b \Sigma_c \mathrm{n}_0 \times \mathrm{P}_a \mathrm{P}_b \mathrm{P}_c)\ X\ T \tag{6}$$

- The value of the fatigue index generated by assumed design life of 25 years

$$D = \sum_{i=1}^{k} \frac{nLi}{NI} \tag{7}$$

- Fatigue Life, F.L

$$fatigue\ life = \frac{design\ life}{D} \tag{8}$$

3. Results and discussion

3.1. Validation of vessel hydrostatic parameters in hull modeler software

Based on the Principal Particular and General Arrangement data from the MT. Gamkonora that has been obtained, modeling can be carried out as the first step of the analysis. The results of modeling FSO GAMKONORA using hull modeler software can be shown in Figure 5.

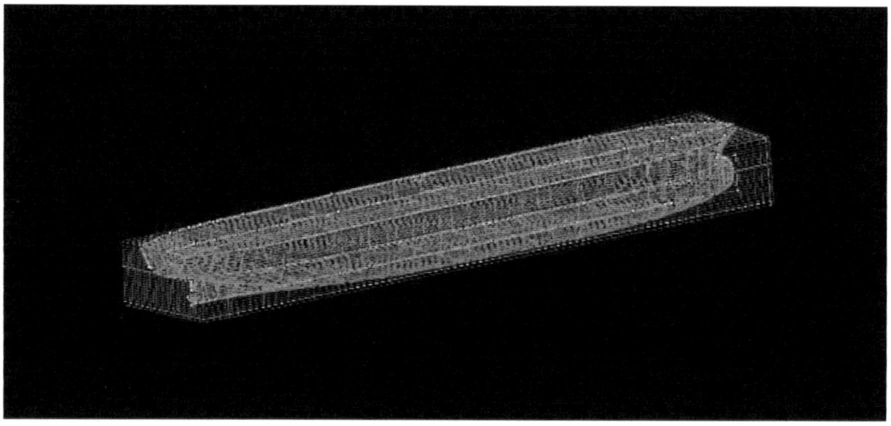

Figure 5. Isometric View of FSO Gamkonora Model on Hull Modeler Software

3.1.1. Full load variation model. Because the full load draught along with the KG points have been obtained in the data, the validation of the displacement on full load draught condition in the hull modeler software must be done by making the displacement data obtained on the software as close as possible to the displacement values obtained in the data as can be seen in Figure 6.

FULL LOAD VALIDATION		
Full Load Draft Based on Data =	12.70	m
Full Load Displacement Based on Data =	109431.30	ton
Full Load Displacement on Hull Modeler Software =	109117.37	ton
Validation of Full Load Displacement =	**0.29%**	**ACCEPTED**
VCG for Full Load Cond. (98%) based on data =	12.5	m

Figure 6. Model Validation Results on Full Load Condition

3.1.2. Light load variation model. The calculation of the center of gravity height (KG) is carried out using the approach (see Figure 7.), and also the calculation of the light load draught by making the displacement obtained in the excel calculation process as close as possible to the draft value that will be obtained in the hull modeler software. Then validation is done by matching it to the displacement value contained in the ship data as shown in Figure 8.

Light Load Condition VCG Calculation		
Tank Height	19.2	m
Double Bottom Height from General Arrangement	2.3	m
Total Height (H)	21.5	m

Construction VCG (0.3*H)	6.45	m
Cargo Oil VCG (98%)	13.82	m

Cargo Oil VCG (20%) without double bottom	3.32	m
Cargo Oil VCG (20%) with double bottom	5.62	m

VCG for Light Load Condition (20%)	**6.0**	**m**

Figure 7. KG Point Calculation on Light Load Condition

LIGHT LOAD VALIDATION		
Full Load Displacement on Hull Modeler Software =	109117.37	ton
Density =	0.8	(Estimation)
100% Data full Include SLOP T =	128090.10	
100% Volume Slop T =	6084.27	
100% Cargo Volume (No Slop T)=	122005.83	
Full (98%) Cargo Volume =	119565.71	
80% Cargo Vol. =	95652.57	
DWT C.O 80% =	76522.06	ton
Light Load Displacement by Calculation (20%) =	**32595.31**	ton

Light Load Draft based on Data According to Calculated Displacement =	**4.20**	m
Light Load Displacement Based on Data =	32627.60	ton
Light Load Displacement on Hull Modeler Software =	32236.99	ton

Validation of Light Load Displacement	**1.20%**	**ACCEPTED**

Figure 8. Draft Calculation and Model Validation Results on Light Load Condition

Based on the results of the validation that has been done, it can be concluded that the modeling that has been made can be used for further analysis. The analysis process carried out requires some additional data, namely the ship's water draft (T) according to existing data or data from validation

IOP Conf. Series: Earth and Environmental Science 1166 (2023) 012029 doi:10.1088/1755-1315/1166/1/012029

results that have been carried out for full load and light load model conditions. After the analysis, the data can be shown in Table 4. and Table 5.

Table 4. MT. GAMKONORA Model Hydrostatic Data on Full Load Condition

Condition	Displacement (ton)	WPA (m^2)	Draft (m)	LCB (m)	TCB (m)	VCB (m)
Full Load	109117.37	9441.28	12.70	7.07	0.00	6.67

Table 5. MT. GAMKONORA Model Hydrostatic Data on Light Load Condition

Condition	Displacement (ton)	WPA (m^2)	Draft (m)	LCB (m)	TCB (m)	VCB (m)
Light Load	32236.99	8192.67	4.20	13.14	0.00	2.20

3.2. Motion response analysis of free-floating conditions

The analysis process carried out requires additional data, namely the ship's water draft (T), the direction of loading, and the ship's Center of Gravity (Xo, Yo, and Zo) as a reference point for RAO. These data have been obtained using the help of calculations using Microsoft Excel and hull modeler software.

Figure 9. And Figure 10. will show the graphs of the structure responses in full load and light load conditions with 5 environmental loading directions aft wave load (heading 0°), aft quarter wave (45° heading), side wave (90° heading), and bow quarter wave (heading 135°) and bow wave (heading 180°).

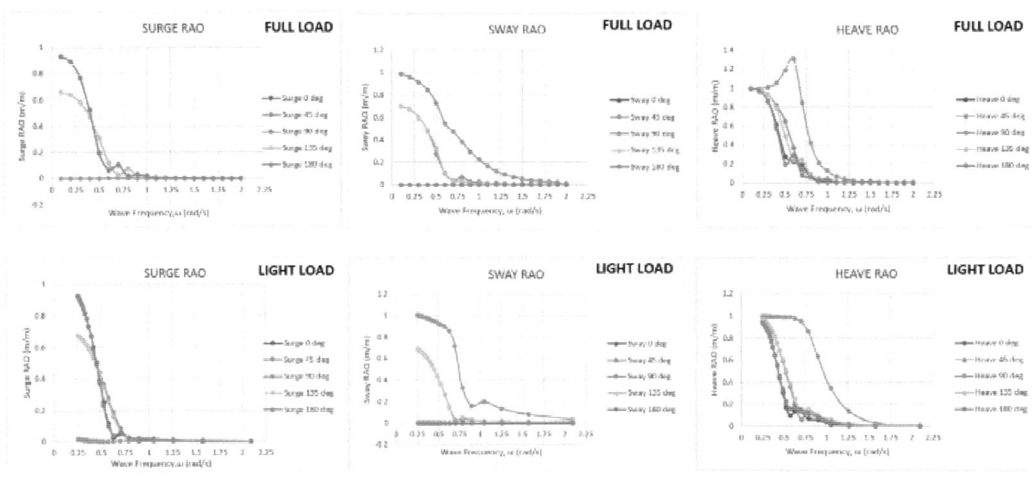

Figure 9. Translational Motion Response of the Structure

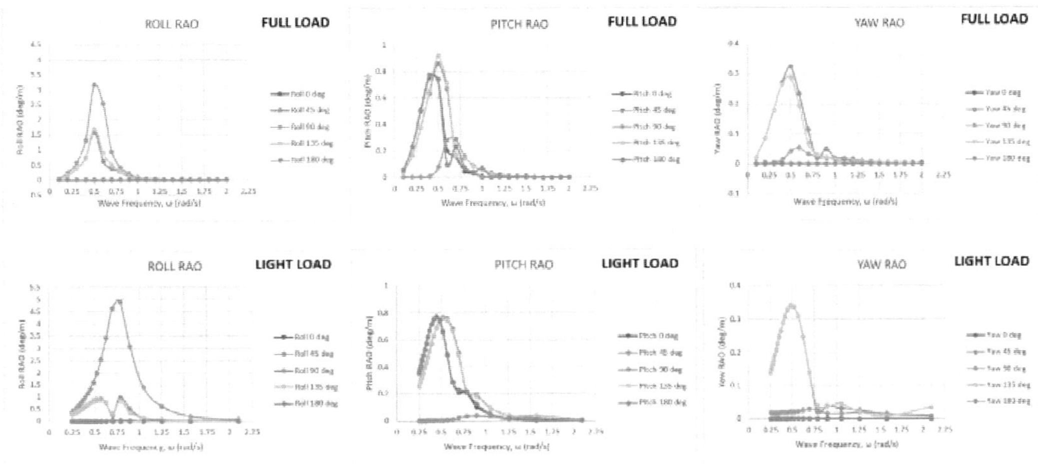

Figure 10. Rotational Motion Response of the Structure

In the 6 types of ship motions, translational motion is represented by Surge, Sway, and Heave RAO charts, while rotational motion is represented by Roll, Pitch, and Yaw RAO charts. Referring to reference [6], on a floating structure the 3 vertical movements that are more dominant than the other 3 movements that move horizontally. The three vertical movements are Heave, Roll, and Pitch.

In Heave RAO, the maximum value achieved at Full Load condition is higher than the maximum value of Heave RAO in Light Load condition. Heave and Pitch movements will mutually affect each other in a floating structure due to the influence of the first and second resonances in the Heave and Pitch movements. That is the cause why the maximum value for the Pitch RAO at Full Load conditions is also greater than in Light Load conditions

Furthermore, because the motion of the Roll and Pitch is very similar from the rotational-vertical mode of motion, it can be seen that the graph trends generated on the two charts are very similar. For Roll movement, it can be seen that the maximum value of Roll RAO in Light Load condition is much greater when compared Roll RAO in Full Load condition, with the maximum difference in the Beam Sea direction (90°) reaching 1.5°/-m, so is the value on the RAO Sway. One of the biggest factors that causes this to happen is the effect of the difference in the draft of the ship which is too far between the full load condition (12.7m) and the light load condition (4.2m). Thus, the area of the ship that exposed as the wind load area is also greater in light-load ships, with a relatively short draft

3.3. Chain stress output on Dynamic Analysis Software

The ship model visualization on dynamic analysis software is made according to the ship's principal particular. The mooring line modeling in FSO refers to a predefined configuration. The mooring system will use a spread mooring with a 4x2 chain line configuration. The eight mooring ropes are homogeneous studlink chains with a diameter of 87 mm and a length of 350 meters each. The direction of the environmental load used in this analysis is in collinear loading condition, where the current, wind, and wave loads come from the same direction. The direction of the incoming environmental load consists of 5 different directions, direction 0°, 45°, 90°, 135°, and 180°. Finally, it is necessary to separate the load case for light load and full load loading conditions, because they have different drafts. The configuration of the mooring line and the environmental load direction scheme used in this analysis can be shown in Figure 11.

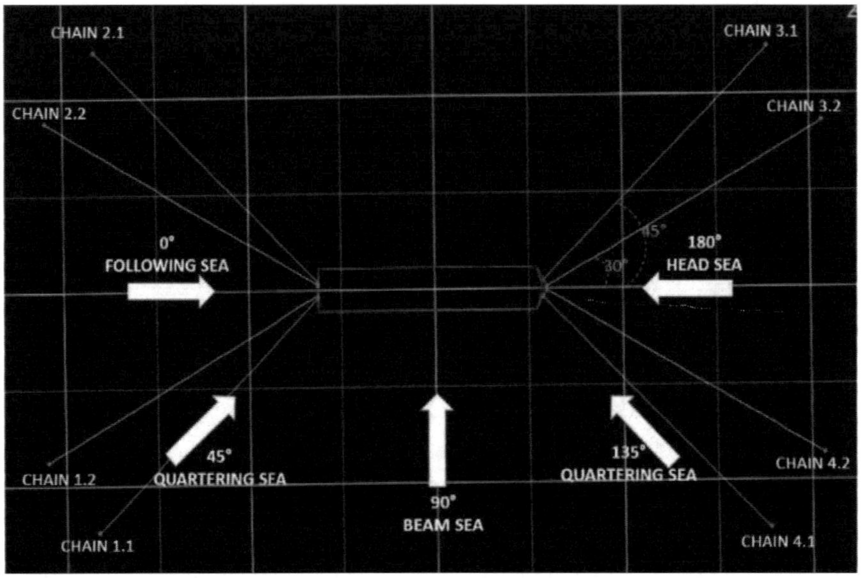

Figure 11. Mooring Configuration and Loading Direction

Analysis of mooring line tension will use the time domain analysis method. According to [7], the minimum duration of time domain simulation to get the peak value of a stable response statistic that can represent the environment is 3 hours (10800 seconds). However, in this analysis, the cut off method is used where the time is taken in the range of 1200 seconds, between the time range of 4500 seconds to 5700 seconds. This time range is considered to be representative considering that the wave crests in the existing wave window occur in that time range.

From the results of the 30 load cases that have been run (6 classes Hs-Tp x 5 variations in wave direction), it was found that the Maximum Tension Value experienced by FSO in Full Load and Light Load conditions mostly occurred in chain 4.1. Therefore, if we look at the results of the tension value trend in chain 4.1 as shown in Figure 12. Below.

Condition		**Maximum Tension Anchor Chain 4.1 (kN)**
Heading	**Loading Condition**	**Hs 2.75**
0°	0°- Full Load (98%)	683.33
	0°-Light Load (20%)	455.57
45°	45°- Full Load (98%)	762.82
	45°-Light Load (20%)	521.19
90°	90°- Full Load (98%)	1258.26
	90°-Light Load (20%)	675.40
135°	135°- Full Load (98%)	729.70
	135°-Light Load (20%)	591.56
180°	180°- Full Load (98%)	384.94
	180°-Light Load (20%)	481.62

Figure 12. Maximum Tension Anchor Chain 4.1 on Hs 2.75m

Then the Maximum Tension Value in order from the largest to the smallest comes from the direction of the Beam Sea, followed by the Quartering Sea, and finally the directions that are parallel to the ship, namely Head Sea and Following Sea. This phenomenon is caused by the area exposed to the wind load and also the current in the direction of the beam sea loading having much larger area than the loading in the Quartering Sea, Head Sea and Following Sea directions.

The second largest tension value is experienced in the quartering direction, because although the area of wind and current loads is not as large as the beam sea direction, the tension in the direction of the quartering load focuses more on 1 pair of mooring lines only. It is different when loading occurs in the Head Sea and Following Sea directions, where the wind and current loads are borne directly by 2 pairs of mooring lines from the portside and starboard side. Coupled with the load area in the current direction of Head Sea and Following Sea which is relatively smaller than the other directions.

In Figure 13 below, it can also be seen the trend of the maximum tension value that will follow the Hs value which is getting bigger from left to right, with the largest tension values experienced by chain 4.1 at FSO full load condition in the beam sea loading direction. In accordance with the results obtained, the input that will be used for the chain stopper loading on the FEA software will use the output we get from the mooring rope tension in full load condition with analysis that will focus on the bow chain stopper starboard 2 as the recipient of the biggest tension load.

Condition		**Individual Maximum Tension Anchor Chain 4.1 (kN)**					
Heading	**Loading Condition**	**Hs 0.50**	**Hs 1.00**	**Hs 1.50**	**Hs 2.00**	**Hs 2.50**	**Hs 2.75**
180°	180°-Full Load (98%)	318.06	322.74	325.78	359.39	368.91	384.94
90°	90°-Full Load (98%)	570.91	593.46	601.64	734.28	952.34	1258.26
45°	45°-Full Load (98%)	412.00	437.55	455.58	569.34	737.71	762.82

Figure 13. Maximum Tension Anchor Chain 4.1 on Full Load Condition

3.4. Ship's bow modeling with finite element analysis software to get the maximum principal stress

First, global coarse meshing will be performed automatically by changing the element size several times so that a rough image of the approximate location of the highest stress will be obtained. The tensile force applied to the chain stopper will make the rear connection between the chain stopper and the deck and supporting structures experience the highest stress. This is also due to the moment force that occurs at the rear, which will be greater than the connection at the front of the chain stopper.

Furthermore, size refinement will be carried out in critical areas where high stress occurs by using the "edge sizing" and "vertice sizing" features so that the results obtained in these areas can be more accurate. Various meshing experiments have been carried out with different numbers of nodes and elements. As seen in Figure 14. below, the meshing sensitivity analysis graph shows that stable results began to be obtained in the experiment with 179237 meshing elements This meshing condition will be used for running each load case using FEA software.

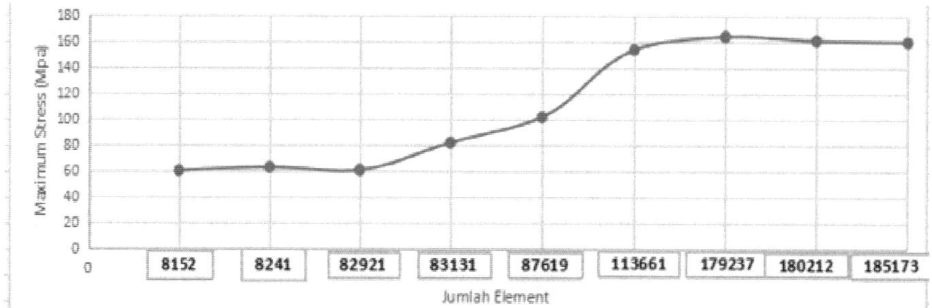

Figure 14. Meshing Sensitivity Analysis

The boundary conditions will be given to the 3D structure of the bow of the FSO as can be seen in Figure 15. And Figure 16., where fixed support is given at the rear of the 3D modeling of the bow of the ship and at the end of the bulwark of the 3D modeling of the bow of the ship. Meanwhile, Roll Support will be given to each face of the construction pillars in the 3D model of the ship's bow.

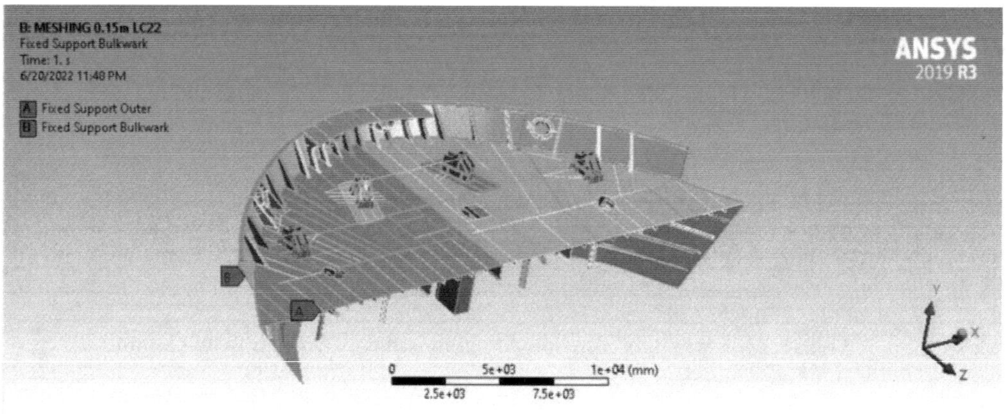

Figure 15. Fixed Support on Bow Model

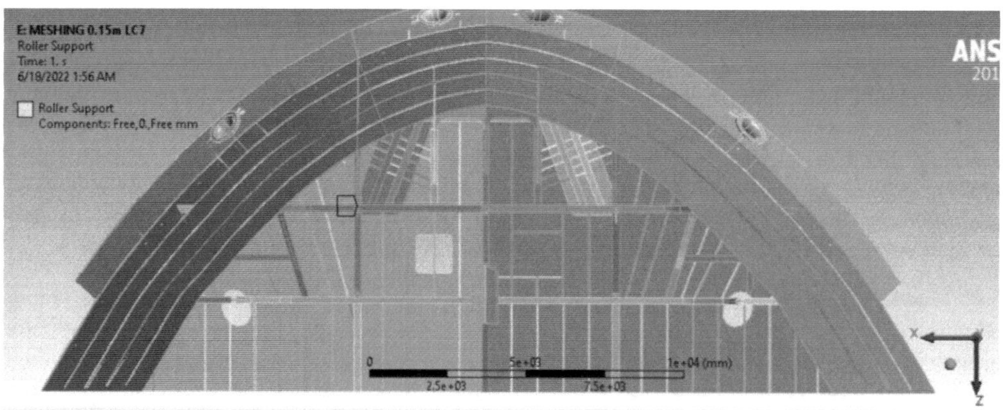

Figure 16. Roll Support on Bow Model

The tensile force will be applied to the chain stopper after first calculating the resultant force that will occur in the pulling direction from the chain stopper. This needs to be done because the force that occurs on the chain line is of course different from the force that occurs on the chain stopper because the direction of the force and the position of the chain stopper is different from the chain line. First, we will convert the force vector that initially points from the anchor to the chain to coincide with the positive Fx - negative Fy plane. Then a resultant force will be created in the direction of the direction of the chain stopper 2 SB (see Figure 17.).

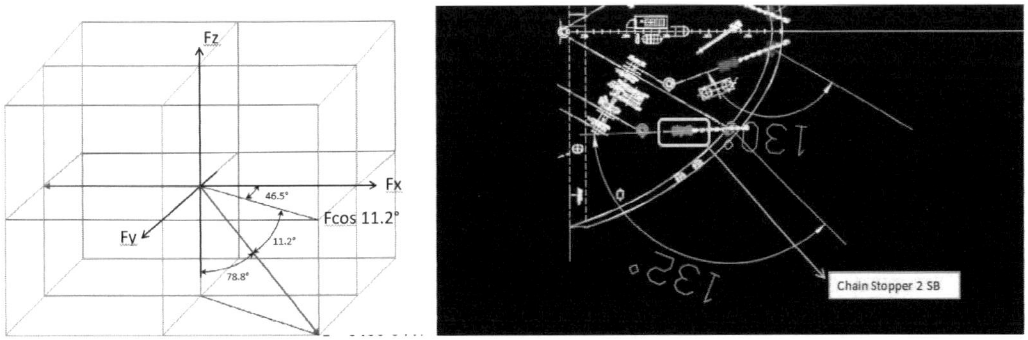

Figure 17. Force Vector Direction on Chain Line

After all the resultant forces are obtained, each load case will be formed with a different amount of loading, then the running process will be carried out using FEA software with a total of 30 load cases. The stress calculation is done through software running by entering the Maximum Principal Stress command solution for each load case. The results that come out will be taken using the probe feature in the area where the greatest stress occurs. This value is already a Hotspot Stress value, without the need to multiply it with another stress concentration factor (SCF).

3.5. Making of stress transfer function

The stress transfer function (Stress RAO) will be obtained by plotting the results of running all 30 load cases that have been obtained previously into 5 different stress RAO graphs according to the variations in the direction of loading as can be seen in Figure 18. The assumption that the value of the stress range will be equal to twice the value of the stress amplitude will be used in this analysis. Where the amplitude stress itself is the hotspot stress obtained from the FEA software output. One of the plots of the graph of the stress transfer function I obtained can be seen in Figure 19.

No	Frequency (rad/s)	Principal Stress Range (Mpa)				
		Following Sea	Quart Sea 45deg	Beam Sea	Quart Sea 135deg	Head Sea
1	0.59	177.78	198.46	327.36	82.09	100.15
2	0.69	129.51	193.65	268.80	179.18	97.08
3	0.80	120.12	155.79	200.96	165.04	93.94
4	0.90	99.13	127.86	167.41	143.20	87.52
5	1.00	88.24	117.14	155.90	135.78	84.52
6	1.11	87.51	114.84	154.86	132.23	84.14
7	1.21	86.21	113.02	153.68	130.07	83.82
8	1.31	84.48	111.56	152.39	129.00	83.55
9	1.41	82.75	110.10	151.10	127.92	83.28
10	1.52	81.02	108.64	149.82	126.85	83.01
11	1.62	79.28	107.19	148.53	125.78	82.75

Figure 18. Summary of Stress Range in 5 Loading Direction

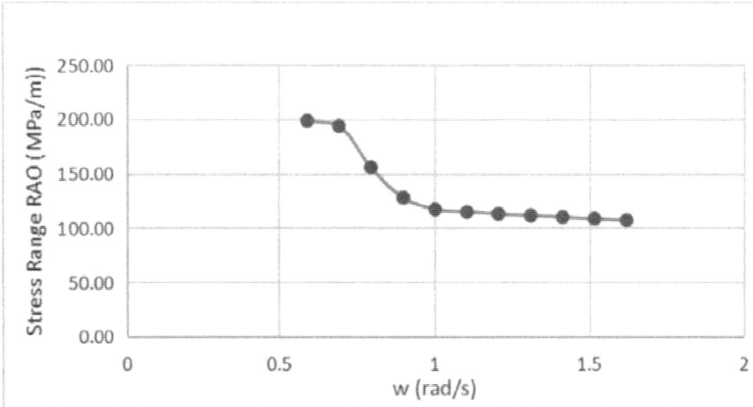

Figure 19. Stress Transfer Function on Loading Direction Quartering Sea (45deg)

3.6. Stress response spectrum calculation

One example of the calculation table for the spectral response of the stress response spectrum graph that has been obtained can be seen in Figure 20. and Figure 21.

1	2	3	4	4*4	2*4*4	2*4*4*3	1*1*2*4*4*3
w (rad/s)	S(w) (m^2/rad/s)	SM	RAO (Mpa/m)	RAO^2 (Mpa^2/m^2)	Sr(w) (Mpa^2/(rad/s))	Sr(w)*SM (Mpa^2)	w^2*Sr(w)*SM (Mpa^2(rad/s))
0.59	1.68E-07	1	198.46	39387.16544	6.63E-03	6.63E-03	2.31E-03
0.69	2.92E-04	4	193.65	37502.19816	1.10E+01	4.38E+01	2.11E+01
0.80	7.41E-03	2	155.79	24270.43063	1.80E+02	3.60E+02	2.28E+02
0.90	3.16E-02	4	127.86	16348.37237	5.17E+02	2.07E+03	1.67E+03
1.00	6.12E-02	2	117.14	13722.09198	8.40E+02	1.68E+03	1.69E+03
1.11	8.75E-02	4	114.84	13188.45528	1.15E+03	4.62E+03	5.64E+03
1.21	8.65E-02	2	113.02	12772.52777	1.11E+03	2.21E+03	3.23E+03
1.31	6.77E-02	4	111.56	12445.34497	8.42E+02	3.37E+03	5.79E+03
1.41	5.31E-02	2	110.10	12122.4073	6.44E+02	1.29E+03	2.58E+03
1.52	4.15E-02	4	108.64	11803.71476	4.90E+02	1.96E+03	4.51E+03
1.62	3.29E-02	1	107.19	11489.26734	3.78E+02	3.78E+02	9.91E+02
	M0 =		617.102013		Σ=	1.80E+04	2.63E+04
	M2 =		904.2488727			Σ0	Σ2

Figure 20. Stress Response Spectra Table with Hs = 1.00 Tp = 5.48s, Loading Direction Quartering Sea (45deg)

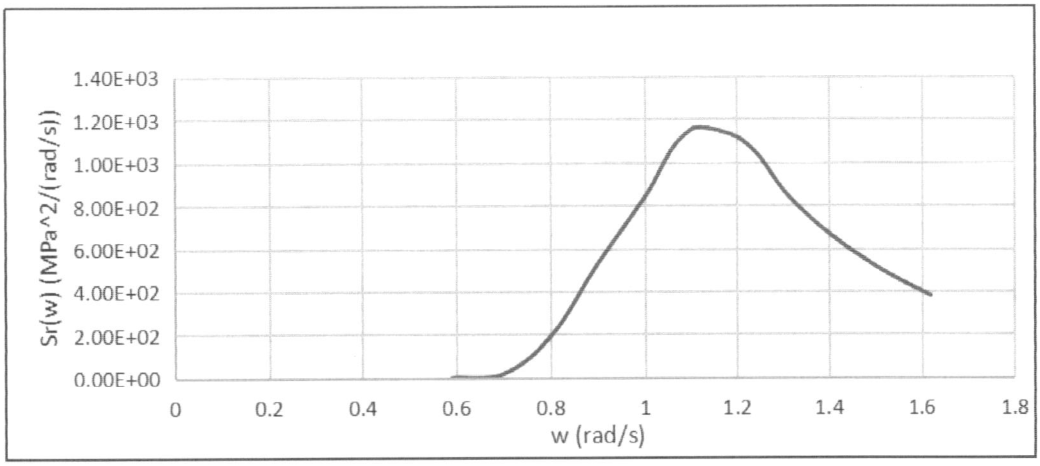

Figure 21. Stress Response Spectra Graph with Hs = 1.00 Tp = 5.48s, Loading Direction Quartering Sea (45deg)

From the example above, for each response spectral calculation table, the stochastic components can be obtained. The area under the spectral curve (m0) and the second moment of the area (m2) can be found using Simpson's formula. Meanwhile, the stress distribution that occurs, when plotted into a curve of the relationship between the stress range on the x axis and the number of events (cycles) that may occur on the y axis will form a Weibull distribution curve with parameter shape = 2, or what is also known as as a Raileigh distribution. In this case, the loading directions of 225°, 270°, and also 315° will be considered equal to the opposite loading directions (45°, 90° and 135°).

3.7. Short-term analysis

In short-term wave analysis, the number of cycles of the stress range (n) per second can be calculated using the moment area under the stress response spectral curve by applying the Raileigh distribution statistic as can be seen in Equation (4). In order to simplify the process of calculating the fatigue life, the stochastic components can be made as can be seen like the example in Figure 22. below.

Notasi	Heading Angle							
	Quarter Sea 135°	Beam Sea 90°	Quarter Sea 45°	Following Sea 0°	Quarter Sea 135° (2)	Beam Sea 90° (2)	Quarter Sea 45° (2)	Head Sea 180°
m_0	817.55	1127.87	617.10	355.46	817.55	1127.87	617.10	332.18
m_2	1208.84	1669.36	904.25	517.84	1208.84	1669.36	904.25	497.65
P_θ	0.70%	0.32%	12.56%	0.70%	0.15%	0.26%	1.68%	4.71%
n	1.94E-01	1.94E-01	1.93E-01	1.92E-01	1.94E-01	1.94E-01	1.93E-01	1.95E-01
n 25 tahun	1.53E+08	1.53E+08	1.52E+08	1.51E+08	1.53E+08	1.53E+08	1.52E+08	1.54E+08
[redacted]								
n (Hs,T 25 tahun)	1.06E+06	4.84E+05	1.91E+07	1.06E+06	2.23E+05	3.90E+05	2.55E+06	7.24E+06

Figure 22. Stochastic Component of Short-Term Stress Range with Hs= 1.00 Tp = 5.48s

3.8. Long-term analysis

In long-term analysis, the value of n obtained as cycles per second must be converted into n25 years by multiplying the value of n by the design life in the same unit. So that every n value that will be obtained from each load case must be multiplied by (60x60x24x365x25 = 788400000) seconds. Then the nL result for each operating mode will be multiplied again by the probability of occurrence of waves according to 8 directions in one year (Figure 23.), whose data is taken from the wave scatter in the operating area.

Direction (From)	North 135°	North East 90°	East 45°	South East 0°	South 135° (2)	South West 90° (2)	West 45° (2)	North West 180°	Percentage (%)
0.00-0.50	5.103%	3.699%	35.815%	7.721%	2.807%	2.209%	5.140%	11.774%	74.267%
0.50-1.00	0.696%	0.317%	12.559%	0.701%	0.146%	0.255%	1.682%	4.714%	21.069%
1.00-1.50	0.058%	0.018%	2.409%	0.040%	0.001%	0.010%	0.241%	1.013%	3.790%
1.50-2.00	0.010%	0.001%	0.429%	0.003%	0.000%	0.001%	0.033%	0.226%	0.704%
2.00-2.50	0.003%	0.000%	0.074%	0.000%	0.000%	0.000%	0.005%	0.054%	0.136%
2.50<	0.001%	0.000%	0.014%	0.000%	0.000%	0.000%	0.001%	0.018%	0.034%
PERCENTAGE (%)	5.871%	4.035%	51.299%	8.465%	2.954%	2.475%	7.101%	17.798%	100.000%

Figure 23. 1-Year Wave Scatter Probability

From the short-term analysis, we have also obtained the probability of each stress range that we determined in each load case, which if totaled, then the total probability in one load case is equal to 1. From the result of multiplying the probability of each stress range with the total number of cycles involved occurs in 25 years (n25), the number of cycles *nL* can be obtained.

3.9. Fatigue life calculation

The fatigue life of the structure near the chain stopper will be obtained from the quotient between the Design Life (which has been set for 25 years) and the Fatigue Ratio (D) that has been obtained. This fatigue index itself is the result of the sum of all the nL/N values generated by each loading direction which has their own allocated cycles according to the probability of the wave scatter which has been discussed in the previous part. The value of N will be obtained from the selected S-N curve. In this case, citing the DNV rule (2005), the D-class S-N curve will be used with parameters Log A = 12,6007 and m = 3. By making a calculation table as shown in Figure 24., the total fatigue index (D) is 0.988 (equals to 25.29 years fatigue life).

This indicates that the structure around the chain stopper can pass the specified design life, but with the recommendation of safety factor (DFF) of 2 times the design life [8], the construction around the bow chain stopper starboard 2 cannot reach the target of 50 years fatigue life.

Stress Range (Mpa)	nL (cycles) Σn Total	N (cycles)	Cumulative Damage n/N
0.0	0	-	0
1.0	2780880	3987493609042	6.97E-07
2.0	5378429	498436701130	1.08E-05
3.0	7631769	147684948483	5.17E-05
4.0	9420802	62304587641	1.51E-04
5.0	10677314	31899948872	3.35E-04
6.0	11387470	18460618560	6.17E-04
7.0	11586212	11625345799	9.97E-04
8.0	11345666	7788073455	1.46E-03
9.0	10760502	5469812907	1.97E-03
10.0	9933259	3987493609	2.49E-03
11.0	8962004	2995862967	2.99E-03
12.0	7931676	2307577320	3.44E-03

-
-
-

374.0	0	76223	3.34E-06
375.0	0	75615	3.24E-06
		D=	0.988625905
	FATIGUE LIFE D.L/D=	25.29	YEARS

Figure 24. Fatigue Index Calculation Using D-Class S-N Curve

4. Conclusions

- The Roll and Sway RAO occurred to the ship in the Light Load condition has a much greater value than the RAO in the Full Load condition with the difference in RAO Roll at the highest value reaching 1.5° per meter.
- The maximum tension value experienced by FSO in Full Load conditions and also in Light Load conditions with variations of 6 Hs classes and 5 loading directions occurs most often in chain 4.1.
- From the results obtained, it can be concluded that the weight of the cargo carried by ship has a greater influence than the characteristics of the ship's movement (ship's RAO) itself.
- Using the D class S-N curve, the fatigue ratio (D) value is 0.988.
- Using the assumption of design life of 25 years, the fatigue life of the structure around the chain stopper in the 4x2 spread mooring chain line configuration FSO PAPA Terminal case study is 25.29 years.

5. References

[1] E. B. Djatmiko, "Analisa Kelelahan Struktur Bangunan Laut," *Pembin. Dasar Eng. Dan Insp. Bangunan Lepas Pantai Terpancang (Fixed Offshore Platform)*, 2006.
[2] P. H. Wirsching dan Y.-N. Chen, "Considerations of probability-based fatigue design for marine structures," *Mar. Struct.*, vol. 1, no. 1, hal. 23–45, Jan 1988, doi: 10.1016/0951-8339(88)90009-3.
[3] DNV, "DNV-RP-C203," *Recomm. Pract. DNVGL-RP-00052014-06*, no. April, hal. 126, 2005, [Daring]. Tersedia pada: ftp://128.84.241.91/tmp/MSE-4020/Fatigue-Design-Offshore.pdf.
[4] Biro Klasifikasi Indonesia, *Guidelines for Floating Production Installations*, vol. 3. Jakarta, 2021.
[5] N. D. P. Barltrop dan A. J. Adams, *Dynamics of Fixed Marine Structures : Third Edition*. Epsom: Atkins Oil & Gas Engineering Limited, 1991.
[6] E. B. Djatmiko, *Perilaku dan Operabilitas Bangunan Laut di Atas Gelombang Acak*, no. September. ITS Press, 2012.
[7] DNV GL, *Offshore Standard - Position mooring (DNVGL-OS-E301)*. Norway, 2017.
[8] DNV GL, "OFFSHORE STANDARD DNVGL-OS-C102- Structural design of offchore ships," no. July, 2015, [Daring]. Tersedia pada: www.dnvgl.com.

Acknowledgments
The authors would like to thank all the lecturers and colleagues in the Department of Ocean Engineering, Faculty of Marine Technology, ITS Surabaya that help us to accomplish this work together, as well as all the directors and employees of PT. CM of useful data and discussions.

IOP Conf. Series: Earth and Environmental Science 1166 (2023) 012030 doi:10.1088/1755-1315/1166/1/012030

Selection of mooring system configuration of a FSO using multi-criteria methods

Murdjito[1*], H M N Shidqi[1] , W Wardhana[1], E B Djatmiko[1], I Prasetyawan[2], and F Rachmadiarto[3]
[1] Department of Ocean Engineering, Institut Teknologi Sepuluh Nopember, Surabaya, Indonesia
[2] Senior Engineer, Lloyds Register of Shipping, Malaysia
[3] Senior Marine Engineer, PT ZEE Indonesia

*murdjito@its.ac.id

Abstract. In order to maintain and restrain the movement of a temporary storage tanker during the loading and offloading process, it is necessary to design a mooring system that is safe and reliable for its operation needs. The main problem that must be solved in this research is choosing the appropriate mooring system configuration for the temporary storage tanker equipped with a catenary mooring configuration that can be used for oil loading operations from the storage tanker to a transport tanker using the ship to ship method. This research focusses in the selection of mooring system configurations based on the strength of the anchor chain. There are nine alternatives configurations and predetermined criteria that were carried out in the study. Each alternative configuration will provide a different value of tension in each mooring line and ship movement. The analysis of the mooring tension is carried out using dynamic method with time domain approach. There are many criteria to be considered in the selection of the alternatives. The analytical hierarchical process (AHP) method is used in the selection of the best configuration based on multi criteria conditions. As case study of the study is a temporary storage tanker at Papa Field in Pertamina (Persero) ONWJ operation area as temporary storage taker is an Aframax tanker size.

1. Introduction

The Special Task Force for upstream oil and gas business activities of the Republic of Indonesia (SKK MIGAS) has a policy to avoid an unplanned shutdown in upstream oil and gas activities [1]. In general, exporting oil and natural gas from offshore field to onshore processing unit can be done in two ways, namely through subsea pipelines or vessel tankers [2]. In Indonesia, pipeline systems are often used instead of using tankers to export the crude oil from the offshore oil fields to the onshore processing units for not too far away distance. On the other hand, the use of subsea pipelines also have a great potential for oil leaks that can pollute the environment and harm the company [3]. In addition, the company will also experience financial losses due to the temporary cessation of the product oil delivery process.

Pertamina ONWJ indicated that there had been a crude oil leak in the subsea pipeline network at the Papa field terminal, which was quite large during 2021, round 6000 BPOD, which has the potential to pollute the environment and have a significant impact on the economic value of the company's business [4]. For this reason, Pertamina ONWJ plans to place a temporary storage tanker (TST) as a crude oil storage and offloading system to the oil tanker vessel carrier during the investigation and repair of the subsea pipelines that have leaked in the Papa Field Terminal's area [5].

A Temporary Storage Tanker (TST) can be a temporary storage and offloading terminal for storing crude oil and loading oil into transport vessels using the tandem or side-by-side offloading methods [6]. In the moored condition, TST will store crude oil and then loads it into a tanker that is moored to the TST. Crude oil will be transferred to the tanker via a floating hose. To limit the movement of TST and export tanker, a mooring system is needed where TST is moored with the spread mooring system while export tanker is moored side by side to TST. The proposed coordinate location of the TST can be seen in Figure 1.

From Figure 1. above, it is known that many seabed facilities are still actively used. In addition, the determination of the mooring system configuration must also consider the length of the floating hose which is limited to 500 m from the processing platform, the restricted area should not be more than 500 m around the TST position as per government regulations, the ease of maneuvering of the export tanker during berthing and unberthing process, as well as operational safety considerations in the event of an emergency. So that in the placement of the mooring system at that location, the facilities around the system must be taken into account, especially in the coordinate placement of anchors. Until now, there has not been much research focus on the selection of mooring system configurations for TST with an export tanker mooring system using a side by side method for offloading.

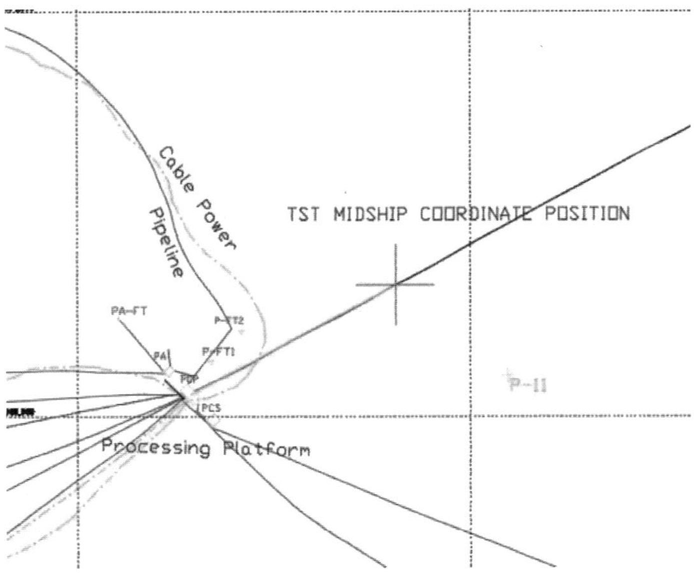

Figure 1. Location of FSO [27]

In this study, mooring system configurations will be selected based on the mooring line strength of the temporary storage tanker (TST) at Papa Field terminal. Nine alternative configurations and six criteria used in determining the most suitable TST mooring system configuration. A dynamic time-domain analysis of the mooring line tension was carried out to determine the mooring line tension. The analysis uses a time integration procedure to produce the mooring line tension response every time [7]. Each mooring system configuration used will be compared in pairs wise using the MCDM approach with the Analytical Hierarchy Process (AHP) [8], so that the best mooring system configuration can be selected for the Temporary Storage Tanker (TST) in this Papa Field study case.

2. Literature Review

2.1. Temporary storage tanker (TST)

Temporary storage tanker (TST) is a tanker that is temporarily converted as an Offshore Floating Storage Terminal and operates as temporary terminal. Yang Shen had mentioned that the oil transfer terminal is an important node for loading, unloading, transhipment and temporary storage of petroleum products, and is an important interface for connecting different oil transportation modes [9]. He studied the operational efficiency indicator system of the oil transfer depot and evaluates its operational efficiency by fuzzy comprehensive evaluation method. In this research he evaluated the operational efficiency of oil transfer terminal.

Determining the design of FSO, several parameters must be analysed such as FSO capacity, type of hull (double hull or single hull), loading and unloading systems, mooring systems, and others [10]. A few studies have examined several aspects of the use and selection of FSO had been done in the past. From the economic aspect, a charter rate using economics calculation, such as IRR (Internal Rate of Return) is usually used [10]. Biasotto examined the selection of oil tankers that will be converted into FPSOs by evaluating aspects of the ship's hull configuration [11]. Jose and Phillipe analysed the structure of the conversion FPSO to ensure the safety of the operation [12]. Neto stated that the critical aspects that need to be explored in the conversion of oil tanker to FSO or FPSO is the selection of oil tankers to be converted and the project start up structural and installation design in conjunction with large production facilities [13]. The methodology for the conversion procedure is based on ABS (2015) ''Rules for building and classing Floating Production Installations'' [14].

2.2. Mooring Configuration

The motion behaviour of a floating unit such as a temporary storage tanker will have an influence on the mooring system capability. The mooring system is used on the floating unit to maintain the position of the floating unit in the desired location. The mooring system is the most critical component because the failure of the mooring system can result in losses in the form of global damage [15]. The performance of the floating unit depends more directly on the performance of the system of lines and anchors rather than on the performance of a single line or anchor [16]. Proper configuration of the mooring system can reduce the tension in the mooring line [17]. The distribution of the mooring line tension and the structure offset are both impacted by the mooring line spread angle, which is another crucial factor to take into account. The length of the mooring line employed can also be impacted by the choice of angle of spread, which has an effect on both the economic and technical aspects. An investigation of the mooring line tension is necessary to determine whether a mooring system can function properly and provide the necessary strength [9]. Analysis can be done by comparing with the applicable criteria, one of which is DNVGL-OS-E301 [18]. An analysis of the mooring system has been carried out by Lim et al. (2016) about the difference between Quasi-static analysis and dynamic simulation on the motion and tension behaviour of mooring ropes for floating structures [19].

Based on research conducted by Cendekiawan (2016), it was found that the use of catenary mooring is suitable to be carried out in shallow water up to deep water [20]. In addition, modelling the spread mooring system with a group configuration/ bundle system with the same number of chains provides better tension and motion than the evenly distributed configuration. Apart from the chain configuration, variations in the length of the mooring line also affect the tension and motion of the moored vessel [21].

2.3. Side by Side offloading

Ship to Ship Transfer (SST) Operations are described with the standards of MARPOL Annex I, Chapter 8, as amended by Resolution MEPC.186(59): Prevention of Pollution during Transfer of Oil Cargo between Oil Tankers at Sea, Regulations 40, 41, and 42 [22]. Oil tankers with a gross tonnage of 150 or more that are involved in STS operations, or the transfer of oil cargo between oil tankers at sea, are restricted by the regulations. Bunkering operations shall not be subject to the regulations. Ship to ship transfer (SST) may fall under the broad category of mooring, and therefore require specialised fittings or equipment in addition to those for mooring the vessel to the terminal. In any of these cases, an efficient mooring system is essential for the safety of the ship, terminal and environment [19,23].

2.4. Analytical Hierarchy Process

Multi Criteria Decision method (MCDM) is a method to select alternatives that accommodate wide information, incomplete, uncertainty, and even contradictive between one criterion to another [29]. Triantaphyllou and Mann argued that MCDM is not a tool that determines the final decision, but rather as a decision support tools [24].

Multi Criteria Decision Method (MCDM) is one of the most widely used methods in the decision-making area. The purpose of MCDM is to choose the best alternative from several mutually beneficial alternatives on the basis of criteria or attributes determined by decision makers. There are 2 basic approaches to MCDM problems, namely Multiple Attribute Decision Making (MADM) and Multiple Objective Decision Making (MODM). MADM makes decisions by taking into account several attributes that sometimes conflict with each other, while in MODM the number of alternatives is infinite and the reciprocity between criteria is described using a continuous function [10].

Analytical Hierarchy Process (AHP) is one of the most popular methods of multiple criteria decision making [25]. This paper utilises AHP technique to select the best alternatives of TST mooring configurations.

3. Methodology

3.1. Data collection

The principle dimensions of TST and shuttle tanker used in the modelling are in the Table 1 and 2 [5].

Table 1. Main Dimension of TST MT Gamkonora

Principal Dimension	Unit	Dimension
Length overall (LOA)	m	244.5
Length between perpendicular (LPP)	m	233
Breadth (B)	m	44
Depth (H)	m	21.5
Draft (T)	m	12.7
DWT	ton	88,258
Displacement	ton	109,431

Table 2. Main Dimension of Export Tanker MT Sanana

Principal Dimension	Unit	Dimension
Length overall (LOA)	m	183
Length between perpendicular (LPP)	m	175.5
Breadth (B)	m	32.5
Depth (H)	m	17.1
Draft (T)	m	10.5
Displacement	ton	40627

3.2. Hull modelling of TST and export tanker

Based on the General Arrangement and main dimension of MT Gamkonora as a temporary storage tanker (TST) and export tanker MT Sanana, the 3D modelling can be carried out using MOSES software. The results of the Temporary Storage Tanker (TST) and export tanker modelling can be seen in Figure 2 and 3. In the modelling, it should consider the validity of the parameter data obtained. According to ABS [26], the data to be confirmed is from its hydrostatic parameters, i.e. the error rate should not exceed 2.0% if compared to the original data.

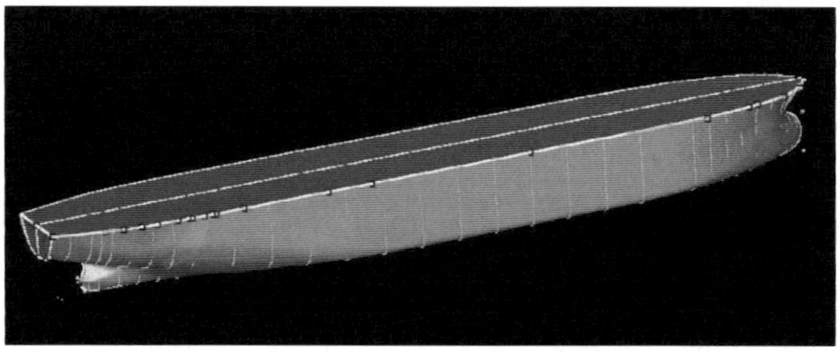

Figure 2. 3-D modelling of TST MT Gamkonora

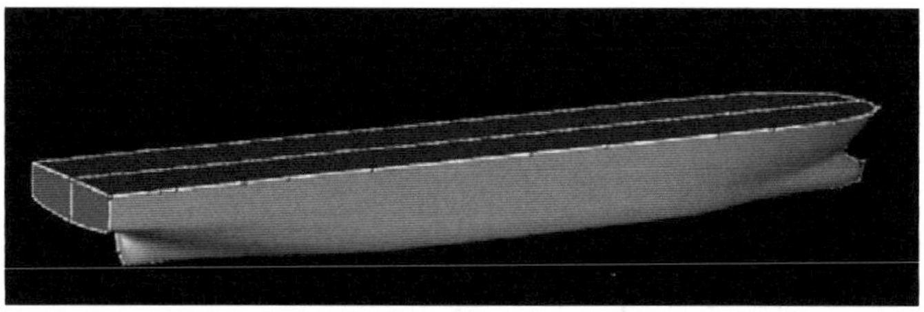

Figure 3. 3-D modelling of export tanker MT Sanana

3.3. Mooring Configurations

There are nine (9) alternatives spread mooring configurations that were simulated in the selection of the optimum mooring configurations, as see in the Figure 4.

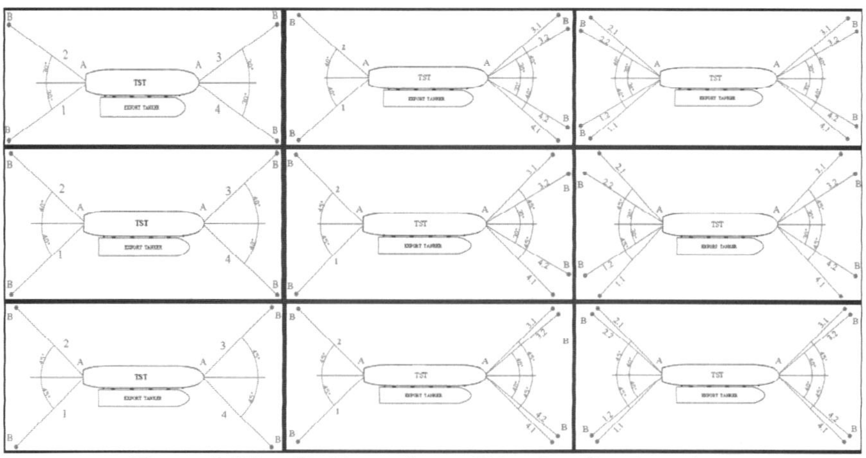

Figure 4. Nine (9) Alternatives of Mooring Configurations

3.4. Mooring analysis

The tension analysis of mooring lines and FSO offsets is carried out regarding the dynamic formulation of floating body motions [19,27,28]. Variables in eq. (2) represent matrices of the 1st-order wave excitation $\mathbf{F}_{j-wv}^{(1)}$, 2nd-order wave excitation $\mathbf{F}_{j-wv}^{(2)}$, current excitation $\mathbf{F}_{j-c}$, wind excitation $\mathbf{F}_{j-w}$, and excitation due to mooring system $\mathbf{F}_{j-m}$.

$$\sum_{j=1}^{6}\left[-\omega^2\left(\mathbf{M}_{jk}+\mathbf{A}_{jk}\right)-i\omega\mathbf{B}_{jk}+\mathbf{C}_{jk}+\mathbf{C}_{jk}^{m}\right]\zeta_j=\mathbf{F}_j \tag{1}$$

j,k	=	1, 2, 3 for translational motions in the direction of x-, y-, and z-axis, namely surge, sway, and heave, while 4, 5, 6 for rotational motions about the x-, y-, and z-axis, namely roll, pitch, and yaw;
ω	=	wave frequency;
M_{jk}	=	matrix of floating structure mass and mass moment of inertia about the reference axes;
A_{jk}	=	matrix of added mass and added mass moment of inertia in accordance with the floating structure mode of motions;
B_{jk}	=	matrix of hydrodynamics damping in accordance with the floating structure mode of motions;
C_{jk}	=	stiffness matrix of floating structure hydrostatic;
$\boldsymbol{A}_j$	=	amplitude of the jth floating structure mode of motion;
F_j	=	matrix of the jth excitation force and moment, which contains of the following elements:

$$\mathbf{F}_j=\mathbf{F}_{j-wv}^{(1)}+\mathbf{F}_{j-wv}^{(2)}+\mathbf{F}_{j-c}+\mathbf{F}_{j-w}+\mathbf{F}_{j-m} \tag{2}$$

In the mooring line analysis, the analysis is carried out using dynamic methods using time domain simulation approach. The time domain analysis is an analysis that is used to predict dynamic responses based on the time domain. The approach taken in this method will use a time integration procedure and generate a time history response based on the time function x(t). The time domain analysis method can generally be used to analyse all mooring line situations under the influence of wave dynamics. The output of the time-domain simulation is:

a. The response spectrum can be calculated from the time series, the data required for simulation is the same as the frequency domain analysis.
b. Extreme responses can be simulated directly from the response peaks during the simulation time.

The advantage of this time domain method comparing to the frequency domain is that all non-linear parameters such as system matrices and external loads can be modelled more precisely. While the disadvantage is that it requires a longer calculation time. According to DNVGL-OS-E301 (2018), the minimum duration of time domain simulation to obtain a peak statistical response value that is stable and can represent the environment is 3 hours (10800 seconds) [29].

3.5. Mooring Line Tension Calculation

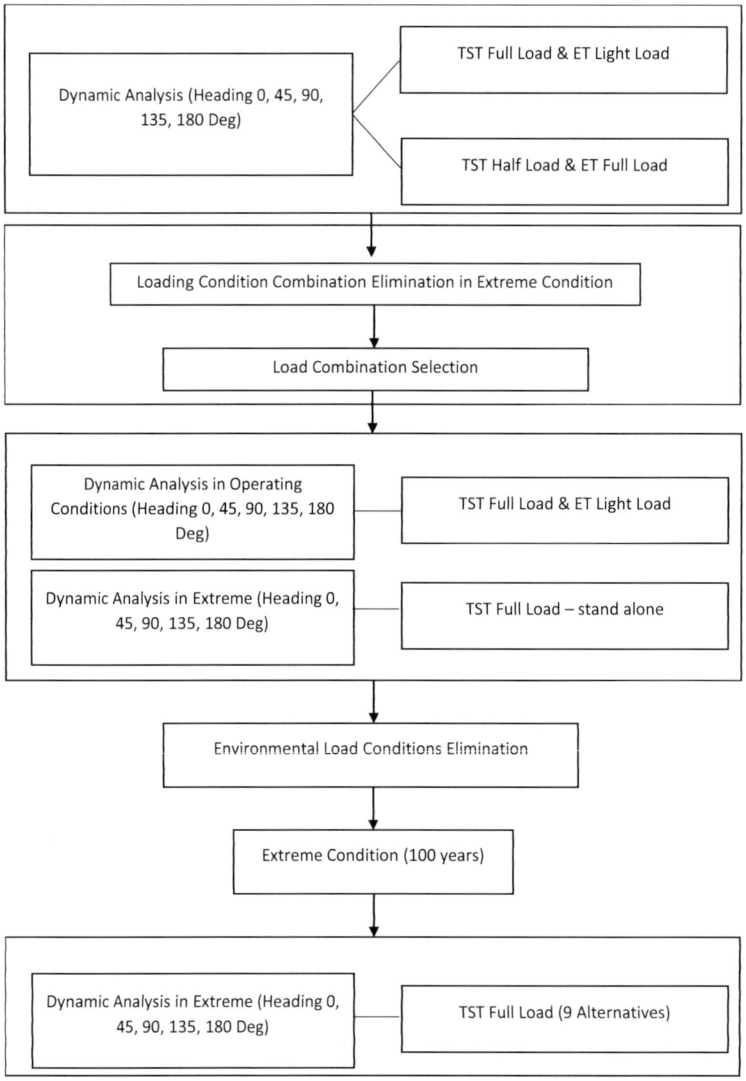

Figure 5. Elimination Procedures of Environmental Loads in Mooring Analysis

The analysis of the maximum tension of each mooring line from each configuration was carried out with the dynamic time domain method. There are five (5) headings of loading, namely 0°, 45°, 90°, 135°, and 180°. In this stage, it used a conservative step to determine the combination of loads and environmental conditions that will be used in dynamic analysis based on the time history domain.

Totally there are 180 model combinations which consist as follows: 2 loading conditions, 2 environmental conditions i.e. operation and extreme condition, 5 headings of loading condition and 9 configuration alternatives. To reduce the simulation time, it is necessary to simplify the analysis procedures in various stages. The procedures in the elimination of the model combinations are presented in the following diagram as seen in Figure 5.

3.6. Analytical Hierarchy Process (AHP)

The method used to conduct an assessment is the Analytical Hierarchy Process (AHP). It was first introduced by Thomas L Saaty in 1980 [25]. To lower the deficiencies in AHP, Belton and Gear proposed the AHP variant by reversing the ranking if there is an addition of one alternative that is the same as one of the existing alternatives [30]. They also propose normalization in the process of determining priority values, the flowchart of the AHP is presented as follows [10].

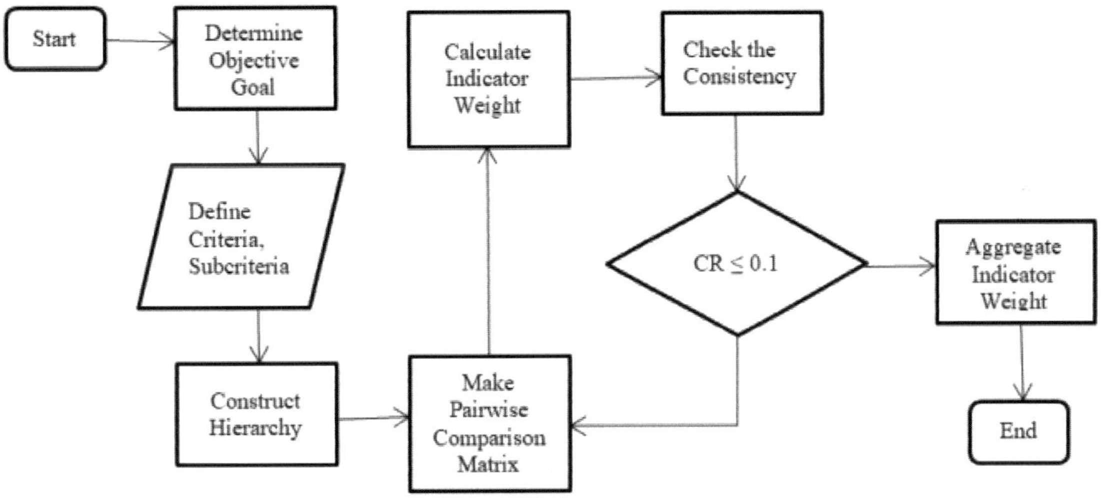

Figure 6. Flowchart of AHP [32]

The goal to be achieved through this MCDM using AHP is to obtain the best configurations form nine alternatives configurations. The assumptions that used are the PAPA's offshore terminal of Pertamina ONWJ located in offshore North West Java Sea, Indonesia. There are six (6) criteria that will be compared through pairwise comparison include Motion, class of mooring chain, length of mooring line, needs of chain stopper and deck modification, berthing ability, and Survivability.

The selection process will involve experts who have experiences in FPSO operation and were classified in two categories, namely experts with field experience more than 3 years and experts with field experience less than 3 years. The evaluation criteria into hierarchy is presented in the following figure

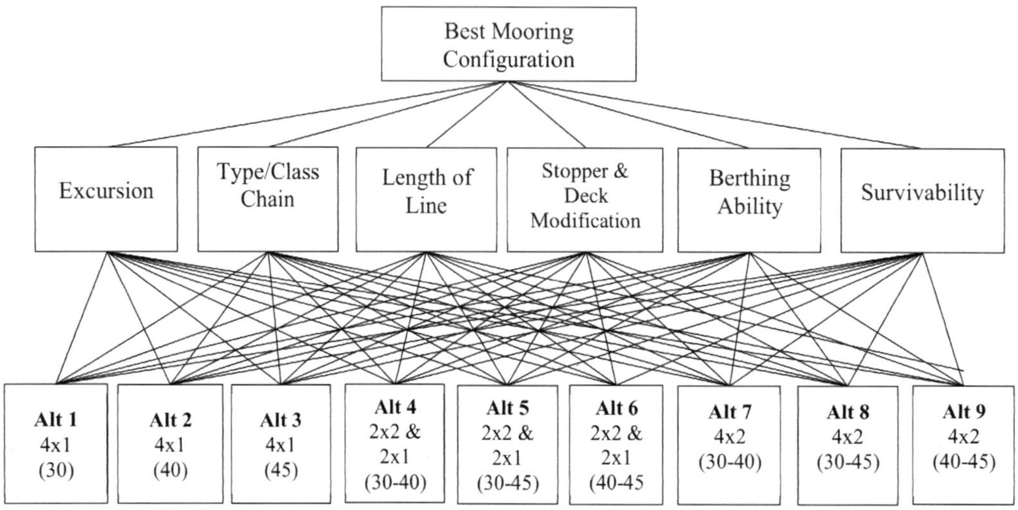

Figure 7. Diagram of the evaluation criteria into hierarchy

The explanation of these criteria is as follows:

1. Excursion
   Excursion criteria of TST in moored condition refer to API RP 2P, which is generally allowed maximum in the range of 8% to 12% of the water depth [31].
2. Type/Class Mooring Chain
   Type or Class of Chain is determined according to the results of chain tension that has met the applicable safety factor requirements, with consideration for the lower type of the chain, are better in used because the unit price is lower and easier to provide (R3, R3S, and R4) [32].
3. Length of mooring line
   The required chain length from each alternative configuration of the mooring system was chosen because the shorter the chain length used, the cheaper the procurement cost and easier installation.
4. Needs of chain stopper and deck modification
   Modifications to the TST deck, especially chain stopper modifications, often have to be made due to adjustments to the needs of the mooring system arrangement. The less that has to be modified in TST the better [33].
5. Berthing Ability
   The export tanker's ability to berth to TST during the berthing process for side-by-side offloading is determined by the maneuver area in front of the TST, which is determined by the outer angle of the Spread mooring System. The smaller the outer angle of the mooring line, the wider the maneuver area will be available and the easier it is for export tankers to berth [34].
6. Survivability
   Survivability is defined as the ability of the mooring system to withstand bad weather in damaged conditions/ ALS condition. So the more mooring lies used, the higher the redundancy of the mooring system and the better it will be to survive in extreme weather [11].

4. Results and Discussions

4.1. Heading and load condition selection

From the selection stage, it is found that the combination of TST Full – ET Light load condition has the highest max tension value in all headings. At the most significant heading selection stage, it was found that the environmental conditions that caused the largest mooring line tension value were extreme conditions from the loading direction of 90 degrees. So from here it is found that the conditions that will be used as a reference for the next analysis are TST conditions only at Full Load loads with extreme environmental conditions. It can be seen in the following Table 3.

Table 3. Maximum Line Tension.

Heading (Deg)	Max Mooring Line Tension (kN)			
	TST Half - ET Full	TST Full - ET Light	Operation Condition	Extreme Condition
0	703.2	1129.8	1129.8	2467.4
45	1174.4	1530.2	1530.2	2916.9
90	1772.0	2040.0	2040.0	4467.5
135	1244.1	1542.2	1542.2	3936.7
180	811.6	925.2	925.2	2017.8

4.2. Mooring Line Configuration Performance

From the analysis of the mooring line performances with time history for nine alternative configurations the summary results are presented in the following Table 4. For the configurations it is found that the configuration with four mooring lines, the mooring lines should comply with API [32] and DnV [18] Standards if R4 class chain type is used as the chain type, R3S chain type for 6 lines of the mooring configurations, and R3 chain type for 8 lines of the mooring configuration.

Table 4. Summary of the performances of mooring lines configurations.

Configuration	**outer angle (deg)**	**berthing length (m)**	**Number of line**	**Line length (m)**	**Excursion max (m)**	**Line Tension max (kN)**	**Chain Type**
Alt 1	30	365	4	1480	4.84	4468	R4
Alt 2	40	330	4	1480	4.69	4364	R4
Alt 3	45	315	4	1480	4.23	4211	R4
Alt 4	40	330	6	2220	3.83	4134	R3S
Alt 5	45	315	6	2220	3.30	4130	R3S
Alt 6	45	315	6	2220	3.21	4094	R3S
Alt 7	40	330	8	2960	3.16	3222	R3
Alt 8	45	315	8	2960	3.03	3122	R3
Alt 9	45	315	8	2960	2.53	2690	R3

4.3. Expert Judgement

In this analysis, expert respondents are divided into two categories of experts, namely, experts with less than 3 years of field experience in FPSO operations and experts with less than 3 years of field experience. The following figures are presented the results of pairwise comparison using AHP for each criterion. The results of pairwise comparison with AHP for expert qualification with less than 3 years of field experience can be seen in the following Table 5. While the results of priority vector of the expert qualification with greater than 3 years of field experience can be seen in the following Table 6. Table 7 consists of all expert

Table 5. Priority vector of expert qualification less than 3 year experience

Criteria	Priority Vector (W)	Rank
Survivability	0.434	1
Berthing Ability	0.143	2
Excursion	0.129	3
Chain Type	0.113	4
Chain length	0.107	5
Deck Modification	0.073	6

Table 6. Priority vector of expert qualification greater than 3 year experience.

Criteria	Priority Vector (W)	Rank
Survivability	0.457	1
Berthing Ability	0.219	2
Excursion	0.127	3
Chain length	0.117	4
Chain Type	0.041	5
Deck Modification	0.039	6

Table 7. Priority vector of all expert

Criteria	Priority Vector (W)	Rank
Survivability	0.448	1
Berthing Ability	0.189	2
Excursion	0.128	3
Chain length	0.113	4
Chain Type	0.070	5
Deck Modification	0.052	6

From these results, it was found that for both expert categories there were differences in the priority weighting on the criteria for the chain length and the chain type. Furthermore, the priority vector results from the two categories of experts are weighted, with a weighting of 40% for experts with qualifications of less than 3 years and 60% for expert qualifications of more than 3 years. The results of the combined priority vector ranking are obtained as follows.

4.4. Configuration Selection

By referring to the results of the dynamic analysis on the mooring lines that have been carried out and then each condition is weighted, a summary of the results for each alternative configuration is obtained based on each criterion as shown in the following table 8.

From the Table 8, it is found that based on the dynamic time domain simulation, pairwise comparison with AHP and the weighting procedure in the selection of alternatives of the mooring line configuration that has been done. The alternative configuration that has the highest priority vector for the configuration of the mooring system at Papa Field is configuration alternative 7 which is 4x2 lines, line angle of 30-40 degree with a priority vector of 0.1436.

Table 8. Summary results of priority vector of alternative configurations

Alternative	Excursion	Chain Class	Chain Length	Deck Modifications	Berthing Ability	Survivability	Total
Alt 1	0.008	0.0039	0.0189	0.004	0.0405	0.0249	0.1001
Alt 2	0.008	0.0039	0.0189	0.004	0.027	0.0249	0.0866
Alt 3	0.008	0.0039	0.0189	0.004	0.0135	0.0249	0.0732
Alt 4	0.016	0.0078	0.0126	0.004	0.027	0.0498	0.1171
Alt 5	0.016	0.0078	0.0126	0.004	0.0135	0.0498	0.1036
Alt 6	0.016	0.0078	0.0126	0.004	0.0135	0.0498	0.1036
Alt 7	0.016	0.0116	0.0063	0.0081	0.027	0.0746	0.1436
Alt 8	0.016	0.0116	0.0063	0.0081	0.0135	0.0746	0.1301
Alt 9	0.024	0.0116	0.0063	0.0121	0.0135	0.0746	0.1421

The advantages and disadvantages of selecting alternative seven as the main priority of the selected configuration can be described as follows:

- If one of the mooring lines is broken due to the force under extreme conditions, it will not cause a failure of the mooring line integrity as a whole.
- If one line chain damage occurs in extreme environmental conditions, then no additional number and capacity of Tugboats are required.
- The berthing and un-berthing processes for the export tanker to TST are relatively easy.
- The operating costs of the mooring system are relatively cheaper than other alternatives.
- Capitan and installation expenditures are relatively expensive.
- The deck modifications for TST are required for additional chain stoppers.

5. Conclusions

This paper aims to find out that the mooring configuration with 4x2 lines and an outer line angle of 30-40 degrees is the best solution for the mooring system for the PAPA field temporary storage terminal (TST). The second best option is a mooring configuration with 4x2 lines and outer line angle of 40-45 degrees.

The selection of the best mooring system configuration can be made using the multi-criteria decision method (MCDM), which is carried out in combination with the technical parameters and expert judgments. The use of MCDM is not the final method of determining best mooring system configuration, but can be used as a means of supporting decision-making at managerial level. Detailed engineering calculations, Hazop, and costs analysis need to be done for follow-up decisions at the managerial level. Determination of the best mooring system configuration still requires many more criteria as the basis for its evaluation. Further research is needed on the optimization of the selected mooring system configuration so that the system is reliable, robust, and cost effective.

Acknowledgements
Firstly, the authors would like to express their deepest gratitude to the officers of Pertamina ONWJ who have given the permission to study the PAPA Terminal's Project of the temporary storage tanker and at the same time provide all the necessary supporting data. Secondly, the authors are very grateful to the senior officers of PT Zee Indonesia for supporting this study by providing the software facilities for modelling.

References

[1] Satuan Kerja Khusus Pelaksana Kegiatan Usaha Hulu Minyak dan Gas Bumi (SKK MIGAS) 2018 *Pedoman Tata Kerja Nomor: PTK-041/SKKMA0000/2022/S0 Tentang Pemeliharaan Fasilitas Produksi Minyak Dan Gas Bumi Revisi-02* (Jakarta)
[2] Havard D 2013 *Oil and Gas Production Handbook - An introduction to oil and gas production, transport, refining and petrochemical industry* (Oslo: ASEA Brown Boveri (ABB))
[3] Agbakwuru J 2011 Pipeline Potential Leak Detection Technologies: Assessment and Perspective in the Nigeria Niger Delta Region *J. Environ. Prot. (Irvine,. Calif).* **02** 1055–61
[4] Umah A 2021 Pipa PHE ONWJ Bocor, Produksi Minyak Hilang 6.000 Barel/Hari *CNBC Indones.*
[5] Pertamina ONWJ 2021 *FEED Study of Temporary Storage Tanker At Papa Station: Final Report, Project FEED West FSO at Papa, Doc No PAPA-G-PHI-002*
[6] The Oil Companies International Marine Forum (OCIMF) 2006 *International Safety Guide for Oil Tankers and Terminals (ISGOTT 6)* (London)
[7] Maharani D F 2017 *Analisa time-domain pengaruh spread mooring dengan variasi jumlah line terhadap tension pada flexible riser* (Institut Teknologi Sepuluh Nopember)
[8] Tzeng G-H and Huang J-J 2011 *Multiple Attribute Decision Making: Methods and Applications* vol 4 (New York: Chapman and Hall/CRC)
[9] Shen Y and Mu D 2018 Evaluation and optimization of operation efficiency of oil transfer depot *AIP Conf. Proc.* **2036**
[10] Susilowati H, Gurning R O S, Artana K B, Dinariyana A A B and Setiyono H A 2020 Analysis of coastal state regulation impact on the capacity optimization of floating storage and offloading using multi criteria decision making *IOP Conf. Ser. Earth Environ. Sci.* **557**
[11] Biasotto P, Bonniol V and Cambos P 2005 Selection Of Trading Tankers For FPSO Conversion Projects *Offshore Technol. Conf.* OTC-17506-MS
[12] Esteve J and Philippe C 2014 Conversion and life extension of floating production unit *Proc. Indon. Petrol. Assoc., 38th Ann. Conv.,2014* ed Indonesian Petroleum Association (Jakarta)
[13] Neto T G and Lima H A de S 2001 Conversion of Tankers into FPSOs and FSOs: Practical Design Experiences *Offshore Technol. Conf.* OTC-13209-MS
[14] American Bureau of Shipping 2015 *Rules for Building and Classing Floating Production Installations* (Houston: American Bureau of Shipping)
[15] Grue I H 2014 *Loads on the Gravity-net-cage from Waves and Currents* (Norwegian University of Science and Technology)
[16] Choi Y J, Gilbert R, Ding Y and Zhang J 2006 *Reliability of Mooring Systems for Floating Production Systems* (Texas)
[17] Putri N D 2018 Analisis Variasi Konfigurasi Sistem Mooring Dan Olah Gerak Struktur Model Collar Cage Floating Offshore Aquaculture Untuk Laut Lepas Indonesia *Tugas Akhir*
[18] Det Norske Veritas 2010 *Environmental conditions and environmental testing* vol 2 (Høvik, Norway)
[19] Murdjito, Rosari M P and Djatmiko E B 2018 Analysis on the Critical Conditions of Side-by-Side Offloading Operation between SSP Type-FPSO and Shuttle Tanker *Appl. Mech. Mater.* **874** 53–63
[20] Cendekiawan V 2016 *Studi Komparasi Perilaku Gerak FLNG Masela Terhadap Sistem Tambat Spread Mooring Dengan Konfigurasi Spread dan Group (Bundle)* (Institut Teknologi Sepuluh Nopember)

[21] Prasodjo B S 2019 Study Optimasi Panjang Mooring Line Tipe Spread Mooring pada F(P)SO *Wave J. Ilm. Teknol. Marit.* **5** 13–9
[22] IBICON Ship to Ship Transfer Operations Plan (STS) *IBICON*
[23] Villa-Caro R, Carral J C, Fraguela J Á, López M and Carral L 2018 A review of ship mooring systems *Brodogradnja* **69** 123–49
[24] Triantaphyllou E and Mann S H 1995 Using the analytic hierarchy process for decision making in engineering applications: some challenges *Int. J. Ind. Eng. Theory, Appl. Pract.* **2** 35–44
[25] Saaty T L 1990 How to make a decision: The analytic hierarchy process *Eur. J. Oper. Res.* **48** 9–26
[26] American Bureau of Shipping 2018 *Rules for Building and Classing Mobile Offshore Drilling Units* (Houston: American Bureau of Shipping)
[27] Chen X 2011 Offshore hydrodynamics and applications *IES J. Part A Civ. Struct. Eng.* **4** 124–42
[28] Van Walree F and Van den Boom H J J 1991 Wind, Wave, and Current Loads on Semisubmersibles *Offshore Technol. Conf.* OTC-6521-MS
[29] Det Norske Veritas 2004 *Offshore Standard DNV-OS-E301: Position Mooring* (Høvik, Norway: Det Norske Veritas)
[30] Belton V and Gear T 1983 On a short-coming of Saaty's method of analytic hierarchies *Omega* **11** 228–30
[31] American Petroleum Institute 1987 Recommended Practice for the Analysis of Spread Mooring Systems for Floating Drilling Units *API RP 2P* (USA)
[32] API-RP2SK 2005 *Design and Analysis of Stationkeeping Systems for Floating Structures* (Washington D. C.: American Petroleum Institute)
[33] International Association of Classification Societies 2017 *Common Structural Rules for Bulk Carriers and Oil Tankers* ed International Association of Classification Societies (International Association of Classification Societies)
[34] Thoresen C A 2003 *Port Designer'S Handbook: Recommendations and Guidelines* (London: Thomas Telford Publishing)

IOP Conf. Series: Earth and Environmental Science 1166 (2023) 012031 doi:10.1088/1755-1315/1166/1/012031

Improving Safety of Marine Cable Survey Operation Through Safety Assessment using the Formal Safety Assessment Method (Case Study RV. Baruna Jaya)

A Muhtadi[1], D H Waskito[2], and D F Prasetyo[3]

[1] Research Center for Transportation Technology, National Research and Innovation Agency, Serpong, Indonesia 15314
[2] Directorate of Research Vessel Fleet Management, National Research and Innovation Agency, Jakarta, Indonesia 10250
[3] Research Center for Hydrodynamic Technology National Research and Innovation Agency, Surabaya, Indonesia 60112

E-mail : ahmad.muhtadi@brin.go.id

Abstract. In the last five years, survey operations for marine cables have become a yearly activity carried out by Indonesia's Research Vessel. Safety in survey operations is one of the biggest concerns in marine cable route surveys. This paper focuses on handling the survey operation risk and excluding the vessel operation risk. To solve the problem, we need a risk analysis methodology with systematic steps to identify, prioritize, and control risks. The International Maritime Organization's (IMO) methodology to deal with them is the Formal Safety Assessment (FSA). Firstly, Hazard Identification (HAZID) is used to identify hazards and prioritize Top Events where Towfish Survey Equipment Lost is found as the Highest Top Event with basic events Bad Communication, Fisheries Aggregating Device, Seabed Collision, and Wire & Cable fastening using the Fault Tree Analysis Method as the second step. Determining a Risk Control Option (RCO) based on the Basic Event in the FTA is the third step, of which Mitigation and Prevention will be the highlight. Lastly, RCO will be assessed according to Cost Benefit Assessment (CBA). The results showed that by using Administrative Risk Control Option , we could reduce the probability of survey equipment being lost by 46.55%.

1. Introduction

In recent years, Indonesia has expanded its underwater network cable to the east. In this case, Palapa Ring Timur started in 2017 and was still expanded with phase 2 and 3[1]. Based on the Indonesia Ministry of Communication, there are 360.000 km of fibre optic cables across land and sea. The current Indonesia Sea Cable Communication system (SKKL) is 115,000 km long, including 55,000 km of cables in the Economic Exclusive Zone (ZEE)[2]. Baruna Jaya Research Vessel is one of the research and survey vessels that were involved during Palapa Ring cable route survey in NTT, Lumori cable route survey, and North Papua Cable Route Survey Phase 2. To increase the safety and reliability of survey operation due to breakdowns and accidents, we approach those topics with the aim to improve the safety of marine cable route survey operations.

In this modern world, infrastructure expansion has been in constant progress. In line with the project's progress, the increased concern regarding safety has been constantly improved. The International Maritime Organization (IMO) introduced Formal Safety Assessment (FSA) as a rational and systematic process for accessing the risk related to maritime safety and the protection of the marine environment and for evaluating the costs and benefits of the IMO's options for reducing these risks[3]. The last amendment came out on April 9th, 2018, in MSC-MEPC.2/circ.12/Rev.2. FSA itself is not exempt from critics. Stated that lack of reliable safety data and lack of confidence in safety assessment has been significant problems in safety analysis which is undoubtedly true in FSA, where the uncertainty levels are high[4]. FSA also relies highly on expert opinion, which could lead to excessive assumptions from questionably qualified experts[3]. However, the IMO FSA guideline has laid out a guideline on concordance matrix to reduce the disagreement between experts. Due to the nature of accident data being either confidential or not published, this research will tackle the lack of data with expert judgment.

FSA itself has been in constant study. At first, it was used as a tool for the IMO decision-making process and not exclusively for ship safety cases[5]. On the other hand, the methodology itself is being tried and succeeded when applied to vessel safety improvement; Safety in Traditional Shipping[6], Safety on Ship Collision[7], Shipping Traffic[8]. However, FSA study regarding safety on marine survey is scarce, especially FSA study that focus on survey or research vessels[9]. Asuelimen et al. study on FSA of Marine Seismic Survey Vessel Operation approaches the risk assessment of research vessels by FSA application incorporating Fault Tree Analysis and Risk Matrix. This paper gave us guidance and methodology to approach safety on marine seismic survey vessels. Thus we take note and modify the approach to the survey operation rather than the vessel as a whole.

This study aims to tackle the accident that will likely occur during survey operations based on systematic and proactive ways that are deployed on FSA Methodology. At the end of the study, we will get the list of most likely accidents based on expert judgment that has been processed by Fault Tree Analysis. The event or accident will then be approached by Risk Control Option that is used with Cost Benefit Assessment, thus inventing a rational and cost-effective decision. It needs to be known that the lack of statistical data and history will be a huge question for an accurate result on decisions issued and carried out. However, due to that question also, we chose to use both qualitative and quantitative approaches using Fault Tree Analysis (FTA) and Hazard Identification (HAZID) to give the most proper result that could be delivered, based on available data.

1.1. Marine Cable Route Survey

Marine cable route survey is one of the primary requirements for the first phase of underwater network optic cable, which is based on the selected route from a desktop study. The survey's primary purpose is to obtain the bathymetric data, topologic and seabed features, and geotechnical properties of the proposed underwater cable area. From the three data mentioned above, the geotechnical, geologist, and oceanographer should be able to identify the geotechnical, geology, and hydrology information that will be used for location selection, construction, and maintenance of underwater network cable[10].

Generally, the survey of marine cable route uses research vessels, which ship's particular vary according to their capabilities. Normally a research vessel has a length of 50-70 meters, a width of 10-12 meters, a draft from 4 to 5 meters, and usually has a spacious deck to accommodate deck activities and research equipment. The vessels are also equipped with A-frame that can lift up to 5-10 tons, Main Crane for lifting activities, and several winches to conduct the CTD and SVP. Conductivity, Temperature, Depth (CTD), and Sound Velocity Profiler (SVP) are activities to acquire data on the seawater column's physical properties. In this case, both are used to ensure the multibeam deviations are limited to the minimum required[11]. Some newly built research vessels are also equipped with a fairing – An additional mini hull that contains research equipment such as a multibeam echosounder, ADCP (Acoustic Doppler Current Profiler), or SBP (Sub Bottom Profiler)-which increases the draft of the vessel that limits the movement of the vessel in the nearshore area, which also will be analyzed in this study.

One of the essential components of surveying is the survey equipment that will be used to obtain data, images, or samples needed for several purposes. Generally, on marine cable route surveys, data acquisition is performed by deploying, towing, or putting down the equipment into the water at a certain depth. With that method, there will be a hazard primarily for the equipment itself, such as loss of equipment, damage to equipment, and hazard for humans that operate the equipment or hazard to the ship as a vessel containing the equipment. Moreover, the hazards are not only in the operational phase, but the preparation and completion phase must be taken into account, such as the recovery of the equipment and lifting of the equipment using a crane during preparation. The lists of survey equipment for marine cable route survey that has the potential as a 'high risk' equipment are listed in table 1.

Table 1. List of Survey Equipment for Marine Cable Route Survey[12].

Equipment	Function	Data Acquisition Method
SVP (Sound Velocity Profiler) / CTD (Conductivity Temperature Depth)	Correction for sound wave propagation	Deployed from the side of the ship to a certain depth up to 4000 m. The ship is stationary
Multibeam Echosounder	Provide depth of the seabed	Mounted : Attached to the ship's bottom hull Portable: Attached to ship's side hull, can be moved
Sub Bottom Profiler	Identify sediment type	Mounted : Attached to the ship's bottom hull Portable: Towed behind the ship
Gravity Corer / Box Corer	Obtain sample of sediment from seabed	The Corer is dropped to the seabed by ship's Gantry/ A Frame
Marine Magnetometer	Detect any pipe or existing cable	Towed behind the Ship when the ship is moving
Side Scan Sonar	Create an image of seafloor	Towed behind the ship when the ship is moving

2. Methods and Formal Safety Assessment

FSA methodology is described as a systematic process when assessing the case. It has five stages of assessment, as shown in Figure 1.

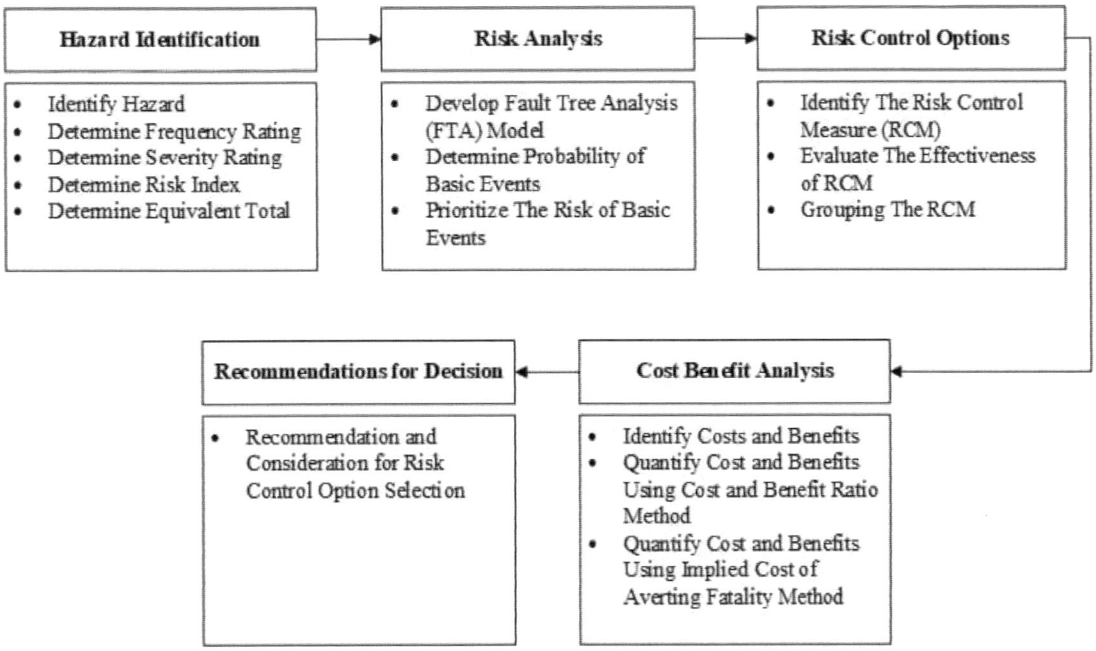

Figure 1. Flowchart of the FSA methodology employed on this study.

2.1. Hazard Identification

2.1.1. List of Hazard

Hazard identification is performed to obtain a list of hazards that may or might cause accidents in marine surveys. In this study, HAZID was obtained solely by expert judgment, which was composed of choosing experts, Elicitation, and Aggregation[13]. The expert group consists of people with experience in marine survey for ten to twenty years as Party Chief, Head of Deck Operation, and Chief Scientists. The list of accidents that can happen during marine cable route survey for each survey equipment is obtained by doing an expert group interview, reviewing a survey plan and HSE report available during marine cable route survey from 2017 to 2021. The hazards associated with accidents are also chosen when doing the interview. The hazard refers to what already happened in the past and predicts relevant events that may occur in the future[14].

2.1.2. Risk Index

The next step is to screen the hazard and its relation to risk. After the identification of the top event (accident) and hazard had been conducted, the experts were asked to fill the Risk Index form, which can be constituted as a function of probability (P) and consequences (C) as denoted in equation (1). The frequency index table is obtained according to IMO, while the severity index is the modified version from IMO that has been adjusted with the condition in marine cable route surveys, such as the addition of operational day delay and damage to equipment index. Since the frequency index is the probability of an accident in a specific period and severity index is the consequences level of a hazard, the two indexes become a logarithmic function as denoted in equation (2)[15]. Therefore, the risk index is the summation of the logarithmic probability index and logarithmic severity index.

$$R = PC \qquad (1)$$

$$Log(R) = Log(P) + Log(C) \quad (2)$$

Table 2. Frequency Index[14].

FI	Frequency	Definition	F (per ship year)
7	Frequent	Likely to occur once per month on one ship	10
5	Reasonably probable	Likely to occur once per year in a fleet of 10 ships, i.e. likely to occur a few times during the ship's life	0.1
3	Remote	Likely to occur once per year in a fleet of 1,000 ships, i.e. likely to occur in the total life of several similar ships	10^{-3}
1	Extremely remote	Likely to occur once in the lifetime (20 years) of a world fleet of 5,000 ships	10^{-5}

Table 3. Severity Index, compiled and adjusted from several standards.

Severity Index	Indicator	Health Safety	Operational delay	Environmental impact	Damage to survey equipment	Damage to vessel
1	Minor	Injury not requiring first aid	Hourly delay	No Environmental impact	No cosmetic survey equipment damage	No cosmetic vessel damage
2	Significant	Injury requiring first aid	Daily delay	Damage to Seabed	Cosmetic damage	Cosmetic damage
3	Severe	Severe injury	Weekly delay	Damage to Coral Reef	Major survey equipment damage	Major vessel damage
4	Catastrophic	Loss of life	Monthly delay	Damage to Marine Pipeline and Cables	Loss of survey equipment	Loss of vessel

After collecting data from the expert, we found the 'critical hazard,' i.e., hazard with the highest risk index, for each top event for all survey equipment by averaging the risk index from different experts. Meanwhile, the value of the 'Highest RI' for each hazard was obtained from the maximum value of the critical hazard.

Table 4. HAZID results for SVP and Portable Multibeam

Equipment : SVP (Sound Velocity Profiler)			Equipment : Portable Multibeam		
Top Event	Hazard with Highest Risk Index	Highest Risk Index (RI)	Top Event	Hazard with Highest Risk Index	Highest Risk Index (RI)
Fire	Electric Shock	7	Fire	Electric Shock	8
Contact	Heavy Weather	8	Contact	Heavy Weather	8
Survey Equipment Lost	Fatigue	7	Survey Equipment Lost	Heavy Weather	7
Survey Equipment Damaged	Unskilled Operator	7	Survey Equipment Damage	Unskilled Operator	9
MOB (Man Overboard)	Heavy Weather	8	MOB (Man Overboard)	Deck Slippery	7
Survey Support Equipment Damaged	Heavy Weather	7	Survey Support Equipment Damage	Winch (Equipment Failure)	9

Table 5. HAZID results for Coring and Hull-Mounted Multibeam

Equipment : Coring (Gravity Core, Box Core)			Equipment : Hull-Mounted Multibeam		
Top Event	Hazard with Highest Risk Index	Highest Risk Index (RI)	Top Event	Hazard with Highest Risk Index	Highest Risk Index (RI)
Fire	Friction of Material	8	Survey Equipment Damaged	Navigation Equipment Failure	7
Man Falling	Personnel Fatigue	7	Contact	Uncharted Object	8
Man Overboard	Lubrication Oil Leak	7			
Survey Equipment Lost	Wear of Equipment Wire	9			

Survey Equipment Damaged	Wear of Equipment Wire	9
Contact	Heavy Weather	8
Survey Support Equipment Damaged	Lubrication Oil Leak	7

Table 6. HAZID results for Portable SBP and Towfish Equipment

Equipment : Portable SBP			Equipment : Towfish Equipment; Side Scan Sonar, Marine Magnetometer, USBL (Ultra Short Baseline) Beacon)		
Top Event	Hazard with Highest Risk Index	Highest Risk Index (RI)	Top Event	Hazard with Highest Risk Index	Highest Risk Index (RI)
Fire	Electric Spark	7	Fire on acquisition Room	Electric Spark	8
Survey Equipment Lost	Fishing Aggregated Devices	8	Survey Equipment Lost	Fishing Aggregated Devices	8
Survey Equipment Damaged	Cargo Handling	7	Survey Equipment Damaged	Fishing Net	8
Man Overboard	Heavy Weather	8	Man Overboard	Heavy Weather	8
Man Falling	Heavy Weather	8	Man Falling	Heavy Weather	8
Survey Support Equipment Damaged	Cargo Handling	7	Survey Support Equipment Damaged	Cargo Handling	7

2.1.3. Equivalent Total

Tables 4-6 are used to analyze and identify the risk index for each survey equipment. To determine the top event with the highest risk for all equipment, the method to integrate the value of Risk Index is requisite. The Equivalent Total Method (ET) is generally used. This method is the logarithmic summation based on the number of times each Risk Index appears[16]. In this calculation, risk Index 4 is used as a base risk. An example of ET calculation for Towfish Equipment from one expert is explained in table 7. The details calculation for the Equivalent Total are:

ET for Fire on acquisition room:

$$4 + Log(0 + 0 + 0 + 19 \times 10^3 + 0 + 0) = 8.28$$

ET for Survey Equipment Lost:

$$4 + Log(7 \times 10^0 + 14 \times 10^1 + 4 \times 10^2 + 47 \times 10^3 + 6 \times 10^4 + 0) = 9.03$$

ET for Survey Equipment Damaged:

$$4 + Log(3 \times 10^0 + 18 \times 10^1 + 2 \times 10^2 + 49 \times 10^3 + 6 \times 10^4 + 0) = 9.04$$

ET for Man Overboard:

$$4 + Log(2 \times 10^0 + 15 \times 10^1 + 3 \times 10^2 + 21 \times 10^3 + 8 \times 10^4 + 0) = 9$$

ET for Man Falling:

$$4 + Log(0 + 0 + 2 \times 10^2 + 28 \times 10^3 + 3 \times 10^4 + 0) = 8.76$$

ET for Survey Support Equipment Damage:

$$4 + Log(0 + 7 \times 10^1 + 2 \times 10^2 + 51 \times 10^3 + 3 \times 10^4 + 0) = 8.90$$

Table 7. Example of Number of Occurrence for each Risk Index.

Risk Index	No. of occurrence for Top Event Category					
	Fire on acquisition Room	Survey Equipment Lost	Survey Equipment Damaged	Man Overboard	Man Falling	Survey Support Equipment Damaged
4	-	7	3	2	-	-
5	-	14	18	5	-	7
6	-	4	2	3	2	2
7	19	47	49	21	28	51
8	-	6	6	8	3	3
9	-	-	-	-	-	-

The results of ET for each survey equipment are averaged with results of ET from another expert, and the results are shown in table 8.

IOP Conf. Series: Earth and Environmental Science 1166 (2023) 012031 doi:10.1088/1755-1315/1166/1/012031

Table 8. Average Equivalent Total for all Survey Equipment.

Average Equivalent Total of Risk of Top Event (from experts)							
	Top Event						
Survey Equipment	Fire	Contact	Survey Equipment Lost	Survey Equipment Damaged	MOB(Man Overboard)	Survey Support Eq. Damage	Man Falling
SVP	3.8	5.26	4.92	6.61	4.49	4.95	-
Portable Multibeam	2.18	5.51	4.32	5.48	3.43	4.87	-
Coring	2.21	5.73	5.88	5.58	4.01	3.89	6.36
Hull-Mounted Multibeam	-	-	-	5.32	4.57	-	-
Portable SBP	4.16	-	4.94	6.45	3.64	5.85	4.43
Towfish Equipment	2.56	-	8.26	7.96	3.64	5.27	5.27

Table 8 shows that the 'Survey Equipment Lost' Top Event on Towfish Equipment is the highest Risk according to ET calculation, followed by 'Survey Equipment Damaged' on Towfish Equipment. Therefore it can be concluded that Towfish Equipment has the most risk on marine cable route surveys and needs to be controlled. A full-scale risk analysis will focus on the 'Survey Equipment Lost' top event.

2.2. *Risk Analysis*

Risk can be defined as a situation involving exposure to hazards, dangerous opportunities, or the possibility that something unexpected or undesirable will happen[15]. Risk analysis aims to investigate the causes and underlying events and risks of the main event scenarios that exist at the HAZID stage[14]. In this study, the investigation of the causes and basic events of the top events that have been obtained at the HAZID stage will use Fault Tree Analysis (FTA) modeling. FTA is a top-down and systematic approach to analyzing a series of events. An FTA chart is constructed using 'AND' or 'OR' gates to illustrate the failure cause of an event. The 'AND' gate is used to relate the failure causes of an event if the causes occur simultaneously, while the 'OR' gate is used if the failure causes of an event occur directly (i.e., not concurrently). Logic gates define the addition or multiplication of probabilities to get the top event value. In this study, the Top Event Survey Equipment Lost is modeled by FTA according to figure 2

From the FTA model created, 24 basic events can potentially cause the loss of survey equipment. To determine the occurrence of the top event survey equipment lost, estimating the probability of the basic event is necessary. Estimation of the basic event probability can be determined using actual data that occurs in the field based on recorded accident data that has occurred or historical data. By accident data that has been recorded, the rate of basic event failure/occurrence in units of time can be obtained[7]. This historical data will provide a more genuine interpretation when used to predict the possibility of top events occurring in the future. Existing historical data related to accidents that have occurred in submarine cable survey operations are incomplete, so this study will use a questionnaire with data from the experience of experts[17],while historical data will be used as a comparison. The following table 9, shows details of basic events and basic event probabilities from several experts.

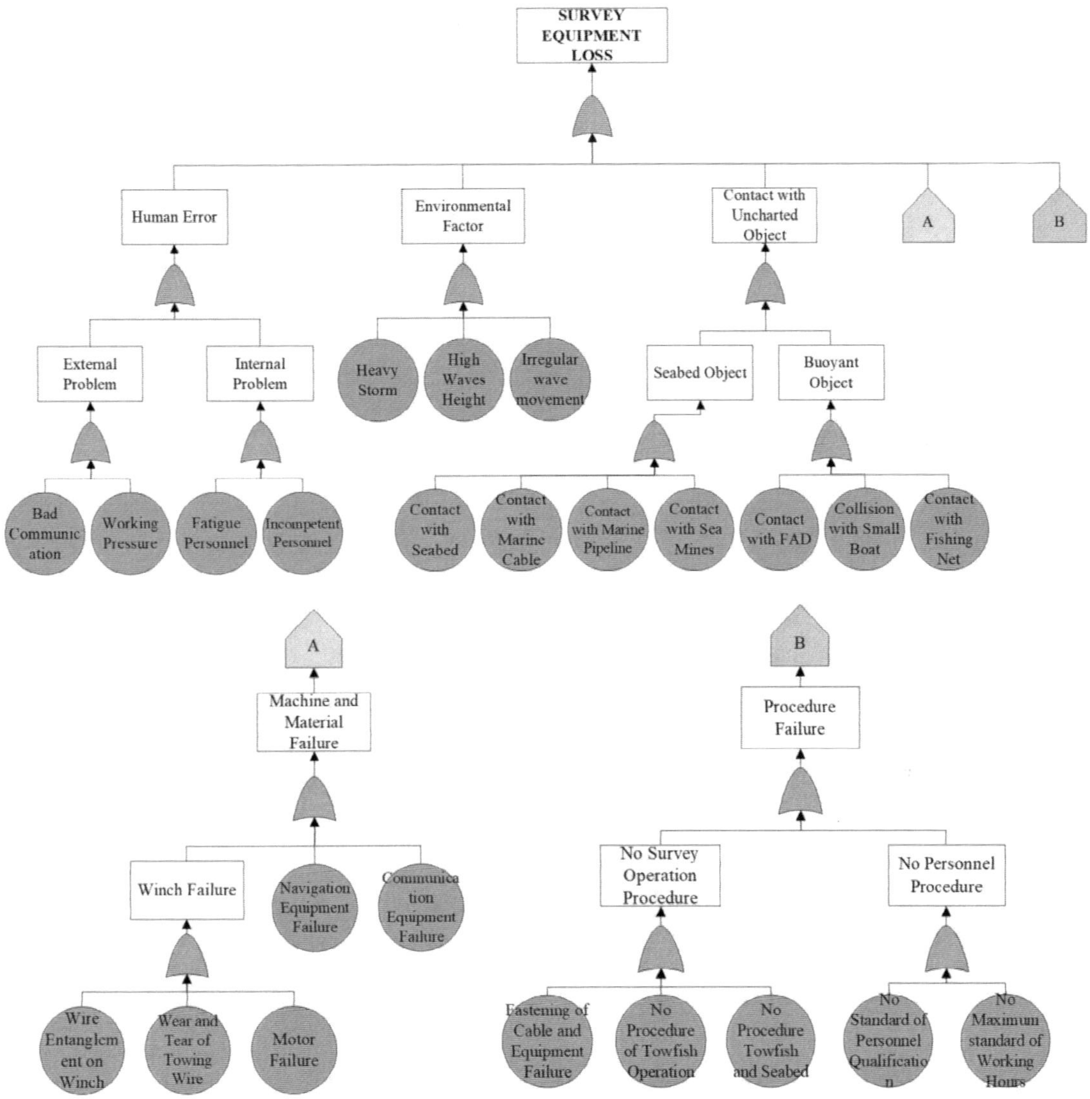

Figure 2. FTA model with top event 'Survey Equipment Lost'.

Table 9. Basic events data that have occurred based on expert judgment.

Cut Set Abbreviation	Basic Event	Occurrence (Average frequency from Expert Judgment from 0 -10 scale)	Probability of basic event happening on one ship in a year
X1	Bad Communication causing Human Error	1,4	3.84×10^{-3}

X2	Working Pressure Causing Human Error	0,6	$1.64x10^{-3}$
X3	Fatigue of Personnel causing human error	0,2	$5.48x10^{-4}$
X4	Incompetent Person causing human Error	0,4	$1.1x10^{-3}$
X5	Heavy Storm	0,4	$1.1x10^{-3}$
X6	High Waves	0,4	$1.1x10^{-3}$
X7	Irregular Wave Movement	0,2	$5.48x10^{-4}$
X8	Towfish Contact with Seabed	1,6	$4.38x10^{-3}$
X9	Towfish Contact with Marine Cable	0	0
X10	Towfish Contact with Marine Pipeline	0	0
X11	Towfish Contact with Sea Mines	0	0
X12	Towfish Contact with FAD (Fisheries Aggregation Devices)	1,4	$3.84x10^{-3}$
X13	Towfish contact with Small Boat	0	0
X14	Towfish contact with Fishing Net	0,6	$1.64x10^{-3}$
X15	Wire Entanglement	0,6	$1.64x10^{-3}$
X16	Wear and Tear of Wire	0,4	$1.1x10^{-3}$
X17	Winch Motor Failure	0,6	$1.64x10^{-3}$
X18	Navigation Equipment Failure	0,4	$1.1x10^{-3}$
X19	Communication Equipment Failure	0,6	$1.64x10^{-3}$
X20	Fastening of Cable and Equipment	1	$2.74x10^{-3}$
X21	No Procedure of Towfish Operation	0	0
X22	Seabed and Towfish distance standards	0,2	$5.48x10^{-4}$
X23	No Personnel Qualification standard	0,4	$1.1x10^{-3}$
X24	No Working Hours Standard	0,2	$5.48x10^{-4}$

With the FTA model generated according to figure 2, the probability value of the top event 'Survey Equipment Lost' with the basic event from table 9 is 0.0318 (summation the probability from each basic events). For comparison, historical data during the voyage is used, obtained from the HSE report during the 2020-2021 time range (118 sailing days). There were 12 basic events that led to the loss of Towfish Equipment twice. First is the loss of USBL Beacon and Side Scan Sonar Towfish in Kendari, Indonesia-2020, and the second is the loss of USBL Beacon, Side Scan Sonar, and Marine Magnetometer in Papua-2021. The occurrence of the basic event that causes the Top Event is shown in table 10.

Table 10. Basic events data that have occurred based on historical data.

Basic Event	Occurrence	Frequency of occurrence (118 days)
Bad Communication Causing Human Error (X1)	1	$8.4x10^{-3}$
Fatigue of Personnel Causing Human Error (X3)	1	$8.4x10^{-3}$
Towfish Contact with Seabed (X8)	1	$8.4x10^{-3}$
Towfish Contact with FAD (X12)	3	$2.5x10^{-2}$
Towfish Contact with Fishing Net (X14)	1	$8.4x10^{-3}$
Wire Entanglement (X15)	1	$8.4x10^{-3}$
Wear and Tear of Wire (X16)	1	$8.4x10^{-3}$
Winch Motor Failure (X17)	2	$1.7x10^{-2}$
No Procedure of Towfish Operation (X21)	3	$2.5x10^{-2}$
Seabed and Towfish distance standard (X22)	3	$2.5x10^{-2}$
No Personnel qualification standard (X23)	2	$1.7x10^{-2}$
No working hours standard (X24)	2	$1.7x10^{-2}$

According to table 10, the probability value of the Top Event by historical data is 0.0219 (summation the probability from each basic events). From the expert judgment and historical data (HSE Report), it is found that the latter data is slightly higher than the former because the data obtained using expert judgment has a higher number of basic events than the historical data in the HSE report. However, expert judgment data is parallel with findings from historical data. Contact with FAD is a high contributor to both data. In addition, the data on expert judgment is comprehensive because it is based on the experience of experts who have almost 15 years of work in the survey field, while historical data in the HSE report only exists for the last two years. For further analysis, the data obtained from the expert will be used.

By using the results from expert judgment, nineteen basic events were obtained while five basic events (Towfish Contact with Marine Cable, Towfish Contact with Marine Pipeline, Towfish Contact with Sea Mines, Towfish Contact with Small boat, and No Procedure of Towfish Equipment) not happened yearly on expert experience, but will potentially happen in the future. Nineteen basic events will be prioritized to determine which one will be risk-controlled and which will be in the ALARP region and considered negligible.

2.3. *Risk Controlling*

It is a crucial stage where we issue various options to make safety improvements to the top events that occur according to the FTA that has been modelled. Risk Control Options (RCO) is a list of options from RCM (Risk Control Measure) that will be carried out to approach the problems found in the FTA model [14]. RCO must overcome existing and emerging risks due to its implementation of RCO itself. Therefore, the technique for determining the RCO must be able to identify the risks that will arise due to the application. Expert opinion is also influential and contributes to the method to overcome the frequency and severity of accidents. An expert opinion provides considerable input on the options for the following approach., RCO should follow four principal stages to provide effective and practical options[15]:

1. Focus on the risk area which needs control. The risk addressed here should come from the FTA model developed in the risk analysis stage. It should have high frequency and severe outcomes.
2. Identifying the Risk Control Measures (RCM).
3. Evaluating the effectiveness of the RCM.
4. Grouping the risk control measure into practical regulatory options.

Four basic events were taken to solve the problem. These four basic events were chosen because their probability, based on the results of expert judgment, can occur once a year. The results were, 'Contact with Seabed' could occur 1.6 times a year, 'Contact With FAD' could occur 1.4 times a year, 'Bad Communication' could occur 1.4 times a year, and 'Fastening of Cable' could occur once a year. Based on the HSE framework for decisions on the tolerability of risk, there were three regions of tolerability: (A) intolerable, (B) as low as is reasonably practicable (ALARP), and (C) broadly acceptable[18]. In this study, the ALARP region is where the event has the probability of occurrence below once a year. It means that when it is likely to occur once a year, the basic event falls in the ALARP region, which is the target of implementing the RCO, to make all the basic events that lead to loss of survey equipment, at least in the ALARP region.

RCO in this study refers to Category A Attributes [14], which are divided into two types, mitigation and prevention control. After the RCM of mitigation and prevention is made, a top-down analysis is made to see the involvement of four types of risk controls: elimination, Substitution, Engineering, and Administrative. These four items are planned to be the backbone of the RCO produced in this study. The RCO presentation that will be performed is explained in table 11 and 12, describing the prevention and mitigation measures and their derivatives.

Table 11. Preventive Risk Control.

	Preventive Risk Control Measurement of selected basic events			
Risk Control Option	Bad Communication causing Human error	Towfish Contact with Seabed	Towfish Contact with FAD	Fastening of Cable and Equipment
RCO1 (Elimination) – Inherent Risk Control	Provide an adequate number of two-way radios	Perform the multibeam echosounder survey as depth verification	Perform the visual scouting of survey area	Using strong clamp and quality wire
RCO2 (Substitution) – Change of material,	Provide marine standard two way radio (waterproof,	-	-	Using excellent casting material

alternative equipment	floatable, shockproof)			
RCO3 (Engineering)- Engineering Risk Control	Using of frequency Booster	-	Adding net cutter on the head of the towfish	Using effective / additional safety wire
RCO4 (Administrative)- Procedural Risk Control	• Creating SOP of minimum working hour • Creating SOP for Toolbox Meeting • Frequent Kick off Meeting • Frequent Communicatio n Training	• Creating SOP of survey operation • Frequent Communication Training • Training Simulation when towfish contacted with seabed	• Creating SOP of survey operation • Frequent Communication Training	• Creating SOP of equipment handling • Training of equipment handling

Table 12. Mitigative Risk Control.

Risk Control Option	Mitigative Risk Control Measurement			
	Bad Communication causing Human error	Towfish Contact with Seabed	Towfish Contact with FAD	Fastening of Cable and Equipment
RCO1 (Elimination) – Inherent Risk Control	Provide a backup communication method	-	-	-
RCO2 (Substitution) – Change of material, alternative equipment	-	Provide decent safety wire	Provide decent safety wire	Replace the cable with marine standard
RCO3 (Engineering)- Engineering Risk Control	Provide additional safety wire	-	-	-
RCO4 (Administrative)- Procedural Risk Control	-	Training Simulation when towfish contacted with seabed	Training Simulation when towfish contacted with FAD	-

IOP Conf. Series: Earth and Environmental Science 1166 (2023) 012031 doi:10.1088/1755-1315/1166/1/012031

2.4. Cost Benefit Assessment (CBA)

In order to evaluate the effectiveness of every RCO, a Cost Benefit Assessment (CBA) is needed to assess whether the cost needed to reduce the risk of basic events with RCO is beneficial compared to the benefit attained. There are several methods to perform CBA; The cost-benefit ratio and ICAF method will be discussed in this study. To conduct CBA with a risk-associated method, the following steps were used[14]:

1. Define the problems and RCO
2. Identify costs and benefits
3. Quantify costs and benefits with table
4. Rank the RCO

2.4.1. Cost and Benefit Ratio Method

In terms of CBA with risk-associated methods, the loss of equipment can also be incorporated as a consequence. In this study, the consequence is associated with the cost of purchasing new Towfish Equipment, which is approximately 1.9 billion Rupiah and includes the cost of Side Scan Sonar, Marine Magnetometer, and USBL beacon. Meanwhile, the cost of RCO consists of Installation, capital cost, training, labour, and administrative cost. The maintenance cost is excluded since the RCO will be assessed only for one year period. The cost of RCO was obtained from several procurement documents, online markets, and expert estimations.

The benefits, in this case, mean reducing the number of probabilities of basic events that can induce the loss of towfish survey equipment. RCOs were used to eliminate four basic events with the highest probabilities. After the RCO is performed, the probability of a basic event becomes 0. This assumption was made because the RCO was deemed effective in solving the problem and made the probability of the basic event into 0.

To calculate the benefit as the function of reduced risk, equation (3) was used as a modified version of equation (1)[19].

$$\Delta R = (P_0 - P_1)C \qquad (3)$$

Where P_0 is the probability of the top event happening before risk control, P_1 is the probability of the top event happening after the risk control, and C is the consequences of the survey equipment lost which also can be assumed as the price of the survey equipment. In this study, the RCO is only applied for one year (the RCO's lifetime did not take into account); therefore, the reduced risk also can be stated as a benefit (B). To calculate the cost-benefit ratio, perform the ratio between the cost of RCO (C_{RCO}) and the benefit for each RCO. In order to achieve the beneficial status, the cost-benefit ratio's results must not be greater than one.

$$\frac{C_{RCO}}{B} < 1 \qquad (4)$$

Table 13. Cost Benefit Analysis of RCO using Cost-Benefit Ratio.

List of RCO	New Probability of Top Event(P_1)	Probability Difference after RCO (ΔP)	Benefit of RCO (B)	Cost of RCO (C_{RCO})	Cost Benefit Ratio	Conclusion
RCO1 (Elimination)	0,017	0,015	Rp28.797.534	Rp131.000.000	4,54	Non-Beneficial

RCO2 (Substitution)	0,017	0,015	Rp28.797.534	Rp45.000.000	1,56	Non-Beneficial
RCO3 (Engineering)	0,021	0,010	Rp20.264.932	Rp9.500.000	0,46	Beneficial
RCO4 (Administrative)	0,017	0,015	Rp28.797.534	Rp13.000.000	0,45	Beneficial

The table 13 shows that RCO3 and RCO4 were found to be more beneficial with the ratio below than one, and RCO4 is the most cost-effective option and can tackle all four basic events with the highest probability, while the former cost the least amount of money but only can reduce the probability of three basic events.

2.4.2. ICAF Method

CBA calculation by using ICAF (Implied Cost of Averting Fatality) is more commonly used to assess the effectiveness level of a risk control measure[19]. Since this study focuses more on survey equipment than a fatality, the fatality means the loss of survey equipment. Therefore, in this study, ICAF can also be stated as the cost to prevent the loss of survey equipment. The results from ICAF calculation of each RCO will be compared to the cost of consequences (C), in this case, the cost of new Towfish Equipment. If the ICAF is higher than C, the RCO is considered non-beneficial and vice versa[20].

$$ICAF = \frac{C_{RCO}}{\Delta P} \qquad (5)$$

Table 14. Cost Benefit Analysis of RCO using ICAF.

List of RCO	Reduction of Probability (ΔP)	Cost of RCO (C_{RCO})	ICAF	Conclusion
RCO1 (Elimination)	0,015	Rp131.000.000	Rp8.854.629.630	Non-Beneficial
RCO2 (Substitution)	0,015	Rp45.000.000	Rp3.041.666.667	Non-Beneficial
RCO3 (Engineering)	0,010	Rp9.500.000	Rp912.500.000	Beneficial
RCO4 (Administrative)	0,015	Rp13.000.000	Rp878.703.704	Beneficial

The table 14 shows that the RCO3 and RCO4 were beneficial since their ICAF cost less than the cost of consequences (1.9 billion Rupiah). The two methods that CBA analyzed had similar results. The cost-benefit ratio method is more suitable if the cost of consequences is known and more utilized as a preliminary approach for the CBA calculation. Meanwhile, the ICAF method is more suitable for calculating CBA to reduce fatality or human approach.

2.5. Recommendation for Decision Making

Decision-making is done to select the most cost-effective RCO based on the CBA by considering all the hazards being addressed. In determining which RCO to take, several provisions serve as a reference:

1. ICAF, is it economically beneficial
2. How many problems did the RCO solve
3. How much the decrease in top event after the RCO is applied
4. Will the implementation of RCO create new problems

It was found that RCO3 Engineering and RCO4 Administrative, in ICAF terms, were beneficial. RCO1 elimination and RCO2 substitution, according to ICAF, are not economically profitable. According to the calculation results, it was found that the ICAF of RCO3 and RCO4 was below the price of equipment lost. It means that the two RCO options can be considered beneficial.

RCO1, RCO2, and RCO4, while resolving the hazards, could address all of the basic events that caused the loss of survey equipment. However, RCO3 only addresses the problem of Bad Communication Causing Human Error, Towfish Contact with FAD, and Fastening of Cable and Equipment. Thus its reduction in probability by running the FTA is less than the rest of the RCO. RCO 1,2 and 4 reduce the probability by 46.55%, but RCO3 only reduces the probability by 34.48%.

By doing focused discussions with experts, we found that several problems would arise when carrying out the RCO. However, in RCO4, the apparent risk can be neglected and will improve as long as the Procedure is carried out regularly.

Table 15. Forecasted new risk arises by applying RCO.

RCO	Basic Event
1	Possibility of an accident when scouting on the visible obstacle at sea by rubber boat or other high-speed craft
	Possibility of an accident when recover of eliminating the obstacle using vessel equipment such as winch and gantry frame
	The additional task would result in personnel tiredness
3	Too much safety wire without proper standard operation procedure could result in entanglement.
4	Struggling with new standard operating procedures, at first, can create confusion and mess. However, the risk would reduce over time

3. Discussion and Recommendation

3.1. Discussion

Employing the Formal safety assessment in marine cable route survey operations reduces the risk of equipment lost by 46.55%. The top events addressed here differ from what Asuelimen et al. have found in their paper. Where the top event was 'Man Overboard.' Interestingly, both papers found that as an initial approach, administrative and operational strategies for mitigation are more cost-effective than the rest of the proposed RCO, where the implementation of operational strategies revealed to reduce the top event probability by 40%.

Historical data can provide more actual and precise data compared to expert judgment. The use of expert judgment in this study has represented historical data with verification that has been done previously. The results of expert judgment are linear with historical data, where the 'Contact with FAD' becomes the basic event with the highest probability. The use of expert judgment to replace incomplete historical data has several drawbacks where the expert judgment may have inaccurate estimates. So, an approach to converting uncertainty into mathematical formulas needs to be done. To achieve its objectives, institutions need to take appropriate steps, especially in taking and managing risks. The institution develops a risk matrix to map the risk to simplify which risks are acceptable (ALARP region) and which are not. In this study, the institution does not yet have a risk matrix, so the ALARP region's determination is based on writers' definition based on HSE Framework.

Much input is required in determining the RCO and the implications of implementing the RCO. In this paper, input was only obtained from the team involved in marine cable route survey for the last five years using R.V. Baruna Jaya. Additional input from other survey teams using other vessels will complement and enhance the recommendations provided by this research. Those inputs are also needed to expand the implication and effectiveness of the RCO generated in this study.

While in the process of the CBA, one variable has not been considered in the calculation, which is the time to implement the RCO. According to table 14 and 15, RCO4 will spend the most time on its implementation, since in order to create a procedure, there are several steps to carried out, such as brainstorming of material, experts' judgment, the making of the flowchart, and the most critical phase is the training and application of the procedure which will be spent around two or three weeks. On the other hand, RCO1 will take the least amount of time since the elimination of hazard (FAD) will be executed in less than a day. Without realizing it, there is an unknown relation between time and cost, in this case, profit. The faster RCO can be applied to survey equipment in this case; therefore, the equipment can also gain more financial or scientific benefits for research purposes.

3.2. Recommendation for further study

Determination of probability based on expert judgment, which has the potential to provide inaccurate estimates, is necessary to use a method to change the uncertainty. Converting uncertainty into mathematical formulas can use fuzzy logic simulations. Fuzzy logic provides a measure for the uncertainty of a vague concept.

Determining which ALARP areas and risks should be controlled would be more appropriate if the institution risk matrix had been developed earlier. For further study, ALARP region using an Institution risk matrix could be a better option to evaluate risk.

To enhance CBA's quality and accuracy of CBA, a detailed study of the RCO is needed in the future. This study assumes the probability of a basic event after RCO as zero, which in some cases, is impossible. A detailed study of RCO means an in-depth analysis regarding the probability of basic events after the implementation of the RCO, whether the RCO will reduce the probability to zero, or whether there is still a chance that the basic event will happen in the future. To carry out the study, firstly, the data of every near miss and accident after the implementation of RCO within one to five years is required. Secondly, additional data from private sectors in marine surveys regarding the safety of Towfish Equipment is also preferred. With the detailed RCO study, the RCO that has the most impact on reducing the probability of the loss of survey equipment will be known. Moreover, the data will be used to evaluate the RCO that has been carried out and to develop a more effective RCO.

4. Conclusion

In this study, the safety of marine cable route surveys operation was assessed using the Formal Safety Assessment. Focusing on the risk associated with survey operation, we exclude all vessel accidents and hazards rooted in vessel operation. After processing data from expert judgment and several historical data ranging from 2017-2021, it is found that the loss of survey equipment is the top event that needs to be controlled. Using the FTA reveals that top events such as 'Bad Communication', 'Contact with Seabed', 'Contact with FAD', and 'Bad Wire/Cable Arrangement' are above the ALARP region. To improve the safety of marine cable route survey, an administrative RCO was chosen out of four RCOs. Initial calculation results in a reduction of Loss of Survey Equipment by 46.55%.

This paper excludes vessel risk, meaning that the study focuses solely on survey operation. Nevertheless, it also means ignoring the risk that comes inevitably with a vessel whose top event could be higher than the survey operation's top event, In this case, the loss of survey equipment. Thus, an additional study to compare the result of this study to survey vessel operation through the Formal Safety Assessment method is desired.

Acknowledgements
Authors would like to say thanks to the marine cable route team of BPPT, EGS, and Telkominfra, and National Research and Innovation Agency republic of Indonesia, formerly Agency for Assessment and Application of Technology - Laboratory Marine Survey. Deepest gratitude to the experts that have contributed to this study.

References

[1] PT Palapa Ring, "Sosialisasi Palapa Ring Paket Timur," 2017. http://palaparing.id/ (accessed Jul. 15, 2022).
[2] Biro Humas Kementerian Kominfo, "Potensial Jadi Titik Penghubung Dunia, Menteri Johnny: Optimalkan Trafik dan Utilisasi Fiber Optik Indonesia," 2022. https://www.kominfo.go.id/content/detail/39793/siaran-pers-no-36hmkominfo022022-tentang-potensial-jadi-titik-penghubung-dunia-menteri-johnny-optimalkan-trafik-dan-utilisasi-fiber-optik-indonesia/0/siaran_pers#:~:text=Indonesia memiliki jaringan nasional k (accessed Aug. 19, 2022).
[3] H N Psaraftis, "Formal safety assessment: An updated review," *J. Mar. Sci. Technol.*, vol. **17**, no. 3, pp. 390–402, 2012, doi: 10.1007/s00773-012-0175-0.
[4] J Wang, "The current status and future aspects in formal ship safety assessment," *Saf. Sci.*, vol. **38**, no. **1**, pp. 19–30, 2001, doi: 10.1016/S0925-7535(00)00052-7.
[5] R Skjong, "Formal Safety Assessment and Goal Based Regulations at IMO- Lessons Learned," in *OMAE2005*, 2005, pp. 1–10.
[6] Sunaryo, A Lubis and M B Ramadhan, "Increasing the safety of traditional shipping using formal safety assessment approach," *AIP Conf. Proc.*, vol. **2227**, no. May, 2020, doi: 10.1063/5.0001027.
[7] P H Purba, A A B Dinariyana, D W Handani and A F Rachman, "Application of Formal Safety Assessment for Ship Collision Risk Analysis in Surabaya West Access Channel," *IOP Conf. Ser. Earth Environ. Sci.*, vol. **557**, no. 1, 2020, doi: 10.1088/1755-1315/557/1/012034.
[8] Ö F Görçün and S Z Burak, "Formal Safety Assessment for Ship Traffic in the Istanbul Straits," *Procedia - Soc. Behav. Sci.*, vol. **207**, pp. 252–261, 2015, doi: 10.1016/j.sbspro.2015.10.094.
[9] G Asuelimen, E Blanco-Davis, J Wang, Z Yang and D B Matellini, "Formal Safety Assessment of a Marine Seismic Survey Vessel Operation, Incorporating Risk Matrix and Fault Tree Analysis," *J. Mar. Sci. Appl.*, vol. **19**, no. 2, pp. 155–172, 2020, doi: 10.1007/s11804-020-00136-4.
[10] J Xinmin, J Wei, C Liang and W Jianping, *Submarine Optical Cable Engineering*. 2018.
[11] A. Y. Herwindya, H. K. Febriawan, A. B. Nugroho, and A. Dannari, "Survei Hidro-Oseanografi Di Perairan Raja Ampat, Papua Barat, Indonesia Hydro-Oceanographic Survey At Raja Ampat Waters, West Papua, Indonesia," *J. Ris. dan Rekayasa Kelaut.*, vol. **1**, pp. 48–64, 2020.
[12] D Haryanto, H K Febriawan, Y Haryadi and D Muljawan, "Prosedur Pemetaan Bawah Laut Untuk Rencana Pemasangan Indonesia Cable Based Tsunameter (Ina-Cbt)," *J. Ris. dan Rekayasa Kelaut.*, vol. **1**, pp. 34–47, 2020.
[13] N Endrina, J C Rasero and D Konovessis, "Risk analysis for RoPax vessels : A case of study for the Strait of Gibraltar," *Ocean Eng.*, vol. **151**, no. July 2016, pp. 141–151, 2018, doi: 10.1016/j.oceaneng.2018.01.038.
[14] IMO, "REVISED GUIDELINES FOR FORMAL SAFETY ASSESSMENT (FSA) FOR USE IN THE IMO RULE-MAKING PROCESS," London, 2018.
[15] E A Kishore, "Formal Safety Assessment in Maritime Industry - Explanation to IMO Guidelines," *Cl. Notes MTech NA&OE 1st Year*, no. September 2013, 2013.
[16] MSC, "FSA"trial application to high speed passenger catamaran vessel"," 1997.

[17] B Satria and D Manfaat, "Aplikasi Formal Safety Assessment untuk Penilaian Risiko Kecelakaan pada Helipad FSO: Studi Kasus FSO Kakap Natuna," *J. Tek. Pomits*, vol. **1**, no. 1, 2012, [Online]. Available: https://core.ac.uk/download/pdf/267876695.pdf.

[18] J Wang, "Offshore safety case approach and formal safety assessment of ships.," *J. Safety Res.*, vol. **33**, no. 1, pp. 81–115, 2002, doi: 10.1016/S0022-4375(02)00005-1.

[19] S Kristiansen, *Maritime Transportation: Safety Management and Risk Analysis*, vol. **1**, no. 1. 2005.

[20] A Hopkins, "How much should be spent to prevent disaster? A critique of consequence times probability.," *J. Pipeline Eng.*, vol. 14, no. 2, pp. 69–78, 2015, [Online]. Available: https://search.ebscohost.com/login.aspx?direct=true&db=bth&AN=108516365&site=ehost-live.

IOP Conf. Series: Earth and Environmental Science 1166 (2023) 012032 doi:10.1088/1755-1315/1166/1/012032

Validation of multiphase water wave propagation using smoothed particle hydrodynamics

M A Salvikran[1], A Trimulyono[1*], E S Hadi[1]

[1]Department of Naval Architecture and Shipbuilding Engineering, Diponegoro University, Semarang, Indonesia

[*]Correspondence: anditrimulyono@lecturer.undip.ac.id

Abstract. Water wave propagation is one of the classical problems in free surface flow that involved water and air. There are a lot of studies have been conducted on both experimental and numerical methods. One numerical method is computational fluid dynamics (CFD), which is divided into two major i.e. CFD mesh-based and meshfree. In this paper, smoothed particle hydrodynamics (SPH) is used to carry out a study of water wave propagation. SPH is a mesh-free CFD and Lagrangian approach that is suitable for free surface flow such as water waves. Two-phase SPH was used to reproduce long-distance propagation using a piston-type wavemaker. The wave probe was set 24.6 m in front of the wavemaker with a water depth is 0.75 m. The wave period is 1.15 s with an amplitude of the wavemaker is 33.0 mm. The simulation is carried out with a water and air phase (multiphase). The results indicate SPH is in reasonable agreement with experimental for the wave height, celerity, and amplitude.

1. Introduction

Water wave propagation is one of the classical problems in free surface flow which is still demanding for the marine or ocean engineering field. There are a lot of studies have been conducted on both experimental and numerical methods. One numerical method is computational fluid dynamics (CFD), which is divided into two major, i.e., CFD mesh-based and meshfree. In this paper, smoothed particle hydrodynamics (SPH) is used to study water wave propagation in long-distance propagation. SPH is a mesh-free CFD and Lagrangian approach suitable for free surface flow, such as the SPH application for sloshing with advanced visualization [1,2]. The results indicate SPH agrees well with the experiment for sloshing without and with baffles. SPH application for water waves has been conducted for long-distance propagation in single-phase [3], and 3D water waves for large free surface deformation [4]. Estimating sea wave impact on the coastal structure was conducted using SPH for tsunami and wave breaking [5]. Later on, the application of SPH modeling was developed for solitary waves and breaking waves as a complete tool in coastal engineering applications [6].

This study conducted long-distance propagation of two-phase SPH based on a previous study [3]. The shorter wave basin was used to reduce computation cost (see Fig. 1), and only a water depth of 0.75 m was used, with the amplitude of the wavemaker is 33.0 mm. The period of wave maker is 1.15 s with piston type wavemaker was used. SPH solver DualSPHysics version 5.0 was used with computation conducted in GPU [7]. It indicates SPH has fairly in agreement with experiment results. Moreover, computation in two-phase flow increases the computation time compared single-phase SPH. In addition, because calculation takes longer compared single-phase, only several parameters are carried out in this study.

2. Methods

Smoothed Particle Hydrodynamic is a Langrangian approach that has been widely used for free surface flow modeling as the advantage for solving large deformation in the free surface flow. The multi-phase numerical method of SPH uses Navier-Stoke as the base equation for Langrangian form governing equations, which are expressed in Equation 2, where g is gravity, τ is diffusive terms, and p is pressure. Equation 3 is the derivative of vector position, and Equation 4 shows the particle approximation in SPH, where a is the particle, r is the spatial position, and W is the kernel function. Equation 1 shows the continuity equation, where v is velocity and ρ density.

The momentum equation base on Colagrossi and Landrini is used for multi-phase SPH [8], where Equation 5 is the momentum equation for the liquid phase and Equation 6 is for the gas phase. The equation state for multi-phase is a development from the equation of state of weakly compressible smoothed particle hydrodynamics (see Equation 7). Before wave propagation simulation is conducted, first and foremost, we must select the parameter that can affect the numerical simulations. In this study, we mainly study three main parameters in this research: artificial viscosity, speed of sound, and particle distance. The numerical parameter setup for SPH computation shows in Table 1. This parameter is based on a previous study [3] and as a basic setup. Fig. 1 indicates the sketch of the numerical domain..

$$\frac{D_\rho}{D_t} = -\rho\nabla\upsilon, \tag{1}$$

$$\frac{D_v}{D_t} = -\frac{1}{\rho}\nabla P + g + \tau \tag{2}$$

$$\frac{D_r}{D_t} = v \tag{3}$$

$$A(r\alpha) \approx \sum\nolimits_b A(rb)W(ra - rb, h)\frac{mb}{\rho b} \tag{4}$$

$$\frac{dv_a}{dt} = -\sum\nolimits_b mb\left(\frac{P_a + P_b}{\rho_a \cdot \rho_b} + \Pi_{ab}\right) + \nabla_a W_{ab} + g, \tag{5}$$

$$\frac{dv_a}{dt} = -\sum\nolimits_b m_b\left(\frac{P_a + P_b}{\rho_a \cdot \rho_b} + \Pi_{ab}\right)\nabla_a W_{ab} - 2a\rho_a^2 \sum_b \frac{m_b}{\rho_b}\nabla_a W_{ab}, \tag{6}$$

$$\text{Where } \Pi_{ab} = \begin{cases} \frac{-\propto \overline{c_{ab}}\mu_{ab}}{\overline{\rho_{ab}}} & \upsilon_{ab}\cdot r_{ab} < 0 \\ 0 & \upsilon_{ab}\cdot r_{ab} > 0 \end{cases}$$

$$P = b\left[\left(\frac{\rho}{\rho_0}\right)^\gamma - 1\right] + X - a\rho^2 \tag{7}$$

Table 1. Numerical Parameter

Parameters setup	
Kernel Function	Wendland
Time Step Algorithm	Symplectic
Dp (Particle Distance)	0.006
CFL Number	0.2
Artificial Viscosity (α)	0.001
Cs (Speed Of Sound) Water	50
Cs (Speed Of Sound) Air	200

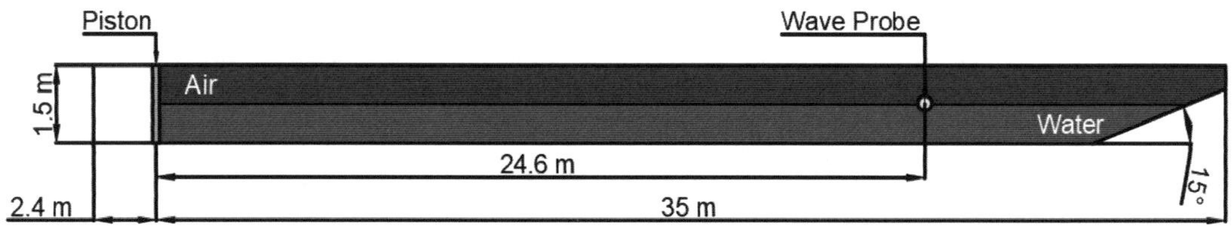

Figure 1. Sketch of numerical domain for SPH computation

3. Results and Discussion

Fig. 2 shows a wave profile with a speed of sound of water at 65 m/s, and the wave profile indicates the accuracy decreases slightly as the results simulation was stopped. In this case, the sensitivity of the speed of sound significantly affects the wave profile. The latest version of DualSPHysics has a significant effect compared to the previous version. Using larger particle distance, the accuracy shows similar results though there is some minor phase with the experiment. Fig. 3 indicates the wave profile with a coefficient of artificial viscosity is 0.0001. The wave profile means the phase difference becomes dominant though only a small change of artificial coefficient viscosity. It was evidence that speed of sound and coefficient of artificial viscosity are tuning parameters, and it significantly affects the accuracy of wave profile.

Based on Table 1, the computation was conducted and shown in Fig. 4. This parameter was based on the work of the sloshing case [9,10]. The study indicates the ratio of the speed of sound in multiphase SPH significantly affects the pressure and accuracy of the fluid phenomenon. Fig. 4 shows the accuracy is relatively in agreement with the experiment, though there is a minor phase with the experiment. However, from the authors' knowledge, the long-distance propagation in two-phase SPH is still rare, and this preliminary study evidences SPH application in ocean engineering. Fig. 5 shows water wave propagation from the wave maker to the end of the tank. Blue indicates the air phase, and red is the water phase, repectively. It showed waves propagate at least twelve waves from the wavemaker to the probe. This was a long distance-propagation for the case of a water wave because the common researcher used two or three waves for a numerical wave tank (NWT).

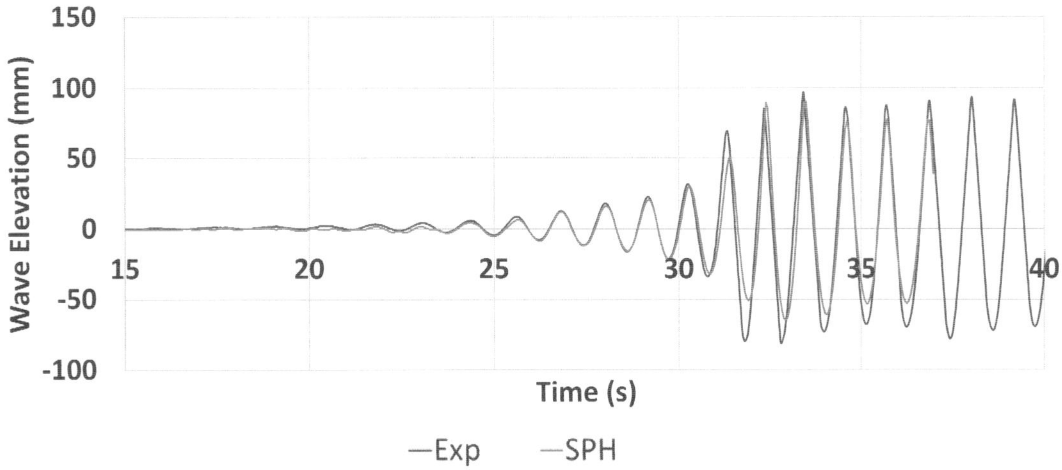

Figure 2. Comparison of wave profile between SPH and Exp with Cs water 65 m/s

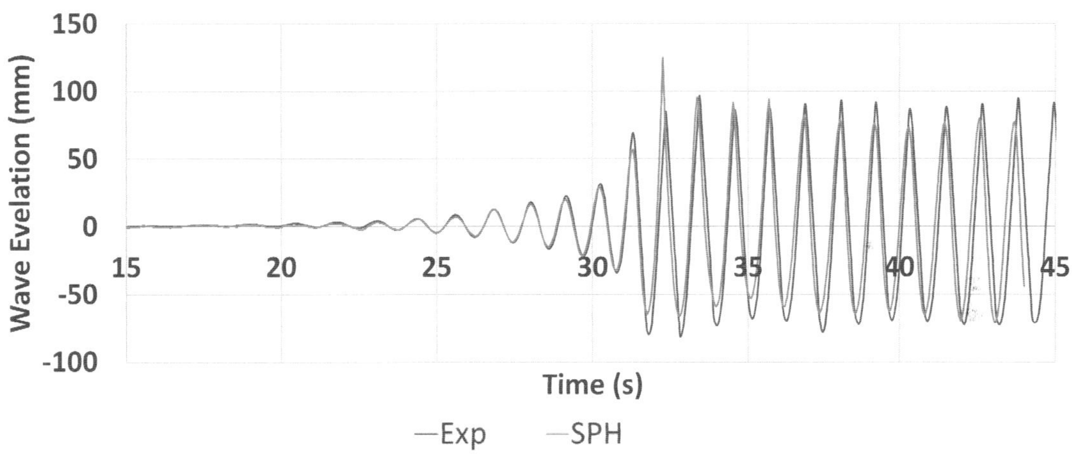

Figure 3. Comparison of wave profile between SPH and Exp with $\alpha = 10^{-4}$

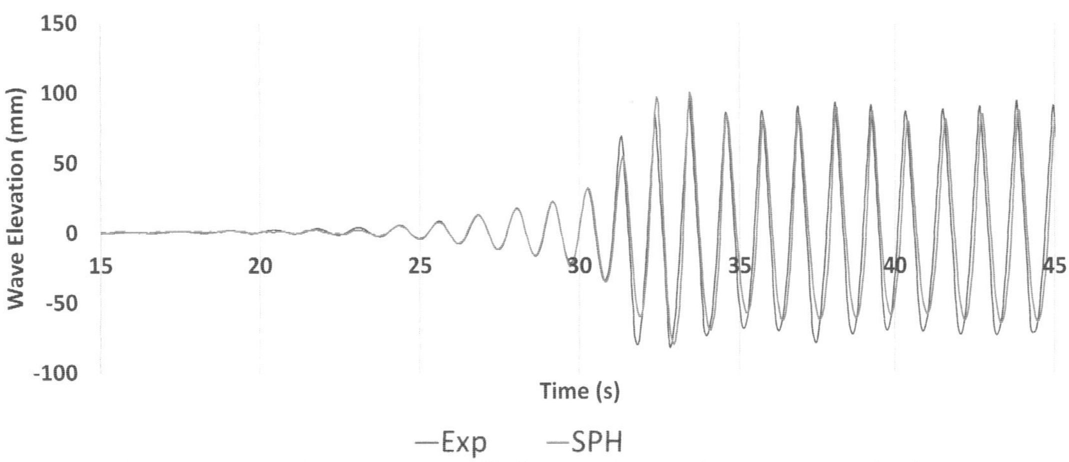

Figure 4. Comparison of wave profile between SPH and Exp Cs water 50 m/s

Figure 5. Wave Propagation Simulation at 10, 25, and 50 second

4. Conclusion

A two-phase water wave simulation was generated using DualSPHysics. The study indicates the speed of sound in water is 50 m/s, and in the air is 200 m/s showing the best value to get a wave profile similar to the experiment. An increase in both speeds of the sound of water and air made noticeable changes in wave elevation. It showed that the coefficient of artificial viscosity significantly affects wave profile. Thus the best artificial viscosity is 0.001. Further study must be carried out for a better understanding of the effect of the parameters in DualSPHysics, such as particle distance, shifting number, etc.

5. Acknowledgments

The authors wish to thank the ship hydrodynamics laboratory, Department of Naval Architecture, for providing computer facilities.

References

[1] A. Trimulyono, D. Chrismianto, H. Atthariq, and Samuel 2022 Numerical Simulation Low Filling Ratio of Sway Sloshing in the Prismatic Tank Using Smoothed Particle Hydrodynamics *CFD Letters* **14**(7) p 113-123

[2] A. Trimulyono, H. Atthariq, D. Chrismianto, and Samuel 2022 Investigation of Sloshing in the Prismatic Tank with Vertical and T-Shape Baffles *Brodogradnja* **73**(2) p 43-58

[3] A. Trimulyono and H. Hashimoto 2019 Experimental Validation of Smoothed Particle Hydrodynamics on Generation and Propagation of Waterwaves *Journal Of Marine Science and Engineering* **7**(1).

[4] A. Trimulyono and A. Wicaksono 2018 Simulasi numerik large-deformation surface wave dengan smoothed particle hydrodynamics *Kapal: Jurnal Ilmu Pengetahuan dan Teknologi Kelautan* **15**(3) p. 102-106.

[5] C. Altomare, A. J. C. Crespo, J. M. Domínguez, M. Gómez-Gesteira, T. Suzuki, T. Verwaest 2015. Applicability of Smoothed Particle Hydrodynamics for estimation of sea wave impact on coastal structures. *Coastal Engineering*, ***96***, p 1-12.

[6] J. M. Domínguez, C. Altomare, J. Gonzalez-Cao and P. Lomonaco 2019 Towards a more complete tool for coastal engineering: solitary wave generation, propagation and breaking in an SPH-based model *Coastal Engineering Journal* **61**(1) p. 15-40.

[7] J.M. Domínguez, G. Fourtakas, C. Altomare, R.B. Canelas, A. Tafuni, O. García-Feal, I. Martínez-Estévez, A. Mokos, R. Vacondio, A.J.C. Crespo, B.D. Rogers, P.K. Stansby, M. Gómez-Gesteira, 2021. DualSPHysics: from fluid dynamics to multiphysics problems. *Computational Particle Mechanics*.

[8] A. Colagrossi and M. Landrini 2003 Numerical simulation of interfacial flows by smoothed particle hydrodynamics *Journal of Computational Physics* **191**(2) p. 448-475.

[9] A. Trimulyono, H. Hashimoto, and A. Matsuda 2019 Experimental validation of single- and two-phase smoothed particle hydrodynamics on sloshing in a prismatic tank *Journal Of Marine Science and Engineering* **7**(8).

[10] A. Mokos, B.D. Rogers, P.K. Stansby, J.M. Domínguez 2015 Multi-phase modelling of violent hydrodynamics on GPUs. *Computer Physics Communications* **196** p. 304–316.

IOP Conf. Series: Earth and Environmental Science 1166 (2023) 012033 doi:10.1088/1755-1315/1166/1/012033

Strength Analysis of Crane Hook Structure with Trapezoidal Cross-Section Based on Rigging Configuration on the Upper Deck Lifting Process

Y S Hadiwidodo [1], R G Permana[1], and Handayanu[1]

[1]Department of Ocean Engineering, Institut Teknologi Sepuluh Nopember, Surabaya, Indonesia

radiktyagp@gmail.com

Abstract. Lifting is one of the most important processes in offshore platform fabricaction. This process involves structures with large loads and dimensions which can lead to structural failure. One of the failures that often occur in lifting operations is the failure of the crane hook structure. In this study, strength analysis of the rigging configuration and crane hook is carried out in the upper deck lifting process at the stacking phase using analytical and numerical methods. The hook crane model used is Ramshorn Forged Hooks with a design based on DIN15402. Furthermore, local stress analysis is carried out on the crane hook structure to determine the deformation, maximum Von Mises stress, and Von Mises stress in cross-section that occurs in the crane hook structure with a trapezoidal cross-section. Simulations are carried out using analytical and finite element method software. The results of the stress analysis on the trapezoidal cross-section will then be validated. Then the calculation of the safety factor (SF) based on the results of the analysis of the maximum Von Mises stress is carried out to determine the failure of the crane hook structure when given a predetermined load.

1. Introduction

Along with the development of technology, oil drilling is not only done on land but also offshore. Basically, offshore drilling has a system that is not much different from the existing system on land, but due to very different environmental conditions, offshore drilling development and operation methods require more advanced technology. There are various types of offshore platforms that are built based on the needs and environmental conditions namely compliant tower, semi-submersible rig, jack up rig, and one of the most commonly used types is the jacket platform.

Jacket platform is a fixed offshore structure which sits on the seabed and is rigidly connected to it by means of foundation piles (e.g. jackets) or under the effects of its weight (e.g. gravity base structure) [1]. The jacket platform is considered stable and able to withstand wind and water movement, but the most optimal use of this platform is when it is installed in shallow waters less than 400 feet (122 m) [2]. This structure has three main components, namely topside, jacket, and pile, each of which has its own function.

At the construction phase of an offshore structure, there are various stages such as the design process, fabrication, transportation and installation process [3]. All of these processes have an important role in building a platform. The work on this project cannot be separated from lifting operations. The lifting operation is one of the most important and high-risk stages in the construction of an offshore structure. This is because lifting operations on offshore structures involve structures with large dimensions with loads of tens to hundreds of tons, therefore it is necessary to pay attention to both structural and environmental loads as well as rigging equipment. Incorrect lifting calculation will result in failure that can cause fatal losses.

One of the failures that often occur in lifting operations is the failure of the crane hook. Crane hook is one of the main components in lifting operations which functions as a hook that connects the sling with the crane. This tool must be designed and simulated properly to avoid failure. To facilitate planning and simulation, it is necessary to use finite element-based software. In this study, the crane hook strength analysis was carried out using an analytical and numerical method. It is necessary to consider several variables that affect the lifting process, such as the static load of the deck structure itself, as well as the dynamic load caused by wind at the fabrication site.

2. Basic Theory

2.1. Lifting

Lifting is an activity necessary to lift or assist a structure by crane(s) [4]. Lifting operations are one of the common stages in fabrication projects and offshore building installations. There are several types of lifting operations during the offshore building fabrication process, such as stacking, roll-up, flipping, etc [5]. Lifting operation is a method of moving a structure by raising or lowering it with the help of a crane [4]. In lifting operations, all lifting equipment must be in good condition to carry out lifting operations to avoid failures and losses.

2.2. Rigging Components

Rigging components are equipment used during lifting operations which include:

a. Lifting Point

Lifting point is the location where intensive sling tensions meet with module structure. Lift points should be properly designed to allow sling tensions smoothly transfer to other strong structural members [5]. These lifting points can be padeyes, trunnions, and/or directly attaching the sling to the module without the aid of a substructure. To determine the lifting point must be calculated properly by determining the right center of gravity of the module so that in the lifting process the structure remains balanced and does not tilted.

b. Hook Point

Hook point is a point that represents the location of the hook on the crane during the lifting process. So that during the lifting process there is no excessive tilt, then the placement of the hook point must be parallel to the CoG (center of gravity) of the module to be lifted. The centre of gravity of the structure is the point around which its entire weight may be concentrated. To make a level lift, the crane hook or point of suspension must be clearly above this point. The CoG of the lifting structure greatly affects the lifting point during the lifting process and also the rigging activity of the lifting process, and the location of the CoG on the structure depends on the shape, dimensions and size of the structure.

c. Sling

Sling is one of the lifting components in the form of a strap that connect lifting point and hook point during lifting. Types of slings can be in the form of wire rope slings, chain slings, webbing slings, and round slings. The sling used in this analysis is a wire rope type.

Wire rope is made from several groups of wires that are threaded and become strands and thread the center of the rope to form a wire rope. The more the number of wires used, the greater the level of flexibility and lifting power, but at the same time, the level of abrasion is also much greater.

d. Crane Hook

A crane hook is a tool for hooking and lifting loads using a device such as a hoist or crane. Lifting hooks are used in material handling and rigging systems together with wire ropes or chains [6]. In the industrial world, crane hooks have an important role in lifting heavy loads. In order to prevent disengagement lifting wire rope sling, chain or rope to which the load is attached, crane hooks usually equipped with a safety latch [7].

There are two types of crane hooks based on their shape, namely single crane hooks and double crane hooks. Double crane hook designed with smaller mount than a single hook with the same lifting capacity, this type of crane hook will be used in this study.

2.3. Curved Beam Theory

Curved beam theory is used to calculate the stress distribution in a curved cross-section such as a crane hook. A curved beam is a rod/beam whose axis is not straight and is curved at its elevation. The distribution of stress in a curved flexural member is determined by using the following assumptions: the cross-section has an axis of symmetry in a plane along the length of the beam, plane cross-sections remain plane after bending, and the modulus of elasticity is the same in tension as in compression [6].

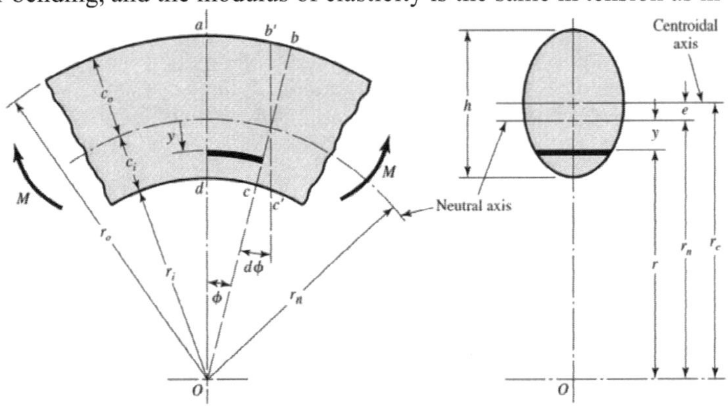

Figure 1 Curved Beam Cross-Section

Figure 1 is a cross-section of a curved beam which is a beam whose axis is not straight and is curved at its elevation. The stress distribution on the curved beam at a distance r (mm) from the center of the bend can be found by balancing the external moment against the internal resisting moment [8] which can be calculated through the following equation:

$$M = P.r_c \tag{1}$$

$$\sigma = \frac{My}{Ae\,(R-y)} \tag{2}$$

where M is bending moment (N.mm), P is the normal load acting on critical cross-section (N), r_c is the radius of the centroidal axis (mm), σ is the normal stress of the curved beam (Pa), y is the length moment arm (mm), e is the distance from the centroidal axis to neutral axis (mm), and R is the is the radius of the neutral axis (mm).

The positive direction of the moment (M) in Figure 1 shows that the stress distribution in the curved member follows a hyperbolic pattern. The maximum stress is always on the inside (concave) of the beam. The critical stresses occur on the inner and outer surfaces where $y = c_i$ and $y = -c_0$, respectively, are described in the following equation:

$$\sigma_i = \frac{Mc_i}{Aer_i} \text{ atau } \sigma_0 = \frac{Mc_0}{Aer_0} \tag{3}$$

where σ_i is the critical stresses occur at the inner surface (Pa), σ_0 is the critical stresses occur at the inner surface (Pa), r_i is the radius inner fiber (mm), r_0 is the radius outer fiber (mm), c_i is the distance from the neutral axis to inner fiber (mm), and c_0 is the distance from the neutral axis to outer fiber (mm).

2.4. Finite Element Method

Finite element method (FEM) is a numerical approach used for engineering problems where exact solutions/analytical solutions cannot solve them [9]. From an engineering point of view, FEM is a method for solving engineering problems such as stress, heat transfer, fluid flow, and electromagnetic analysis performed by computer simulation. The essence of FEM is to divide an object to be analyzed into several parts with a finite number. These parts are called elements where each element is connected to each other by nodes (nodes). Then a mathematical equation is built which represents the object. The process of dividing an object into several parts is called meshing.

3. Result and Discussion

3.1. Rigging Configuration Modelling

Rigging configuration modeling is a global modeling using numerical model software which is carried out to obtain dead load values and to determine the coordinates of the center of gravity (CoG) of the upper deck structure. The deadload of the upper deck structure is then multiplied by the load factor, namely the contingency factor and DAF as shown in Table 1. Dynamic amplification factor (DAF) is a factor by which the gross weight is multiplied, to account for accelerations and impacts during the lifting operation [10], and the contingency factor is a factor applied to the weight of an object, when an object is not to be weighed, to account for uncertainties related to the design and fabrication of the object [11].

Table 1. Upper Deck Structure Weight by Factoring

Description	Weight (MT)
Dead Load	155.15
Contingency Factor (Dead Load x 1,03)	159.805
DAF (Onshore) (Dead Load x 1,03 x 1,15)	183.776

After getting the deadload value, then we can determine the coordinates of the center of gravity of the upper deck structure. The coordinates of the center of gravity of the upper deck structure are shown in Table 2 and in Figure 2.

Table 2. Center of Gravity on Upper Deck Structure

Axis	Coordinate (mm)
X	-84.35
Y	-64.34

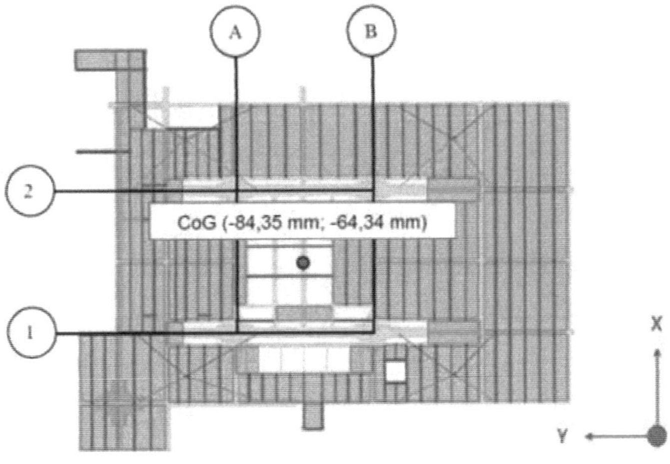

Figure 2. Center of Gravity Coordinate on Upper Deck Structure

After getting the center of gravity of the upper deck structure, then the rigging configuration modeling is carried out. The lifting configuration is also determined based on the angle on the sling, where the smaller the angle of the sling, the smaller the stress on the sling along with the safety factor [6]. This upper deck lifting process uses 4 cranes with 4 slings in each crane, and hook point coordinates can bee seen in Figure 3.

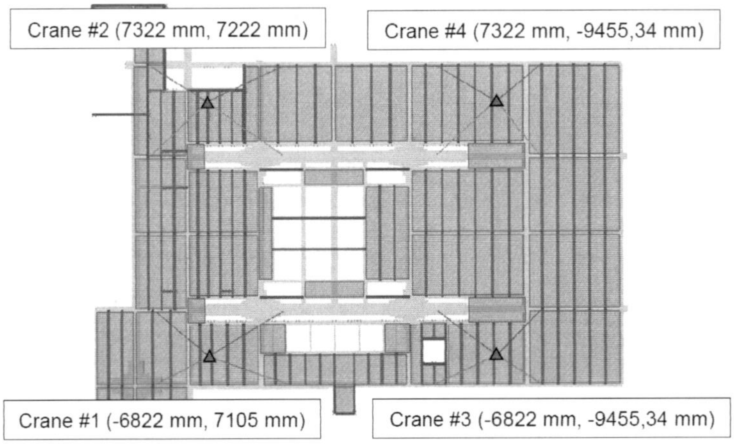

Figure 3. Hook Point Coordinates on Upper Deck Structure

Sling angle less than 45° is usually avoided because the lower the sling angle, the higher the tension factor on the sling [12]. The rigging configuration that has been carried out with a large angle on each sling with the smallest angle is found in sling 2.1, sling 2.2, sling 2.3, and sling 2.4 of 60.25° can be seen in Figure 4.

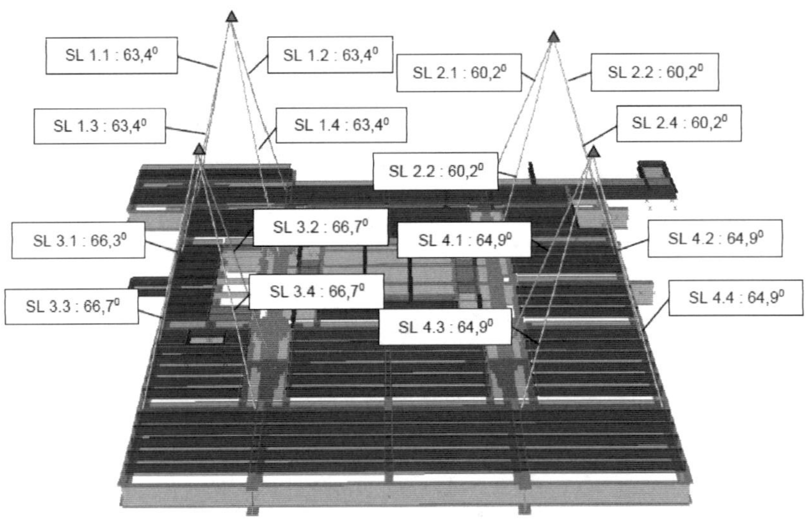

Figure 4. Lifting Upper Deck Configuration and Angle Size on Each Sling

3.2. Analysis of Upper Deck Member

Analysis of the strength of the deck member is the maximum UC (Unity Check) value of the member deck multiplied by the consequence factor. Consequence factors are factor applied in the design of critical components to ensure that these components have an increased factor of safety in relation to the consequence of their failure. Consequence factor of main elements supporting the lift point is 1.35 and for other member is 1.15 [11]. The UC values in the member deck around lift points can be seen in

Table 3 and the five highest UC values in other member deck can be seen in Table 4. The critical member deck with the highest UC occurred in member 7743-7718 is less than 1 with a value of 0.32 as can be seen in Figure 5.

Table 3. UC Value for Member Deck at Lifting Point

Lifting Point	Member ID	Profile	Length (mm)	Unity Check	Cons Factor	Factored UC	Remarks
SL 1.1	7429 - 7487	W 920x420x344	416	0,03	1,35	0,04	OK
SL 1.2	7292 - 7289	W 920x420x344	566	0,06	1,35	0,07	OK
SL 1.3	7515 - 7472	W 920x420x344	914	0,07	1,35	0,09	OK
SL 1.4	7101 - 7184	W 920x420x344	112	0,02	1,35	0,02	OK
SL 2.1	7293 - 7242	W 920x420x344	800	0,04	1,35	0,05	OK
SL 2.2	7761 - 7785	W 920x420x344	1050	0,03	1,35	0,05	OK
SL 2.3	7211 - 7185	W 920x420x344	112	0,03	1,35	0,04	OK
SL 2.4	0011 - 7754	W 920x420x344	457	0,02	1,35	0,03	OK
SL 3.1	7444 - 7446	W 760x265x147	750	0,15	1,35	0,20	OK
SL 3.2	7143 - 7352	W 920x420x344	400	0,02	1,35	0,03	OK
SL 3.3	7414 - 7402	W 760x265x147	967	0,11	1,35	0,15	OK
SL 3.4	7121 - 7125	W 920x420x344	967	0,08	1,35	0,10	OK
SL 4.1	7144 - 7360	W 920x420x344	400	0,02	1,35	0,03	OK
SL 4.2	7742 - 0013	W 760x265x147	457	0,24	1,35	0,32	OK
SL 4.3	7205 - 7132	W 920x420x345	967	0,08	1,35	0,10	OK
SL 4.4	7706 - 7713	W 760x265x147	967	0,16	1,35	0,22	OK

Table 4. Five Highest UC on Member Deck

Member ID	Profile	Length (mm)	Factored UC	Remarks
7743-7718	W 760x265x147	1029	0.32	OK
0011-7754	W 760x265x147	457	0.28	OK
7742-0013	W 760x265x147	457	0.27	OK
7753-0011	W 760x265x147	686	0.26	OK
0013-7743	W 760x265x147	571	0.26	OK

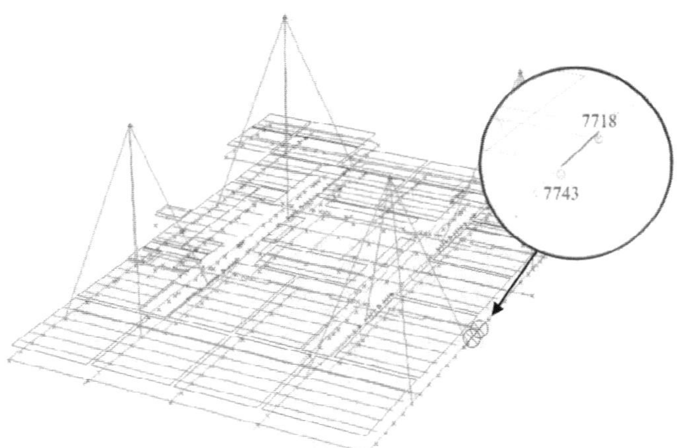

Figure 5. Critical Member Location on Upper Deck Structure during Lifting Procedure

The amount of deflection on the deck is the joint displacement value obtained from the results of static analysis. The calculation of the deflection limit is the length of the deflected beam divided by 360 and the deflection value must not exceed the result of the deflection limit. The five highest deflections on the member deck can be seen in Table 5 and their location in Figure 6 with the largest deflection on the member deck occurs at the 7204 joint of 0.945, less than the deflection limit value of 0.999.

Table 5. Five Highest Deflections on Member Deck

Joint ID	Length (cm)	Limit Type	Limit Value (cm)	Deflection (cm)	Remarks
7204	359.6	L/360	0.999	0.945	OK
7256	570	L/360	1.583	0.943	OK
7250	570	L/360	1.583	0.937	OK
7255	570	L/360	1.583	0.936	OK
7231	775	L/360	2.153	0.930	OK

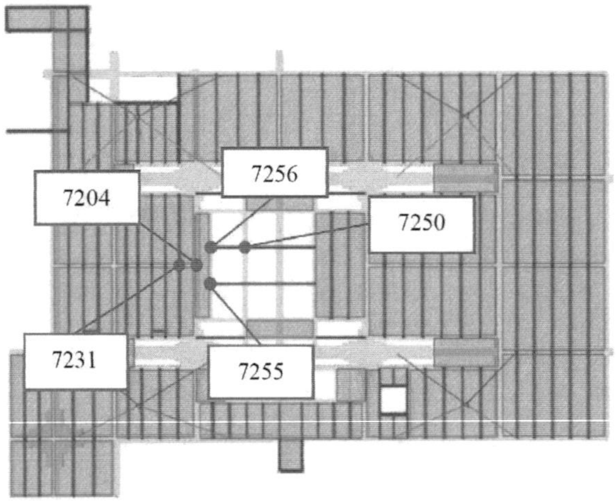

Figure 6. Location of Five Joints with Highest Deflection

3.3. Analysis of Sling

Analysis of sling strength begins with determining the value of unity check (UC). Sling strength analysis is very crucial in lifting operations, this is because the sling is a connecting component between the crane and the structure to be lifted, if the sling fails, the lifting operation will automatically fail too. Based on the results of static analysis, the sling member with the highest UC occurred in member 7184-HK01 is less than 1 with a value of 0.328 as shown in Figure 7.

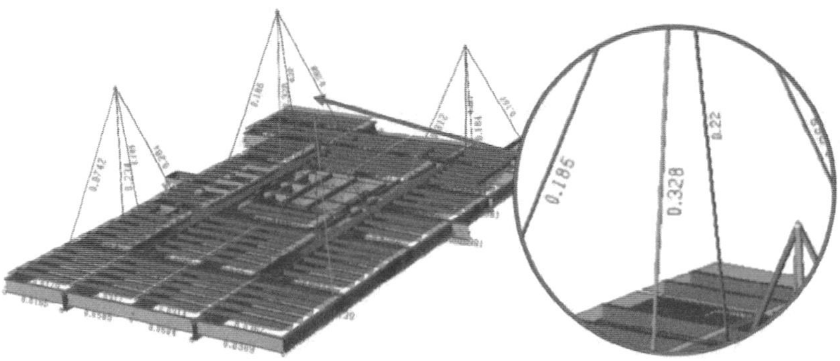

Figure 7. Highest UC Member Sling Value in Rigging Configuration

After the UC sling is considered safe, the next step is to calculate the maximum tension that occurs in the sling. The sling stress value which is the output of the static analysis is multiplied by the load factor, namely the contingency factor, DAF, and SKL. Skew load factor (SKL) is a factor to account for additional loading caused by rigging fabrication tolerances, fabrication tolerances of the lifted structure and other uncertainties with respect to asymmetry and associated force distribution in the rigging arrangement [10]. The results of checking the sling tension can be seen in Table 6.

Table 6. Sling Tension Check

Member	Tension (cm)	Contigency Factor	DAF	SKL	Factory Tension (MT)
	A	B	C	D	E = A x B x C x D
SL 1.1	15.162	1.1	1.05	1.25	21.890
SL 1.2	18.433	1.1	1.05	1.25	26.613
SL 1.3	12.754	1.1	1.05	1.25	18.413
SL 1.4	22.582	1.1	1.05	1.25	32.602
SL 2.1	15.198	1.1	1.05	1.25	21.942
SL 2.2	10.825	1.1	1.05	1.25	15.629
SL 2.3	21.426	1.1	1.05	1.25	30.934
SL 2.4	12.638	1.1	1.05	1.25	18.247
SL 3.1	7.255	1.1	1.05	1.25	10.475
SL 3.2	19.502	1.1	1.05	1.25	28.156
SL 3.3	5.103	1.1	1.05	1.25	7.367
SL 3.4	16.097	1.1	1.05	1.25	23.240
SL 4.1	18.496	1.1	1.05	1.25	26.703
SL 4.2	9.369	1.1	1.05	1.25	13.526
SL 4.3	13.777	1.1	1.05	1.25	19.891
SL 4.4	6.828	1.1	1.05	1.25	9.858

After getting the maximum stress value of the sling, then proceed with calculating the safety of the rigging configuration based on the maximum tension of the sling. In this analysis, sling configurations with lengths of 11 m, 10.2 m, and 10.1 m are used with each having a diameter of 2.5 inches. For its strength, the wire rope sling has a SWL (Safety Working Load) value of 50 tons. The SWL value is then divided by the maximum sling stress to obtain the safety factor as shown in Table 7.

Table 7. Sling Capacity Check

Member	Dimension	Tension (MT)	SWL (ton)	Safety Factor	Remark
		A	B	C = B / A	OK = C > 1
SL 1.1	2½" x 11 m	21.890	50	2.284	OK
SL 1.2	2½" x 11 m	26.613	50	1.879	OK
SL 1.3	2½" x 11 m	18.413	50	2.715	OK
SL 1.4	2½" x 11 m	32.602	50	1.534	OK
SL 2.1	2½" x 10.2 m	21.942	50	2.279	OK
SL 2.2	2½" x 10.2 m	15.629	50	3.199	OK
SL 2.3	2½" x 10.2 m	30.934	50	1.616	OK

Member	Dimension	Tension (MT)	SWL (ton)	Safety Factor	Remark
		A	B	C = B / A	OK = C > 1
SL 2.4	2½" x 10.2 m	18.247	50	2.740	OK
SL 3.1	2½" x 10.1 m	10.475	50	4.773	OK
SL 3.2	2½" x 10.1 m	28.156	50	1.776	OK
SL 3.3	2½" x 10.1 m	7.367	50	6.787	OK
SL 3.4	2½" x 10.1 m	23.240	50	2.151	OK
SL 4.1	2½" x 10.1 m	26.703	50	1.872	OK
SL 4.2	2½" x 10.1 m	13.526	50	3.697	OK
SL 4.3	2½" x 10.1 m	19.891	50	2.514	OK
SL 4.4	2½" x 10.1 m	9.858	50	5.072	OK

3.4. Crane Hook Modelling
In this study, a crane hook with a trapezoidal cross-section was modeled in three dimensions using computer-aided design (CAD) based software. The crane hook model used is Ramshorn Forged Hooks with a design based on DIN15402 as shown in Figure 8. Ramshorn hook is a category of crane hook with double slings and is mounted in one shank [13]. The crane hook used has a working load limit (WLL) capacity of 100 tons. Based on DIN15402 the material used in the crane hook model is structural steel with material properties as shown in Table 8.

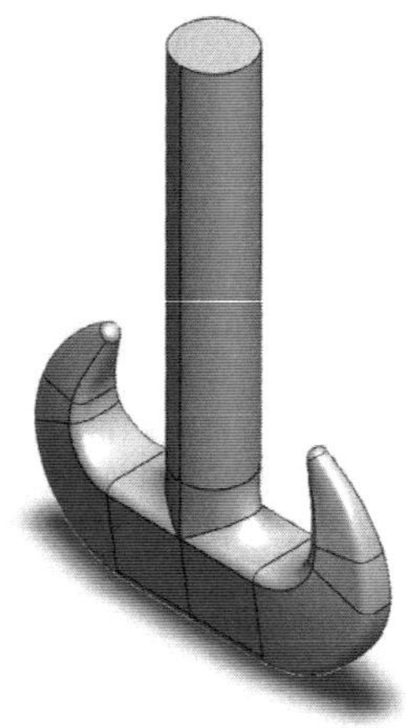

Figure 8. Crane Hook Model with Trapezoidal Cross-Section

Table 8. Structural Steel Properties

Description	Value
Density (kg/m^3)	7850
Elasticity modulus (MPa)	203156.33
Poisson ratio	0.30
Tensile yield strength (MPa)	250
Compressive yield strength (MPa)	250
Tensile ultimate strength (MPa)	460

The crane hook model has dimensions as shown in Table 9. The cross-sectional area is the result of drawing using CAD software and the weight of the structure is the output of the same software.

Table 9. Crane Hook Model Dimension

Sectional Properties (mm)	Area of Cross-Section (mm)	Weight (kg)
b = 196	58,027.43	1027.7

3.5. Analytical Investigation of Crane Hook

Analytical investigation is calculated using curved beam theory. Because the crane hook model is a double hook that has two cross-sections, then there is load and tension in each cross-section. The force used is the tension of the sling with the smallest sling angle, namely the sling 2.1, sling 2.2, sling 2.3, and sling 2.4 with a sling angle of 60.25°. The load acting on a sling hook (P) with Q is the load of the sling and is the sling angle [14] with the following equation:

$$P = Q/2\cos\alpha \tag{4}$$

where α is the angle of the sling, then the load at section 1 is 522490.935 N and section 2 is 334739.849 N. Then the manual Von Mises calculation is carried out using the curved beam theory with the results of the Von Mises stress in section 1 is 83 MPa and at section 2 the Von Mises stress value is 53 MPa with a total Von Mises stress is 136 MPa.

3.6. Local Analysis of Crane Hook

3.6.1. Boundary Conditions

After modeling, the next step is to input the force that occurs on the crane hook cross-section using finite element method based software. Same as the analytical method, the force used in this analysis is the tension of the sling with the smallest sling angle, namely the sling 2.1, sling 2.2, sling 2.3, and sling 2.4 with a sling angle of 60.25°. Figure 9 is the input force on the model shown on the label A-D with the force as follows:

- SL 2.1 = 215176.5 N
- SL 2.2 = 153268.5 N
- SL 2.3 = 303360.75 N
- SL 2.4 = 178938.375 N

with the tensions of sling SL 2.1 (label B) and SL 2.3 (label C) applied to section 1 and the tensions of sling SL 2.2 (label A) and SL 2.4 (label D) applied to section 2, then on label E is the location of the fixed support on the crane hook model, which means that the component is fixed and does not deform.

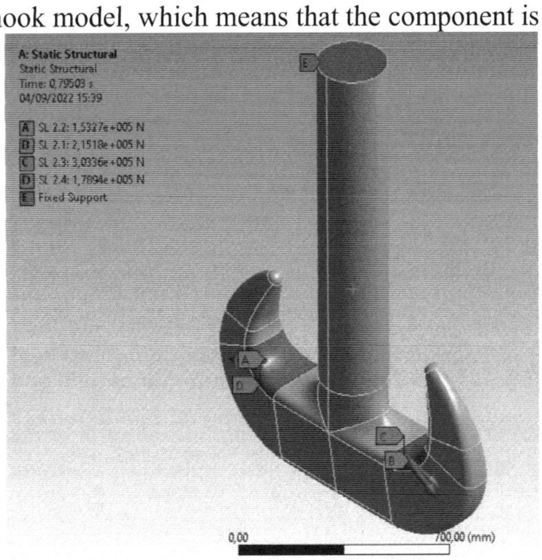

Figure 9. Force and Fixed Support on Crane Hook Model

3.6.2. Mesh Sensitivity

Local stress analysis begins with performing a mesh sensitivity analysis. Mesh sensitivity is a variation of meshing density to obtain a constant stress at the same point [15]. Meshing sensitivity is carried out at a point located in the middle of the section that receives the greatest load, in this case namely at section 1 as shown in Figure 10.

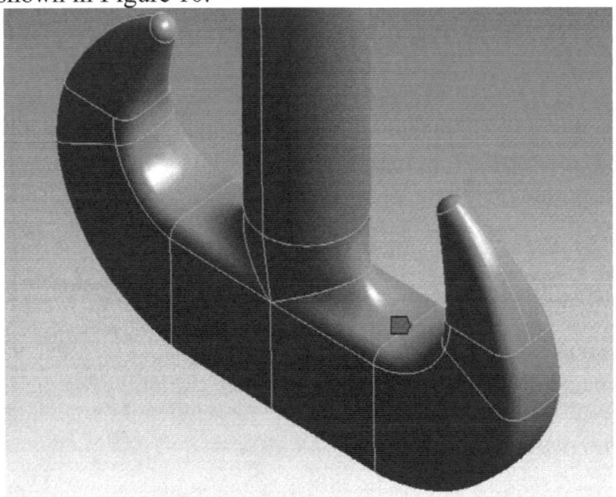

Figure 10. Location of Mesh Sensitivity Analysis on Crane Hook Cross-section

In this study, mesh sensitivity is carried out by determining and changing the size of element until a constant Von Mises stress is obtained. Mesh sensitivity analysis starts at a mesh size of 20 mm to 10 mm with meshing shape used is tetrahedron. The results of the mesh sensitivity obtained constant Von Mises stress values (stress difference not more than 3%) at mesh size 17 mm to 10 mm as shown in Table 10 and Figure 11.

Table 10. Mesh Sensitivity Analysis Results

Mesh Size (mm)	Nodes	Element	Von Mises Stress (MPa)	Total Deformation (mm)	Diff
10	1505600	1090196	79.586	5.4816	-
11	1136231	819487	78.361	5.4815	1.54%
12	876792	630516	77.634	5.4814	0.93%
13	693298	496751	77.61	5.4813	0.03%
14	557761	398268	77.013	5.4812	0.77%
15	455333	324017	76.993	5.4811	0.03%
16	379569	268664	77.159	5.4811	0.22%
17	314652	222428	77.021	5.4806	0.18%
18	267346	187996	74.323	5.4806	3.50%
19	226369	159119	76.704	5.4803	3.20%
20	196663	137426	74.072	5.4806	3.43%

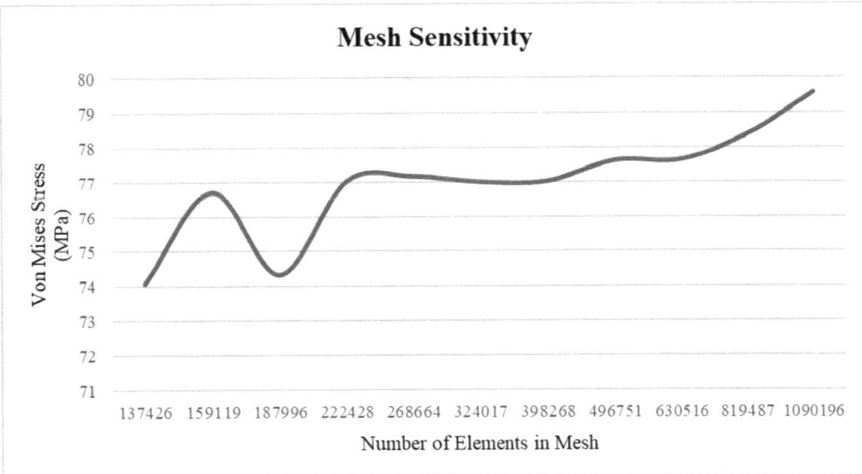

Figure 11. Graph of Meshing Sensitivity Analysis Results

It can be seen in the Table 10 and Figure 11 that the Von Mises stress does not seem constant up to a mesh size of 17 mm, then the mesh results are constant at a mesh size of 17 mm to 14 mm. Meanwhile, the total deformation there is no significant difference with values ranging from 5.48 mm. Therefore, in this analysis a mesh size of 16 mm is used because the meshing density value is constant. The mesh on the crane hook model with a trapezoidal cross section with a mesh size of 16 mm can be seen in Figure 12. The meshing configuration used has 384562 nodes and 272281 elements.

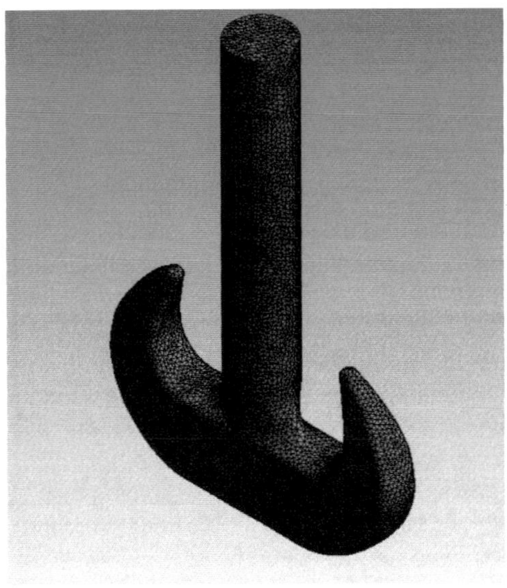

Figure 12. Crane Hook Model with Mesh Size 16 mm

3.6.3. Linear Static Analysis Result

Figure 13 is a crane hook model with a trapezoidal cross section with a mesh density of 16 mm along with the Von Mises stress that occurs. The results of the analysis showed that the maximum Von Mises stress was 191.51 MPa, with the red color being the maximum stress value and the blue color being the minimum stress value. The total deformation that occurs on the crane hook can be seen in Figure 14 with a maximum total deformation value of 5.4811 mm. Furthermore, in Figure 15 is a trapezoidal cross section crabe hook along with the Von Mises stress that occurs. The maximum stress on the crane hook piece with a trapezoidal cross section is 77.166 MPa and the minimum Von Mises stress is 9.4399 MPa.

IOP Conf. Series: Earth and Environmental Science 1166 (2023) 012033 doi:10.1088/1755-1315/1166/1/012033

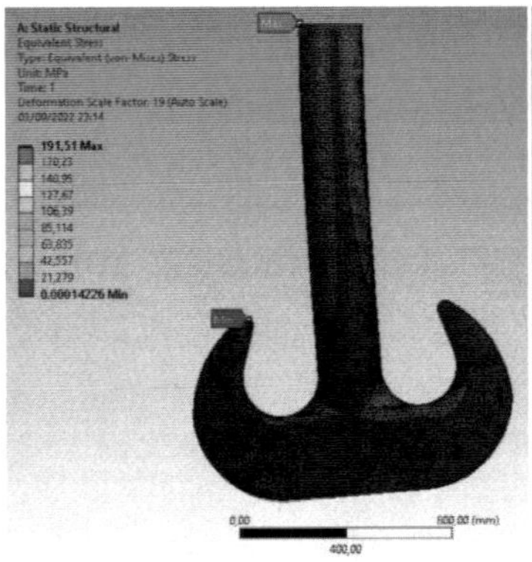

Figure 13. Maximum Von Mises Stress on Crane Hook.

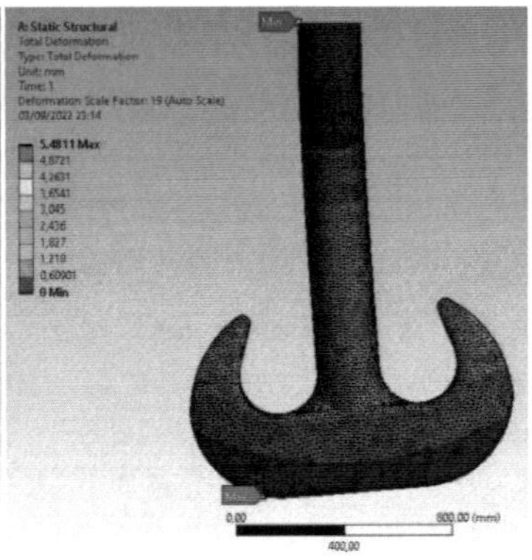

Figure 14. Total Deformation in the Crane Hook.

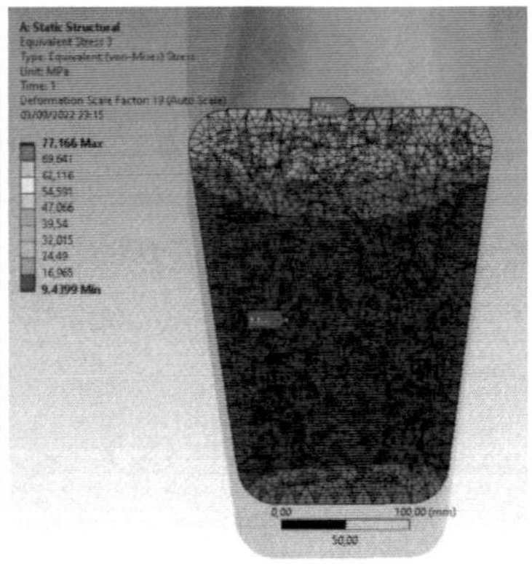

Figure 15. Von Mises Stress in Crane Hook Section.

3.7. Validation Between Analytical and Numerical Analysis

After obtaining the results of the stress, the next step is to validate the results of the stress. Validation is done by adjusting the results of the manual calculation of the Von Mises stress with the results of the analysis using numerical analysis. The results of manual calculations and the results of numerical analysis can be seen in Table 11.

Table 11. Validation Stress Result Between Analytical and Numerical Analysis Results

Curved Beam Theory (MPa)	Numerical Analysis (MPa)	Error
73.195	77.159	4.3%

After validation is done, the next step is to calculate the factor of safety to determine the failure of the crane hook structure when carrying out lifting operations. The safety factor is a factor used to evaluate so that the planning of machine elements is guaranteed to be safe with minimum dimensions [16]. The equation of the safety factor is as follows:

$$F_s = \frac{Sy}{\sigma} \tag{5}$$

where F_s is the factor of safety, S_y is the yield strength material (MPa), and σ‘ is the Von Mises stress (MPa). Based on the above equation, the safety factor of the crane hook structure can be calculated. The results of the calculation of the safety factor on the crane hook structure with a trapezoidal cross-section can be seen in Table 12.

Table 12. Safety Factor Calculation Results

Curved Beam Theory	Numerical Analysis
2.060	1.305

4. Conclusion

Based on the analysis and discussion that has been carried out previouly, the conclusions are:

- The member of the upper deck structure does not fail with the application of the rigging configuration model for lifting. This is because critical members or members that have the largest unity check are found in members 7742 - 0013 with a unity check value of 0.32, less than 1. And for the largest deflection on the member deck occurs at the 7204 joint of 0.945, less than the deflection limit value of 0.999.
- Based on the results of stress validation between manual calculations and numerical analysis results, an error of 4.3% was obtained. Because the error is less than 5%, it can be concluded that manual calculations with numerical analysis results are in accordance with actual conditions.

- The smallest value of safety factor is found in the numerical analysis of crane hook stress on a trapezoidal cross-section with a safety factor value of 1.305. The finite element analysis results show that the material has a safety factor higher than 1.25, it means that the material can withstand static loads [17], so it can be concluded that the crane hook design does not fail.

5. References

[1] Chakrabarti SK. Handbook of Offshore Engineering. Vol. II. Amsterdam: Elsevier Ltd.; 2005. 1127–1143 p.
[2] API RP 2A-WSD 22nd Ed. Planning, Designing, and Constructing Fixed Offshore Platforms - Working Stress Design. API Recommended Practice. Washington DC: American Petroleum Institute; 2014. 324 p.
[3] Pramastyo N, Sujantoko. Analisis Respon Struktur pada Proses Lifting. In: Prosiding 4th Seminar Nasional Penelitian & Pengabdian Kepada Masyarakat 2020. Surabaya: Institut Teknologi Sepuluh Nopember; 2020. p. 40–5.
[4] DNV-OS-H205. Lifting Operations (VMO Standard - Part 2-5). Norway: Det Norske Veritas AS; 2014..
[5] Liang L. Heavy Lift Installation Study of Offshore Structures. National University of Singapore; 2004.
[6] Onur YA. Investigation of the effect of the sling angle and size on the reliability of lifting hooks. Simul Trans Soc Model Simul Int [Internet]. 2018;94(10):931–42. Available from: https://doi.org/10.1177/0037549717744646.
[7] Shrestha B, Bhandari A, Poudel S, Rao MK V. Crane Hook Analysis for different Cross-Section using ANSYS. Int J Adv Sci Res Eng. 2020;05(12):67–73..
[8] Popov EP. Mechanics of Materials. 2nd ed. Berkeley: University of California; 1984..
[9] Fish J, Belytschko T. A first course in finite elements. Vol. 45. Chicester: John Wiley & Sons, Ltd; 2007. 45-3218-45–3218 p.
[10] GL Noble Denton. Guidelines for Marine Lifting & Lowering Operations 0027/ND. Norway: Noble Denton Group Limited.; 2015. 1–84 p.
[11] DNV GL Noble Denton. Noble Denton Marine Services - Marine Warranty Wizard. Norway: DNV GL Noble Denton; 2016. 1–57 p.
[12] Clement C. Offshore Heavy Lift Design and Operations: Sling Tension in Heavy Lifts with Quadruple Sling Arrangement. 2016;(August):1–28.
[13] Singh B, Singh P. Stress and Deformation Analysis of Ramshorn Hook with Various Cross Stress and Deformation Analysis of Ramshorn Hook with Various Cross Sections. Int J Adv Sci Technol. 2020;29(10S):4915–22.
[14] 1. Singh B, Singh P. Finite Element Analysis of Ramshorn Hook Using CAE Tools. Int J Adv Sci Technol. 2020;29(10S):4860–6.
[15] Yansah AR. Analisis Konfigurasi Rigging Dan Padeye Pada Saat Proses Installation Deck Structure UL.A Platform Dengan Cara Lifting. Departemen Teknik Kelautan, Fakultas Teknologi Kelautan, Institut Teknologi Sepuluh Nopember; 2016.
[16] 1. Awali JA. Analisa Kegagalan Poros Dengan Pendekatan Metode Elemen Hingga. Turbo. 2014;2(2):39–44.
[17] 1. Wibawa LAN. Numerical Study of the Effect of Wall Thickness and Internal Pressure on Von Mises Stress and Safety Factor of Thin-Walled Cylinder for Rocket Motor Case. JST (Jurnal Sains dan Teknol. 2020;9(1):30–8.

Experimental and Numerical Studies on Motion Response of Side-by-Side Two Floating Vessels in Waves

H M Bafadhal, R W Prastianto, Murdjito, D M Rosyid, Y Mulyadi and F Ardhiansyah

Department of Ocean Engineering, Institut Teknologi Sepuluh Nopember, Surabaya, East Java 60111, Indonesia.
rudiwp@oe.its.ac.id

Abstract. In the world of oil and gas, there will definitely be a structure for either the production process, exploration and so on. The structure consists of three types, namely the fixed structure, the floating structure and the moored structure. Floating structures can be the best option in terms of cost and time. Things that can interfere with floating structure when operating are environmental loads. One example of its operation is offloading, it can be offloading side by side or tandem configurations. The main focus of this study is the motion response both when single body and side by side configuration. The result of experimental studies obtained the maximum RAO (Response Amplitude Operator) for heave, surge, and pitch motion modes occurred when the model was configured side by side with a distance of 0.5 from the width. For numerical studies, the maximum RAO of heave motion when single body is 0.69 cm/cm, the maximum RAO of surge motion occurs when side by side configuration with a gap of 10 cm which is 0.71 cm/cm, and the maximum RAO of pitch motion occurs when the model is configured side by side with a gap of 10 cm which is 1.54 rad/rad.

Keywords. ***Keywords.*** *Floating Structure, RAO, Side-by-Side*

1. Introduction

In the world of oil and gas, there will definitely be a structure for the production process, exploration and many others. The structure consists of three types which are more specifically called the fixed building structure, the floating building structure and the moored building structure. For a fixed building structure, the load received by the structure is transformed by the construction of its legs through the foundation to the seabed. for the bonded structure has the advantage that the conductor pipes are installed on the side of the structure, it can eliminate the vertical movement of the structure and the position of the deck will always be above the water surface. If the moored structure is an offshore structure that is supported by steel ropes connected to the seabed. While the floating building structure has the advantage of being very easy to move, it only needs to build one structure but can be used in many wells[1].

Just like any other type of structure, this floating structure has many things to pay attention to. These include environmental loads, the response of the structure to these loads, related matters, the type of building that will be used, and the system that will be used to help the building itself. This floating structure is a very inexpensive project and requires sophisticated technology so it must be really mature in terms of design to maintenance. Environmental loads that must be considered are currents, winds, waves, tides, waves and others [2]. Responses made after being exposed to environmental loads are sway, heave, surge, roll, yaw and pitch [3].

When at the installation location while operating, an analysis of the motion response of the floating building is urgently needed. Both in its own condition and when offloading with a tandem system or side by side. The analysis is carried out in order to obtain the stability value and the response in order to get an overview to be able to get optimal results. In addition to smooth operation process, this motion response analysis is also to plan things that need further analysis such as risk analysis, safety analysis and so on. After obtaining the motion response, it can later be predicted and correlated the results of the motion response with the actual movement at sea. This usually includes the procedures applied to evaluate probability [4]. For this study, the focus is to find out how the difference in the effect of the distance between the gaps from the two floating buildings both experimentally and numerically as a research which in my opinion is a continuation of research on [5] side by side numerically only.

2. Research Method

2.1. Collecting Data

2.1.1. Structure Model

In preparing experimental models as well as floating structures, it is done as much as possible in order to be able to present the original model. geometry of the physical model and the full scale model and its scaling as shown in table 1

Table 1. Principal dimension

Skala	Lenght *L*	Beam *B*	Height *D*	Draught *T*	*Displacement*
Full Scale (1:1)	120 m	24 m	18 m	6 m	1.52x 10^7 kg
Model (1:80)	150 cm	30 cm	23 cm	8 cm	29.8 kg

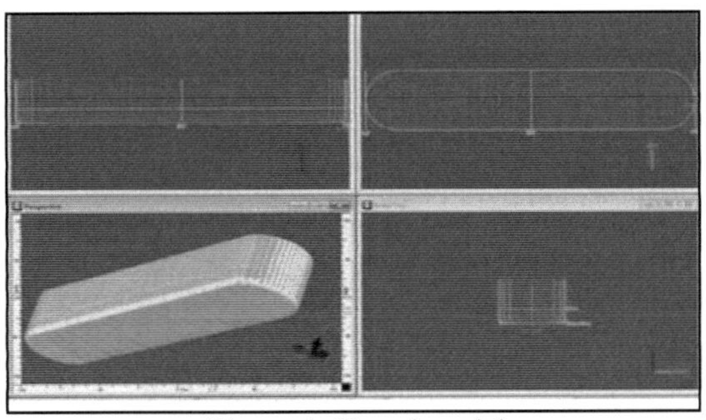

Figure 1. Vessel model

2.1.2. Test Data

The table below is a test matrix for both experimental and numerical wave inputs after scaling from the reference[6] &[7] for single body and side by side configurations.

Table 2. Matrix test data

Test no-*	H (cm)	T (sec.)
1		1.40
2		1.27
3	5	1.17
4		1.10
5		1.07

6	1.02
7	0.98
8	0.95
9	0.92
10	0.89
11	0.87
12	0.85
13	0.83
14	0.82
15	0.80
16	0.79
17	0.77
18	0.76

2.1.3. Data Layout

The layout of the side by side configuration modeling can be seen in figure 2.

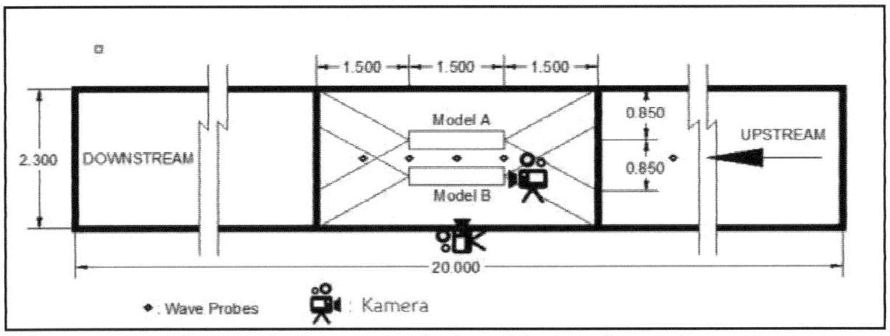

Figure 2. Lay out side-by-side configuration

In accordance with the layout above, the gap distance if the side by side configuration that can be applied without being affected by the wall effect is 1 times the width or 30 cm. When the gap distance is more than 30 cm, the floating structure model will be wider and closer to the laboratory wall so that the working waves cannot be said to be regular because they are already affected by the wall effect[8] From the picture above, the two models are tied with spread mooring but are still said to be free floating because the natural period of the building system and the binding wire is outside the natural period of the incident wave generated[9]. In this study, the motion response recorded was only the motion response in model B. This research applies two ships with side-by-side configuration when in deep sea [6] & [10]. Last, to get the data as we expected, this research must follow the existing guidelines, especially experimental studies. For this experimental study using guidelines from ITTC [9], [11], [12], and [13].

2.1.4. Material test equipment and its support

a. Wave Probe

An instrument for measuring the elevation of a wave. The measuring instrument works when the string attached to the wave probe has been submerged in water and the calculation of the wave probe with the voltage level of the string itself. The higher the depth, the higher the yield. The wave probe needs to be calibrated first in order to get optimal results.

b. Computer

Three devices are needed to carry out this experimental test. two computers belong to the laboratory and one is private. Computer facilities in the laboratory are used to control the wave generator as a tool to enter the period and wave height to be entered and another computer to obtain wave elevation data from the wave probe. A personal computer is used to control the camera captures that record the movement response of the tested model.

c. Camera

The camera is used to record video of the movement of the ball on top of the model. It is used to obtain a translational motion response. It takes two cameras to get the data and they are placed in front and beside the model as shown in the layout. Then the processing of the video recordings is processed using tracker software

d. Gyroscope

Motion sensor / gyroscope is used to obtain and record the motion response of the model when it has been subjected to waves. The motion response obtained from this tool is a rotational movement response, namely, roll, pitch, and yaw. This research uses the WIT Motion trademark as a gyrometer instrument.

2.1.5. Laboratory

The experiment was carried out in the laboratory of the wave flume tank, Department of Ocean Engineering, Faculty of Marine Technology, Sepuluh Nopember Institute of Technology, Surabaya. The dimensions of the flume tank were 20 m x 2.3 m x 2 m. wave generator is a device for generating waves located at the front end of the flume tank. The wave generator is a plunger type with a range of periods that can be generated from 0.5 to 3 seconds. For the water depth so that waves can still be generated is 0.8 m and the maximum height of the wave generator is 0.3 m. Before carrying out the test, make sure to calibrate the wave generator.

2.2. Calibrate and Validation

Calibrate the test equipment described above. Calibration in order to get data and results that are in accordance with what researchers want. So that the variations that will be tested by researchers have a very high similarity with the scenarios produced by several instruments. For validation, it is used to see how the difference is when modeling from experimental to numerical. To model this using the IACS Req reference. 2004/Rev.3 2017[14]. Its validation is related to its geometric size.

2.3. Simulation of each variation

After the installation of the experimental model, it is continued with simulation in each variation. The simulated experimental method is:

1. Regular waves that will be used for running with a direction of 180 degrees or headseas.
2. One time simulation if at least 20 wave cycles have been achieved
3. Every time you do a simulation, you have to keep it in the stilling basin
4. The estimated time to do one run is 6 minutes (2 minutes for simulation and 4 minutes to ensure the water conditions are calm)
5. The amount of time required under ideal conditions in each experimental component is about 2160 minutes or about 36 hours (18 variations in wave height and period times 5 variations in gap distance between floating structures times 6 minutes times 2 repetitions)

Likewise for numerical, simulations are carried out with regular waves from the direction of 180 degrees or headseas. With the same period range as the experiment so that the results of the motion response can be compared. The simulation was carried out for 2 minutes using the ANSYS AQWA software. There are 5 variations for side-by-side configuration. 0.3 of the ship's width or 10 cm, 0.5 of the ship's width or 15cm, 0.67 of the ship's width or 20 cm, 0.83 of the ship's width or 25 cm and the same as the width of the ship or 30 cm. This research using frequency domain analysis [15].

2.4. Result and Discussion

After performing both single and side-by-side simulations experimentally and numerically, the RAO (Response Amplitude Operator) is obtained with a calculation formula that fits the context of this research. The calculation results will be explained in the form of tables, graphs and descriptions related to the tables and graphs.

2.5. Conclusion

From the results of the analysis and discussion, it will be obtained how the RAO (Response Amplitude Operator) of each scenario. have a great result if the RAO (Response Amplitude Operator) from numerical studies and experiments has a very close resemblance. This chapter will also explain about suggestions on how to get more optimal results, perform more complex variations and things that can support this research to gain more knowledge.

3. Analysis and Discussion

3.1. Calibrate and Validation

3.1.1. Wave Probe Calibration

The results obtained from the wave probe calibration test are the voltage at each depth of the wave probe is immersed in water. The data is then regressed by making a graph with the abscissa axis in the form of depth in centimeters where the wave probe is immersed and the ordinate axis in the form of stress in millivolts. regression graph of wave probe calibration as shown as figure 3.

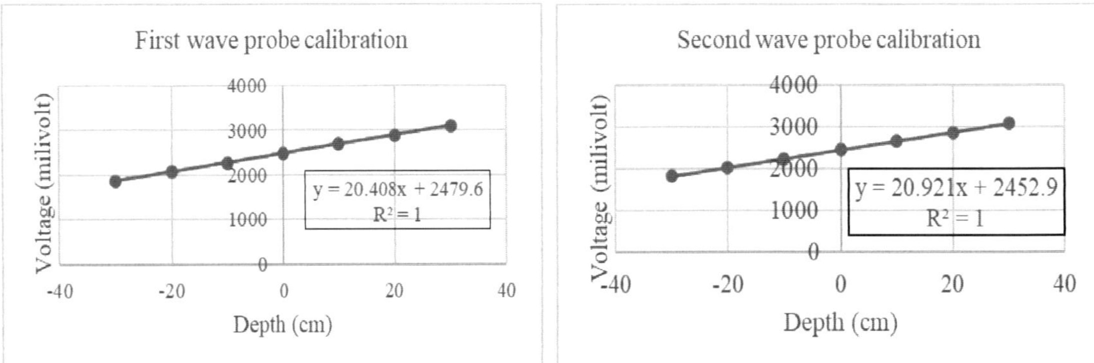

Figure 3 Graphic of wave probe calibration

In terms of the results of the table above, from two times calibration the wave probe, the regression value of the two experiments is worth 1. It means that the wave probe has a high level of accuracy.

3.1.2. Gyroscope Calibration

Manually calibrate the gyroscope with the help of an arc ruler. This is so that the data obtained from the rotational motion response is precise and accurate. Here are the calibration results:

Table 3. Gyroscope calibration

Angle X (Roll)			**Angle Y (Pitch)**		
Angle	Output Gyro	Error	**Angle**	Output Gyro	Error
0	0.01	0.01%	**0**	0.02	0.02%
20	20.05	0.25%	**20**	20.1	0.50%
40	40.08	0.20%	**40**	39.94	-0.15%
60	59.98	-0.03%	**60**	60.04	0.07%
90	90.02	0.02%	**90**	89.52	-0.53%

3.1.3. Validation

Modeling using ANSYS software with its sub applications, namely SPACECLAIM and DESIGNMODELER. With these two sub-applications, a model is made in such a way that it is validated according to the standard and also the experimental model. The results of the experimental and numerical model validation based on IACS Req. 2004/Rev.3 2017[14] as shown as table 4:

Dimension	**Experiment**	**Numerical**	**Error**
LOA	1.5	1.5	0%
B	0.3	0.3	0%
D	0.225	0.225	0%
T	0.068	0.0688	-1%
Displacement	29.8	29.163	2%

3.2. Motion Response Data

3.2.1. Experiment

After calibrating each instrument, it is continued with scenario testing for each variation. The first with one floating structure then followed by two floating structures. For side-by-side configuration, perform the gap distance according to the scenario described in chapter 2.3. To get RAO (Response Amplitude Operator) data, the translation mode is assisted by the camera to record motion of the ball on the floating structure and the processing of recording data from the camera is assisted by using the TRACKER software. As for the rotation mode using a Gyroscope meter. There are only 3 degrees of freedom will be analyzed in this study.

Figure 4. Front and side view of floating structure of experiment study

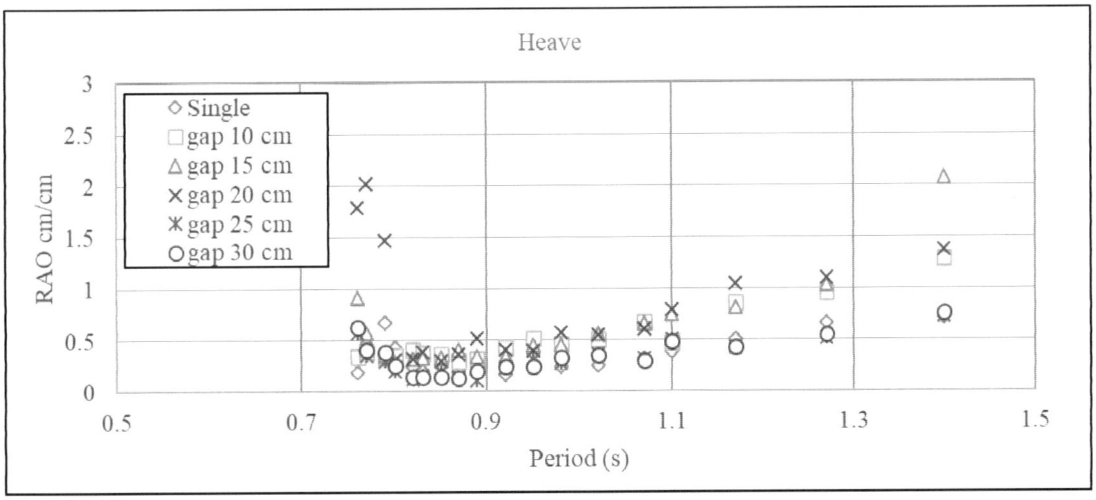

Figure 5. Heave motion graph for single body and side by side configuration with variation of gap distance

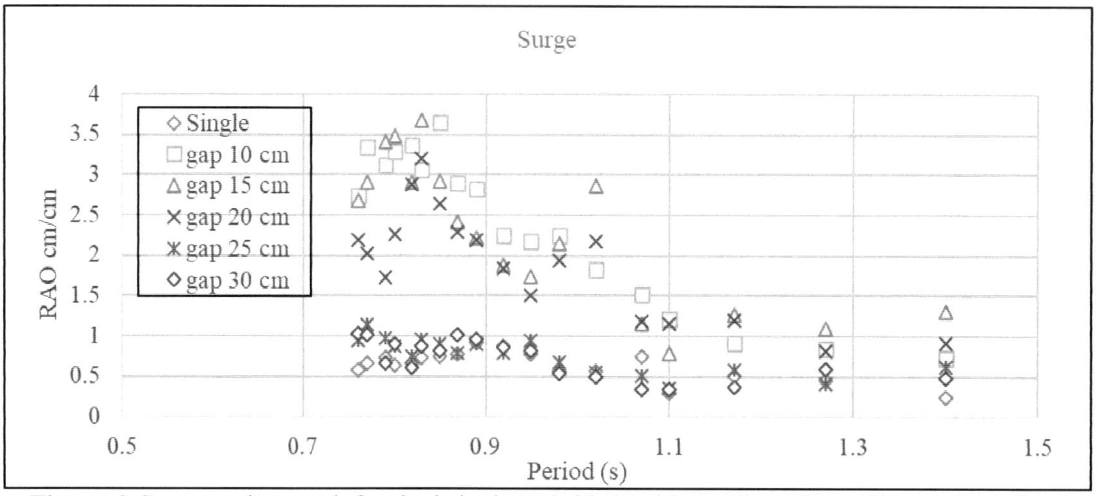

Figure 6. Surge motion graph for single body and side by side configuration with variation of gap distance

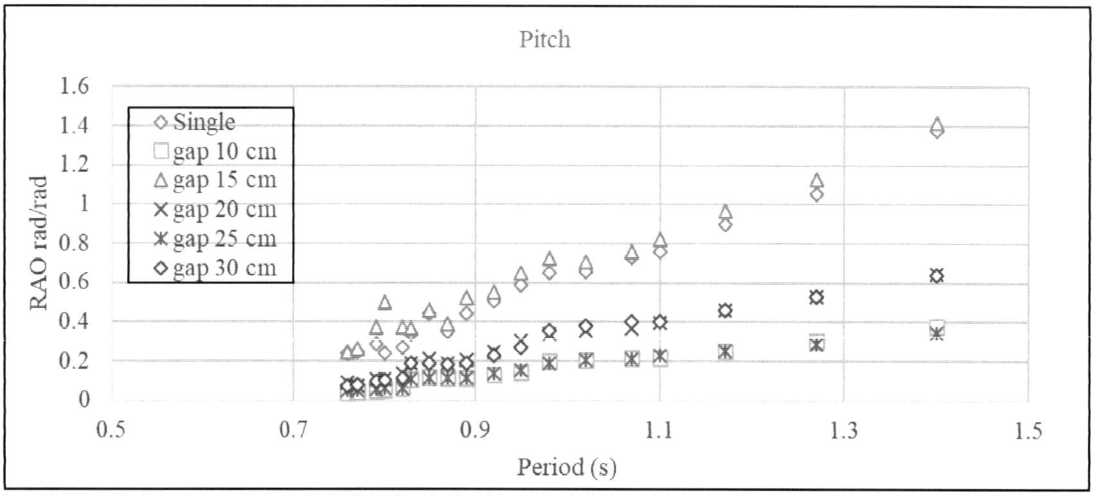

Figure 7. Pitch motion graph for single body and side by side configuration with variation of gap distance

The table below is a summary of the maximum RAO values for experimental studies from each mode of motion, both single body and side by side configurations with variations in the gap distance.

Table 4. Maximum RAO for each variation

Maximum	Heave (cm/cm)	Surge (cm/cm)	Pitch (rad/rad)
Single	0.713806	0.915163	1.378435
Gap 10 cm	1.281352	3.648384	0.37635
Gap 15 cm	2.072126	3.677945	1.413899
Gap 20 cm	2.022944	3.200988	0.641581
Gap 25 cm	0.7248	1.137821	0.341804
Gap 30 cm	0.760483	1.033997	0.444189

From the motion heave graph, it can be concluded that experimental studies, both single and side by side configurations with all variations of the gap distance have a trendline similar to the upward trend, but for those with a gap of 15 cm during a period of 1.5 seconds, the RAO value increases significantly. For surge motion with side-by-side configuration with a gap of 15 cm, the trendline and values are similar to the side-by-side configuration with gaps of 10 cm and 20 cm. In addition, the single body has similarities with the side by side configuration with a gap of 25 cm and 30 cm. For pitch movements, there are three pairs that have the same RAO value, namely single body with side by side configuration with a gap of 15 cm, side by side with a gap of 30 cm equals a gap of 20 cm and side by side with a gap of 10 cm has an RAO value. which is equal to the slit distance of 25 cm. For the large and small effect of the slit distance, the side by side configuration model does not have a conclusion that can describe how the slit distance affects the motion response.

The summary table above contains the maximum RAO (Response Amplitude Operator) value for each variation in each model configuration. The maximum heave, surge, and pitch values occur when the side by side configuration model has a gap distance of 0.5 from the model width or can be said to be 15 cm apart. Heave is at 2,072 cm/cm, surge is at 3,678 cm/cm, and pitch is 1,414 rad/rad. For the surge motion why has a very random value, it might happen because the measurement of the motion response reference point is above the model by 24.4 cm. Which means, when the model is pitched, the surge that is read is quite far because of the moment arm.

3.2.2. Numeric

Like an experiment, also testing the floating structure in each variation after modeling which is in accordance with the conditions during the experiment. Here are the results of running ANSYS AQWA in the form of RAO (Response Amplitude Operator):

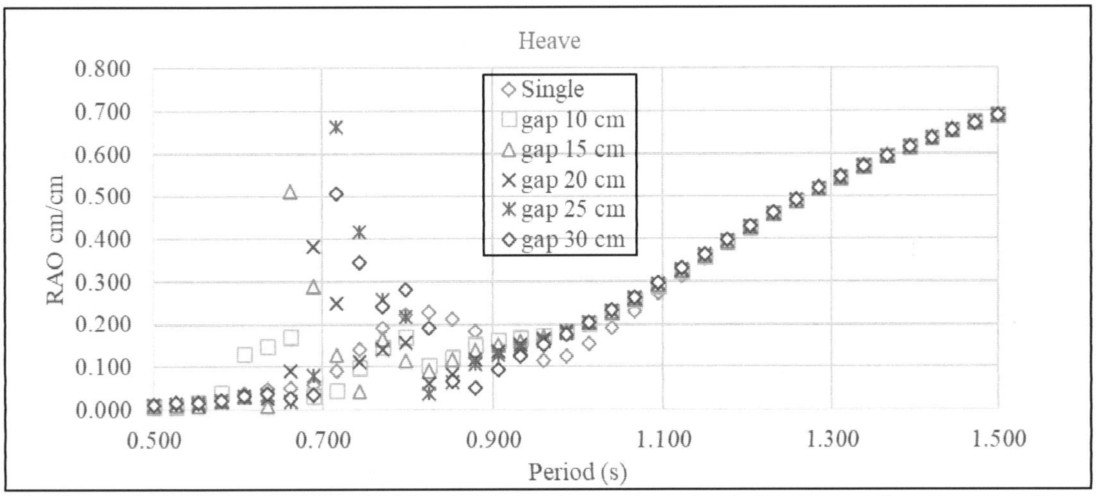

Figure 8. RAO graph of heave single body motion mode and side by side configuration for all variations of gap distance

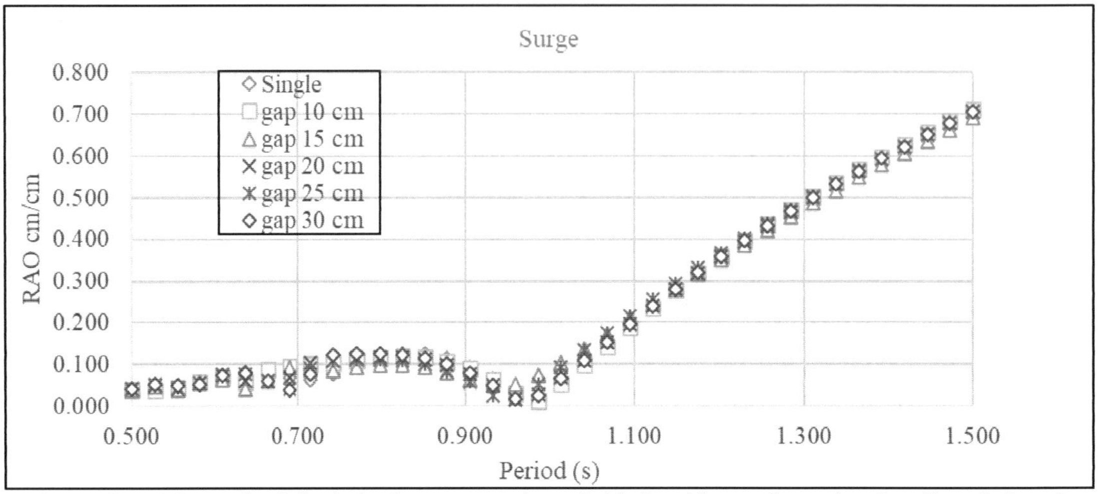

Figure 9. RAO graph of single body surge mode and side by side configuration for all variations in gap distance

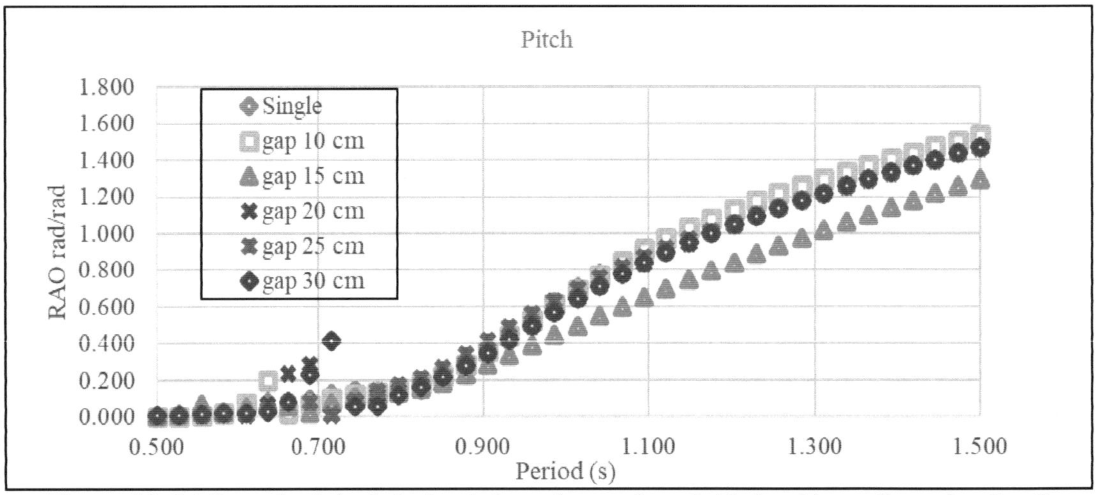

Figure 10. RAO graph of single body pitch motion mode and side by side configuration for all variations of slit distance

The table below is a summary of the maximum RAO values for numerical studies from each mode of motion, both single body and side by side configurations with variations in the gap distance.

Table 5. Maximum RAO for each variation

Maksimum	***Heave* (cm/cm)**	***Surge* (cm/cm)**	***Pitch* (rad/rad)**
Single	0.690859	0.710513	1.526713
Gap 10 cm	0.690622	0.711559	1.536315
Gap 15 cm	0.689938	0.691492	1.295771
Gap 20 cm	0.689628	0.706871	1.470763
Gap 25 cm	0.689199	0.705193	1.462809
Gap 30 cm	0.688838	0.707308	1.470405

From the motion heave graph, it can be concluded that numerical studies, both single and side by side configurations with all variations of gap distance have a trendline similar to an upward trend when the period has reached 0.9 seconds. Before the 0.9 second period there was a significant difference in the RAO value for the side by side configuration with a gap of 15 - 30 cm. For motion surges for single or all side by side configurations, the graphs have a fairly high degree of similarity. For motion pitch for single or all side by side configurations, the graphics have a fairly high level of similarity as well. However, there is a slight difference in the RAO value for the side by side configuration with a gap of 20 and 30 cm when the period is 0.7 seconds. From some of the explanations above, it can be concluded that the gap distance of 10 - 30 cm does not have a significant effect on motion heave, surge, and pitch.

The summary table above contains the maximum RAO (Response Amplitude Operator) value for each variation in each model configuration. The maximum heave value occurs when the model is not in a side by side configuration but in a single body. The heave value is 0.691 cm/cm. The maximum surge and pitch values are 0.712 and 1.536 when the model is configured side by side with a gap of 10 cm.

3.2.3. Experimental and Numerical Comparison

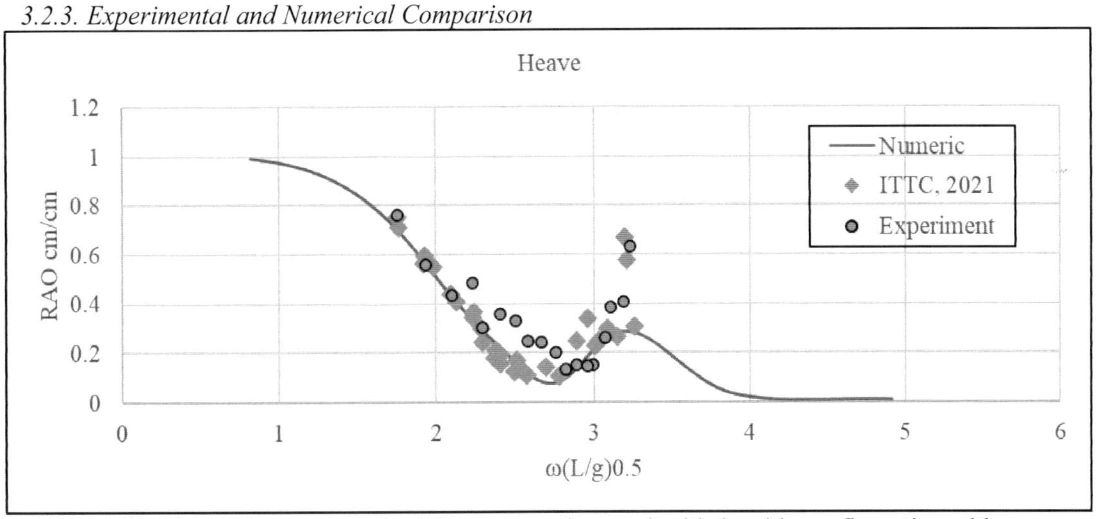

Figure 11. Comparison graph of RAO heave motion mode side by side configuration with a gap distance of 30 cm

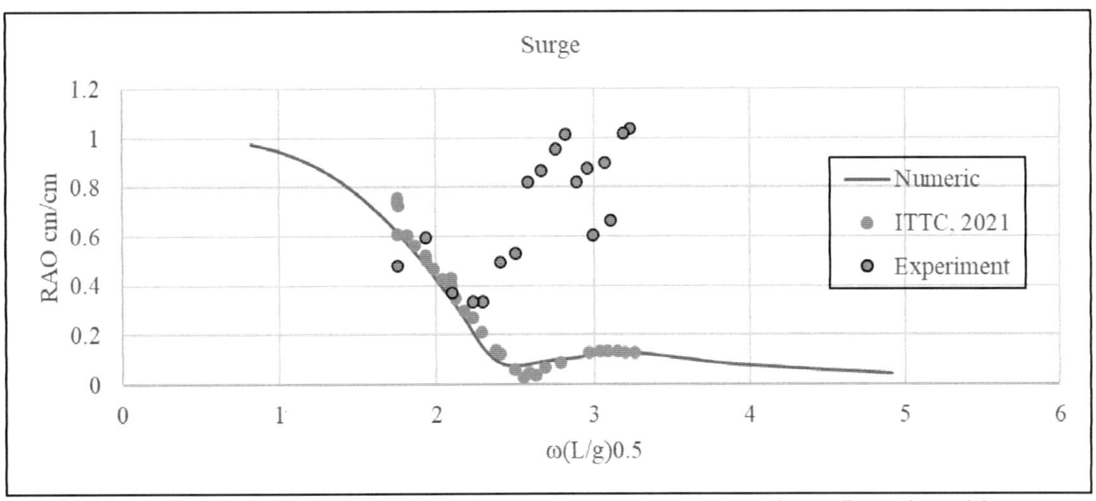

Figure 12. Comparison graph of RAO surge motion mode side by side configuration with a gap distance of 30 cm

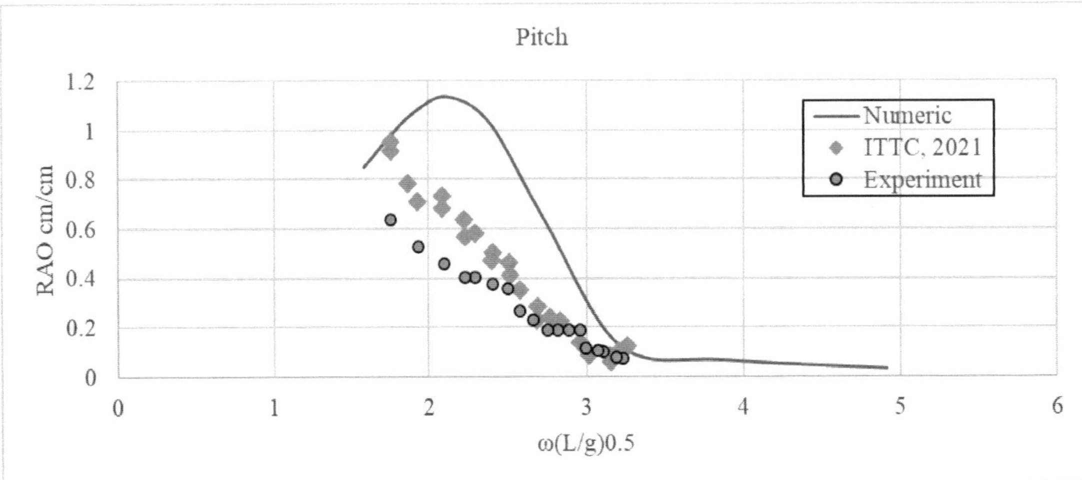

Figure 13. Comparison graph of RAO pitch motion mode side by side configuration with a gap distance of 30 cm

From the side-by-side comparison chart with a gap of 30 cm, in addition to the comparison between numerical and experimental, it is also compared with references originating from[16]. The motion heave has a trendline that has a very high similarity between the three studies. Likewise, the motion pitch for the experimental and ITTC is quite close and for the numerical it is not how far the value and the trendline of the RAO are. For the ITTC and numerical surges, the similarity is very high, but for the experiment it is still random because it is influenced by different reference points.

4. Conclusion

From the analysis of motion response between numeric and experiment the conclusion that can be taken are:

1. For this comparison case, it only explains when the model is configured side by side with a gap of 30 cm because of the difficulty of finding references as validation material for this research. For variations other than above, if there are differences, it can be chosen for numerical studies as reference material due to the many limitations of experimental studies as one example is very old from the wave generator engine so that it cannot work perfectly and also human error from the researchers themselves.
2. From the comparison, the heave value which produces the graph tends to be the same from the three research results.
3. For the pitch motion, the graph from the reference with the experiment from the researcher is similar, but for the numerical study, the researcher has a trendline that is almost the same but the RAO value is slightly larger.
4. For motion surge, as explained in chapter 3.2.1, the experimental results of the researchers are very random. However, for numerical studies, researchers and references from ITTC have almost the same RAO values.

References

[1] Pratama, N. Syahroni, and M. Murtedjo, "ANALISIS UMUR KELELAHAN STRUKTUR CATENARY ANCHORED LEG MOORING BUOY PADA SINGLE POINT MOORING FSO ARCO ARDJUNA SAAT KONDISI TERTAMBAT."

[2] Arifannisa, "Analisis Umur Kelelahan Pada Anchor Chain Single Point Mooring 3," Institut Teknologi Sepuluh Nopember Surabaya, 2016.

[3] R. A. Prasiwi and I. Rochani, "ANALISIS PERILAKU FPSO (FLOATING PRODUCTION STORAGE AND OFFLOADING) TERHADAP INTERNAL TURRET MOORING SYSTEM

BERBASIS SIMULASI TIME DOMAIN," *J. Tek. ITS*, 2014, [Online]. Available: 10.12962/j23373539.v3i2.7551.

[4] A. Dianiswara, "OPERABILITAS OFFLOADING PADA FLOATING STORAGE REGASIFICATION UNIT (FSRU)," 2020.

[5] Y. A. Wibowo and A. N. Firdaus, "ANALISIS NUMERIK PENGARUH MULTIBODY PADA KONFIGURASI TRANSFER LNG SECARA SIDE-BY-SIDE DENGAN VARIASI JARAK," *MARLIN*, 2020.

[6] E. B. Djatmiko, *Perilaku dan Operabilitas Bangunan Laut di Atas Gelombang Acak*. Surabaya: ITS-Press, 2012.

[7] ITTC, "The Ocean Engineering Committee Final Report and Recommendations to the 25th ITTC," vol. I, pp. 263–324, 2021.

[8] S. Chakrabarti, "Offshore structure modelling," *Appl. Ocean Res.*, vol. 17, no. 6, pp. 391–392, 1995, doi: 10.1016/s0141-1187(96)00012-0.

[9] ITTC, "Guideline for Model Tests of Multi-Bodies in Close Proximity," 2017.

[10] T. Tezdogan, A. Incecik, and O. Turan, "Full-scale unsteady RANS simulations of vertical ship motions in shallow water," *Ocean Eng.*, vol. 123, pp. 131–145, 2016, doi: 10.1016/j.oceaneng.2016.06.047.

[11] ITTC, "Analysis Procedure for Model Tests in Regular Waves," 2017.

[12] ITTC, "Floating Offshore Platform Experiments," 2017.

[13] ITTC, "Seakeeping Experiments," 2017.

[14] IACS, "Computer Software for Onboard Stability Calculations," 2017.

[15] DNV, "DNVGL-OS-E301 Position Mooring," 2017.

[16] ITTC, "The Ocean Engineering Committee Final Report and Recommendations," 2021.

Mooring Analysis of Ocean FarmITS (OFITS)

Y Mulyadi[1*], Y N Purnawanti[2], N Huda[2], N Syahroni[1], Wahyudi[1], K Sambodho[1], and A P Sulaksono[1]

[1] Ocean Engineering Department, Sepuluh Nopember Institute of Technology Surabaya
[2] Fisheries Mechanization Department, Marine and Fisheries Polytechnic Sorong

*yeyesmulyadi@gmail.com

Abstract. Ocean Farm ITS (OFITS) is a pilot project of offshore aquaculture designed by Institut Teknologi Sepuluh Nopember (ITS) which will be installed at Sidoasri Beach, Malang Districts. OFITS structure is an integration of fish cages with ecotourism facilities. This paper describes the mooring analysis for OFITS installation. The mooring configuration uses a rectangle grid consisting of 4 mooring buoys, 8 top inner mooring grids, and 14 mooring lines arranged symmetrically. The mooring line uses a type of nylon line with a size of 42 mm in diameter and 70 m in length. Five mooring lines on the front side and two mooring lines on the back side secure the moored buoy to the seafloor. The maximum capacity of buoy is 3 tons. The anchor for mooring uses an anchor with a weight of 0.25 tons for each anchor. The purpose of this analysis is to obtain the maximum tension of the mooring rope at OFITS based on extreme environmental loads. This analysis uses the time domain analysis method.

1. Introduction

Nowadays, the demand for the supply of fishery commodities is increasing in the world market. However, this is not balanced with good fisheries management, thus encouraging the proliferation of overfishing activities. In 2015, it was recorded that 33.1% of the world's marine fishery stocks were met by overfishing [1]. So, to restore and improve the condition of marine fisheries due to overfishing, it is necessary to have marine aquaculture efforts in Indonesia. Talking about marine aquaculture, in 2017, Indonesia has a potential area of marine cultivation reaching 12,123,383 Ha, with a utilization rate of only 325,825 Ha or equal to 2.7% [2]. So it is necessary to optimize the marine aquaculture area to increase marine fishery production and Indonesian food security.
Now Center for Marine Studies at the Institute for Research and Community Service (LPPM) of the Sepuluh Nopember Institute of Technology, Surabaya starting to develop modern fisheries that switch to open fish farming systems (floating net cages) off the coast, often known as offshore aquaculture.

This technology has been developed in several developed countries, such as Norway, Japan, and America. Offshore aquaculture has many advantages, including better water quality, more available locations, and higher economic potential. In addition, other advantages are that it can reduce overfishing, reduce pollution of the coastal environment, and improve the welfare of coastal fishing communities [3].

The Center for Marine Studies of the Institute for Research and Community Service (LPPM) at the Sepuluh Nopember Institute of Technology, Surabaya, is also conducting research on offshore aquaculture. In 2017, the Ocean Farm ITS floating hotel and structural design new innovation developed for tuna fish farming were released (OFITS), and this design is very suitable for Indonesian waters [4] [8].

As an offshore aquaculture, Ocean Farm ITS must have the ability to anticipate movements due to waves, currents, and winds. So it is necessary to have a mooring system to keep the structure in position in good condition. Research in the design of mooring systems and analysis of the structure of floating net cages in the Sea Station model for the Indonesian high seas has been carried out [5]. Based on these results, there is one configuration of the mooring layout that produces the smallest mooring tension and off set, namely the rectangular array type. In addition, similar research on the Collar Cage model for the Indonesian high seas has also been carried out [6]. In line with the offshore aquaculture research, research on the creation of the ITS Ocean Farm design innovation concept, both its structure and mooring system, has also been carried out [7]. Analysis of the structural strength of OFITS has also been carried out by for the conditions of the southern waters of Java Island [8]. However, this model has undergone many changes along with the research and development carried out by LPPM ITS so it is necessary to re-do the mooring system analysis on the OFITS model.

In its design, Ocean Farm ITS is applied for aquaculture at a depth of 20 m and has a hotel for marine tourism. So it is necessary to consider the factors of safety and comfort in it. There is one type of mooring system that has not been researched and analyzed to be applied to the latest model of Ocean Farm ITS, namely the catenary mooring system with rectangular array configuration. In this mooring system, the wire rope on OFITS is not tethered directly to the anchor but through the buoy. This is done to reduce the stress on the mooring system, as research has been done by [9]. The purpose of this study was to determine the maximum tension on the mooring rope attached to OFITS due to extreme environmental loads.

Ocean Farm ITS as shown in Figure 1 is an offshore aquaculture model developed by the Institut Teknologi Sepuluh Nopember through the Center for Marine Studies LPPM-ITS in 2017. At the beginning of its design, several types of offshore marine cages that have been developed in the world were selected for analysis first. , namely Seastation, Farm Ocean and Collar Cage. The behavior of the existing offshore marine cage is used as a benchmark for the innovative design of Ocean Farm ITS. The structure is planned to be applied in the waters of Sidoasri, Malang, East Java (Figure 2). In its design, ecosystems in aquaculture are made in such a way as to resemble the native habitat of marine biota cultivated and consider local commodities, availability of materials, environmental conditions, and added value that can provide many benefits for the community in the location [4].

Figure 1. Ocean Farm ITS (OFITS) Design

Figure 2. Location of OFITS Installation

2. Methodology

The analysis of the maximum tension on the OFITS mooring line is carried out in several stages, namely modeling ofits building surface, mooring system modeling, motion response analysis, and maximum tension analysis. Motion response analysis can be seen from the Response Amplitude Operator (RAO) frequency domain. The equations of motion in the frequency domain representing surge, heave and pitch degrees of freedom can be written as:

$$\sum_{i=1}^{3}[-\omega^2\{M_i + A_i\,(\omega)\} + i\omega C_i(\omega) + K_i]X_i(\omega)e^{i\omega t} = F_i\,(\omega) \quad (1)$$

where [M_i], ω, [$A_i(\omega)$], [$C_i(\omega)$] , [K_i] , [$X_i(\omega)$], and [$F_i\,(\omega)$] are mass matrix, wave frequency, added mass matrix, damping matrix, stiffness matrix, displacement amplitude and wave force for degrees of freedom.

Response Amplitude Operator (RAO) is the ratio of the response amplitude to the wave amplitude for linear systems that is plotted versus wave frequency or wave period:

$$x = X_i(\omega)e^{i\omega t}$$
$$RAOx = \frac{X_i(\omega)}{\varsigma_a} \quad (2)$$

where $X_i(\omega)$and $\varsigma_a$are response and wave amplitude of i-th motion, respectively.
RAOs of velocity and acceleration are expressed as follows:

$$\dot{x} = X_i(\omega)i\omega e^{i\omega t}$$
$$RAOx = \frac{\dot{X}_i\,(\omega)}{\omega\varsigma_a}$$

$$\ddot{x} = X_i(\omega)\omega^2\omega e^{i\omega t}$$
$$RAOx = \frac{\ddot{X}_i\,(\omega)}{\omega^2\varsigma_a} \quad (3)$$

where, $\dot{X}_i(\omega)$ and $\ddot{X}_i\,(\omega)$are velocity and acceleration amplitude of the response.

2.1 Environmental Load

Environmental load, wave load formulated as wave spectrum. The wave spectrum in the southern waters of Java Island was simulated using the JONSWAP wave spectrum. The selection of the JONSWAP spectrum states that closed waters can use the JONSWAP spectra [10]. The JONSWAP spectrum equation is as follows :

$$S(\omega) = \alpha g^2 \omega^{-5} exp\left[-1.25\left(\frac{\omega}{\omega_0}\right)^{-4}\right]\gamma^{exp\left[\frac{(\omega-\omega_0)^2}{2\tau\omega_0^2}\right]} \quad (4)$$

where :

α : 0,076 $(X_0)^{-0.22}$
X_0 : gX/Uw^2
X : fetch length
Uw : wind velocity
α : 0.0081 if X not determined
Υ : peak parameter
$\gamma = 5$ for $T_p/\sqrt{H_s} \leq 3.6$; and
$\exp\left(5.75 - 1.15\frac{Tp}{\sqrt{H_s}}\right)$ for $Tp/\sqrt{H_s} > 3.6$
where :
Tp : spectrum peak period $(\frac{2\pi}{\omega_0})$
Hs : significant wave height
τ : shape parameter (for $\omega \leq \omega_o = 0{,}07$ and $\omega \geq \omega_o = 0{,}09$)
ω_o : wave peak frequency $2\pi(\frac{g}{U_\omega})$ $(Xo)^{-0,33}$

Based on the calculation through Eq. 4, it is found that the Tp for the southern waters of the island of Java is 7.73 s, so that $\frac{Tp}{\sqrt{Hs}}$ = 5.25. The following is the JONSWAP spectrum curve for the southern waters of Java Island (Figure 3).

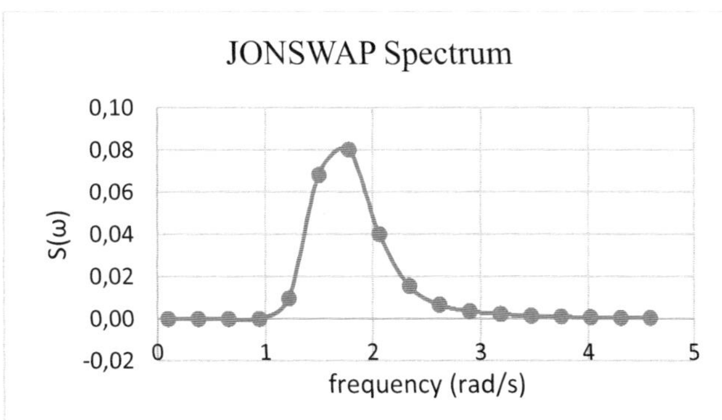

Figure 3. JONSWAP wave spectrum (Hs = 2.19m , Tp = 1.78s)

2.2 Time Domain Analysis

The time domain analysis method is based on an approach that uses time integration procedures and produces a time history response based on the time function x(t). Common time domain analysis methods, such as computer programs, can be used to analyze all mooring rope situations under the influence of wave frequency dynamics. However, this method requires a more complex and time-consuming process requires (time history simulation). Time history provides maximum tension results, anchor loads, and others [11]. The advantage of this method over the frequency domain is that all nonlinear types (system matrices and external loads) can be modeled more precisely. While the disadvantage is that it requires a longer calculation time. According to DNV OS E301, the minimum time domain simulation is 3 hours [12].

2.3 OFITS Modeling

Ocean Farm ITS is an *offshore aquaculture* design not like most *aquaculture.* There is a uniqueness in the ITS Ocean Farm model. This is, one of which is the existence of a hotel above this floating net cage structure. In addition, Ocean Farm ITS consists of several main components, namely; Hotel, *Top Deck, Lower Deck* (consisting of *Main Frame* and *Buoyancy Frame*), *Ball Joints*, as well as *Floaters* .

There is some assumptions in loading this, that is; burden man assumed originated from an amount 10 people with mass total eachof 100 kg. Meanwhile, the mass of fish is based on each person it can accommodate about 20 kg of fish. More detailed information regarding OFITS specifications can be seen in Table 1.

Table 1. OFITS Spesification

Structural Type	Ocean Farm (steel)
Dimensions	Diameter 19.8mx Height 5m
Weight	5.03 tons
Draft	1.6 m
Displacement	32.89 tons
Center of gravity coordinates (Xg; Yg; Zg)	0.09 m, 0.00 m, -1.55 m
Center of Buoyancy coordinates (Xb; Yb; Zb)	(0.0m; 0.0m; -4.2m)

The mooring configuration uses a rectangle grid consisting of 4 mooring buoys, 8 top inner mooring grids, and 14 mooring lines arranged symmetrically (Figure 4, Figure 5.). The mooring line uses a type of nylon line of 42 mm in diameter and 70 m in length, with properties in Table 2. Five mooring lines on the front side and two mooring lines on the back side secure the moored buoy to the seafloor. The maximum capacity of a buoy is 3 tons. The anchor for mooring uses an anchor with a weight of 0.25 tons for each anchor. Modelling the mooring line should be accommodated safety factor 1,67. [12]

Table 2. *Mooring* Property

Mooring Line	Weight (KN)	MBL/SF (KN)
Nylon 42 mm (ML1)	0.4	38.4
Nylon 42 mm (Mooring Grid)	0.4	38.4

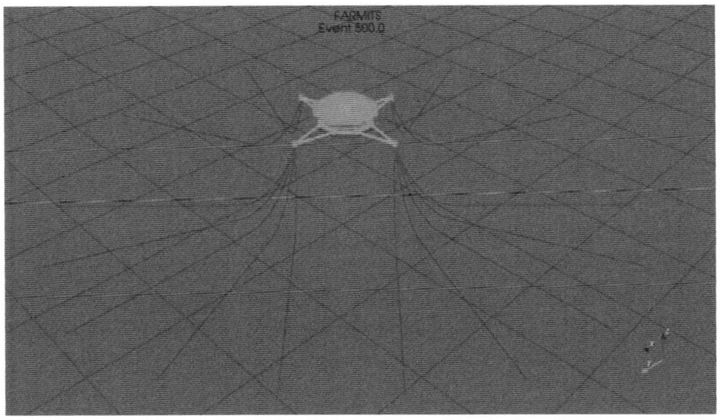

Figure 4. OFITS Mooring Configuration

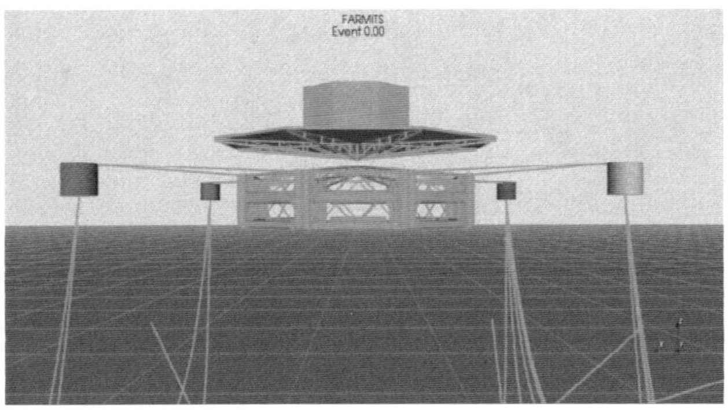

Figure 5. Isometric View of OFITS Mooring Configuration

3. Result and Discussion

3.1 OFITS Motion Responses

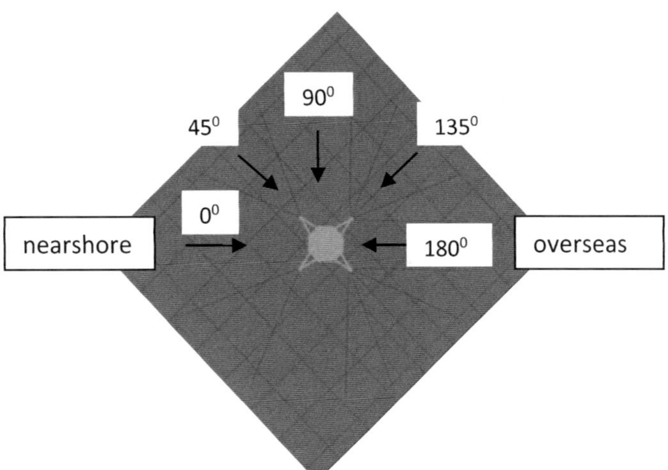

Figure 6. Wave directions

The computational results of the motion response of the OFITS on regular waves in free floating conditions can be seen through the RAO curves. Computing is done in 5 wave directions, namely following seas (0^0), quarter sea (45^0, 135^0), beam seas (90^0), and head seas (180^0) as shown in Figure 6. In the following sea and head seas, the mode of motion is dominated by surge, heave, and pitch motions because the structure is driven by waves from the front or back. In heading beam seas, the motion is dominated by sway and roll, because the waves hit the sides of the structure. Waves from the oblique seas, also called quarter seas, contribute to the sway, roll motion because the waves strike from the cross or diagonal direction to the longitudinal axis of the structure. In the free-floating condition, the vertical motions, heave, roll, and pitch are more dominant than horizontal motions, surge, sway, and yaw. [13][14]

3.2 Heave, Roll, Pitch Motions

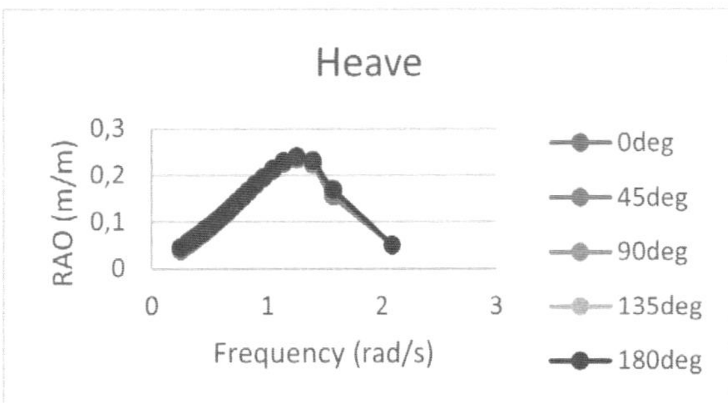

Figure 7. Heave Motion of OFITS structure

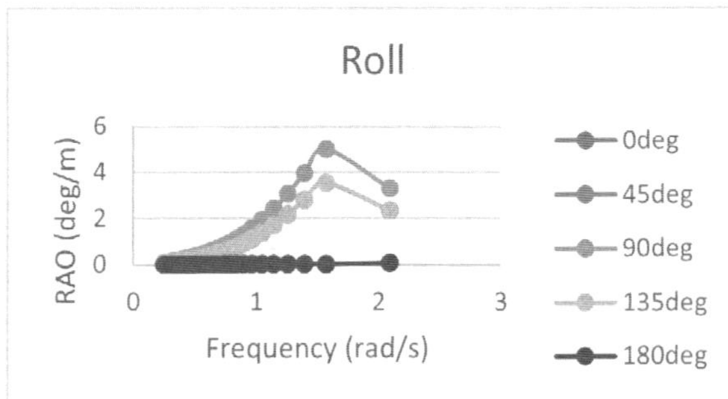

Figure 8. Heave Motion of OFITS structure

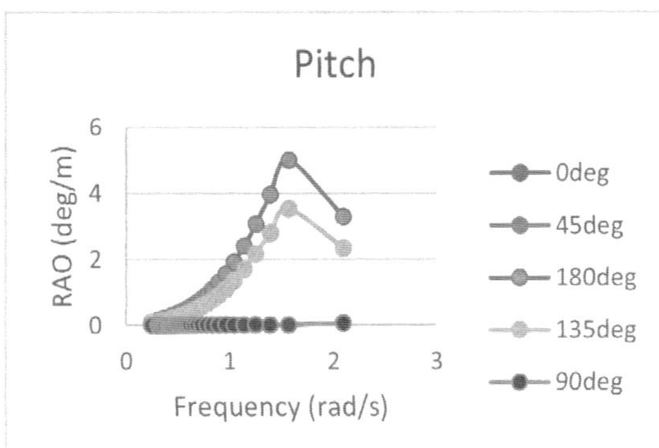

Figure 9. Heave Motion of OFITS structure

Figure 7 presents the heave motion RAO on OF ITS a ta frequency of 0-2.09 rad/s. The heave motion rises sharply to 0.23 m/m at peak frequency 1.14 rad/s for all heading seas , then drops to 0.05 m/m at a frequency 2 rad/s. In the subcritical region of OFITS, the roll motion (Figure 8) rises slowly until it rises sharply in the resonance region, reaching 4.9 deg/m and then drops sharply to close to 0 deg/m in the supercritical region. Dominant roll motion occurs when getting a 90 deg heading. The characteristics of the pitch motion of RAO in Figure 9 are similar to the roll motion, which are both rotational motions. However, the dominant pitch motion is from the direction of wave 180.

3.3 Mooring Tension

Analysis of the mooring rope tension on the OFITS mooring system was carried out using the time domain method. Simulations were carried out under extreme wave conditions for 3 hours, as recommended by DNV [12]. This study has two mooring line tension outputs, namely the mooring grid and the mooring line. mooring grid is the rope that is tied from the ofits to the buoy, while the mooring line is the rope that is tied from the buoy to the anchor. Figure 10 is the result of the running of the mooring grid. The figure shows that the maximum tension is 18.55 tons on the L7 mooring grid (Figure 10). It happens because the rope is exposed to a direct load from the direction of 135deg.

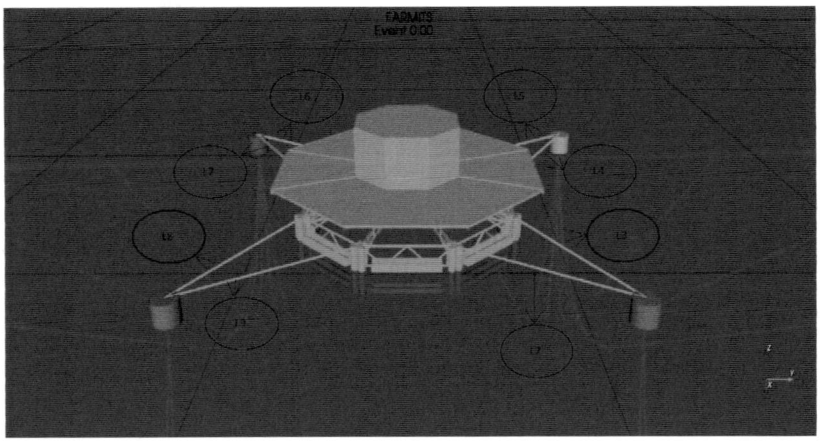

Figure 10. Location of Mooring Grid Maximum Tension

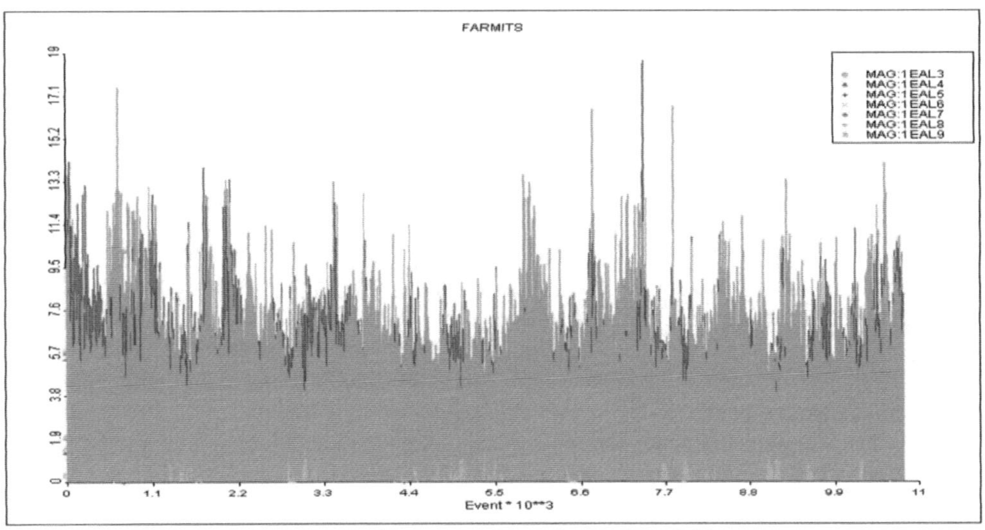

Figure 11. Maximum Tension of Mooring Grid

In the Figure 12 is the result of the running mooring line. It can be seen that the maximum tension on Line RCHB2 is 3.45 tons. This happened because the line was carrying a direct wave load from overseas in the direction of 135deg.

IOP Conf. Series: Earth and Environmental Science 1166 (2023) 012035 doi:10.1088/1755-1315/1166/1/012035

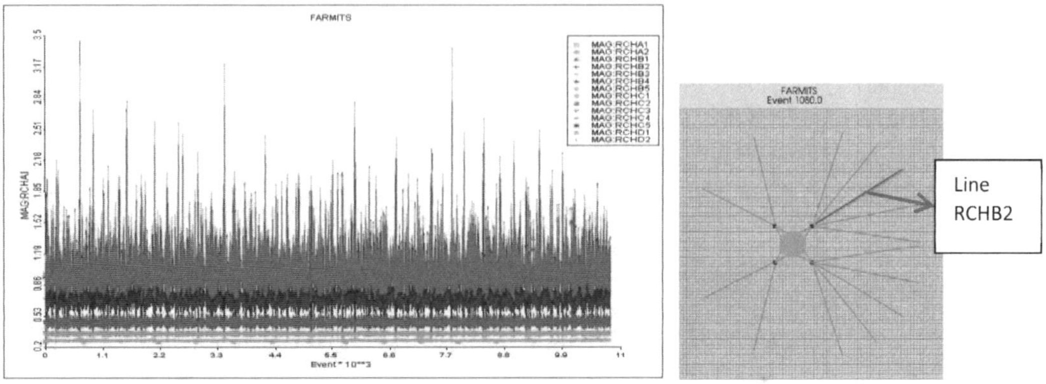

Figure 12. Location of Mooring Line Maximum Tension

4. Conclusion

In this study, time domain analysis was used to determine the maximum tension mooring line of the Ocean Farm ITS (OFITS). The mooring line is made of nylon and has a diameter of 42 mm and a minimum breaking load (MBL) of 38.4 tonnes. The tension mooring line results for each mooring line from OFITS to buoy and buoy to anchor are 18.55 tons and 3.45 tons, respectively. The safety factors for each mooring line are 2.04 and 1.5, respectively. This safety factor still meets the DNVGL-OS-minimum E301's requirement of 1.45.

In order to make operations more secure, it is necessary to conduct a risk assessment study that allows the most severe structures from each group to be assessed further for finite element analysis [15]. It is necessary to conduct studies on fish conditions due to stress during the acclimation process, handling and transporting cultured fish, which can cause metabolic disorders [16].

5. References

[1] FAO, *World Fisheries and Aquaculture, FAO:Rome,2020*. 2020.
[2] KKP, "Rencana Strategis Kementrian Kelautan dan Perikanan," 2017.
[3] M. H. Aziz, "Desain Aquaculture dengan Menggunakan Pipa HDPE untuk Budidaya Ikan Baronang," Institut Teknologi Sepuluh Nopember, 2015.
[4] Y. Mulyadi, N. Syahroni, K. Sambodho, M. Zikra, H. B. P. Adia, and others, "Mooring Design Selection of Aquaculture Cage for Indonesian Ocean," in *IOP Conference Series: Earth and Environmental Science*, 2018, vol. 135, no. 1, p. 12022.
[5] H. B. Prasetyo Adi, "Perancangan Sistem Mooring dan Analisa Olah Gerak Struktur Aquaculture Model SeaStation untuk Laut Lepas Indonesia," Institut Teknologi Sepuluh Nopember, 2017.
[6] N. D. Putri, "Analisis Variasi Konfigurasi Sistem Mooring dan Olah Gerak Struktur Model Collar Cage FLoating Offshore Aquaculture Untuk Laut Lepas Indonesia," Institut Teknologi Sepuluh Nopember, 2018.
[7] M. R. Athoillah, "Konsep Inovasi Desain Aquaculture System Untuk Ikan Tuna Di Laut Indonesia," Institut Teknologi Sepuluh Nopember, 2017.
[8] Y. Purnawanti, N. Syahroni, and Y. Mulyadi, "Analysis of static structural of ocean FARMITS offshore cage due to environmental load," in *MATEC Web of Conferences*, 2018, vol. 177, p. 1023.
[9] O. Mursid, K. R. Malau, N. Huda, A. M. A. Abidin, and G. Sutarno, "The Effect of Additional Subsea Floater on Mooring Line Tension and Motion of CBM (Conventional Buoy Mooring) Ø4.5 m," in *IOP Conference Series: Earth and Environmental Science*, 2021, vol. 698, no. 1, p. 12034.
[10] S. K. Chakrabarti, *Hydrodynamics of offshore structures*. WIT press, 1987.

[11] D. F. Mahanani, “Analisa time-Domain Pengaruh Spread Mooring dengan Variasi Jumlah Line terhadap Tension pada Flexible Riser,” Institut Teknologi Sepuluh Nopember, 2017.
[12] D. N. V Gl, “Position mooring,” no. July, 2015.
[13] O. Yilmaz and A. Incecik, “Hydrodynamic design of moored floating platforms,” *Mar. Struct.*, vol. 9, no. 5, pp. 545–575, 1996.
[14] E. B. Djatmiko, “Perilaku dan Operabilitas Bangunan Laut di Atas Gelombang Acak,” *ITS-Press. Surabaya. Inonesia*, 2012.
[15] R.D. Riyanto, M. Resiman, N. Syahroni, "Risk Assessment of Aging Offshore Jacket Platform Group: A Case Study on “B” Field Platforms," in IOP Conf. Series: Earth and Environmental Science 649 (2021) 012068, doi:10.1088/1755-1315/649/1/012068.
[16] Hidayati, D., Nurindra, A.R., Abdulgani, N., Syahroni, N., Mulyadi, Y, “Fish disease study of asian seabass (Lates calcarifer) in a floating marine cage: Endoparasite and blood profile,” in Biodiversitas, Volume 22, Number 10, October 2021, Pages: 4505-4511, DOI: 10.13057/biodiv/d221044.

IOP Conf. Series: Earth and Environmental Science 1166 (2023) 012036 doi:10.1088/1755-1315/1166/1/012036

Hot Spot Stress Analysis of a Two-Planar DTY Tubular Joint Subjected to Axial Loading

R W Prastianto[1], I A Adha[1], Y S Hadiwidodo[1], D M Rosyid[1], and Y Mulyadi[1]

Department of Ocean Engineering, Institut Teknologi Sepuluh Nopember, Surabaya, Indonesia

rudiwp@oe.its.ac.id

Abstract. Air Cooler Support Frame (ACSF) is one of the supporting structures in oil and gas field that consist of tubular and beam members. Currently, the existing common ACSF has a conventional design and could be overdesign. Therefore, the ACSF structure must be designed by considering local stress characteristics to obtain sustainable innovative design. Significant structural failure usually occurs in tubular joints because of the hot spot stress in the joint area. In this study, the ACSF structure was globally analysed to determine the critical joint which was a two-planar DTY joint, then it was locally modelled by using finite element method-based of ANSYS software for local stress analysis. The tubular joint has been modelled using solid elements including weld model and by creating fine meshing in the vicinity of the brace-chord intersection area for all braces. The axial loading is applied to investigate the local stress distribution along the brace-chord intersection line of the tubular joint for each brace. The result show that the hot spot stress of about 145 MPa for the DTY joint occurred in Brace 1 around the saddle area on the chord side under the axial tension loading.

1. Introduction

In the production of oil and gas, it is not only drilling facilities needed in the processing stage, but there are also other necessary supporting facilities needed, such as facilities at the gathering station. The gathering station works as a place to collect the fluid produced from oil wells in the field and then the fluid will be separated according to its purposes [1].

Air Cooler Support Frame (ACSF) is a supporting platform installed at the gathering station in the form of a space frame structure composed of open profile-beams and tubular members that functions as a support structure for the air cooler facilities. This structure has a fairly large in shape and placed in a narrow place such as a river. The ACSF can experience damage or failure due to both material degradation and variations in the load imposed to the structure. Failure tends to occur in components that are old due to natural processes even though there has never been a failure before.

The hot spot stress (HSS) of tubular joints directly affected to the accuracy of the fatigue life calculation by using the HSS approach. The HSS is the maximum geometric stress occurred along the weld toe of the chord/brace intersection. The load imposed simultaneously to braces of multiplanar tubular joints will create geometric stress in the all braces. Thus, the induced HSS is a function of all nominal stresses in all the brace/chord members due to an interaction effect among the braces of the multi-planar tubular joints that different than that occurred in common uniplanar joints.

Study on the structural static strength has been carried out by [2] in term of a local analysis of the tubular K-joint and obtained the maximum stress on the joint. Local stress analysis on multiplanar tubular double KT (DKT) joint was performed by [3] and obtained stress concentration factor distribution of the inclined brace of the joint.

This present study focused on the analysis of local stresses in a two-planar tubular joint of DTY joint as a critical joint of the ACSF structure by using the finite element method software that can represent detailed structural component with much greater accuracy better that simplified hand calculations [4]. The analysis produced stress distribution characteristics along the weld toe of brace-chord intersection of the tubular joint and finally identified the hot spot stress on the joint.

2. Structure Modelling

2.1 Structure and Environment Data

The ACSF structure has 8 legs that support 18 air cooler units with the structural data as presented in Table 1. The dominant environmental load affected the ACSF structure are only wind and current because the structure is located on a river. Table 2 and 3 show the detail data of the wind load, and current load.

Table 1. The ACSF Structure Data.

Description	Appurtenance
No of legs	8 (Un-grouted Legs)
Deck Leg Diameter	30" (Row 1) & 30" (Row 2)
Deck Leg Spacing	16.25 cm × 21.5 cm (Center to center of leg)
Leg Batter	No Batter
Working Point	(+) 4.80 m
Splash Zone	Between ± 1 m from MSL
Deck Elevation	Upper Deck – EL. + 12.72 m
	Lower Deck – EL. + 6.22 m

Table 2. Wind Load Data.

Wind Criteria	Wind Speed Operational Condition (m/s)
3 second gust	29.3
1 minute mean	23.2

Table 3. Current Load Data.

Elevation (m)	Current Speed Operational Condition (m/s)
0	2
100	2

2.2 Global Modelling

Global structural modeling and inplace analysis of the ACSF structure were carried out using SACS 12.0 software. In the inplace analysis, the working loads were applied to the structure as well as the environmental loads affected the structure. Figure 1 shows the global model of the ACSF structure. Results of the inplace analysis are presented in the form of unity check values for the main part of the structural members such as the legs and related members attached to the legs of the structure, as shown in Table 4.

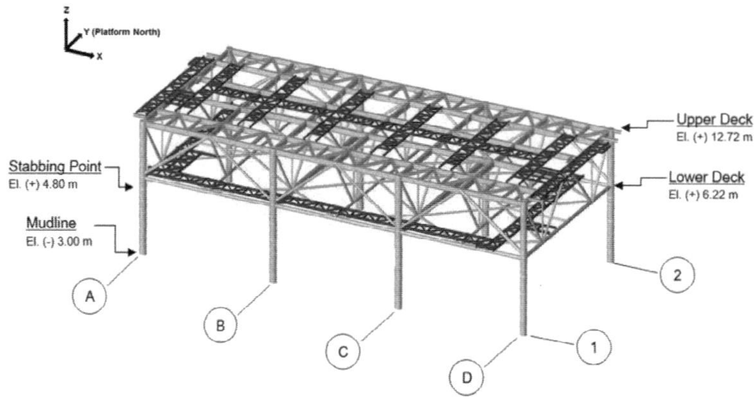

Figure 1. The Air Cooler Support Frame Structure.

Table 4. Unity Check Member Deck Leg ACSF.

Group	Member	Max. UC	Critical Condition	Load Condition
PA1	101L-0010	0.14	SHEAR	1006
PB1	102L-0011	0.2	CM+BN	1007
PC1	103L-0014	0.19	CM+BN	1007
PD1	104L-0015	0.13	SHEAR	1008
PA2	105L-0106	0.13	SHEAR	1003
PB2	106L-0021	0.2	CM+BN	1003
PC2	107L-0019	0.2	SHEAR	1003
PD2	108L-0018	0.14	SHEAR	1002
DLG	201L-301L	0.14	SHEAR	1002
DLG	202L-302L	0.16	SHEAR	1003
DLG	203L-303L	0.27	SHEAR	1003
DLG	204L-304L	0.13	SHEAR	1004
DLG	205L-305L	0.14	SHEAR	1008
DLG	206L-306L	0.16	SHEAR	1007
DLG	207L-307L	0.15	SHEAR	1007
DLG	208L-308L	0.14	SHEAR	1006
DL	0010-201L	0.14	SHEAR	1006
DL	0011-202L	0.21	SHEAR	1007
DL	0014-203L	0.19	SHEAR	1007
DL	0015-204L	013	SHEAR	1008
DL	0018-208L	0.13	SHEAR	1002

DL	0019-207L	0.2	SHEAR	1003
DL	0021-206L	0.2	SHEAR	1003
DL	0106-205L	0.13	SHEAR	1003
16A	201L-0000	0.38	LOCAP	1005
16A	201L-0029	0.27	CM+BN	1005
16A	201L-0926	0.1	LOCAP	1005
16A	0926-0927	0.08	LOCAP	1005
16A	0000-202L	0.36	LOCAP	1001
16A	202L-0001	0.3	LOCAP	1005
12B	202L-0030	0.3	CM+BN	1005
12B	0029-202L	0.42	CM+BN	1001
WPG	202L-0941	0.19	LOCAP	1007
16A	0001-203L	0.29	LOCAP	1001
16A	203L-0002	0.37	LOCAP	1005
12B	0030-203L	0.3	CM+BN	1001
12B	203L-0031	0.4	CM+BN	1005
WPG	203L-0788	0.2	SHEAR	1007
16A	**0838-204L**	**0.46**	**LOCAP**	**1001**
16A	204L-0768	0.16	LOCAP	1003
16A	0031-204L	0.25	CM+BN	1001
16A	205L-0032	0.28	CM+BN	1005
16A	205L-0459	0.25	LOCAP	1005
16A	0459-0460	0.21	LOCAP	1005
16A	0003-206L	0.19	LOCAP	1001
16A	206L-0004	0.15	LOCAP	1005
12B	0032-206L	0.43	CM+BN	1001
12B	206L-0033	0.31	CM+BN	1005
WPG	0775-206L	0.18	LOCAP	1003
16A	0004-207L	0.14	LOCAP	1001
16A	207L-0005	0.23	LOCAP	1005
12B	0033-207L	0.29	CM+BN	1001
12B	207L-0034	0.4	CM+BN	1005
WPG	0777-207L	0.18	SHEAR	1003
16A	0942-208L	0.32	LOCAP	1001
16A	0034-208L	0.25	CM+BN	1001
16A	0770-208L	0.15	LOCAP	1007

The modeled ACSF structure is then validated using the MAPE parameter criteria as formulated in Equation 1 which is stated that the prediction is highly accurate if the MAPE value is below 10% [5].

$$MAPE = \frac{1}{n}\sum_{t=1}^{n} \left|\frac{A_t - F_t}{A_t}\right| \times 100 \quad (1)$$

Where, A_t and F_t are real and model values at time t. In the present study, the validation compares the weight of the modeled structure to the weight of the structure based on the original the ACSF data. Table 5 shows the results of the MAPE validation with the MAPE value of 4.31% which is under the 10%.

Table 5. The ACSF Model Weight Validation with MAPE Criteria.

Weight Component	Base Weight (ton)	Error (%)
Modelling	2,963.565	4.31
Real data	2,835.690	

2.3 Local Structure Modelling

For local analysis the critical joint to be determined is based on the member that has largest unity check on the ACSF structure. In the UC results shown in Table 4, member 0838-204L is selected as a critical joint to be locally modeled and analyzed for local stress distribution. Local modeling of the tubular joint was carried out using the ANSYS Static Structural 2020 R2 software by using the Solid187 element type that can capture both linear and non-linear behaviors [6].

Chord length of the tubular joint model was based on [7], in which the chord length must greater than six times of the chord diameters in order to ensure that the stress occurred at the brace-to-chord intersection of the model was not affected by the support conditions at the both end of the chord. Figure 2 shows the joint model as a two planar DTY joint with 90-degree angle between the joint planes. Material and section properties data of the tubular joint are listed in Table 6.

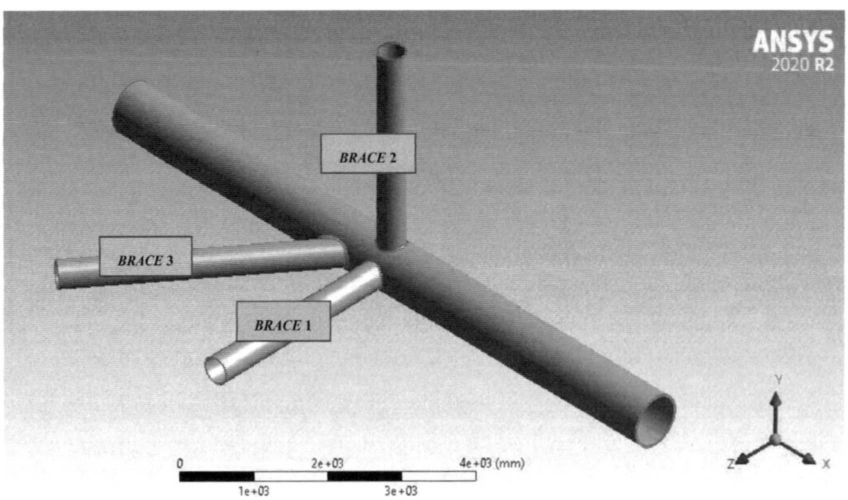

Figure 2. A Two Planar DTY Joint (Isometric View).

Table 6. Material and Section Properties Data of the Two Planar DTY Joint.

Element	Length (m)	Young Modulus (kN/cm^2)	Yield Strength (kN/cm^2)	Density (tonne/m^3)	Outer Diameter (OD) (cm)	Wall thickness (Wt) (cm)
Chord 1	5.00	20000	24.8	7.849	76.2	3.175
Chord 2	1.42	20000	24.8	7.849	76.2	3.175
Chord 3	3.58	20000	24.8	7.849	76.2	3.175
Brace 1	3.00	20000	24.8	7.849	40.64	1.905
Brace 2	3.00	20000	24.8	7.849	40.64	1.905
Brace 3	4.30	20000	24.8	7.849	40.64	1.905

The modeling of the weld on the tubular joint used weld size which is calculated based on [8] with calculated dimensions of weld length (Lw) and weld height (Hw) as shown in Table 7. Material of the weld metal and base metal of the joint are assumed same which is the ASTM A36 steel material. The detail nomenclature of the weld dimension for T-joint part and Y-joint part of the DTY joint are shown in Figures 3 and 4.

Tubular joints in offshore platforms, the chord support behaviors ranges from fixed to pinned conditions, and is generally considered closer to fixed support [7]. In several previous studies on tubular joints stress analysis, both ends of the chord have fixed supports [9]. So, in the present study, fixed supports at the ends of the chords are used. Fixed support is defined as a limitation of movements in all directions, both translationally and rotationally.

Table 7. Weld Geometry and Dimension of the DTY Joint.

Joint Type	Hw (mm)	Lw (mm)
T joint	14	10
Y joint	48	40

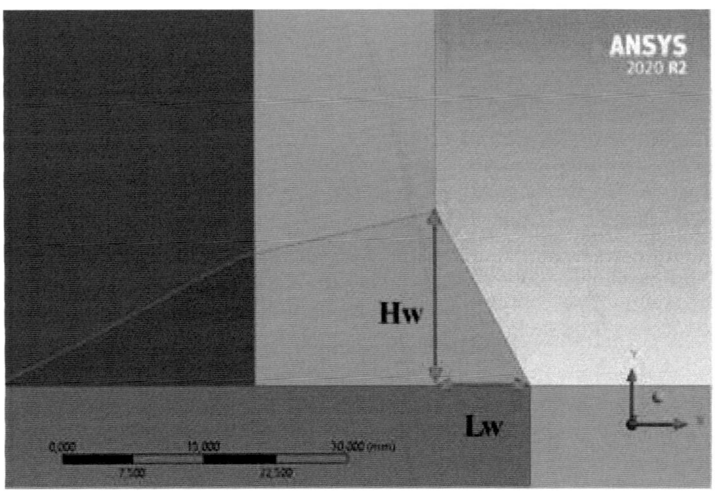

Figure 3. Weld Dimension for the T Joint Part.

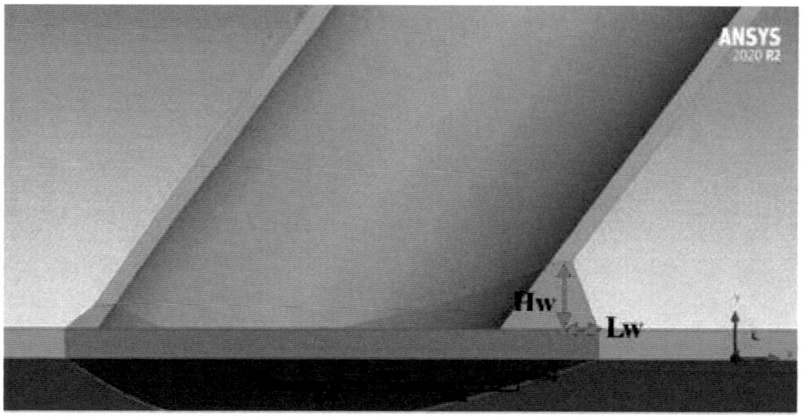

Figure 4. Weld Dimension for the Y Joint Part.

2.4 Critical Joint Loading

For the local analysis the axial load was applied to each brace of the tubular joint model. The magnitude and directions of the axial load is obtained from the global inplace analysis. Among the other output, the analysis produced member end forces for all structural members in both tension or compression condition. Table 8 shows the magnitude of the axial load for each brace, and the direction of each axial load is illustrated in Figure 5.

Table 8. Axial Loading on DTY Joint.

Element	Member	Axial Force (kN)
Brace 1	0838-204L	356.751 (tension)
Brace 2	204L-0768	31.2638 (tension)
Brace 3	0031-204L	-439.819 (compression)

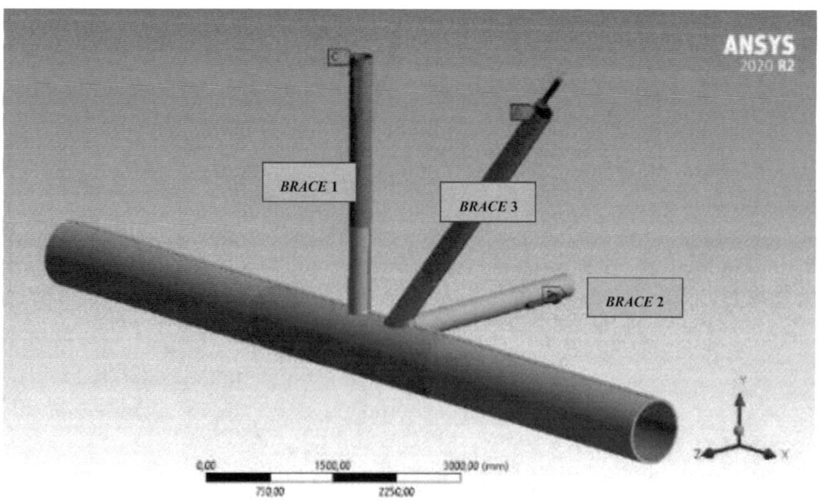

Figure 5. Axial Loading Directions applied on the DTY Joint Model.

3. Results and Discussion

3.1 Meshing Sensitivity Analysis

Meshing sensitivity analysis is carried out for the model in order to obtain a stable stress result. The meshing performed in this local modeling is arranged so that the fine meshing is created in area approaches the brace-chord intersection of the tubular joint, and gradually the meshing size gets larger for the area far away from the joint intersection [10]. The shape of the meshing used is hexahedral in the region far from the brace-chord intersection, and tetrahedron form in the area close to the brace-chord intersection of the joint. A good quality meshing of the tubular model will ensure to produce good and valid stresses output. Complate meshing of the model is shown in Figure 7 including the detail meshing in the region close to the weld area of the tubular joint.

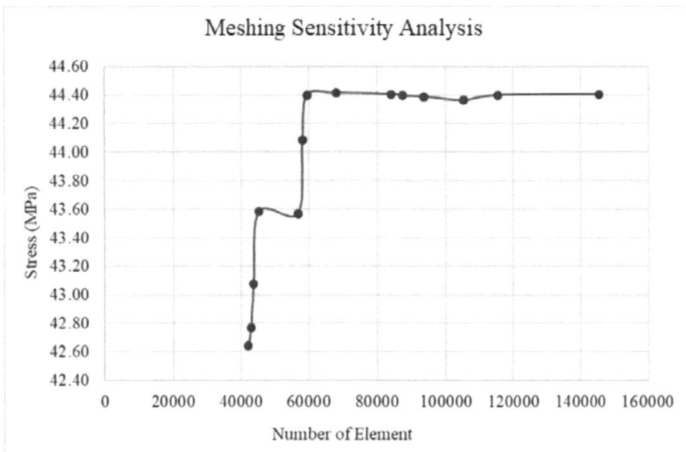

Figure 6. Meshing Sensitivity Analysis of the DTY Joint.

The meshing sensitivity analysis was done by mean that the number of meshing is increased slowly until a stable stress probe value is obtained in the location where the stress is being evaluated for this analysis. It can be seen from the graph in Figure 6 that the stationary stress of around 44.4 MPa occurred after reached the elements number of 60,000 until 140,000 and this study used 115,436 elements for the shake of computing efficiency.

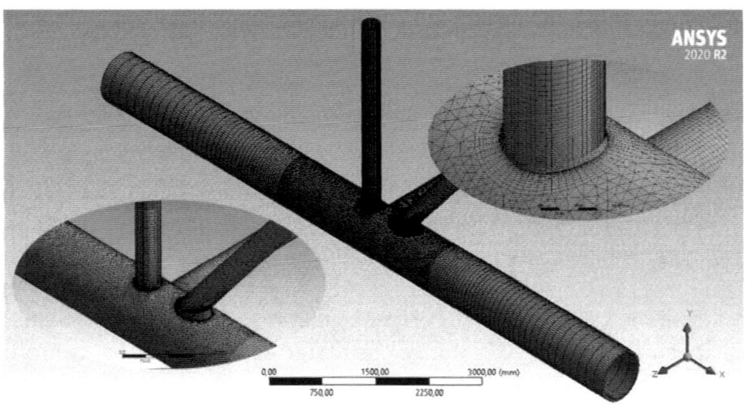

Figure 7. Meshing Detail of the DTY Joint.

3.2 Stress Distribution in the Brace-Chord Intersection Area

The stress distribution analysis uses von Mises stress or equivalent stress which is this stress allows each variable of three-dimensional stress state to be represented as a single positive stress value [11]. The combined stress between shear stress and axial stress in all directions of the element plane with matrix approximation which can be calculated by the equation below.

$$\sigma_e = \frac{1}{2}\sqrt{\left[\left(\sigma_x - \sigma_y\right)^2 + \left(\sigma_y - \sigma_z\right)^2 + (\sigma_z - \sigma_x)^2 + 6(\tau_{xy}^2 + \tau_{yz}^2 + \tau_{xz}^2)^2\right]} \quad (2)$$

In Eq. 2, σ_e is von mises stress, σ_x is x axis normal stress, σ_y is y axis normal stress, σ_z is z axis normal stress, τ_{xy} is xy plane shear stress, τ_{yz} is yz plane shear stress, and τ_{zx} is zx plane shear stress.

As it was described in the previous section, this local analysis is carried out by imposing axial loads on each brace based on the member end force results obtained from the global inplace analysis that has been performed in the previous global modeling. The locations reviewed to determine the stress distribution of the tubular joint are the weld toe area along the chord side and the weld toe area along the brace side in each brace of the DTY joint. Figures 8 – 10 show the stress distribution around the weld area of the Brace 1, Brace 2, and Brace 3, respectively.

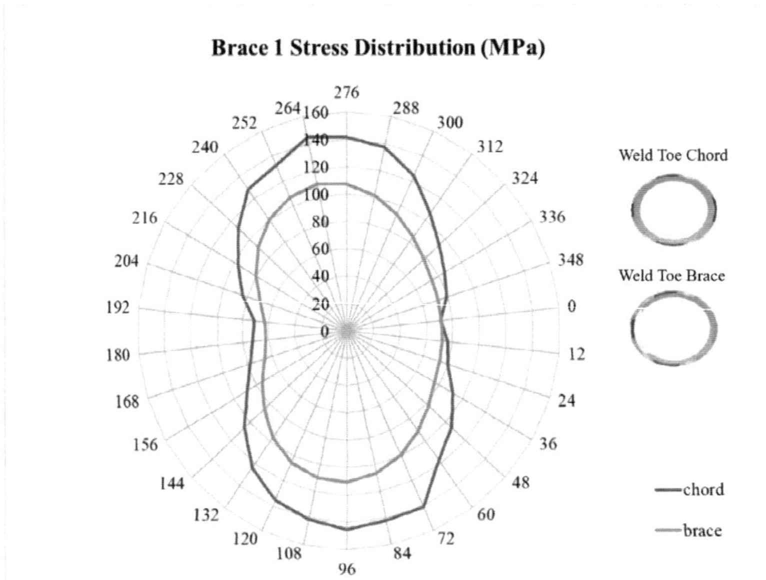

Figure 8. Stress Distribution on the Brace 1 of the DTY Joint.

Figure 2 and Figure 5 illustrated that Brace 1 and Brace 3 are two braces that form a TY joint in one plane of the DTY joint. Meanwhile, Brace 2 is a brace that forms a T joint in another plane of the DTY joint. Brace 1 and 2 are loaded by axial tension forces, while Brace 3 is loaded by compression force. Figure 8 shows stress distribution around the Brace 1 along the weld toe for both on the chord and the brace sides. The maximum stress generally occurred around the saddle area which are around 145.70 MPa on the chord side and 110 MPa on the brace side. This Brace 1 is exposed to a fairly large axial force.

Meanwhile, stress distribution around Brace 2 along the weld toe on the chord side as well as on the brace side are depicted in Figure 9. This Brace 2 is loaded by a relatively small axial force.

The maximum stress occurred at a position of 60 degrees that slightly shifted from the saddle area. The stresses are around 23 MPa on the chord side and 14 MPa on the brace side.

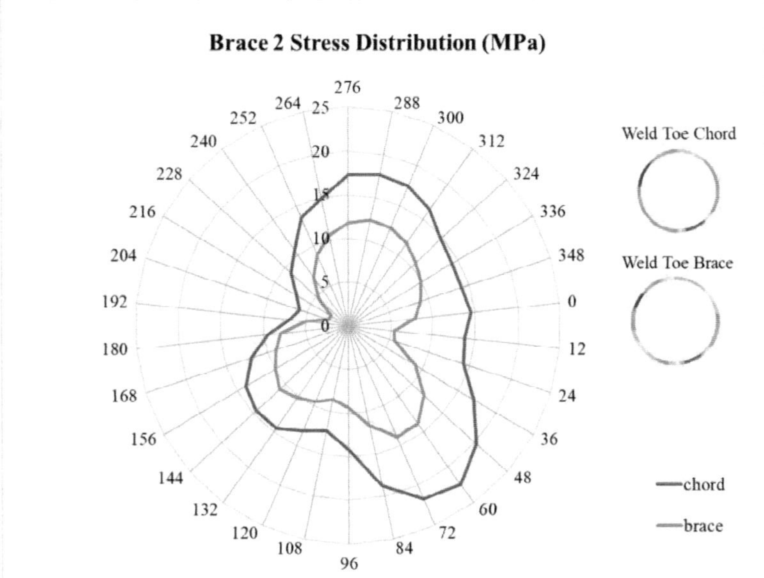

Figure 9. Stress Distribution on the Brace 2 of the DTY Joint.

For Brace 3, stress distribution that occurred along the weld toe on the chord side as well as on the brace side are illustrated in Figure 10. This Brace 3 is loaded by a largest axial force than that experienced by Brace 1 and 2, but in compression. The maximum stress occurred at a crown position of 180 degrees which are about 123 MPa on the chord side and 80 MPa on the brace side.

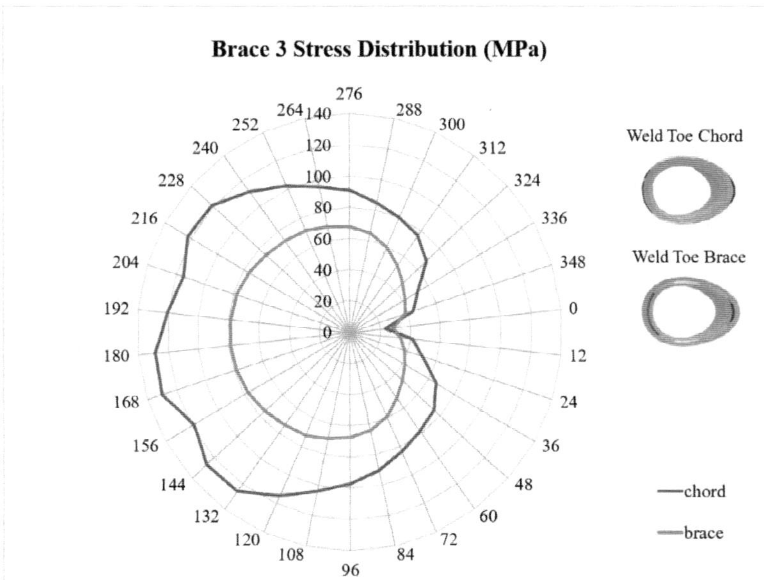

Figure 10. Stress Distribution on the Brace 3 of the DTY Joint.

The maximum stress or hot spot stress (HSS) in the DTY joint occurred in the brace 1 weld toe along the chord which is 145.7 MPa. This HSS does not exceed the yield stress of ASTM A36 steel material, which is 250 MPa. From the graph, it can be seen that the HSS occurred in brace 1 is at the upper and lower saddles positions, both on the weld toe of the chord side and on the weld toe of the brace side. In brace 2 the maximum stress occurred at the lower saddle for the weld toe attached to the chord and also the weld toe attached to the brace. Meanwhile, for brace 3 the maximum stress occurred at the left crown position for both on the weld toe attached to the chord and to the brace.

4. Conclusion

From the results of the structure modelling, analysis, and discussion that have been done, the conclusions of the study are:

- The HSS of about 145.70 MPa occurred around the saddle area of the Brace 1 which is exposed to a fairly large axial tension force.
- This Brace 2 is loaded by a relatively small axial tension force. The maximum stress of about 23 MPa occurred at a position of 60 degrees that slightly shifted from the crown area.
- Among all braces, Brace 3 is loaded by a largest axial compression force. The maximum stress of about 123 MPa occurred at a crown position of 180 degrees on the chord side.

References

[1] Herawati I 2016 *Teknik Produksi Minyak dan Gas* (Medan)

[2] Chandran A and S A 2016 Static Strength Analysis of a Tubular K- Joint of an Offshore Jacket Structure *Int. J. Sci. Res.* **5** 708–13

[3] Prastianto R W, Hadiwidodo Y S, Fuadi I F, Prawira Naibaho M B Y and Rahardianto P N 2018 Stress Concentration Factor Distribution of Inclined Brace in Multiplanar Offshore Tubular Double KT Joints *IOP Conf. Ser. Earth Environ. Sci.* **162** 012049

[4] Kautsar R Y G, Prastianto R W, Rifqy M J and Farhana R F 2022 Analysis of a Lower Deck Rigging Configuration and Reusable Padeye Design based on Finite Element Model for Improving Stacking Operation Efficiency *IOP Conf. Ser. Earth Environ. Sci.* **1081** 012043

[5] Bird E and Lewis N 2015 *Beach Renourishment* (Melbourne: Springer)

[6] ANSYS Inc 2020 *Mechanical User ' s Guide* vol 15317

[7] Efthymiou M 1988 Development of SCF formulae and generalized influence functions for use in fatigue analysis *OTJ'88* (Surrey (UK)) p 46

[8] AWS D1.1 2010 *Structural Welding Code Steel*

[9] Ahmadi H and Alizadeh Atalo A 2021 Geometrical effects on the degree of bending (DoB) of multi-planar tubular KK-joints in jacket substructure of offshore wind turbines *Appl. Ocean Res.* **111** 102678

[10] Syalsabila F, Prastianto R W, Hadiwidodo Y S, Adaalah A H and Syarifudin M R 2022 Structural Analysis of Floating Net Cage Bracket in Current and Wave *IOP Conf. Ser. Earth Environ. Sci.* **972** 012017

[11] Logan D L 1987 *A First Course in the Finite Element Method* vol 3

IOP Conf. Series: Earth and Environmental Science 1166 (2023) 012037 doi:10.1088/1755-1315/1166/1/012037

Strength analysis of air cooler support frame (ACSF) platform due to corrosion

D M Rosyid[1], R W Prastianto[1], E L Wahda[1], Murdjito[1], K Sambodho[1], and M R Syarifudin[1]
[1]Department of Ocean Engineering, Institut Teknologi Sepuluh Nopember, Surabaya, Indonesia

rudiwp@oe.its.ac.id

Abstract. Design of an Air Cooler Support Frame (ACSF) platform installed at waters area must be sustainable and capable to withstand the environmental loads. In the estuary, corrosion due to seawater accumulates around the platform legs leads to reduce the strength of the structure. Therefore, this study was conducted to predict the platform's strength under critical corrosion conditions during its working life. The analysis was carried out using the Finite Element Method. In the global analysis, the operating and environmental loads were applied. Then, various patterns in term of length and intensity of corrosion were adjusted to the corrosion zone's elevation based on calculations from inspection data to determine the pattern that has the most significant effect to the platform's strength. The most critical pattern of the corrosion was used in the model for local stress analysis. The results indicated that the appropriate corrosion model was applied to the ACSF Platform has a corrosion length pattern of about half of the leg length with an intensity of 40%. Also, the strength of the ACSF platform decreases after corrosion occurs. The stress in the corroded leg has exceeded the material allowable stress but still undergoes acceptable deformation.

1. Introduction

A fixed offshore platform is designed with an operating life according to the purpose and function of the platform. Generally, the operating life of the platform is approximately 20 years. However, some platforms have passed their operating life and have extended their lifespan. Before the platform is decided to extend its operating period, it is necessary to carry out inspection and analysis to determine the condition and feasibility of the platform qualitatively and quantitatively.

During the operation period, naturally, the platform will experience a decrease in performance to support the loads it receives. The decline in platform performance is caused by an anomaly or damage to the platform structure. Such damage can occur due to corrosion, environmental conditions, fatigue, impact, or heavy objects falling. Therefore, an important inspection is carried out in order to qualitatively determine whether there is any damage that affects the feasibility of the platform for an extended operating life. Then from the inspection results, further analysis is carried out to review the strength of the platform quantitatively due to the damage.

A decrease in the performance or strength of the platform can be quantified by stress. A structure fails to meet the strength criteria when the stress induced by loading is greater than the capacity of the structural material to withstand the load without breaking. Stresses are analyzed to ensure that the satisfactory structural performance of the components under operating conditions can be estimated. In global analysis platform strength is measured by the ratio of actual stress per allowable stress or called UC (Unity Check). In local analysis, it can be seen the stress distribution that occurs in the structural members specifically, so that it can be seen the points on the structure that have the potential to be more severely damaged. Degradation will lead to a reduction in the performance of the structure so that the structure cannot meet basic serviceability and safety requirements before the design life ends. If analyzed by dynamic load, the ratio of residual strength of the structure that has been corroded for 3 years after inspection is 2.962 [1].

Figure 1. The ACSF Platform

The Air Cooler Support Frame (ACSF) functions to support the Air Cooled-Heat Exchanger (ACHE) facility or also called fin-fan cooler for processing oil and gas processing activities as shown in Figure 1. Located in the Mahakam River Delta, Samarinda, East Kalimantan, the ACSF is an unmanned platform. Usually, the leg of platform consists of tubular members and the deck consists of beam girders [2]. While at the legs of ACSF, the beam girder is attached for the lowest deck. Design of an air cooler support frame platform installed at waters area must be sustainable and capable to withstand the environmental loads. The inspection carried out on ACSF in 2020 provided information that there was corrosion on the main structural members in the splash zone area, namely the legs and girders. However, from these inspections no further analysis has been carried out to determine the strength of the platform due to corrosion, especially local analysis. Based on the severity and intensity of corrosion, a decrease in structural strength can occur as a result of the resulting material degradation [3].

In this paper, ultimate strength of locally pitted tubular members under axial compression is numerically studied, based on novel models constructed via a modelling method [4]. Localized corrosion pits are randomly introduced on the local surface of the model of each pitted member, forming patch corrosion with various corrosion features varied in the location, size (length and width) and shape. Before performing extensive nonlinear analyses, these models are validated against benchmark column tests. Stochastic simulation is utilized to simulate the random nature of localized pitting. Taking into account the variation of corrosion features, localized pitting effect on compressive behaviour and ultimate strength is systematically studied. A unified empirical formula is proposed to predict the ultimate strength of locally pitted tubular members.

2. Theory

2.1. Uniform corrosion

Corrosion is the degradation that occurs on the surface of a material due to a chemical or electrochemical reaction. Corrosion of materials consists of corrosion of metals and non-metals, which can be defined as damage, deterioration, and degradation of materials due to the effects of chemical or electrochemical reactions in the environment. Based on the form of damage from metal corrosion, corrosion can be classified into two categories, namely general corrosion and local corrosion. General corrosion is corrosion that occurs on all surfaces of metal components and is connected together. Local corrosion is corrosion that occurs only on a small part of the metal surface, while other parts are not corroded [5]. Uniform corrosion is one of the most characteristic corrosions that occurs in metals. Uniform corrosion can result in massive metal material loss. The destructive behavior of this corrosion is the gradual thinning of the metal material until it finally fails.

2.1.1. Corrosion zone

The corrosion zone is divided into four zones, including the atmospheric zone, splash zone, submerged zone, and mud zone (buried in soil) with different corrosion rate standards as shown in Table 1 [6]. The mud zone is the part of the structure that is below the lowest end of the immersed zone and is buried in the ground or seabed. Submerged zone is the part of the structure that is below the lowest end of the splash zone and is always below sea level. Splash zone is part of the structure that intermittently enters or leaves sea water during the service life of the structure. Tides and winds are often the cause of wetting this section. The range of the zone varies with sea level rise and fall due to daily tidal cycles in a particular geographic area. The atmospheric zone is the part of the structure above the splash zone, and by definition this zone is not wetted or affected by rising tidal waves.

Table 1. Corrosion rate in each zone [6]

Zone	Corrosion Rate (mm/yr)
Mud zone	0.06 – 0.10
Submerged zone	0.10 – 0.20
Splash zone	0.20 – 0.40
Atmospheric zone	0.05 – 0.08

2.1.2. Corrosion rate

Corrosion rate is the amount of corrosion that occurs in a unit time. The corrosion rate is expressed as a decrease in material weight (weight loss) per unit area per unit time [7], where k is corrosion rate, (mm/yr), w is weight loss (g), A is surface area (mm^2), and ρ is density (g/cm^3).

$$k = \left(\frac{w \times 10^4}{A}\right)\left(\frac{0.365}{\rho}\right) \sigma_{\perp W} = 1.4\sigma_{\perp E1} - 0.4\sigma_{\perp E2} \quad (1)$$

The corrosion rate can be used to estimate the operating life by dividing it by the wall thickness of the material/equipment [8].

$$Estimated\ service\ lifetime = \frac{Equipment\ wall\ thickness}{Corrosion\ rate} \sigma_{\perp W} = 1.4\sigma_{\perp E1} - 0.4\sigma_{\perp E2} \quad (2)$$

2.2. Allowable Stress

Allowable stress is the maximum limit stress that is the reference for designing structural members. The allowable stresses shall not be exceeded by any part of the structure. The allowable stress is made significantly lower than the ultimate strength obtained from the static test. Based on API RP 2A-WSD 22nd Ed [9] the allowable stresses for cylindrical members as show in Eq. 3 – Eq.5, where σ_y is yield stress (MPa), E is Modulus Young (MPa), K is effective length factor, l is unbraced length (mm), and r is radius of gyration (mm).

Allowable Stresses for Axial Tension:

$$\sigma_t = 0.6\sigma_y \sigma_{\perp W} = 1.4\sigma_{\perp E1} - 0.4\sigma_{\perp E2} \quad (3)$$

Allowable Stresses for Axial Compression:

$$\sigma_a = \frac{\left(1 - \frac{(Kl/r)^2}{2C_c^2}\right)\sigma_y}{\frac{5}{3} + \frac{3(Kl/r)}{8C_c} - \frac{(Kl/r)^3}{8C_c^3}} \text{ for } Kl/r < C_c \sigma_{\perp W} = 1.4\sigma_{\perp E1} - 0.4\sigma_{\perp E2} \quad (4)$$

$$\sigma_a = \frac{12\pi^2 E}{23(Kl/r)^2} \text{ for } Kl/r \geq C_c \sigma_{\perp W} = 1.4\sigma_{\perp E1} - 0.4\sigma_{\perp E2} \quad (5)$$

$$C_c = \left(\frac{2\pi^2 E}{\sigma_y}\right)^{1/2} \sigma_{\perp W} = 1.4\sigma_{\perp E1} - 0.4\sigma_{\perp E2} \quad (6)$$

2.3. Equivalent Stress

The stress state can be comprehensively described by the 3×3 tensor. The way that we can use a single value to represent the state of a material is with the equivalent stress or Von Mises stress. In evaluating the stress results, 9 contour plots are needed to visualize the stress tensor, the equivalent stress allows one to view the structural stresses with a single plot as shown in Figure 2. Equivalent stresses are widely used to represent material states for ductile materials. Engineers use these simple scalar values to determine whether a material has yielded or failed.

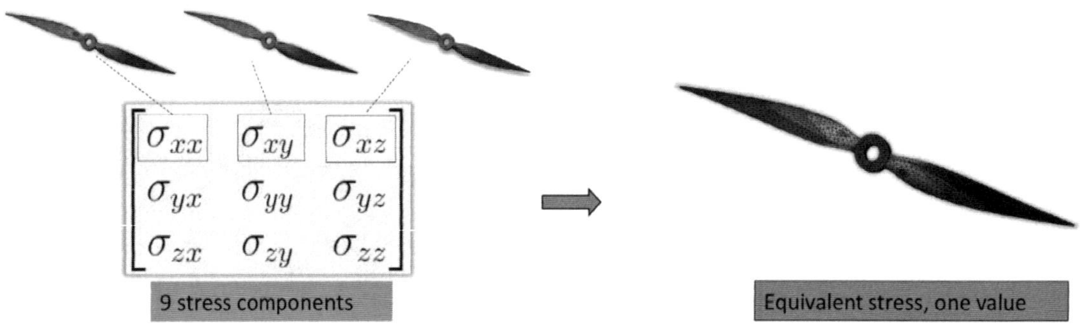

Figure 2. Equivalent stress

Von Mises stresses are best described in a principal coordinate system, due to easier visualization. Another important feature of the Von Mises surface is that the cylinder height is not limited, which means that both ends extend indefinitely. This means that no matter how large the values of the three principal voltages are, the Von Mises voltage can remain the same. On one Von Mises surface, all points have the same Von Mises stress value.

3. Methodology

3.1. Data collection

3.1.1. Structural Data

The ACSF platform used for this study has 47 m × 21.5 m in dimension, has 2 decks and 8 legs with different pile penetration for each leg. The structural data is summarized in Table 2. The platform legs have ASTM A36 material type with material properties is presented in Table 3. The platform experiences structural loading that includes equipment loads.

Table 2. The ACSF platform description

Description	Proportions	
Number of Legs	8 (Ungrouted Legs)	
Deck Leg Diameter	30" (Row 1) & 30" (Row 2)	
	Row 1	Row 2
	46.6 m	47.5 m
Pile Penetration	48.4 m	53.2 m
	45.5 m	47.3 m
	45.0 m	46.5 m
Deck Leg Spacing	16.25 m × 21.5 m (Center to center of leg)	
Leg Batter	No Batter	
	(+) 12.72 m	
Pile Plan Elevation	(+) 6.22 m	
	(+) 4.80 m (Stabbing point)	
	(-) 3.00 m (Mudline)	
Deck Elevation	Upper Deck - EL. + 12.72 m	
	Lower Deck - EL. + 6.22 m	
Working Point	(+) 4.80 m	
Splash Zone	Between ±1 m from MSL	

Table 3. Properties of material

Material	ASTM A36
Yield Strength (σ_y)	250 MPa
Ultimate Strength (σ_{ult})	460 MPa
Young Modulus (E)	200 GPa
Poisson's Ratio (U)	0,3
Fracture Toughness (K_{IC})	90 ksi.in^{0,5}
Density (ρ)	7800 kg/m^3

3.1.2. Environmental data

The ACSF platform is located at a depth of 6 meters in the Mahakam River Delta, Samarinda, East Kalimantan at latitude of 0°34'45.38"LS and longitude of 117°22'54.85"E. Because it is located in a river, the dominant environmental load are only current and wind loads. Table 4 and Table 5 show the current and wind data. Current and wind conditions are steady, and its profile is assumed shear.

Table 4. Current speed data

Elevation (m)	Current Speed (m/sec)	
	1 year	100 years
0	2.0	3.0
100	2.0	3.0

Table 5. Wind speed data

Wind associated with wave velocity	Wind Speed (m/sec)	
	1 year	100 years
3 second gust	29.3	36.3
1 minute mean	23.2	20.1

3.1.3. Inspection data

Table 6 shows Inspection data in 2020 with total of 5% corrosion at the legs. Corrosion is assumed to occur on all legs and is uniform with an unknown location specifically. Figure 3 is an example of the condition of the leg based on inspection.

Table 6. Inspection data of the ACSF platform.

Above Water Visual Inspection of Main Structure		
Main Structure Component	Inspection Findings at Zone	
	Level-1	Level-2
Frames (Column, Girder & Beam)	Light corr. on leg (5%)	Light corr. on girder (3%)
Deck (Plate & Grating)	Good condition	Rusty on grating (2%)
Stairs & Ladder	Rusty on stairway (5%)	Light corr. on stairway (2%)
Handrail	Rusty on handrail (5%)	Rusty on handrail (2%)

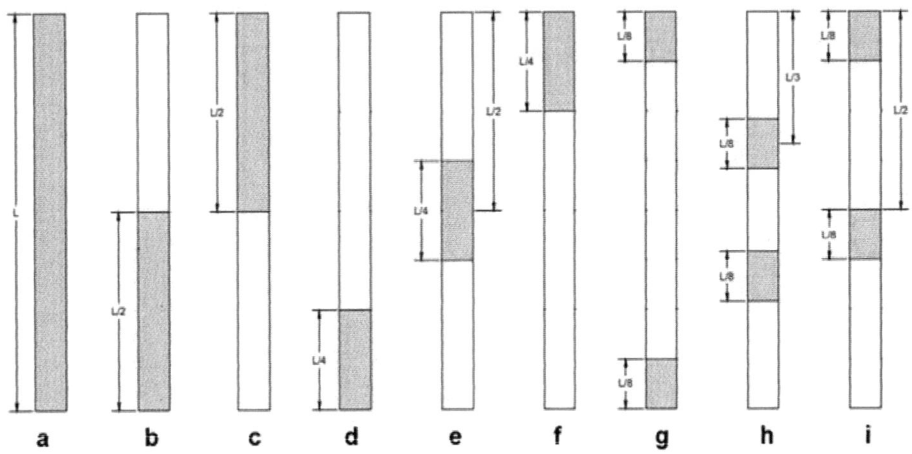

Figure 3. Corrosion length patterns

3.2. Global Simulation

The ACSF platform is modeled using finite element software. The platform model that has been made is validated by comparing the weight of the original structure (from the data) and the weight of the modeled structure with the error criterion that occurs is 5%. If the error is more than 5% it must be remodeled. This is so that the modeled platform structure can represent the actual structure. After the platform model meets the validation criteria, In-place static analysis is then carried out. In-place analysis was carried out for 2 conditions, namely:

a. No Corrosion

The platform is modeled without any corrosion. In-place analysis aims to determine whether the UC (Unity Check) of the structure already meets the criteria of API-RP-2A-WSD [9]. The parameter that determines the security of the platform is the stress ratio or commonly referred to as UC (Unity Check). The stress ratio or UC is defined as the actual stress per allowable stress. If the UC value > 1 then the member is declared unsafe. Through the UC value the critical members of the platform structure are known.

b. With various corrosion

In-place analysis was performed on the platform leg for nine variations of the corrosion length pattern by Wang [4] (Figure 3) with seven variations of corrosion intensity (5%, 20%, 30%, 40%, 50%, 60%, 70%) to determine the most critical corrosion length pattern. The leg with the most critical pattern will then proceed to the local analysis stage. For this strength analysis of existing fixed platforms, corrosion will be modeled by the decrease of thickness and the size of corroded area on the structural members

[10]. Corrosion is modeled adjusted to the corrosion zone: atmospheric zone, splash zone, submerged zone and mud zone. Mass loss and corrosion rate are assumed to be solely under the influence of uniform corrosion, thus providing insight into the harshness of the metal environment [6].

Corrosion rate is calculated based on inspection data as shown in Table 7. The corrosion rate is assumed to be the corrosion rate in the splash zone. Corrosion rates were then interpolated for each corrosion zone at seven variations of corrosion intensity.

Table 7. ACSF *Platform* Corrosion Rate

Orig. outer diameter	762.00	mm
Orig. wall thickness	31.75	mm
Corrosion allowance	6.40	mm
Corrosion in 2020 (5%)	1.59	mm
Corrosion rate	0.08	mm/yr

3.3. Local Simulation

The critical leg obtained from inplace analysis for structural conditions in the presence of corrosion was then modeled locally by numerical modeling using finite element software. The geometry dimension is shown in Table 8 while the boundary conditions or applied loads as shown in Table 9. The legs without corrosion and the legs with corrosion were modeled with Solid 187 elements. Solid elements provide a more accurate and detailed stress distribution around chord-brace junctions [11]. The legs with corrosion were modeled with the pattern of the critical length of corrosion and the intensity of corrosion which still met the corrosion allowance.

Table 8. Dimension of Local Model

Dimensions	Chord (mm)	Brace A (mm)	Brace B (mm)	Beam (mm)
Outer diameter	762.00	406.40	304.80	
Wall thickness	31.75	19.05	19.05	
Length	1,272.00	6,000.00	3,850.00	4,654.00

The analyzed output of the platform leg local analysis is based on the equivalent stress (Von Mises stress). The equivalent stress output is obtained from running finite element software with the boundary conditions applied to the specimen parts of the platform legs in the form of axial force obtained from global analysis. Fixed supports are applied to the upper and lower ends of the chord. Mesh sensitivity analysis aims to get the right meshing size in numerical modelling so that the resulting output is accurate.

Table 9. Local model boundary condition

Member	Axial Force (kN)		Behavior
	Corrosionless	Due to Corrosion	
A-left	280,95	282,32	Tension
A-right	139,25	143,55	Tension
B-left	-591,70	-589,14	Compression
B-right	-429,56	-429,13	Compression
Rear beam	23,97	27,41	Tension

4. Results and Discussion

4.1. Validation

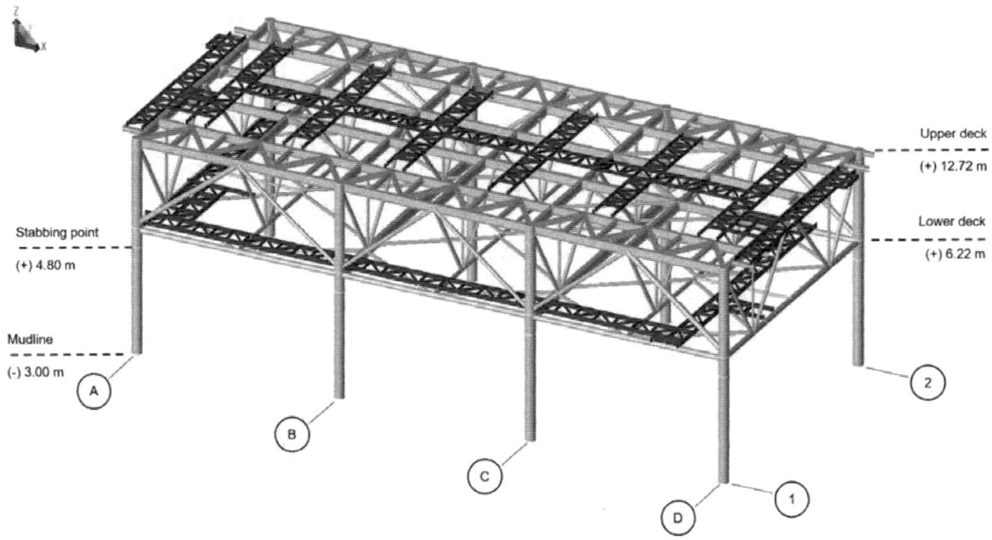

Figure 4. Global model of the platform

Global modelling for the validation is shown in Figure 4. The error value is obtained by calculating the difference in the weight of the structure and then compared with the weight of the structure in the data. Platform weight on report is 2835.69 kN, while the weight of the modeled platform is 2963.57 kN. Table 10 shows the model error of 4.51% which does not exceed the 5% validation criteria., then the model is declared valid and represents the real platform. After the platform model under normal conditions has been declared valid, the model can be modified by applying corrosion variations.

Table 10. Model validation

Structure Weight (kN)		Error (%)
Data	Model	
2835.69	2963.70	4.51%

4.2. Global Analysis

4.2.1. No corrosion

Table 11 presents the maximum UC of ACSF Platform. Based on the in-place analysis that has been carried out, the maximum UC is found in the W8 beam member, which is 0.74, then the platform is declared safe. Meanwhile, Table 12 presents the UC of the platform legs. Based on the results, the maximum UC value for the platform leg is 0.21 for the leg B1. Then based on Table 13 joint maximum deflection of the leg still meets the criteria D H/200.

Table 11. Maximum UC

Group	Member	Max. UC	Check UC < 1,00
W8	0608-0209	0.74	OK
10A	0482-0057	0.71	OK
MC8	0386-0237	0.65	OK

Table 12. UC of Legs

Member Group	Platform Leg UC							
	A1	B1	C1	D1	A2	B2	C2	D2
DL	0.14	0.21	0.19	0.13	0.13	0.20	0.20	0.13

Table 13. Joint Maximum Deflection of Legs

Joint	Defl (X) (cm)	Defl (Y) (cm)	Defl (Z) (cm)	H/200 (cm)	Check D ≤ H/200
205L	-0.626			7.86	OK
306L		-0.752		7.86	OK
302L			-0.285	7.86	OK

4.2.2. With various corrosion

Different zone in a leg as shown in Figure 5 modeled with various corrosion pattern. Stress ratio in in place platform analysis without and in the presence of corrosion shows different results. However, the majority of the maximum UC is on the leg B1, so it can be concluded that the leg B1 is the most critical leg and will proceed to local analysis. Based on the graph shown Figure 6, the corrosion patterns a, c, f, g, and i tend to form an exponential curve. This shows that as the intensity of corrosion increases, the UC value of the platform foot will increase. Meanwhile, the corrosion patterns b, d, e, and h curves tend to be stable, not showing any significant changes. This shows that regardless of the intensity of corrosion that is applied to the pattern will not increase or affect the UC of the platform legs.

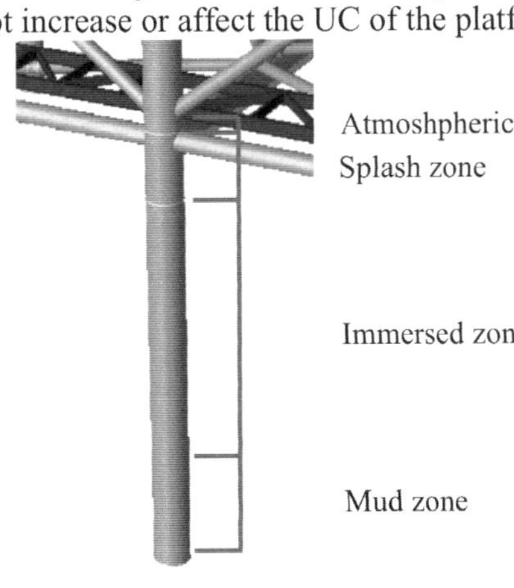

Figure 5. Corrosion zone

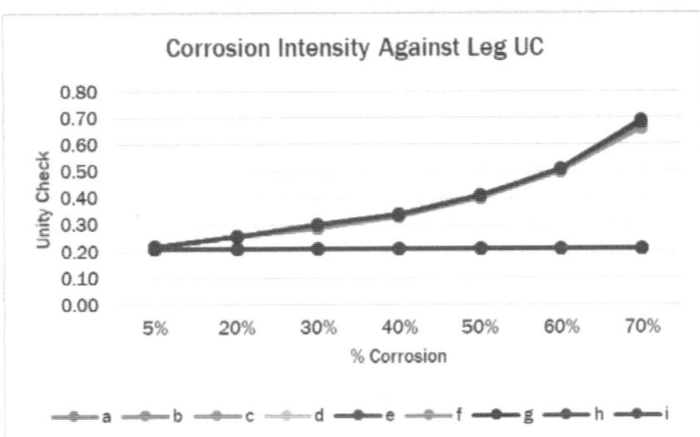

Figure 6. Chart of corrosion intensity against leg UC

The most extreme curve of the nine corrosion patterns is the pattern c curve, so with this pattern c will be chosen as the corrosion length pattern to be applied to the local geometry of the leg. Corrosion length pattern c is the most critical pattern affecting leg strength compared to the other eight patterns. Due to the chosen pattern c, the calculation of the accumulation of corrosion that occurs at each corrosion intensity is carried out. Based on Table 14 selected corrosion intensity 40%. The submerged zone is a zone that is difficult to protect from the risk of corrosion [12]. Therefore, an intensity of 40% was chosen because the corrosion in the immersed zone is still not far from the corrosion allowance limit, which is 6.65 mm. In addition, the 40% intensity is still below the corrosion allowance limit when pattern c is applied, which is 4.20 mm.

Table 14. Accumulation of corrosion along the leg

Zone	Leg Length (mm)	5% (mm)	20% (mm)	30% (mm)	40% (mm)	50% (mm)	60% (mm)	70% (mm)
Atmospheric zone	200	0.31	1.26	1.89	2.52	3.15	3.78	4.41
Splash zone	1520	1.59	6.35	9.53	12.70	15.88	19.05	22.23
Submerged zone	4800	0.83	3.33	4.99	6.65	8.32	9.98	11.64
Mud zone	3000	0.48	1.92	2.87	3.83	4.79	5.75	6.70
Accumulation along leg (mm)		0.83	3.32	4.98	6.64	8.30	9.96	11.62
Accumulation along pattern-c (mm)		0.53	2.10	3.15	4.20	5.26	6.31	7.36

4.3. Local Analysis

4.3.1. Meshing sensitivity analysis

The leg B1 as a chord is modeled with a T brace (Brace A), Y brace (Brace B), and a beam connected to the 202L joint. Leg B1 member cutting refers to with 6 x OD [13]. Both Brace A, Brace B, and Beam are modeled to surface on chords.

Figure 7 shows that the probe stress fluctuates from an increase of 128000 to 158000 elements with the highest error occurring at 2.23%. The probe stress when the number of elements 138000 and 148000 is small, it can occur because the density of the elements in the area around the probe point is too tenuous and the quality of the meshing of the elements is not good, causing the results to be much different from the others. The graph starts to hit when it rises from 158000 to 217000 with the highest error of 0.14%, then one of the criteria for the number of meshing elements among them can be used. After going through the considerations, the model with the number of elements is 207196. Meanwhile, in Figure 8, the stress probe fluctuates from an increase of 78000 to 98000 elements with the highest error occurring at 2.11%. The chart starts to slope as it rises from 98000 to 168000 with the highest error of 0.24%. After going through the considerations, the model with the number of elements is 148592.

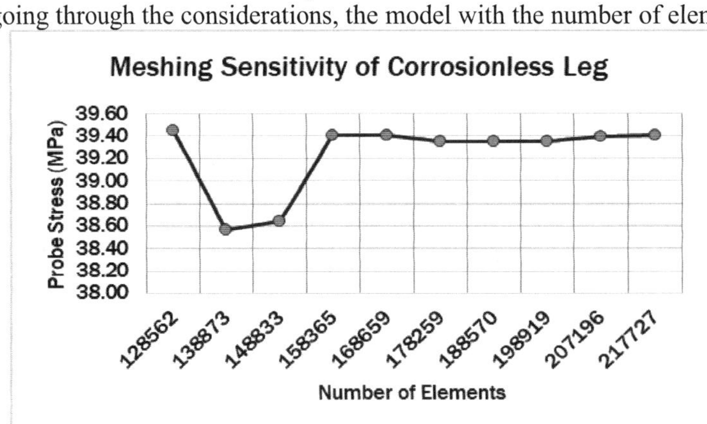

Figure 7. Meshing sensitivity chart of corrosionless leg

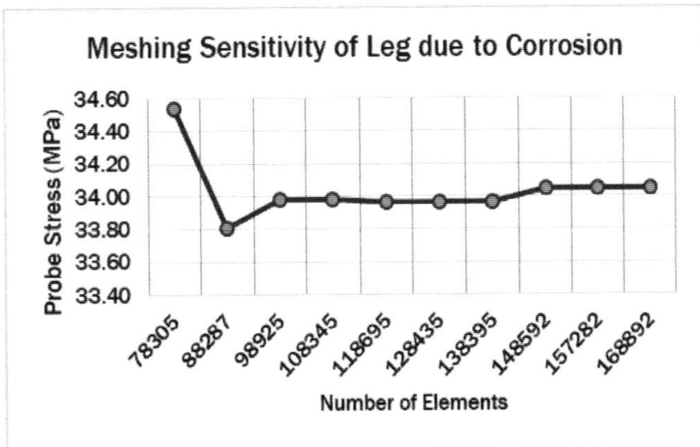

Figure 8. Meshing sensitivity chart of leg due to corrosion

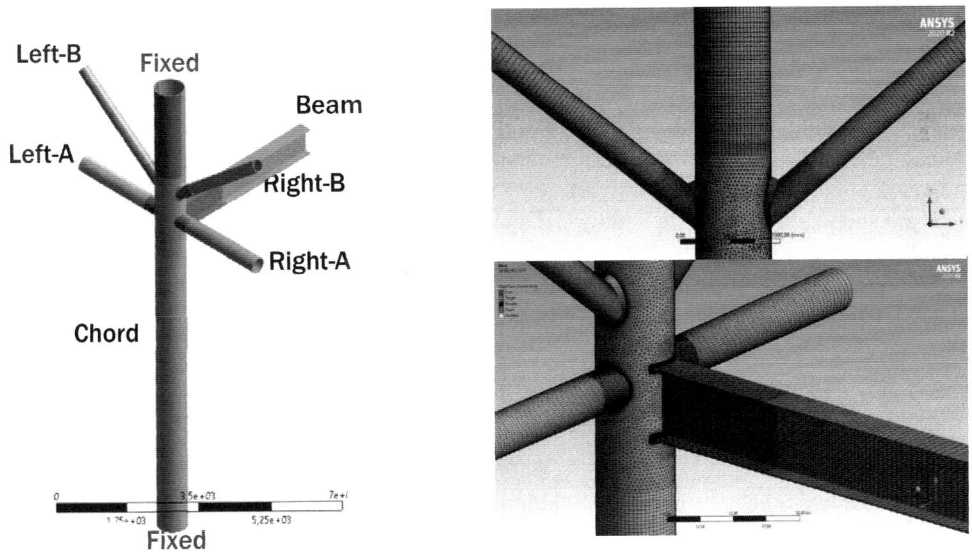

(a) (b)

Figure 9. Model (a) Local model of the leg (b) Leg meshing

On the criteria for the number of elements, the leg model is meshed with the largest element size being 35 mm, while the welded part is meshed with the largest element size being 10 mm. Meshing on the leg with Brace A is a tetrahedron, while the other part is a hexahedron. On the leg where there is corrosion, meshing tetrahedron is applied to the splash zone to the atmospheric zone. The mesh size is made bigger the farther away from the weld area, as shown in Figure 9.

4.3.2. Stress distribution

Based on Figure 10, the highest maximum stress on corrosionless leg is found in the Left-Brace B specimen, which is 153.88 MPa with the location shown in Figure 12. The tension experienced by the Left-Brace B is the highest that can occur because the axial load given to it is the largest among other leg specimens. Then Right-Brace B with 117.67 MPa, Beam with 83.86 MPa, Left-Brace A with 66.74 MPa, and Right-Brace A with the lowest maximum stress of 44.51 MPa. However, the focus of this analysis is on the Chord or the foot of the platform itself. The stress that occurs in the Chord is 119.67 MPa.

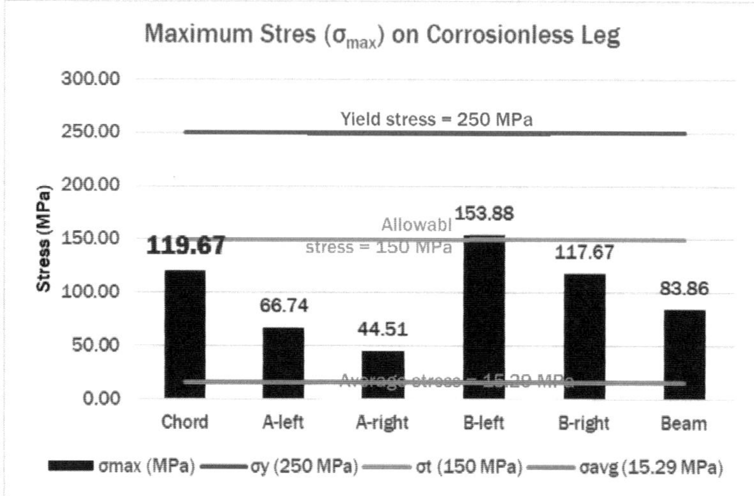

Figure 10. Chart of maximum stress on corrosionless leg

The stress that occurs on the leg (Chord) when there is no corrosion is still safe. This is because the stress that occurs is still below the allowable stress limit of 150 MPa based on the criteria of API RP 2A-WSD 22nd Ed [9]. with a safety factor of 0.6 yield stress. In addition, the stress that occurs on the Chord without corrosion is still below the ASTM A36 material yield stress limit of 250 MPa. So that the legs are still in a proportional condition, have not experienced elastic deformation. Chord specimens are not applied axial loads, but fixed support at both ends. However, Chord (leg) receives axial loads from other specimens with tension and compression behavior to cause deformation as shown in Figure 13. The maximum deformation at the foot of the platform without corrosion and cracking occurs in the Left Brace A which is 1.79 mm.

Based on Figure 11, the highest maximum stress in the leg with the presence of corrosion is found in the Left-Brace B specimen, which is 197.61 MPa, with the same location as the corrosionless leg. Then Right-Brace B with 146.00 MPa, Right-Brace A with 130.03 MPa, Left-Brace A with 91.57 MPa, and Beam with the lowest maximum stress of 74.72 MPa. The stress that occurs in the Chord is 191.09 MPa. These results indicate that the maximum stress that occurs in the foot with corrosion is higher than the foot without corrosion. The leg stress is increased by 1.6 times the corrosionless leg stress. Therefore, it can be stated that corrosion affects the stresses experienced by the platform legs.

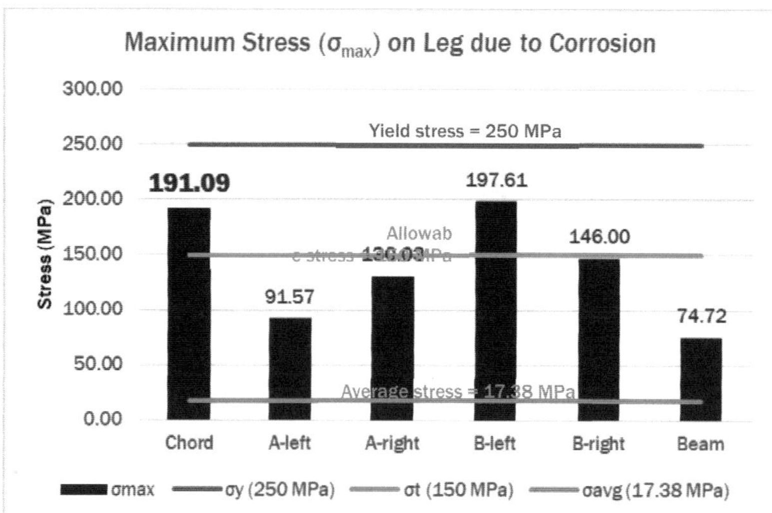

Figure 11. Chart of maximum stress on leg due to corrosion

The stress that occurs in the leg (Chord) after corrosion is not safe. This is because the stress that occurs has exceeded the allowable stress limit of 150 MPa based on the criteria of API RP 2A-WSD 22nd Ed [9]. with a safety factor of 0.6 yield stress. However, the stress that occurs in the leg (Chord) due to corrosion is still below the yield stress limit of ASTM A36 material, which is 250 MPa, so that the deformation experienced is still in the form of elastic deformation. Elastic deformation is a deformation or change in shape caused by the application of a load, where if the load is removed, the shape and size will return to its original shape or the deformation that occurs will be lost. The maximum deformation that occurs after corrosion is 2.58 mm on the Left-Brace A.

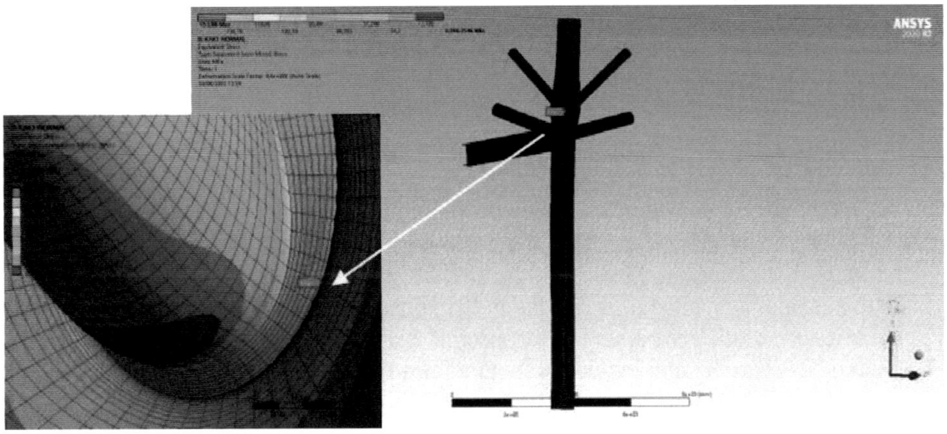

Figure 12. Maximum stress location

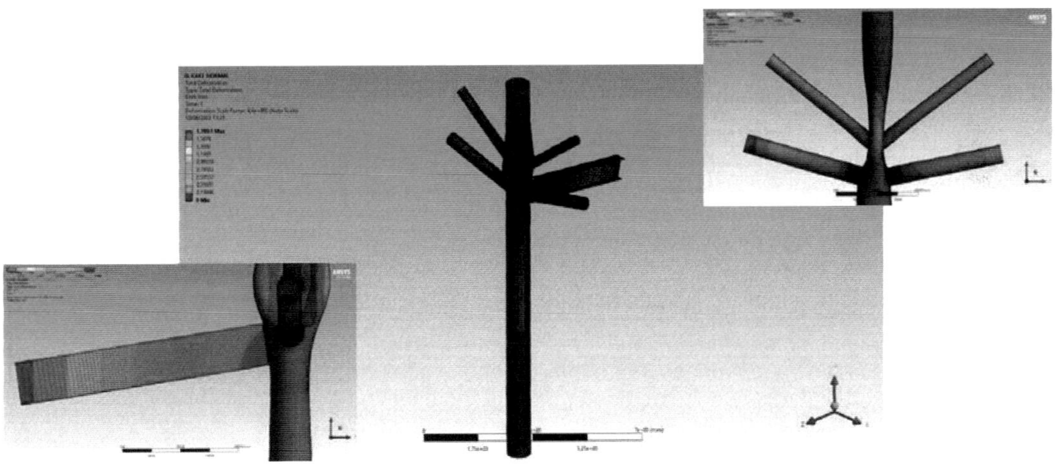

Figure 13. Leg deformation

5. Conclusion

The correct corrosion length pattern modeling based on inspection data to be applied to the ACSF Platform is a corrosion length pattern of c or half of the leg length with an intensity of 40%. Corrosion length pattern c is the most critical pattern affecting leg strength compared to the other eight patterns. While the 40% intensity was chosen because the corrosion in the submerged zone which is a zone that is difficult to protect from corrosion risk is within the corrosion allowance limit, besides that the 40% intensity is still below the corrosion allowance limit when pattern c is applied. The strength of the ACSF Platform is reduced after corrosion occurs. The stress occurred in the corroded leg has increased 1.6 times has exceeded the material allowable stress but still undergoes elastic deformation.

References

[1] Deviany M A D, Hadiwidodo Y S and Wisudawan A 2018 *Analisis Kekuatan Sisa Struktur Bangunan Lepas Pantai Terpancang yang Terkorosi* (Institut Teknologi Sepuluh Nopember)

[2] Giri Kautsar R Y, Prastianto R W, Rifqy M J and Farhana R F 2022 Analysis of a Lower Deck Rigging Configuration and Reusable Padeye Design based on Finite Element Model for Improving Stacking Operation Efficiency *IOP Conf. Ser. Earth Environ. Sci.* **1081** 012043

[3] Yu W 2009 Corrosion Effects on the Ductile Fracture , Strength and Reliability of Membranes ,Plates and Shells

[4] Wang R, Guo H and Shenoi R A 2020 Compressive strength of tubular members with localized pitting damage considering variation of corrosion features *Mar. Struct.* **73** 102805

[5] Huang Y and Zhang J 2018 *Materials Corrosion and Protection* (Shanghai: Shanghai Jiao Tong University Press)

[6] Khodabux W, Causon P and Brennan F 2020 Profiling corrosion rates for offshorewind turbines with depth in the North Sea *Energies* **13**

[7] Uhlig H H 2011 *Uhlig's Corrosion Handbook* (United Kingdom: Wiley)

[8] Tait W S 2018 *Controlling corrosion of chemical processing equipment* (Elsevier Inc.)

[9] API RP 2A-WSD 22nd Ed. 2014 Planning, Designing, and Constructing Fixed Offshore Platforms - Working Stress Design *API Recomm. Pract.* 324

[10] Hadiwidodo Y S and Prastianto R W 2018 Structural corrosion modeling for reserve strength assessment of fixed jacket platform *Int. J. Civ. Eng. Technol.* **9** 1006–12

[11] Jiang Y, Yuan K and Cui H 2018 Prediction of stress concentration factor distribution for multi-planar tubular DT-joints under axial loads *Mar. Struct.* **61** 434–51

[12] Derrington J 2014 Offshore Structures *Corrosion Control for Offshore Structures* pp 57–88

[13] Efthymiou M 1988 Development of SCF formulae and generalized influence functions for use in fatigue analysis *Recent Developments in Tubular Joints Technology*

A study into the effect of tip clearance on the performance of b-series and kaplan-series ducted propellers

B A Adietya[1], W D Aryawan[1], I K A P Utama[1]

[1]Department of Naval Architecture, Institut Teknologi Sepuluh Nopember, Surabaya, 60111, Indonesia

kutama@na.its.ac.id

Abstract. A study into the effect of tip clearance on ducted propellers was carried out. Two series of propellers, namely B-series and Kaplan-series, were used. The gap between tip of propeller and ducted was varied from 2mm to 5mm. The study was conducted numerically using computational fluid dynamics (CFD) approach. The solver is based on the Reynolds-Averaged Navier-Stokes (RANS) solutions and two turbulence modelling, i.e. k-ω SST by Menter and explicit algebraic stress model (EASM), were applied in order to find out the optimum results. This paper objective to discuss energy efficiency on ships by investigating variations of propeller tip clearances. Overall results showed that EASM demonstrated better score (particularly in term of thrust and efficiency) than the k-ω SST. The highest thrust, torque and efficiency were achieved at 3mm tip clearance for both series. Moreover, Kaplan-series showed better results than the B-series propeller indicating that the use of ducted propeller using Kaplan-series is more appropriate.

Keywords: tip clearance; ducted propeller; B-series; Kaplan-Series; turbulence modelling; energy efficiency.

1. Introduction

Following the previous research on ducted propeller compared with the conventional propeller, ducted propellers are very useful to provide advanced operability and higher bollard thrust to support marine vessels (1). B-series propeller is much better as open propeller (both default and with PBCF), whereas Kaplan series is excellent when fitted with ducted (2). An important factor influencing the accurate prediction of ducted propeller performance is the tip clearance or gap. The blade tip clearance between a blade and the duct has been the major source of many unfavourable flow phenomena. The influence of the tip clearance of a rigid ducted propeller was investigated by the incompressible Navier-Stokes equations with a Multiple Reference Frame (MRF) method (3). The existence of tip clearance has been found to have a significant influence on the efficiency of the machinery (4).

Based on the above information, two types of propellers were investigated, namely the common B4-70 and the Ka4-70 propellers. Both propellers have the same size and variation of tip clearance, i.e.,

IOP Conf. Series: Earth and Environmental Science 1166 (2023) 012038 doi:10.1088/1755-1315/1166/1/012038

2mm, 3mm, 4mm, and 5mm. The type of ducted uses the 19A duct with diameter (D) = 300mm and a pitch-diameter ratio (P/D) = 1.2. The investigation was carried out numerically using computational fluid dynamics (CFD) technique together with the use of Reynolds-Averaged Navier-Stokes Equations (RANSE) and two turbulence modelling, namely k-ω SST from Menter (5) and Explicit Algebraic Stress Model (EASM) (6).

2. Method

2.1. Geometry

Two types of propellers using ducted were shown in Figure 1 and the meshing of the two models were shown in Figure 2. CFD simulation is used to evaluate the propeller performance and the measured in terms included thrust coefficient (K_T), torque coefficient (K_Q), and propeller's efficiency (η).

a) Propeller B4 – 70 (b) Propeller Ka4-70

Figure 1. Propeller with Nozzle 19 A

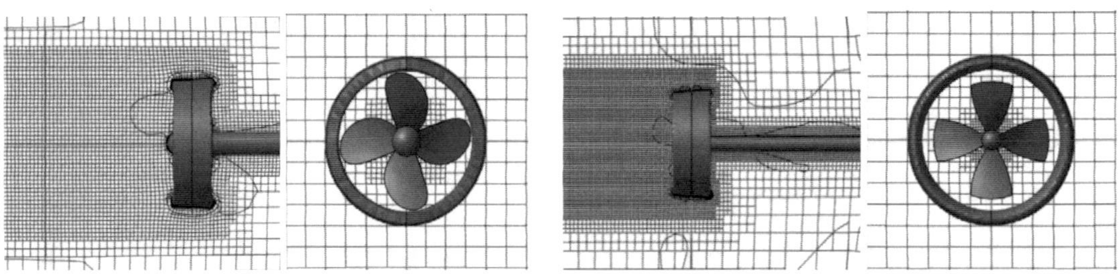

a) Propeller B4 – 70 b) Propeller Ka4-70

Figure 2. Meshing of the models

2.2. Numerical Analysis

The numerical simulation is carried out using continuity equation, Reynolds-Averaged Navier-Stokes Equation (RANSE), and turbulence k-ω SST and EASM to solve the turbulence phenomena. The equations for those three are shown in Equations (1) to (4). Furthermore, the CFD setup and variation of numerical test using CFD are provided in Tables 3 and 4, respectively.

$$\frac{\partial U_i}{\partial x_i} = 0 \quad (1)$$

Where: ρ is fluid density, t is time, U_j is the flow velocity vector field.

$$\frac{\partial U_i}{\partial t} + U_j \frac{\partial U_j}{\partial x_j} = \frac{\partial p}{\partial x_i} + \frac{\partial}{\partial x_j}\left[Re_{eff}^{-1}\left(\frac{\partial U_i}{\partial x_j} + \frac{\partial U_j}{\partial x_i}\right)\right] + S_i \quad (2)$$

Where U_i = (u, v, w) represented Reynolds average velocity components; x_i = (x, y, z) signified the independent coordinate direction; the term Si denoted the mean strain-rate tensor for a body force, the piezometric pressure p, and the Re_{eff} were effective Reynolds numbers.

The Menter's SST model combines the advantages of the *k-ω* model to achieve an optimal model formulation for a wide range of applications. For this, a blending function *F1* is introduced which is equal to one near the solid surface and equal to zero for the flow domain away from the wall. It activates the *k-ω* wall region and the *k-ε* model for residual flow. By this approach, the attractive near-wall performance of the *k-ω* model can be used for the free stream sensitivity.

$$\frac{\gamma}{v_t} P - \beta\rho\omega^2 + \frac{\partial}{\partial x_j}\left[(\mu + \sigma_\omega \mu_t)\frac{\partial \omega}{\partial x_j}\right] + 2(1 - F_1)2\rho_{\omega 2}\frac{1}{\omega}\frac{\partial k}{\partial x_j}\frac{\partial \omega}{\partial x_j} - \left(\frac{\partial(\rho\omega)}{\partial t} + \frac{\partial(\rho u_j \omega)}{\partial x_j}\right) = 0 \quad (3)$$

In term of single-point closures, the Reynolds Stress Transport Model (RSTM) provides the best representation of flow physics. However, the reduction of the tensor basis requires an enormous mathematical effort. Originally most of the explicit algebraic stress model (EASM) are formulated using a 10-term basis:

$$b_{ij} = \sum_{\lambda=1}^{10} G^{(\lambda)} T_{ij}^{(\lambda)} \quad (4)$$

The outputs of the numerical estimation are thrust coefficient (K_T), torque coefficient (K_Q), and hull efficiency (η_0) and the formulations of those items are given on Equations (5), (6), and (7).

$$K_T = \frac{T}{\rho D^4 n^2} \quad (5)$$

$$K_Q = \frac{Q}{\rho D^5 n^2} \quad (6)$$

$$\eta_0 = \frac{T V_a}{2\pi Q n} \quad (7)$$

3. Results and Discussion

Table 1 and 2 showed the results of the B4-70 Propeller CFD calculations. There are four variations of tip clearances, i.e., 2mm to 5mm.

Table 1. CFD Calculation for B4-70 Series using EASM

No	Model	Tip Clearance (mm)	K_T	$10K_Q$	η_0
1	A	2	0.125	0.301	0.53
2	B	3	0.151	0.344	0.56
3	C	4	0.145	0.330	0.56
4	D	5	0.140	0.319	0.56

Table 2. CFD Calculation for B4-70 Series using k-ω SST

No	Model	Tip Clearance (mm)	K_T	$10K_Q$	ηo
1	A	2	0.123	0.295	0.53
2	B	3	0.149	0.339	0.56
3	C	4	0.142	0.324	0.56
4	D	5	0.136	0.311	0.56

In the case B4-70 propeller (see Tables 1 and 2), both turbulence modelling (EASM and k-ω SST) showed that the tip clearance of 3mm possess the highest values of thrust, torque and efficiency. The EASM had higher results than the k-ω SST; the discrepancy between thrust and efficiency is less than 2%.

Table 3 and 4 demonstrated the results of the Ka4-70 CFD calculations. There are four variations of tip clearances, i.e., 2mm to 5mm.

Table 3. CFD Calculation for Kaplan-Series using EASM

No	Model	Tip Clearance (mm)	K_T	$10K_Q$	ηo
1	A	2	0.086	0.257	0.43
2	B	3	0.157	0.377	0.53
3	C	4	0.101	0.284	0.45
4	D	5	0.090	0.634	0.18

Table 4. CFD Calculation for Kaplan-Series using k-ω SST

No	Model	Tip Clearance (mm)	K_T	$10K_Q$	ηo
1	A	2	0.086	0.249	0.44
2	B	3	0.154	0.367	0.53
3	C	4	0.101	0.281	0.46
4	D	5	0.089	0.622	0.18

In the case of Ka4-70 propeller (see Table 3 and 4), the values of thrust, and torque using EASM are higher than k-ω SST with the difference values of thrust and efficiency are less than 2%. The highest value occurs at 3mm tip clearance.

IOP Conf. Series: Earth and Environmental Science 1166 (2023) 012038 doi:10.1088/1755-1315/1166/1/012038

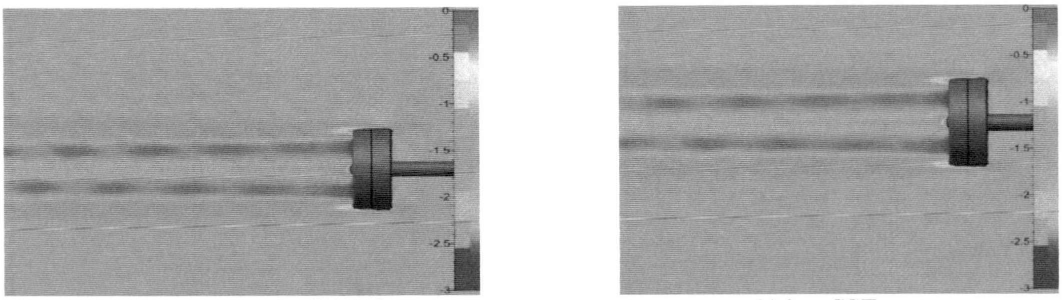

a) EASM b) k-ω SST

Figure 3. B4-70 tip clearance 2mm

It can be seen in Figure 3 for B4-70 propeller, when the tip clearance is 2mm, that the stronger flow behind the propeller is demonstrated by EASM (blue darker), whilst the k-ω SST showed lighter blue.

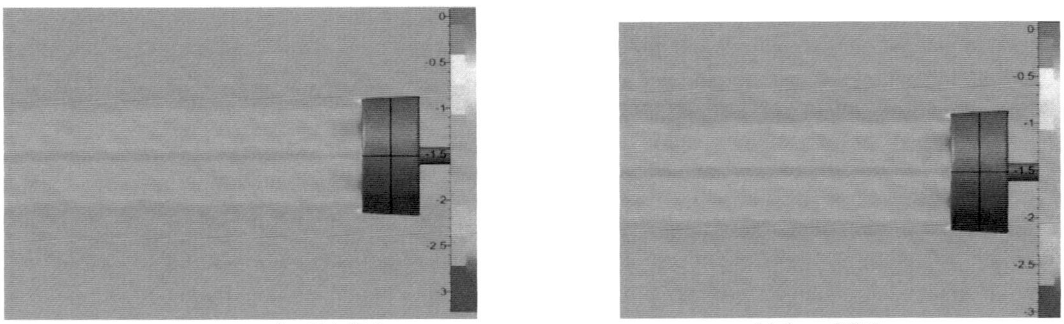

a) EASM b) k-ω SST

Figure 4. Ka4-70 tip clearance 2mm

Figure 4 depicts the Ka4-70 propeller at a tip clearance of 2mm, revealing EASM's demonstration of the stronger flow behind the propeller (blue darker, denser) in contrast to the k-ω SST's lighter blue. Nevertheless, the two-turbulence technique results in a flow of fluid that is comparable after the propeller.

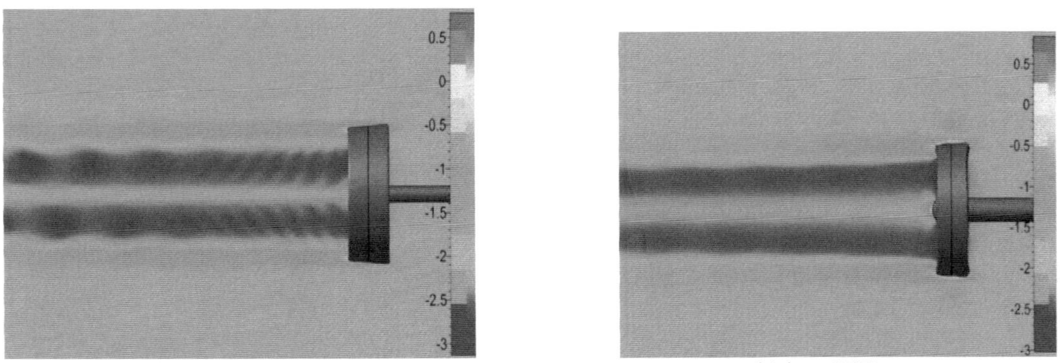

a) EASM b) k-ω SST

Figure 5. B4-70 tip clearance 3mm

Figure 5 demonstrates for a B4-70 propeller that when the tip clearance is 3mm, the k-ω SST shows a stronger and heavier flow behind the propeller, but the EASM shows a less dense flow behind the propeller. This can be observed when the propeller has a tip clearance of 3mm. Furthermore, the flow pattern of the k-ω SST approach seems smooth, but the EASM method exhibits a pattern of previous fluid flow cycles.

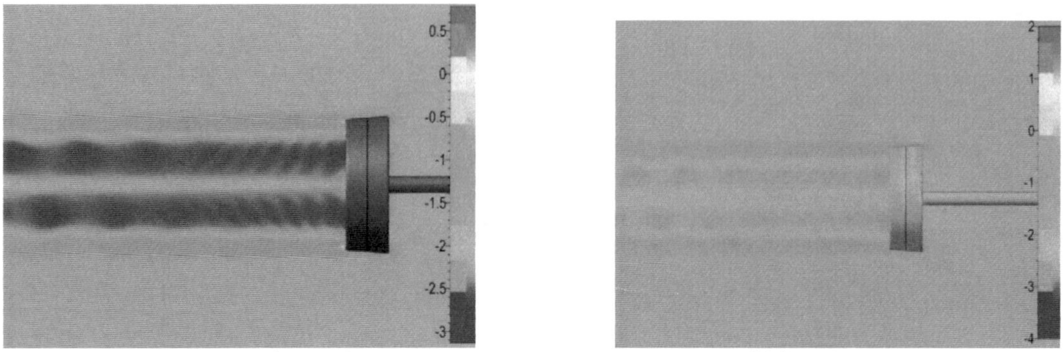

a) EASM b) k-ω SST

Figure 6. Ka4-70 tip clearance 3mm

Both turbulence methods produce fluid flow patterns that are visually identical after the propeller type, as shown in Figure 6. However, When the tip clearance is 3mm on a Ka4-70 propeller, as shown in Figure 6, EASM shows a stronger and heavier flow behind the propeller (darker blue) while k-ω SST shows a lighter blue.

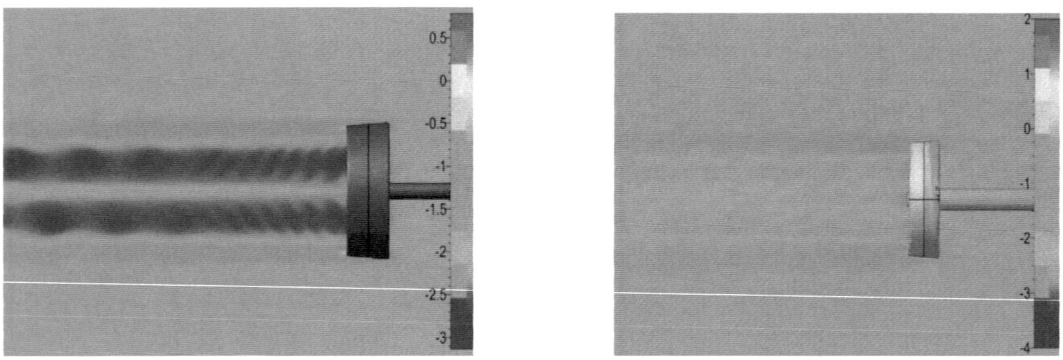

a) EASM b) k-ω SST

Figure 7. B4-70 tip clearance 4mm

Following Figure 7 shows, for the B4-70 propeller, that when the tip clearance is 4mm, there is a stronger and denser flow behind the propeller, which is proven by EASM (blue that is darker), but the k-ω SST revealed lighter blue. This can be observed when the propeller is given a tip clearance of 4mm.

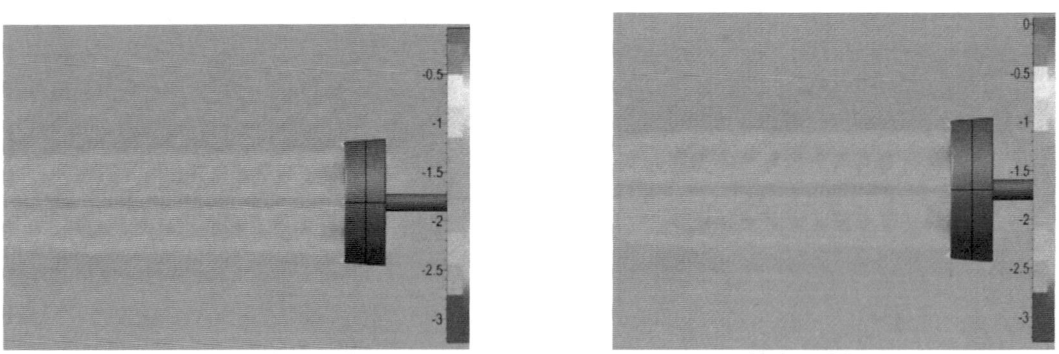

a) EASM b) k-ω SST

Figure 8. Ka4-70 tip clearance 4mm

It can be observed in Figure 8 that for the Ka4-70 propeller, when the tip clearance is 2mm, both the EASM and the k-ω SST turbulence models revealed a pattern of flow behind the propeller that was quite comparable.

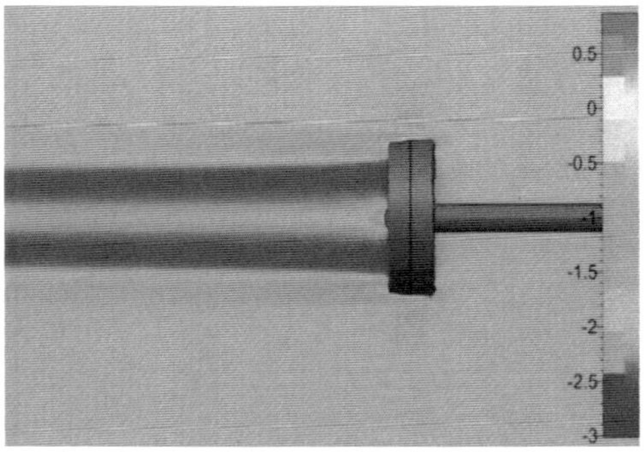

a) EASM

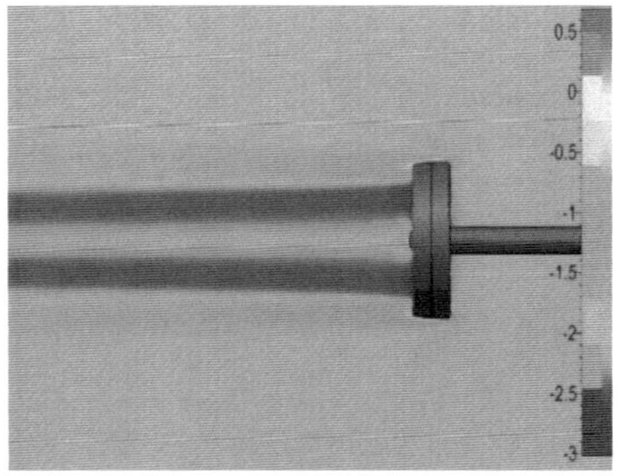

b) k-ω SST

Figure 9. B4-70 tip clearance 5mm

The using of tip clearance is 5mm, it is important to identify in Figure 9 for the B4-70 propeller that the higher flow behind the propeller is proven by EASM (blue that is darker), although the k-ω SST exhibited less dark blue. This occurs when the tip clearance is 5mm.

a) EASM

b) k-ω SST

Figure 10. Ka4-70 tip clearance 5mm

Following that, the choice of tip clearance is 5mm, it is possible to observe in Figure 10 for the Ka4-70 propeller that the higher flow behind the propeller is exhibited by EASM (blue that is darker), but the k- SST revealed lighter blue. This is the case when the tip clearance is 5mm.

4. Conclusion

A study into the effect of tip clearance or gap on ducted propeller has been conducted using B4-70 series and Ka4-70 series of propellers. The work was carried out numerically using CFD technique together with tip clearance from 2mm to 5mm, and the EASM turbulence model demonstrated better results than the k-ω SST model. Both propeller types indicated that the gap of 3mm was optimum in terms of thrust, and torque. Furthermore, the Ka4-70 has better performance than the B4-70 as ducted propellers and the EASM turbulence model is preferable to k-ω SST model.

Acknowledgement

The authors wished to thank the Institut Teknologi Sepuluh Nopember for financing the research under a research scheme called "Research for Doctoral Dissertation" under contract number: 1566/PKS/ITS/2022.

References

1. Funeno I. Influence of Hydrodynamic Interaction between Ducted Propellers and Struts on Performance of Azimuth Thrusters. In: Fifth International Symposium on Marine Propulsors SMP'17. 2017. Available from: https://www.marinepropulsors.com/proceedings/2017/THA2-1.pdf
2. Adietya BA, Utama IKAP, Aryawan WD, Sutiyo. CFD Analysis into the Effect of using Propeller Boss Cap Fins (PBCF) on Open and Ducted Propellers, Case Study with Propeller B-Series and Kaplan-Series. CFD Letter. 2022 May 6;14(4):32–42. Available from: https://doi.org/10.37934/cfdl.14.4.3242
3. Yongle D, Baowei S, Peng W. Numerical investigation of tip clearance effects on the performance of ducted propeller. International Journal Naval Architecture and Ocean Engineering . 2015;7(5):795–804. Available from: https://doi.org/10.1515/ijnaoe-2015-0056
4. Liu Y, Tan L, Wang B. A Review of Tip Clearance in Propeller, Pump and Turbine. Energies. 2018 Aug 22;11(9):2202. Available from: https://doi.org/10.3390/en11092202
5. Menter FR, Kuntz M, Langtry R. Ten Years of Industrial Experience with the SST Turbulence Model. In: 4th Internal Syposium, Turbulence, Heat and Mass Transfer. New York, Wallingford: Begell House; 2003. p. 625–32.
6. Gatski TB, Speziale CG. On explicit algebraic stress models for complex turbulent flows. Journal of Fluid Mechanic . 2006/04/26. 1993;254:59–78. Available from: https://doi.org/10.1017/S0022112093002034

The Effect of geometric parameters on stress concentration factor of a multiplanar DKT tubular joint through finite element model

R W Prastianto[1], F A Admiral[1], D M Rosyid[1], Y S Hadiwidodo[1], Handayanu[1], and F Syalsabila[1]
[1]Department of Ocean Engineering, Institut Teknologi Sepuluh Nopember, Surabaya, Indonesia

rudiwp@oe.its.ac.id

Abstract. It is important to consider geometric parameters on Stress Concentration Factor (SCF) at the tubular joint, especially multi-planar joint, when designing a jacket platform. Thus, a jacket platform capable to withstand the extreme environmental loads. Many researchers use the Finite Element (FE) method to analyze stress at joints and estimate the SCF with the variation of geometric parameters. That studies obtained the result that geometric parameters have significant effect on decreasing or increasing the SCF. Therefore, this study will delve the effect of geometric parameters (β, τ, and γ) on maximum SCF caused by axial loading, in-plane bending moment, and out-of-plane bending moment at DKT tubular joint. The increase of β results in the decrease in maximum SCF at brace-side and chord-side weld toe, respectively. The increase of τ and γ results in the increase in maximum SCF at brace-side and chord-side weld toe, respectively.

1. Introduction

Tubular joints are the main configuration of the jacket structure. Tubular joints are divided into two types based on the number of planes, namely uni-planar and multi-planar joints. A uni-planar connection has a branching configuration in only one plane. While the multi-planar connection has a branching configuration in more than one plane (usually consisting of two to three planes). The two types of connections have various configurations, which consist of variations in the configuration of Y, T, X, K, and KT.

Tubular joints are the most frequent part to the occurrence of stress concentration. That is, the tubular joints are the component that is very vulnerable to the occurrence of hotspot stresses and has a significant effect on the service life of a jacket structure.

The finite element method is one of the most widely used methods in the calculation of hotspot stress because it is able to model complex tubular joints simply with considerable results. The hotspot stress results obtained from the finite element method can be used to calculate the stress concentration factor (SCF).

SCF on the tubular joint weld toe can be known after knowing the stress distribution. The effect of multi-planar joints plays an important role in the stress distribution that occurs on weld toes of the joint [1]. Thus, the stress distribution studied in uni-planar connections is not accurate in determining the stress distribution in multi-planar connections which is present in the majority of jacket structure joints. Then, some studies show that the variation of geometric parameters have influences to hotspot stress, which later influence the SCF [2,3]. The amount of SCF depends on the geometry of the joint, the type of loading, the size, and the type of weld [4].

Based on the previous statement, an accurate SCF is required specifically for the tubular joint under study, namely the DKT multi-planar tubular joint. This research studies the effect of geometric characteristics on SCF by determining several variations of geometric parameters (β, τ, and γ). The loads applied to the DKT multi-planar tubular joints are axial loads, in-plane bending moment, and out-of-plane bending moment.

2. Theory

2.1. Multi-planar DKT tubular joint

A multi-planar tubular joint is a joint with a brace in more than one chord plane, usually 2 to 3 planes (or more). Multi-planar joints are widely used in platform jackets, as well as being the main connection of the strength of the structure. Hotspot stress in multi-planar tubular junctions can have a greater or lesser value when compared to uni-planar tubular junctions. This fact makes the parametric formula for uni-planar tubular junctions inaccurate for use in multi-planar tubular junctions [4]. Figure 1 is an example of a multi-planar tubular joint in a DKT (Double KT) configuration:

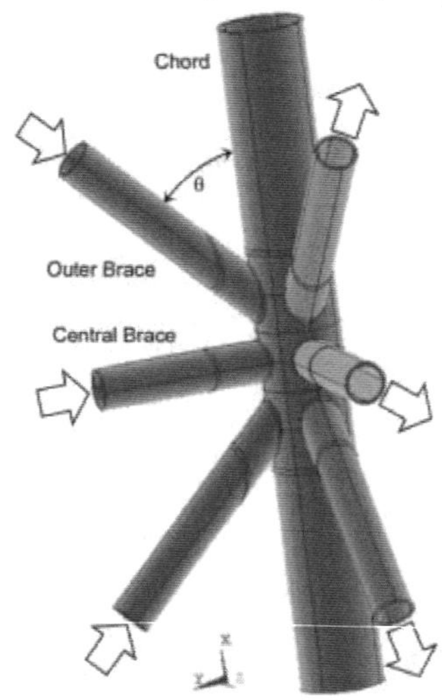

Figure 1. Multi-planar DKT tubular joint [4]

2.2. Geometric parameter

The geometric parameters of the tubular joint that will be used in this research are:

1. Beta (β)
   Beta (β) is the ratio of the diameter of the brace and chord. This parameter is related to the ovalization of the chord and affects the size of the SCF. The smaller this parameter, the larger the SCF that occurs (inversely proportional). Although it has an effect on SCF, it does not have a significant impact [4,5]. In addition, this parameter also affects the stress distribution that occurs around the joint.
2. Tau (τ)
   Tau (τ) is the ratio between the thickness of the brace and the thickness of the chord. The increase in the value of this parameter is related to the increase in SCF (directly proportional). This parameter is very influential on SCF, so it needs to be a major consideration [4,5]. The main effect of this parameter is on the bending moment brace at the joint.
3. Gamma (γ)
   Gamma (γ) is the ratio of the chord diameter to the chord thickness. Based on the comparison, the greater the gamma parameter, the larger the hollow section chord will be. This parameter affects the size of the SCF with increasing this parameter making the SCF even larger (directly proportional). This parameter includes an effect on SCF, but the magnitude of this parameter is strongly influenced by the magnitude of the parameter τ [5].

Determination of geometric parameters is carried out at the design stage by referring to existing fatigue design guides, such as [6]. This determination is important because the size of the geometric parameters can determine the size of the SCF. In determining geometric parameters, a range is determined for each parameter which is called the validity range. The following geometric parameter validity ranges are given in Table 1.

Table 1. Geometric parameter validity range

Geometric parameter	Validity range	
	API RP 2A-WSD 22nd Ed. [6]	Oshogbunu (2019) [7]
β	$0.2 \leq \beta \leq 1.0$	$0.4 \leq \beta \leq 0.6$
τ	-	$0.25 \leq \tau \leq 1.0$
γ	$10 \leq \gamma \leq 50$	$12 \leq \gamma \leq 24$

2.3. Hot spot stress

Hotspot stress is the maximum stress that occurs at a point (in this case, the tubular connection) with a value greater than the nominal stress. These stresses are distributed unevenly along the joint, resulting in no uniformity of values, and cannot be solved by just one equation. For analyze the critical tubular member, local analysis is needed. The analyses focus on the stress distribution along the weld toe [8].

A popular method for determining the distribution and magnitude of the hotspot stress value is using the numerical-based finite element method. Linear extrapolation of stress needs to be carried out around the connection to determine the hotspot stress value based on the recommendations from [6]. There are various references that provide linear extrapolation length calculations. The minimum and maximum distance from the extrapolation area to the weld toe on the brace and chord with the illustration given in Figure 2 [9]. In the brace using 0.4t and 1.0t, while the chord uses 0.4T and 1.0T (t: brace wall thickness; T: chord wall thickness). The following Eq. 1 is the critical stress equation based on IIW-XV-E which was reformulated by Nassiraei and Rezadoost [10], where $\sigma_{\perp W}$ is hotspot stress along the extrapolation points (ksi), $\sigma_{\perp E1}$ is stress on the first extrapolation point (ksi), and $\sigma_{\perp E2}$ is stress on the second extrapolation point (ksi):

$$\sigma_{\perp W} = 1.4\sigma_{\perp E1} - 0.4\sigma_{\perp E2} \quad (1)$$

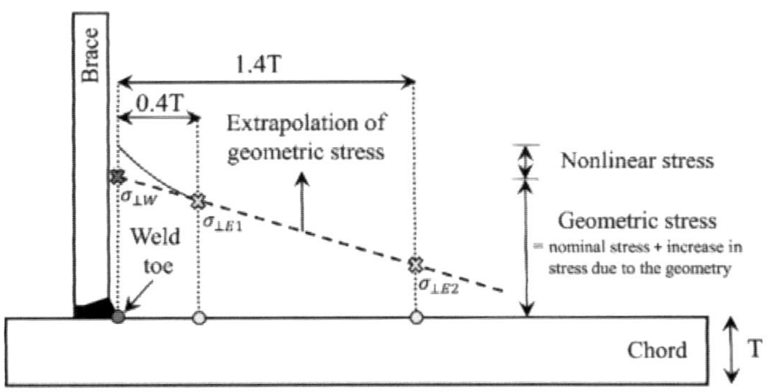

Figure 2. Linear extrapolation based on IIW-XV-E [9]

2.4. Stress Concentration Factor (SCF)

Stress concentration factor (SCF) is the ratio between the stress in the critical region and the nominal stress in the cross section. The influencing factors are the size of the connection, and the connection configuration. The following Eq. 2 is the stress concentration factor equation [10], where SCF is stress concentration factor (ksi) and σ_n is nominal stress (ksi):

$$SCF = \sigma_{\perp W}/\sigma_n \quad (2)$$

Stress concentration factor varies in each joint geometry configuration. This factor is an indicator of the strength of the joint, especially in determining the service life of a structure. This very crucial role makes the accuracy of the stress concentration factor in a tubular joint which is required to be high.

3. Methodology

3.1. Data

Global analysis is conducted by applying modified Morison's equation for calculating the waves and current load on fixed offshore platform [11]. Then, the result from the global analysis is extracted, namely platform data, geometry data, and material properties of double KT tubular joints. The global analysis data used from the previous study by Noviyanti [2,12] with the detail DKT connection shown in Figure 3 and Figure 4.

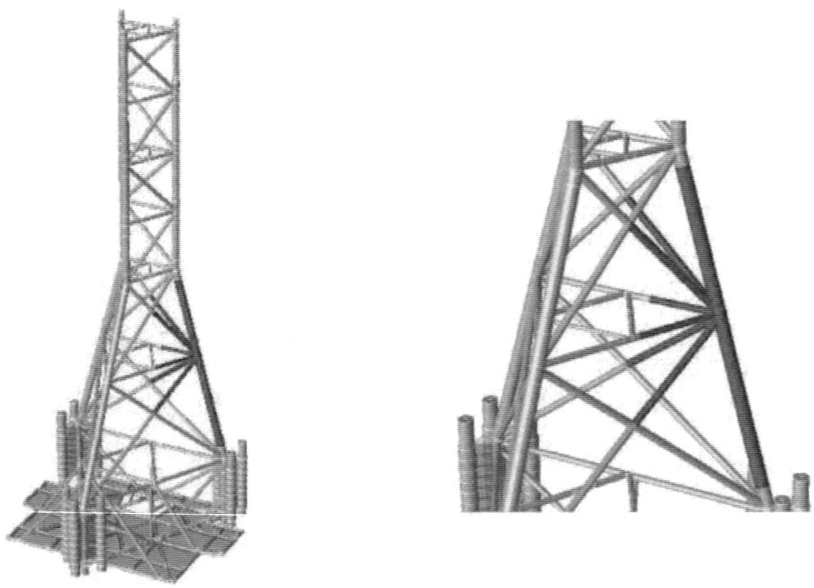

(a) (b)

Figure 3. (a) Minimalist jacket platform (b) Multi-planar DKT joint

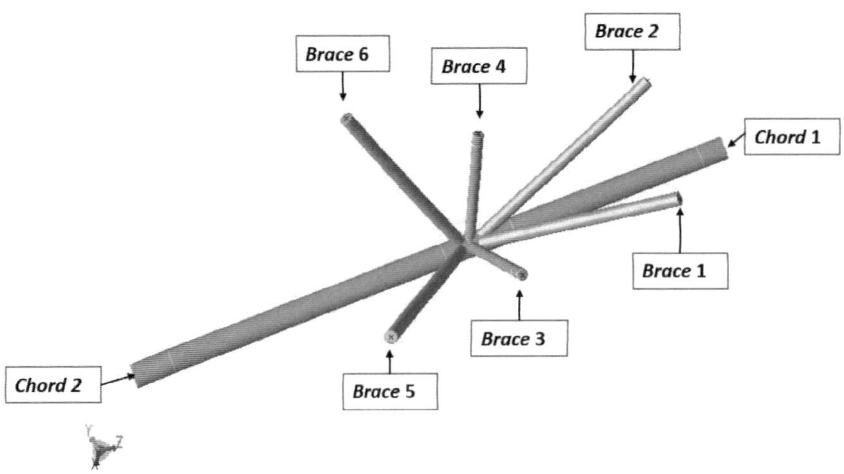

Figure 4. Multi-planar DKT joint details

1. Geometry and material properties of DKT tubular joints are shown in Table 2 and Table 3:

Table 2. Geometry data sets on the reviewed multi-planar DKT joint

Member	Thickness *(t)* (in)	Outside diameter (OD) (in)	Inside diameter (ID) (in)	Length (L) (ft)
Chord 1	2,5	54	49	52
Chord 2	2,5	54	49	64
Brace 1	1,25	28	25	43
Brace 2	1,25	28	25	45
Brace 3	0,75	24	22,5	32,5
Brace 4	0,75	24	22,5	25
Brace 5	1,25	28	25	43
Brace 6	1,25	28	25	40

Table 3. Material property data sets on the reviewed multi-planar DKT joint

Material Properties	Value
Density, ρ (lb/ft3)	490
Yield Strength, σ_y (ksi)	36
Modulus Young, E (ksi)	29000
Shear Modulus, G (ksi)	11200
Poisson's Rasio, υ	0.3

2. Types of loading on DKT tubular joint
The load that will be given to the local modelling of the DKT tubular connection is the axial force, IPB, and OPB based on Member End Force data from global analysis. The difference in the direction of stress orientation and bending rotation between global analysis and local analysis should be considered.
Table 4 below is the joint load data gathered from global analysis that has been adjusted with the direction of the local analysis's direction and rotation:

Table 4. Axial forces and bending moment data

Member		Axial Forces (kips)	Bending Moment (in-kips)		
		fx	Mx	My	Mz
Brace 1	0016-0050	225.58	14.04	345.77	437.75
Brace 2	0050-0022	135.86	-34.08	502.83	35.08
Brace 3	0056-0050	-73.13	-61.41	-1015.78	25.07
Brace 4	0050-0104	-49.52	53.86	-881.28	68.05
Brace 5	0068-0050	-48.40	143.02	931.57	645.36
Brace 6	0072-0050	24.39	-7.13	877.52	75.68

3.2. Geometric parameter variations

The variations of geometric parameters in this research are beta (β) in Eq. 3, tau (τ) in Eq. 4, and gamma (γ) in Eq. 5, where d is brace diameter, D is chord diameter, t is brace thickness, and T is chord thickness:

$$\beta = \frac{d}{D} \tag{3}$$

$$\tau = \frac{t}{T} \tag{4}$$

$$\gamma = \frac{D}{2T} \tag{5}$$

The variety of the three parameters are obtained with a validity range that refers to [7] as shown in Table 1. Then, seven variations are made based on the validity range to be as in Table 5 below:

Table 5. Geometric parameters variations of multi-planar DKT joint

Variation	D (in)	T (in)	d (in)	t (in)	β	τ	γ
Original	54.00	2.50	28.00	1.25	0.52	0.50	10.80
1	54.00	2.25	24.00	1.20	0.40	0.50	12.00
2	54.00	1.50	24.00	0.80	0.40	0.50	18.00
3	54.00	1.13	24.00	0.60	0.40	0.50	24.00
4	54.00	2.25	24.00	1.68	0.40	0.75	12.00
5	54.00	2.25	24.00	2.40	0.40	1.00	12.00
6	54.00	2.25	27.00	1.20	0.50	0.50	12.00
7	50.00	2.00	30.00	1.00	0.60	0.50	12.50

3.3. Modeling the multi-planar DKT joint

Double KT tubular joint obtained from the variation was then modelled using finite element software. The main geometries and their variations to loading using the data in Table 2 to Table 5.

3.4. Meshing sensitivity

The meshing sensitivity method aims to analyze the sensitivity of the mesh on the KT double tubular joint, especially in the stress hotspot area (weld toe). This analysis produces the most optimal number and size of elements. The optimal mesh provides a high level of accuracy in describing every stress that occurs in the model. This stage is done by changing the size and increasing the number of mesh with a certain range to get a mesh configuration with a constant stress. The location under review must be the same when performing meshing sensitivity.

3.5. Calculating hotspot stress

Hotspot stress on the weld toe is obtained by first extrapolating the weld toe linearly. In this research, the point that is used as a reference to perform linear extrapolation is the location point that experiences the maximum equivalent stress. After getting the extrapolation point, the hotspot stress can be calculated using Equation 1.

3.6. Calculating SCF

The stress concentration factor is obtained by dividing the hotspot stress by the nominal stress. Equation 2. The hotspot stress is obtained by linear extrapolation, and the nominal stress is obtained by using a constant stress around the weld toe. The greater the value of SCF, the more concentrated a stress that occurs at the location under consideration.

3.7. Study of the effect of geometric parameters to SCF

The study of the influence of geometric parameters aims to obtain what parameters have a crucial, moderate or no effect on the SCF from finite element method. The results of this study show the effect of each variable (β, τ, and γ) on SCF. The aim of this study is to determine the tubular geometry at the design stage.

4. Result and Discussion

4.1. Meshing sensitivity analysis

IOP Conf. Series: Earth and Environmental Science 1166 (2023) 012039 doi:10.1088/1755-1315/1166/1/012039

Meshing sensitivity was conducted to specify the model which has the best consistency of describing the stress. This analysis was conducted with increasing the amount and density of mesh until the stress occurred along the weld toe become constant. Figure 5 below shows the result of meshing sensitivity that applied with axial load, IPB, and OPB respectively:

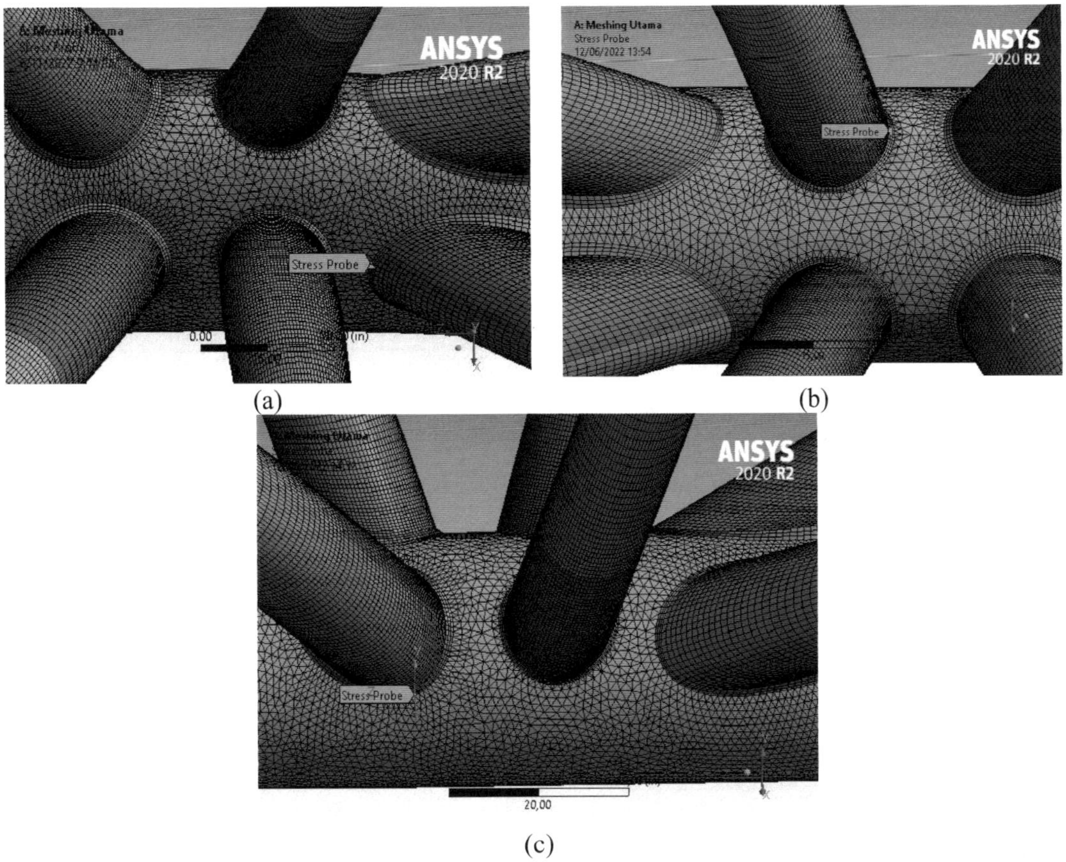

Figure 5. Meshing sensitivity caused by (a) Axial (b) IPB (c) OPB

The results of this analysis for all the applied load (axial, IPB, and OPB) is 810201 meshes. In the sake of brevity, this research does not show the process of meshing sensitivity.

4.2. Effect of parameter geometric to SCF study
This study aims to determine the effect of varying geometric parameters (β, τ, and γ) on the SCF value of the brace and chord-side weld toe due to axial loading, IPB, and OPB. The results of the effect of geometric parameters on the SCF can be used at the design stage of the tubular joint.

4.2.1. Effect of β to SCF.
The increase in the geometric parameter which is the ratio of the brace diameter and chord diameter to the chord diameter size is made constant (constant) making the brace diameter size increase, and vice versa. The parameter effect study gave three different β (0.4; 0.5; 0.6) with the parameters τ and γ kept constant. Based on the study, the variation model used is variation model 1 (β = 0.4; τ = 0.5; γ = 12), variation model 6 (β = 0.5; τ = 0.5; γ = 12), and the variation model 7 (β = 0.6; τ = 0.5; γ = 12). Figure 6 is a graph of the effect of parameter β on SCF:

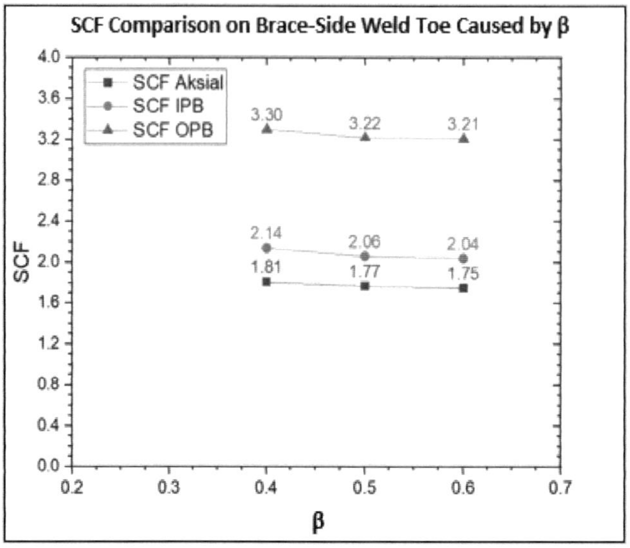

(a)

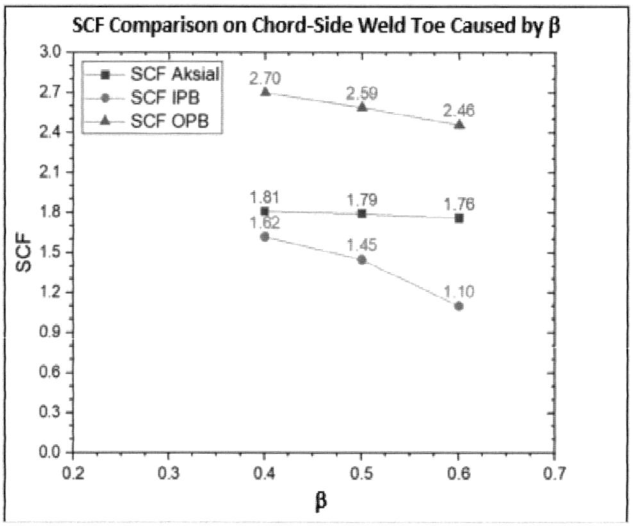

(b)

Figure 6. SCF comparison due to β (0,4; 0,5; 0,6)
(a) Brace-Side Weld Toe (b) Chord-Side Weld Toe

Based on the study results shown in Figure 6, the larger the parameter β, the smaller the SCF in the brace and chord-side weld toe. On the other hand, the smaller the parameter β, the larger the SCF in the brace and chord-side weld toe.

4.2.2. Effect of τ to SCF

The increase in the geometric parameter which is the ratio of the thickness of the brace and the thickness of the chord to the size of the thickness of the chord is made constant (constant) makes the size of the thickness of the brace increase, and vice versa. The parameter effect study gave three different τ (0.5; 0.75; 1.0) with the parameters β and γ kept constant. Based on the study, the variation model used is variation model 1 ($\beta = 0.4$; $\tau = 0.5$; $\gamma = 12$), variation model 4 ($\beta = 0.4$; $= \tau\ 0.75$; $\gamma = 12$), and the variation model 5 ($\beta = 0.4$; $\tau = 1.0$; $\gamma = 12$). Figure 7 is a graph of the effect of parameter τ on SCF:

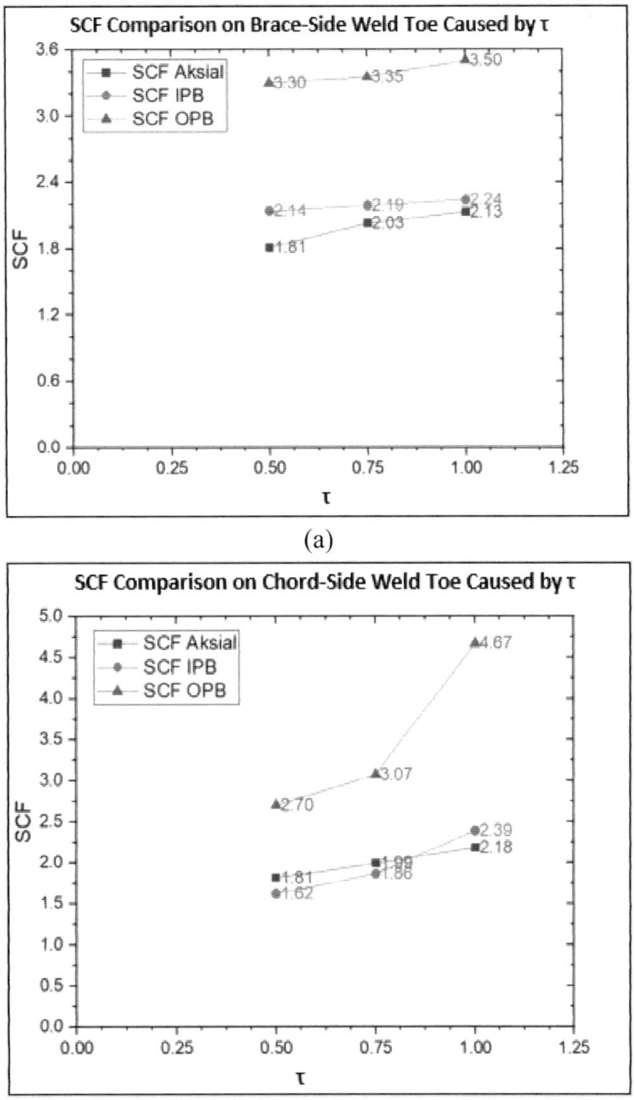

Figure 7. SCF comparison due to τ (0,5; 0,75; 1,0)
(a) Brace-Side Weld Toe (b) Chord-Side Weld Toe

Based on the study results shown in Figure 7, the greater the parameter τ, the greater the SCF in the brace and chord-side weld toe. On the other hand, the smaller the parameter τ, the smaller the SCF in the brace and chord-side weld toe.

4.2.3. Effect of γ to SCF

The increase in the geometric parameter which is the ratio of the chord diameter and chord thickness to the chord diameter size is made constant (constant) making the chord thickness size decrease, and vice versa. The parameter effect study gave three different γ (12; 18; 24) with parameters β and τ kept constant. Based on the study, the variation model used is variation model 1 ($\beta = 0.4$; $\tau = 0.5$; $\gamma = 12$), variation model 2 ($\beta = 0.4$; $\tau = 0.5$; $\gamma = 18$), and the variation model 3 ($\beta = 0.4$; $\tau = 0.5$; $\gamma = 24$). Figure 8 is a graph of the effect of parameter γ on SCF:

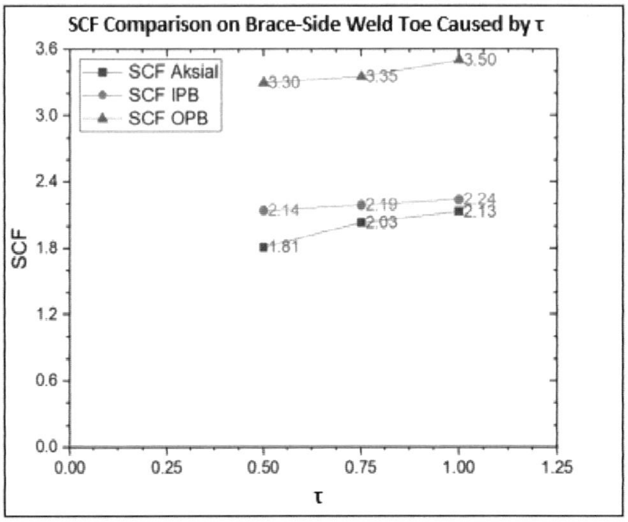

(a)

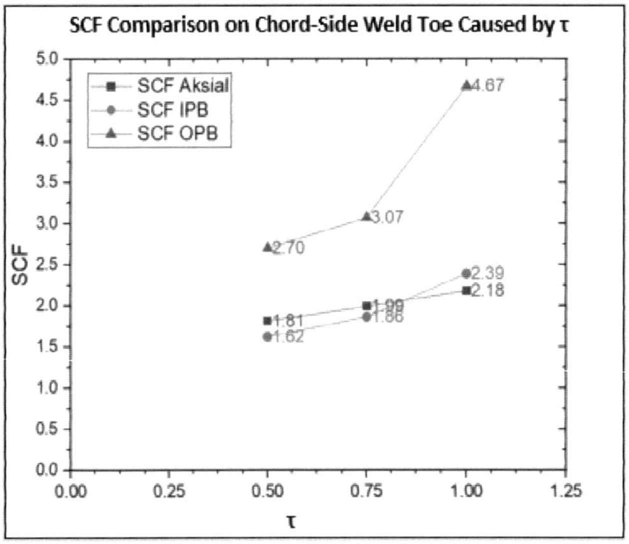

(b)

Figure 8. SCF comparison due to γ (12; 18; 24)
(a) Brace-Side Weld Toe (b) Chord-Side Weld Toe

Based on the study results shown in Figure 8, the greater the parameter γ, the greater the SCF in the brace and chord-side weld toe. On the other hand, the smaller the parameter γ, the smaller the SCF in the brace and chord-side weld toe.

5. Conclusion

From the process of study results, calculations, simulations, and discussions that have been carried out, the following conclusions can be summarized from this research:

1. SCF in double KT tubular joints can be obtained using the finite element method from variations in geometric parameters due to axial loading, in-plane bending moment, and out-of-plane bending moment.
2. The study of geometric parameters shows that the smaller the parameter β, the larger the SCF obtained. Meanwhile, the smaller the parameters τ and γ, the smaller the SCF obtained. The effect of this parameter is suitable for axial loading, IPB, and OPB.

References

[1] Prastianto R W, Hadiwidodo Y S, Widhestomo S W and Yazhahir R 2018 Stress Distribution along the Weld Toes of Tubular KT and KDT Joints under Balance Axial Loads and In-Plane-Bending Moments *Ocean and Coastal Engineering, Environtmental and Natural Disaster Management* ed Suntoyo and Silvianita pp 170–5

[2] Noviyanti I, Prastianto R W and Murdjito M 2021 Stress Distribution Analysis of DKDT Tubular Joint in a Minimum Offshore Structure *Rekayasa* **14** 191–9

[3] Prastianto R W, Hadiwidodo Y S, Fuadi I F, Naibaho M B Y P and Rahardianto P N 2018 Stress concentration factor distribution of inclined brace in multiplanar offshore tubular kt joints *IOP Conf. Ser. Earth Environ. Sci.* **162** 011001

[4] Ahmadi H, Lotfollahi-Yaghin M A and Aminfar M H 2011 Distribution of weld toe stress concentration factors on the central brace in two-planar CHS DKT-connections of steel offshore structures *Thin-Walled Struct.* **49** 1225–36

[5] Zavvar E, Hectors K and De Waele W 2021 Stress concentration factors of multi-planar tubular KT-joints subjected to in-plane bending moments *Mar. Struct.* **78** 103000

[6] API RP 2A-WSD 22nd Ed. 2014 Planning, Designing, and Constructing Fixed Offshore Platforms - Working Stress Design *API Recomm. Pract.* 324

[7] Oshogbunu E O 2019 *Optimisation of Offshore Wind Farm Maintenance* (The University of Manchester)

[8] Hendrawan M F, Prastianto R W and Rosyid D M 2022 Local Stress Analysis at Critical Tubular Joint of Offshore Wind Turbine Jacket Structure *Recent Advances in Mechanical Engineering* (Singapore: Springer) pp 238–246

[9] IIW-XV-E 2000 *Fatigue Design Procedure for Welded Hollow Section Joints* (Cambridge: Abington Publishing)

[10] Nassiraei H and Rezadoost P 2022 Probabilistic analysis of the SCFs in tubular T/Y-joints reinforced with FRP under axial, in-plane bending, and out-of-plane bending loads *Structures* **35** 1078–97

[11] Syarifudin M R, Prastianto R W, Hadiwidodo Y S, Adaalah A H and Syalsabila F 2022 Strength Analysis of Offshore Aquaculture Net Structure under Wave and Current Loads *IOP Conf. Ser. Earth Environ. Sci.* **972**

[12] Noviyanti I and Prastianto R W 2022 Numerical Study on the Stress Distribution Analysis of Two-Planar DKDT Tubular Joint Under Variation of Axial Loading Conditions *Recent Advances in Mechanical Engineering* (Singapore: Springer) pp 19–26

Exploring the Impact of Covid 19 on the Maritime Transport Sector

Nowar Deeb[1], Adam Leonardo[2]

[1,2] Department of Marine Engineering, Institut Teknologi Sepuluh Nopember

Email: nowardeeb12@gmail.com

Abstract. In the maritime industry, The unanticipated COVID-19 viral epidemic is an unforeseeable circumstance, and other nations implemented enormous containment measures to stop the Coronavirus epidemic from spreading around the world. Thus, directly affecting the maritime shipping sector. This paper will discuss the current problems facing the shipping industry, taking into account the congestion problems, delays, and uncertainty timeframes, using the Los Angeles port as a case study. These problems and more were addressed directly by increasing the operating hours, workload, and available staff, and indirectly by looking for alternatives for shipping goods, and creating more cargo space, furthermore, this study will use Monte Carlo simulation to predict the effectiveness of these solutions on the congestion at the port.

1. Introduction

The COVID-19 pandemic is spreading across the globe, affecting every element of modern life [1], Following the initial encounter, the virus rapidly migrated from China to other nations and regions throughout the globe. The COVID-19 pandemic has been deemed a public health emergency by the World Health Organization (WHO) that is primarily affecting human activities and economic development, and it has been categorized as a quarantine scenario by the WHO [2].

Activities involving the sea are crucial to the world economy and to societal progress. The maritime and shipping industry is one of the industries that has been severely damaged by the growth of COVID-19. Initially numerous countries ceased to operate their seaports and restricted the majority of their export and import activities since the spread of COVID-19. Significant lockdowns and social distancing measures have been implemented in several countries, resulting in immediate reductions in productivity and consumption [3]. Moreover, Maritime ports have stayed open for business in the aftermath of the global epidemic, at the expense of seafarers' health and safety. Seafarers faced major travel challenges to and from their ships due to the continued worldwide reliance on marine trade while other countries' borders remained closed. This has resulted in a humanitarian "crew change problem" at sea, in which labor contracts are routinely broken, resulting in forced labor at sea [4].

The global "crew change crisis" at sea is characterized by persistent labor contract violations that result in forced labor at sea. The Wakashio, an oil tanker that broke in half after striking aground in July 2020, had been at sea for almost a year and was understaffed. This has resulted in a loss of happiness and mental health, as well as an increase in the likelihood of accidents. The approaches of international organizations to the global "Crew Change Crisis" at Sea are examined in this article [5].

On September 24th, World Maritime Day, the global humanitarian crew change crisis reached six months, and the number of stranded seafarers reached 400.000. By May 2021, the number had dropped to around 200,000, which is still far too high. The failure to end this long-running issue indicates that much more is in danger [6]. Furthermore, this shows the importance to assess the impact of COVID-19 clearly on maritime shipping.
Countries have partially or closed their borders to stop the spread of COVID-19, causing delays in the flow of products, capital, and people and disrupting global supply chains, which in turn has led several industries to shut or slow down the manufacturing processes since there was no contact with the global market to distribute the commodities.

For we can see its negative effects on port throughput including the shutdown of shipping lines, disruptions in the transportation sector, and an increase in the health risk of imported products [7]. Enhanced connection and better efficiency of international trade have become a double-edged sword as a result of increased mobility. As a result, increased maritime mobility has allowed more items to be transported to more and more remote ports, increasing the potential of pandemic transmission.

1.1. Port Performance

To prevent the spread of largescale cross-infection, governments around the world have implemented a variety of prevention and control measures, including work stoppages and travel restrictions [8] [9].

The global economy was unavoidably shattered, resulting in a significant decline in international trade. The shipping industry, as the primary carrier of international trade, suffered the brunt of this health crisis. Due to low demand for transportation, liner firms decided to stop some routes to save money [9]. The efficiency of cargo handling is reduced, and freight prices have risen significantly, due to a loss in efficient transport capacity, port quarantine, and a shortage of staff.

Nonetheless, to guarantee that maritime activities run well during COVID-19, some governments took some extra measures to ensure that the ports can still work during the pandemic such as Waiver of the penalty fee; digitalization activities; force majeure on works; port safety measures; waiver of all charges for vessels carrying oxygen and oxygen-related cargo at important Indian ports; During a pandemic, seaport CSR monies were used to support medical and community development programs, as well as the repatriation of seafarers [8].

1.2. Consumers' Behavior during the Pandemic

Consumers have been stockpiling since the start of the COVID-19 epidemic (early 2020), which is a substantial departure from their regular buying habits. An instant surge in demand for alternate distribution channels has resulted from the limited accessibility of shop premises, paired with consumers' increased health concerns. Unexpected rules causing social separation are also having a significant impact on customers' preferred buying channels. For example, whereas online grocery purchasing has grown steadily but slowly over the last decade [10], it grew dramatically during the COVID-19 pandemic outbreak [11]. In addition, older and less digitally adept consumers have begun to find and enjoy online shopping, embracing the security that technology provides. Other distribution techniques that do not require direct human interaction, in addition to online retailing, may develop in popularity [12].

The spike in demand for commodities happened as a result of outside factors that weren't included before within the freight shipping industry, or it was just overlooked as an insignificant probability. However, with the COVID restrictions taking place and the consumers starting to start spending on goods other than services rapidly increased the demand and the supply started to fall behind, afterward, and the ports reduced working capacity to a minimum due to the pandemic circumstances also played an important part with the supply chain delay.

The daily pricing data submissions, summed over all georeferenced sites in the study area, reveal continuously higher grain prices in 2020 than the previous year, according to [13] research. Retail prices of maize (white and yellow) and rice (local and imported) were 26 percent and 44 percent higher in 2020 than in 2019, respectively, across the eleven-week data collection period. For all commodities, the price index has been trending upward since week 24 (2 weeks before the full lifting of lockdown restrictions).

1.3. Supply Chain

Throughout the pandemic, people are buying a lot more things and a lot of times buying them online, and that has led to the increased spending on commodities, which meant more demand for shipping that has reflected lower capacities inside the shipping sector because the demand exceeds the supply capabilities, therefore some commodities couldn't be shipped upon demand resulting in the long waiting period and delivery times. As an example, the United States demand and supply pattern over the pandemic had almost a peak level volumes, therefore, demands outstripped supply and ocean freight costs have spiked up, now it's extremely expensive to ship a container from Asia to the United States west coast as the price is almost triple what it was before.

One of the most important elements impacting port throughput is the volume of international trade. As a result of the epidemic, China's foreign trade has been seriously hindered, and the container scale has substantially decreased [14]. However, during the following period, we find out that the container scale reported a significant improvement of 23.84 percent reaching a new record of 294.32 Billion USD. The pandemic recession and recovery have resulted in a huge shipping crisis, with public health regulations hampering operations in ports and industries, even as demand for durable goods has fluctuated dramatically, first dropping then skyrocketing, causing an impact across the economy. Just-in-time techniques have increased delays by keeping stocks and expenses low while leaving retailers and producers with empty hands if there is a big stoppage [15].

Recently all-time high numbers for ship congestion were recorded in the ports of the United States. The number of container ships anchored or drifting in San Pedro Bay, just off the coasts of Los Angeles and Long Beach, has shattered all prior records. According to the Marine Exchange of Southern California, the most recent peak occurred in September 2021, when there was an all-time high of 73 container ships in the queue at San Pedro Bay. Because anchorages were filled, 36 ships were forced to wander offshore [16].

The Los Angeles/Long Beach port complex's constraints are shown by statistics from the Marine Exchange. The overall number of container ships at anchor or dock has grown and declined since the congestion began, reaching an all-time high of 100 in September, more than five times pre-COVID levels. However, one statistic has remained amazingly consistent: the number of container ships berthing at Los Angeles/Long Beach berths has stayed in a tight band of roughly 27-31 per day — that is, what the land side of the metaphorical funnel can handle. All ship arrivals beyond that threshold have overflowed into the anchorages and drift zones throughout 2021.

When post-lockdown import demand soared in August 2020, there were almost no ships anchored off the coast of Southern California. According to Marine Exchange data, there were an average of 36 ships at anchor per day in August. In August, the Port of Los Angeles handled 485,672 twenty-foot equivalent units of imports, down 5.9% from the previous month. As Port of Los Angeles Executive Director Gene Seroka pointed out, 26 ships were anchored in Los Angeles on the last day of August 2021, waiting for berths. These ships carried 205,000 TEUs of cargo, which was scheduled to arrive in September 2021. (Just as delayed cargo from this September will be pushed into October 2021) [16].

The port's overall throughput decreased by 0.8 percent in August to 954,377 TEUs, year over year, as a result of fewer imports and a 23 percent decline in loaded exports, which were somewhat offset by a 17 percent rise in outgoing empty containers. The port anticipates volumes will decline over the following two months from their peak in August to 930,000 TEUs in September and 950,000 TEUs in October. For the entire year, it expects a throughput of 10.8 million TEUs.

2. Methodology

The shipping cost has increased drastically throughout the pandemic. Every year during peak season when there's a rush to ocean freight and volumes are elevated some containers will be rolled as there's not enough space on the ship, therefore, the container will get delayed and has to wait for a longer period inside the port before it gets picked up by the ship.

In the shipping industry suppliers could also pay a premium as a method to prioritize their containers and make sure the shipment will arrive the soonest. Furthermore, a large proportion of containers need to have these additional costs just to get on the ship because it is oversold, and this is not just the situation for small shippers but even very large shippers as they don't have the guarantee, even though they have contracts and better cost.

These problems are not only faced by small suppliers, even big shippers are facing the same problems, for example, it was reported recently that Home Depot which is the third-largest United States retailing supplier, had chartered their container ship, as they had enough with delays and not having reliability and being able to control their supply chain so it was a very unusual and radical step to contract their ship.

Other large companies that are relying on ocean freight to deliver supplies have decided to switch focus to airway shipping instead as the increased demand for ocean freights has created more and more delays for these companies, as the case Amazon has contracted new airplanes for use. In order to continue providing customers with service, the company expanded its fleet by making its first-ever acquisition of a Boeing 767-300 aircraft. 11 aircraft are included in the deal and will be added to the network by 2022. The growth of the Air fleet comes at a time when customers are depending more than ever on quick, free deliveries.

2.1. Satellite-Assisted Data Collection

A lack of data has been one cause of uncertainty for those responsible for forecasting supply chain outcomes. Human data entry is commonly used by shipping businesses and ports, which are prone to errors and delays.

One option to prevent such errors is to employ satellite data, which depends on orbiting sensors to gather information on the activity on Earth, including gathering ship locations that are communicated via onboard radio transponders. When things are normal, the data helps companies place orders more effectively and can spare them from paying penalties for keeping cargo containers at ports after ships have left port and headed back out to sea.

The shipping industry is motivated to develop more accurate risk-based models of how weather impacts marine supply chains, and satellites that collect weather data could help the industry, which includes logistics businesses, wholesalers, and manufacturers, provide more accurate ETAs. Currently, satellites can assist in the documentation of traffic data. The carrier may receive notification that a ship carrying a specific container has arrived at a port, such as the Port of Los Angeles.

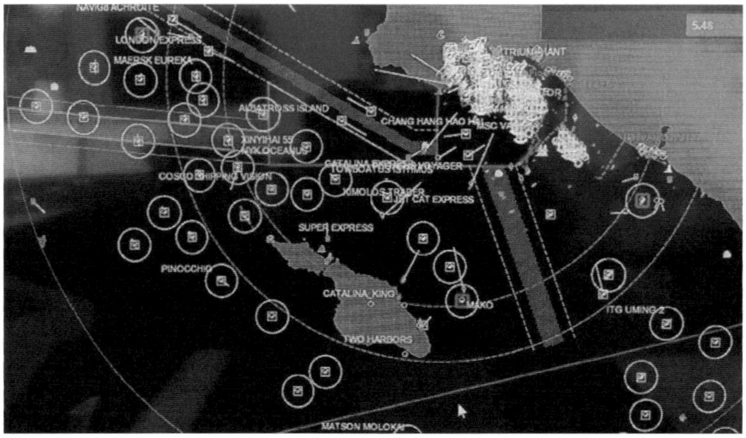

Fig. 1, Screenshot of a vessel traffic system showing where drifting ships are located on November 18. (Photo: Marine Exchange of Southern California).

2.2. *Operating Hours at Ports*

Some governments have already started addressing the maritime traffic at the largest container ports, as the United States has started efforts in September 2021 to expand the operating strategy for 90 days.

The ports of Los Angeles and Long Beach took the lead by extending gate hours and declaring their readiness to keep accepting incoming shipments. More steps were taken to improve the other parts of the supply chain, including the railroads, trucking, and the major importers in the United States — the Walmarts, FedExes, Home Depots, and others — and set commitments to operate more, 24 hours a day, seven days a week, because off-peak capacity is currently necessary to address this major problem [17].

While extended port gate hours are being used to assist flush out loaded containers of finished retail products waiting to be moved inland, off-peak-hour operations also give up prospects for agricultural exports. The Georgia Ports Authority (GPA) is attempting to establish the South Atlantic Supply Chain Relief program.

To move the import-filled containers taking up space at the port, off-dock sites, which are made up of inland areas with access to trucks or rail, are being built. This might have a big influence on exports as well as imports, clearing the yard and opening up the field for exports, and restarting some of the longer [receiving gate] periods.

Table.1 Port Of L.A. Operations Report (November 9, 2021).

Date		9-Nov-21					
Container Vessels in Port (POLA)		Containers	Tanker/ATB	General/Bulk	Vehicle	Cruse	Total
		17	3	0	0	1	4
Vessels at Anchore/Drifting		Containers	Tanker/ATB	General/Bulk	Vehicle	Cruse	Total
	San Pedro Bay	78	10	17	0	1	106
		Containers	Tanker/ATB	General/Bulk	Vehicle	Cruse	Total
	Port of Los Angeles	44	1	15	0	0	60
	Container Vessels enroute	Port of Los Angeles		Port of Long Beach			Total
	(3 days out forecast)	14		6			20
	Average POLA days at Anchor	15.5 days					
Terminal Update (POLA)	POLA Dewll time on terminal (As of Oct 28 - Last 30 days)	Local			Peak		
		9.7 days			9.7 days		
	POLA Dewll on-dock rail (As of Oct 28 - Last 30 days)	Local			Peak		
		4.5 days			13.4 days		

By comparison between the most recent reports from the port of Los Angeles, we can find out the improvement in the number of waiting for ships as it has decreased from an all-time high of 73 vessels at anchor/drifting in September 2021 to just 60 in November 2021. However, the average time the vessels spend at anchor is 15.5 days, which has increased. Meanwhile, the port of San Pedro has an increased number of vessels at anchor/drifting. This highlights the results of extended work hours and full personnel policy used by the port of Los Angeles, for the readiness levels and longer more efficient working hours reduced the number of vessels waiting at the port efficiently.

2.3. *Port Automation*

The employment of mechanical, hydraulic, pneumatic, electric, electronic, and computerized elements or systems to control equipment is known as industrial automation. It allows for less human engagement in industrial tasks while increasing control over the equipment and processes involved. Port container terminals are the biggest proponents of automation in the port industry. The automation of these facilities is the subject of this paper.

A port terminal is a modal interchange facility with an inland storage area that is used to coordinate the movement of products by sea and land. Port container terminals (PCTs) have some characteristics that allow them to achieve a far higher level of systematization than other types of freight terminals. The great degree of automation of equipment and operations in these types of port facilities is due to this level of standardization and specialization [18].

A modern corporate idea centered on long-term activity development requires three strategic needs, and PCT automation is a strategic initiative that meets those objectives. Among these objectives are commitments to environmental sustainability, greater safety and security, and improvements in operational performance. Increased quay and yard densities made possible by automated terminals' higher productivity translate into better facility capacity and more effective use of available space. Automation techniques reduce human error in operation, which enhances safety, but they also lower the chance of accidents by erecting a physical barrier between people and the area where operations are carried out.

Automation processes not only improve safety by minimizing human errors in operation, but they also lessen the effect of potential accidents by separating humans from the area where activities are physically carried out. Because minor automation standardizes the way activities are carried out while reducing the importance of operators, they reduce the number of accidents. Automation aids in the optimization of operations in all aspects, reducing equipment travels, empty runs, container shuffles, and so on, resulting in a reduction in energy consumption.

IOP Conf. Series: Earth and Environmental Science 1166 (2023) 012040 doi:10.1088/1755-1315/1166/1/012040

2.4. *Monte Carlo Simulation*

Monte Carlo simulation estimates mathematical functions and simulates the activities of complicated systems using random sampling and statistical modeling. This research takes accumulated data over an extended period between the 9th of November 2021, and until the 7th of June 2022. Moreover, the simulation iterations were multiplied by 5000 repetitions for each of the selected values for; The total number of vessels in the Port of Los Angeles while neglecting any negative values; the Total number of container ships in San Pedro Bay only considering the number of ships under 200 ships at a time; furthermore, the average number of days within the distance of 40NM awaiting berthing, while also neglecting the number of days greater than 24 days of wait. Furthermore, all the calculations for this simulation were completed using the Monte Carlo Simulation on the Microsoft Excel program.

3. Results

Using the End of year analytics the port of Los Angeles provides, we can monitor the improvements to the congestion traffic and waiting times for the shipping vessels entering the port. As the number of vessels in port on December 24 is showing as 17 containers and 1 General cargo ship.

Improved tracking of the incoming vessels has been added to the report, with 23 incoming vessels within 40 NM headed to San Pedro Bay. Moreover, the number of vessels loitering, drifting, or slow steaming heading to San Pedro Bay is 66 vessels. This brings the total to 89 vessels.

Table.2 Port Of L.A. Operations Report (December 24, 2021).

Date		24-Dec-21					
Vessels in Port (POLA)		Containers	Tanker/ATB	General/Bulk	Vehicle	Cruse	Total
		17	0	1	0	0	18
Container Vessel Update	(A) Vessels within 40 NM	Port of Los Angeles		Port of Long Beach			Total
	Heading to San Pedro Bay	4		19			23
	(B) Vessels that are loitering/drifting/slow	Port of Los Angeles		Port of Long Beach			Total
	streaming heading to San Pedro Bay	34		32			66
	Total (A) + (B)	Port of Los Angeles		Port of Long Beach			Total
		38		51			89
	Average days within 40 NM awaiting berthing			18.9 days			
Terminal Update (POLA)	POLA Dewll time on terminal (As of Dec 24 - Last 30 days)	Local			Peak		
		6 days			11 days		
	POLA Dewll on-dock rail (As of Dec 24 - Last 30 days)	Local			Peak		
		2 days			13.4 days		

In accordance, also the time spent inside the port has been improved going down to an average of 7 days per vessel. However, for more details about the different types of vessels, we can find out that the dry bulk ships on week 47 of the year 2021 had the longest time in the port of 14 days, while Ro/Ro ships kept the smallest wait time in port, which is around 1 day.

The container ship waiting time fluctuated and has reached the maximum waiting time in week 47 of around 8 days, however, we can see the waiting time steadily decreasing and on week 51 it has reached around 5 days of waiting, with an improvement of 3 days.

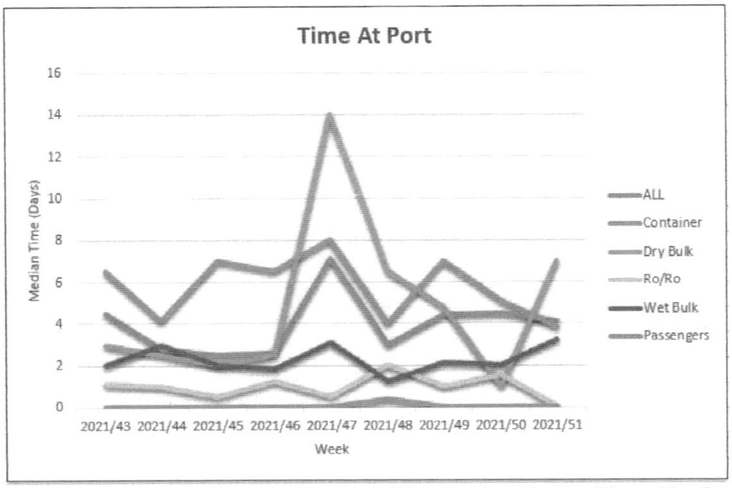

Fig .2 shows the median time spent at port of LA during weeks 43 to 51 of year 2021

The visible results on the ground have shown improvement in the congestion problem and promise significant progress for the future. Moreover, Monte Carlo simulation results suggested the same for the investigated values.

The total number of vessels in the Port of Los Angeles has shown a 39% chance to decrease from the maximum value of 21 ships, while also showing a 50% chance to increase in the future, while the remaining 11% was neglected due to exclusion.

The total number of container ships in San Pedro Bay has shown a 44% chance to decrease from the maximum value of 132, while also showing an 29% chance to increase in the future, while the remaining 27% was neglected due to exclusion.

The average number of days within the distance of 40NM awaiting berthing has shown a 45% chance to decrease from the maximum value of 132, while also showing a 17% chance to increase in the future, while the remaining 38% was neglected due to exclusion.

			Date			Max	Min	Ave
			9-Nov-21	24-Dec-21	7-Jun-22			
Vessels in Port POLA		Containers	17	17	14	17	14	16
		Tanker/ATB	3	0	0	3	0	1
		General/Bulk	0	1	2	2	0	1
		Vehicle	0	0	0	0	0	0
		Cruse	1	0	0	1	0	0.333333
		Total	21	18	16	21	16	18.33333
Container Vessel Update	(A)	Port of Los Angeles	20	4	0	20	0	8
		Port of Long Beach	6	19	0	19	0	8.333333
		Total	26	23	0	26	0	16.33333
	(B)	Port of Los Angeles	60	34	17	60	17	37
		Port of Long Beach	40	32	10	40	10	27.33333
		Total	106	66	27	106	27	66.33333
	(C)		132	89	27	132	27	82.66667
	(D)		15.5	18.9	0	18.9	0	11.46667
Terminal Update (POLA)	(E)	Local	9.7	6	4.3	9.7	4.3	6.666667
		Peak	9.7	11	11	11	9.7	10.56667
	(F)	Local	4.5	2	7.1	7.1	2	4.533333
		Peak	13.4	13.4	13.4	13.4	13.4	13.4

(A) Vessels within 40 NM Heading to San Pedro Bay
(B) Vessels that are loitering/drifting/slow streaming heading to San Pedro Bay
(C) Total A+B
(D) Average days within 40 NM awaiting berthing
(E) POLA Dewll time on terminal (As of Date - Last 30 days)
(F) POLA Dewll on-dock rail (As of Date - Last 30 days)

Total (POLA)	Chance of decrease	39%
	Chance of increase	50%
Container Total (SPB)	Chance of decrease	44%
	Chance of increase	29%
Average waiting days	Chance of decrease	45%
	Chance of increase	17%

Fig .3 Monte Carlo Simulation Results.

4. Discussion

The current methodology implicated by direct suppliers to the local markets, which only tackles their problems personally such as paying an extra fee or chartering their ships to deliver goods does not affect resolving the source problem of current maritime instability. Moreover, the direct involvement of using airline shipping or building an own air supply chain may seem like a radical solution, however, airway shipping will deliver fewer goods as per the difference in cargo volumes between the maritime and airway supply.

The satellite tracking technology at the time is giving a good amount of information on the maritime shipping vessels position and heading, however, adding some more live tracking elements to the requested shipment specifically for the importer, will reduce the uncertainty element with arrival dates, therefore, the stakeholder could have the inland shipper be available at the exact time of arrival to collect the shipment and move it inland towards the final destination, which in turn will reduce the need to stack and keep the containers waiting inside port after arrival, and avoid unnecessary work of restacking multiple times until the shipment is collected.

Currently, most ports are still facing lower capacity due to the pandemic spread, and to avoid any unnecessary infections that may happen at the terminals, therefore, they also reduced working hours creating the gap between the required hours and capacity to efficiently move with the process and the current capabilities. However, the success of working with full capacity and longer hours at the designated port has led to reduced congestion, and the port management should make it a priority to keep all the employees safe and able to fulfill their roles correctly to avoid more traffic ant the port. This alongside pushing the inland network to also coordinate and work at full capacity also reduces the container build-up at the port yard and allows new supplies to enter and be restacked in the available space.

Furthermore, creating a new inland space to collect shipments outside the ports and directly connected by railroad is a revolutionary method to allow the containers to stay waiting for collection, while at the same time creating more space in the port that will allow the exports to build up and wait to be shipped outwards, especially agricultural exports that need to be prioritized to avoid and deterioration.

Finally, the port automation will greatly increase the ability to meet the required work hours based on the current supply chain requirements, as minimal human interaction is required to keep the port running at full efficiency, also is safer for the humans working due to limited contact with the outside elements.

5. Conclusion

The maritime industry had difficulty adapting during the COVID-19 scenario because operations were not prepared, as seen by low percentage outcomes. As a result, maritime officials should be concerned about such a low proportion of complete preparedness and investigate why there was a lack of planning and trouble adapting. An attempt should be made to create a framework and guidelines for maritime stakeholders, a business continuity plan, a national logistics policy, a consolidated national institutional framework, and a disaster management plan for emergencies in order to lessen the effects of a pandemic.

Authorities and politicians must place a high priority on the safety of their personnel and other players in the marine chain. They should put in place the appropriate emergency response processes to keep ports and shipping operations operational. Thus, keeping the shipping sector operating at a suitable capacity is more concerning the workload expected and required.

The pandemic drives the sector to enhance its capacity for responding to disruptions, as well as to build better long-term plans for dealing with uncertainty, and to adopt more sustainable operations, despite the fact that it hurts the economy and the well-being of people. For instance, it is crucial that shipping firms continue to deliver food and medical supplies without hesitation. In order to respond to the changing environment, the shipping industry will need to become more agile and resilient. To quickly recover from setbacks and reestablish themselves stronger than before, they must focus on creating efficient response strategies and execution plans.

On the other side, the Covid-19 epidemic is likely to be a defining moment for the use of remote technology and automation in transportation. For example, an autonomous port may be able to manage staff changes with minimal interactions, reducing the risk of virus transmission. The pandemic might be a major catalyst for new technology adoption, collaborative solutions, and more efficient use of space and resources. In terms of avoiding pollution and optimizing energy efficiency, the innovation in the shipping system appears to be more sustainable, safe, efficient, and reliable.

References

[1] Corlett, R.T., Primack, R.B., Devictor, V., Maas, B., Goswami, V.R., Bates, A.E., et al., 2020. Impacts of the coronavirus pandemic on biodiversity conservation. Biol. Conserv. 246, 108571.

[2] Uddin, Md Bashir, Hasan, Mahmudul, Ahmed, Harun-Al-Rashid, et al., 23 March 2020. Novel coronavirus (COVID-19): Molecular evolutionary analysis, global Burden and possible threat to Bangladesh. PREPRINT (Version 1), Available at: Research Square. https://doi.org/10.21203/rs.3.rs-18985/v1.

[3] Liu, L., Miller, H.J., Scheff, J., 2020. The impacts of COVID-19 pandemic on public transit demand in the United States. PloS One 15 (11), e0242476. https://doi.org/ 10.1371/journal.pone.0242476

[4] UNCTAD, U., 2020. COVID-19: Shipping data hints to some recovery in global trade | UNCTAD [WWW Document]. UNCTAD. URL https://unctad.org/news/covid-19-shipping-data-hints-some-recovery-global-trade (accessed 6.8.22).

[5] De Beukelaer, C. (2021). COVID-19 border closures cause humanitarian crew change crisis at sea. *Marine Policy*, *132*(October 2020), 104661.

[6] IMO, I., 2020. 400,000 seafarers stuck at sea as crew change crisis deepens [WWW Document]. 400,000 seafarers stuck at sea as crew change crisis deepens. URL https://www.imo.org/en/MediaCentre/PressBriefings/Pages/32-crew-change-UNGA.aspx (accessed 6.8.22).

[7] Campbell, S.J., Jakub, R., Valdivia, A., Setiawan, H., Setiawan, A., Cox, C., Box, S., 1920. Immediate impact of COVID-19 across tropical small-scale fishing communities. Ocean Coast Manag. 200, 105485.

[8] Narasimha, P. T., Jena, P. R., & Majhi, R. (2021). Impact of COVID-19 on the Indian seaport transportation and maritime supply chain. Transport Policy, 110(May), 191–203. https://doi.org/10.1016/j.tranpol.2021.05.011

[9] Xu, L., Shi, J., Chen, J., & Li, L. (2021). Estimating the effect of COVID-19 epidemic on shipping trade: An empirical analysis using panel data. Marine Policy, 133(August), 104768. https://doi.org/10.1016/j.marpol.2021.104768

[10] Harris, P., Dall'Olmo Riley, F., Riley, D., Hand, C., 2017. Online and store patronage: a typology of grocery shoppers. Int. J. Retail Distrib. Manag. 45 (4), 419–445. https:// doi.org/10.1108/IJRDM-06-2016-0103.

[11] Pantano, E., Pizzi, G., Scarpi, D., Dennis, Ch, 2020. Competing during a pandemic? Retailers' ups and downs during the COVID19 outbreak. J. Bus. Res. 116, 209–213. https://doi.org/10.1016/j.jbusres.2020.05.036.

[12] Kirk, C.P., Rifkin, L.S., 2020. I'll trade you diamonds for toilet paper: consumer reacting, coping and adapting behaviors in the COVID-19 pandemic. J. Bus. Res. 117, 124–131. https://doi.org/10.1016/j.jbusres.2020.05.028.

[13] Adewopo, J. B., Solano-Hermosilla, G., Colen, L., & Micale, F. (2021). Using crowd-sourced data for real-time monitoring of food prices during the COVID-19 pandemic: Insights from a pilot project in northern Nigeria. *Global Food Security*, *29*, 100523. https://doi.org/10.1016/j.gfs.2021.100523

[14] Xu, L., Yang, S., Chen, J., & Shi, J. (2021). The effect of COVID-19 pandemic on port performance: Evidence from China. Ocean and Coastal Management, 209(February), 105660. https://doi.org/10.1016/j.ocecoaman.2021.105660

[15] Fernholz, T., 2021. What the shipping supply-chain crunch looks like from space. [online] Quartz. Available at: <https://qz.com/2067501/what-the-shipping-supply-chain-issues-look-like-from-space/> [Accessed 11 October 2021].

[16] Miller, G., 2021. Record shattered: 73 container ships stuck waiting off California. [online] FreightWaves. Available at: <https://www.freightwaves.com/news/record-shattered-61-container-ships-stuck-waiting-off-california?fw > [Accessed 11 October 2021].

[17] Gallagher, J., 2021. Biden administration pushing shippers, carriers to expand operating hours. [online] FreightWaves. Available at: <https://www.freightwaves.com/news/biden-administration-pushing-shippers-carriers-to-expand-operating-hours> [Accessed 11 October 2021].

[18] Martín-soberón, A. M., Monfort, A., Sapiña, R., Monterde, N., & Calduch, D. (2014). Automation in port container terminals. Procedia - Social and Behavioral Sciences, 160(Cit), 195–204. https://doi.org/10.1016/j.sbspro.2014.12.131

[19] General Administration of Customs, P.R.China.

[20] Indian Ports Association Report, 2021. http://www.ipa.nic.in/ accessed on 26.09.2021.

IOP Conf. Series: Earth and Environmental Science 1166 (2023) 012041 doi:10.1088/1755-1315/1166/1/012041

Risk Accident Analysis Due to the Effect of Wrong Loading on Ship Stability

S S Dewi[1], M Zaky[2], A Wahidin[1], and I P S Asmara[1]

[1]Department of Naval Architecture, Faculty of Engineering, Shipbuilding Institute of Polytechnic Surabaya, Surabaya, East Java, Indonesia

[2]Research and Development Division, Biro Klasifikasi Indonesia, Tg. Priok, Jakarta, Indonesia

ssinta@student.ppns.ac.id, zaky@bki.co.id, aangwahidin@ppns.ac.id, putusindhu@ppns.ac.id

Abstract. In this study, the effect of laying and shifting load is considered as one of the main factors of ship safety. Ferry Ro-Ro that carry vehicles and passengers were chosen as the object of study with the assumption that the vehicles are not lashed. Safety measurement is carried out through static stability analysis with criteria limits IMO Intact Stability Code 2008 and PM44 2021 and dynamic stability with roll motion limitations due to beam sea to result quantitative risk assessment data. The wave height is taken from the waters of the Bali strait with a Beaufort scale conversion of BF1 to BF6. Because of the vehicle is not lashed, the vehicle is simulated to shift horizontally as far as 10% to stack onto the starboard, while the assumption of passenger movement is vertical and horizontal. The biggest heel angle due to the shifting load is 9 degrees when all loads are stacked on the starboard (G1). Analysis of the calculation of static stability shows that several conditions do not meet the criteria with the greatest Roll Amplitude value of 39.56 deg at BF6 4 m. In dynamic stability conditions through maxsurf motion, the safe wave height for Ferry Ro-Ro operation is up to BF4 1.14 m. The highest risk of ship capsizing occurs when operating on BF5 with wave heights above 2 m.

1. Introduction

Based on the investigation report of National Transportation Safety Committee (KNKT), more than 120 cases of ship accidents occurred from 2010 to 2020. Ferry Ro-Ro is one type of ship that often got sea accidents. The waves are closely related to the ship's performance when operating considering that several routes in Indonesia have high waves due to the influence of seasonal winds. In June 2014 the wave height in the Bali Strait was between 1.5-2 meters and the highest wave occurred in the East monsoon with a height of 2.1 meters (Dikatama, 2019). Even the height of the waves in these position can reach a height of up to 4 meters. Conditions like this are very risky, especially for passenger ships. In some cases, the vehicle on the ferry ro-ro is not lashed. Whereas lashing is closely related so that the load does not shift while operating. The shift of load to one side of the ship causes a heel which affects the change in the center of gravity of the ship. The waves that come from the side of the ship (beam sea) will affect the ship's motion (Seakeeping), especially on the roll motion. Because the laying and shifting of load is very influential on the stability, it is necessary to do research by simulating the laying and shifting of ferry ro-ro load due to the effect of Beam Sea. The vehicle is simulated to shift horizontally to starboard, while the assumption of passenger movement is vertical and horizontal. This is to determine the fulfillment of the static stability criteria at each loading condition. For dynamic stability analysis carried out with variations in wave height of the Bali Strait conversion of the Beaufort scale. After doing the simulation, the maximum wave height that is suitable for the Ro-Ro Ferry will be obtained. Risk assessment will be carried out using quantitative methods, taking into account the results of meeting the stability criteria in each condition.

2. Shift Load Angle

The angle limits are obtained from calculations using empirical formulas and KNKT final report. The following limits the angle of shifting charge according to several sources.

Table 1. Shift Load Angle Limit Criteria

reference	µs	*θ (deg)*	Descripsion
KNKT	-	10 - 15	-
IMO CSS CODE	0,3	16,7	Steel-Rubber
J. Jia (Chalmer)	0,65	33	Dry Area
	0,37	20	Wet Area

2.1. Calculation With Empirical Formula

The calculation of the angle is obtained using the Pythagorean triangle formula with the assumption that the height used is the freeboard Ferry Ro-Ro height of 1.15m and the shifting distance of 13.8m. The following are the calculation steps:

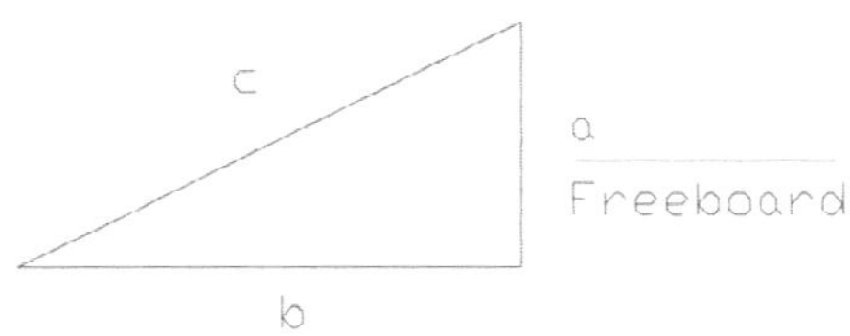

Figure 1. Freeboard Ferry Ro-Ro

$b^2 = c^2 - a^2$

$b^2 = (13.8)^2 - (1.15)^2$

$b^2 = 189{,}117$

$b^2 = \sqrt{189{,}117}$

$b = 13{,}752$

Based on Figure 1, it can be seen that (a) represents the freeboard height of ferry ro ro, (c) represents the possible distance the vehicle may shift. Next calculate the normal force acting on the vehicle. Through calculations with equation 1 assuming the weight of the car is 2.3 tons and the truck is 20 tons, the results of the calculation are as follows:

- **Car**

N = m x g cos α

N = Wcos α

N = 2300 x 10 x (b/c)

N = 22.916,66 N

- Truck

N = Wcos α

N = 20000 x 10 x (b/c)

N = 199.275,36 N

Then calculate the frictional force of the vehicle assuming **µs =0,325**. The following calculation results are obtained:

- **Car**

W sin $\alpha - f = m \times a$

W sin α = **μs.N**

23000 sin α = 0,325 x **22.916,66**

sin α = 0,324

- **Truck**

W sin $\alpha - f = m \times a$

W sin α = **μs.N**

200000 sin α = 0,325 x **199.275,36**

sin α = 0,324

While the calculation using the empirical formula for shifting objects the limit is 18 deg. After getting an angle of 18 deg, it can be made a limit when calculating the seakeeping, it will be known the size of the roll response motion with a certain wave height starting from BF1 to BF6, if the roll response amplitude (deg) > heel angle (18 deg), then the assumed load shift and the possibility of the Ferry Ro-Ro vessel capsizing will be even greater (high risk level category). In addition, this angle limitation will be used as a consideration in calculating the capsize probability of a Ro-Ro Ferry ship.

3. Static Stability Analysis

Ferry Ro-Ro hull modeling is the initial stage of this study. The modeling was carried out using 3D modeling of surface types in Maxsurf Modeler Advanced Software. The modeling was carried out with the aim of obtaining a ferry roro model used for the Maxsurf Stability analysis stage.

3.1. Loading Condition

The author uses several variations of vehicle loads on the main deck which are divided into three variations, namely Simulation of 1 vehicle 100% Truck, 2 vehicle simulation 50% Truck 50% Car and 3 vehicle simulation 100% Car. Based on these variations, with the same area, a variety of vehicles with different locations and numbers will be obtained following the deck layout of the Ro-Ro Ferry vehicle. As a support, a simulation of placing passengers vertically on each deck will be carried out to determine the effect on the stability condition of ferry roro.

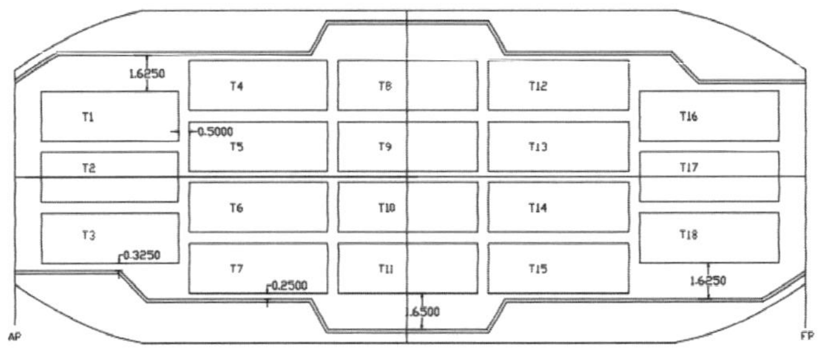

Figure 2. Initial Condition Simulation 1 (100% Truck)

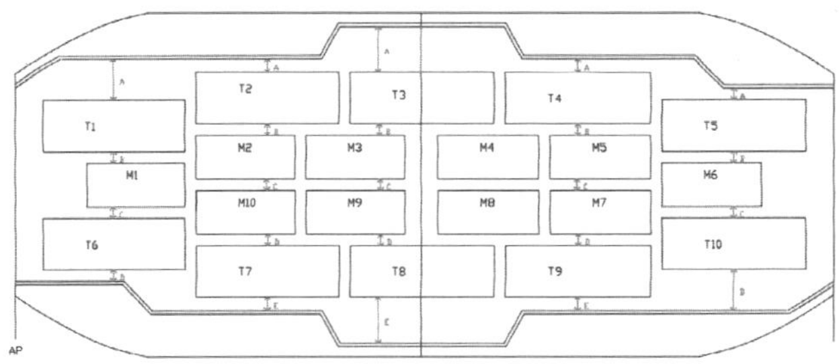

Figure 3. *Initial Condition Simulation 2 (50% Truck 50% Car)*

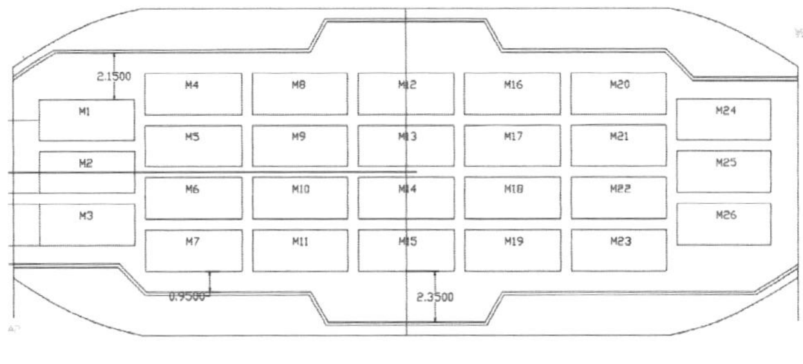

Figure 4. Initial Condition Simulation 3 (100% Car)

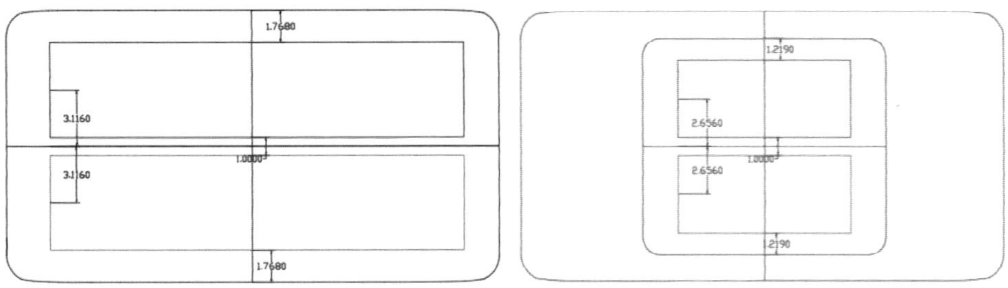

Figure 5. (a) Initial Condition Passenger Deck 1, **(b)** Initial Condition Passenger Deck 1,

Table 2. Simulation Matrix

Loading Condition		100 % Truck	50 % Truck 50 % Car	100 % Car
		1	2	3
Initial Condition	A	A1	A2	A3
Shift 10%	B	B1	B2	B3
Shift 20%	C	C1	C2	C3
Shift 30%	D	D1	D2	D3
Shift 40%	E	E1	E2	E3
Shift 50%	F	F1	F2	F3
Load Piled Up to Starboard	G	G1	G2	G3
100% of passengers on deck Passenger 1	H	H1	H2	H3
100% of passengers on deck Passenger 2	I	I1	I2	I3
Overlap	J	J1	J2	J3

Based on Table 2, it is known that due to variations in the types of vehicles on the main deck and the accumulation of passengers on one of the decks, a total of 30 conditions were obtained which were divided into conditions A1 to J1, conditions A2 to J2 and conditions A3 to J3.

3.2. The Effect of Loading Condition

Due to the horizontal shift of load (vehicles and passengers) simulation, different heel angle values are obtained in each simulation. In addition to horizontal shifts, a simulation of passenger stacking is carried out on one of the decks so that a difference in center of gravity (KG) will be obtained for each simulation of laying vertically. The following is a recap of the simulation results of horizontal load shift and vertical placement of passenger.

Table 3. Result of Loading Condition

Loading Condition	100 % Truck				50% Truck 50% Car				100% Car			
	Draught		KG (m)	HEEL (deg)	Draught		KG (m)	HEEL (deg)	Draught		KG (m)	HEEL (deg)
	Ap	Fp			Ap	Fp			Ap	Fp		
Initial Condition	1.888	1.867	4.617	0	1.573	1.547	4.622	0	1.170	1.139	4.615	0
Shift 10%	1.887	1.866	4.617	1,7	1.571	1.546	4.622	1,4	1.170	1.139	4.615	0,5
Shift 20%	1.879	1.857	4.617	4,5	1.569	1.543	4.622	2,6	1.169	1.138	4.615	1,1
Shift 30%	1.872	1.851	4.617	5,7	1.565	1.539	4.622	3,6	1.167	1.136	4.615	1,8
Shift 40%	1.872	1.850	4.617	5,8	1.560	1.534	4.622	4,6	1.166	1.135	4.615	2,1
Shift 50%	1.872	1.850	4.617	6,8	1.555	1.529	4.622	5,4	1.165	1.133	4.615	2,5
Load Piled Up to Starboard	1.845	1.824	4.617	9	1.546	1.520	4.622	6,6	`1,159	1.130	4.615	3,4
Overlap	1.876	1.854	4.781	5,2	1.563	1.537	4.761	4,2	1.165	1.133	4.669	2,4
100% of passengers on deck Passenger 1	1.888	1.867	4.585	0	1.573	1.547	4.580	0	1.170	1.139	4.554	0
100% of passengers on deck Passenger 2	1.888	1.867	4.759	0	1.573	1.547	4.802	0	1.170	1.139	4.882	0

In Table 3 above, it can be seen that the largest heel angle is in condition G in all simulations. With a 100% truck condition, especially in the G1 condition, which is 9.0 deg. The largest KG was obtained at 4.882m in condition I3 or the accumulation of all passengers on deck 2 with a vehicle variation of 100% cars.

3.3. Static Stability Calculation Analysis

After planning the laying conditions and load shifting, the longitudinal (LCG), transverse (TCG) and vertical (VCG) positions of the center of gravity are obtained in each condition. The result is used as a reference to create a loadcase which will be analyzed for its effect on static stability. In this study, the only condition considered is full loaded departure. The criteria used in this analysis refer to the Regulation of the Minister of Transportation of the Republic of Indonesia number 44 of 2021 (PM 44 2021) and IMO Intact Stability Code 2008. Furthermore, an analysis of the 30 conditions above was carried out against each of the existing criteria.

3.3.1. Criterion 1 : Area 0 to 30 >= 0,0550 m.rad

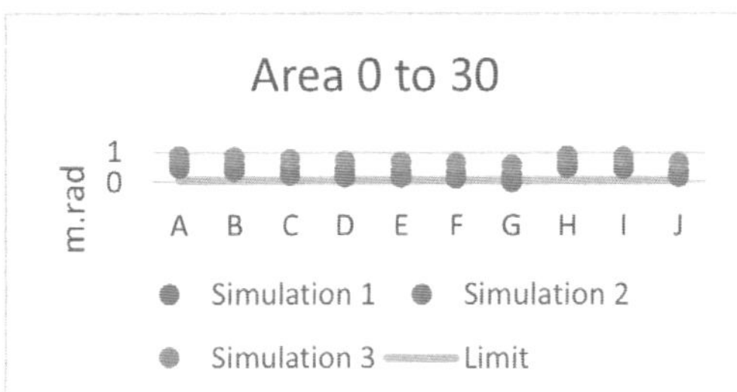

Figure 6. Graph of Results Criterion 1

From figure 6, it can be seen that almost all conditions meet the criteria for the 0 to 30 area except for the G1 condition because it has a value of 0.0057 m.rad where the value is less than the criteria limit of 0.055 m.rad

3.3.2. Criterion 2 : Area 0 to 40 >= 0,09 m.rad

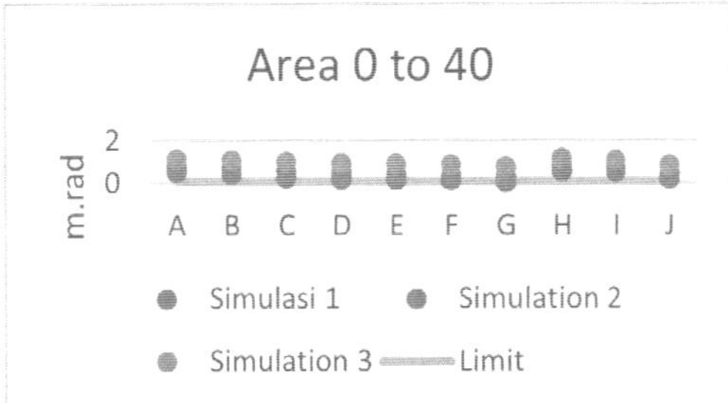

Figure 7. Graph of Results Criterion 2

From Figure 7. it can be seen that almost all conditions meet the criteria except in the G1 condition because it has a value of 0.0057 m.rad where this value is less than the criterion limit of 0.09 m.rad.

3.3.3. Criterion 3 : Area 30 to 40 >= 0,03 m.rad

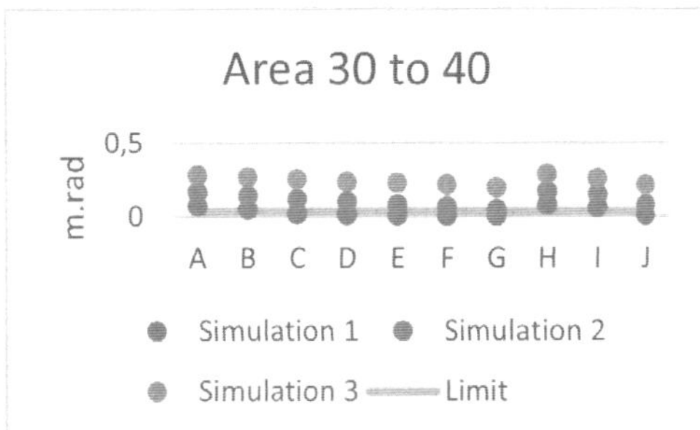

Figure 8. Graph of Results Criterion 3

From Figure 8. it can be seen that some of the conditions do not meet the criteria limits. There are 6 simulations that do not meet the criteria for criteria 3, namely conditions C1, D1, E1, F1, G1 and I1 because they have a value of less than 0.03m as the specified limit.

3.3.4. Criterion 4 : Max. GZ at 30 or Greater >= 0,2 m.

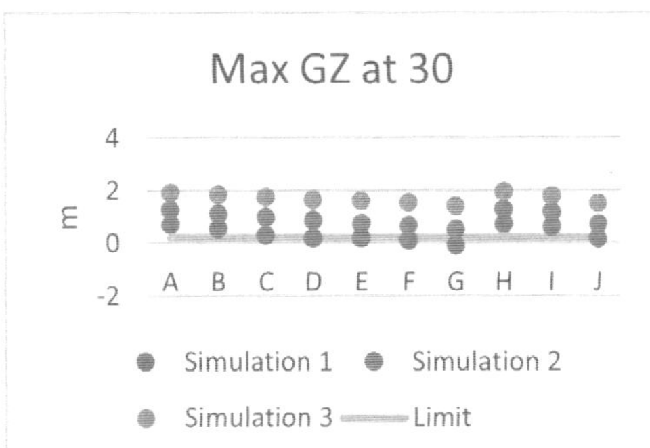

Figure 9. Graph of Results Criterion 4

From Figure 9. it can be seen that some conditions do not meet the criteria for the GZ value at an angle of 30. Conditions that do not meet the stability criteria are conditions D1, E1, F1, G1, and J1 because they have values below 0.2m so they do not meet the criteria limits. which is determined.

3.3.5. Criterion 5 : Initial GMt >= 0,2 m.

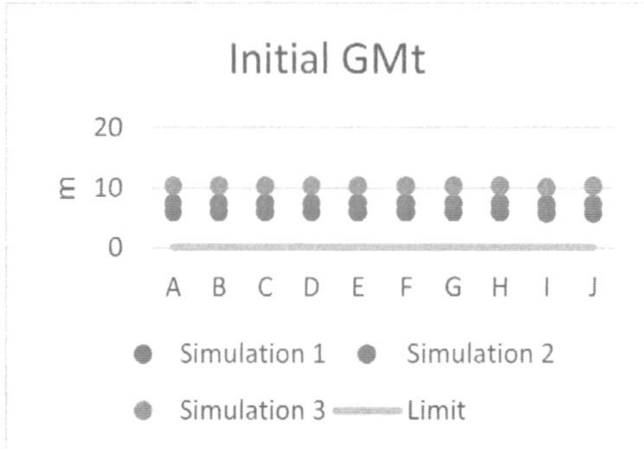

Figure 10. Graph of Results Criterion 5

From Figure 10. it can be seen that all conditions meet the criteria for GM values which should not be more than 0.15 m.

3.3.6. Criterion 6 : Angle of Max GZ >= 15 deg.

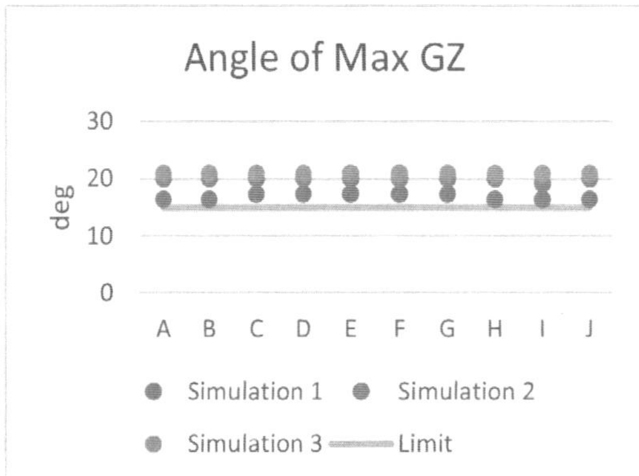

Figure 11. Graph of Results Criterion 6

From Figure 11. it can be seen that all conditions meet the criteria for the GM value which should not be less than 15 deg.

3.3.7. Criterion 7 : Area 0 to 30 or GZ Max >= 0,055-0,07 m.rad..

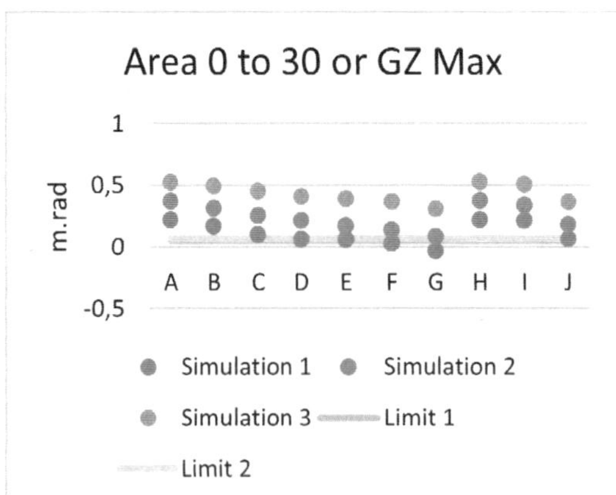

Figure 12. Graph of Results Criterion 7

From Figure 12, it can be seen that some of the conditions do not meet the GZ value limit in the area 0 to 30. There are 5 simulations that do not meet the 7 criteria limits, namely conditions D1, E1, F1, G1, and J1 because they have values less than 0.055-0.07 m.rad so that it does not meet the specified limit criteria.

3.3.8. Criterion 8 : Weather Criterion

1. *Angle of steady heel* ≤ 16°
2. *Angle of steady heel / Deck edge immersion angle* ≤ 80%
3. Area1 / Area2 ≥ 100%

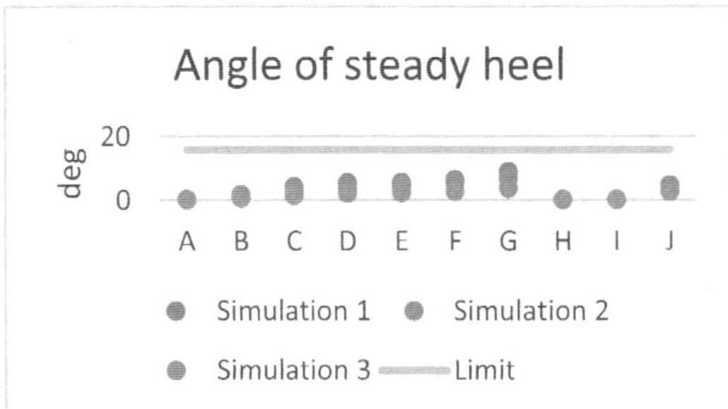

Figure 13. Graph of Results Criterion 8

From Figure 13, it can be seen that all conditions meet the criteria limits because the analysis value is not more than 16 degrees.

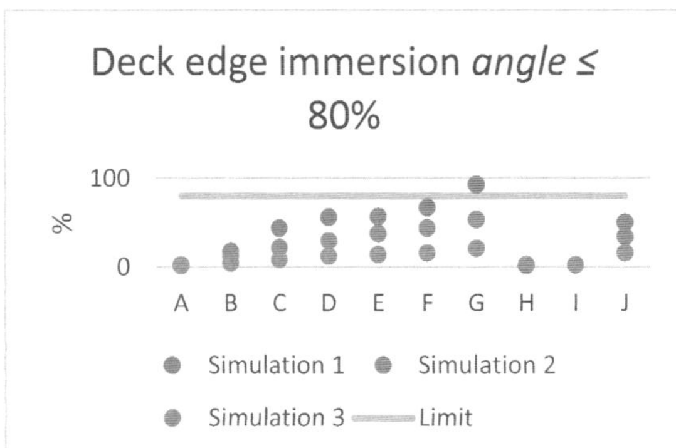

Figure 14. Graph of Results Criterion 8

From Figure 14, it can be seen that almost all conditions meet the criteria except for the G1 condition because it has an analysis value of more than 80%.

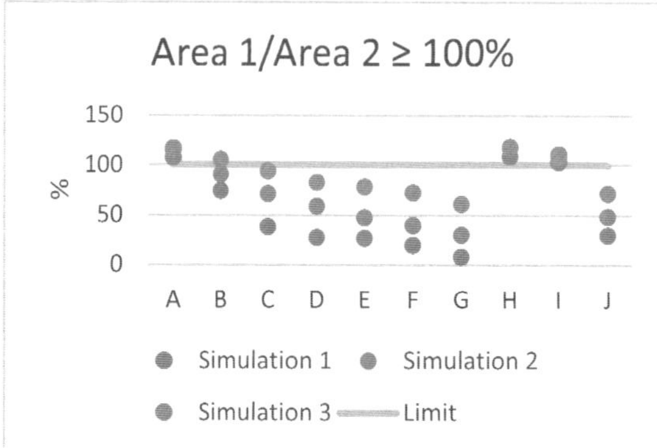

Figure 15. Graph of Results Criterion 8

From Figure 15, it can be seen that almost all conditions do not meet the criteria except for conditions A1, H1, I1, A2, H2, I2, A3, B3, H3 and I3 because they have an analysis value of more than 100%.

3.3.9. Criterion 9 : Passenger Crowding : Angle of Equilibrum <=10 deg

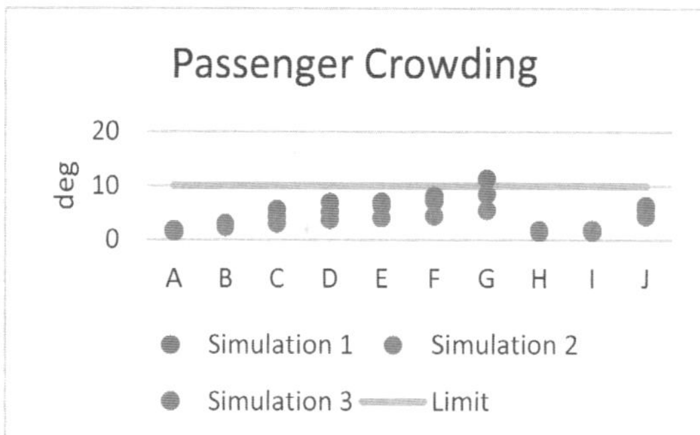

Figure 16. Graph of Results Criterion 9

From Figure 16, it can be seen that almost all conditions meet the slope angle limitation due to the gathering of passengers except for the G1 condition which has a value of 11.4 deg so it does not meet the criteria limit of 10 deg.

3.3.10. Criterion 10 : Turn Angle of Equilibrum <=10 deg

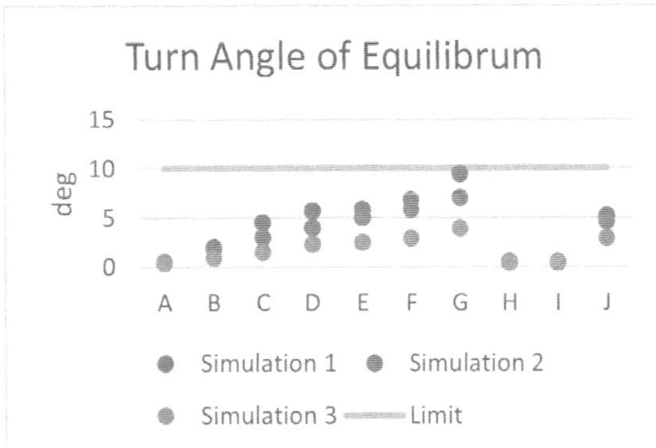

Figure 17. Graph of Results Criterion 10

From figure 17, it can be seen that all conditions meet the angle limit not exceeding 10 deg.

4. Dinamic Stability Analysis

At this stage, dynamic stability analysis begins with maxsurf motion software. The analysis was carried out using the conditions of each loadcase at the planned wave height, namely Beaufort 1 to Beaufort 6. Here are the wave heights for each Beaufort.

1. Beaufort 1 (BF1) 0.24 m
2. Beaufort 2 (BF2) 0.56 m
3. Beaufort 3 (BF3) 1.05 m
4. Beaufort 4 (BF4) 1.14 m
5. Beaufort 5 (BF5) 2.03 m
6. Beaufort 6 (BF6) 4.00 m

4.1. Criterion 11 : Angle of Roll Amplitude ≤ Angle of Deck Edge Immersion
4.1.1 Simulation 1 : 100% Truck

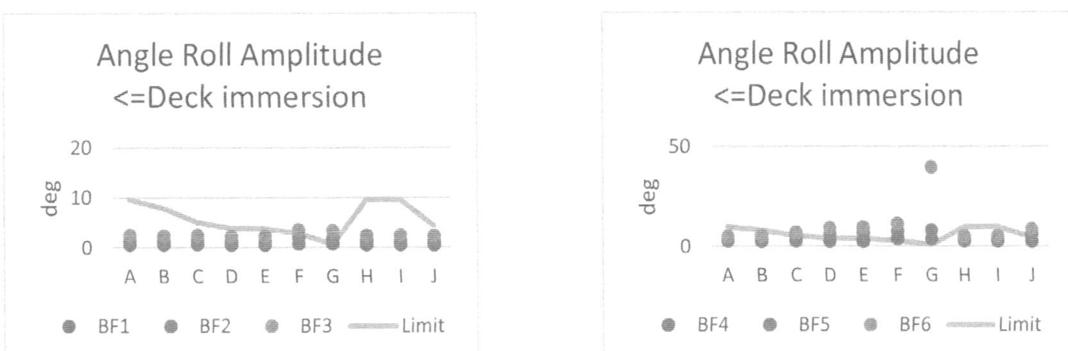

Figure 18. Graph of Result Criterion 11 in simulation 1

From figure 18, it can be seen that some of the conditions do not meet criteria 11, namely the G1 condition at BF1 0.24m, the G2 condition at BF2 0.56m and the F3 G3 condition at BF3 1.05m. Then the failure

IOP Conf. Series: Earth and Environmental Science 1166 (2023) 012041 doi:10.1088/1755-1315/1166/1/012041

occurs in conditions F1 and G1 on BF4 1.14m, conditions C1 D1 E1 F1 G1 and J1 on BF5 2.03m, and conditions C1 D1 E1 F1 G1 and J1 on BF6 4m because these conditions have a greater Angle Roll Amplitude value. than deck edge immersion

4.1.2 Simulation 2 : 50% Truck 50% Car

Figure 19. Graph of Result Criterion 11 in simulation 2

From Figure 19, it can be seen that in BF1 to BF3 all conditions still meet the expected criteria. However, the failure occurred in conditions G2 on BF5 2.03m, condition E1 F1 G1 on BF6 4m because these conditions had an Angle Roll Amplitude value greater than deck edge immersion.

4.1.3 Simulation 3 : 100% Car

Figure 20. Graph of Result Criterion 11 in simulation 3

From Figure 20, it can be seen that all conditions starting from BF1 to BF6 meet the criteria.

4.2. Criterion 12 : Probability Capsize ≤ 5%

Furthermore, the calculation for the probability of the capsize of the Ro-Ro Ferry ship is carried out with the following formula:

$$Pc = \exp\left(\frac{-\theta^2 \mathrm{DF}}{2m0c}\right)$$

To calculate the Pc (Probability Capsize) value, the maximum sinking angle value is limited by the minimum leakage angle or minDF obtained from the previous analysis. The angles considered are Vanishing Stability, Downfloading Point, Deck Edge Immersion, and the calculation of angle limits using empirical formulas. Because the smallest angle value is found in the Deck Edge Immersion, then that angle is used as a reference in calculating the capsize probability of a Ro-Ro Ferry ship.

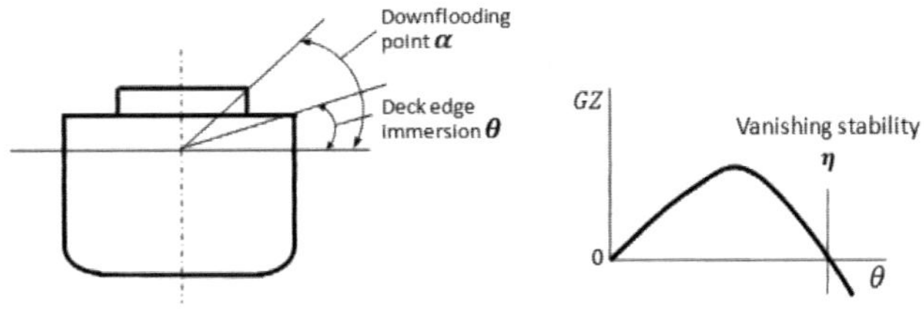

Figure 21 Downflooding Angle

(Zaky et al,2019)

4.2.1 Simulation 1 : 100% Truck

Figure 22. Graph of Result Criterion 12 in simulation 1

From Figure 22, it can be seen that some of the conditions in simulation 1, especially the conditions in BF4 to BF6, do not meet criteria 12 because they have a probability value of above 5%. Most failures occur in BF6 as the highest wave is 4 m.

4.2.2 Simulation 2 : 50% Truck 50% Car

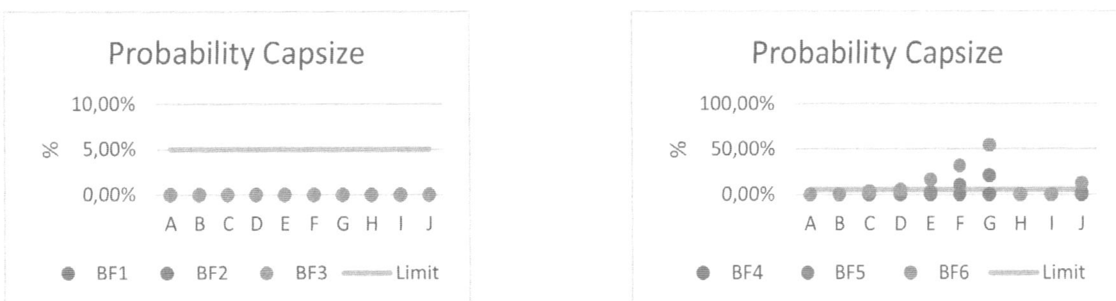

Figure 23. Graph of Result Criterion 12 in simulation 2

From Figure 23, it can be seen that all conditions in BF1 to BF3 meet criteria 12 with a capsize probability value of not more than 5%. In almost all wave heights, all conditions have a capsize probability of 0% except for BF3 with a probability of 0.01% in F1 conditions, 0.01% in G1 conditions. While some conditions in BF4 to BF6 do not meet criteria 12 because they have a capsize probability value above 5%. Most failures occur in BF6 as the highest wave is 4m. In BF4 all conditions meet with almost all conditions having a capsize probability except for F1 of 0.03% and G1 of 0.04%.

4.2.3 Simulation 3 : 100% Car

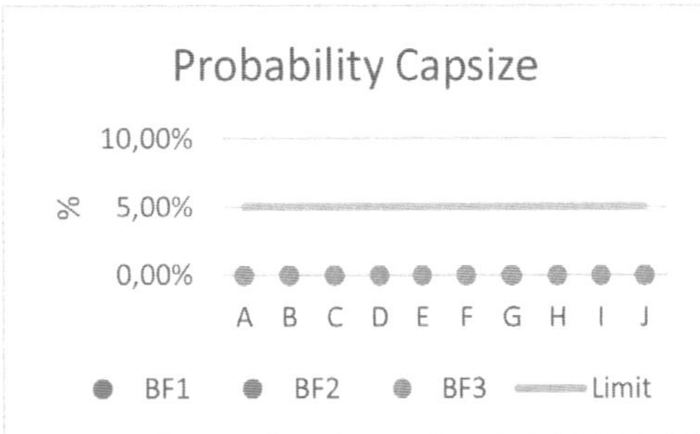

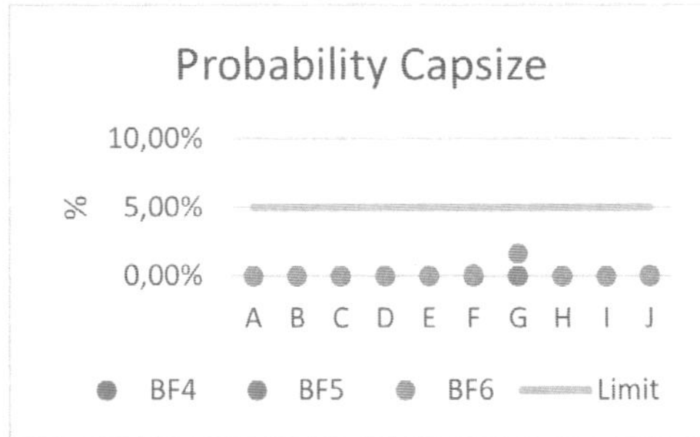

Figure 24. Graph of Result Criterion 12 in simulation

From Figure 24, it can be seen that all conditions meet criteria 12. In almost all wave heights, all conditions have a capsize probability of 0% except for BF6 as the most extreme wave height in this simulation. The probability is 0.02% in E1 condition, 0.1% in F1 condition and 1.64% in G1 condition.

5. Risk Assessment

The results of all conditions are taken into consideration for the Ferry Ro-Ro risk analysis. With quantitative methods, it is possible to know the probability and impact that will occur based on the conditions that have been planned so that a risk matrix can be made that maps the level of risk for each condition.

5.1. Likehood and Impact

Based on the results of fulfilling the static and dynamic stability criteria, a rating of the likelihood and risk impact of each condition at each wave height can be made from BF1 to BF6. The following rating table the likelihood and impact of the risk of a ship accident.

Table 4. Likehood Rating

Rating	Description	Criteria
1	Extremely remote	complete all stability criteria
2	Remote	complete almost all stability criteria except weather criteria
3	Resonably probable	partially meet the stability criteria (minimum initial positive GM)
4	frequent	all stability criteria are not met

Table 5. Impact Rating

Impact	minor	significant	Severe	catastrophic
Safety	One person injured or minor accident	More than one person is injured or seriously injured	One person dies or several people are seriously injured	More than one person died
Asset	Insignificant damage (occurs only to local equipment), no damage	Damage to the ship is not severe, minor damage to essential equipment (essential), damage to non-essential equipment	Severe damage, major damage to critical equipment (essential)	Loss of assets and total damage (total loss)
Environment	No spills, minor spills, no response required	Medium spill, needs limited response in short duration	Serious spills/requires significant commitment of resources	large spill/needs significant cleanup/full response scale

After knowing the likelihood and impact of the risk, the risk probability and impact rating (Rating) is determined for each wave height from BF1 to BF6. Determining the ranking of possible risks using the approach of many stability criteria that meet and do not meet each condition. As for the impact of an accident, it can be known through the impact that occurs as a result of these conditions. The impact in question is viewed from several aspects ranging from ship stability to asset damage due to shifting and stacking of cargo horizontally to the side of the starboard. The following is an analysis of the possibility and impact of each wave height.

5.2.Risk Matrix

From the probability and impact data, it is determined the tolerance limit or which requires further prevention. The following is a table of the risk matrix that has been determined in Table 6.

Table 6. Risk Matrix

	Possible		Impact			
			A	B	C	D
BF1	frequent	$F \ge 10$				
	Moderat (Resonably probable)	$0.1 \le F < 10$	D1; E1	F1;J1	G1	
	Remote	$10^{-3} \le F < 0.1$	B1; C1; H1, I1 B2; C2; H2; I2; D2; E2, C3, D3, E3; H3; I3	F2,J2, F3	G2, G3	
	Extremely remote	$10^{-5} \le F < 10^{-3}$	A1,A2, A3, B3			
			minor	significant	Severe	catastrophic

	Possible		Impact			
			A	B	C	D
BF2	frequent	$F \ge 10$				
	Moderat (Resonably probable)	$0.1 \le F < 10$	D1; E1	F1;J1	G1	
	Remote	$10^{-3} \le F < 0.1$	B1; C1; H1, I1 B2; C2; H2; I2; D2; E2, C3, D3, E3; H3; I3	F2,J2, F3	G2, G3	
	Extremely remote	$10^{-5} \le F < 10^{-3}$	A1,A2, A3, B3			
			minor	significant	Severe	catastrophic

	Possible		Impact A	Impact B	Impact C	Impact D
BF3	frequent	$F \geq 10$				
	Moderat (Resonably probable)	$0.1 \leq F < 10$	D1; E1	F1;J1	G1	
	Remote	$10^{-3} \leq F < 0.1$	B1; C1; H1, I1 B2; C2; H2; I2; D2; E2, C3, D3, E3; H3; I3	F2,J2, F3	G2, G3	
	Extremely remote	$10^{-5} \leq F < 10^{-3}$	A1,A2, A3, B3			
			minor	significant	Severe	catastrophic
BF4	frequent	$F \geq 10$				
	Moderat (Resonably probable)	$0.1 \leq F < 10$		D1; E1; F1;J1	G1	
	Remote	$10^{-3} \leq F < 0.1$	B1; C1; H1, I1 B2; C2; D2; E2, H2; I2; C3, D3, E3; H3; I3	F2,J2, F3; J3	G2, G3	
	Extremely remote	$10^{-5} \leq F < 10^{-3}$	A1,A2, A3, B3			
			minor	significant	Severe	catastrophic
BF5	frequent	$F \geq 10$				
	Moderat (Resonably probable)	$0.1 \leq F < 10$		C1; D1; E1	F1; J1	G1
	Remote	$10^{-3} \leq F < 0.1$	B1; H1, I1 B2; C2; H2; I2; C3, D3, E3; H3; I3	D2; E2; F2,J2, F3; J3	G2, G3	
	Extremely remote	$10^{-5} \leq F < 10^{-3}$	A1,A2, A3, B3			
			minor	significant	Severe	catastrophic
BF6	frequent	$F \geq 10$				
	Moderat (Resonably probable)	$0.1 \leq F < 10$		C1	D1; E1; F1; J1; F2; G2	G1
	Remote	$10^{-3} \leq F < 0.1$	B1; H1, I1 B2; C2; H2; I2; C3, D3, E3; H3; I3	D2; F3; J3	E2; J2, G3	
	Extremely remote	$10^{-5} \leq F < 10^{-3}$	A1,A2, A3, B3			
			minor	significant	Severe	catastrophic

Negligible, no need for further prevention
As Low As Resonably Practicable
Intolerable, risk can't be tolerated except in extraordinary circumstances

Based on this matrix, it is possible to map the conditions that should be watched out for in the voyage of the Ro-Ro Ferry. Therefore, a limitation is made that the color yellow is a condition that can still be tolerated in a risk. So that conditions that have a large and very high risk (high risk) or that are included in the red category in the matrix must be handled (treatment). The handling aims to minimize the impact of a condition so that the risk can be accepted.

IOP Conf. Series: Earth and Environmental Science 1166 (2023) 012041 doi:10.1088/1755-1315/1166/1/012041

5.3. Mitigation or Recommendation

Based on the results of the risk matrix, there are mitigation or preventive measures to be more alert when the ship sails. The following is the mitigation of the risk matrix in Table 7.

Table 7. Risk Mitigation

Number	Initial Risk Level	Mitigation
1	Minor	Vehicle's lashing
2	Significant	Vehicle lashing and fender placement on the side of the ship
3	Severe	Need to be careful not to operate in wave conditions above BF5 2 m
4	Castatrophic	Need to be careful not to operate in wave conditions above BF5 2m

After a treatment has been carried out as described in table 4.18, conditions that need to be watched out for, especially for conditions with a red rating on the risk matrix, the condition that originally received a red rating turns into a yellow rating with a much lower impact than before. However, in conditions that are located at high BF still require regular monitoring. The risk matrix is obtained as follows:

Table 8. Risk Matrix after Treatment

	Possible		Impact			
			A	B	C	D
BF1	frequent	$F \geq 10$				
	Moderat (Resonably probable)	$0.1 \leq F < 10$	D1; E1	F1;J1		
	Remote	$10^{3} \leq F < 0.1$	B1; C1; H1, I1 B2; C2; H2; I2; D2; E2, C3, D3, E3; H3; I3	F2,J2, F3	G1, G2, G3	
	Extremely remote	$10^{5} \leq F < 10^{3}$	A1,A2, A3, B3			
			minor	significant	Severe	catastrophic

	Possible		Impact			
			A	B	C	D
BF2	frequent	$F \geq 10$				
	Moderat (Resonably probable)	$0.1 \leq F < 10$	D1; E1	F1;J1		
	Remote	$10^{3} \leq F < 0.1$	B1; C1; H1, I1 B2; C2; H2; I2; D2; E2, C3, D3, E3; H3; I3	F2,J2, F3	G1, G2, G3	
	Extremely remote	$10^{5} \leq F < 10^{3}$	A1,A2, A3, B3			
			minor	significant	Severe	catastrophic

IOP Conf. Series: Earth and Environmental Science 1166 (2023) 012041 doi:10.1088/1755-1315/1166/1/012041

	Possible		Impact			
			A	B	C	D
BF3	frequent	$F \geq 10$				
	Moderat (Resonably probable)	$0.1 \leq F < 10$	D1; E1	F1;J1		
	Remote	$10^{-3} \leq F < 0.1$	B1; C1; H1, I1 B2; C2; H2; I2; D2; E2, C3, D3, E3; H3; I3	F2,J2, F3	G1, G2, G3	
	Extremely remote	$10^{-5} \leq F < 10^{-3}$	A1,A2, A3, B3			
			minor	significant	Severe	catastrophic

	Possible		Impact			
			A	B	C	D
BF4	frequent	$F \geq 10$				
	Moderat (Resonably probable)	$0.1 \leq F < 10$		D1; E1; F1;J1		
	Remote	$10^{-3} \leq F < 0.1$	B1; C1; H1, I1 B2; C2; D2; E2, H2; I2; C3, D3, E3; H3; I3	F2,J2, F3; J3	G1, G2, G3	
	Extremely remote	$10^{-5} \leq F < 10^{-3}$	A1,A2, A3, B3			
			minor	significant	Severe	catastrophic

	Possible		Impact			
			A	B	C	D
BF5	frequent	$F \geq 10$				
	Moderat (Resonably probable)	$0.1 \leq F < 10$		C1; D1; E1		
	Remote	$10^{-3} \leq F < 0.1$	B1; H1, I1 B2; C2; H2; I2; C3, D3, E3; H3; I3	D2; E2; F2,J2, F3; J3	F1, G1, J1, G2, G3	
	Extremely remote	$10^{-5} \leq F < 10^{-3}$	A1,A2, A3, B3			
			minor	significant	Severe	catastrophic

	Possible		Impact			
			A	B	C	D
BF6	frequent	$F \geq 10$				
	Moderat (Resonably probable)	$0.1 \leq F < 10$		C1		
	Remote	$10^{-3} \leq F < 0.1$	B1; H1, I1 B2; C2; H2; I2; C3, D3, E3; H3; I3	D2; F3; J3	D1, F1, G1, J1, E2; F2; G2; J2, G3	
	Extremely remote	$10^{-5} \leq F < 10^{-3}$	A1,A2, A3, B3			
			minor	significant	Severe	catastrophic

6. Conclusion

Based on the results of the analysis that has been done, it can be said that:

1. Horizontal load shift affects:
- The largest heel angle occurs due to the entire load piling onto the starboard, which is 9.0 degrees.
- Referring to the stability criteria according to the IMO Intact Stability Code 2008 and PM 44 of 2021 and the dynamic stability criteria, it is known that some conditions do not meet the criteria. In the analysis that has been done, simulation 1 becomes a simulation that has several conditions that do not meet the stability criteria, especially in the conditions of F1 and G1.

2. The effect of vertical passenger position on each deck affects:
- There was a difference in center of gravity (KG) in conditions H and I, where the accumulation of passengers on deck 2 (condition I) resulted in a KG greater than the accumulation on passenger deck 1 (condition H).
- Referring to the stability criteria according to the IMO Intact Stability Code 2008 and PM 44 of 2021 as well as dynamic stability criteria, the stacking conditions of all passengers on one of the decks still meet all the specified stability criteria. It can be seen that the stability of the stacking conditions on passenger deck 1 (condition H) is better than the stacking conditions on passenger deck 2 (condition I).

3. The biggest roll amplitude is in the G condition. However, the most extreme condition is the G1 condition at a BF6 4 m roll amplitude of 39.59 degrees
4. Simulation 1 has many unsafe conditions at several wave heights. As the most extreme condition, the condition of G1 (100% truck load and close to the starboard) so that there is no wave height that meets these conditions. For F1 conditions, the wave height that is still safe is BF2 as high as 0.56m. Apart from F1 and G1 conditions, most of the conditions meet up to the BF4 wave height as high as 1.14m. For Simulation 2, most of the conditions meet the criteria up to the BF4 wave height. As for simulation 3, all conditions are met for all wave heights from BF1 to BF6.
5. The highest risk of ship capsizing occurs when operating on BF5 with wave heights above 2 m.

Acknowledgments

The authors would like to thank at Biro Klasifikasi Indonesia and Department of Naval Architecture, Faculty of Engineering, Shipbuilding Institute of Polytechnic Surabaya for their support and valuable comments on this research.

References

[1] Azhar Ali, & Kristiyono Tri Agung. (2019). Small Ship Stability (Vol. 53, Issue 9).

[2] Dikatama, M. (2018). Analysis of Extreme Wave Heights With Return Periods Related to Shipping Safety in Bali Strait Waters.

[3] Education, Minister. (2015). Fundamentals of Handling of Merchandise and Electronic Merchandise Payload Arrangements 1.

[4] IMO. (2008). Intact Stability Code.

[5] Kusuma, P. (2008). "Lashing System Analysis on Ferry". University of Indonesia.

[6] Transporatations, Minister. (2008). Ship stability. Minister of Transportation Regulation PM.44/2021

[7] Zaky, et., al. (2019). The Development of Freeboard Formula for Open Top Vessels in Indonesian Waterways, Proceeding of ICEST conference 2019. Makassar.

Effects of adding sideboard and longitudinal bulkhead construction on ship stability of open-top cargo ship

Amron Rosyada[1], R Dimas Endro W[2], and M Arif Kurniawan[3]

[1]D4 Marine Engineering Study Program, Shipbuilding Institute of Polytechnic Surabaya, Indonesia
[2]Piping Engineering Department, Shipbuilding Institute of Polytechnic Surabaya, Indonesia
[3]Research and Development Division of Biro Klasifikasi Indonesia, Indonesia

[1]rosyadaamron@gmail.com, [2]dimasend@ppns.ac.id, [3]arif.kurniawan@bki.co.id

Abstract. The cargo cover removal on cargo ship will surely affect the technical condition of ship accomplishment that is regulated by rule. In this study, the author would like to evaluate the down-flooding point and free surface effect on open-top cargo ship to the stability acceptance criteria using IMO and DNV test procedures. The simulation result shows that the ship stability value is not in the range of the value required with IMO A.749(18) stability criteria. The failures are caused by lower down-flooding point and free surface effect. The Construction modifications effect by adding sideboard and longitudinal bulkhead is believed can increase the stability value. The simulation result using modified construction shows that the stability value is getting better. Using IMO MSC.Circular608 test procedure, the green-water is still allowed to enter cargo room of modified ship, 633,56 m3 at 25% load, 428,87 m3 at 50% load, and 194.94 m3 at 75% load. It is better than the previous ship that the green-water is not allowed at all. Using DNV test procedure, all of the modified ship stability values are accepted by IMO A.749(18). The value of area 30-40° is 21.06~21.00 m.deg and the value of maximum GZ angle is 28.2°.

1. Introduction

One of the solutions to increase the effectivity in ship loading and unloading process is by removing the cargo cover. However, that modification surely changes the safety rule to the ship. The opened cargo room will cause the new down flooding points appearance which become the way green-water entering the ship hull during heeling or because of weather. Due to its entry, there will be free surface effect to the ship which is greatly affect stability value of the ship [1]. The lower down-flooding point causes the poor down-flooding angle. So, in order to fix it, the down-flooding point height must be increased. For the free surface effect problem, the best way to overcome it is by adding longitudinal bulkhead into the cargo room [2]. Based on those problems, the Author would like to identify a coal carrier cargo ship which was modified by removing its cargo cover. Then, the ship will be recalculated its stability value using testing procedure by IMO MSC Circular608 about Open-Top Container Ship and by DNV about Hatch-coverless Ship. There will be construction engineering to prevent poor down-flooding angle and to overcome free surface effect by adding sideboard and longitudinal bulkhead. The stability value of previous model and this construction modifications will be compared and must be accepted by stability criteria of IMO A.749(18).

IOP Conf. Series: Earth and Environmental Science 1166 (2023) 012042 doi:10.1088/1755-1315/1166/1/012042

2. Method

In this section, there are several methods which are used to calculated the stability value of the ship started from the ship modelling based on its principal dimension to the stability simulation.

2.1. Ship Dimension

The principal data of the ship and its general arrangement based on its stability booklet are shown below and in Figure 1.

LOA	= 68,50 m
LWL	= 67,46 m
Breadth	= 13,00 m
Depth	= 4,200 m
Draft	= 3,125 m
Displacement	= 2489.485 ton
LWT	= 496,535 ton

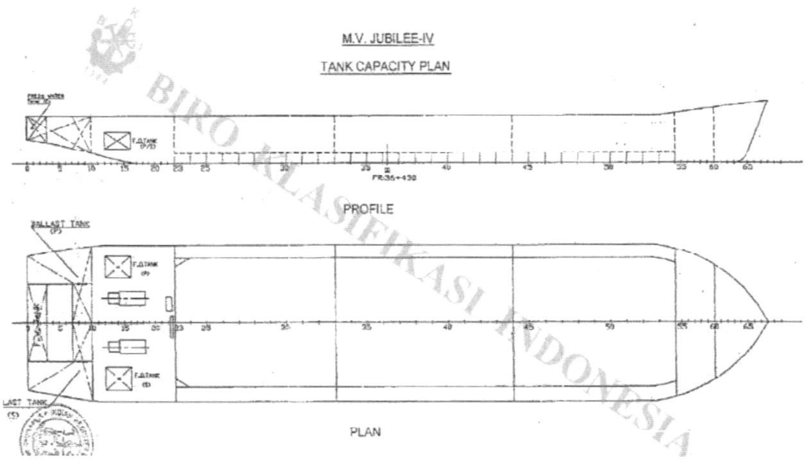

Figure 1. General Arrangement

2.2. Ship Modelling

The ship modelling process is based on its general arrangement based on its stability booklet. From that general arrangement, then draw the 3D model using Autocad and Maxsurf Modeller software. The model must be validated by DNV Rules about Loading Computer Systems (Lcs) For Stability and Longitudinal Strength [3]. The Acceptable Tolerance for stability Aspect is shown in Table 1 below.

Table 1. Acceptable Tolerance for stability Aspect.

Hull Form Dependent Parameters		Tolerance
Displacement	(t)	2%
LCB	(m)	1%/50 cm
LCF	(m)	1%/50 cm
KB	(m)	1%/5 cm
KMt	(m)	1%/5 cm
KML	(m)	1%/50 cm
MTc	(MT-m/cm)	2%

The next step after getting validated model is determining room definition. It is like making some tanks and compartments into the ship by inputting the coordinates based on data of tanks and room positions written in stability booklet. Because there will be flooding simulations, the cargo room is defined as tank.

2.3. Defining The Loading Condition (Load-case)

The load on the ship is regulated by IMO MSC Circular.608 and DNV part 6 chapter 5. Based on IMO MSC Circular.608 test procedure, there are several points must be noticed [4].

1. All of cargo holds are treated as flooded by seawater (70% permeability) in variations of loads
2. The effect of free surface caused by flooding water must be considered
3. The ship must have residual stability

From those points, the load-case for IMO test procedure is determined as shown in Table 2.

Table 2. IMO Load-case

Load-case	Consumable tanks	Cargo Load	Fluid in Cargo room
25% Loaded	100%	25%	xx%[a]
50% Loaded	100%	50%	xx%[a]
75% Loaded	100%	75%	xx%[a]
100% Loaded	100%	100%	xx%[a]

[a]The variable of flooding water is simulated up to its maximum limit

The IMO simulation is intended to obtain information about how much the maximum flooding water allowed in cargo room with variations of ship load is. On the other hand, the stability value must be in acceptance of stability criteria. IMO MSC Circular608 basically discusses about container ship. However, this research is using cargo ship. So, the result of using this test procedure will be used to compare the stability performance between previous construction and modified ship with additional construction.

Turning into DNV part 6 chapter 5, this chapter discusses about Hatch-coverless cargo ship. In the testing process, there are several points that must be noticed [5].

1.Cargo holds are treated as 2 m height flooded
2.There is no load at cargo room
3.The flooding water doesn't pour out during heeling

From the first point, there is a conversion from 2 m height to percentage volume (%Vfw). The conversion is calculated below.

$$\%\text{Vfw} = \frac{t_{fw}}{t_{cr}}\, x\ 100\% \qquad (1)$$

Where,

tfw = Height of flooding water
= 2 m

tcr = Height of cargo room from tank-top
= 3.25 m

So,

$\%V_{fw}$ = 61,54%

From those points, the load-case for DNV test procedure is determined as shown in Table 3.

Table 3. DNV Load-case

Load-case	Consumable tanks	Cargo Load	Fluid in Cargo room
Ballast Condition Departure	100%	0%	61,54%
Ballast Condition Arrival	10%	0%	61,54%

2.4. *Construction Modifications*

The modifications on the ship that will be simulated are by adding sideboard and longitudinal bulkhead. The sideboard construction is to increase down-flooding point. So, the down-flooding angle can be expected to be increased too. The sideboard construction is planned to surround above cargo room with 2.4 m height. This refers to the sideboard height on the 220-240 ft barge [6]. Because of this construction, the height of down-flooding point is set as 6.6 m from the previous as 4.2 m. the calculation of additional weight due to this construction is shown in Table 4.

Table 4. Sideboard Construction

Item	Dimension	Density (ton/m^3)	Quantity	Weight (ton)
Sideboard Plate (P & S)	46.5 m x 2.4 m x 8 mm	7.85	2	14.02
Sideboard Stiffener (L.125 x 75 x 8 mm)	46.5 m	7.85	8	4.48
Stanchion (HB.200 x 200 x 8 x12 mm)	2.4 m	7.85	24	2.80
Support Stanchion (HB.200 x 200 x 8 x12 mm)	2.4 m	7.85	24	2.80
Sideboard Plate (F & A)	10.6 m x 2.4 m x 8 mm	7.85	2	3.20
Sideboard Stiffener (L.125 x 75 x 8 mm)	10.6 m	7.85	8	4.06
Stanchion (HB.200 x 200 x 8 x12 mm)	2.4 m	7.85	8	0.94
Support Stanchion (HB.200 x 200 x 8 x12 mm)	2.4 m	7.85	8	0,94
Total				33.24

So, there must be recalculating the percentage of flooding water when using DNV test procedure. The calculation is based on Formula 1 with a change on t_{cr} from 3.25 m to 5.65 m height. Then, the percentage of fluid in cargo room of modified ship becomes 35,4 %.

The second modification is adding longitudinal bulkhead construction. This modification is to overcome free surface effect caused by flooding treatment in cargo room. The construction is planned as long as cargo room length (46.5 m) with the height from tank-top to the top of cargo room (5.65 m). The calculation of additional weight caused by this construction is shown in Table 5 below. Due to this construction addition, the vertical centre of gravity of the light ship was changed become 2.875 m from the previous as 2.65 m.

Table 5. Longitudinal Bulkhead Construction

Item	Dimension	Density (ton/m^3)	Quantity	Weight (ton)
Bulkhead Plate	46.5 m x 5.65 m x 10 mm	7.85	-	20.62
Longitudinal Ordinary Siffener (L.80 x 80 x 8)	46.5 m	7.85	6	2.66
Bulkhead Vertical Web Stiffener (L.300 x 100 x 10)	5.65 m	7.85	37	6.40
Senta (L.300 x 100 x 10)	46.5 m	7.85	1	1.42
Total				31.1

2.5. Stability Simulation
The stability value must be accepted by IMO A.749 -stability criteria [7] as shown in Table 6.

Table 6. Stability Criteria of IMO A.749(18)

Criteria	Value	Units
Area 0-30°	3.151	m.deg
Area 0-40°	5.157	m.deg
Area 30-40°	1.179	m.deg
Max GZ at 30°	0.200	m
Angle of Max GZ	25.0	deg
Initial GMt	0.150	m

The simulation is carried out using Maxsurf Stability Advance software. The ship stability value of previous model and modified construction model will be compared and analysed.

3. Result and Analysis

3.1 Ship Model
The ship's 3D hull model from the previous ship construction is shown in Figure 2. This model was validated by DNV rules (see Table 1). All hydrostatic data from Maxsurf's model are compared with hydrostatic data from stability booklet. The result shows that the differences (corrections) between both data are still below the maximum allowed tolerance for the parameters. The validation result can be seen in Table 7.

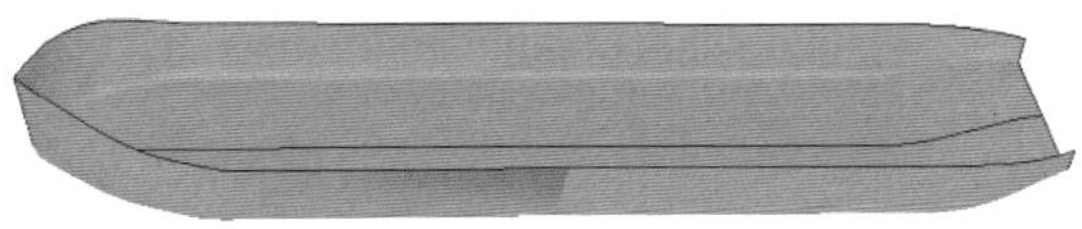

Figure 2. Previous Ship 3D Model

The 3D hull model of ship with additional constructions (Sideboard and Longitudinal Bulkhead) is shown in Figure 3.

Figure 3. Modified Ship 3D Model

These two models which both are made from Maxsurf Modeller Advanced software will be processed into Maxsurft Stability Advanced software to determine room definitions, input load-cases, and calculate stability value.

Table 7. Result of Model Validation

Hull Form Dependant Parameters	Data of Model	Data of Stability Booklet	Correction	Tolerance	Status
Displacement (t)	2489.00	2489.49	0.02%	2%	Pass
LCB (m)	31.92	31.98	0.21%	1%/50 cm	Pass
LCF (m)	30.68	30.80	0.39%	1%/50 cm	Pass
KB (m)	1.62	1.62	0.22%	1%/5 cm	Pass
KMt (m)	6.20	6.23	0.49%	1%/5 cm	Pass

Hull Form Dependant Parameters	Data of Model	Data of Stability Booklet	Correction	Tolerance	Status
KML (m)	119.32	120.31	0.83%	1%/50 cm	Pass
MTc (MT-m/cm)	45.22	44.40	1.86%	2%	Pass

3.2 Room Definition

The result of defining rooms in ship hull of previous ship and modified ship using Maxsurf Stability Advance software is shown in Figure 4 and Figure 5 below.

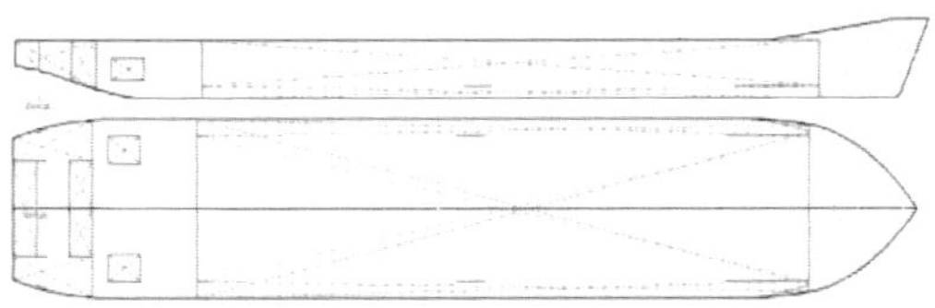

Figure 4. Previous Ship Room Definitions

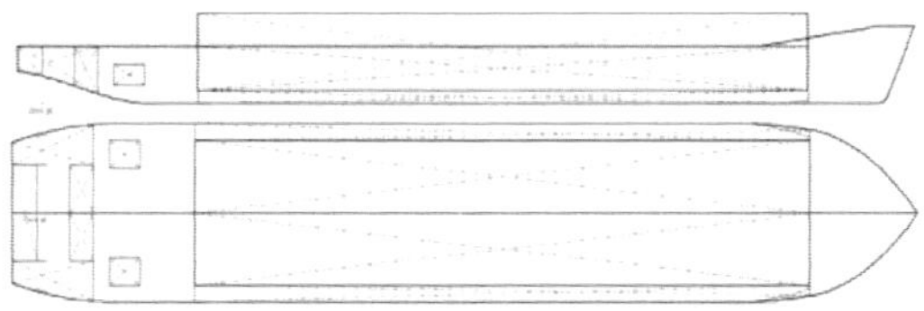

Figure 5. Modified Ship Room Definitions

3.3 Stability Simulation Result

1. IMO Test Procedure

This test is obtained to find out the maximum volume of flooding water allowed in cargo room. The result of the simulation is shown in Table 8.

Table 8. Simulation Result using IMO Test Procedure

Load-case	Fluid in Cargo room (m^3)	
	Previous	Modified
25% Loaded	0	633.56
50% Loaded	0	428.87
75% Loaded	0	194.94
100% Loaded	0	0

From Table 8, the result shows that the ship with additional construction of sideboard and longitudinal bulkhead is better than the previous construction. At 25% load, 633.56 m^3 of flooding water is still allowed to fill the cargo room from the previously not at all. At 50% load, 428.87m^3 of flooding water is still allowed to fill the cargo room from the previously not at all. At 75% load, 194.94 m^3 of flooding water is still allowed to fill the cargo room from the previously not at all.

2. DNV Test Procedure

This test is obtained to find out the ship stability value. The result of the simulation is shown by graphs in several points based on item of stability criteria below.

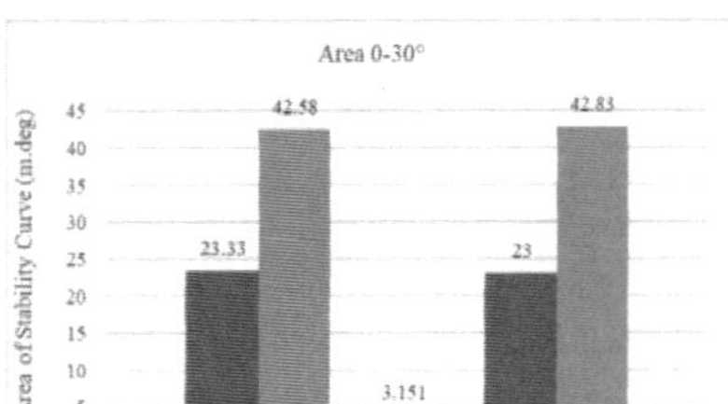

Figure 6. Area 0-30°

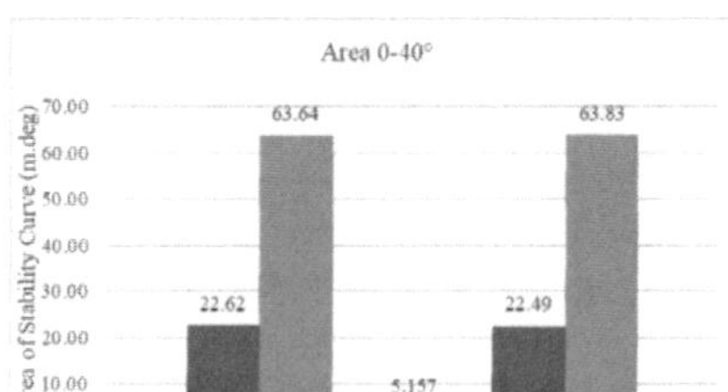

Figure 7. Area 0-40°

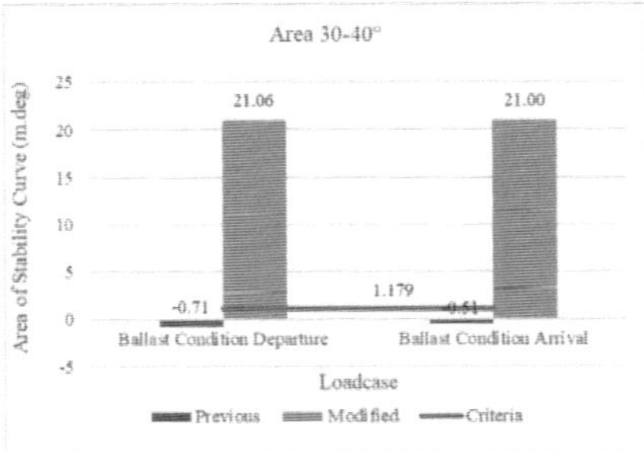

Figure 8. Area 30-40°

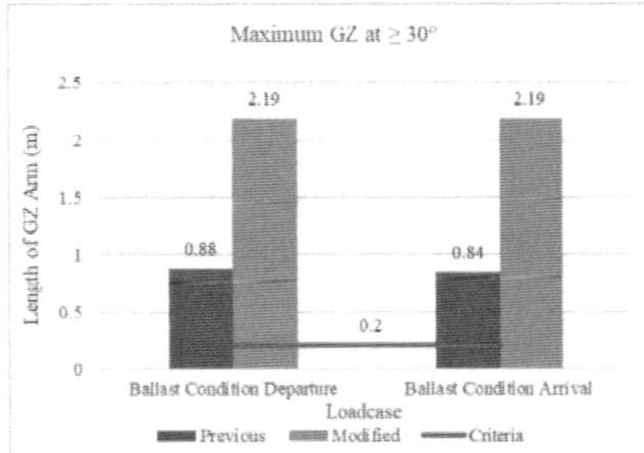

Figure 9. Maximum GZ at ≥ 30°

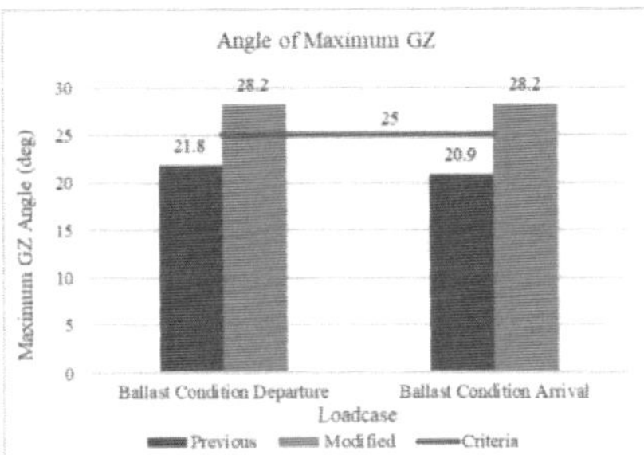

Figure 10. Angle of Maximum GZ

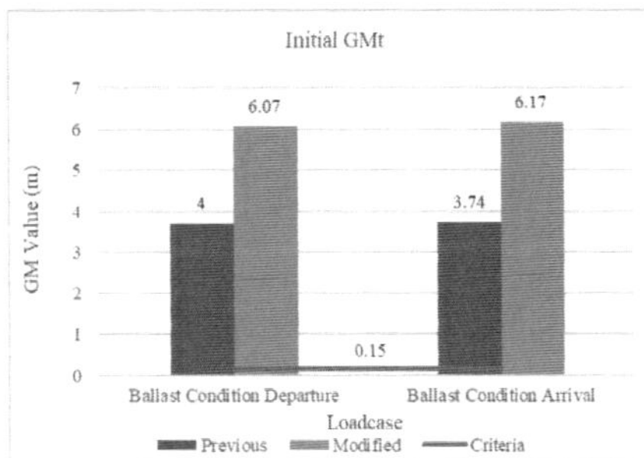

Figure 11. Initial GM

- Area 0-30°

Figure 6 below shows that the value of area 0-30° for previous ship is 23.33 m.deg in ballast condition departure and 23 m.deg in ballast condition arrival. The value of area 0-30° for modified construction ship is 42.58 m.deg in ballast condition departure and 42.83 m.deg in ballast condition arrival. Both of the area value of previous and modified ship are accepted or exceeding the stability criteria (3.151 m.deg) by IMO A.749(18). But the modified ship has better value than the previous ship.

- Area 0-40°

Figure 7 shows that the value of area 0-40° for previous ship is 22.62 m.deg in ballast condition departure and 22.49 m.deg in ballast condition arrival. The value of area 0-40° for modified construction ship is 63.64 m.deg in ballast condition departure and 63.83 m.deg in ballast condition arrival. Both of the area value of previous and modified ship are accepted or exceeding the stability criteria (5.157 m.deg) by IMO A.749(18). But the modified ship has better value than the previous ship.

- Area 30-40°

Figure 8 shows that the value of area 30-40° for previous ship is -0.71 m.deg in ballast condition departure and -0.51 m.deg in ballast condition arrival. The value of area 30-40° for modified construction ship is 21.06 m.deg in ballast condition departure and 21.00 m.deg in ballast condition arrival. The area value of modified ship is accepted or exceeding the stability criteria (3.151 m.deg) by IMO A.749(18). But the previous ship is not accepted by the criteria. The failure is caused by the lower of down-flooding point. Due to that point, the maximum heeling of ship is decreased.

- Maximum GZ Arm at ≥ 30°

Figure 9 shows that the value of maximum GZ arm above 30° for previous ship is 0.88 m in ballast condition departure and 0.84 m in ballast condition arrival. The value of maximum GZ arm above 30° for modified construction ship is 2.19 m in ballast condition departure and 2.19 m in ballast condition arrival. Both of the maximum GZ of previous and modified ship are accepted or exceeding the stability criteria (0.2 m) by IMO A.749(18). But the modified ship has better value than the previous.

- Angle of Maximum GZ

Figure 10 shows that the maximum GZ arm of previous ship happens at 21.8° in ballast condition departure and at 20.9° in ballast condition arrival. the maximum GZ arm of modified ship happens at 28.2° in ballast condition departure and at 28.2° in ballast condition arrival. The value of modified ship is accepted or exceeding the stability criteria (25°) by IMO A.749(18). But the previous ship is not accepted by the criteria. The failure is caused by green-water movement due to free surface in cargo room during heeling. It is called free surface effect.

- Initial GM

Figure 9 shows that the GM value of previous ship is 4 m in ballast condition departure and 3.74 m in ballast condition arrival. The GM value of modified ship is 6.07 m in ballast condition departure and 6.17 m in ballast condition arrival. Both of the GM value of previous and modified ship are accepted or exceeding the stability criteria (0.15 m) by IMO A.749(18). But the modified ship has better value than the previous.

4. Conclusion

The simulation of stability analysis shows that the ship with construction modification by adding sideboard and longitudinal bulkhead has better stability value than the previous construction. At IMO MSC Circular608 test procedure, the green-water is still allowed to enter cargo room of modified ship with certain limits. 633,56 m^3 at 25% load, 428,87 m^3 at 50% load, and 194.94 m^3 at 75% load. It is better than the previous ship that the green-water is not allowed at all to enter cargo room in any load-case. Using DNV test procedure, all of the modified ship stability values are accepted by IMO A.749(18) stability criteria at all load-case. But, the ship with previous construction failed in criteria items of area 30-40° and of angle of maximum GZ. Those failure are caused by lower down-flooding point and free surface effect. Based on the result, it can be concluded that adding sideboard and longitudinal bulkhead are effective to increase stability value of Open-Top cargo ship.

5. Suggestion

There must be further research on other options in terms of construction engineering applied to the ships that may be able to increase the value of ship stability and to be more effective and efficient in loading and unloading process.

6. References

[1] D. P. Yuwandana, Y. Novita, and B. H. Iskandar, "Efektivitas Sirip Peredam Dalam Meredam Efek Free Surface Yang Mempengaruhi Gerakan Rolling Kapal Model," *1*, vol. 6, no. 2, Art. no. 2, 2015, doi: 10.24319/jtpk.6.187-197.

[2] A. Abdurrahman, "Pengaruh Variasi Sekat Pada Ruang Muat Kapal Pengangkut Ikan Hidup Terhadap Stabilitas Kapal," vol. 9, no. 2, p. 8, 2019.

[3] DNV-GL, *Special Equipment And Systems Additional Class Part 6 Chapter 9*. 2005.

[4] IMO MSC/Circular.608/Rev.1, "5 Procedure of model tests," 2022. https://www.imorules.com/GUID-7B0FC6B4-AE73-4046-8539-2B65D1777273.html (accessed Jul. 03, 2022).

[5] DNV-GL, *DNV-GL Rules for Classification Part 6 Chapter 5*. 2020.

[6] Smith22, "Belajar Mengenai Kapal: Konstruksi Tongkang," *Belajar Mengenal Kapal*, Jul. 04, 2015. https://smithship.blogspot.com/2015/07/konstruksi-tongkang.html (accessed Jun. 30, 2022).

[7] IMO, "International Code on Intact Stability, 2008 Criteria regarding righting lever curve properties," 2022. https://www.imorules.com/GUID-CC04FC48-9A3D-487D-BB76-7B6064FA75CD.html (accessed Apr. 29, 2022).

Conceptual Design and Operation Plan of Floating Medical Wastes Treatment Facility for Archipelagic Area: A Case Study of Maluku and Papua

S D Lazuardi[1], E W Ardhi[2], S Nugroho[3], F Hadi[4], H I Nur[5], D V Devintasari[6], Y R Izza[7]

[1,3,4,6,7] Department of Marine Transportation Engineering, Institut Teknologi Sepuluh Nopember, Surabaya, Indonesia
[2] University of Strathclyde, Glasgow, UK
[5] University of Auckland, New Zealand

corresponding author's email: dwilaz@seatrans.its.ac.id

Abstract. Indonesia has been facing considerable challenges in medical wastes management. In 2020, the medical wastes reached 294.66 tons/day, but the total capacity of the aggregate wastes treatment facilities nationally was only 224.23 tons/day. Maluku and Papua are Indonesia's largest archipelagic region consisting of four provinces, which are included in the region with the lowest index of medical wastes management as well as the minimum number of medical wastes treatment facilities. The existing wastes treatment in Maluku and Papua does not meet the standards, it is carried out by open-air burning and open dumping. Therefore, the objective of this study is to create the conceptual design and operation plan of floating medical wastes treatment for archipelagic area. By using a multi-port relay operating pattern for each province, the medical wastes transport started from the hub point which was determined by using the centre of gravity method. Considering the number of medical wastes in each province so that the nonlinear programming method was used to determine the most optimum main dimensions of floating facility. The result obtained that the number of floating facilities required was four SPCB-type with costs overall by 63.14 billion IDR/year.

Keywords: Medical wastes treatment, floating facility, conceptual design, ship optimization

1. Introduction

Maluku and Papua are Indonesia's largest archipelagic regions, which consists of four provinces, namely Maluku, North Maluku, Papua, and West Papua, cover approximately 26.1% or more than a quarter of Indonesia's territory. These regions own 111 general hospitals, three specialty hospitals, and 218 floating public health centres. In 2020, The Directorate of Environmental Health of the Ministry of Health stated that aggregately, of 1,055 healthcare facilities in these regions were only 14 facilities or 1.33% that met the medical wastes treatment standards. Provincially, the highest percentage among these regions was reached by Maluku around 4.12%. [1]

The existing medical wastes treatment in Maluku and Papua is conducted by each healthcare facility, carried out by manual shredding process for the sharp wastes (used syringes and other needles), followed by open-air burning and open dumping (Figure 1).

Table 1. Number of healthcare facilities that meet the medical wastes treatment standards.

Items	Registered Healthcare Facilities	Healthcare Facilities that Meet the Medical Wastes Treatment Standards	Percentage (%)
Aggregate	1,055	14	1.33
Maluku Province	243	10	4.12
North Maluku Province	160	1	0.63
West Papua Province	181	2	1.10
Papua Province	471	1	0.21

Those processes have not met the standard for medical wastes treatment and potentially harm human health. The open-air burning process produces some toxic compounds, for instance it usually forms carbon monoxide (CO), which can bind to form carboxy haemoglobin so that the blood can no longer supply sufficient oxygen to body tissues. Hence, burning the infectious medical wastes will be more dangerous. [2] Meanwhile, the World Health Organization (WHO) states that open dumping can harm humans and other living things through the food chain. [3]

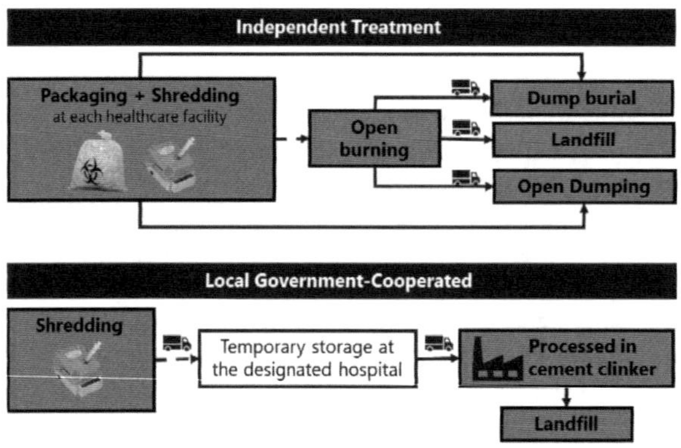

Figure 1. Existing medical wastes management flow in Maluku and Papua

According to the Directorate of Environmental Health in 2020, the total capacity of medical wastes treatment facilities in Indonesia reached 224.23 tons/day (hospital-based wastes treatment capacity was 53.12 tons/day, and the third party-based capacity reached 171.11 tons/day). However, this number does not describe the actual capacity and condition. The medical wastes treatment facility mentioned above is centralized in Java, with a total of 29.44 tons/day of hospital-based capacity and 133.308 tons/day of third party-based capacity (aggregately 72.56% of the national capacity). Ironically, there is no medical wastes treatment facility in Maluku and Papua. The medical wastes treatment facilities in some public hospitals in Maluku and Papua no longer operate, and there is no third-party provider in charge to build those facilities.

The gap between the number of medical wastes and its treatment capacity and the unevenly distributed capacity results in a large pile of unprocessed medical wastes. Moreover, as the impact of the Covid-19 pandemic, the number of medical wastes continuously rises. Responding those problems, along with several private companies, Indonesia's government through the Ministry of Environment and Forestry has been planning to establish some medical wastes treatment facilities in several regions. Unfortunately, this establishment plan has been delaying due to no private companies/ third-party wastes treatment providers who plan to build the facilities in Maluku and Papua. Therefore, an appropriate medical wastes treatment facility is required to accommodate the medical wastes in Maluku and Papua.

2. Determining medical wastes

Due to lack of historical data to predict the number of medical wastes that could be generated in Maluku and Papua, the following step is required to identify the number of wastes in those areas. The first step is to estimate the number of patients in the hospital per day. It is assumed to be 1.23% of the population in the area covered by each hospital [4]. Furthermore, based on the equation obtained from the previous research [5], the number of medical wastes for each hospital is determined as follows:

$$GW = -15.76 + 1.21(PAT) + 0.714(BED) + 10.74(TYPE) \quad (1)$$

Where:

GW = The number of medical wastes (Kg/day)
PAT = Estimated number of patients (people/day)
BED = Number of hospital beds (units)
TYPE = dummy variable; 0 for private hospitals, 1 for government hospitals

Therefore, the number of medical wastes in each pick-up point (the nearest port of the health facilities on each island) can be seen in Table 2 as below:

Table 2. The number of medical wastes at each pick-up point

Port Name	Port Code	Medical Wastes (Kg/day)	Port Name	Port Code	Medical Wastes (Kg/day)
Namlea	NAM	93.52	Sofifi	SFI	34.90
Bula	BLA	64.68	Nabire	NBX	99.95
Gorom/Gorong	GMR	21.91	Serui	ZRI	65.94
Banda Naira	NDA	30.92	Biak	BIK	145.79
Amahai	AHI	152.87	Korido	IKU	28.87
Haria/Saparua	RUH	43.90	Sarmi	ZRM	12.51
Ambon	AMQ	1,173.82	Jayapura	DJJ	870.08
Tual	TUA	171.92	Amamapare	AMA	211.40
Dobo	DOB	38.21	Agats	AGS	47.89
Saumlaki	SXK	66.00	Moor	MMR	44.26
Moa	KWU	12.09	Boven Digoel	BDL	93.74
Bobong	BOG	14.91	Merauke	MKQ	236.42
Sanana	SQN	33.14	Kaimana	KNG	30.09
Laiwui	LWI	15.19	FakFak	FKQ	59.17
Babang/Labuha	BBG	67.56	Bintuni	NTI	23.86
Weda	WED	24.62	Teminabuan	NAB	34.25
Buli	BLI	81.46	Sorong	SOQ	337.76
Tobelo	TBO	95.38	Waisai	WSA	49.15
Daruba (Morotai)	DRA	36.11	Manokwari	MKW	283.00
Jailolo	JIO	61.91	Wasior	WSR	52.30
Ternate	TEI	311.38			
Trikora (Soasio)	SIO	74.87			

3. Operation plan

3.1. Packaging and ship type

According to the Regulation of Ministry of Environment and Forestry Number P.56/Menlhk-Setjen/2015 regarding Procedures and Technical Requirements for the Management of Hazardous and Toxic Wastes from Healthcare Facilities, medical wastes can be stored in an enclosed space for maximum of 90 days in low temperature (below 0^0C). [6]

Utilizing the sea modes, where the medical wastes were transported by ship, the most appropriate storage is a reefer container – using cold storage system. A container vessel is needed to transport that wastes by using reefer container. Considering the vessel compatibility with the operational area, the floating medical wastes treatment facility should have shallow draft that allows the vessel to operate in a shallow water. Therefore, a self-propelled container barge is the most suitable vessels to deal with this operation.

3.2. Hub point and Operating Pattern

Hub point is determined as a port near the toxic and hazardous wastes landfill sites planned for each province. The floating facility in this hub point is anchored after picking up the medical wastes from each pick-up point to the processed point. Since there are more than two pick-up points in each province, a multi-port relay operating pattern is used under this operation. Hereby, the process used to compute the hub point of each province:

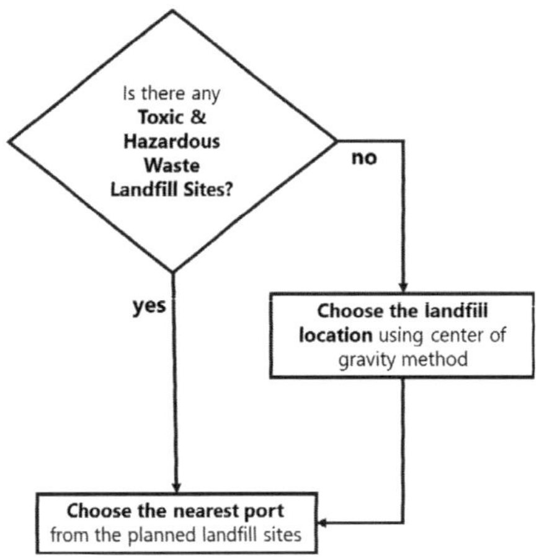

Figure 2. Determining the hub point

Centre of gravity method is applied to determine the hub point for each province by using the following formulation [7]:

$$C_x = \frac{\sum Dl_x W_l}{\sum W_l} \quad (2)$$

$$C_y = \frac{\sum Dl_y W_l}{\sum W_l} \quad (3)$$

Where:

Cx = Longitude of the landfill sites
Cy = Latitude of the landfill sites
Dlx = Longitude of the pick-up point
Dly = Latitude of the pick-up point

Wl = The number of medical wastes at pick-up point (Kg)

The following figures are the results of centre of gravity method to find out the hub point as well as to create the operating routes of the floating medical wastes treatment facilities.

3.2.1 Maluku Province

According to integrated maps of toxic and hazardous wastes in Indonesia, Maluku Province has one landfill site located in Buru (Figure 3). Hence, the hub point location is in the port of Buru Island, which is Namlea Port.

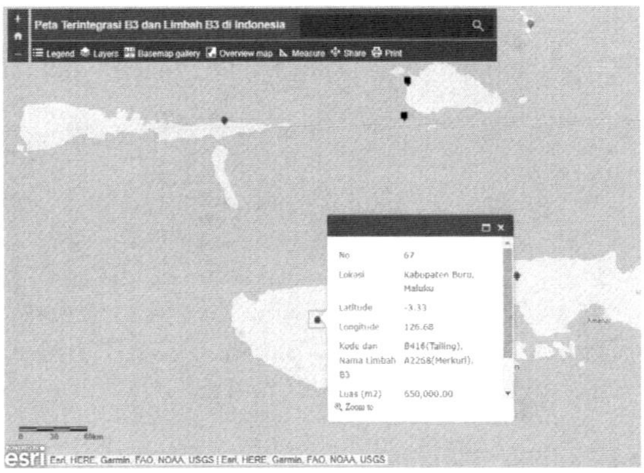

Figure 3. Toxic and hazardous landfill sites in Maluku Province

Namlea Port would be port of origin as well as port of the last destination for ship that assigned to load all medical wastes around Maluku Province. The route is started from Namlea-Bula-Gorom-Tual-Dobo-Saumlaki-Moa-Banda Naira-Haria-Amahai-Ambon-Namlea. The following Figure 4 depicts the operating route of ship assigned in this province:

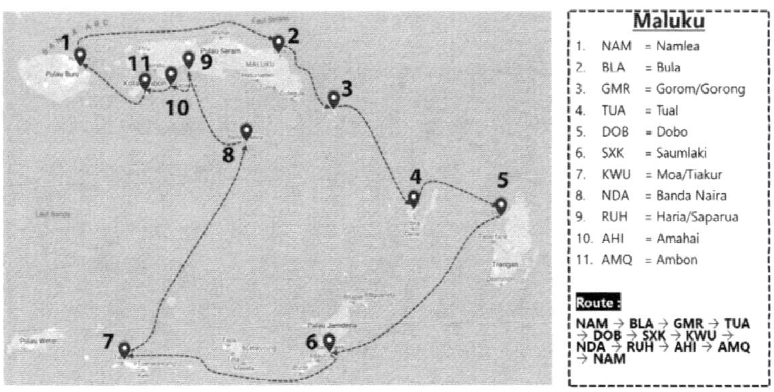

Figure 4. The operating route in Maluku Province

3.2.2 North Maluku Province

According to the Integrated maps of the toxic and hazardous wastes in Indonesia, there are no landfill sites in North Maluku. Thus, the hub point would be determined by using centre of gravity method as shown in Figure 5. As a result, the weighted centre of gravity that would be selected as a hub point for North Maluku Province is in Ternate.

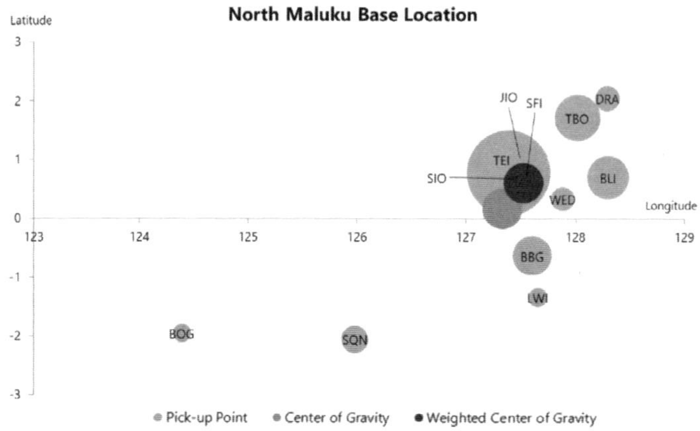

Figure 5. Centre of gravity diagram for hub point in North Maluku Province

In Ternate Port, the number of ships assigned would have an operational pattern as shown in Figure 6, where they pick-up all medical wastes around North Maluku and send them back to Ternate. The route is started from Ternate-Bobong-Sanana-Laiwui-Babang-Weda-Buli-Daruba-Tobelo-Jailolo-Sofifi-Soasio-Ternate. Afterwards, the ship will be anchored, and the medical wastes will be processed.

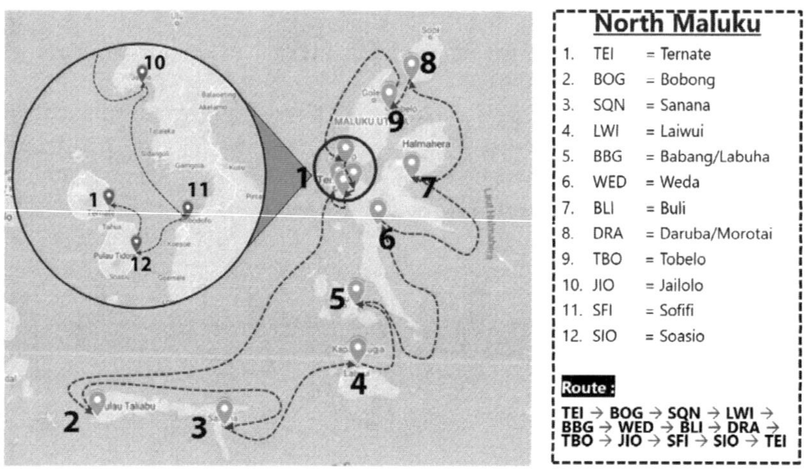

Figure 6. The operating route in North Maluku Province

3.2.3 Papua Province

The condition in Papua Province is the same as North Maluku, which is unavailability of landfill sites. Therefore, centre of gravity method was applied to find out the hub point for Papua Province. According to Figure 7, Jayapura was selected to be hub point for collecting the medical wastes around Papua Province.

As a hub point, Jayapura Port will handle the number of ships assigned to compile the medical wastes from each pick-up point around Papua Province to be returned to Jayapura for processing the wastes. The route is started from Jayapura-Sarmi-Serui-Nabire-Biak-Korido-Amamapare-Agats-Moor-Boven Digoel-Merauke-Jayapura. The detail for ship operation route in Papua Province can be seen in Figure 8.

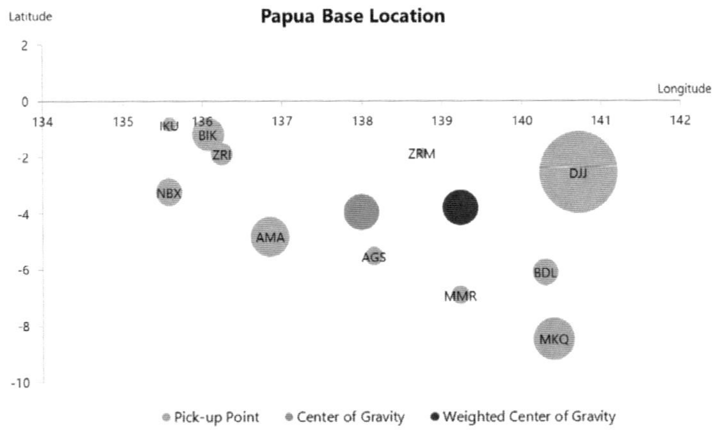

Figure 7. Centre of gravity diagram for hub point in Papua Province

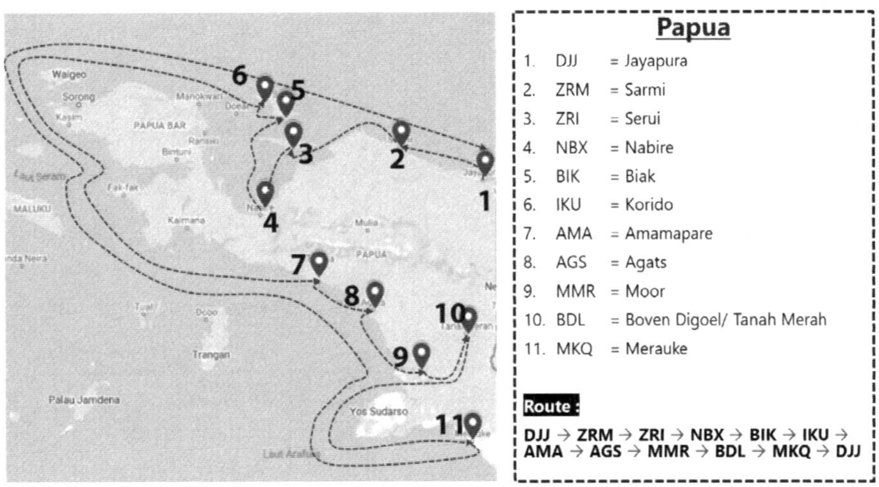

Figure 8. The operating route in Papua Province

3.2.4 West Papua Province

Similar to the previous province, West Papua Province does not have landfill sites as well. Therefore, centre of gravity method was applied to find out the hub point for this Province. According to Figure 9, Fak-fak was selected to be hub point for collecting the medical wastes around West Papua Province.

In Fak-fak Port, the number of ships assigned would have an operational pattern as follows: Fakfak-Kaimana-Wasior-Manokwari-Waisai-Sorong-Teminabuan-Bintuni-Fakfak. They pick-up all medical wastes around West Papua Province and send them back to Fak-fak. Afterwards, the ship will be anchored, and the medical wastes will be processed. The detail pattern for ship operation route in West Papua Province can be seen in Figure 10.

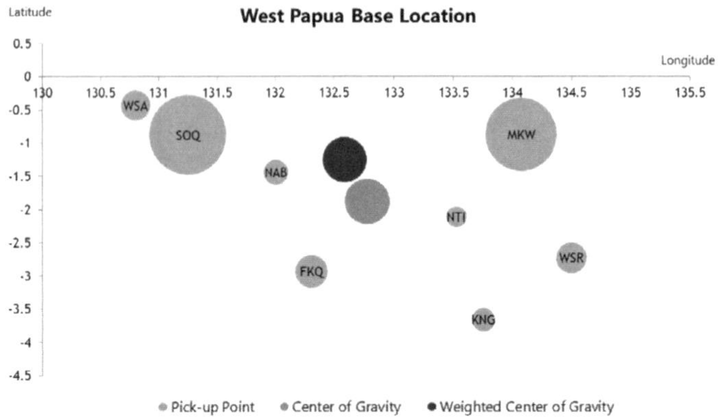

Figure 9. Centre of gravity diagram for hub point in West Papua Province

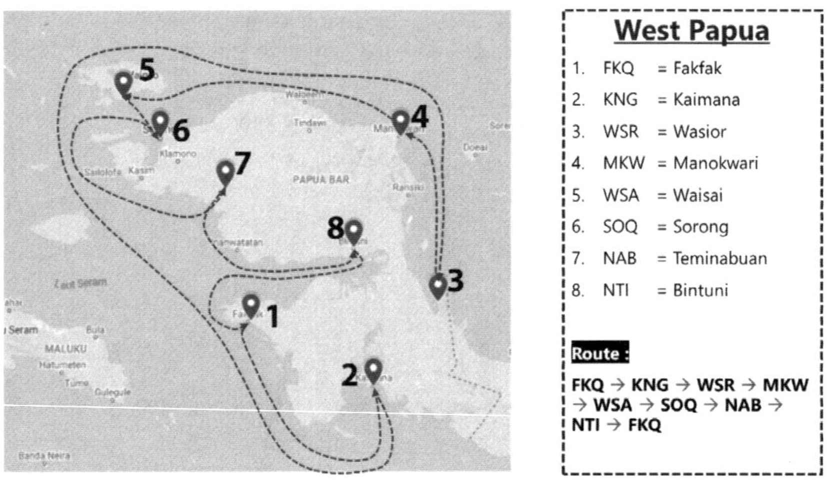

Figure 10. The operating route in West Papua Province

4. Fleet optimization and design

4.1. Fleet optimization

Nonlinear programming method was used to determine the optimum main dimension of the floating medical wastes treatment facility [9]. The objective function of the optimization model is to minimize the total production cost of ship [10] [11]. The formulation to determine the minimum total production costs is described in equation (4) to equation (6). There are three main constraints applied in this model such as main dimensions of ship from equation (7) to equation (10), main dimension ratio from equation (11) to equation (13) and ship stability requirements from equation (14) to equation (19).

The first constraint is obtained from the identification of space requirements on the ship as well as limitation to geographical conditions (wave height). The second constraint is applied to ensure the ship dimension has proportional shape. The last constraint is required to comply with the IMO stability standards.

$$\boldsymbol{Min\ TPC} = W_{ST}.C_{ST} + W_{E\&O}.C_{E\&O} + W_{ME}.C_{ME} \tag{4}$$

$$\boldsymbol{W_{ST}} = Lpp\,.B.D_A.C_s \tag{5}$$

$$\boldsymbol{W}_{\boldsymbol{E\&O}} = L_{E\&O} \, . \, B_{E\&O} \, . \, H_{E\&O} \quad (6)$$

Subject to:

1. Main dimension

$$Lpp \geq 64{,}90 \quad (7)$$

$$B \geq 12{,}75 \quad (8)$$

$$2{,}48 < T \leq 5{,}88 \quad (9)$$

$$H \geq 5{,}48 \quad (10)$$

2. Main dimension ratio

$$3.13 \leq Lpp/B \leq 5.29 \quad (11)$$

$$2.49 \leq B/T \leq 8.87 \quad (12)$$

$$10.57 \leq Lpp/T \leq 30.48 \quad (13)$$

3. International Maritime Organization (IMO) stability standards

$$e_{30^\circ} \geq 0.555 \quad (14)$$

$$e_{40^\circ} \geq 0.09 \quad (15)$$

$$e_{30\text{-}40^\circ} \geq 0.03 \quad (16)$$

$$h_{30^\circ} \geq 0.2 \quad (17)$$

$$\Theta_{max} \geq 25 \quad (18)$$

$$GM_0 \geq 0.15 \quad (19)$$

Where:

TPC	= Total production cost of ship	(Rp)
W_{ST}	= Steel weight	(ton)
C_{ST}	= Steel price coefficient	(USD/ton)
$W_{E\&O}$	= Equipment and outfitting weight	(ton)
$C_{E\&O}$	= Equipment and outfitting price coefficient	(USD/ton)
W_{ME}	= Machinery weight	(ton)
C_{ME}	= Machinery price coefficient	(USD/ton)
Lpp	= Length of perpendicular	(m)
B	= Breadth	(m)
D_A	= Steel construction depth, defined as correction of depth (H)	(m)
C_S	= Steel coefficient	
$L_{E\&O}$	= Length of Equipment and Outfitting	(m)
$B_{E\&O}$	= Breadth of Equipment and Outfitting	(m)
$H_{E\&O}$	= Depth of Equipment and Outfitting	(m)
Lpp	= Length between Perpendicular	(m)
B	= Breadth	(m)
T	= Draught	(m)
H	= Depth	(m)

The optimization results depicts that the number of ships required are four units under the Self-Propelled Container Barge (SPCB) type. These ships will be used as floating medical wastes treatment facilities to cover the whole area in Maluku and Papua. The main dimensions of this floating facility are: LPP

IOP Conf. Series: Earth and Environmental Science 1166 (2023) 012043 doi:10.1088/1755-1315/1166/1/012043

(length of ship) = 70,99 m; B (width of ship) = 13,80 m; T (draft) = 2,48 m and H (height of ship) = 5,60 m. The total production costs accrued IDR 84,95 billion.

Table 3. Optimization results

Item	Value	Unit
Lpp	70.99	m
B	13.80	m
T	2.48	m
H	5.60	m
Ship production cost	84.95	billion IDR
Number of ships built	4	unit

4.2. Fleet design

Hereby the General Arrangement of floating medical wastes treatment facility based on the calculation result:

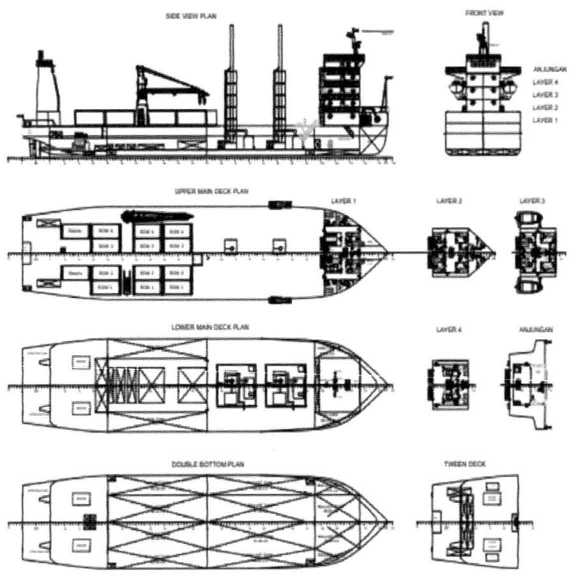

Figure 11. General arrangement of the floating medical wastes treatment facility

The layout and operation process on board can be seen in Figure 12. Part "A" as shown in Figure 12 is storage zone for medical wastes with a capacity of 12 TEUs. Part "B" is the medical wastes treatment zone, which consists of two incinerators and two medical wastes destroyers. Meanwhile, part "C" is the liquid wastes and residue storage zone of processed medical wastes. In part "C", there are five main rooms namely: (1) residue storage room with size 6 m x 6 m x 1.5 m); (2) initial liquid wastes storage tank with size 6 m x 6 m x 1 m; (3) stabilization tank with size 1.8 m x 6 m x 1 m; (4) aeration tank with size similar to stabilization tank and (5) maturation tank with size 1.2 m x 6 m x 1 m.

Meanwhile, the process of medical wastes treatment on board can be seen through the notation of numbers in the side view plan in Figure 12, the detail as follows:

1. Wastes taken out from Storage Zone (A);
2. Wastes transferred from inside the container to the elevator using a forklift;
3. Wastes shredded in Medical Wastes Destroyer (B) and transferred into the incinerator (B);

4. The residue (fly ash) accommodated in the Residue Storage Room (C);
5. Liquid Wastes from treatment stored in the Initial Liquid Wastes Storage Tank (C);
6. Liquid Wastes in the Initial Liquid Wastes Storage Tank is processed in the Stabilization Tank, Aeration Tank, and Maturation Tank (C), and;
7. The wastes from the Residue Storage Room transferred into the Container (the transferred to the landfill sites inland), while the liquid wastes taken out by pumping it through the sea chest.

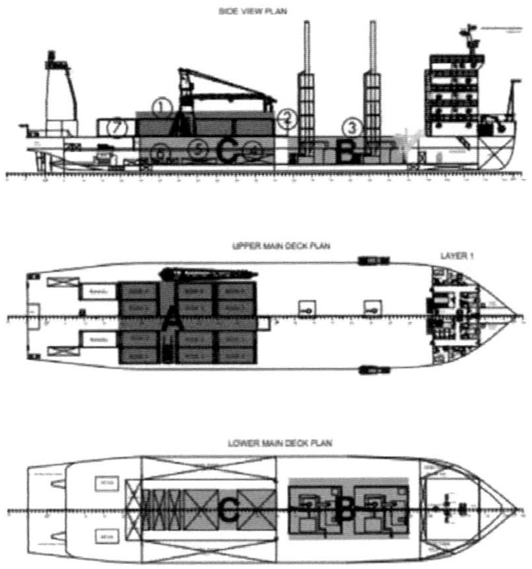

Figure 12. Layout and operation process on board of floating medical wastes treatment facility

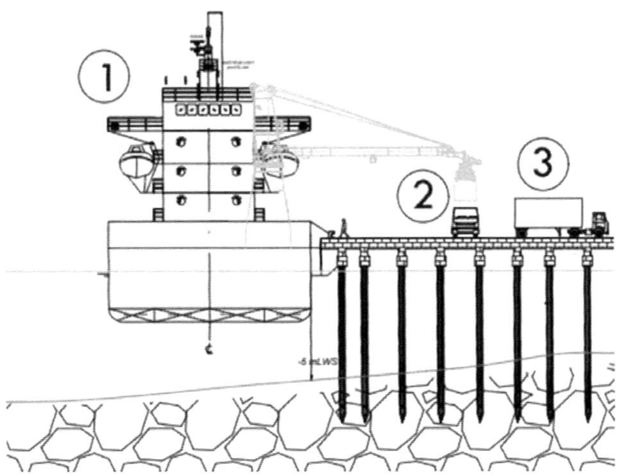

Figure 13. Side view of pick-up point berth

The process of loading and unloading medical wastes from and to the ship can be seen through the number notation in Figure 13, in detail as follows:
1. Ship is berthing in the jetty;
2. The container crane took the empty containers from the ship onto the truck or took the full containers from the truck onto the ship;

3. The truck carried out the receiving/delivery process to bring empty containers back to each healthcare facilities to store the medical wastes and/or deliver the full containers to the pick-up point.

5. Determining total costs

Total costs of ship are all costs incurred by each ship for one year operation, consist of fixed costs and variable costs. The fixed cost component consists of capital costs and operational costs, while variable costs are costs incurred by the ship when the ship operates a voyage such as fuel costs and port charges. Hereby the detail of determining total cost of floating medical wastes treatment in four provinces:

Table 4. Determining the total cost per each province

Item	Unit	Maluku	North Maluku	Papua	West Papua
Capital cost	billion IDR/year	7.31	6.99	7.67	6.44
Operating cost	billion IDR/year	3.25	3.11	3.41	2.86
Voyage cost	billion IDR/year	4.87	4.66	5.12	4.29
Cargo handling cost	billion IDR/year	0.49	0.47	0.51	0.43
Additional cost	billion IDR/year	0.32	0.31	0.34	0.29
Annual Total Cost	billion IDR/year	**16.24**	**15.54**	**17.05**	**14.31**

Additional cost is incurred by electricity, handling wastes equipment and reefer container during the operational of this floating facility. Furthermore, the result of annual total cost for each province can be seen in Table 5 as below:

Table 5. Total cost

Province	Value	Unit
Maluku	16.24	billion IDR/year
North Maluku	15.54	billion IDR/year
Papua	17.05	billion IDR/year
West Papua	14.31	billion IDR/year
Annual Total Cost	**63.14**	billion IDR/year

The total costs of floating medical wastes treatment facilities in Maluku, North Maluku, Papua and West Papua is 63.14 billion IDR/year. The production cost was included in these total costs under the capital costs.

6. Conclusion

- Medical wastes management in Maluku and Papua have not met the standard due to the lack of hospitals with licensed incinerators, unavailability of third-party company that planned to operate an incinerator in Maluku and Papua until the year of 2022 and the most appropriate wastes processing in these regions were carried out by conducting wastes shredding, open burning and landfilled.
- Maluku produced about 56,08 tons/month of medical wastes and its hub point is in Namlea, while North Maluku gained about 25,32 tons/month and its hub point is in Ternate. On the other hand,

Papua and West Papua produced 26,90 tons/month and 55,70 tons/month of medical wastes, simultaneously and those hub points are in Jayapura and in Fak-fak.

- Considering the number of medical wastes in each province so that the nonlinear programming method was used to determine the most optimum main dimensions of floating facility. The result obtained that the number of floating facilities required was four SPCB-type with costs overall by 63.14 billion IDR/year. Furthermore, the main dimensions of this floating facility are: Lpp=70,99 m, B=13,80 m, T=2,48 m, and H=5,60 m. These floating facilities would be operated by a multi-port relay pattern starting from hub point in each province.

References

[1] Direktorat Kesehatan Lingkungan. (2020, January 15). *Pantau Data Pengelolaan Limbah Media (PLM)*. Opgehaald van Direktorat Kesehatan Lingkungan, Direktorat Jenderal Kesehatan Masyarakat, Kementerian Kesehatan Republik Indonesia: http://kesling.kesmas.kemkes.go.id/limbahfasyankes

[2] Malayadi, A. F. (2017). *Karakteristik dan Sistem Pengelolaan Limbah B3 Laboratorium Universitas Hasanuddin Kota Makassar.* Makassar: Fakultas Kesehatan Masyarakat Universitas Hasanuddin.

[3] WHO. (2014). Safe management of wastes from health-care activities: 2nd Edition. Malta: World Health Organization.

[4] Huang, C.-H. Chen, W.-J., Ma, M. H. -M., Lai, C.-L., Lin. F.-Y., & Lee, a. Y.-T. (2001, September). *Ambulance Utilization in Metropolitan and Rural Areas in Taiwan.* Journal of the Formosan Medical Association. Dipetik Juli 14, 2021, dari https://bit.ly/3g3bNcO.

[5] Bdour, A., Altrabsheh, B., Hadadin, N., & Al-Shareif, M. (2007). Assessment of medical wastes management practice: A case study of the northern part of Jordan. Waste Management, XXVII, 746-759.

[6] KLHK RI. (2015). *Peraturan Menteri Lingkungan Hidup dan Kehutanan RI No. PermenLHK/56/2015 tentang Tata Cara Dan Persyaratan Teknis Pengelolaan Limbah Bahan Berbahaya Dan Beracun Dari Fasilitas Pelayanan Kesehatan.* Jakarta: Kementerian Lingkungan Hidup dan Kehutanan RI.

[7] Firmansyah, Y. (2020). *'Penentuan Lokasi External Warehouse Dengan Menggunakan Metode Center of Gravity (Studi Kasus Di PT. RPZ Surabaya)'.* Jurnal Manajemen Industri dan Logistik, 4(1), pp. 58-66

[8] Craig B. Smith. (2007). *Extreme Waves and Ship Design, 10th International Symposium on Practical Design of Ships and Other Floating Structures,* Texas,USA.

[9] Lazuardi, S. D., van Riessen, B., Achmadi, T., Hadi, I., & Mustakim, A. (2017). Analyzing the national logistics system through integrated and efficient logistics networks: A case study of container shipping connectivity in Indonesia. In Applied Mechanics and Materials (Vol. 862, pp. 238-243). Trans Tech Publications Ltd.

[10] Yunianto, I. T., Lazuardi, S. D., & Hadi, F. (2018, March). Freight calculation model: a case study of coal distribution. In IOP Conference Series: Earth and Environmental Science (Vol. 135, No. 1, p. 012013). IOP Publishing.

[11] Watson, DGM. (1998). *Practical Ship Design.* Oxford: Elsevier Science.

Acknowledgments

Authors would like to be obliged to the Directorate of Environmental Health of the Ministry of Health for providing data and information.

Performance Simulation of Hybrid Refrigeration System Using Eutectic Salt PCM For Eco Reefer Container

D J Ditya[1] **and S P Fitri**[1]
[1]Departement of Marine Engineering, Institut Teknologi Sepuluh Nopember, Surabaya, Indonesia

*djditya16@gmail.com, sutopopf@its.ac.id

Abstract. Today, fresh, chilled, or frozen products usually use reefer containers. The main problem with using reefer containers is the large energy consumption of 3.6kW/TEUS and the continuous supply of electric power. One solution available is using Phase Change Material (PCM) in the hybrid refrigeration system combined with a vapor compression system. Using PCM, the temperature of the reefer container can be maintained for several hours of cooling with the compressor off. This research aims to simulate the performance of a hybrid refrigeration system using PCM and the distribution of cooling air temperature in a laboratory-scale eco-reefer container. A simulation was performed using numerical software and CFD software. From the simulation analysis, the power value of the compressor at -20 °C is 0.3794 kW for the conventional system, 0.463 kW for the 22% NaCl PCM hybrid system, and 0.411 kW for the 20% Propylene Glycol PCM hybrid one. The COP value is 2,560 for the conventional system, 2,572 for the 22% NaCl PCM hybrid system, and 2,569 for the 20% Propylene Glycol PCM hybrid one. The CFD simulation results that PCM NaCl and Propylene Glycol can maintain the reefer container room temperature and product load.

Keywords: phase change material, reefer container, refrigeration system.

1. Introduction

Reefer container is a form of cold storage used in the refrigeration chain logistics industry for the distribution and transportation of perishable product commodities such as fish, meat, fruit and vegetable products, and medicines. Reefer container transportation can be done by land transportation modes such as trains and trucks as well as sea transportation modes, namely ships. Reefer containers are equipped with integrated refrigeration units that depend on the supply of electricity at locations on land, ships, or ports. Refrigeration units can regulate and maintain the cold air temperature inside the refrigerated container and are used to maintain product quality.

When operating, it is very important to maintain the specified temperature inside the reefer container. This will later result in large energy consumption and may also have an impact on the environment because of the emissions it produces. To maintain a predetermined temperature, reefer containers generally require the energy of around 3.6kW/TEUS [1]. In addition to energy problems, the main problem faced by reefer containers is the electricity cut-off due to blackouts, accidents, or other damage so that the reefer containers cannot operate, and the cargo is damaged due to inappropriate temperatures inside.

One solution is to use Phase Change Material (PCM) which is applied to a hybrid refrigeration system that functions to maintain a cold temperature in the container when the compressor in the refrigeration system stops/turns off or there is no electricity. PCM is a material that can change its phase by melting or freezing. At the time of phase change, the PCM will release or absorb the heat close to a constant temperature, so it is very effective to use to maintain hot or cold conditions.

In this paper, the hybrid refrigeration system used is a combination of a vapor compression refrigeration system and PCM. In this system, the PCM charging/freezing process is carried out by the vapor compression refrigeration system, and after the PCM temperature is met and then the cooling temperature in the container room is replaced by PCM so that the vapor compression system is turned off. Therefore, in this paper, the problem to be solved is to simulate the hybrid refrigeration systems' performance due to the charging/discharging process and the distribution process of cooling airflow in the container room with the hybrid method. The simulations are performed on a laboratory-scale 10-ft container model.

2. Literature Review

Reefer Container is a refrigerated container used in intermodal goods transportation to support transportation activities on certain cargoes sensitive to ambient temperature changes in general, a reefer container has a data logger to record temperature changes that occur in the container. Reefer containers work by conditioning the temperature in the container that is adjusted to the temperature requirements of the transported commodity. Each commodity has different transport temperature requirements according to its characteristics to maintain quality so as not to degrade when arriving at the destination.

The intake air is cooled through the evaporator coil in the container refrigeration unit. The cold air is then exhaled by the evaporator fan through the T-Bar Floor at the bottom of the container. Furthermore, hot air with a density higher than cold air will move toward the roof of the container which will then be reabsorbed by the evaporator to continue the refrigeration cycle. In general, the average reefer containers available are 20 Feet (1 TEU) and 40 Feet (2 TEU).

The reefer container refrigeration system works with a vapor compression refrigeration system by using a compressor to flow high-pressure air to the condenser, which in turn will flow to the evaporator before the conditioning air enters the reefer container. The work of the refrigeration unit depends on the power source on the ship and the dock.

A compressor is used as the refrigerant pumping device in vapor compression refrigeration. Low-pressure refrigerant vapor entering the suction side is pushed into the compressor, becomes high-pressure refrigerant vapor, and is released downstream (discharge). The high-pressure and low-pressure sides are known in this process. [2].

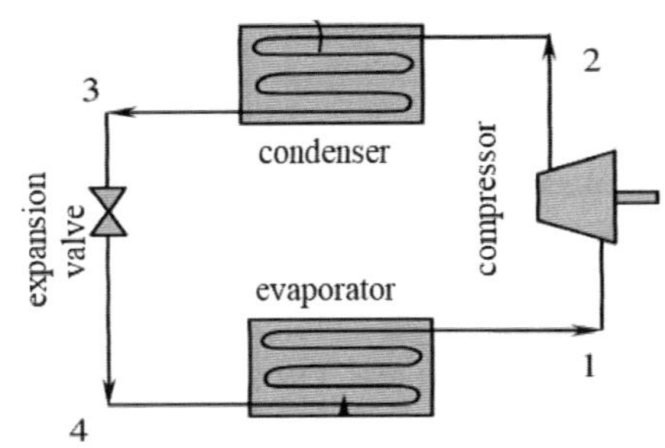

Figure 1. Vapor compression refrigeration system cycle

Refrigerant vapor leaves the evaporator and enters the compressor, increasing the pressure and temperature of the superheated vapor. After the refrigerant vapor is compressed by the compressor, it enters the condenser for the condensation process, turning the refrigerant vapor or gas back into liquid. The hot and high-pressure liquid refrigerant enters the expansion device where the pressure and temperature are reduced to the point required for the evaporation process. In this state, the saturation temperature of the refrigerant is lower than the ambient temperature or the temperature of the medium being cooled, and the system absorbs heat from the refrigerant to the surroundings. [3].

The thermal performance of the refrigeration process is determined by the coefficient of performance (COP) formula.

$$COP=\frac{Q_L}{W_C}=\frac{\dot{m}_{ref}\times(h_1-h_4)}{\dot{m}_{ref}\times(h_2-h_1)}=\frac{(h_1-h_4)}{(h_2-h_1)} \tag{1}$$

Which:

- W_C : compressor power (kW)
- Q_L : Q evaporator (kJ/s)
- $\dot{m}_{ref}$: mass flow rate of refrigerant (kg/s)
- h : enthalpy of refrigerant (kJ/kg)

PCM is a material that can change its phase by melting or freezing. At the time of phase change, the PCM will release or absorb the heat close to a constant temperature [4]. PCM could store and release a significant amount of heat at a steady temperature, as well as a very high heat storage density. As a result, PCM can be used as an alternate heat storage medium in a range of applications [5].

With the use of engineering numerical software, Varen (2021) tested the performance of a prototype 10-foot eco-reefer container equipped with a VC-PCM hybrid refrigeration system. Only 18% NaCl + H_2O and 30% Propylene Glycol + H_2O were allowed in the PCM. According to the analysis of the prototype's performance, variations in the surrounding temperature have an impact on the prototype's performance. As the temperature rises, the cooling load and compressor work increase while the Coefficient of Performance (COP) is decreased. The lowest COP value and the highest cooling load occurred at 10.00 with a value of 2,594 and 2574 W for the hybrid system, while for the conventional system, the lowest COP value and the highest cooling load occurred at 12.00 with a value of 2,542 and 948 W for the evaporator temperature setting -20 °C, as well as 3,044 and 649.6 W for setting the evaporator temperature to -10 °C [6]

Gin (2011) tried to combine PCM panels on the walls with the 3-dimensional transient simulation of cold storage using CFD. For freezers and cold stores, the impact of changes in air temperature during a loss of electrical power (and loss of cooling) was subsequently examined and validated. The thermal load that enters the cold storage is absorbed by the PCM as it melts, lowering the cold storage's temperature rise. The temperature rise during a power outage is minimized by using PCM in the freezer. Without PCM, the air temperature steadily increases to over -8°C in just one hour, whereas it was maintained at a steady temperature of -8°C for seven hours. The ambient temperature keeps rising in the absence of PCM. As a point of comparison, the air inside the cold store equipped with PCM got as cold as -7 °C and maintained that temperature for 60 hours. Over the period of 60 hours, the PCM melts gradually raising the temperature. When the PCM has fully melted, both the air and PCM's temperature quickly rise. [7].

Elsony (2021) examined the temperature distribution capabilities of PCM materials for applications involving hybrid eco-reefer container cooling systems. Propylene glycol (PG) PCM containers' room temperature increased more quickly than NaCl. While the PG could only keep the loading chamber temperature at -10°C for 8 hours, PCM NaCl was able to do it for a full 14 hours. In containers that use PG, the temperature of the loading area is higher than NaCl due to the phase change temperature of PG being higher than NaCl. [8].

3. System Setup

Software for numerical computation and CFD simulation were used. A hybrid refrigeration system's performance is simulated using numerical software. CFD software is used to simulate the distribution of cooling air temperature in laboratory-scale eco-reefer containers. In the CFD simulation, the required data are the dimensions of the container space, the properties of the materials used as the container structure, and the load to be loaded into the eco-reefer container. After all, data is

IOP Conf. Series: Earth and Environmental Science 1166 (2023) 012044 doi:10.1088/1755-1315/1166/1/012044

collected, CFD modeling and simulation are continued. A laboratory-scale of 10 ft reefer container with a length of 1,2 m, width of 1,2 m, height of 1,2 m, and a ¾ HP compressor were used in this paper. The refrigerant used in this system is R404A. The PCM used in this paper is NaCl with a concentration of 22% in water and Propylene Glycol with a concentration of 20% in water. The cargo or cooling load is in the form of fresh fish which is placed in a container/pallet before being put into the prototype eco-reefer container. In the design of this eco-reefer container, a pallet with basket material measuring 28 × 16.8 × 12 cm will be used with a total load of 57,6 kg of fish or 15 pallets.

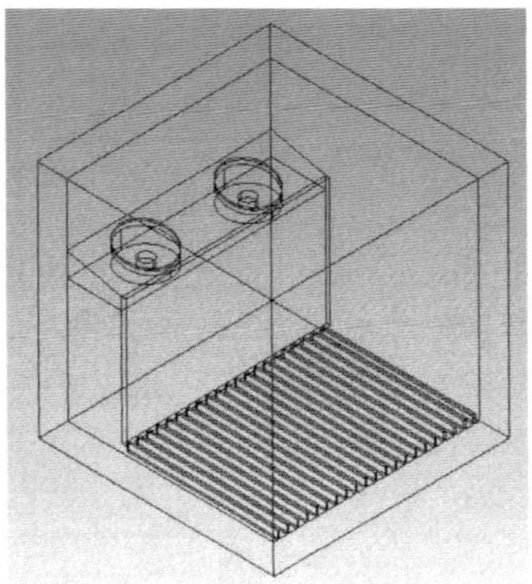

Figure 2. 3D modeling of conventional systems

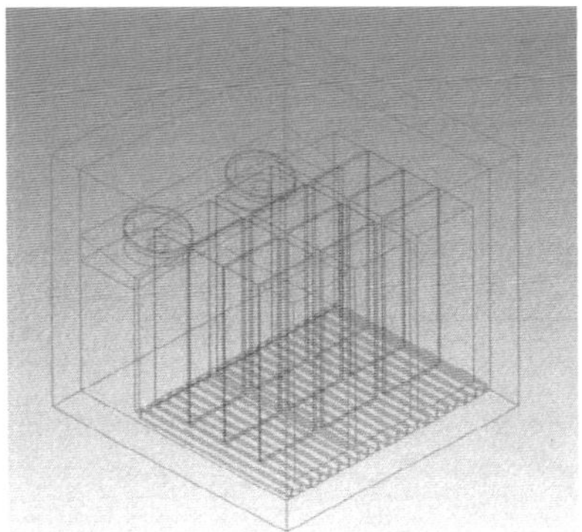

Figure 3. 3D modeling of conventional systems with cooling loads

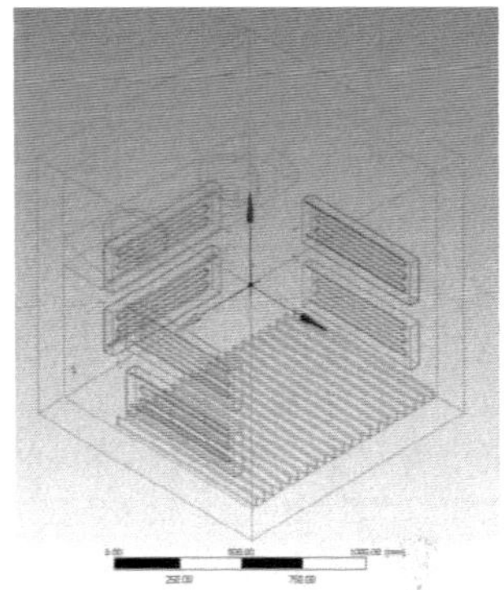

Figure 4. 3D modeling of hybrid systems without cooling loads

4. Result and Discussion

4.1. Performance of Refrigeration System

In this study, numerical software was used to simulate the refrigeration system's performance. A programming language must first be created before any simulation can be carried out. The simulation results for conventional and hybrid systems are shown below.

Table 1. Performance of the refrigeration system simulation result

System	Total	m_{fr}	condenser	$Q_{evaporator}$	$W_{compressor}$	COP
Conventional	986,3	0,007285	1,37	0,9863	0,384	2,56
Hybrid NaCl 22%	1205	0,008899	1,205	1,674	0,4691	2,572
Hybrid PG 20%	1061	0,007838	1,474	1,061	0,4132	2,569

Table 1 shows that compared to the other systems, the conventional system uses the lowest power for the compressor. This is so that the PCM can be charged, which the conventional system lacks. The temperature in the evaporator is altered at numerous sub-zero temperatures to discover its relationship to the mass flow rate of the refrigerant. For lower temperatures, the mass flow rate value will be higher, or the mass flow rate will be faster. This is because, at a given compressor pressure, the enthalpy value at the evaporator output differs more from the constant enthalpy value at the evaporator input. The mass flow rate on PCM NaCl 22% has the highest value compared to other conditions because PCM NaCl 22% has the highest charging load.

The pressure of the compressor is varied to see its relationship to the power of the compressor. The result is the higher the working pressure of the compressor used at the same evaporator temperature, the higher the input power value of the compressor used will be. This is because when the compressor working pressure is higher, the value of the intersection between the entropy line and the pressure on the P – H diagram of the refrigerant is further away from the gas saturation line, so the enthalpy value will be greater.

IOP Conf. Series: Earth and Environmental Science 1166 (2023) 012044 doi:10.1088/1755-1315/1166/1/012044

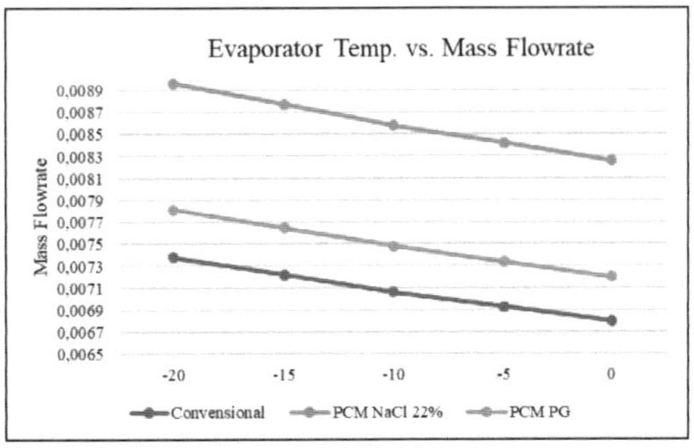

Figure 5. Comparison between evaporator temperature and mass flowrate of refrigerant

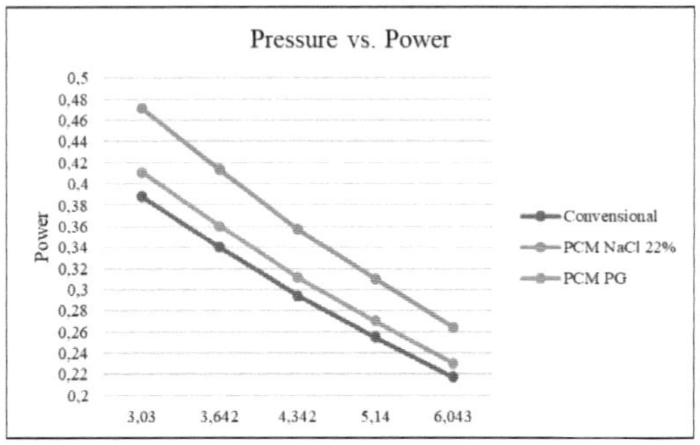

Figure 6. Comparison between compressor pressure and compressor power comparison

4.2. Distribution of cooling air temperature in Hybrid System

In this section, a simulation is carried out using the CFD method with the initial temperature in the container -20°C, PCM initial temperature -20°C, and the fan from the evaporator turned on at -20°C within 200 s.

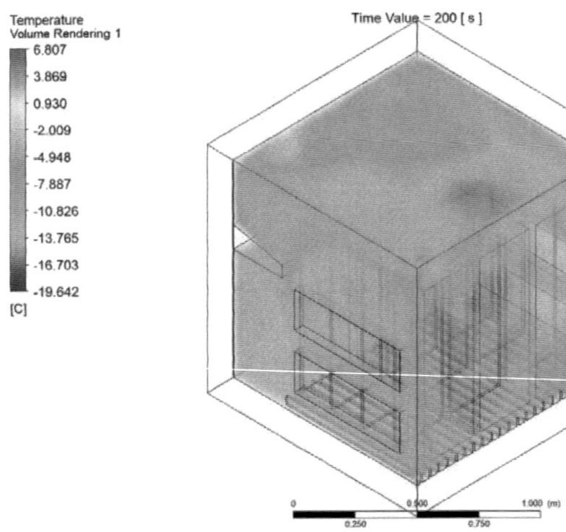

Figure 7. Air temperature distribution in PCM NaCl 22% hybrid system

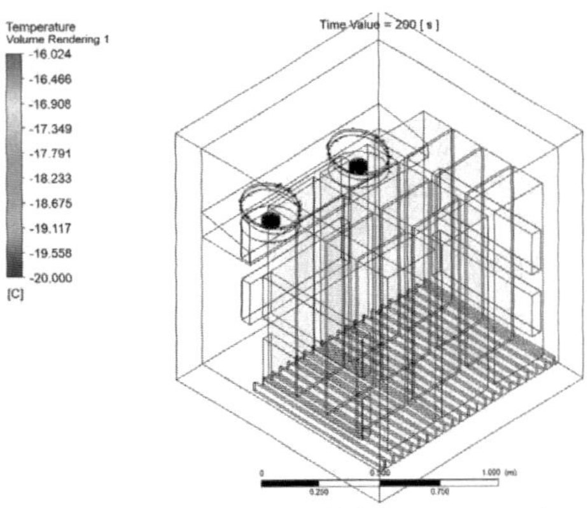

Figure 8. Cooling load distribution in PCM NaCl 22% hybrid system

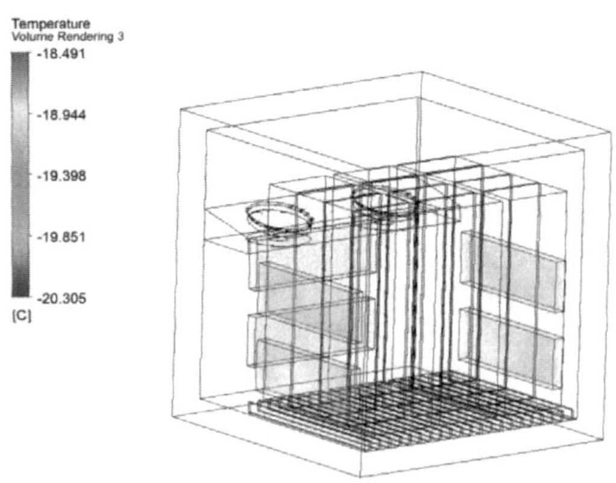

Figure 9. PCM distribution in PCM NaCl 22% hybrid system

According to figure 7, after 200 seconds, the NaCl 22% hybrid system's maximum air temperature was 6°C on the container's top side due to convection, and the lowest temperature value was measured at -19.6°C with an average temperature of -12.163°C. Since the air absorbs heat from the load, the temperature of the air appears to rapidly decrease. Figure 8 shows that the fish were kept at a temperature of -20°C, with a maximum value of -16°C and an average temperature of -19.788°C. The PCM boxes' average temperature was measured at -19.85°C, with a minimum temperature of -20°C in the middle and a maximum of -18.4°C at each box's corners, as shown in figure 9.

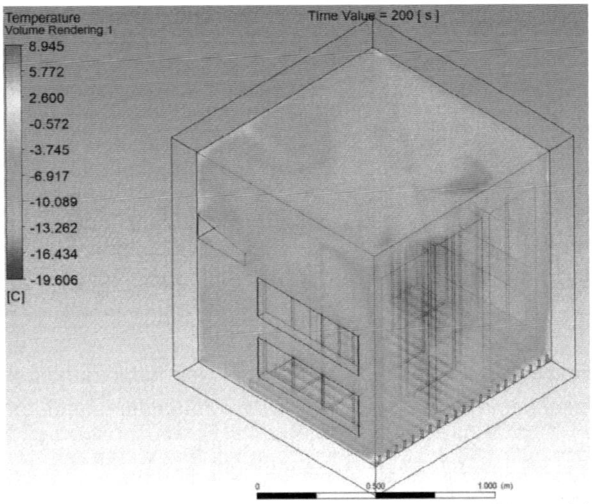

Figure 10. Air temperature distribution in PCM Propylene Glycol 20% hybrid system

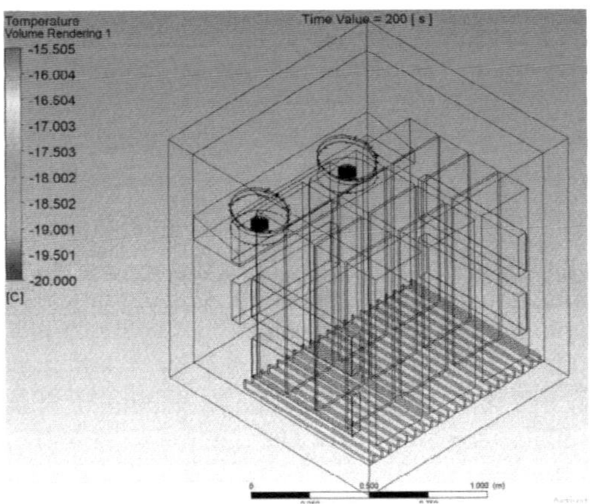

Figure 11. Cooling load distribution in PCM Propylene Glycol 20% hybrid system

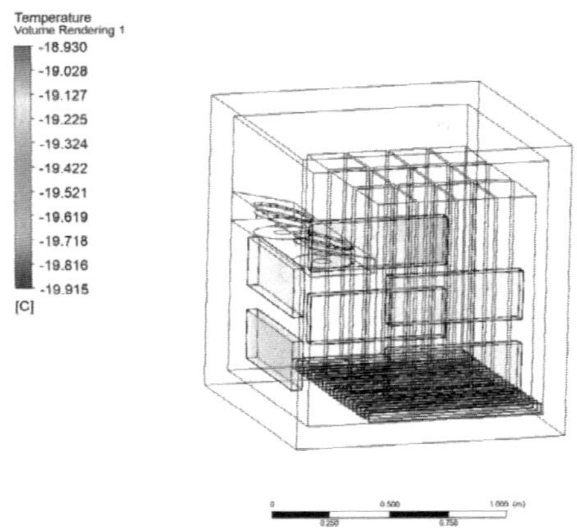

Figure 12. PCM distribution in PCM Propylene Glycol 20% hybrid system

The maximum air temperature for the propylene glycol 20% hybrid system was 8°C on the top side of the container after 200 seconds, and the minimum temperature value was recorded at -19,8°C and the average temperature was -11.47°C as shown in figure 10. So, because air absorbs heat from the load, the temperature of the air appears to rapidly decrease. Figure 11 shows that the fish were kept at a temperature of -20°C, with a maximum value of -15.5°C and an average temperature of -19.788°C. Due to the PCM's position, the temperature on the PCM can be seen differently. The PCM under the evaporator has a minimum temperature of -19.6°C and a maximum temperature of -19.3°C. The PCM's lowest and maximum temperatures for the right and left sides of the container wall are -19°C and -18.9°C, respectively. This occurs as a result of the load heating the PCM, which is situated on the side of the container wall as illustrated in figure 12.

5. Conclusion

Numerical software was used to simulate the performance of the refrigeration system in three different configurations: conventional system, hybrid 22% NaCl PCM system, and hybrid 20% Propylene Glycol PCM system. The power value of the compressor at -20°C is 0.3794 kW for the conventional system, 0.463 kW for the 22% NaCl PCM hybrid system, and 0.411 kW for the 20% Propylene Glycol PCM hybrid system, according to the simulation of the performance of the refrigeration system analysis. For the conventional system, the COP value is 2.560, for the 22% NaCl PCM hybrid system is 2.572, and for the 20% Propylene Glycol PCM hybrid system is 2.569.

The CFD simulation results that PCM NaCl and Propylene Glycol can maintain the reefer container room temperature and product load. PCM material only drops in the range of -19°C and -20°C. so that it can maintain the temperature of the fish at the freezing point in the temperature range of -19°C and -20°C.

Reference

[1] Wild Y. Refrigerated Containers and CA Technology, Container Handbook, The German Insurance Association. Berlin, Germany. 2009

[2] Budiarto, U. and Amiruddin, W. (2017) ‘Analisa Unjuk Kerja Desain Sistem Refrigerasi Kompresi Uap Pada Kapal Ikan Ukuran 5 Gt Di Wilayah Rembang’, Jurnal Teknik Perkapalan, 4(4), pp. 768–778.

[3] Muhsin, Z. and Rasyid, A. R. (2017) ‘Analisis Unjuk Kerja (COP) Mesin Pendingin Hibrid dengan Menggunakan Refrigeran R-22’, Teknologi, 17(Oktober), pp. 49–58.

[4] Mofijur, M. et al. (2019) ‘Phase change materials (PCM) for solar energy usages and storage: An overview’, Energies, 12(16), pp. 1–20. doi: 10.3390/en12163167.

[5] Nurfi, H. and Saktioto (2020) ‘Eco-Green Air Cooler System: Pemanfaatan Phase Change Material Sebagai Alternatif Pengganti Air Conditioning di Kota Pekanbaru’, 5(September), pp. 15–17.

[6] Denaor, A. K. V. (2021) TUGAS AKHIR: RANCANG BANGUN PROTOTIPE ECO-REEFER CONTAINER DENGAN MENGGUNAKAN SISTEM REFRIGERASI HYBRID VAPOUR COMPRESSION – PHASE CHANGE MATERIAL (VC-PCM). Surabaya: Institut Teknologi Sepuluh Nopember.

[7] Elsony, F. (2021) TUGAS AKHIR: KINERJA TERMAL PROTOTIPE SISTEM PENDINGIN HYBRID MENGGUNAKAN PHASE CHANGE MATERIAL (PCM) UNTUK ECO REEFER CONTAINER. Surabaya: Institut Teknologi Sepuluh Nopember.

[8] Gin, B., Farid, M. M. and Bansal, P. (2011) ‘Modeling of phase change material implemented into cold storage application’, HVAC and R Research, 17(3), pp. 257–267. DOI: 10.1080/10789669.2011.572222.

.

Using genetic algorithm for fleet assignment of small-scale LNG supply chain

Firmanto Hadi[1], Heri Supomo[2], Tri Achmadi[1]

[1]Dept. of Marine Transportation Engineering, Institut Teknologi Sepuluh Nopember
[2]Dept. of Naval Architecture, Institut Teknologi Sepuluh Nopember

firmanto@seatrans.its.ac.id

Abstract. This paper presents the development of a model of how to assign LNG carriers from a single LNG source to a number of scattered small-scale power plants in a milk-run operation strategy. Typically, researchers assume that ports/terminals to be visited are as given parameters. In this paper, the specifications of ports/terminals (infrastructures) are considered in the model as a consequence of the selected LNG carriers. The objective is to determine the optimum size and number of LNG carriers for a particular route by minimizing the unit logistics costs which include shipping costs and required infrastructures (ports and facilities) costs. The proposed approach of this paper is a metaheuristic by using genetic algorithm. The computational results shows that the optimal solution can be achieved in a relatively short period of time and the optimum cuve satisfies the law of deminishing returns. Finally, some encouraging computational results and suggestions for further study are presented.

Keyword: Fleet assignment, genetic algorithm, milk-run operations, optimization

1. Introduction

In contrast to other industries where the levels of planning – strategic, tactical and operational – can be clearly distinguished, it is not easy to differentiate among the planning levels of maritime transportation system, due to interplay between them. On the one hand, lower-level or short-term decisions are strongly influenced by the higher-level ones. On the other hand, in order to make a good planning at a higher-level, details regarding short-term decisions are needed. However, some researchers have tried to classify the planning levels of maritime transportation planning as can be cound in [1], [2], [3], [4], [5] and [6] whose classify that fleet assignment falls into the tactical planning.

Fleet assignment also known as fleet deployment problem in maritime transportation basically dealing with how to assign ships efficiently to predetermined route to meet demand [2] and sometime is also the basis of scheduling [3]. The objective function, can be seen from two different perspectives depending on the problem, where in liner and tramp shipping, the objective function is in form of maximizing profit while minimizing costs is commonly used in industrial shipping [7]. In addition, the problem dealing with how to allocate the fleet in such a way so that utility of fleet is maximum is part of the fleet assignment problem [8]. This paper is focus on the industrial shipping fleet assignment problem.

As a plan at the tactical level, several parameters on a fleet assignment problem, especially demand and route are given before the decision on fleet assignment or deployment is made on both deterministic

([9], [10], [11], [12]) and stochastic ([13], [14], [15]) problem. In addition, most studies on fleet assignment problem, design of the port/ terminal where ship will berth is generally known or given in advance. In this paper, the design of port/ terminal will be considered as a consequencies of fleet deployed in certain routes. Therefore, the design of port/ terminal will vary depending on the optimal ships selected.

This paper aims to develop a ship assignment or deployment from a particular set of ships in a milk-run operation using a genetic algorithm with the objective function to minimize logistics costs. The milk-run operation is a logistic strategy where vehicles (ships) depart from one location to another followong a predetermined route to collect/deliver good to be distributed [16]. The logistics costs considered in this paper consist of transportation and infrastructures costs both for capital expenditures and operating expenditures. The reason that infrastructure costs are included in the objective function is because the model was developed based on the existing condition where LNG receiving infrastructures are not available yet. In other words, this fleet assignment model takes into account the consequences of the selected ships on the required infrastructures specifications (jetty/platform, facilities and equipments).

This paper is structured with the following framework. Section 2 discusses the literature study on the fleet assignment or deployment that has been carried out by other researchers. The description of the real problem used as a case study is described in Section 3. Section 4 presents a mathematical model developed. Section 5 presents computation result and sensitivity analysis. Finally, conclusions and suggestion for furhter research are presented in Section 6.

2. Literature review

Distribution management is a major decision-making problem in most companies or organizations due to the distribution costs to the total cost is quite significant, especially if fuel price are getting higher. Therefore, fleet assignment and route planning are major issues in almost all distribution management. The number and size of vehicles required to accommodate demand at minimum cost is a natural question that arises in this distribution management issues which is difficult to answer because of combinatorial characteristics [17]. Therefore, this issues has attracted many researchers to conduct a research related to network planning, fleet sizing and fleet assignment – which is belongs to Vehicle Routing Problem (VRP) - since more than 60 years ago [18].

The VRP was first introduced as a real-world application by Dantzig and Ramser concerning the delivery of gasoline by trucks [19]. This paper based on generalization of the Traveling Salesman Problem solved by using Linear Programming (LP) for obtaining a near optimal solution. Another monumental work in fleet allocation problem using LP when Laderman produced their paper that solved the minimum number of vessels required and allocate them to the various routes [20]. Five years later, Clarke and Wright proposed a heuristic method to solve problems such as Dantzig and Ramser problem with a more effective solution [21]. Following this brilliant pioneer, many papers have been published to present mathematical models, exact and heuristic algorithms.

Perakis and Jaramillo developed a model of fleet deployment optimization for liner shipping on various routes by using a LP algorithm under deterministic consideration [22]. Their model fall in the category of strategic planning rather than operational level with the resulting output consisting of: number and type of ships to be operated, ships to be allocated, and which owned-ships to lay-up. This research was continued by Powel and Perkins who developed model for the same case as that carried out by Perakis and Jaramillo by using an Integer Programming (IP) approach [23]. With this model they can determine the optimal ships assignment on a particular routes, levels of service, charter rates, and compatibility constraints.

Another method which are widely used in fleet assignment problems are mixed-integer linear programming (MILP) and mixed-integer non linear progamming (MINLP). Fleet assignment problem which not only determines the ship assigned but also determines how much cargo to be transported per shipment by considering inventory level constraints at the port is done by [9] using MINLP method. While [24] applying MILP method in developing an LNG carrier assignment model to determine the optimum route with the objective function of minizing cost. The application of mixed-integer programming method was also proposed by [25] to solve fleet assignment problem in hub-and-spoke networks conditions considering realistic factors from handling and holding cost, traveling time to the late delivery penalties.

Milk-run is a logistics strategy where vehicles depart from one location to another following a predetermined route to collect/ deliver goods to be distributed [16]. According to Sadjadi et al [26], this milk-run operation provides more economical transportation costs since it allows small and frequent demand aggregations. Related to this operation strategy, Adriano et al [27] developed a fleet assignment and routing problem in dynamic milk-run by using vehicular ad hoc networks.

The focus of most studies presented above is only on the fleet deployment for the given route of an existing port or terminal. The purpose of this paper is to propose a fleet assignment model where the receiving infrastructures (port/terminal) are not available yet and is taken into account in the model as a consequence of the ship assigned with the objective function of minimizing unit logistics costs.

3. Problem description

The power plant gasification program of Indonesian Government motivates this research where the problem is how to transport LNG from the source to meet the demand for scattered small-scale power plants with the minimum unit logistics costs. This program is planned to replace diesesl oil (HSD) – which is currently used as primary energy for small-scaled power plants – due to rising fuel prices. There are several small-scale power plants targeted by this program with location spread mainly in Eastern part of Indonesia. The real case study used in this paper are power plants in the North Papua Cluster, which consists of 5 power plants on the Northern part of Papua with LNG source from LNG-Tangguh where the demand for each power plants as shown in **Table 1**.

Table 1. Demand of each power plants

Power plants	Demand (BBTUD)
Jayapura	9,22
Serui	0,92
Manokwari	2,49
Nabire	2,00
Biak	2,21

The model was developed to select the optimal ship (from the set of available ships) and the consequences for the infrastructure required on a specific route with the objective function to minimize the total logistics costs. Optimal route selection is carried out using a separate model and is beyond the scope of this paper. The logistics costs consist of 2 (two) main component costs as follows:

- Shipping costs, consist of fixed costs (capital, crew wages, supplies, docking repair, running repair and sertification) and variable costs (port charges, and fuel costs).
- Infrastructure costs consist of capital expenditures (platform/ jetty, regasification system, piping system, and storage tank) and operating expenditures which is assumed to be 4% of capital expenditures.

The fleet consists of a set of LNG carriers with capacities ranging from 1.000 m^3 to 65.000 m^3 with different technical specifications (capacity, deadweight, length overall, draugh, speed laden and ballast, building cost) as shown in **Table 2**. The voyages of ships are not pre-assigned to a specific routes, thus the assignment of each voyages to meet the demand is also part of the optimization problem.

Table 2. LNG carriers data

Ship name	Capacity	DWT	LOA	T	Vs laden	Vs ballast	Ship price
	(m3)	(ton)	(m)	(m)	(knot)	(knot)	(mio USD)
v-1.000	1.000	1.564	70,78	3,82	7,00	8,00	13,00
v-2.000	2.000	2.531	83,01	4,42	7,70	8,80	20,00
v-3.000	3.000	3.511	91,12	4,82	7,70	8,80	25,00
v-3.500	3.500	3.643	94,40	4,98	8,40	9,60	28,00
v-4.000	4.000	4.052	97,36	5,12	8,40	9,60	33,00
v-6.000	6.000	5.532	106,87	5,57	8,40	9,60	38,00
v-8.000	8.000	6.404	114,18	5,92	9,10	10,40	40,00
v-8.500	8.500	6.715	115,79	6,00	9,10	10,40	42,00
v-10.000	10.000	7.589	120,19	6,21	9,10	10,40	44,00
v-12.000	12.000	8.676	125,35	6,45	9,10	10,40	49,00
v-15.000	15.000	10.192	131,94	6,76	9,10	10,40	56,00
v-16.500	16.500	10.919	134,86	6,90	9,10	10,40	60,00
v-18.000	18.000	11.613	137,59	7,03	9,10	10,40	62,00
v-30.000	30.000	18.000	180,00	7,30	11,00	12,50	70,00
v-45.000	45.000	27.000	196,00	10,30	13,00	14,50	85,00
v-65.000	65.000	35.000	200,00	11,00	14,00	15,50	98,00

In addition, since receiving infrastructures of each power plants is not available yet, the specification of the receiving infrastructures is also part of the optimization problem as a consequences of the selected LNG carrier. Therefore, costs or infrastructures required (both capital expenditures and operating expenditures) are included in the objective function of the model. The route applied in this model is a milk-run system where the route combinations are developed in a separate model, therefore, the route in this model is a given parameter. The milk-run route from the LNG source to each power plant site consisting of 6 legs is as shown in **Figure 1**. Leg_1 to leg_5 are laden voyages since in this legs ships on these legs is in a loaded condition, while leg_6 is ballast voyage because the ship is empty.

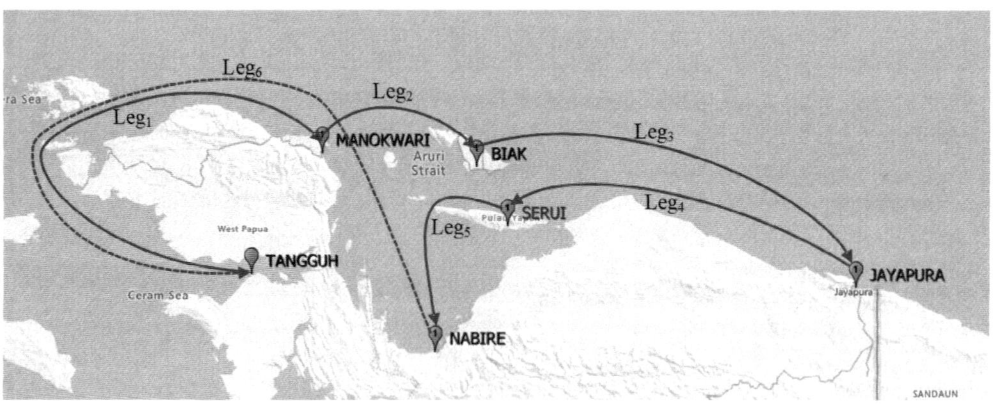

Figure 1. The given milk-run route

As previously mentioned, the focus of this paper is industrial shipping, therefore, the time horizon of the calculation in this model is carried out over the ship's economic life which is assumed to be 20 years. In order to simplify the problem, the receiving infrastructures are also assumed to have the same economic life as the ships. Thus, the calculation in this model are carried out with the same time horizon of 20 years. Both capex and opex costs are calculated annually and are assumed to increase every year. Furthermore, to get a uniform value (for the sake of the calculation in the model) for 20 years, the annuity method is used.

If more than one ship is required to meet the demand, it is assumed that these ships are identical, where the fixed cost is calculated as a function of the number of ships while the variable cost is a function of the number of trips (voyages) required per ship. In addition, the demand for each power plants is assumed to have a certain value (deterministic) and constant for a periode of 20 years.

4. Mathematical model

Since the operation strategy used in this LNG distribution is a milk-run operation, the first step is to determine the volume of LNG (transported by the selected ship) that must be discharged at each power plants visited according to a predetermined route. This volume is calculated as a ratio between the demand of each power plant to total demand of all power plants in this cluster multiplied by the net capacity of the ship by the following equation:

$$Vd_{k,n} = \frac{Q_n}{\sum_{n=1}^{N} D_n} P_k, \qquad \forall_{k \in K} \tag{1}$$

Where:

N = number of nodes (power plant site) excluding LNG source, $n = \{1,2,\dots,N\}$

K = number of difference size of ships, $k = \{1,2,\dots,K\}$

$Vd_{k,n}$ = the discharged volume of the kth ship at the nth node (m^3/trip)

Q_n = demand of LNG at nth node per year (m^3/year)

P_k = the net capacity of the kth ship which is assumed to be 95% of total capacity (m^3)

The next step is to calculate the roundtrip days required for the ship to complete one voyage from loading at origin, discharging at each power plants and return to origin. Basically, the duration of the roundtrip days is the sum of sea time and port time. The sea time is grouped into laden voyages and ballast voyage and therefore, sea time is a function of laden speed and ballast speed of the ship. Formula to calculate sea time is as follows:

$$Ts_k = \left(\sum_{n=1}^{N} \frac{d_n}{vladen_k.24}\right) + \frac{d_{N+1}}{vballast_k.24}, \qquad \forall_{k \in K} \tag{2}$$

Where:

Ts_k = sea time of the kth ship to complete one voyage

d_n = The distance from source to the first site for $n = 1$ and distance between sites for $n = 2,3,\dots N$ (nm)

d_{N+1} = distance from last site to the source (nm)

$vladen_k$ = laden speed of the kth ship (knot)

$vballast_k$ = ballast speed of the kth ship (knot)

Meanwhile, the formula for calculating port time both for loading and dischaging is shown in Equation (3) below, where the first part is the port time at each power plants site and the second part is the port time at the LNG source. Therefore, the formula to calculate the total roundtrip days required by the ship is as shown in Equation (4)

$$Tp_k = \left(\sum_{n=1}^{N} \frac{Vd_{k,n}}{Rd_{k,n}}\right) + \frac{Pk}{Rs}, \qquad \forall_{k\in K} \tag{3}$$

$$RTD_k = Ts_k + Tp_k \tag{4}$$

Where:
Tp_k = port time of the kth ship at each nodes
$Rd_{k,n}$ = discharging rate of the kth ship at the nth nodes (m^3/hour)
Rs = loading rate at the source node (m^3/hour)
RTD_k = Rountrip days of the kth ship (days)

The number of ships needed to meet the demands of all power plants is calculated by comparing the RTD of the ship with the duration of consumption of LNG discharged at each port. This duration is calculated by dividing volume $Vd_{k,n}$ with the daily demand $D_n/365$. Therefore, the number of ships required is calculated by the following equation:

$$nShip_k = Round' \frac{RTD_k}{\left(\frac{Vd_{k,n}}{Q_n/365}\right)}, \qquad \forall_{k\in K, n\in N} \tag{5}$$

Where:
$nShip_k$ = number of the kth ship required (unit)

Fixed costs component of shipping costs consist of capital and operating costs, where annual capital costs is calculated from ship price S_p by using annuity formula as shown in Equation (6), while the operating costs are obtained from the given data or assumption. Operating costs considered in this model include manning, docking repair, running repair and ship sertification. Therefore, the total fixed costs calculated by the following equation:

$$CC_k = Sp_k \left[\frac{i(1+i)^{Per}}{(1+i)^{Per} - 1}\right] \tag{6}$$

$$TFCs_k = nShip_k(CC_k + OC_k) \tag{7}$$

Where:
Sp_k = total price of the kth ship (\$)
i = weighted average of capital cost (%)
Per = contract period (years)
$TFCs_k$ = total annual fixed costs of the kth ship (\$)
CC_k = annual capital cost of the kth ship (\$)
OC_k = annual operating costs of the kth ship (\$)

Variable costs consist of fuel costs and port charges, where the formula to calculate fuel cost is shown in Equation (8) and formula to calculate port charges is shown in Equation (9), therefore, the total variable costs is as shown in Formula (10) and total shipping costs as shown in Equation (11):

$$CF_k = Ts_k FOC_k Round'\left(\frac{\sum_{n=1}^{N} Q_n}{nShip_k}\right) PO, \qquad \forall_{k\in K} \tag{8}$$

$$CP_k = \prod_{n=1}^{N} Round'\left(\frac{\sum_{n=1}^{N} Q_n}{nShip_k}\right) PC_n, \qquad \forall_{k\in K} \tag{9}$$

$$TVCs_k = CF_k + CP_k \tag{10}$$

$$TCs_k = TFCs_k + TVCs_k \tag{11}$$

Where:
CF_k = annual fuel oil cost of the kth ship ($)
FOC_k = fuel oil consumption rate of the kth ship (MT/hour)
PO = fuel oil price ($/MT)
CP_k = annual port costs of the kth ship ($)
PC_n = port charges of the nth node ($)
$TVCs_k$ = total annual variable costs of the kth ship ($)
TCs_k = total annual shipping cost of the kth ship ($)

The infrastructure costs are calculated as a consequence of the selected ship, which is as in shipping cost, consists of fixed and variable costs. The formulas to calculate infrastructures capital expenditures are obtained by applying the regression method to the real data, while the operating expenditures as mentioned are assumed to be 4% of capital expenditures. The storage tank is assumed to be of the bullet type with a volume of 500 m3 per unit, therefore, the number of tanks required at each node is a multiple of this volume. The formula related to the capital expenditures of receiving infrastructures at each node is shown in the following equation:

$$FCj_n = (100{,}70\ DWT_k + 5{,}9\ .\ 10^6)\left[\frac{i(1+i)^{Per}}{(1+i)^{Per}-1}\right] \tag{12}$$

$$FCr_n = (2{,}25\ .\ 10^6 Q_n + 1{,}5\ .\ 10^6)\left[\frac{i(1+i)^{Per}}{(1+i)^{Per}-1}\right] \tag{13}$$

$$FCt_n = \left(Round'\left(\frac{Vd_{k,n}(1+10\%)}{vSt}\right)Pt\right)\left[\frac{i(1+i)^{Per}}{(1+i)^{Per}-1}\right] \tag{14}$$

$$FCp_n = \left((Lpg_n + Lpv_n).Pp\right)\left[\frac{i(1+i)^{Per}}{(1+i)^{Per}-1}\right] \tag{15}$$

$$OCinf_n = \left((100{,}70\ DWT_k + 5{,}9\ .\ 10^6) + (2{,}25\ .\ 10^6 Q_n + 1{,}5\ .\ 10^6) + \left(Round'\left(\frac{Vd_{k,n}(1+10\%)}{vSt}\right)Pt\right)\right).4\% \tag{16}$$

$$TCinf_n = FCj_n + FCr_n + FCt_n + OCinf_n \tag{17}$$

Where:
FCj_n = annual fixed cost of jetty of the nth node ($)
FCr_n = annual fixed cost of regas unit of the nth node ($)
FCt_n = annual fixed cost of storage tank of the nth node ($)
vSt = volume of bullet tank (m^3/unit)
Pt = unit price of bullet tank ($/unit)
FCp_n = annual fixed cost of piping system of the nth node ($)
Lpg_n = length of LNG line of the nth node (m)
Lpv_n = length of Vapor line of the nth node (m)
Pp = unit price of cryogenic pipe ($/m)
$OCinf_n$ = annual operating cost of infrastructures of the nth node ($)
$TCinf_n$ = Total annual cost of infrastructures of the nth node ($)

By using the formulas as previously mentioned, the optimization equation can be expressed as follows:

Decision variable:

$$x_k = \begin{cases} 1, & \text{if ship assigned} \\ 0, & \text{otherwise} \end{cases} \tag{18}$$

Objective function:

$$Min \sum_{k-1}^{K} \sum_{n=1}^{N} x_k \frac{(TCs_n + TCinf_n)}{Q_n} \tag{19}$$

Subject to:

$$Vd_{k,n} Round' \left(\frac{\sum_{n=1}^{N} Q_n}{nShip_k} \right) \geq Q_n \tag{20}$$

5. Computational result

5.1. Optimum results

With the input parameters of demand for each power plants and available alternative ships as shown in **Table 1** and **Table 2** respectively, the model shows that optimal solution (minimum unit logistics costs) are achieved by assigning ship with a capacity of 10.000 m^3 as shown in **Table 3**.

Table 3. The results of running model

Trial	Ship name	No. of ships required (unit)	Utility of ships (%)	Shipment frequencies (times)	Rountrip days (days)	Unit logistics cost ($/mmbtu)
1	v-30.000	1	29,82%	11	9,63	8,29
2	v-1.000	67	18,66%	335	13,25	56,76
3	v-18.000	1	50,70%	17	10,59	7,14
4	v-2.000	9	58,15%	153	12,14	13,78
5	v-15.000	1	63,99%	21	10,82	6,60
6	v-10.000	1	93,77%	31	10,74	5,89
7	v-16.500	1	57,10%	19	10,67	6,83
8	v-8.500	2	53,58%	36	10,57	7,46
9	v-12.000	1	78,59%	26	10,73	6,23
10	v-45.000	1	16,20%	7	8,22	10,01
11	v-65.000	1	11,27%	5	8,00	12,18
12	v-4.000	3	82,31%	78	11,24	8,20
13	v-3.000	5	70,75%	105	11,96	10,45
14	v-3.500	4	68,98%	88	11,13	9,26
15	v-8.000	2	59,20%	40	10,51	7,40
16	v-6.000	2	81,09%	52	11,07	7,34

The results show that there is a high relationship between demands and the capacity of the selected ships. For small demands, the use of small ships requires a larger number of ships due to cargo discharged at each power plant is also small. The less cargo recived, the shorter the inter-arrival time between ships required, so that more ships are needed since the rountrip days of the ship is greater than the required inter-arrival time. For example, the use of 10.000 m^3 ships (see **Table 3**.), to meet the demand for each power plant, 67 ships are needed with a total number of shipment frequencies of 335 time/year with an average utility of each ships quite low (18,66%).

On the other hand, the use of large ships requires a larger tank storage capacity at each location since the cargo discharged per shipment is also large which has an impact on the higher costs of the infrastructures. The more cargo received, the longer the inter-arrival time of the ship, so the utility of the ship is also low which has an impact on the higher shipping costs. For example, in order to meet the demands of each power plant, the use of ships with 65.000 m3 capacity, the number of ships required is only 1, but with a frequency of only 5 times/year, so the utility of ships is very low (11,27%). If the relationship between the capacity of the selected ship and the unit logistics costs are plotted, it will look like in **Figure 2**. From this figure, it can be seen tha the large capacity of the ship to a certain value is able to reduce unit logistics costs, but after reaching the optimum value, the larger the ship will cause the unit logistics costs to be even higher. This satisfies the law of diminishing returns.

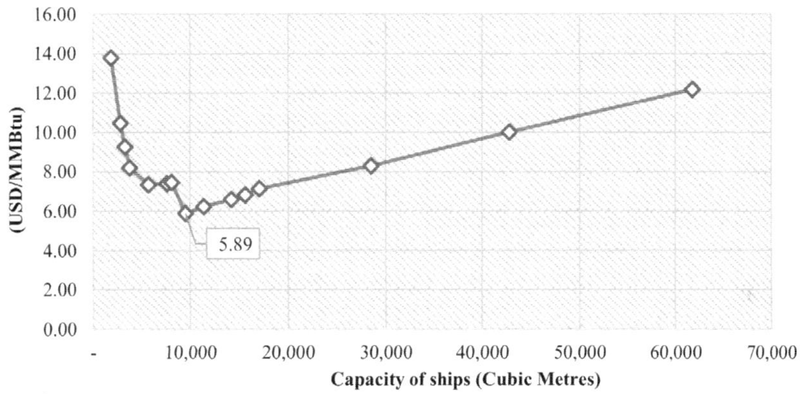

Figure 2. The minimum unit cost of LNG distribution for North Papua Cluster

The optimum size of the fleet assigned is an LNG Carrier of 10,000 cubic metres with a minimum unit logistics cost is 5.89 USD/MMBtu. The number of shipment frequeuncies is 31 times per year with a round trip days of 10,74 days. The total capital expenditures (CAPEX) and operating expenses (OPEX) for both shipping and infrastructure required are presented in **Table 4**. Meanwhile, the specification for the receiving infrastructures at each power plants site, especially for the number of storage tanks required and the length of the trestle, can be seen in **Table 5**.

Table 4. Capex and opex for both shipping and infrastructures

Item	Capex		Opex		Unit logistcs cost	
	(mio USD)	(%)	(mio USD)	(%)	(USD/mmbtu)	(%)
Shipping	44,00	24,69%	8,26	60,61%	2,31	39,26%
Infrastructures	134,23	75,31%	5,37	39,39%	3,57	60,74%
Total	178,23	100%	13,63	100%	5,89	100%

IOP Conf. Series: Earth and Environmental Science 1166 (2023) 012045 doi:10.1088/1755-1315/1166/1/012045

Table 5. Infrastructures needs

Infrastructures	Units	Manokwari	Nabire	Serui	Jayapura	Biak
Gross storage	(m^3)	1.545	1.241	571	5.721	1.371
Number of tank	(unit)	4	3	2	12	3
Regasification unit	(m^2)	14 x 6,5	14 x 6,5	14 x 6,5	14 x 6,5	14 x 6,5
Pipeline & MRS	(unit)	2 line LNG 1 line vapour	2 line LNG 1 line vapour	2 line LNG 1 line vapour	2 line LNG 1 line vapour	2 line LNG 1 line vapour
Jetty head	(m^2)	20 x 40	20 x 40	20 x 40	20 x 40	20 x 40
Trestle length	(m)	250	280	205	250	150

5.2. Sensitivity analysis

To investigate whether the optimal solution is sufficiently robust, it is usefull to perform a what-if analysis – also known as sensitivity analysis – to see how optimal solution changes when some parameters vary. In this paper, sensitivity analysis is performed by varying the demands for each power plant and investigating the consequences for the objectiv function with the results as shown **Figure 3**. This sensitivity analysis resulted in a downward sloping unit logistics cost curve that satisfies the law of economies of scale, where the larger the ship size the lower unit cost.

Nevertheless, from the **Figure 3**, it can be seen that when the demand is less than 90% or greater than 105% from the given data, the ship with 10.000 m3 capacity is no longer valid as the optimum decision. Since the assumption used in this paper is that the demand for each power plant is constant for 20 years, which is mostlikely to change, this sensitivity analysis is very important and can provide important information for decision maker related to the size of the ships that mus be built.

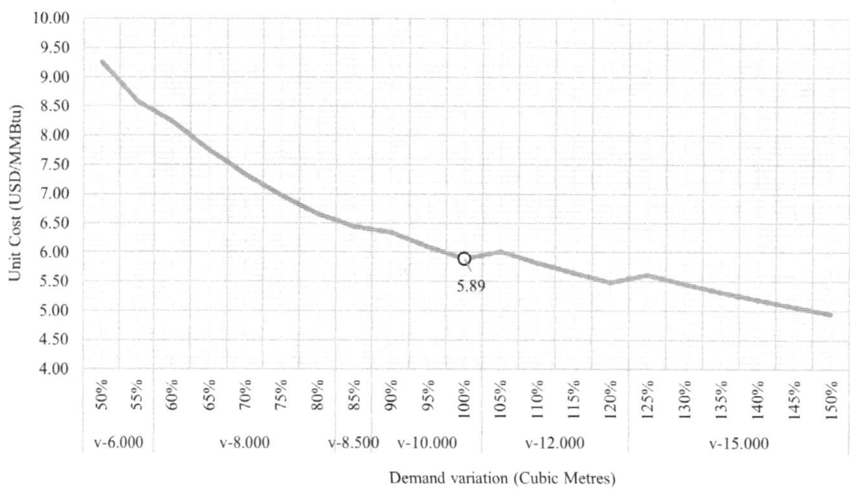

Figure 3. Sensitivity analysis of demand and unit logistics cost

6. Conclusion

This paper presents a fleet assignment model from a real-life problem of LNG small-scale supply chain by considering the receiving infrastructures required as a consequences of the selected ship and included in the objective function. Even the demand is small, the use of ships that are too small is not the best solution because of large number of ships are needed to meet the inter-arrival time needs at each power plant site. On the other hand, the use of large ships is also not the best solution not only because the utility of the ship is very low, but also has an impact on higher infrastructures costs.

The application of the genetic algorithm method in this case study is proven to be able to reach the optimum point in a relatively short time. In addition, the developed mode in this paper is able to produce an optimum curve that satisfies the law of deminishing return. The result of the sensitivity analysis show that the variation in demand of each power plant around 15% will not affect the optimum solution.

Further research needs to be carried out in the following directions: (1) Consider the problems with the uncertain (fluctuating) demand for a long time horizon. (2) Extend the model with the hub-and-spoke networks where transshipment must be involved. (3) combination of options (1) dan (2).

7. Acknowledgment

The support of all colleagues in the marine transportation and logistics laboratory, department of marine transportation, Insitut Teknologi Sepuluh Nopember is highly appreciated for all their contributions.

References

[1] M. Ball, C. Barnhart, G. Nemhauser, and A. Odoni, "Chapter 4: Maritime Transportation," in *Handbooks in Operations Research and Management Science*, vol. 14, no. C, C. Barnhart and G. Laporte, Eds. 2007, pp. 1–67. doi: 10.1016/S0927-0507(06)14001-3.

[2] X. Lai, L. Wu, K. Wang, and F. Wang, "Robust ship fleet deployment with shipping revenue management," *Transportation Research Part B: Methodological*, vol. 161, pp. 169–196, Jul. 2022, doi: 10.1016/j.trb.2022.05.005.

[3] A. J. Su, W. D. Yang, C. Zhang, and M. X. Kong, "Robust modeling for fleet assignment problem based on GASVR forecast," *J Phys Conf Ser*, vol. 1187, no. 4, May 2019, doi: 10.1088/1742-6596/1187/4/042062.

[4] K. Fagerholt, M. Christiansen, L. Magnus Hvattum, T. A. V. Johnsen, and T. J. Vabø, "A decision support methodology for strategic planning in maritime transportation," *Omega (Westport)*, vol. 38, no. 6, pp. 465–474, Dec. 2010, doi: 10.1016/j.omega.2009.12.003.

[5] B. Dong, T. Bektaş, S. Chandra, M. Christiansen, and K. Fagerholt, "A new formulation for the combined maritime fleet deployment and inventory management problem," in *Lecture Notes in Computer Science (including subseries Lecture Notes in Artificial Intelligence and Lecture Notes in Bioinformatics)*, 2017, vol. 10572 LNCS, pp. 321–335. doi: 10.1007/978-3-319-68496-3_22.

[6] T. L. Magnanti and R. T. Wong, "Network design and transportation planning: Models and algoritms," *Transportation Science*, vol. 18, no. 1, pp. 1–55, 1984, doi: 10.1287/trsc.18.1.1.

[7] V. Windeck, "A liner shipping network design," Dissertation, University of Hamburg, Hamburg, 2012.

[8] S. C. Parikh, "On a Fleet Sizing and Allocation Problem," *Source: Management Science*, vol. 23, no. 9, pp. 972–977, May 1977, [Online]. Available: http://www.jstor.orgURL:http://www.jstor.org/stable/2630778

[9] S. Gelareh and Q. Meng, "A novel modeling approach for the fleet deployment problem within a short-term planning horizon," *Transp Res E Logist Transp Rev*, vol. 46, no. 1, pp. 76–89, 2010, doi: 10.1016/j.tre.2009.06.004.

[10] Q. Meng, T. Wang, and S. Wang, "Short-term liner ship fleet planning with container transshipment and uncertain container shipment demand," *Eur J Oper Res*, vol. 223, no. 1, pp. 96–105, Nov. 2012, doi: 10.1016/j.ejor.2012.06.025.

[11] J. Xia, K. X. Li, H. Ma, and Z. Xu, "Joint planning of fleet deployment, speed optimization, and cargo allocation for liner shipping," *Transportation Science*, vol. 49, no. 4, pp. 922–938, Nov. 2015, doi: 10.1287/trsc.2015.0625.

[12] Y. Wang, Q. Meng, and Y. Du, "Liner container seasonal shipping revenue management," *Transportation Research Part B: Methodological*, vol. 82, pp. 141–161, Dec. 2015, doi: 10.1016/j.trb.2015.10.003.

[13] Q. Meng and S. Wang, "Liner ship fleet deployment with week-dependent container shipment demand," *Eur J Oper Res*, vol. 222, no. 2, pp. 241–252, Oct. 2012, doi: 10.1016/j.ejor.2012.05.006.
[14] Q. Meng, T. Wang, and S. Wang, "Multi-period liner ship fleet planning with dependent uncertain container shipment demand," *Maritime Policy and Management*, vol. 42, no. 1, pp. 43–67, Jan. 2015, doi: 10.1080/03088839.2013.865848.
[15] S. Wang, Z. Liu, and X. Qu, "Minimax regret model for liner shipping fleet deployment with uncertain demand," *Transp Res Rec*, vol. 2549, pp. 45–53, 2016, doi: 10.3141/2549-06.
[16] D. D. Adriano, C. Montez, A. G. N. Novaes, and M. Wangham, "Dynamic milk-run vehicle routing solution using fog-based vehicular ad hoc networks," *Electronics (Switzerland)*, vol. 9, no. 12, pp. 1–24, Dec. 2020, doi: 10.3390/electronics9122010.
[17] B. Golden, A. Assad, L. Levy, and F. Gheysens, "The fleet size and mix vehicle routing problem," *Computer & Operations Research*, vol. 11, no. 1, pp. 49–66, 1984.
[18] P. Toth, D. Vigo, and Society for Industrial and Applied Mathematics, *Vehicle routing : problems, methods, and applications*. 2014.
[19] G. B. Dantzig and J. H. Ramser, "The Truck Dispatching Problem," *Manage Sci*, vol. 6, no. 1, 1959.
[20] J. Laderman, L. Gleiberman, and J. F. Egan, "Vessel allocation by linear programming," *Naval Research Logistics Quarterly*, vol. 13, no. 3, pp. 315–320, Sep. 1966, doi: 10.1002/nav.3800130307.
[21] G. Clarke and J. W. Wright, "Scheduling of Vehicles from a Central Depot to a Number of Delivery Points," *Oper Res*, vol. 12, no. 4, pp. 568–581, Aug. 1964, doi: 10.1287/opre.12.4.568.
[22] A. N. Perakis and D. I. Jaramillo, "Fleet deployment optimization for liner shipping part 1. background, problem formulation and solution approaches," *Maritime Policy and Management*, vol. 18, no. 3, pp. 183–200, 1991, doi: 10.1080/03088839100000022.
[23] B. J. Powell and A. N. Perkins, "Fleet deployment optimization for liner shipping: An integer programming model," *Maritime Policy and Management*, vol. 24, no. 2, pp. 183–192, 1997, doi: 10.1080/03088839700000069.
[24] A. Zetos, C. Papaleonidas, and D. v. Lyridis, "Developing a comprehensive fleet deployment model for LNG shipping companies," 2020.
[25] L. Zhen, S. Wang, G. Laporte, and Y. Hu, "Integrated planning of ship deployment, service schedule and container routing," *Comput Oper Res*, vol. 104, pp. 304–318, Apr. 2019, doi: 10.1016/j.cor.2018.12.022.
[26] S. J. Sadjadi, M. Jafari, and T. Amini, "A new mathematical modeling and a genetic algorithm search for milk run problem (an auto industry supply chain case study)," *International Journal of Advanced Manufacturing Technology*, vol. 44, no. 1–2, pp. 194–200, Sep. 2009, doi: 10.1007/s00170-008-1648-5.
[27] D. D. Adriano, A. G. N. Novaes, and M. S. Wangham, "Providing a dynamic milk-run vehicle routing using vehicular ad hoc networks," in *DIVANet 2019 - Proceedings of the 9th ACM Symposium on Design and Analysis of Intelligent Vehicular Networks and Applications*, Nov. 2019, pp. 31–38. doi: 10.1145/3345838.3356006.

Analysis of hull performance on fast patrol boat with an extended study of survivability under damaged conditions

A S Pratama[1], A R Prabowo[1,*], N Muhayat[1], T Putranto[2,3] and T Tuswan[4]

1 Department of Mechanical Engineering, Universitas Sebelas Maret, Surakarta 57126, Indonesia
2 Department of Civil Engineering and Architecture, Tallinn University of Technology, Tallinn 19086, Estonia
3 Department of Naval Architecture, Institut Teknologi Sepuluh Nopember, Surabaya 60111, Indonesia
4 Department of Naval Architecture, Universitas Diponegoro, Semarang 50275, Indonesia

Corresponding author: aditya@ft.uns.ac.id

Abstract. The prevalence of maritime accidents that occur due to ship leakage causes a significant risk to the safety of passengers and crew. Therefore, it is important to estimate stability during critical conditions, including characterizing the hull performance in sailing conditions. This research analyzes the effect of compartments and bulkheads on the ship's stability when a leak occurs. The research method was carried out using the numerical approach to obtain resistance, stability, and large-angle stability results. The study also uses floodable length simulations to predict ship leakage and the distance between compartments. The variations were composed, including the location of the compartments and the distance between the partitions. Based on the obtained results, it can be seen that the addition of the compartment will affect the intensity of the ship to sink completely. Furthermore, the effect of leakage in each compartment will contribute to the stability state of the ship.

Keywords: Resistance, Stability, Floodable length, Large-angle stability, Compartment.

1. Introduction

Ships are a means of water transportation (which concept of advanced material and propulsion are considered in the ship [1,2]) needed and used in various water sectors. Shipbuilding is adapted to its functions and missions, including transporting natural resources or goods, transporting people, tourism facilities, and war defense equipment. Characteristics of waters, such as rivers, lakes, or oceans, determine the dimensions and shape of the ship to be used so that it is effective and increases safety during the voyage. Therefore, it is necessary to perform a study before designing a ship to get a ship that is multipurpose and optimal to support ship performance [3].

Designing a ship must also consider the ship's dimensions, the shape of the hull, the power required, the arrangement of engines, and the primary structure. According to the required characteristics, the proper determination will support the ship's performance in terms of resistance, ship stability, good maneuverability and seakeeping, speed, cargo capacity, and deadweight [4-8]. According to the International Maritime Organization (IMO) regulations, a ship must also meet the requirements of a unique agency for ship safety and security. However, in realization, some factors can cause the ship to have an accident and sink. Ship accidents can be caused by collisions, weather, terrain, and being shot during the war, so a study is needed to prevent ships when they experience accidents or leaks to increase the ship's survival power for the safety of passengers and crew.

Prevention of the ship when it leaks is by adding bulkheads or compartments on the ship. The hull is divided into several compartments so that when a leak occurs on one side, it does not immediately fill the entire hull but only the Compartment where the leak occurs. The flood margin line is measured from the bulkhead deck with a minimum distance of 76 millimeters (3 inches). The bulkhead deck adjusts from the position of the highest bulkhead point of the transverse bulkhead. [3,5]. Vessels with potholes and flooding will have a poor stability margin, making the ship unstable. When the equilibrium state is stable, the center of buoyancy and gravity are in one direction, and the longitudinal and transverse lines are aligned with the trim angle. Flooding on the ship will also affect the trim angle, dynamic stability, roll period, and pitch period, and these factors contribute significantly to the ship's sinking [9-13].

This study aims to analyze the initial design of the ship to test its feasibility when it is realized. The analysis used is floodable length, one of the factors that must be met in designing a ship to analyze when the boat leaks when sailing. This study also investigates the effect of adding bulkheads to the ship. This research was done because of the prevalence of ship accidents, which makes prevention needed to improve ship survival so that passengers and crew can be saved before the ship sinks.

2. Methodology

In this research, simulations are carried out to obtain the main factors of ship design, such as resistance and stability, and simulation of additional compartments to support the ship's resilience when it experiences a leak. Ship design is carried out by collecting data by looking for original ship references to get the primary size of the ship. A simulation will be carried out from the design obtained to get some data and processed to get the best results. The analysis results were identified to determine the feasibility of the ship from the resistance and stability factors, and additional compartments were also identified to determine the optimal distance and position on the designed ship.

2.1 Ship Data

Hull modeling is started with designing a line plan according to the shape of the ship type and by predetermined criteria, as shown in Table 1. Variations of ship dimensions are chosen to determine which dimensions are the most optimal in the simulation of resistance and stability. The design refers to the original ship for the main sizes of ships with different hull shapes, so there are differences in the main sizes and variations in the hull. There are three designs: Design 1 concerning the NSW WATER POLICE - CLASS 1 ship, Design 2 concerning the Austal INSHORE Patrol 21 ship, and Design 3 concerning the DASTOOR AUSTAL Patrol 22 ship.

Table 1. Principal dimension of patrol boat designs

Parameters	Value		
	Design 1 - NSW WATER POLICE - CLASS 1	Design 2 - Austal INSHORE Patrol 21	Design 3 - DASTOOR AUSTAL Patrol 22
Length Overall (m)	21.2	21.2	21.6
Length Waterline (m)	19.761	20.137	19.929
Beam (m)	5.5	5.5	6
Depth (m)	3.3	2.8	3.6
Draft (m)	1.700	1.83	1.500
Cb (-)	0.346	0.401	0.339
Displacement (t)	56.83	79.02	42.66
Length: Beam Ratio	4.025	3.812	4.580
Beam: Draft Ratio	2.888	2.910	2.858

Additional compartments are also made in each ship design to identify how to react when the ship leaks. Therefore, it is known how the condition of the ship when it leaks when using a compartment and without a compartment. In addition to compartments, the permeability value will also affect how the volume of the compartment can accommodate water and keep the ship from sinking. From the designs that have been made, the lines plan design is obtained as shown in Figures 1 to 3, and the 3D design is shown in Figure 4.

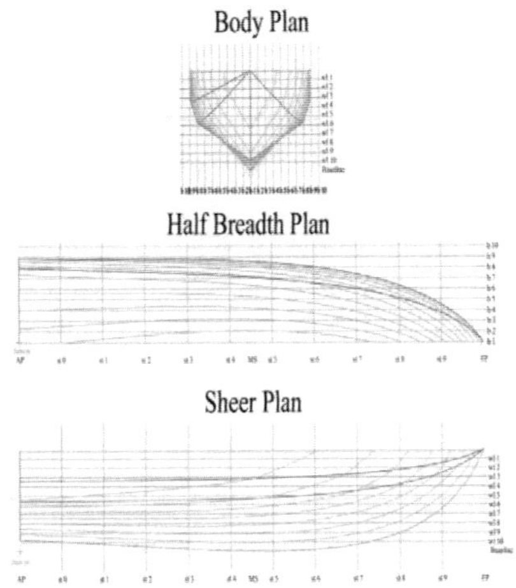

Figure 1. Linesplan Design 1

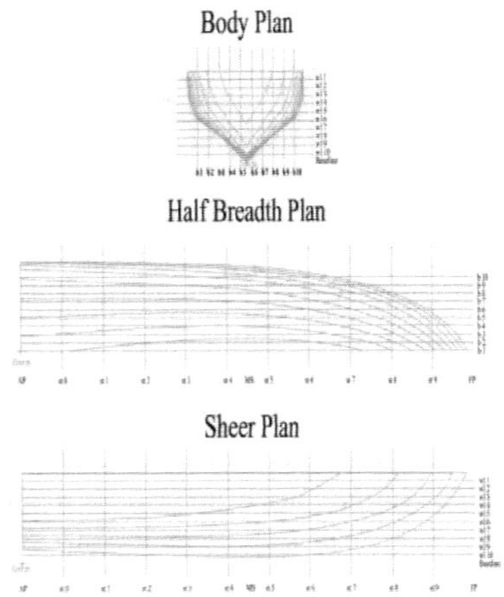

Figure 2. Linesplan Design 2

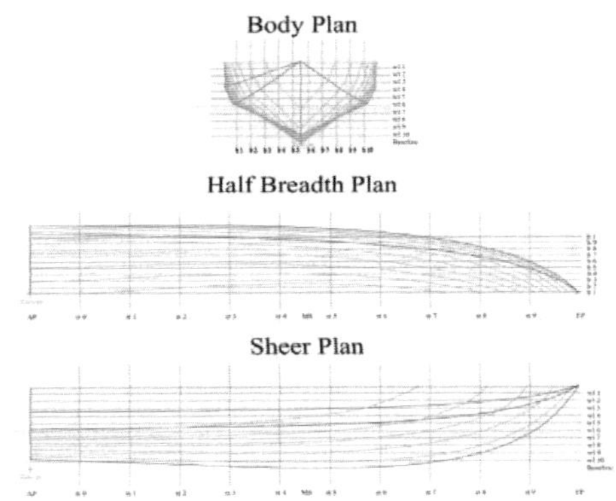

Figure 3. Linesplan Design 3

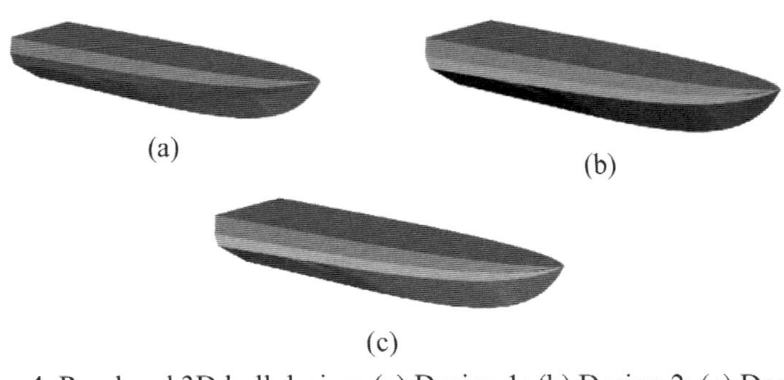

Figure 4. Rendered 3D hull design: (a) Design 1; (b) Design 2; (c) Design 3.

2.2 Research Parameter

The analysis used includes simulation of hull resistance and floodable length. Each simulation uses a control variable to control the results and minimize other influences. Froude number is used as a control variable in the simulation of resistance; the value of the Froude number is based on the length of the ship, submerged point, and immersion depth [15]. In the calculation of resistance, essential variables can determine the final result, including speed and displacement, overall and beam length, deadrise angle, and longitudinal center of gravity (LCG).

The displacement of the shipload can determine the allowable flood length curve with compartment permeability (μ) which also plays a significant role in calculating the flood length value [3]. Permeability is the ratio of the volume that can be submerged to the total volume of the Compartment. The selected permeability ranges from 85% to 95% according to the conditions of the room used on the ship. In addition to permeability, variations in the bulkhead location are also used to obtain efficient values when a leak occurs, as shown in Table 2 and the permeability of each bulkhead in Table 3.

Table 2. Permeability and Bulkhead variations

Model	Variation	Bulkhead				
		Range 1	Range 2	Range 3	Range 4	Range 5
Model 1	Variation 1	3.7 m	5 m	8.6 m	13.5 m	17.7 m
	Variation 2	3.8 m	8.5 m	13.4 m	15.3 m	17.7 m
	Variation 3	2.2 m	3.4 m	8.1 m	15.8 m	18.2 m
Model 2	Variation 1	3.7 m	5 m	8.6 m	13.5 m	17.7 m
	Variation 2	3.8 m	8.5 m	13.4 m	15.3 m	17.7 m
	Variation 3	2.2 m	3.4 m	8.1 m	15.8 m	18.2 m
Model 3	Variation 1	3.7 m	5 m	8.6 m	13.5 m	17.7 m
	Variation 2	3.8 m	8.5 m	13.4 m	15.3 m	17.7 m
	Variation 3	2.2 m	3.4 m	8.1 m	15.8 m	18.2 m

Table 3. Permeability of bulkhead

Bulkhead	Value Permeability (%)
1	95
2	85
3	98
4	95
5	95
6	95

2.3 Hull Resistance

The method for resistance prediction is the Savitsky method which is one of the reliable approaches to calculating resistance and predicting power requirements. The Savitsky method can be calculated manually or using software with several input data such as speed, mass, deadrise angle, and center of gravity. The simulation results will be total resistance, lift force, trim and sinkage, and wetted surface area [16,17]. The speed coefficient by Savitsky is stated in Equation 1.

$$C_v = \frac{v}{\sqrt{g.b}} \tag{1}$$

where, C_v is the velocity coefficient; v is the speed of the ship; g is the center of gravity; and b is the maximum beam over chine. The Savitsky method uses the equation of the Reynolds number formula which is stated as follows Equation 2.

$$R_n = \frac{v_s.Lwt}{v} \tag{2}$$

where, V_s is the service speed; *Lwl* is the length of waterline; υ is the viscosity of seawater. Coefficient that occurs between the hull and the fluid can use the Equation 3.

$$D_f = \frac{C_f \rho v_1^2 \left(\lambda b^2\right)}{2\cos\beta} \tag{3}$$

where, C_f is the coefficient of friction resistance; ρ is the water density; V_1 is the average bottom velocity; *b* is the maximum beam over chine; β is the deadrise angle; λ is the wetted length-beam ratio. To find the Froude number value, Equation 4 is used in calculation.

$$Fn = \frac{V}{\sqrt{gL}} \text{ or } Fn = \frac{V}{\sqrt{gD}} \tag{4}$$

where, *Fn* is Froude number; *V* is the average flow velocity; *L* is the characteristic length; *D* is the ratio of the surface area to the surface width; where *D* is *A*/*T*.

2.4 Ship Stability

Based on the International Maritime Organization (IMO) standard, the ship must have a good stability value by the established criteria. The maximum righting lever is one of the factors that represent the level of the ship's ability to survive when the ship shakes. The righting lever's maximum value on the heel's angle is not less than 25° with the following Equation 5 [6].

$$\frac{dGZ}{d\varphi}\left(\varphi \geq 25°\right) = 0 \tag{5}$$

where, φ is the angel of heel and GZ is righting lever.

The area under the righting lever curve (GZ curve) at an angle of 0° to 30° should not be less than 0.055 m.rad. For heel angles at a radius of 0° to 40°, the area under the curve should not be less than 0.09 m.rad, and at a heel angle between 30° to 40°, the area under lighting lever GZ should not be less than 0.03 m.rad. Righting lever is shown to be calculated by Equation 6.

$$GZ = GM_0 \sin\phi + M_0S \tag{6}$$

where M_0S is the residual stability lever.

2.5 Floodable Length

The floodable length is the length of standing water in the compartment that can cause the ship to tilt and the vessel to descend until the ship's boundary line reaches the waterline. [3]. There are two methods for analyzing leaks. The first is the added weight method, which is a method that adds incoming water as the weight of the ship or adds displacement. Second, the lost buoyancy method is a method that assumes that the incoming water is not part of the ship and is still part of the seawater. Hence, the ship only loses buoyancy and does not increase the ship displacement. Permeability value is one of the factors that can affect the results of floodable length. With permeability, the volume of the compartment can be determined using the following Equation 7.

$$V = \frac{Vc}{\mu} \tag{7}$$

where (μ) is the permeability of the compartment, (υ) is the compartment volume, (V_c) is the permeability ratio of the floodable. The volume of incoming water can be calculated by Equation 8.

$$Vc = V_{\varphi} - v \tag{8}$$

where V_{φ} is the volume at which the vessel leaks, and v is the original volume of the vessel.

With the method of adding weight to the floodable length analysis, the ship's displacement is calculated based on the amount of intact displacement and the mass of the flooding water so that the floodwater becomes part of the ship and produces a free surface that must be calculated and considered. At minimum trim changes and negligible heel angles, Equation 9 is obtained.

$$\Delta_F = \Delta_I + \rho v$$

$$LCG_F.\Delta_F = LCG_I.\Delta_I + lcg.\rho.v$$

$$TCG_F.\Delta_F = tcg.\rho.v \tag{9}$$

In this equation, the lower script F represents differences like the ship when it is inundated, the lower script I represents those of the intact ship; lcg is the longitudinal center of gravity of the flooding; v is the volume of incoming water; TCG is the transverse center of gravity.

3. Result and Discussion

3.1 Ship Resistance Calculation

Calculating the value of ship resistance on fast boats is a complex hydrodynamic problem. One of the calculation methods is the Savitsky method [16]. This calculation is to find out the resistance value of the design that has been made. From the three designs that have been made, it will be known which design has excellent and effective resistance and power values. The following is the result of data from the resistance simulation of three designs (summarized in Table 4, and a comparison between resistance and a graph of power and Froude number is presented in Figure 5.

Tabel 4. Test result of hull resistance and hull power

No	Forude Number LWL	Savitsky Planing Resistance					
		Design 1		Design 2		Design 3	
		Resistance (N)	Power (W)	Resistance (N)	Power (W)	Resistance (N)	Power (W)
1	0	0	0	0	0	0	0
2	0.554	56817.81	487160.15	89474.34	777132.69	36009.66	307576.22
3	0.739	73833.56	844072.53	113786.21	1318376.2	47566.68	541965.31
4	0.924	90284.68	1290179.24	135131.22	1957612.95	59778.02	851604.22
5	1.109	101135.12	1734280.04	145414.64	2528335.85	70079.41	1198207.08
No	Forude Number LWL	Savitsky Planing Resistance					
		Design 1		Design 2		Design 3	
		Resistance (N)	Power (W)	Resistance (N)	Power (W)	Resistance (N)	Power (W)
6	1.293	107168.02	2144021.85	147976.48	2999774.56	77617.86	1547290.51
7	1.478	111694.4	2553802.87	148585.27	3443061.43	83871.96	1911192.86
8	1.663	116727.47	3002489.83	150142.5	3914607.73	90190.42	2312422.84
9	1.848	123084.1	3517773.86	153678.48	4452605.16	97302.89	2772319.68
10	2.033	131014.98	4118884.37	159423.63	5081442.44	105505.84	3306970.5
11	2.217	140544.68	4820161.84	167275.43	5814280.75	114836.4	3925205.1

The results of the resistance simulation show the differences between the three designs made from both resistance and power. This compares the effects of the main ship sizes to the value of the ship's resistance and power (see Figure 5). The simulation uses the Froude number reference as the main variable, with a minimum value of 0.554 and a maximum value of 2.217. Design 1 on the Froude number 0.554 has a resistance value of 56817.81 N with the required power of 487160.15 W, and the Froude number value of 2.217 Design 1 has a resistance value of 140544.68 N with the required power of 4820161.84 W. In Design 2, with a Froude number value of 0.554, it has a resistance value of 89474.34 N with a required power of 777132.69 W and a Froude number value of 2.217 Design 2 has a resistance value of 167275.43 N with a required power of 5814280.75 W. Design 3 on the Froude number 0.554 has a resistance value of 36009.66 N with the required power of 307576.22 W, and the Froude number value of 2.033 Design 3 has a resistance value of 114836.4 N with the required power of 3925205.1 W.

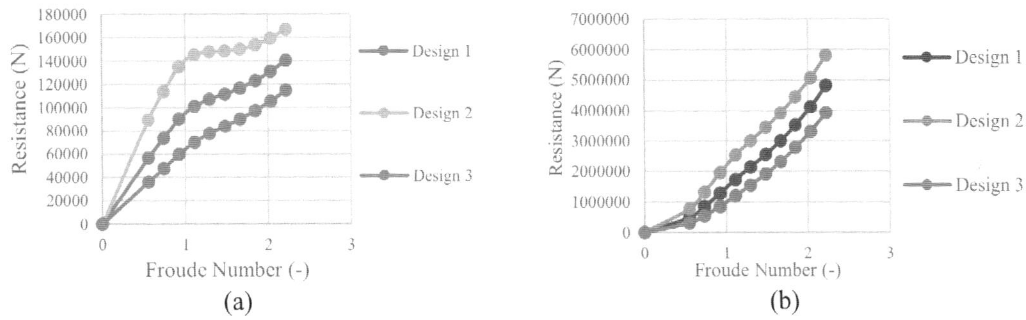

Figure 5. Resistance simulation: (a) Froude Number and Resistance; (b) Froude Number and Power.

The simulation results show that the smallest resistance value is in Design 3 of 114836.44 N, the second is Design 1 with a minimum resistance of 140544.68 N, and the third is Design 2 with a resistance value of 167275.43 N. From the simulation results, it is also found that the required power value (power) with the most undersized required power is found in Design 3 with a value of 3925205.18 W, the second is in Design 1 with the required power value of 4820161.84 W, and the third is in Design 2 with a power value of 5814280.75 W. Based on the simulation, it is also known that the increase in resistance and power is directly proportional to the increase in the value of the Froude number, the greater the value of the Froude number, the greater the value of resistance and power required. The difference in the main size of the ship used is known to also affect the results of the ship's resistance and power, even though the shape of the hull used is the same. This difference is, of course, influenced by the area of the hull exposed to water, so there are differences in the results obtained.

3.2 Large Angle Stability

One of the factors of ship safety in sailing is the stability of the ship, which has been determined by IMO. Stability is a factor that affects the stability of the ship when sailing both in calm conditions and under external forces. The following are the results of the large angle stability test on the three designs that have been simulated and are presented in Table 5. The stability simulation results show that Design 1 on the criteria Area 0° to 30° has a value of 0.09 m.rad where the value exceeds the required value of 0.055 m.rad, so that it passes this criterion. On the criteria, 0° to 40° Design 1 also meets the requirements which have a value of 0.156 m.rad exceeding the required value of 0.09 m.rad. Design 1 also meets the criteria for an area angle of 30° to 40° which has a value of 0.066 m.rad and exceeds the required value of 0.03 m.rad. The last criterion is Max GZ at 30° or greater where Design 1 also meets the criteria with a value of 0.487 m from the 0.2 m requirement, on this criterion it can also be seen the Angle of maximum GZ which in Design 1 has a value of 51.8 deg.

Table 5. Test result of hull large angle stability

Criteria	Design 1	Design 2	Design 3
Area 0° to 30° (m.rad)	0.09	0.16	0.056
Area 0° to 40° (m.rad)	0.16	0.25	0.113
Area 30° to 40° (m.rad)	0.07	0.09	0.057
Max GZ (m)	0.49	0.52	0.725
Angle of Maximum GZ (deg)	51.8	33.6	60.9

Testing Design 2 on the stability test meets the specified criteria starting from the criteria Area 0° to 30° Design 2 has a value of 0.156 m.rad exceeding the required value of 0.055 m.rad. On the criterion of the area at 0° to 40°, Design 2 has a value of 0.246 m.rad, which passes the required value of 0.09 m.rad. Design 2 also passed the criterion of the area at 30° to 40° with a value of 0.089 m.rad (surpassing the required value of 0.03 m.rad). Then, there is the maximum GZ criterion at 30° or greater where design 2 has a value of 0.517 m. This exceeds the required value of 0.2 m and has an angle of maximum GZ which is 33.6 deg.

Stability in Design 3 is also feasible and passes in this simulation where the criterion of the area at 0° to 30° has a value of 0.0559 m.rad passes the required value of 0.055 m.rad, although the difference in value is not too significant. In the criterion of the area at 0° to 40°, the value from Design 3 is 0.1133 m.rad which also meets the criteria of a required value of 0.09 m.rad. The value in the 30° to 40° Design 3 criteria is 0.0574 m.rad where this value passes the requirements of 0.03 m.rad. Design 3 has a value of 0.725 which exceeds the required value, namely 0.02 on the criterion of maximum GZ at 30° or greater, and has an Angle of maximum GZ value of 60.9 deg.

Based on the stability simulation results from the three designs, Design 3 has the best stability of the three designs with the largest maximum GZ at 30° or greater value of 0.725 m.rad, and an angle of maximum GZ of 60.9 deg. Summary of the analysis results indicated that the differences in beam, depth and draft will also affect the length of the stability arm and the maximum angle of inclination of the ship. The greater the value of maximum GZ at 30° or greater and the angle of maximum GZ, the better the level of stability of the ship.

3.3 Floodable Length Curve Analisys

A ship with stability following the specified standards will be safe to sail. However, this condition changes when a compartment is damaged. The damage allows the incoming liquid to fill the ship and affect the ship's draft, trim, and heel. To increase survivability, the hull is divided into several watertight compartments. The inundation length can be calculated, and the results can be plotted in a floodable length diagram where the x-axis is the longitudinal length of the ship. The y-axis is the allowable flood length. The triangle on the graph represents flooding in the compartment, the bottom triangle represents the compartment distance, and the top of the triangle represents the actual flood.

The floodable length simulations of various watertight compartments are presented in Figures 6 to 14. The floodable length with Design 1 - variation 1 graph shows that there is no triangle or the actual flood length exceeds the allowable flood length. Almost all triangles have the same vertex except the second triangle of the Machinery room. The triangle or bulkhead close to the allowable flood limit is the afterpeak room, while the triangle that is still quite far from the allowable flood limit is the Machinery room. Design 1 and bulkhead variation 1 (Figure 6) show that the ship is safe when it leaks because no bulkhead exceeds the flood limit and makes it sink.

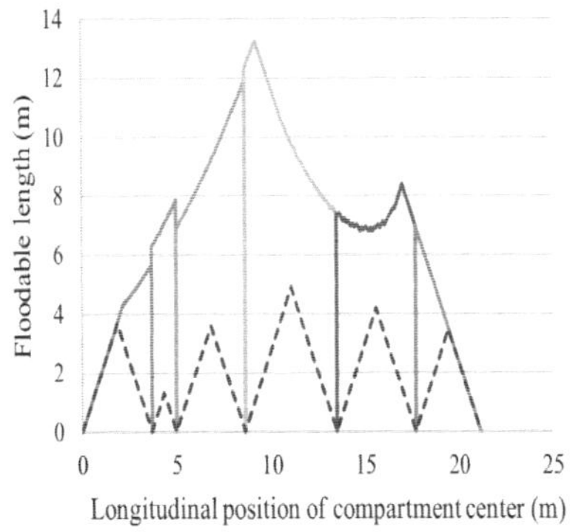

Figure 6. Graph of floodable length design 1-variation 1

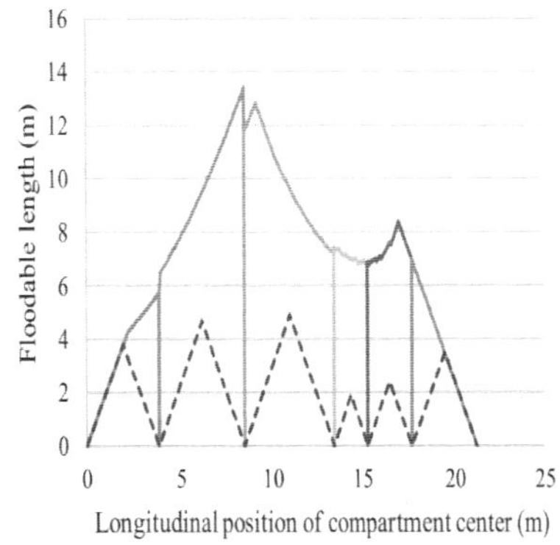

Figure 7. Graph of floodable length design 1-variation 2

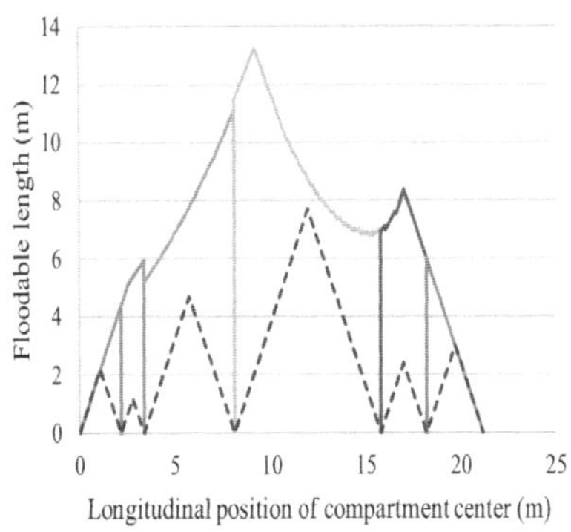

Figure 8. Graph of floodable length design 1-variation 3

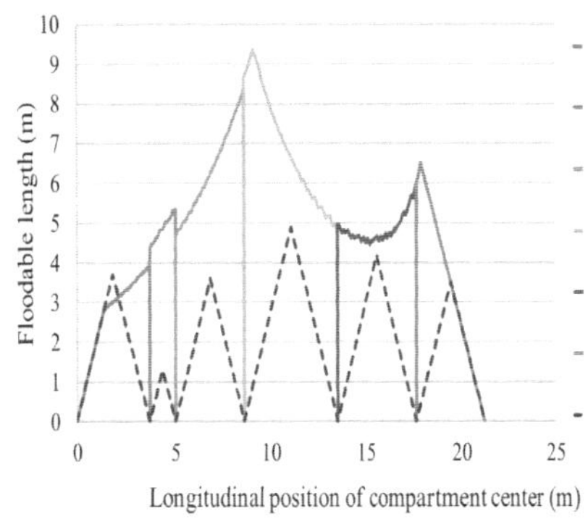

Figure 9. Graph of floodable length design 2-variations 1

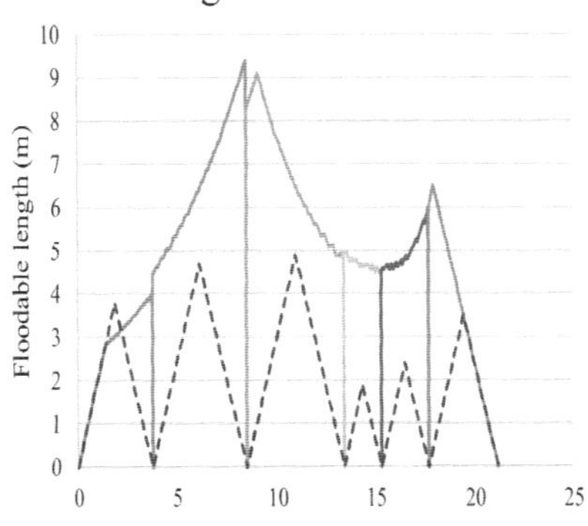

Figure 10. Graph of floodable length design 2-variations 2

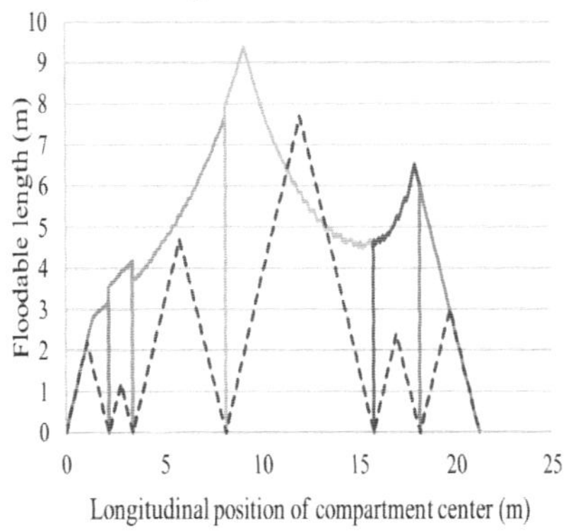

Figure 11. Graph of floodable length design 2-variations 3

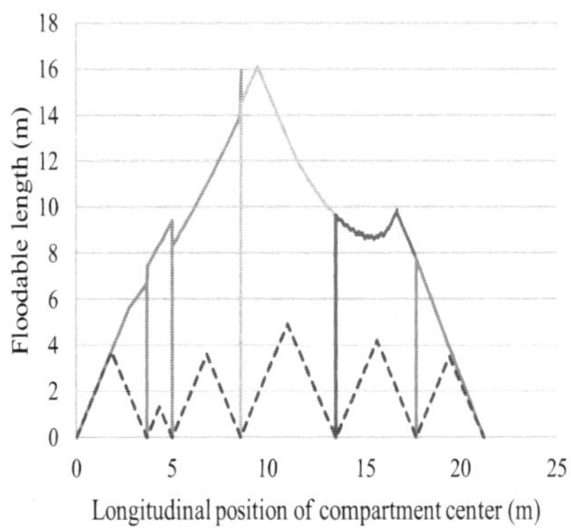

Figure 12. Graph of floodable length design 3-variations 1

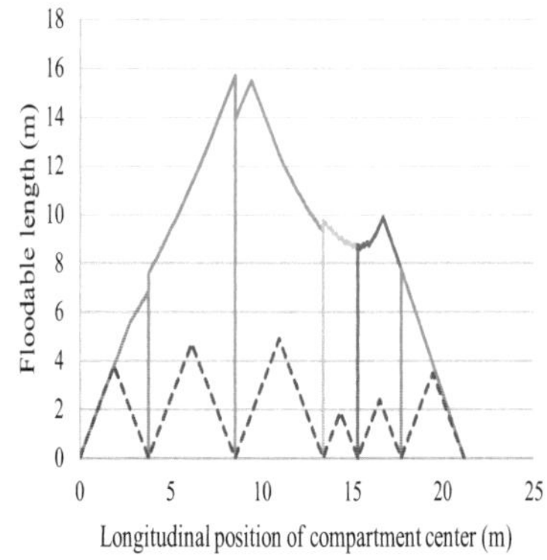

Figure 13. Graph of floodable length design 3-variations 2

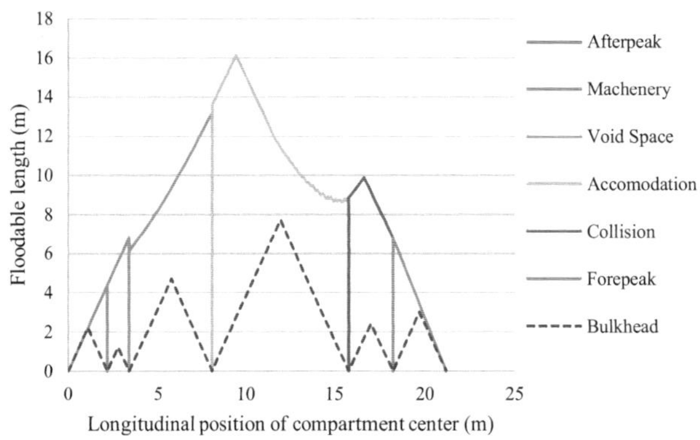

Figure 14. Graph of floodable length design 3-variations 3

The three variations of the bulkhead arrangement show that variations 1 and 2 are more dangerous in the afterpeak room, and variation 3 is more dangerous in room accommodation, judging by the several designs that have triangles crossing or approaching the allowable flood line. These results indicate that ships with higher water drafts and larger displacements tend to have lower allowable flood boundaries. On the other hand, if the displacement is smaller, it will increase the height of the allowable flood boundary line, which means more water volume can reach the waterline.

4. Conclusions

Based on the conducted research, it is found that the difference in the ship's main dimensions will affect the value of the resistance and stability of the ship. Design 3 is the best compared to Designs 1 and 2. Design 3 has a resistance value of 2.217 Froude number of 114836.4 N with the required power of 3925205.1 W. This value is better than design 1, which has a resistance of 140544.68 N with the required power of 4820161.84 W, and design 2 which has a resistance of 167275.43 N with the required power of 5814280.75 W. The angle stability simulation shows that the three feasible designs pass the stability test. In the floodable length test, several graphs indicate a dangerous compartment to flooding. The compartment is in Design 2, wherein in variations 1 and 2, the first compartment triangle crosses the flood boundary line. In contrast, in Design 2 - variation 3, the fourth compartment triangle exceeds the flood line limit. Whereas in Designs 1 and 3, no compartment triangle crosses the boundary line, it is feasible in Designs 1 and 3. It passes with the compartment arrangement determined in the floodable length simulation.

References

[1] Yuniwantoro C, Surojo E, Smaradhana D F and Imaduddin F 2020 Effect of Filler on Shear Strength and Electrical Conductivity in Epoxy Based Aluminium Bonded. *Mekanika: Majalah Ilmiah Mekanika.* **19** 1-6.

[2] Wicaksono H, Susilo S H, Pranoto B and Fakhruddin M 2021 Initial Rotation Characteristic Investigation of a Hybrid Savonius - Darrieus Wind Turbine using 6 DOF Computational Fluid Dynamics. *Mekanika: Majalah Ilmiah Mekanika.* **20** 9-17.

[3] Elwira K 2017 Floodable length of a bulk carrier. *Sci. J. Maritime Univ. Szczecin.* **51** 71–77.

[4] Wu Y H and Shaw H J 2011 Document based knowledge base engineering method for ship basic design. *Ocean Eng.* **38** 1508–1521.

[5] Dabit A S, Lianto A E, Branta S A, Laksono F B, Prabowo A R and Muhayat N 2020 Design of unmanned ships for distributing fish feed in coastal areas based on arduino microcontroller. *Mekanika: Majalah Ilmiah Mekanika.* **19** 74-82.

[6] Prabowo A R, Martono E, Muttaqie T, Tuswan T and Bae D M 2022 Effect of hull design variations on the resistance profile and wave pattern: A case study of the patrol boat vessel. *J. Eng. Sci. Tech.* **17** 293-312.

[7] Yusvika M, Fajri A, Tuswan T, Prabowo A R, Hadi S, Yanigsih I, Muttaqie T and Laksono F B 2022 Numerical prediction of cavitation phenomena on marine vessel: Effect of the water environment profile on the propulsion performance. *Open Eng.* **12** 293-312.

[8] Julianto R I, Prabowo A R, Muhayat N, Putranto T and Adiputra R 2021 Investigation of hull design to quantify resistance criteria using holtrop's regression-based method and savitsky's mathematical model: A study case of fishing vessels. *J. Eng. Sci. Tech.* **16** 1426-1443.

[9] Julianto R I, Muttaqie T, Adiputra R, Hadi S, Hidayat R L L G and Prabowo A R 2020 Hydrodynamic and structural investigations of catamaran design. *Proc. Struct. Integ.* **27** 93-100.

[10] Dabit A S, Lianto A E, Branta S A, Laksono F B, Prabowo A R and Muhayat N 2020 Finite Element Analysis (FEA) on autonomous unmanned surface vehicle feeder boat subjected to static loads. *Proc. Struct. Integ.* **27** 163-170.

[11] Febrianto R A, Prabowo A R, Baek S J and Adiputra R 2021 Analysis of monohull design characteristics as supporting vessel for the COVID-19 medical treatment and logistic. *Trans. Res. Proc.* **55** 699-706.

[12] Yusvika M, Prabowo A R, Baek S J and Tjahjana D D D P 2020 Achievements in observation and prediction of cavitation: effect and damage on the ship propellers. *Proc. Struct. Integ.* **27** 109-116.

[13] Mallat C K, Rousset J M and Ferrant Pierre 2011 The Transient and Progressive Flooding Stages of Damaged Ro-Ro Vessels: A Systematic Review of Entailed Factors. *Offshore Mech. Arctic Eng.* **133** 031105.

[14] Wu H, He J, Liang H and Noblesse F 2019 Influence of Froude number and submergence depth on wave patterns. *European J. Mech.-B/Fluids.* **75** 258-270.

[15] Khai D Q and Quynh N T T 2021 A comparison study on the resistance calculation of planning boat. *Int. J. Eng. Appl. Sci. Tech.* **5** 23-36.

[16] Savitsky D and Brown P W 1976 Procedures for hydrodynamic evaluation of planing hulls in smooth and rough water. *Marine Tech. SNAME News.* **13** 381-400.

[17] Söding H 2002 Flow computations for ship safety problems. *Ocean Eng.* **29** 721–738.

Fast ship prototype design simulation with fin stabilizer on hydrodynamic characteristics for ship realization planning

H Diatmaja[1], A R Prabowo[1,*], N Muhayat[1], T Tuswan[2], T Putranto[3,4]

[1] Department of Mechanical Engineering, Universitas Sebelas Maret, Surakarta 57126, Indonesia
[2] Department of Naval Architecture, Universitas Diponegoro, Semarang 50275, Indonesia
[3] Department of Civil Engineering and Architecture, Tallinn University of Technology, Tallinn 19086, Estonia
[4] Department of Naval Architecture, Institut Teknologi Sepuluh Nopember, Surabaya 60111, Indonesia

*Corresponding author: aditya@ft.uns.ac.id

Abstract. This study aims to determine the effect of adding variations of fin stabilizers on the hydrodynamic characteristics of the ship design. It compares the hull without modification with the hull modified by a fin stabilizer on a prototype scale. The results of this study is assessed to obtain the best performance of ships with various models to be used in the design of ship realization. The research method is conducted by numerical approach, including simulation of resistance, stability, and seakeeping. The hull design references chosen are the Fast Police Boat and SMIT Patrol Boat with a Length Overall value of 1 m. The fin stabilizer uses hydrofoil type NACA-0015 with an aspect ratio of 0.7. Then the length variations are 3 cm, 4 cm, and 5 cm, with angle variations being 25°, 30°, and 35°. The results showed that adding a fin stabilizer could increase the ship's stability. The best fin stabilizers variations are fins with a length of 5 cm and an angle of 30° for the prototype design of the Fast Police Boat hull and fins with a length of 5 cm and an angle of 35° for the prototype design of the SMIT Patrol Boat hull.

Keywords: Fast boat design; Fin stabilizer; Resistance; Stability; Seakeeping.

1. Introduction

An archipelagic country has a transportation sector which is a substantial aspect of the country's development. Indonesia is the largest archipelagic country globally that has the potential to become the world's maritime axis [1]. As a maritime country, it must have sea transportation which plays a vital role in defense and economy. Currently, ships as sea transportation with various hull shapes and hull types adjusting to their needs. Therefore, it the importance to plan the ship design as needed.

Ship design considers the shape, dimensions, and weight that affects the hydrostatic characteristics. One of the primary problems that ships constantly face when sailing in the sea is resistance [2,3], especially on fast patrol boats with high speed to support their missions. If the ship's resistance can be reduced, the power needed to push the ship to reach the planned speed is also reduced, or with the same power, the ship sails faster [4]. One way that can be done to overcome these problems is by making modifications to the shape of the ship [5,6]. Modification of the hull shapes also considers the stability of the ship.

One ship's safety is determined by the ship's stability [7]. Stability is the ability of the ship to return to its original position after experiencing forces from outside and from inside the ship that caused the ship to tilt [8]. Gravity and buoyancy work in opposite directions, thus causing the ship to float on the water's surface. The ship's stability is affected by the rolling motion on the starboard and the port. The ship is stable. If it stays on the right or left side, it can return to a vertical position. The modification of adding a fin stabilizer can increase stability without increasing the resistance value significantly. The fin stabilizer is the most effective device for vessels sailing at medium-high velocity [9]. Most of the latest research on traditional fin stabilizers has focused on introducing new control strategies and utilizing lift feedback to improve control [10]. The highest rolling motion generally occurs when the ship has a low speed. In previous studies, fin stabilizers overcome the rolling motion at low speed. Based on these data, the ship was tested by adding various fin stabilizers at low speed to high speed in this study. The ship used in the test is a patrol-type fast boat that uses a V-hull type.

2. Methodology

One of the methods that can be used in ship design is the comparison method, in which a sailing ship is used as a reference for designing and developing [11]. The ship data using reference ships designed by the company is based on experts in the maritime industry [12]. This method does not require complicated calculations according to actual ship calculations. Then the ship model is made according to the selected reference ship using the SolidWorks software. The ship model is calculated based on resistance, stability, and seakeeping by a numerical approach. Furthermore, an analysis of the simulation results compares the hull without a fin stabilizer with the hull using a variety of fin stabilizers.

2.1. Ship Data.

This research uses ship models based on reference ships, namely Fast Police Boat and SMIT Patrol Boat. The ship model in the prototype scale with a Length of Overall (LOA) of 1 m. The principal data for the ships are presented in Table 1.

Table 1. Principal dimension of the fast boat prototype designs.

Parameters	**Values**	
	Model I (Fast Police Boat)	Model II (SMIT Patrol Boat)
Length of overall (m)	1.000	1.000
Length of waterline (m)	0.914	0.869
Draft (m)	0.057	0.053
Beam (m)	0.273	0.311
Depth (m)	0.129	0.121
Block coefficient (-)	0.424	0.530
Displacement (kg)	5.675	7.500

The ship design process begins with the design of each type's ship prototype lines plan and it is followed by managing the design grid in the form of sections, buttocks, and waterlines. For each model, i.e., Model I (Fast Police Boat) and Model II (SMIT Patrol Boat) designs, a lines plan was made to determine the shape of the ship. The ship is then modeled into a three-dimensional (3D) design with isometric views with an adjusted scale. The modelled designs are shown in Figure 1.

(a) (b)

Figure 1. Rendered 3D hull designs: (a) Model I; (b) Model II.

2.2. Research Parameters

Parameters as fixed variables include the Length of Overall (LOA) of the two ship models and the waveform criteria are determined as a sinusoidal with a wavelength of 0.15 m and a wave height of 0.0114 m. This waveform has been measured by direct field observations.

2.2.1. Fin stabilizer variations.

There are two designed fin stabilizers on the ship's starboard and portside which are placed in the longitudinal center of gravity (LCG) area on the ship model. The specifications and designs for the fin stabilizer are presented in Table 2 and Figure 2.

Table 2. Specification of fin stabilizer.

Items	Values
Hydrofoil series	NACA-0015
Profile	Trapezoid
Aspect ratio	0.7

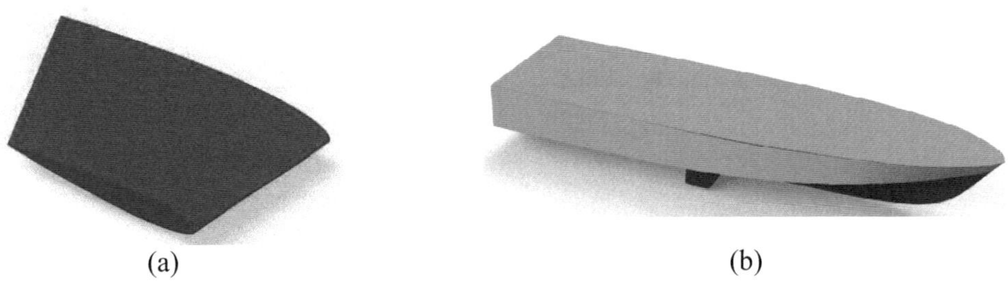

Figure 2. Rendered 3D fin stabilizer design: (a) Fin stabilizer; (b) Fin stabilizer at boat

The stabilizer fin variations consist of fin lengths 3 cm, 4 cm, and 5 cm, while the angle variations are 25°, 30°, and 35°. Therefore, nine ship models with different variations of fin stabilizer are obtained. The nine fin models are summarized in Table 3.

Table 3. Fin stabilizer variations at boat models.

Items	Model									
	0	1	2	3	4	5	6	7	8	9
Length fins (m)	0	0.03	0.03	0.03	0.04	0.04	0.04	0.05	0.05	0.05
Fins angle (°)	0	25	30	35	25	30	35	25	30	35

2.2.2. Hull resistance analysis

The resistance analysis uses Savitsky's method. Several factors influence ship resistance, including keel condition, sub-merged hull form at the bow and stern, wetted surface area, and length of water line (LWL) [13]. The discussion of ship hull with the hydrodynamic characteristics in numerical planning and calculation to find wet areas, drag, pressure centers, stability, and resistance that can function at speed, deadrise angle, and trim [14]. In the ship's speed coefficient, a ship can perform good procedure levels in the center of gravity, trim, and bow height by having a dynamic lift on the ship [15]. The resistance calculation consists of the frictional, viscous, wave, Reynolds, Froude, and wet surface area. The total resistance is shown in Equation (1) [16].

$$R_T = R_F + R_{VP} + R_W \quad (1)$$

where, R_T is the total resistance, R_F is the friction resistance, R_{VP} is the viscous pressure resistance, and R_W is the wave resistance.

Therefore, the total resistance equation is obtained based on the Savitsky method. Daniel Savitsky's formula was obtained from empirical data on prismatic planning hulls [17]. The formula for Savitsky's total resistance is given by Equation (2).

$$R_T = \Delta \tan \tau + \frac{\frac{1}{2}\rho v^2 \lambda b^2 cf}{\cos \tau} \tag{2}$$

where *RT* is the total resistance and τ is the slight angle.

2.2.3. Ship stability assessment.

The ship must have complete stability as the ship operation requires good conditions to increase the ship's stability so that the ship's cargo will be safe in all conditions. Based on the IS Code, the integrated area under the righting lever curve must reach a minimum value [18]. Several criteria are summarized as follows:

a) the area under the righting lever curve shall not be less than 0.055 m.rad up to an angle of 30°. The formula is given by Equation (3);

$$\int_0^{30} GZ(\varphi)\, d(\varphi) \geq 0.055m.rad \tag{3}$$

b) the area under the righting lever curve shall not be less than 0.090 m.rad up to an angle of 40. The formula is given by Equation (4);

$$\int_0^{40} GZ(\varphi)\, d(\varphi) \geq 0.090m.rad \tag{4}$$

c) the area under the righting lever curve between the angle of 30° to 40° shall not be less than 0.03 m.rad. The formula is given by Equation (5);

$$\int_{30}^{40} GZ(\varphi)\, d(\varphi) \geq 0.03m.rad \tag{5}$$

d) the righting lever GZ shall be at least 0.2 m at a heeling angle equal or greater than 30° (see Equation (6));

$$GZ(\varphi = 30°) \geq 0.20m \tag{6}$$

e) the maximum righting lever shall occur at an angle of the heel not less than 25° (see Equation (7));

$$\frac{dGZ}{d\varphi}(\varphi \geq 25^0) = 0 \tag{7}$$

f) the initial metacentric height GM0 shall not be less than 0.15 m (see Equation (8)).

$$GM_0 \geq 0.15m \tag{8}$$

2.2.4. Seakeeping characteristic.

The oscillatory motion is assumed to be linear and hormonal, which the motion equation based on six degrees of freedom is written as follows [19,20].

$$\sum_{n=1}^{6} [(Mjk + Ajk)\zeta k + Bjk\zeta k + Kjk\zeta k] = Fje^{iwt}; j,k = 1,\dots,6 \tag{9}$$

where *Mjk* is the mass matrix and moment of inertia of the marine building (there is a matrix of hydrodynamic added mass coefficients and the number of hydrodynamic damping coefficients); *Kjk* is the matrix of strength coefficients or hydrostatic force and moment; *Fj* is the matrix of the excitation force (*F1*, *F2*, *F3*) and the moment of excitation (*F4*, *F5*, *F6*) in the complex function (expressed by eiwt); *F1* is the excitation force which causes the motion surge; *F2* is the excitation force causing the sway motion; *F3* is the excitation force causing the heave motion; *F4* is the moment of excitation which causes the roll motion; *F5* is the moment of excitation causing pitch motion, and *F6* is the moment of excitation causing yaw motion [19,21].

3. Result and discussion

3.1. Ship resistance.

The resistance calculation method used in this study is the Savitsky method. In the research of Prabowo et al., the comparison between numerical simulations and mathematical calculations using the Savitsky method shows that the difference in error is below 10% [19,22]. The ship model selects the resistance test with low resistance and low power. The simulation results on Model I (Fast Police Boat) are presented in Tables 4 and 5. At the same time, the simulation results on Model II (SMIT Patrol Boat) are presented in Tables 6 and 7.

Table 4. Results of hull resistance of Model I (Fast Police Boat).

Speed (kts)	Resistance (N)										Coefficient Variation (%)
	0	1	2	3	4	5	6	7	8	9	
10	14.17	14.10	14.21	14.21	14.12	14.12	14.27	14.13	14.25	14.25	0.004219238
15	24.29	24.23	24.33	24.33	24.24	24.24	24.37	24.25	24.35	24.37	0.002208462
20	39.81	39.73	39.86	39.85	39.75	39.75	39.90	39.76	39.86	39.90	0.001556516
25	60.18	60.08	60.24	60.23	60.09	60.09	60.29	60.11	60.22	60.28	0.001296623
30	85.16	85.02	85.23	85.23	85.04	85.04	85.30	85.07	85.20	85.29	0.001199614

Table 5. Results of hull power of Model I (Fast Police Boat).

Speed (kts)	Power (W)										Coefficient Variation (%)
	0	1	2	3	4	5	6	7	8	9	
10	85.73	85.37	86.02	86	85.43	85.45	86.34	85.54	86.27	86.27	0.004233943
15	220.5	219.96	220.89	220.86	220.05	220.07	221.27	220.2	221.04	221.2	0.003934545
20	481.88	480.97	482.46	482.43	481.11	481.14	482.99	481.34	482.46	482.92	0.003608477
25	910.53	908.98	911.44	911.38	909.2	909.25	912.23	909.57	911.21	912.14	0.003360667
30	1,546.15	1,543.62	1,547.55	1,547.46	1,543.97	1,544.04	1,548.73	1,544.56	1,546.99	1,548.62	0.003128616

Table 6. Results of hull resistance of Model II (SMIT Patrol Boat).

Speed (kts)	Resistance (N)										Coefficient Variation (%)
	0	1	2	3	4	5	6	7	8	9	
10	16.11	16.14	16.14	16.14	16.15	16.16	16.15	16.16	16.17	16.16	0.000990835
15	16.88	16.92	16.92	16.92	16.93	16.93	16.93	16.94	16.94	16.93	0.000960061
20	17.74	17.77	17.77	17.77	17.78	17.78	17.78	17.79	17.79	17.79	0.000803491
25	18.67	18.7	18.7	18.7	18.71	18.71	18.71	18.72	18.72	18.72	0.000763544
30	19.68	19.71	19.71	19.7	19.71	19.72	19.71	19.72	19.73	19.72	0.00065953

Table 7. Results of hull power of Model II (SMIT Patrol Boat).

Speed (kts)	Power (W)										Coefficient Variation (%)
	0	1	2	3	4	5	6	7	8	9	
10	97.51	97.7	97.7	97.7	97.77	97.79	97.77	97.83	97.85	97.81	0.000954331
15	107.3	107.5	107.5	107.49	107.57	107.58	107.57	107.63	107.65	107.61	0.000890946
20	118.1	118.3	118.3	118.29	118.37	118.38	118.37	118.43	118.45	118.41	0.000809636
25	129.9	130.15	130.15	130.15	130.22	130.24	130.22	130.29	130.3	130.26	0.000743145
30	142.9	143.11	143.11	143.11	143.18	143.2	143.18	143.25	143.26	143.23	0.000681004

The results of the simulation data get a coefficient of variation close to 0 so that it has no significant effect. The lowest resistance and power values in the Fast Police Boat ship model are in Model 1. Model 1 has a resistance value of 85.02 N and a power of 1,543.62 W at the speed of 30 knots. At the same time, the lowest resistance and power values on the SMIT Patrol Boat model are in Model 0. Model 0 is a ship model without adding of a fin stabilizer with a resistance value of 19.68 N and a power of 142.91 W at a speed of 30 knots. The difference in simulation results between Model I and Model II is caused by differences in the shape and dimensions of the ship's hull. The difference in comparison between Length of Overall and Beam (*L*/*B*), Length of Overall and Depth (*L*/*D*), and Beam and Depth (*B*/*D*).

3.2. *Ship stability.*

The purpose of studying ship stability is to determine the behavior of the ship's motion based on the characteristics of the waters to achieve ship safety [7]. Models I and II were tested for stability with variations of adding a fin stabilizer. The simulation results of Model I (Fast Police Boat) and Model II (SMIT Patrol Boat) are presented in Tables 8 and 9.

Table 8. Results of ship stability of Model I (Fast Police Boat).

Model	Stability			
	GZ (cm)	α (deg)	*A* (cm.deg)	The angle of vanishing of point (deg)
0	3.59	43.6	213.6	92.385
1	3.61	45.5	216.8	93.138
2	3.61	45.5	216.6	93.138
3	3.61	45.5	216.9	93.138
4	3.63	45.5	217.9	93.264
5	3.65	45.5	219.5	93.389
6	3.62	45.5	218.0	93.389
7	3.65	45.5	219.6	93.389
8	3.64	45.5	219.2	93.515
9	3.66	45.5	220.2	93.640

Table 9. Results of ship stability of Model II (SMIT Patrol Boat).

Model	Stability			
	GZ (cm)	α (deg)	*A* (cm.deg)	The angle of vanishing of point (deg)
0	5.63	36.4	326.5	90.453
1	5.66	36.4	329.5	90.841
2	5.66	36.4	330	90.971
3	5.67	36.4	330.2	90.971
4	5.68	36.4	331.1	90.971
5	5.68	36.4	331.4	91.004
6	5.69	36.4	331.6	91.130
7	5.69	36.4	332.2	91.130
8	5.70	36.4	333.0	92.248
9	5.70	36.4	332.5	91.248

The results of the ship stability test found that in Model I (Fast Police Boat) was a significant change between the hull without a fin stabilizer and the hull using a fin stabilizer. Model 0 has a *GZ* value of 3.59 cm and α of 43.6°. Meanwhile, Models 1 to 9 increase in the value of *GZ*, and the α is greater than Model 0. The best variation of the Model I (Fast Police Boat) hull based on the ship's stability is found in Model 9. Model 9 is the hull with the addition of a fin stabilizer. Model 9 has a GZ value of 3.66 cm, an α value of 45.5°, and an area under the curve of 220.2 cm.deg.

Model II (SMIT Patrol Boat) did not change significantly between the hull model without a fin stabilizer and the hull using a fin stabilizer. It is because the shape of the hull of the SMIT Patrol Boat is different from the hull of the Fast Police Boat. This difference in stability value is based on the suitability of the used water criteria. The best variation of the Model II hull (SMIT Patrol Boat) based on ship stability is the Model 8, which has a stabilizer fin length of 5 cm and a fin angle of 30°. Model 8 has a *GZ* value of 5.70 cm, an α value of 36.4°, and an area under the curve of 333.0 cm.deg.

3.3. Ship seakeeping.

Based on the results of the ship stability obtained data that has not seen the difference between several variations of the ship. Therefore, the following simulation is needed, namely seakeeping testing. Seakeeping theory studies the motion of surface vessels in waves [23]. Seakeeping analysis is a procedure for computing vessel performance metrics related to the envelope of missions and environmental conditions in which the vessel is to operate [11]. Seakeeping tests get response amplitude operator (RAO) graphs to consist of heave motion and roll motion. The parameter used is the ship's speed of 10 knots. The higher roll motion occurs when the ship's speed is low. While the heading parameters used are 90° and 135°. The results of the seakeeping test on Model I are presented in Table 10 and graphic images are presented in Figures 3 and 4.

Table 10. Seakeeping results of Model I.

Variations	**Seakeeping**			
	Heading 90°		Heading 135°	
	Heaving (m/m)	Rolling (rad/rad)	Heaving (m/m)	Rolling (rad/rad)
0	0.993	6.658	3.587	4.716
1	0.925	6.637	3.372	4.696
2	0.924	6.629	3.394	4.687
3	0.922	6.629	3.354	4.691
4	0.925	6.659	3.363	4.699
5	0.924	6.623	3.339	4.683
6	0.899	6.666	3.311	4.713
7	0.925	6.625	3.394	4.685
8	0.899	6.676	3.274	4.711
9	0.899	6.666	3.295	4.713

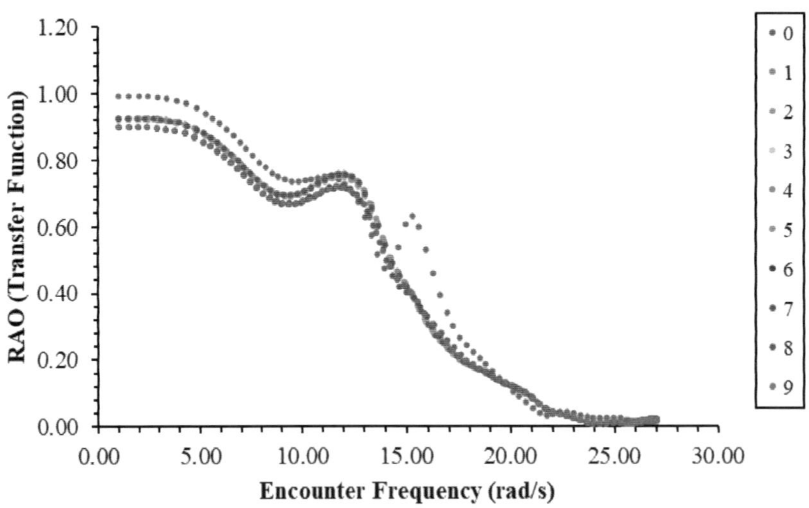

(a)

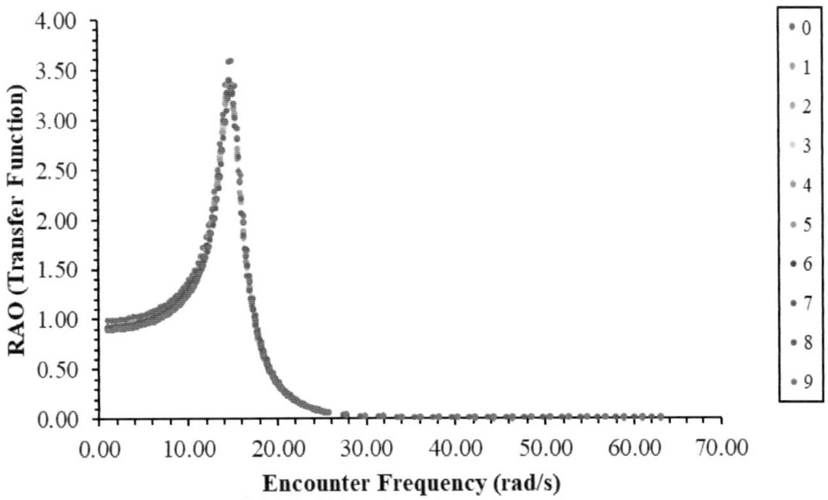

(b)

Figure 3. RAO heave motion of Model I: (a) Heading 90°; (b) Heading 135°.

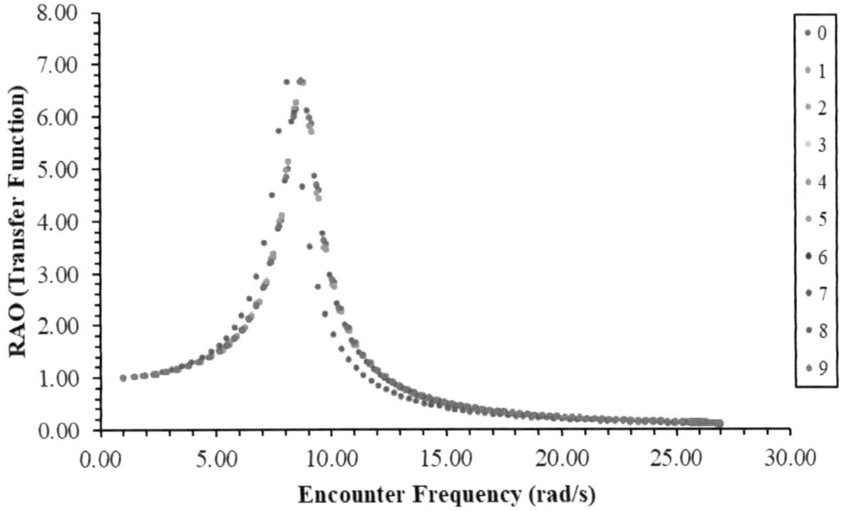

(a)

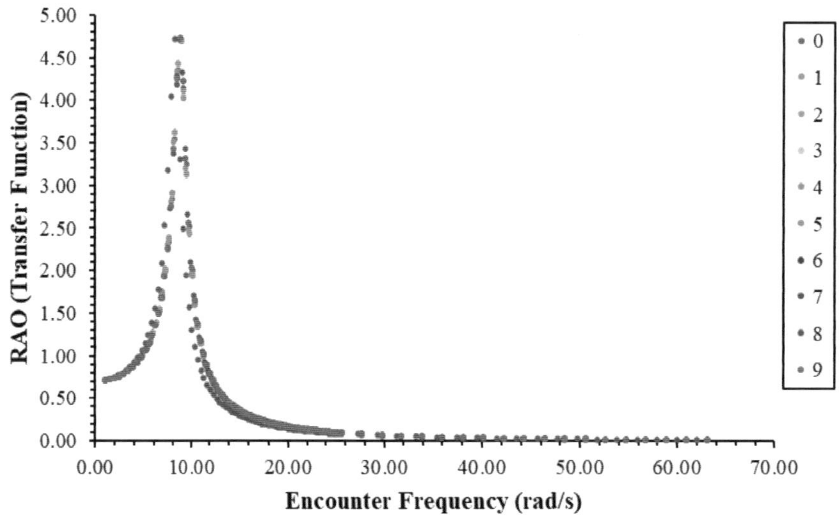

(b)

Figure 4. RAO roll motion of Model I: (a) Heading 90°; (b) Heading 135°.

Based on RAO of the heave and roll motion graphs on Model I (Fast Police Boat) was a significant difference between the hull without a fin stabilizer and the addition of a fin stabilizer on the hull. The hull without a fin stabilizer has the highest heaving of 0.993 m/m and the highest rolling of 6.658 rad/rad at 90° heading. While at heading 135°, the highest heaving is 3.587 m/m and the highest rolling is 4.716 rad/rad. The best variation of Model I is Model 5 has a stabilizer fin length of 4 cm with an angle of 30°—the selection of the best hull variations based on the lowest rolling with low heaving. The results of the seakeeping test on Model II are presented in Table 11, and graphical images are presented in Figures 5 and 6.

Table 11. Seakeeping results of Model II.

Variations	Seakeeping			
	Heading 90°		Heading 135°	
	Heaving (m/m)	Rolling (rad/rad)	Heaving (m/m)	Rolling (rad/rad)
0	0.987	6.624	3.268	4.684
1	0.903	6.602	2.988	4.668
2	0.959	6.577	3.107	4.619
3	0.959	6.601	3.104	4.667
4	0.903	6.594	2.991	4.662
5	0.959	6.680	3.124	4.724
6	0.959	6.680	3.137	4.723
7	0.902	6.639	3.001	4.695
8	0.959	6.678	3.115	4.722
9	0.948	6.640	3.159	4.695

IOP Conf. Series: Earth and Environmental Science 1166 (2023) 012047 doi:10.1088/1755-1315/1166/1/012047

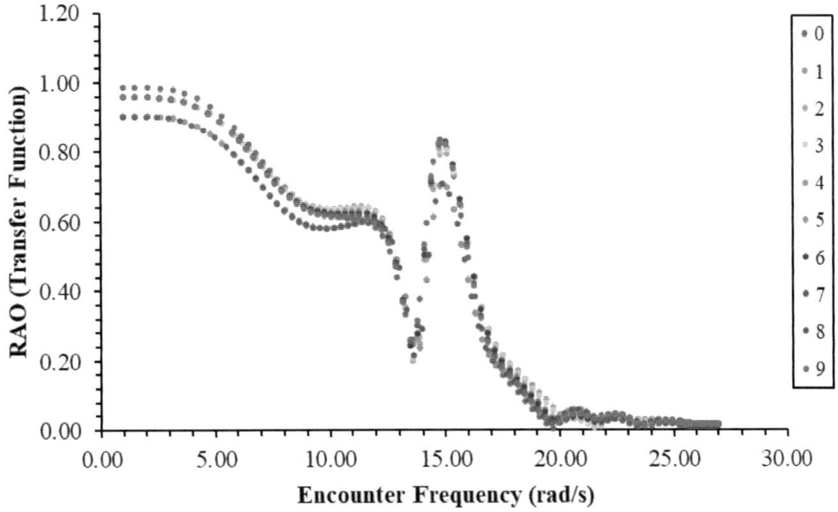

(a)

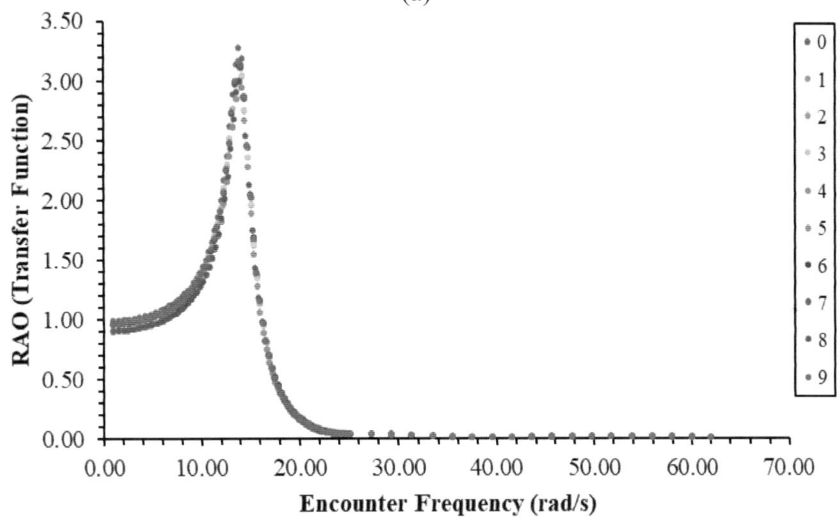

(b)

Figure 5. RAO heave motion of Model II: (a) Heading 90°; and (b) Heading 135°.

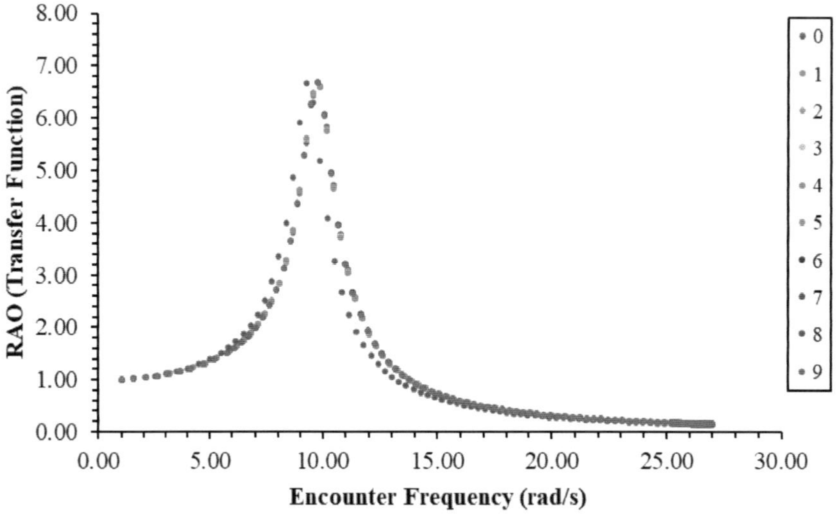

(a)

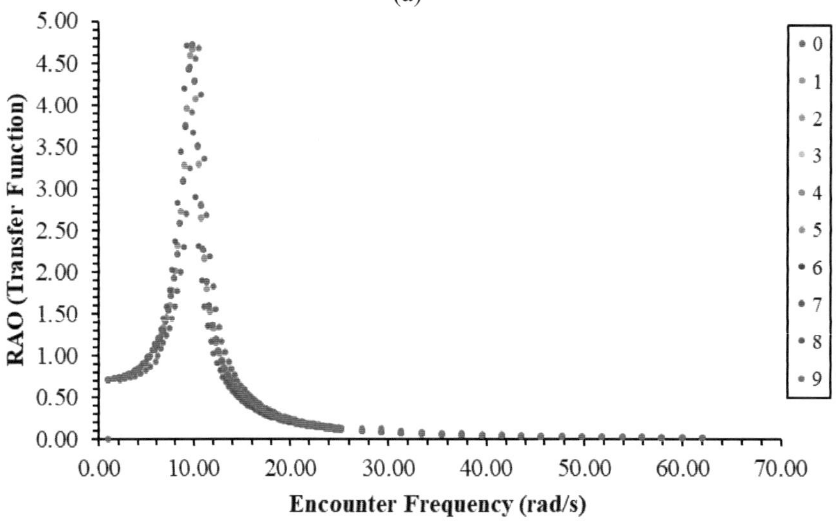

(b)

Figure 6. RAO roll motion of Model II: (a) Heading 90°; and (b) Heading 135°.

The hull without fin stabilizer on the Model II (SMIT Patrol Boat) has the highest heaving of 0.987 m/m and 6.624 rad/rad rolling. The best Model II variation is Model 2 has a stabilizer fin length of 3 cm with an angle of 30°. Therefore, a good stabilizer fin angle is obtained based on the seakeeping test, which is 30°. The difference between the best Model I and the best Model II variation is due to the different hull shapes. In Model I, it is better to use a fin stabilizer with a length of 4 cm, while in Model II, it is better to use a fin stabilizer with a length of 3 cm.

4. Conclusions

Based on the obtained results, the hull with the addition of a fin stabilizer has less influence on the resistance value, evidenced by the coefficient of variation being close to zero. It can be used as a consideration in modifying the hull. With the addition of a fin stabilizer, there is no significant increase in the resistance value with the hull without a fin stabilizer. The hull which has a small resistance value is the Fast Police Boat model, which is Model 1, while the SMIT Patrol Boat model is

Model 0. In the SMIT Patrol Boat, there is no decrease in the resistance value because the shape of the hull is more in line with the without fin stabilizer based on the analysis of the resistance.

Regarding the ship stability criteria, the hull on the Fast Police Boat increases the stability value between the hull without a fin stabilizer and the hull using a fin stabilizer. It is evidenced by the increasing value of GZ, the value of α, and the area under the curve. The best variation of the Model I (Fast Police Boat) based on stability criteria is Model 9, which has a stabilizer fin length of 5 cm and a fin angle of 35°. At the same time, the best variation of the Model II (SMIT Patrol Boat) based on stability criteria is Model 8, which has a stabilizer fin length of 5 cm and a fin angle of 30°.

Finally, the seakeeping test found that the difference was more significant than the ship's resistance and stability test. The best variation of Model I (Fast Police Boat) is Model 5 has a stabilizer fin length of 4 cm with an angle of 30°. On the other hand, the best variation of Model 2 that has a stabilizer fin length of 3 cm with an angle of 30°. A fin stabilizer is generally an anti-rolling on a ship. It can be concluded based on the usefulness of the fin stabilizer by considering the result of roll motion. The angle of a good fin stabilizer is 30°. A clear difference was obtained in several tests, namely in the seakeeping test. Therefore, seakeeping testing is used to select the best variations of ship models.

References

[1] Nugroho O S, Suharyo O S and Rahman A 2020 The development of Indonesian maritime potential and prospects towards a world maritime axis. *STTALPostgraduate-International Conf.* **4** 1-8.

[2] Buterworth J, Atlar M and Shi W 2015 Experimental analysis of an air cavity concept applied on a ship hull to improve the hull resistance. *Ocean Eng.* **110** 2-10.

[3] Prabowo A R, Febrianto R A, Tuswan T and Tjahjana D D D P 2022 Performance evaluation on the designed v-shaped monohull ship models. *J. Appl. Eng. Sci.* **20** 609-623.

[4] Ozdemir Y H, Cosgun T, Dogrul A and Barlas B A 2016 Numerical application to predict the resistance and wave pattern of KRISO container ship, *Brodogr. Teor. i Praksa Brodogr. i Pomor. Teh.* **67** 47–65.

[5] Nazemian A and Ghadimi P 2021 CFD-based optimization of a displacement trimaran hull for improving its calm water and wavy condition resistance. *Appl. Ocean Res.* **113** 102729.

[6] Dabit A S, Lianto A E, Branta S A, Laksono F B, Prabowo A R and Muhayat N 2020 Finite Element Analysis (FEA) on Autonomous Unmanned Surface Vehicle Feeder Boat subjected to Static Loads. *Proc. Struct. Integ.* **27** 163-170.

[7] Mauro F, Braidotti L and Trincas G 2019 A model for intact and damage stability evaluation of CNG ships during the concept design stage. *J. Mar. Sci. Eng.* **7** 450.

[8] Bačkalov I, Bulian G, Cichowicz J, Eliopoulou E, Konovessis D, Leguen J F, Rosén A and Themelis N 2016 Ship stability, dynamics and safety: Status and perspectives from a review of recent STAB conferences and ISSW events. *Ocean Eng.* **116** 312-349.

[9] Crossland P 2003 The effect of roll-stabilisation controllers on warship operational performance. *Control Eng. Pract.* **11** 423–431.

[10] Song J, Zhao P, Liang L and Ji M 2020 Force modeling of zero/low-velocity fin stabilizer and hydrofoil profile optimization. *Ocean Eng.* **213** 107635.

[11] Febrianto R A, Prabowo A R, Baek S J and Adiputra R 2021 Analysis of monohull design characteristics as supporting vessel for the COVID-19 medical treatment and logistic. *Trans. Res. Proc.* **55** 699-706.

[12] Spanos D and Papanikolaou A 2001 Numerical study of the damage stability of ships in intermediate stages of flooding. *Proceedings, 5th Inter. Work. Stab. Oper. Saf. Ships, Trieste.*

[13] Le T H, Vu M T, Bich V N, Phuong N K, Ha N T H, Chuan T Q and Tu T N 2021 Numerical investigation on the effect of trim on ship resistance by RANSE method. *Appl. Ocean Res.* **111** 102642.

[14] Savitsky D and Brown P W 1976 Procedures for hydrodynamic evaluation of planing hulls in smooth and rough water. *Mar. Technol. SNAME News.* **13** 381-400.

[15] Faltinsen O M 2005 *Hydrodynamics of high-speed marine vehicles*. Cambridge University Press.
[16] Demirel Y K, Turan O and Incecik A 2017 Predicting the effect of biofouling on ship resistance using CFD. *Appl. Ocean Res*. **62** 100-118.
[17] Kim D J, Kim S Y, You Y J, Rhee K P, Kim S H and Kim Y G 2013 Design of high-speed planing hulls for the improvement of resistance and seakeeping performance. *Int. J. Nav. Archit. Ocean Eng.* **5** 161-177.
[18] Hanzu-Pazara R, Varsami C, Andrei C and Dumitrache R 2016 The influence of ship's stability on safety of navigation. *IOP Conf. Ser. Mater. Sci. Eng.* **145** 082019.
[19] Prabowo A R, Martono E, Muttaqie T, Tuswan T and Bae D M 2022 Effect of hull design variations on the resistance profile and wave pattern: a case study of the patrol boat vessel. *J. Eng. Sci. Technol.* **17** 106–126.
[20] Dabit A S, Lianto A E, Branta S A, Laksono F B, Prabowo A R and Muhayat N 2020 Design of unmanned ships for distributing fish feed in coastal areas based on arduino microcontroller. *Mekanika*. **19** 74-82 (In Indonesian).
[21] Julianto R I, Prabowo A R, Muhayat N, Putranto T and Adiputra R 2021 Investigation of hull design to quantify resistance criteria using holtrop's regression-based method and savitsky's mathematical model: A study case of fishing vessels. *J. Eng. Sci. Tech*. **16** 1426-1443.
[22] Julianto R I, Muttaqie T, Adiputra R, Hadi S, Hidajat R L L G and Prabowo A R 2020 Hydrodynamic and structural investigations of catamaran design. *Proc. Struct. Integ*. **27** 93-100.
[23] Perez T 2015 Ship seakeeping operability, motion control, and Autonomy-A Bayesian Perspective. *IFAC-PapersOnLine*. **48** 217–222.

Effect of bilge keel variations on the designed fast patrol boats to hull resistance and stability behaviours

T Rahmaji[1], A R Prabowo[1,*], N Muhayat[1], T Tuswan[2], T Putranto[3,4]

[1] Department of Mechanical Engineering, Universitas Sebelas Maret, Surakarta 57126, Indonesia
[2] Department of Naval Architecture, Universitas Diponegoro, Semarang 50275, Indonesia
[3] Department of Civil Engineering and Architecture, Tallinn University of Technology, Tallinn 19086, Estonia
[4] Department of Naval Architecture, Institut Teknologi Sepuluh Nopember, Surabaya 60111, Indonesia

Corresponding author: aditya@ft.uns.ac.id

Abstract. This study aims to analyse the effect of variations in the addition of bilge keel on the value of resistance, ship stability, and seakeeping performances. The research method is carried out by designing and analysing the hull design of the fast patrol boat type with variations in the addition of a bilge keel. Variations made on the bilge keel include length and width. The length variations are considered by applying 4 m, 5 m, and 6 m, while the variations in width include 0.3 m, 0.4 m, and 0.5 m. This study indicates that the nine variations of the bilge keels model with a combination of length and width do not show a significant difference in the resistance value on each ship. Then, for the stability criteria, the bilge keel model IX, with a length of 6 m and a width of 0.5 m, has the best stability value compared to other variations. Meanwhile, the addition of bilge keel in the seakeeping simulation did not show a significant difference. The parameters of this analysis consider the primary needs of fast patrol boats to support the development of shipping technology.

Keywords: Patrol boat; Bilge Keel; Resistance; Stability; Seakeeping

1. Introduction

Indonesia is the largest archipelagic country globally that has the potential to become the world's maritime axis [1]. The dynamic global political situation poses a security threat to Indonesia's maritime sector. Threats in the maritime sector include global terrorism, illegal fishing, illegal trade routes, human trafficking, and military invasions due to a lack of supervision in constrained areas such as inter-island borders. Therefore, Indonesia must have a robust system and infrastructure to overcome various threats in the maritime sector.

One way to strengthen defence in Indonesia's maritime sector is to conduct surveillance in Indonesia's marine areas. The patrol boat is one of the fleets used to protect Indonesia's maritime territory from various threats. This ship has a relatively small size but has a high enough speed so that it can increase time efficiency [2]. Given the very high sea activity, Indonesia needs patrol boats that function optimally to protect its territorial waters. Patrol boats must have good stability when excessive rolling motion occurs when operating at sea.

Stability is the tendency or nature of the ship to return to its original position after getting a slope due to an external force [3]. The ship's stability while operating at sea depends on the rolling motion experienced by the ship [4]. When experiencing a rolling motion, the ship's starboard side will move to the left side longitudinally and repeat alternately [5,6]. If the ship experiences continuous rolling motion, it will cause instability and risk turning it upside down [7].

A bilge keel is a tool that increases the ship's stability. This tool can reduce the ship's rolling motion by 27% when moving at Fr 1.185 [8]. The roll damping produced by the bilge keel is more significant than other roll devices, has an easy installation, and is a passive system. In previous studies, adding a flat bilge keel on the ship prototype scale can reduce resistance by 20.8% [9]. Another study also stated that the geometry of the bilge keel with an angle of 45° and a thickness of 6 mm is an effective bilge keel measure [10–12].

Based on these problems, this research aims to determine the effect of adding variations in the width and length of the bilge keel to the total resistance value and stability of the patrol boat. This research is based on providing comfort for passengers or crew members and the safety of the equipment in it, and increasing the accuracy of weapons systems on warships, especially on fast boats and patrol boats. This analysis parameter considers the primary needs of fast patrol boats to support the development of shipping technology.

2. Methodology

In this study, the methods used in the study include data collection, data processing, and data simulation. Data collection is done by looking for references to patrol boats and the geometry of the bilge keel so that the primary size of the reference vessel and the shape of the bilge keel variation will be analysed. Then, the patrol boat model and the bilge keel model were made using Solidwork Software according to the data that had been collected. The analysis results are expected to identify the effect of adding variations in the width and length of the bilge keel to the total resistance value of the ship, ship stability, and seakeeping performance.

2.1. Ship data

The design process begins with making a lines plan for each type of ship to determine the ship's shape. Then proceed with defining lines, surfaces, and volumes according to the primary size of each ship, as shown in Table 1. The ship is modeled in three dimensions, as shown in Figure 1. After the main design is completed, it is continued with the addition of variations of the bilge keel on the hull, as shown in Figure 2.

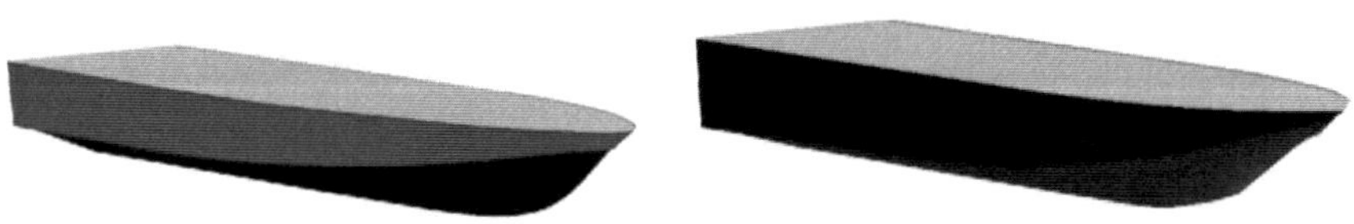

(a) (b)

Figure 1. Rendered 3D hull design: (a) Fast Police Boat; (b) SMIT Patrol Boat.

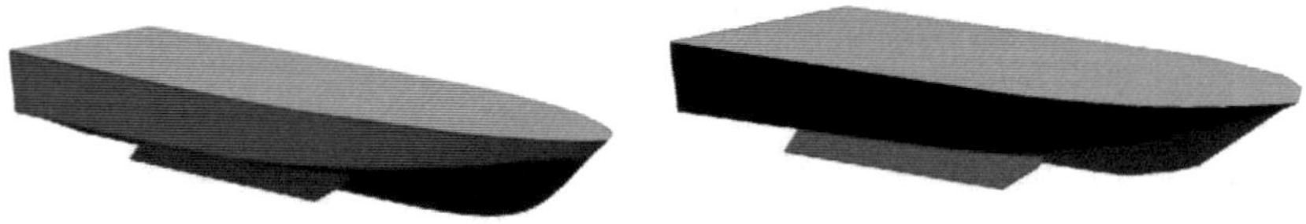

(a) (b)

Figure 2. Rendered 3D hull design with bilge keel: (a) Fast Police Boat; (b) SMIT Patrol Boat.

Table 1. Principal dimension of patrol boat designs.

Parameters	Values	
	Fast Police Boat	SMIT Patrol Boat
Length of Overall (m)	14.95	13.2
Length of Waterline (m)	13.702	11.49
Draft (m)	0.85	0.7
Beam (m)	4.1	4.1
Depth (m)	1.94	1.6
Block Coefficient (-)	0.421	0.531
Displacement (t)	18.93	17.26
Dead Weight Tonnage (kg)	2500	2300

2.2. Research parameters

The fixed parameters used as reference data for this study include the primary size of the ship, the geometry of the flat bilge keel, the thickness of the bilge keel is 6 mm, and the angle of the bilge keel is 45°. At the same time, the parameters are not fixed as the variation of the Model to be analysed, including variations in the width of the bilge keel (0.3 m, 0.4 m, and 0.5 m), as well as variations in the length of the bilge keel (4 m, 5 m, and 6 m). Stability analysis uses moderate wave characteristics with a height of 1.875 m. The bilge keel design is shown in Figure 3.

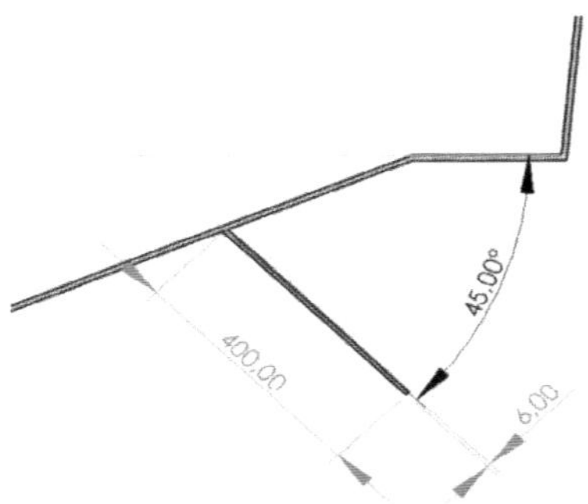

Figure 3. Bilge keel design in millimetre.

2.3. Bilge keel variations.

A bilge keel is installed in the middle of the ship with a length ranging from 25-50% of the ship's length. The geometry of the bilge keel is flat according to a predetermined reference with a bilge keel angle of 45° and a thickness of 6 mm. The width and length of the bilge keel were varied to obtain nine models of bilge keel configuration and 1 model without bilge keel, as shown in Table 2.

Table 2. Bilge Keel Length and Width Variations

Items	Model									
	I	II	III	IV	V	VI	VII	VIII	IX	X
Length Bilge Keel (m)	4	4	4	5	5	5	6	6	6	0
Width Bilge keel (m)	0.3	0.4	0.5	0.3	0.4	0.5	0.3	0.4	0.5	0

2.4. Hull resistance analysis.

One of the critical factors that must be considered in building a ship's hull is the ship's resistance. Resistance is a characteristic parameter of the hydrodynamics of the hull [13]. The level of ship resistance is strongly influenced by the speed of the ship (V_s), displacement (Δ), and the shape of the hull form [14]. The value of the ship's resistance is directly proportional to the speed of the ship. The faster the ship goes, the greater the resistance, so the energy required for the ship to move is also higher. One of the methods for calculating resistance is Savitsky which is stated in Equation 1.

$$D = \Delta \tan + \frac{D_f}{\mathrm{Cos}\,\tau} \tag{1}$$

D is the total drag/resistance, τ is the slight angle, and Df is the friction component.

2.5. Ship stability assessment.

Stability testing is the ship's ability to be upright again after experiencing a certain roll/tilt [15-17]. In general, stability is divided into static stability (Equation 2) and dynamic stability (Equation 3).

$$P.h = P.FS - P.GF\,\mathrm{Sin}\,\gamma \tag{2}$$

$$ED = \int_{\gamma=0}^{\gamma=\gamma'} S3dy \tag{3}$$

Where S3 is $P.h$ in static stability, G is the center of gravity.

2.6. Seakeeping performance analysis.

A measure of how well a ship sails with certain marine environmental conditions is called seakeeping ability [18]. This capability is an important aspect in terms of ship design. In the seakeeping simulation using the Joint North Sea Wave Project (JONSWAP) wave spectrum. The JONSWAP spectrum equation is presented in Equation 4 [19].

$$S^+(\omega) = \frac{\alpha g^2}{\omega^5} \exp\left[-\beta \frac{\omega_p^4}{\omega^4}\right] \gamma^{\exp\left[-\frac{(\omega-\omega_p)^2}{2\omega_p^2\sigma^2}\right]} \tag{4}$$

where, $S^+(\omega)$ is wave spectrum; α is the intensity of the spectra; g is the gravity force; ω is the wave frequency; ω_p is the peak wave frequency; β is a shape factor; γ is peak enhancement factor; and σ is 0.07 if $\omega \le \omega_p$ and 0.09 if $\omega > \omega_p$.

3. Result and discussion

3.1. Ship resistance

In calculating ship resistance, there are various calculation methods, including Savitsky. The research of Julianto et al. [9] stated that the comparison of the Savitsky method approach with numerical calculations is 14%. This research was conducted for the variation of the bilge keel model with the slightest resistance. The following is the resistance simulation data on ten design models on the Fast police boat hull, shown in Tables 3 and 4, as well as the graphs, which are shown in Figure 4. At the same time, the SMIT Patrol Boat hull is shown in Tables 5 and 6, as well as the graph in Figure 5.

Table 3. Test results of hull resistance Fast police boat.

Speed (kts)	Resistance (N)										Coefficient Variation (%)
	I	II	III	IV	V	VI	VII	VIII	IX	X	
15	21037.79	21043.26	21048.67	21042.38	21053.24	21060.3	21045.79	21053.99	21066.27	21025.01	0.000528956
20	26387.43	26394.27	26401.02	26393.52	26408.91	26417.69	26398.36	26408.64	26425.79	26372.73	0.000549593
25	28923.39	28930.78	28938.03	28930.47	28951.36	28960.76	28936.52	28947.64	28970.42	28911.09	0.000583061
30	30179.9	30186.94	30193.82	30186.98	30209.52	30218.4	30193.28	30203.88	30228.21	30170.37	0.000560337
35	31735.23	31741.65	31747.92	31741.94	31764.5	31772.57	31748.08	31757.76	31781.96	31728.18	0.000507741
40	34087.14	34092.99	34098.68	34093.46	34116.04	34123.35	34099.4	34108.22	34132.29	34082.46	0.00045084
45	37305.45	37310.84	37316.08	37311.48	37334.62	37341.32	37317.29	37325.43	37349.93	37303.17	0.00040132
50	41344.94	41349.99	41354.89	41350.79	41375.13	41381.39	41356.57	41364.22	41389.83	41345.14	0.000362496

Table 4. Test results power of hull resistance Fast police boat.

Speed (kts)	Power (kW)										Coefficient Variation (%)
	I	II	III	IV	V	VI	VII	VIII	IX	X	
15	162.342	162.384	162.426	162.377	162.461	162.515	162.403	162.467	162.561	162.243	0.000528333
20	271.497	271.568	271.637	271.56	271.718	271.809	271.61	271.716	271.892	271.346	0.000549852
25	371.987	372.082	372.175	372.078	372.347	372.468	372.156	372.299	372.592	371.829	0.000583247
30	465.777	465.885	465.991	465.886	466.234	466.371	465.983	466.146	466.522	465.629	0.000560434
35	571.411	571.526	571.639	571.531	571.937	572.083	571.642	571.816	572.252	571.284	0.000507566
40	701.438	701.558	701.675	701.568	702.032	702.183	701.69	701.871	702.367	701.341	0.000450885
45	863.621	863.746	863.867	863.761	864.296	864.452	863.895	864.084	864.651	863.568	0.000401449
50	1063.484	1063.614	1063.74	1063.634	1064.26	1064.421	1063.783	1063.98	1064.638	1063.489	0.000362298

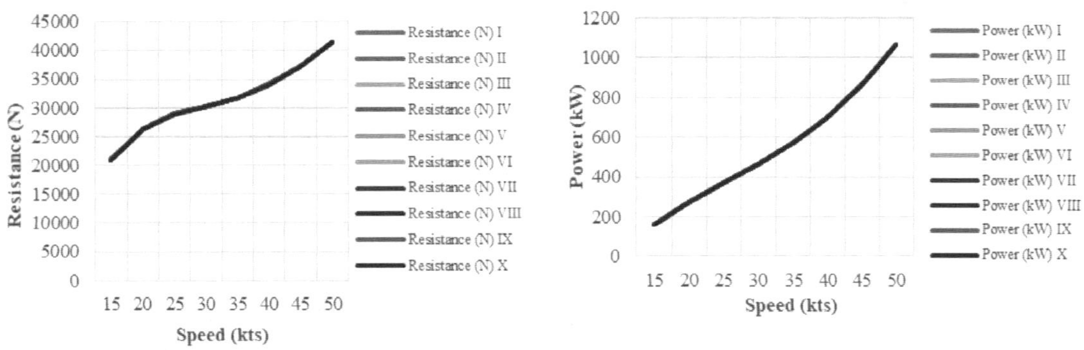

(a) (b)

Figure 4. Fast police boat resistance simulation results graph: (a) Speed vs Resistance; (b) Speed vs Power.

Table 5. Test results of hull resistance SMIT Patrol Boat.

Speed (kts)	Resistance (N)										Variation Coef (%)
	I	II	III	IV	V	VI	VII	VIII	IX	X	
15	22701.95	22715.97	22723.61	22709.71	22725.84	22729.17	22715.86	22727.55	22745.6	22674.47	0.000798451
20	25542.15	25557.3	25566.15	25551.14	25568.4	25572.94	25559.07	25572.25	25591.6	25510.22	0.000811319
25	25505.25	25518.1	25526.44	25513.73	25528.34	25533.78	25521.72	25533.9	25550.41	25475.1	0.000753082
30	25529.61	25539.15	25546.45	25537.04	25547.98	25554.3	25544.35	25554.86	25567.43	25503.18	0.000645186
35	26507.39	26513.98	26520.33	26513.86	26521.57	26528.66	26520.47	26529.51	26538.6	26484.34	0.000529754
40	28496.44	28500.66	28506.29	28502.17	28507.29	28515.07	28508.26	28516.17	28522.46	28475.99	0.000428252
45	31389.85	31392.18	31397.29	31395.05	31398.1	31406.54	31400.81	31407.88	31411.96	31371.26	0.000347388
50	35075.53	35076.35	35081.1	35080.38	35081.76	35090.86	35085.97	35092.44	35094.75	35058.2	0.000286304

Table 6. Test results power of hull resistance SMIT Patrol Boat.

Speed (kts)	Power (kW)										Variation Coef (%)
	I	II	III	IV	V	VI	VII	VIII	IX	X	
15	175.183	175.292	175.351	175.243	175.368	175.393	175.291	175.381	175.52	174.971	0.000798948
20	262.8	262.956	263.047	262.893	263.07	263.117	262.974	263.11	263.309	262.472	0.000811019
25	328.026	328.191	328.298	328.135	328.323	328.393	328.238	328.394	328.607	327.638	0.000753238
30	394.007	394.154	394.267	394.122	394.291	394.388	394.234	394.397	394.591	393.599	0.000645432
35	477.28	477.399	477.513	477.397	477.536	477.663	477.516	477.679	477.842	476.865	0.00052989
40	586.393	586.48	586.596	586.511	586.617	586.777	586.637	586.799	586.929	585.973	0.000428196
45	726.675	726.729	726.847	726.795	726.866	727.061	726.929	727.092	727.187	726.245	0.00034722
50	902.221	902.242	902.364	902.345	902.381	902.615	902.489	902.656	902.715	901.775	0.000286275

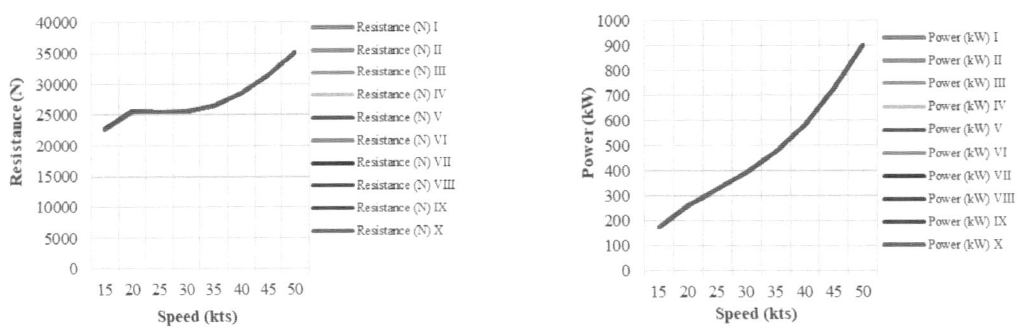

(a) (b)

Figure 5. SMIT patrol boat resistance simulation results graph: (a) Speed vs Resistance; (b) Speed vs Power.

Based on the data above, the simulation results of the two types of ships show that in each Model, the bilge keel variation does not significantly affect the value of the ship's resistance (see Figures 4 and 5). In the graphic image, there is only 1 line that represents the data. This is due to the overlapping values in each model variation, which can be seen from the coefficient of variation close to 0. The data above shows that both ships have the most significant resistance in the variation of Model IX and the slightest resistance in Model I. At a speed of 25 knots for the variation of Model IX, the Fast police boat has a resistance value of 28970.42 N, and the required power is 372.592 kW, while the SMIT patrol boat has a resistance value of 25550.41 N and the required power is 328.607 kW.

Then, at a speed of 25 knots for the variation of Model I, the Fast police boat has a resistance value of 28923.39 N, and the required power is 371.987 kW. On the other hand, the SMIT patrol boat has a resistance value of 25505.25 N, and the required power is 328.026 kW. Model IX has the most significant resistance value. However, the difference is minimal against Model I, only 47.03 N on the Fast police boat and 45.16 N on the SMIT patrol boat. Based on these data, it was also found that the faster the ship's speed, the greater the resistance value.

3.2. Ship stability

The calculation of the ship's stability is used to find the value of stability in each model. Stability refers to the tendency of an object or system to return to its original state after experiencing minor disturbances, so that it can meet the needs of patrol boats. The following are the results of stability tests for each model of the bilge keel variation of the Fast police boat and SMIT patrol boat, which are presented in Tables 7 and 8, and the graph of the GZ curve in Figure 6.

Table 7. Fast police boat stability test data.

Model	Stability				Resistance (V = 25 kts)	
	GZ (m)	α (deg)	A (m.deg)	Angle of vanishing point (deg)	Resistance (N)	Power (kW)
I	0.769	52.7	39.9	101.005	28923.39	371.987
II	0.779	52.7	40.79	101.017	28930.78	372.082
III	0.786	52.7	41.41	102.106	28938.03	372.175
IV	0.78	52.7	40.86	101.638	28930.47	372.078
V	0.788	52.7	41.55	102.106	28951.36	372.347
VI	0.796	53.6	42.31	102.595	28960.76	372.468
VII	0.787	52.7	41.48	102.106	28936.52	372.156
VIII	0.796	53.6	42.31	102.595	28947.64	372.299
IX	0.806	53.6	43.22	103.27	28970.42	372.592
X	0.743	52.7	37.61	99.538	28911.09	371.829

Table 8. SMIT patrol boat stability test data.

Model	Stability				Resistance (V = 25 kts)	
	GZ (m)	α (deg)	A (m.deg)	Angle of vanishing point (deg)	Resistance (N)	Power (kW)
I	1.064	34.5	60.84	96.114	25505.25	328.026
II	1.071	34.5	61.65	96.603	25518.1	328.191
III	1.078	34.5	62.46	97.092	25526.44	328.298
IV	1.069	34.5	61.46	96.481	25513.73	328.135
V	1.078	34.5	62.47	97.092	25528.34	328.323
VI	1.085	34.5	63.22	97.582	25533.78	328.393
VII	1.077	34.5	62.35	97.092	25521.72	328.238
VIII	1.09	34.5	63.82	97.982	25533.9	328.394
IX	1.094	34.5	64.31	98.315	25550.41	328.607
X	1.042	34.5	58.36	94.402	25475.1	327.638

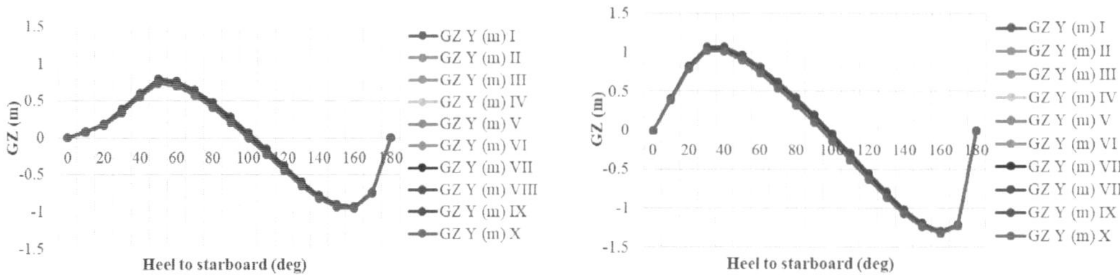

(a) (b)

Figure 6. GZ curve stability graph in each model: (a) Fast police boat; (b) SMIT patrol boat.

Based on the results of stability testing for each model, it can be seen that both ships experienced an increase in the value of stability after the addition of a bilge keel. Fast police boats without a bilge keel have a GZ value of 0.743 m and a maximum boat tilt angle of 52.7°. It means that the ship has a maximum slope to return to its original position, which is 52.7˚. If it exceeds this angle, the ship will capsize. While the Fast police boat with a bilge keel on Model IX, the GZ value is 0.806 and the maximum inclination angle is 53.6°, this seems to have increased the GZ value and the maximum tilt angle. Then, the SMIT patrol boat without a bilge keel has a GZ value of 1.042 m with a maximum tilt angle of 34.5°. At the same time, the SMIT patrol boat with a bilge keel on Model IX has a GZ value of 1.094 m with an angle of 34.5°. It also shows an increase in ship stability after adding the bilge keel. Figure 6 compares the relationship between the maximum ship slope and the GZ curve in each model.

Furthermore, it can be concluded that adding a bilge keel on both types of ships (fast police boat and SMIT patrol boat) can increase ship stability. The stability of the smallest ship with a bilge keel on both ships is shown in Model I. The fast police boat with 52.7° and a maximum GZ value of 0.769 m, while the SMIT patrol boat with 34.5° includes a GZ value of 1.064 m. The most excellent stability is shown in Model IX with 53.6° and a GZ value of 0.806 m (fast police boat), while in the SMIT patrol boat with 34.5° and a maximum GZ value of 1.094 m. In Model IX, both have a wider curve area under the stability graph than the other models. Therefore, the bilge keel Model IX variation is better, so the ship will be more stable and safer when operating at sea.

3.3. Seakeeping analysis

In the calculation of seakeeping using the JONSWAP wave spectrum with wave height and period referring to sea state 4 (Moderate Water) contained in the World Meteorological Organization (WMO). In this study, the results of the seakeeping analysis present heaving and rolling data presented in the Response Amplitude Operators (RAO) Graph. Models X and IX on both types of ships were analysed to find out how much influence the addition of a bilge keel had on seakeeping performance. The response amplitude of the operator on the heaving motion and rolling motion of the Fast police boat is shown in Figure 7. Then, the response amplitude of the operator on the heaving motion and rolling motion of the SMIT patrol boat is shown in Figure 8.

The analysis used a wave heading of 90° (beam seas) from the side of the ship with a ship speed of 20 knots. In the Fast police boat and SMIT patrol boat for heaving criteria, Model X (without bilge keel) shows a relatively more stable value than Model IX (with bilge keel), as shown in Figures 7(a) and 8(a). Meanwhile, in the rolling criteria for both types of ships, Model IX shows better results, but the difference is only minimal, even relatively the same, as seen in Figures 7(b) and 8(b). However, the Model IX and X for both types of ships still have the same trend for heaving and rolling. Each model's transfer function (RAO) will be compared with the wave encounter frequency at an angle of 90°. The highest wave encounter frequency in wave spectral must not superposition with the RAO transfer function of heaving and rolling because the hull design will experience extreme response (heaving or rolling). In both types of ships for all model variations, there is no superposition between the RAO transfer function and the frequency of spectral wave encounters, so there is no wave superposition. From the data presented, adding a bilge keel on Model IX for the two ships tends to be less effective because it lacks significant differences in heaving and rolling criteria.

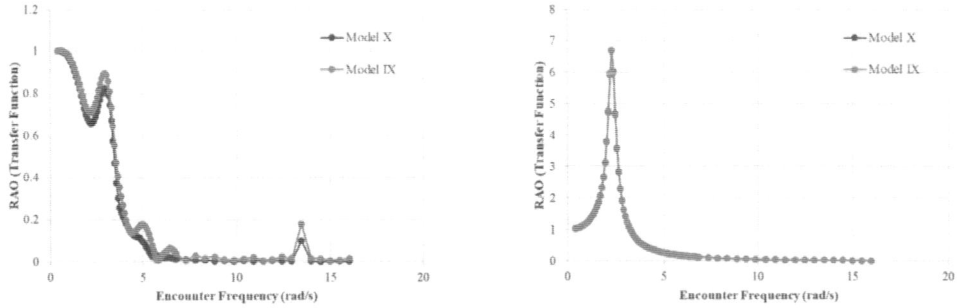

(a) (b)

Figure 7. RAO transfer function Fast police boat with encounter angle 90° and speed 20 kts: (a) Heaving motion; (b) Rolling motion.

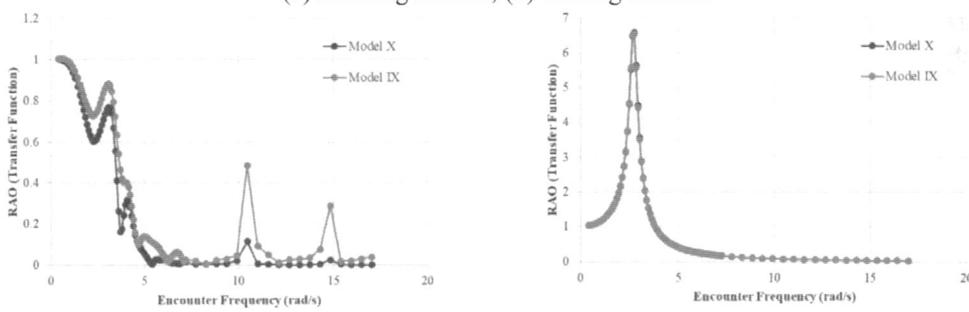

(a) (b)

Figure 8. RAO transfer function SMIT patrol boat with encounter angle 90° and speed 20 kts: (a) Heaving motion; (b) Rolling motion.

4. Conclusions

Based on the results of the study, it can be concluded that the addition of variations in width and length on the bilge keel shown in Model I-Model IX on both ships (fast police boat and SMIT patrol boat) has a more negligible effect on the ship's resistance value, which is indicated by the coefficient of variation of resistance which is close to 0, on each Model. In the resistance test results, Model I have the slightest resistance with a value of 28923.39 N and power of 371.987 kW (fast police boat) and 25505.25 N and power of 328.026 kW (SMIT patrol boat) when compared to other models.

In stability testing, the addition of variations in the width and length of the bilge keel affects the stability value. Model IX has a value of 53.6° and a GZ value of 0.806 m (fast police boat), while the SMIT patrol boat has 34.5° and a maximum GZ value of 1.094 m. Thus, it can be concluded that Model IX, with a configuration of 0.5 m wide and 6 m long, has the best characteristics compared to other models. However, the seakeeping simulation comparing Model IX with Model X (without bilge keel) on heaving and rolling criteria did not show significant results.

Finally, it can be concluded that adding a bilge keel can increase the ship's stability in the water. However, variations in the width and length of the bilge keel have less effect on the ship's resistance value and seakeeping performance. Adding the width and length of the bilge keel tends to increase the ship's resistance so that the energy required will be greater. So the addition of bilge is still not effectively applied to the V-hull patrol boat.

References

[1] Nugroho A, Suharyo O S and Rahman A 2020 The development of indonesian maritime potential and prospects towards a world maritime axis. *STTAL Postgraduate-International Conf.* **4** 1-8.

[2] Prabowo A R, Martono E, Muttaqie T, Tuswan T and Bae D M 2022 Effect of hull design variations on the resistance profile and wave pattern: a case study of the patrol boat vessel. *J Eng Sci Technol*. **17** 293–312

[3] Barrass B and Derrett C D R 2011 *Ship stability for masters and mates*. Elsevier

[4] Mahfouz A B 2004 Identification of the nonlinear ship rolling motion equation using the measured response at sea. *Ocean Eng*. **31** 2139–2156

[5] Wang X, Liu Z, Loughney S, Yang Z, Wang Y and Wang J 2021 An experimental analysis of evacuees' walking speeds under different rolling conditions of a ship. *Ocean Eng.*. **233** 108997.

[6] Gao Z and Tian X 2021 Numerical study on the wave-induced roll motion of a damaged ship in head seas. *Appl. Ocean Res.* **114** 102805.

[7] Liu Y, Han F and Lu Y 2016 Stability and capsizing analysis of nonlinear ship rolling in wind and stochastic beam seas. *Appl Ocean Res.* **57** 52–63

[8] Saputra A, Budiarto U and Rindo G 2019 Analysis of the effect of bulb type bilge keel on V and U hulls on patrol boats using the computational fluid dynamic method (CFD). *J Tek Perkapalan*. **7** 20-28 (In Indonesian).

[9] Julianto R I, Prabowo A R, Muhayat N, Putranto T and Adiputra R 2021 Investigation of hull design to quantify resistance criteria using Holtrop's regression based method and Savitsky's mathematical model: a study case of fishing vessels. *J Eng Sci Technol* **16** 1426–1443

[10] Irkal M A R, Nallayarasu S and Bhattacharyya S K 2016 CFD approach to roll damping of ship with bilge keel with experimental validation. *Appl Ocean Res* **55** 1–17

[11] Watson D G M 1998 *Practical ship design*. Elsevier.

[12] Rui D, Wen-yang D and Shan M A 2013 Investigation of the damping moment induced by a forced rolling cylinder with bilge keels. *Proc. Eng*. **61** 347-355.

[13] Molland A F, Turnock S R and Hudson D A 2017 *Ship resistance and propulsion*. Cambridge University Press.

[14] Dabit A S, Lianto A E, Branta S A, Laksono F B, Prabowo A R and Muhayat N 2020 Finite Element Analysis (FEA) on autonomous unmanned surface vehicle feeder boat subjected to static loads. *Proc. Struct. Integ*. **27** 163-170.

[15] Febrianto R A, Prabowo A R, Baek S J and Adiputra R 2021 *Analysis of Monohull Design Characteristics as Supporting Vessel for the COVID-19 Medical Treatment and Logistic.* Transp Res Procedia. **55** 699–706

[16] Julianto R I, Muttaqie T, Adiputra R, Hadi S, Hidajat R L L G and Prabowo A R 2020 Hydrodynamic and structural investigations of catamaran design. *Proc. Struct. Integ*. **27** 93-100.

[17] Dabit A S, Lianto A E, Branta S A, Laksono F B, Prabowo A R and Muhayat N 2020 Design of unmanned ships for distributing fish feed in coastal areas based on arduino microcontroller, *Mekanika: Majalah Ilmiah Mekanika*. **19** 74-82 (In Indonesian).

[18] Prabowo A R, Febrianto R A, Tuswan T and Tjahjana D D D P 2022 Performance evaluation on the designed v-shaped monohull ship models. *J. Appl. Eng. Sci.* **20** 609-623.

[19] Codecogs 2014 The JONSWAP Spectra In The Wave-Frequency Domain. http://www.codecogs.com/library/engineering/fluid_mechanics/waves/spectra/jonswap.php. Retrived October 9th, 2014

Application of virtual reality as an alternative training media as complement of ship bridge simulator (case study: port of tanjung emas semarang)

Trika Pitana[1], Manungku Trinata Pramudhita[2], Nurhadi Siswantoro[1], Josua Ricky Cristian Napitupulu[1]

[1]Department of Marine Engineering, Institut Teknologi Sepuluh Nopember (ITS), Surabaya, Indonesia
[2]Politeknik Pelayaran, Surabaya, Indonesia

corresponding author's e-mail: trika@its.ac.id

Abstract. The application of computer technology such as 3D modeling, simulators, virtual reality, and augmented reality, will always provide benefits to the maritime sector. The object of study used in this research is the physical bridge simulator at Poltekpel Surabaya which is used as a virtual simulator using virtual reality. Through the application of virtual reality technology, it can be developed into an educational facility for the cadets there which does not only rely on 2D or 3D forms but can be used as a simulator where users can interact directly in the application. With this simulator, it can save in terms of manufacturing costs and in terms of usage it becomes more effective and practical. In this simulator application the user can participate in a virtual 3D environment that is simulated like a real environment. With a simulator using virtual reality, users can learn how to operate a ship like a captain without having to visit the ship's bridge directly. Simulators with virtual reality will provide a virtual learning environment that can be accessed easily anywhere and anytime. Based on the results of the cadet assessment test questionnaire consisting of 7 questions with 30 respondents, it can be concluded that the VR application for the Bridge Simulator is in the very good category because the application is easy to understand by the respondents and the 3D results from the ship's bridge at Poltekpel are the same with those in the simulator application and for the results of the questionnaire calculations have a total value of 908 which is between 840 to 1050, or 87% in the percentage calculation.

Keywords: virtual reality, bridge simulator, simulator, maritime, technology.

1. Introduction

Virtual reality is a technology that can imitate environmental situations and involve the individual minds using a computer system and highlight the interaction between the user and the system. Virtual Reality can create an artificial environment by utilizing unity 3D software. Virtual reality has long been touted as a tool for improving education by providing immersive and interactive experiences in fields ranging from science and engineering to foreign languages and social sciences[1].

Virtual reality can also make it easier to test different scenarios because the environment can be designed to prevent extraneous variables from interfering with test results and can be more controlled [2]. Virtual reality has existed in many forms since the 1960s when the first digital flight simulators were developed and used by the world's major airlines and air forces [3]. As technology develops, "virtual reality" has become an expression to represent devices that create immersive interactive environments with visual realism [4].

Metaverse is a virtual world where you use a VR headset to enter a space that connects all kinds of digital environments. You can shop around, play, study, work and buy. In this space you can have a 3D avatar. It is through this digital representation that you can interact with other avatars [5]. Metaverse is a post-reality world, a persistent and multiuser environment that combines physical reality with digital virtuality. It is based on the convergence of technologies that enable multisensory interactions with virtual environments, digital objects and people as in the use of virtual reality (VR) and augmented reality (AR). Currently there are many ways of education or learning for us to learn about something, an example is a simulator. Simulators as a learning tool is one of the best learning methods to find out the components, functions, and workings of the tool or unit [6].

Politeknik Pelayaran (Poltekpel) Surabaya as one of the maritime academy under the auspices of the Ministry of Transportation in Indonesia that has received approval, cannot be separated from the use of simulators. There are many types of simulators installed in Poltekpel, including ship bridge simulators, ship engine simulators and several other navigation equipment simulators.

The use of simulators as a means of education is deemed necessary to support the learning of the cadets here, and will certainly make it easier for the cadets to face the real world while sailing. In this case, what will be used as research material specifically is the application of Virtual Reality technology on the ship's bridge simulator as a means of education for cadets. It is hoped that the existence of this ship bridge simulator can save more in terms of the cost of making the simulator itself and is expected to facilitate learning for cadets. VR has a strong potential for students to have immersive learning experiences to increase their learning effectiveness and motivation [7]. Making the bridge simulator application is assisted by a VR headset as an intermediary while running the simulator application, the Bridge simulator at Poltekpel Surabaya will be converted into a VR simulator application version which can be used anywhere and anytime.

This simulator application will be created and intended by cadets at Poltekpel Surabaya as educational and learning needs. The deck used in the simulator is the ship's bridge, the cadets can simulate how to drive the ship through the bridge, using VR as a tool to support the simulator. Certainly learning will become more efficient, the use of a bridge simulator will be an effective method.

The application has limited features, only the throttle and steering wheel that can be used. On the steering throttle, there are four speed stages, namely dead slow, half, slow, and full speed, and there is also a speed indicator (knot) on the left side of the steering throttle. Then on the steering wheel there are several indicators, such as the bow degree, the degree of rotation of the rudder, and the speed of the ship (knots).

The components used in virtual reality are divided into two components, there are Hardware and Software:

1. Head Mounted Display (Oculus Quest 2)
   The head-mounted display puts the screen in front of eyes.
2. Blender
   It has functions to define the visual properties of these objects and construct the geometry of objects in the virtual world.
3. Unity 3D
   It is used to define the rules to be followed in the virtual world and to program how these objects are defined. In this final project I use the Unity application to create a VR application.

2. Methods

2.1. Problem identification

Here is explained regarding the problem identification, purpose, research limitation and benefits of the research. The problem that will raised is how to apply virtual reality technology to bridge simulator application for educational facilities for cadets at Politeknik Pelayaran (Poltekpel) Surabaya.

IOP Conf. Series: Earth and Environmental Science 1166 (2023) 012049 doi:10.1088/1755-1315/1166/1/012049

2.2. Data collection
This chapter describes the data collection on the bridge simulator at the Poltekpel Surabaya campus, starting from the view of the port from the Tanjung Emas port of Semarang, the navigation tools on the ship's bridge, and also the ship that will be used as a 3D view. As for the items that will be made into 3D, there are telephone, navtex receiver, navigation aid, ECDISS, CONNING, radar, VHF, steering wheel, and compass.

2.3. Object creation
After the data is complete, then the items are made into 3D form using the application called Blender as in Figure 1.

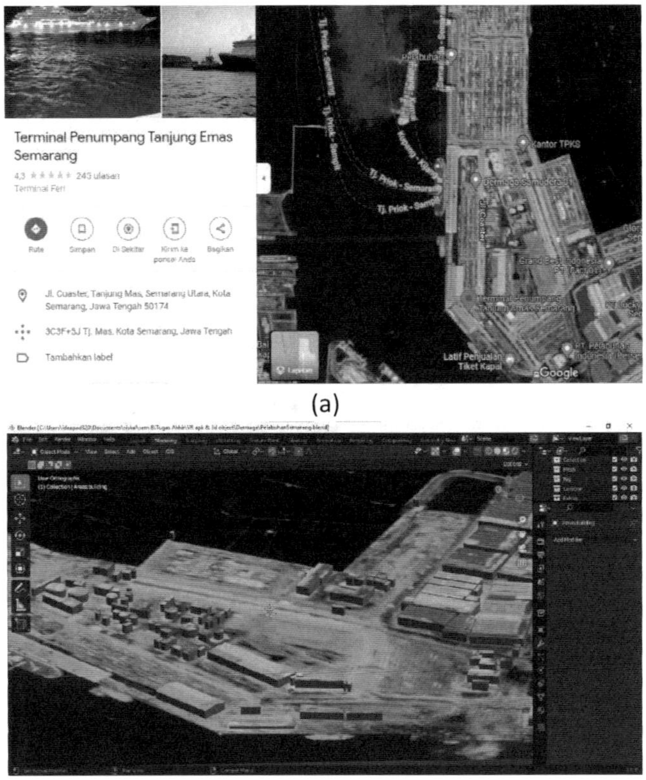

Figure 1. (a) View from Google Maps (b) 3D Object Maker using Blender of Tanjung Emas Port of Semarang

In making 3D of the port, the first stage uses the help of Google Maps to take the layout of the port. Then after taking the layout, the layout results are exported into the blender application. after that do the detailed steps on the blender application.

(a)

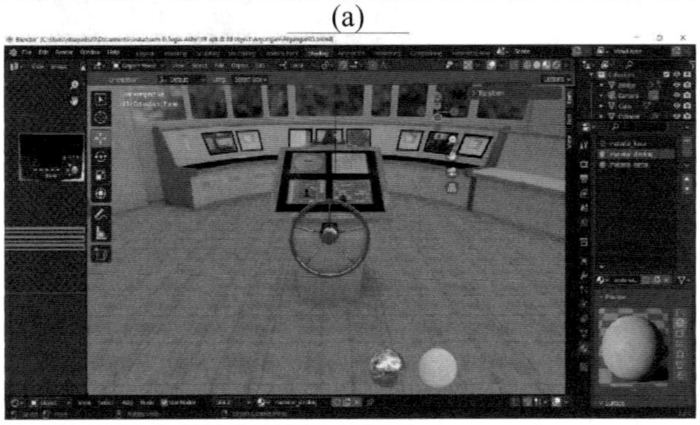

(b)

Figure 2. (a) Real Picture of bridge simulator in Poltekpel Surabaya, (b) 3D Object of bridge simulator using Blender

By using a 3D maker application, called blender, real items will be designed and made as similar as possible to real items as in Figure 2.

2.4. VR application creation

After all items have been made into 3D, then the next step is that the 3D items are inserted into the virtual reality maker application, Unity application. This process takes a long time and must be done carefully.

2.5. Application testing by cadets in poltekpel surabaya

At this stage, testing is carried out to determine whether the Virtual Reality application that has been made in the previous stage is ready to be implemented or not. It is necessary to check several times so that the results of the Virtual Reality application are successful and can run properly.

2.6. Testing analysis

This stage is a comprehensive testing stage and is immediately tested on the Oculus Quest 2 VR headset. All Virtual Reality features must function properly. If the error is still found or does not match the desired function, repair is needed. If all the features and functions of the application run smoothly and no errors are found, then you can proceed with the questionnaire. Total of the respondents are 30, there are cadets from Poltekpel. Questionnaires are used to assess feedback from VR application users with certain parameters.

3. Result and discussion

Table 1. Port area specification

Water Area	400 ha
Land Area	500 ha

(a)

(b)

(c)

Figure 3. (a) (b) (c) 3D Object of Tanjung Emas Port of Semarang

(a)

(b)

(c)

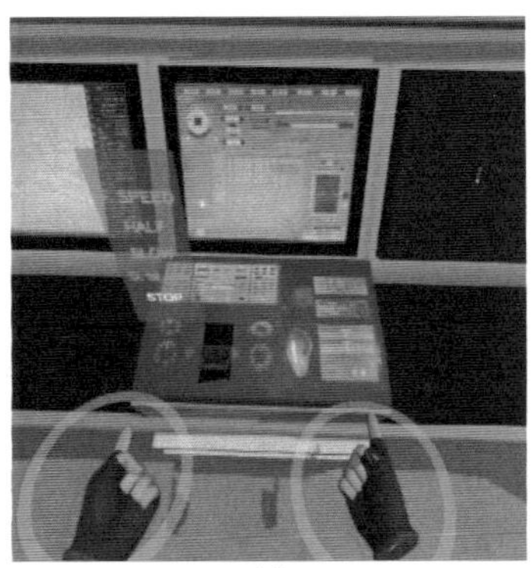

(d)

Figure 4. (a) 3D Object of bridge simulator, (b) 3D Object of steering wheel, (c) 3D Object of compass, (d) throttle

This chapter describes the data collection on the ship's bridge simulator at the Poltekpel Surabaya campus, starting from the view of the port from the Tanjung Emas port of Semarang, the navigation tools on the ship's bridge, and also the ship that will be used as a 3D view. This chapter will also show some equipment / objects that have become 3D models.

The following is a view from various sides of the 3D results of Tanjung Emas Port of Semarang as shown in Figures 3 (a)(b)(c), and 5(a)(b). The results of the 3D modeling of the tools on the ship's bridge simulator as shown in Figure 4 (a)(b)(c)(d).

(a) (b)

Figure 5. (a) Steering throttle with 4 stages speed, (b) Steering wheel with 3 indicators

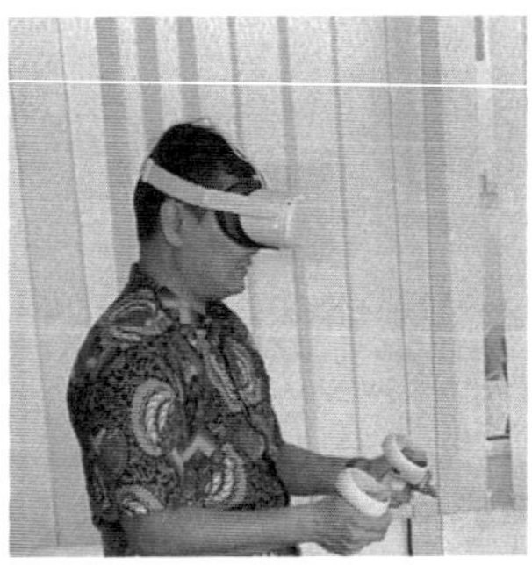

Figure 6. How to use VR Headset

Figure 7. How to hold the joystick

After carrying out the 3D formation process on the ship's bridge tools, the next stage is making a virtual reality simulator application. Making virtual reality applications is done using the Unity application. On the steering throttle, there are four speed stages, namely dead slow, half, slow, and full speed, and there is also a speed indicator (knot) on the left side of the steering throttle. Then on the steering wheel there are several indicators, such as the bow degree, the degree of rotation of the rudder, and the speed of the ship (knots). Here are some views that appear in the virtual reality simulator application as in Figure 6. The first step is to use the Oculus Quest 2 VR headset, the VR headset can be directly put on and the size is adjusted to the user's head as shown in Figure 6. After the VR headset is plugged in, the user can then directly hold both the left and right joysticks of the Oculus Quest 2. The correct way to hold the two joysticks can be seen from figure 7.

Then after Oculus Quest 2 has been installed correctly the application can be run and played immediately. This VR simulator application needs to be tested with the aim of finding flaws and errors in the application. The test was carried out directly at Poltekpel Surabaya by Poltekpel cadets, after testing the cadets filled out a questionnaire and from the questionnaire, the shortcomings and errors found in the simulator application were found.

The questionnaire that will be distributed to the cadets at Poltekpel consists of seven questions. Respondents must give a value that is stated in a number range from one (1) to one (5). The number five (5) indicates a very good score, while the number one (1) indicates a very bad score. To calculate the percentage of each question using the Likert scale formula, the Likert scale is a psychometric scale that is commonly used in questionnaires, and is the most widely used scale in survey research [8]. This scale is named after Rensis Likert, an American educator and psychologist. Rensis Likert has developed a scale to measure people's attitudes in 1932. The Likert scale is used to measure attitudes, opinions, and perceptions of a person or group of people about social phenomena. With a Likert Scale, the variables to be measured are translated into variable indicators [9]. So after doing the Likert Scale calculations, the final results of the questionnaire calculations will be found and conclusions can be drawn from the results of these calculations whether the application is good or there are still many shortcomings. Here is the formula :

$$P = \frac{Total\ Score}{y}\ x\ 100\% \quad (1)$$

Table 2. Formula Description

Symbol	Description
P	Percentage Score
Total Score	The sum of the results of the multiplication of the total respondents with the rating scale.
y	The highest scale times the number of respondents
Highest Scale	5
Respondents	30
y	5 x 30 = 150

The question is as follows:

1. Is the simulation display in VR easy to understand?
2. Is learning to use VR more interesting, easier to understand and more practical?
3. Can learning using VR simulations fulfill functions such as direct simulations in the PoltekPel bridge simulator?
4. Does the 3D Model of the ship's bridge match the original?

5. Does the 3D Model of the Port match the original?
6. Can learning through 3D simulation help indirect learning (do it by yourself)?
7. Are the features in the VR simulation application as expected?

From the description of the formula above, it can be calculated the percentage of each questionnaire question, for example question number 1:

1. Is the simulation display in VR easy to understand?

Table 3. Questionnaire Number 1

Score	Respondents	Total Score
5	23	115
4	7	28
3		
2		
1		
Total	30	143
Average		4,76

Putting it into the Likert percentage formula:

$$P = \frac{143}{150}\ x\ 100\% = 95{,}3\ \% \tag{2}$$

Then, for the percentage result in questionnaire number 1 is 95.3%, this means it is excellent.

2. Is learning to use VR more interesting, easier to understand and more practical?

Table 4. Questionnaire Number 2

Score	Respondents	Total Score
5	15	75
4	13	52
3	2	6
2		
1		
Total	30	133
Average		4,43

Putting it into the Likert percentage formula:

$$P = \frac{133}{150}\ x\ 100\% = 88{,}6\ \% \tag{3}$$

Then, for the percentage result in questionnaire number 1 is 88,6%, this means it is excellent.

3. Can learning using VR simulations fulfill functions such as direct simulations in the PoltekPel bridge simulator?

Table 5. Questionnaire Number 3

Score	Respondents	Total Score
5	5	25
4	18	72
3	7	21
2		
1		
Total	30	118
Average		3,93

Putting it into the Likert percentage formula:

$$P = \frac{118}{150}\ x\ 100\% = 78{,}6\ \% \tag{4}$$

Then, for the percentage result in questionnaire number 1 is 78,6%, this means that it is sufficient but still needs to be improved.

4. Does the 3D Model of the ship's bridge match the original?

Table 6. Questionnaire Number 4

Score	Respondents	Total Score
5	17	85
4	12	48
3	1	3
2		
1		
Total	30	136
Average		4,53

Putting it into the Likert percentage formula:

$$P = \frac{136}{150} x\ 100\% = 90{,}6\ \% \quad (5)$$

Then, for the percentage result in questionnaire number 1 is 90,6%, this means excellent.

5. Does the 3D Model of the Port match the original?

Table 7. Questionnaire Number 5

Score	Respondents	Total Score
5	10	50
4	15	60
3	5	15
2		
1		
Total	30	125
Average		4,16

Putting it into the Likert percentage formula:

$$P = \frac{125}{150} x\ 100\% = 83{,}3\ \% \quad (6)$$

Then, for the percentage result in questionnaire number 1 is 83.3%, this means good.

6. Can learning through 3D simulation help indirect learning (do it by yourself)?

Table 8. Questionnaire Number 6

Score	Respondents	Total Score
5	18	90
4	11	44
3	1	4
2		
1		
Total	30	138
Average		4,6

Putting it into the Likert percentage formula:

$$P = \frac{138}{150} x\ 100\% = 92\ \% \quad (7)$$

Then, for the percentage result in questionnaire number 1 is 92%, this means excellent.

7. Are the features in the VR simulation application as expected?

Table 9. Questionnaire Number 7

Score	Respondents	Total Score
5	7	35
4	14	56
3	7	21
2	2	4
1		
Total	30	116
Average		3,86

Putting it into the Likert percentage formula:

$$P = \frac{116}{150} x\ 100\% = 77{,}3\ \% \quad (8)$$

Then, for the percentage result in questionnaire number 1 is 77.3%, this percentage is small because the bridge simulator project is still in the early stages and in the future there are still many features that will be developed. Here are the total results and percentages for each question:

Table 10. Questionnaire Score Result

Question Number	Respondent 1	2	3	4	5	Score	Percentage
1				7	23	143	95,3 %
2			2	13	15	133	88,6 %
3			7	18	5	118	78,6%
4			1	12	17	136	90,6%
5			5	15	10	125	83,3%
6			1	11	18	138	92%
7		2	7	14	7	116	77,3%
Total Score & Percentage						**909**	**87%**

Based on the results of the cadet assessment questionnaire consisting of 7 questions with 30 respondents, it can be concluded that the VR application for Bridge Simulator is included in the very good category because it has a total score of 908 which is between 840 to 1050, or 87% in percentage calculations.

Previously, must determine the limitations or determinants to see which scale will judge the quality of the application. Here are the steps to determine the determinant:

1. Determine the maximum value, by multiplying the value of the largest answer by the total number of questions and multiplying by the total number of respondents.
   5 x 7 x (30 responden) = 1050
2. Determine the minimum value, by multiplying the smallest answer value by the total number of questions and multiplying by the total number of respondents.
   1 x 7 x (30 responden) = 210
3. Determine the median or quartile II value, by adding up the largest value with the smallest value and then dividing by two.
   ((1050+210))/2 = 630
4. Determine the value of the first quartile, by adding up the smallest value with the median value and dividing by two.
   ((210+630))/2 = 420
5. Determine the value of the third quartile, by adding up the largest value with the median value and dividing by two.
   ((1050+630))/2 = 840

For an explanation of each determine, see Table 11 below.

Table 11. App Rating Category

Excellent	Quartile III ≤ x ≤ Maximum Value	840 – 1.050
Good	Median ≤ x < Quartile III	630 - 840
Enough	Quartile I ≤ x ≤ Median	420 - 630
Bad	Minimum Value ≤ x ≤ Quartile I	210 - 420

VR application for Bridge Simulator is included in the excellent category because it has a total score of 908 which is between 840 to 1050.

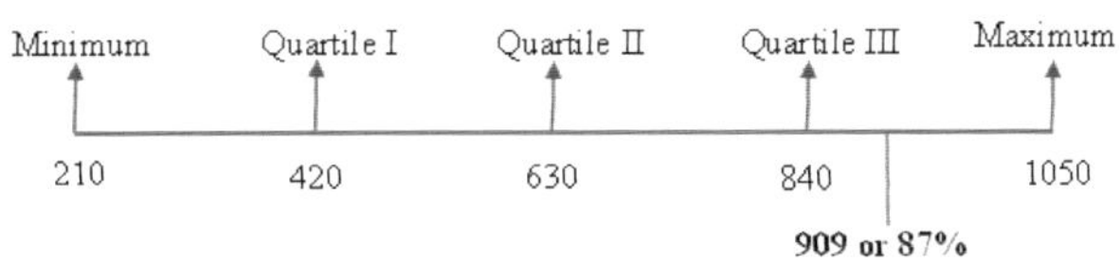

Figure 8. Final Score Result of the Questionnaire

4. Conclusion

The conclusion drawn from the ship bridge simulator application's results is that it is very good, the result provides ship bridge items and looks at the environment around the port of Tanjung Emas Semarang. In this ship bridge simulator application, respondents or users can steer the ship although it is only limited to the forward and backward of the ship and left and right maneuvering using the steering wheel.

Based on the purpose of this study, it can be concluded that the application of this ship bridge simulator can provide education for Surabaya Poltekpel cadets how to steer a ship. With the ship's bridge simulator, it becomes more practical and flexible because it can be used in various places. And based on the results of the user questionnaire, this application is in the very good category because the total score is 908, which is between 840 (quartile III) to 1050 (maximum), which is included in the very good category. However, there must be further development of the application with the reduction of bugs that still sometimes happened.

Acknowledgements

The authors thank to Digital Marine Operation and Maintenance Laboratory, Department of Marine Engineering, and Institut Teknologi Sepuluh Nopember for supporting facilities this research.

References

[1] Blazauskas, T., & Gudoniene, D. (2020). Virtual reality and augmented reality in educational programs. New Perspectives on Virtual and Augmented Reality: Finding New Ways to Teach in a Transformed Learning Environment, 82–94. https://doi.org/10.4324/9781003001874-6

[2] Budi, A. S., Sumardani, D., Muliyati, D., Bakri, F., Chiu, P.-S., Mutoharoh, M., & Siahaan, M. (2021). Virtual Reality Technology in Physics Learning: Possibility, Trend, and Tools. Jurnal Penelitian & Pengembangan Pendidikan Fisika, 7(1), 23–34. https://doi.org/10.21009/1.07103

[3] Damar, M. (2021). Metaverse Shape of Your Life for Future : A bibliometric snapshot. December.

[4] Gouveia, L. B., & Pessoa, U. F. (2014). A Visualisation Design for Sharing Knowledge Luís Manuel Borges Gouveia Submitted for the degree of. December 2001.

[5] Gugenheimer, J., Dobbelstein, D., Winkler, C., Haas, G., & Rukzio, E. (2016). FaceTouch: Touch interaction for mobile virtual reality. Conference on Human Factors in Computing Systems - Proceedings, 07-12-May-2016(September 2018), 3679–3682. https://doi.org/10.1145/2851581.2890242
[6] Hita, I putu agus dharma, D. (2020). Jurnal Menssana, 5 No.2, 146–156.
[7] Janssen, J., Erkens, G., Kanselaar, G., & Jaspers, J. (2007). Visualization of participation: Does it contribute to successful computer-supported collaborative learning? Computers and Education, 49(4), 1037–1065. https://doi.org/10.1016/j.compedu.2006.01.004
[8] Kinateder, M., Ronchi, E., Nilsson, D., Kobes, M., Müller, M., Pauli, P., & Mühlberger, A. (2014). Virtual reality for fire evacuation research. 2014 Federated Conference on Computer Science and Information Systems, FedCSIS 2014, 2014-Janua, 313–321. https://doi.org/10.15439/2014F94
[9] Layanan, P., Yang, S., & Ke, B. (n.d.). 1), 2), 3).
[10] Mayssara A. Abo Hassanin Supervised, A. (2014). Tesis Bab 1. Paper Knowledge. Toward a Media History of Documents, 1–15.
[11] Mystakidis, S. (2022). Metaverse. Encyclopedia, 2(1), 486–497. https://doi.org/10.3390/encyclopedia2010031
[12] Nisa, K., Triwoelandari, R., & Kosim, A. M. (2018). Upaya peningkatan hasil belajar siswa pada mata pelajaran pai di kelas vii smp islam al qudwah dengan model pembelajaran inkuiri. Jurnal Mitra Pendidikan, 2(10), 1063–1077. https://doi.org/10.13140/RG.2.2.28873.49762
[13] Olivan, J. (2022). TRENDING SCIENCE : What is the metaverse , and why is Facebook planning to hire 10 000 in the EU to build it ? 1–3.
[14] Pantelidis, V. S. (2010). Reasons to Use Virtual Reality in Education and Training Courses and a Model to Determine When to Use Virtual Reality. Themes in Science and Technology Education, 2(1–2), 59–70.
[15] Pitana, T., Prastowo, H., & R, B. D. (2020). The Application of Virtual Reality Technology for Introduction to Deck Machineries and Navigation Deck. x(x), 0–5.
[16] Rahm, L. (2021). Education, automation and AI: a genealogy of alternative futures. Learning, Media and Technology, 0(0), 1–19. https://doi.org/10.1080/17439884.2021.1977948
[17] Sellberg, C. (2017). Simulators in bridge operations training and assessment: a systematic review and qualitative synthesis. WMU Journal of Maritime Affairs, 16(2), 247–263. https://doi.org/10.1007/s13437-016-0114-8
[18] Szklarski, A., & Sniegocki, H. (2014). The implementation of a bridge simulator for students training Implementation of bridge simulator in students training. October.
[19] Taluke, D., Lakat, R. S. M., Sembel, A., Mangrove, E., & Bahwa, M. (2019). Analisis Preferensi Masyarakat Dalam Pengelolaan Ekosistem Mangrove Di Pesisir Pantai Kecamatan Loloda Kabupaten Halmahera Barat. Spasial, 6(2), 531–540.
[20] Zghyer, R., & Ostnes, R. (2019). Opportunities and Challenges in Using Ship-Bridge Simulators in Maritime Research. Proceedings of Ergoship 2019, 0, 119–131.

IOP Conf. Series: Earth and Environmental Science 1166 (2023) 012050 doi:10.1088/1755-1315/1166/1/012050

Session Assessment of Seaplane Operations: The Case of Marine Tourism Development on Gili Iyang Island, Indonesia

N Shabrina[1], D Arianto[1], K S Wardani[2], Z Qonita[3], A R Ispandiari[1], Hariyanto[1], N I Gutami[1] and M F Manti[1]

[1] Research Center for Transportation Technology, BRIN Serpong, Banten 15314, Indonesia
[2] Research Center for Hydrodynamic Technology, BRIN Mlati, Yogyakarta 55284, Indonesia
[3] Research Center for Geological Disaster, BRIN Bandung, Jawa Barat 40135, Indonesia

Corresponding author: nurul.shabrina@brin.go.id

Abstract. The development of marine tourism in Indonesia aims to promote economic growth, enhance the well-being of the citizenry, and conserve the environment. Gili Iyang Island in East Java Province has the second highest oxygen concentration in the world and is a popular health tourism destination. The transportation industry plays a crucial role in facilitating access and connectivity between islands and other regions. Seaplanes are an alternative mode of transportation for reaching tourist destinations on small islands with limited land area that are surrounded by water. This study aims to determine the optimal conditions for reaching the island and the number of effective days for seaplane operations. In assessing seaplane operating factors, water conditions such as current speed (1), wind speed (2), waves height (3), and tides (4) have a significant impact. Therefore, seaplanes cannot effectively operate for an entire year. Using a comparative methodology, the characteristics of the seaplane mode of transportation in the waters of Gili Iyang will be examined to determine the optimal operational time. The results of the study indicate that the 19-seat seaplane can operate 258 days per year and recommend the use of air-sea transportation modes for the growth of tourism in Gili Iyang Island.

1. Introduction

Gili Iyang Island is one of the tourist destinations in Sumenep Regency, East Java Province (Indonesia) that has not been fully developed as of yet (Figure 1.). The oxygen content of this island is the second highest in the world. Oxygen is the most abundant element in our bodies, as is well known. Due to this element's abundance in the island's environment, a great number of its inhabitants enjoy a long life span. LAPAN (2006)'s field measurements and analysis indicate that oxygen levels in Gili Iyang are normal at certain times, particularly in February, when concentrations of 20.9% are observed. [1]. In 2011, the Environment Agency and Development Planning Agency at Sub-National Level (Bappeda) of Sumenep Regency discovered that Gili I contained no more than 26.5 percent carbon dioxide and had a noise level of only 36.5 decibels [2]. The Indonesian government will provide special treatment for health tourism on this island in 2022 by enhancing its accessibility and connectivity to Gili Iyang. [3].

Small island marine tourism and connection are frequently distinguished by a constrained selection of services, dependence on basic transportation infrastructure, and resource constraint. The essence of islands is defined by the junction of the landscape and seascape [4][5]. On small islands, though, there are clear seasonal differences in the tourism industry. Whether a small island is in a warm or cold ocean environment affects its seasonal fluctuations. The warmer months, when access to small islands is less restricted by the weather, the environment, and the availability of transportation services, are likely to see seasonal maxima in both scenarios. Visitation patterns in warm-water tropical destinations are influenced by seasonal meteorological conditions, which are at their height outside of hurricane or cyclone season [6][7].

Figure 1. Gili Iyang Island and surrounding

In this paper, we provide recommendations for alternative modes of transportation that would improve the accessibility and connectivity of Gili Iyang island. The chosen mode of transportation is a seaplane or amphibious vehicle because it can reach small islands, as the Indonesian government has planned. Accordingly, information regarding the effective operational time and performance of the seaplane must be compiled in accordance with the necessary standards (safety, environmentally friendly, comfort, security, and health).

2. Method

Comparative analysis is conducted primarily to explain and gain a better understanding of the causal processes underlying the emergence of an event, characteristic, or relationship, typically by comparing variations in the explanatory variable or variables. [8]. Comparative research has a long history and has received increased attention in contemporary research due to globalization, technological advancements, etc. on cross-national platforms. [9]. Therefore, comparative research is a type of research that compares two or more groups of a particular variable. In this study, the characteristics of certain groups of amphibious aircraft are compared to the water conditions on the island of Gili Iyang. Whereas the relationship between the two groups' variables will determine the optimal operational session for Seaplane transportation.

According to [10], comparative research is a subtype of descriptive research that focuses on elements that contribute to the occurrence or emergence of a specific event in order to find fundamental explanations for cause and effect.

2.1. The characteristics of seaplane or amphibious aircraft

There are distinctions between seaplanes and amphibious aircraft. Amphibious aircraft are aircraft that can take off from a runway on land and land on the water surface (sea/lake/river) or vice versa. This aircraft type features a float and landing gear. In contrast, Seaplane is an aircraft that can only take off and land on the water's surface because it lacks landing gear. [11].

Figure 2. Aircraft N219A (PT. Dirgantara Indonesia)

Figure 3. DHC-6-300 (Amphibious) (AIRFAST Indonesia)

Through a national research priority program, Indonesia is currently developing the 19-seat N219A amphibious aircraft (see Figure 2) and DHC-6-300 (see Figure 3). It is hoped that this program will advance the domestic aircraft industry and support remote tourism and logistics distribution.

Several areas in Indonesia are already served by Seaplane, including the Pulau Bawah Resort (Riau Islands) and Sumbawa Island (West Nusa Tenggara). In Indonesia, several operators of this type of aircraft include AIRFAST Indonesia and Trivia-Air.

a. Dimension and Capacity

Table 1 shows the dimensions and capacities of the N219A and Twin Otter Series 400 aircraft.

Table 1. Specifications of existing amphibious aircraft in Indonesia

Name of aircraft	DHC-6-300 (Amphibious)	CESSNA 208 CARAVAN
Operator	AIRFAST Indonesia	Trivia-Air
Type of aircraft	Amphibious	Amphibious
Dimensi	https://www.airfastindonesia.com/ amphibious	(https://www.globalair.com/aircraft-for-sale/specifications?specid=911)
Length	15.77 m (51 ft 9 in)	11.46 m (37 ft 7 in)
Wingspan	19.8 m (65 ft 0 in)	15.87 m (52 ft 1 in)
Height	5,94 m (19 ft 6 in)	36.02 m (14 ft 2 in)
Capacity	https://www.airfastindonesia.com/ amphibious	https://travira-air.com/page/fleet/fix-wing/9/4
Crew	2	2
Seat Capacity	19	8
Empty weight	5,670 kg (12,500 lbs)	1750 kg (4,640 lbs)
Max take-off weight	5.670 kg (12,500 lbs)	8,500 lbs
Max Landing weight	5,579 kg (12,300 lbs)	

b. Performance

Only aircraft with large dimensions, specifically the DHC-6-300, are evaluated for their performance (Amphibious). Table 2 displays the aircraft's performance.

Table 2. Performance

DHC-6-300 (Amphibious)	
Take off Distance	406 m (1,333 ft) run (land) 562 m (1,843 ft) Take off over 50 ft Obstacle (Land) 374 m (1,227 ft) Take off run (water) 599 m (1,965 ft) Take off Over 50 ft obstacle (water)
Landing Distance	320 m (Landing distance from 50 ft)

Figures 4 and 5 illustrate some of the performance characteristics of similar aircraft.

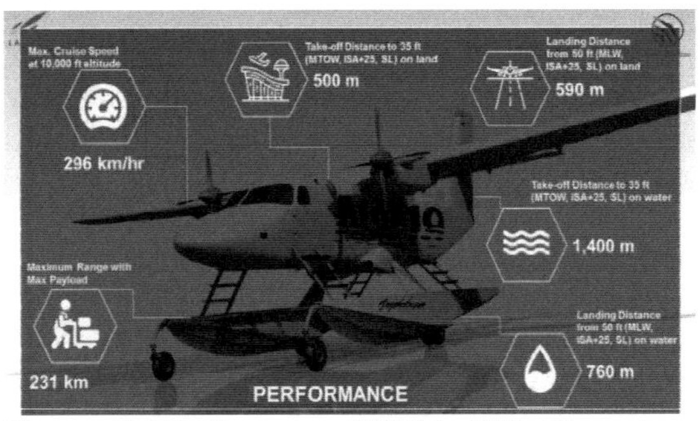

Figure 4. Performance N219A (PT. Dirgantara Indonesia)

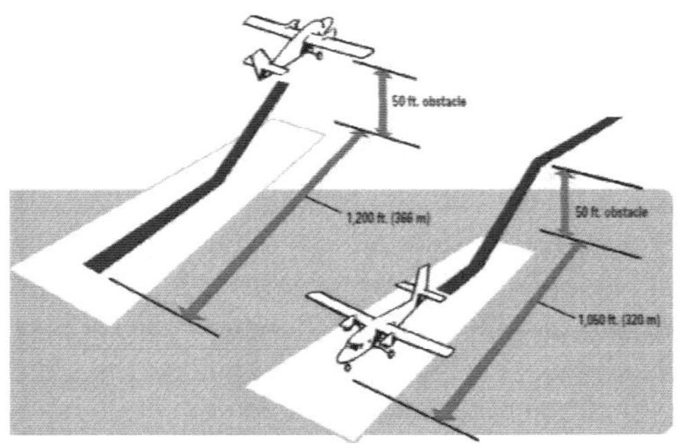

Figure 5. Performance Viking Air Twin Otter DHC 400 Canada

2.2. Gili Iyang Waters

a. Hydrology

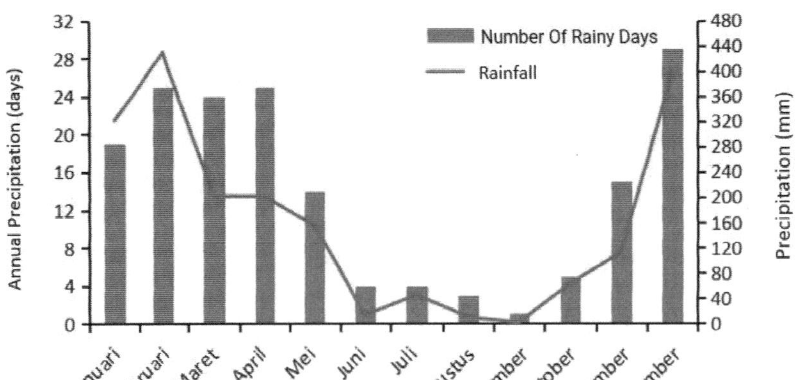

Figure 6. Rainy days and average precipitation in the Dungke District (Source: Sumenep Regency in Figures, 2021)

According to Figure 6, the rainiest days occur between December and April, with an average rainfall of 240 to 400 mm. During the middle of the year, between June and September, precipitation is typically low.

b. Wind

In coastal engineering, wind is the primary source of waves that influence coastal processes. It is difficult to obtain wave data over an extended period of time. In addition, wave measurements for the design of coastal structures are ineffective. In terms of both time and money, its inefficiency is observed. Long-term wave information is required for planning purposes (at least 5-10 years). This is difficult because measuring waves over an extended period of time is not cost-effective and may also be difficult. Because the wind is the source of waves, wave data is calculated using wind data for design purposes. The wind data used in this study are the closest available to the study site. The collected wind data is represented as wind roses.

The wind data is categorized according to the west monsoon (December to February), the transition I monsoon (March to May), the east monsoon (June to August), and the transition II monsoon (September to November) (September – November).

Based on the hourly wind speed and direction gathered by BMKG Kalianget from 2013 to September 2021. The data is categorized by west season (December - February), transition season I (March-May), east season (June - August), and transitional season II (September - November) (September – November). Listed below are the wind roses for each season.

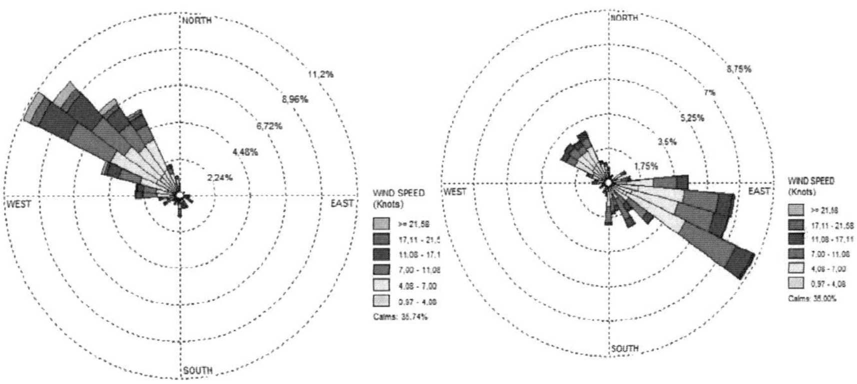

Figure 7. West and Transitional I monsoon, analysis results for the year 2022.

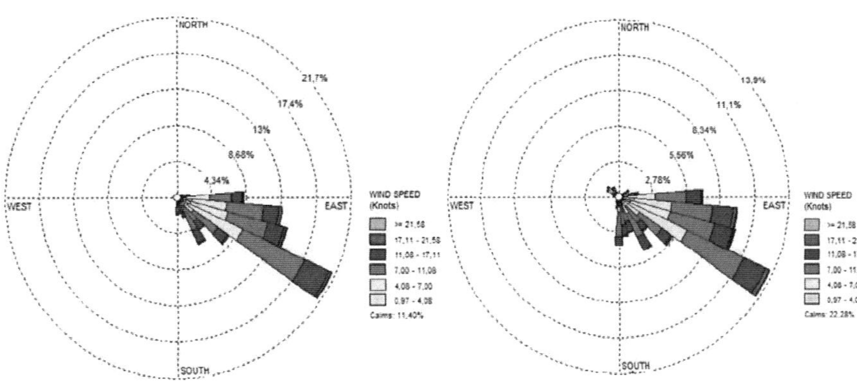

Figure 8. East and Transitional II monsoon, analysis results for the year 2022.

The wind direction is generally from the southeast with an average speed of 5.48 knots, as depicted in the Figure 7 and 8.

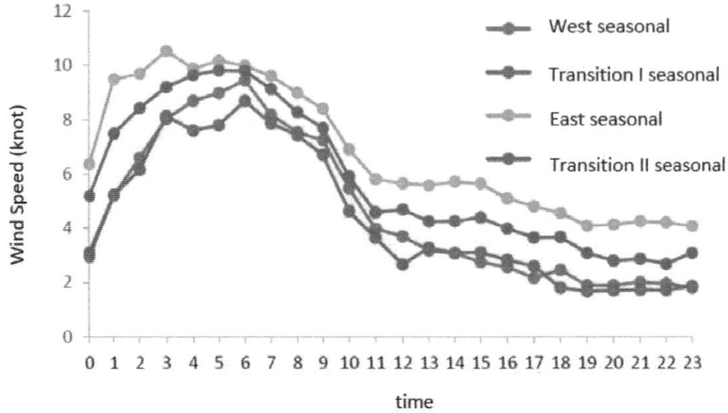

Figure 9. Daily Wind Speed on an Hourly Basis. Analysis Results 2022.

According to weather and wind conditions and a graph of hourly wind speeds over the course of a day (Figure 9), the waters surrounding Gili Iyang have low wind characteristics. According to the data, wind speeds between 6,952 and 10,428 knots are optimal for seaplane and amphibious aircraft operations.

c. Waves

Wave information was retrieved from European Centre for Medium Range Weather Forecasts Reanalysis v5 (ERA5). The data used spans the years 2013 to 2021. The classification of wave data follows the same method as wind data grouping. Figures 10, 11, 12, and 13 display seasonal wave roses.

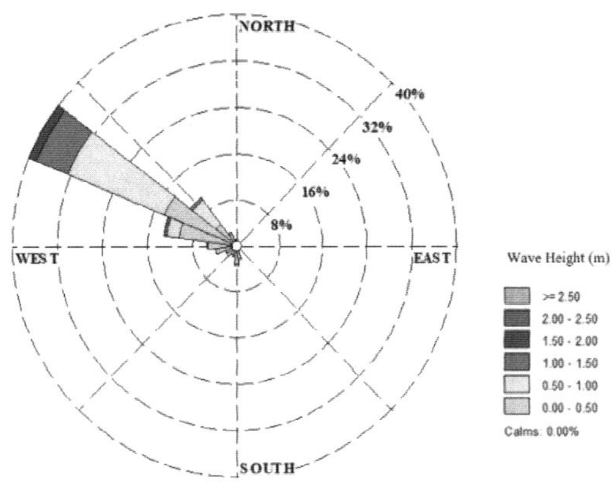

Figure 10. West monsoon

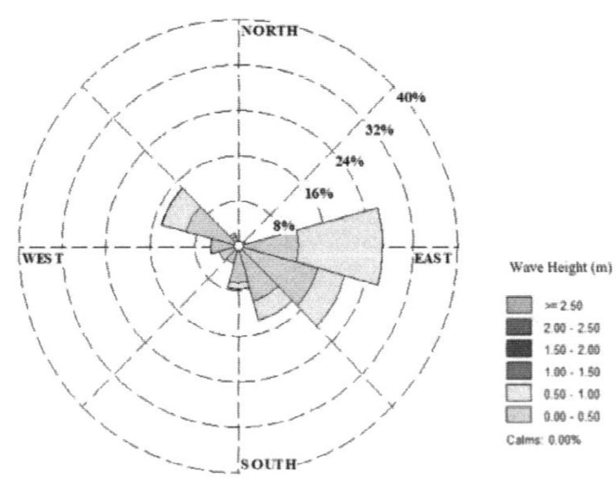

Figure 11. Transitional I monsoon

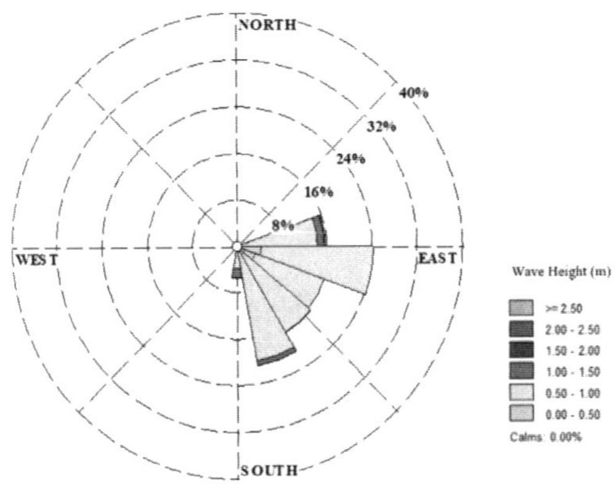

Figure 12. East monsoon

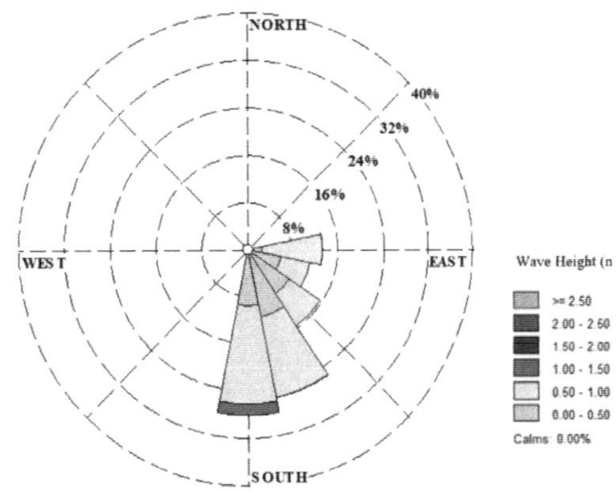

Figure 13. Transitional I monsoon

d. Tides

Tides data in Gili Iyang waters were obtained from the Surabaya Perak Maritime Meteorological Station, BMKG. Gili Iyang's waters are characterized by mixed tides (mixed tides prevailing semidiurnal). In general, tidal fluctuations occur from June to September and from December to March, with the highest tide reaching 2 meters and the lowest low tide reaching -1 meters. In April to May and October to November, tidal fluctuations are significantly lower, with a maximum tidal value of 0.9 meters and a minimum tidal value of -0.8 meters at low tide. Similarly, the beginning of 2016 exhibited a downward trend relative to prior years (see Figure 14).

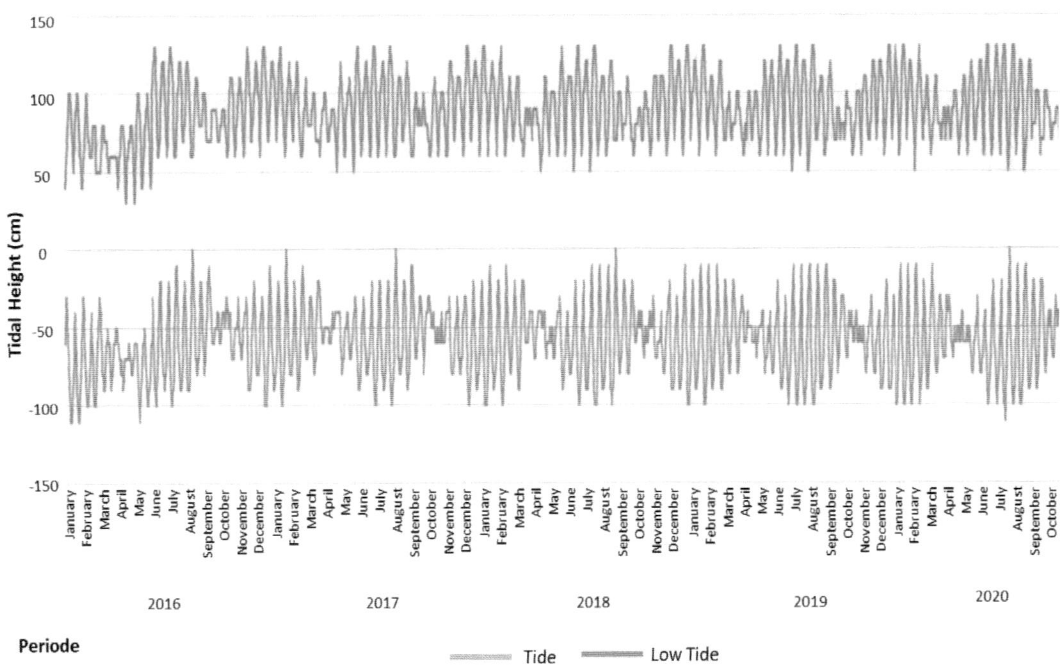

Figure 14. Tides in Gili Iyang Waters (2016-2020)

2.3. Seaplane operational infrastructure and supporting requirements

a. Landing

There are recommendations for infrastructures that accomodate the landing operation. These are shown in Table 3.

Table 3 . Recommendations for Waterway Size, Water Depth, and Safe Operational Distance (meters). Seaplane Bases. 150/5395-1B

Minimum Length (*, **)	Minimum Width	Minimum Water Depth (***)	Explanation
2.500	200	3 (****)	Minimum dimensions for water-based aircraft operations. Minimum approach distance of approximately 20:1 for at least two miles.
3.500	300	4 (****)	Minimum dimensions for commercial aircraft operation are restricted. The minimum approach distance for a minimum departure ratio of 40:1 is 2 miles.
5.000	500	10	Minimum dimensions for more commercial aircraft operations. Minimum approach distance for a minimum departure ratio of 40:1 is 2 miles.
10.000	700	15	Limitless dimensions for operations. Minimum departure approach distance of 50:1 for a minimum of two miles

Notes:
(*) The written length dimensions are applicable when the water is calm, there is no wind, and the temperature at sea level is 15 degrees Celsius.
(**) Temperatures over 15°C during the hottest month on average over the course of a year can cause a 1.5-percentage-point increase in the rate of length change.
(***) The specified dimensions must be increased by 7% for every 1,000-foot (300-meter) increase in elevation above sea level to account for changes in density.
(****) The minimum required water channel depth is 3 feet (1 m), but 6 feet (1.8 m) is preferable (1.8m).

b. Take off

Information/handbook on pilot operations for determining the optimal distance required for seaplane landing and take-off operations is to provide a distance of approximately 50 feet above the water surface from any objects/obstacles that may exist in the Maximum Take-Off Weight (MTOW) for aircraft that are used (see Figure 15). The 50-foot distance from the nearest object (50-foot obstacle) will vary depending on the method of approach, whether by sight or by equipment. (Author is referring to AC 150/5300-13).

- If the seaplane landing is based solely on visual factors, then the distance of 50 feet from the nearest object is 1,000 feet, with a ratio of 20:1 between the viewing distance and landing conditions.
- If the seaplane landing is equipped according to procedures, the obstacle distance is assumed to be approximately 2,000 feet from the landing site's boundary, with a landing gear ratio of 40:1.
- Shorter and closer objects will exceed the 20:1 or 40:1 ratio and require additional calculations.

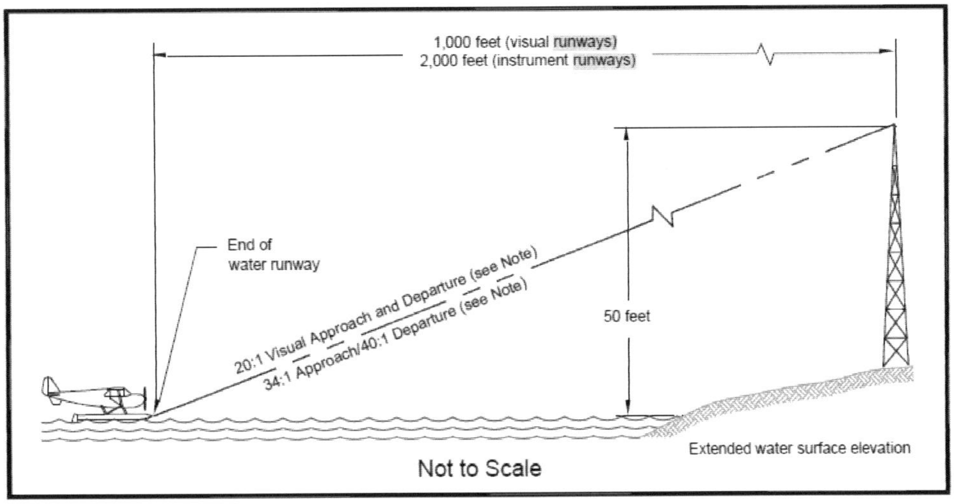

Figure 15. Clearance of Obstacles on Water Runways. (FAA) Document No. 150/5395-1

c. *Minimum size specifications for waterways and a turning basin*

The US Department of Transportation (FAA) 150/5395-1A [11] recommends the following based on the operational characteristics of seaplanes that can operate on land and in water (see Table 4):

Table 4 . The dimensions of the waterway, the water depth, and the safe landing distance with a sloping approach are as follows:

Minimum Seaplane Dimensions	Minimum Dimensions of Waterways	Minimum Turning Basin Dimensions
Length 2500 feet (750m) Minimum Width 200 feet (60m)	Minimum Width 100 feet (30m) Depth (Seaplane single-engine) minimum 3 feet (1 m) or better 6 feet (1.8m)	Diameter 200 feet (60m)

d. *Operational Requirements for Seaplanes Based on Wave, Current, Wind, and Tidal Conditions [11]*

d.1 Water Surface Conditions.

Wave height and the presence of floating debris should be considered in any assessment of the water's surface characteristics. Investigating the various water conditions at the intended site is necessary since open or unprotected water operating regions may become so rough as to be inoperable under specific wind and current conditions. The ideal conditions are when there is a modest amount of surface disturbance, with waves or ripples that are between 3 and 6 inches tall. With twin floats, the typical light seaplane (3,000 pounds or less) can typically operate safely in seas that are 15 inches from crest to trough, whereas seas that are 18 inches will impede normally safe operations. Larger, hull-type, float-equipped aircraft that weigh between 3,000 and 15,000 pounds can typically fly safely in waves up to 2 feet from crest to trough. The difficulties encountered when lifting the floats or drawing off the water during take-off makes smooth or dead calm water the reverse of what is desired. The last stage is to determine whether or not there is floating debris. For extended periods of time, stay away from regions with an offensive amount of debris, or get the debris cleared.

d.2 Currents flow.

Landing and takeoff zones should be located when current speeds are less than 3 knots (5 mph). However, any taxiing operation between the sea lane (or water operating area) and the shoreline facilities would normally require the help of a surface boat. Landing and take-off operations can be carried out in currents of more than 6 knots. Currents greater than three knots can be challenging to handle when seaplanes are in slow taxiing mode while approaching piers or floating docks, or when executing beaching operations such as ramps. The current should therefore ideally flow away from the dock or floating docks in this situation. In some cases, pilots might be able to use favourable prevailing winds to combat unfavourable currents. Locations of the following kinds should be avoided:

1. Where currents are more than 6 knots (7.0 mph);
2. Where extraordinary water turbulence is brought on by a river's acute bend, the meeting of two currents, or an area where tide rips are common.

d.3 Water Levels.

The handling of seaplanes at the coast or waterfront will typically need the usage of floating structures or moderately sloped beaching accommodations if the change in water levels exceeds 18 inches. In locations where water level differences are greater than 6 feet, special or extended developments must be built to support seaplanes. These projects can require a dredging channel, lengthened piers, or specialized lifting apparatus, depending on the slope of the shore. The need for facilities will increase as water variability increases. It is advised that designers reference a list of tide ranges that can be expected at various coastal areas throughout the United States to handle these considerations.

3. Result and discussion

The Seaplane's operational characteristics differ from those of conventional aircraft, including the absence of brakes. In order to ensure the safety and accuracy of the landing area, the Seaplane pilot evaluates characteristics such as the water surface that will be used as the landing area (wind direction and speed and landing direction) as well as other uses of the water area. Consequently, water characteristics such as water depth, wind, and tidal conditions will significantly influence the seaplane's operational capability.

Table 5. The wave height for each season of the year. Analysis results for 2022

Season	Maximum waves height (m)	Minimum waves height (m)
West	2.064	0.138
Transitional I	1.150	0.130
East	1.783	0.235
Transitional II	1.321	0.145

The maximum wave height in the west season is 2,064 meters and the minimum wave height is 0.138 meters, according to Table 5. The maximum wave height in the transitional season I is 1.15 meters and the minimum wave height is 0.13 meters, and in the east season the maximum wave height is 1,783 meters and the minimum wave height is 1.783 meters. The wave height is 0.235 meters, while the transition season II maximum wave height is 1.32 meters and the minimum wave height is 0.145 meters.

Table 6. Percentage of wave occurrence proportional to wave height

Interval (m)	Data	Percentage (%)	Days
0.0 – 0.6	55064	70.46	258
0.6 – 1.0	20857	26.7	97
1.0 – 1.5	1974	2.52	9
>1.5	249	0.32	1
Sum	78144	100	365

According to Table 6, the dominant wave height is with an interval of 0 to 0.6 meters by 70.46 percent, followed by the wave height with an interval of 0.6 to 1.0 meters by 26.7 percent, then the wave height with an interval of 1.0 to 1.5 meters by 2.52 percent, and finally the wave height with an interval of 1.0 to 1.5 meters by 2.52 percent with an interval of more than 1.5 meters by 0.32 percent.

Using previously analyzed data on wave height, the number of days based on wave height can be determined. On the basis of the interval, the number of days the wave height occurs can be calculated as follows:

Number of occurrences of wave height interval = 55064 data

Number of wave height data = 78144

Percentage of occurrence of wave height $= \frac{55064}{78144} \times 100\% = 70.46\%$

Number of days of the occurrence of wave height $= 70.46\ \% \times 365 = 258$ days

Based on an analysis of wave data, 55,064 out of 78,144 waves are safe for seaplane operations, equating to 258 days of safe seaplane operations per year. Detail results for comparative analysis in Table 7.

Table 7. Comparative analysis

Data	**Comparative analysis**		**Correlation**
	Seaplane specifications	**Waters condition in Gili Iyang**	
Wind	6.952 – 10.428 knot	5.48 knots. Wind speeds of > 5 knots are common from May to October, whereas wind speeds of 5 knots prevail from November to April.	
Waves	Max 0.6 m (2ft)	West monsoon max 2.064 min 0.138 Transitional I max 1.15 min 0.13 East monsoon max 1.783 min 0.235 Transitional II max 1.321 min 0.145	Min 0.13m – max 2.064m. The dominant wave height is between 0 and 0.6 meters, which is 70.46 percent optimal for Seaplane flights annually.
Tides	18 inches – 6 feet with certain condition	Min 0.13m – max 2.064m. The dominant wave height is between 0 and 0.6 meters, which is 70.46 percent optimal for Seaplane flights in a given year. June-September and December-March have maximum and	Whereas if change in water levels exceeds 18 inches (0.45 m), floating structures or moderately sloping beaching will be necessary. Where water-level variations exceed 6 feet (1.82 m), seaplanes must be accommodated

Data	Comparative analysis: Seaplane specifications	Comparative analysis: Waters condition in Gili Iyang	Correlation
		minimum values of 2 m and -1 m, respectively. Maximum value of 0.9 m and minimum value of -0.8 m in April-May and October-November.	through the construction of specialized or extended infrastructure.
Current flow	> 6 knot (7.0 mph)	0-20cm/s at the surface and 60cm/s at the seafloor; additionally, it is known that the direction of the current changes at different depths and rotates between 7.5 and 10 meters below sea level.	

4. Conclusion

The potential of water areas for the development of marine tourism in Gili Iyang with the operation of amphibious/seaplane aircraft is high, particularly during times when the water conditions allow for the safe take-off and landing of amphibious/seaplane aircraft. Twin otter models, with a head-to-head length of N219A, a width of 16.49 meters, a length of 19.5 meters, a height of 6.18 meters, and a weight of 4,309 kilograms, are the type of seaplanes that are suitable and commonly used in Indonesia.

Water conditions such as current speed (1), wind speed (2), wave height (3), and tide (4) have a significant impact on the evaluation of seaplane operating factors. Therefore, seaplanes cannot operate effectively, but they can operate for up to 258 days per year outside of their annual service time. As long as the amphibious aircraft is unable to operate optimally, other sea transportation options can be utilized.

For the provision of port water areas as well as infrastructure and superstructure facilities for the development of a waterbase/harbour water/water aerodrome in Gili Iyang, specifically for: Runway for take-off and landing purposes; taxi lanes; movement room; dock; anchoring/mooring places; access; passenger waiting facilities; etc., the area available is sufficient to meet the requirements. Therefore, there is a great deal of potential in the Gili Iyang region for the development of infrastructure for public interest and tourism. A region that is integrated with the nearest seaports on Gili Iyang Island, namely Kalianget Port, Dungkek Port, Ju'aji Port, and the port on the island of Bali.

Acknowledgement

We are grateful to the Ministry of Transportation's Research and Development Agency for allowing us to investigate the findings of the "Waterbase/Water Harbour Survey in Giliyang for Amphibi/Seaplane Aircraft Operation" conducted in 2021. We would also like to thank our colleagues from the National Research and Innovation Agency's Research and Development Center for Sea, River, Lake, and Ferry Transportation, despite our disagreements regarding the interpretation/conclusion of this paper, for their invaluable assistance. We are grateful to the Regional Resource Development Technology Center (PTPSW) BRIN for their assistance with the survey of data collection CATs (Coastal Acoustic Tomography System), which were vital to the research analysis process.

5. Reference

[1] S. Pusat, S. Dan, and T. Atmosfer, "Study of Oxygen Health Tourism Potential in Gili Iyang," Kaji. Potensi Wisata Kesehat. Oksigen ..., pp. 83–90, 2006. (In Indonesian)
[2] Indonesian Information Portal, "Breathe the Best Oxygen to your heart's content on Gili

Iyang," 2021. https://www.indonesia.go.id/kategori/pariwisata/2768/menghirup-oksigen-terbaik-sepuasnya-di-gili-iyang? (accessed Sep. 02, 2022). (In Indonesian)

[3] Detik, "Sandiaga Promises To Develop Gili Iyang By Turning On Sumenep Flights," 2022. https://www.detik.com/jatim/wisata/d-6012520/sandiaga-janji-kembangkan-gili-iyang-dengan-hidupkan-penerbangan-sumenep (accessed Sep. 02, 2022). (In Indonesian)

[4] J. M. Cheer, S. Cole, K. J. Reeves, and K. Kato, "Tourism and islandscapes: Cultural realignment, social-ecological resilience and change," Shima, vol. 11, no. 1, pp. 40–54, 2017, doi: 10.21463/shima.11.1.07.

[5] M. Cheer, "The urgency for sustainability indicators," no. July, 2020.

[6] K. Andriotis, "Seasonality in Crete: Problem or a Way of Life?," Tour. Econ. 11(2)207-224, vol. 9, pp. 298–316, 2001.

[7] T. Cuccia and I. Rizzo, "Tourism seasonality in cultural destinations: Empirical evidence from Sicily," Tour. Manag., vol. 32, no. 3, pp. 589–595, 2011, doi: 10.1016/j.tourman.2010.05.008.

[8] C. Pickvance, "The four varieties of comparative analysis: the case of environmental regulation," Pap. Conf. Small large-N Comp. Solut. Univ. Sussex, no. September, pp. 22–23, 2005, [Online]. Available: http://eprints.ncrm.ac.uk/57/1/chrispickvance.pdf

[9] R. Azarian, "Potentials and Limitations of Comparative Method in Social Science," Int. J. Humanit. Soc. Sci., vol. 1, no. 4, pp. 113–125, 2011.

[10] M. Nazir, Research Methodology. Jakarta, Indonesia: Ghalia Indonesia, 2003. [Online]. Available: https://lib.ui.ac.id/detail?id=20336654. (In Indonesian)

[11] U.S. Department of Transportation and Federal Aviation Administration, "Advisory Circular: Seaplane Bases," no. Chapter 1, 2018.

Analysis of Location and Wave Characteristics on Capsizing of RoPax Opened-Car-Deck

Hasanudin[1], Achmad Zubaydi[1*], Wasis Dwi Aryawan[1], Aleik Nurwahyudi[3]

[1]Department of Ocean Engineering, Institut Teknologi Sepuluh Nopember (ITS), Keputih, Sukolilo, Surabaya City, East Java 60117, Indonesia

[3]National Transportation Safety Committee (NTSC) of Indonesia, Medan Merdeka Timur No. 5, Transportation Ministry Building, 3th Floor, Jakarta, Indonesia

*Correspondent Author: Zubaydi@na.its.ac.id

Abstract. RoPax is a ship that carries passengers, cargo, and cars simultaneously by waterway. RoPax with Opened-Car-Deck (OCD) type is the most popular and dominant population in archipelagos developing countries to operate: between islands, strait, and river banks. Nevertheless, the RoPax has the most contributions in casualties, so the ship needs serious intention to reduce the ship accident. Internal ship problems, lousy weather, and human errorsc contribute to shipping accidents. A capsizing ship occurs for single or multiple combination reasons. The paper discusses locations measured from the coastline, significant wave height, and period waves when a capsizing ship. The paper's research methodology: collecting data on some capsizing RoPax OCD with a range of 25 years in Indonesia water, searching wave characteristics on Copernicus Marine Service, analyzing wave characteristics with Ocean Data View (ODV), and making results. From the location of the case study investigating all capsizing RoPax near the coastline with a range of 0.12-24 miles, two capsizing RoPax OCD occurred on the river. This case suits the IMO study that almost various ships happen in the coastal area . The result of wave height characteristics shows that RoPax OCD capsizes in the low significant wave height of 0.2-1.75 meters, with nine cases in the low wave periods, two in the extended period, and a reasonable period not be founded. In conclusion, stability problems and small size cause the capsizing of RoPax OCD. Crews have more intention in onshore areas and do not carry overload.

Keywords: RoPax, Open Car Deck, Location, Wave Characteristics

1. Introduction

A Ship operates in uncertain, calm, and sometimes in bad weather. The ship must sail under all conditions: weather, area operation, and capable load suitable determined. Choosing an operation route and the weather forecast is one way to reduce ship accidents when shipping operations [1]. RoPax is the world's most famous ship that simultaneously carries miscellaneous passengers, cargo, and cars that loading-unloading use ramp doors [2]. Most RoPax are operated in developing archipelagos countries to cross the water between river banks, straits, and islands [3]. However, the RoPax is the ship that causes the most prominent casualties in the world compared to other types of vessels, so it needs serious attention to reduce fatal accident risk [4] [5]. Ship accidents will affect economic losses, marine pollution, and human casualties. Most RoPax accidents are experienced in developing countries: Indonesia, the Philippines, and Bangladesh [6]. One of the biggest types of the shipwreck is caused by stability problems that make the ship capsize at sea. For small RoPax < 60 meters, the dominant problem is related to intact stability defects, while in large RoPax > 60 meters, the dominant problem is related to damaged stability. A small

RoPax generally has an open vehicle deck (OCD) shape, and the goal is to minimize GT, minimize the ship's weight and not need the air ventilation system. The shape of the open deck of the vehicle reduces the freeboard and the high risk of water on the deck so that it has a high risk of overturning [7]. On the other hand, one of the main problems of RoPax safety is that there is no specific intact stability criterion that addresses the problem of the stability of this ship, so it needs to be investigated more intensively and regulations are made so that victims of this ship accident can be minimized.

Ship accidents are generally caused by three main factors: human error, the internal factor of the ship's feasibility conditions, and extreme weather external conditions [6]. At the time of an accident, these three factors can coincide as a single cause or in combination. Human error is the dominant factor causing ship accidents: work overload, fatigue, lack of discipline, communication errors, lack of operator knowledge, and others [8] [9]. Ship internal factors cause such as worthy ship, overload, damage, and design do not comply with rules and regulations. Factor external can be caused by heavy rain, tornado, storm, worse wave, etcetera [10]. Global warming is the present increase and difficult to predict effects in high-risk ship accidents [11]. Sea waves have random patterns but can be modeled with regular and irregular waves, characteristic of waves commonly determined by significant wave height and wave period [12]. Wave-spectra draws wave energy effects from wave height and wave period [13]. Indonesia wave characteristics had to be mapped to analyse the structure and regulation of the ship [14]. Some literature had been studied the dangerous location sea of ship accidents [13] [15] [16].

This paper studies the location of RoPax OCD capsizing and its wave characteristics so vulnerable areas, and the same incident can be ovoid for other ships. The first collection date of RoPax OCD capsizing, location, and measure the location from the coastline where the data sources get from NTSC [17]. Then analyzing, wave characteristics such as significant wave height and wave period were drawn in maps and graphs using the free software Ocen Data View [18]. Each incident is gathered, so the pattern of location and characteristics of RoPax is known. The research results are expected to be used as additional material to support the manufacture of specific criteria and regulations for RoPax OCD vessels. They can be applied in the field, so that ship crews are careful when operating in these vulnerable areas.

2. Sea Wave Algorithm

Wave characteristics are essential in the design and construction of ports, rivers, bridges, oil platforms, pipelines, coastal projects, and shipping routes [19]. Characteristics of ocean waves, especially significant wave height and wave period, are very influential on the overturning of the ship. Ocean wave patterns can be approximated by mathematical algorithms using Fourier Series Equations and Ocean Wave Spectra, which have been validated using experiments in the laboratory and records of natural waves at sea. Ocean waves occur randomly $h(\bar{r},t)$ is the sum of a collection of time-dependent sinusional waves whose amplitudes are defined as follows [20]:

$$h(\bar{r},t) = \sum_{k} \tilde{h}\left(\bar{k},t\right)e^{i\bar{k}\bar{r}} \tag{1}$$

Where $\bar{k} = (k_x, k_y)$ is 2 (two) dimensions of wave amplitude vector amplitude, $k = \sqrt{k_x^2 + k_y^2}$ is a representation of the number of waves $\bar{r} = (x,y)$ at one point and time t. A set of complex Fourier series amplitude domains and initial phase for elevation at time 0 (zero) is defined as follows:

$$\tilde{h}_0\left(\bar{k}\right) = \frac{1}{\sqrt{2}}(\xi_{Re} + i\xi_{Im})\sqrt{S_{PM}(k)} \tag{2}$$

Where ξ_{Re} and $i\xi_{\mathrm{Im}}$ are independent derivatives of generating many random generators, with average 0 and standard. S_{PM} *(k)* is an ocean wave spectra that can be modeled using the Pierson Maskovits method. In the analysis of ocean wave excitation, it is generally not taken at maximum conditions but at significant wave height conditions $(h_{1/3})$ that is 1/3 of the maximum wave height [21]. Assuming that wave is linear, sea surface waves are dispersive, and the propagation mechanism is regulated, then the relationship can be written in the form of wave frequency as follows [20]:

$$\omega\left(\bar{k}\right) = \sqrt{gk\tanh(kd)} + \bar{k}\cdot\bar{U} \tag{3}$$

Where ω is angular wave frequency, d is seawater depth, $U = \left(U_x, U_y\right)$ is a sea surface velocity vector. The frequency characteristic of an angular wave can be converted into a wave period $T\left(\bar{k}\right)$ Which is formulated as follows:

IOP Conf. Series: Earth and Environmental Science 1166 (2023) 012051 doi:10.1088/1755-1315/1166/1/012051

$$T(\bar{k}) = \frac{2\pi}{\omega(\bar{k})} \qquad (4)$$

The location, area, and characteristics of the islands affect the wave character. Islands provide more excellent protection than the open sea influence wave characteristics. Based on a study of significant wave height, islands have 6 %-9 % lower wave heights compared to the open sea [21].

3. Research Methodology

This paper uses a case study capsizing RoPax OCD due to the problem of intact stability between 1996-2021 in Indonesian water. This country is an archipelagic developing country with the largest RoPax fleet in the world and a significant contribution to ship accidents, making it ideal for research case studies. The research methodology is made in a hierarchical and sequential process to facilitate the analysis process. The methodology in this paper is made with steps as shown in Figure 1 as follows:

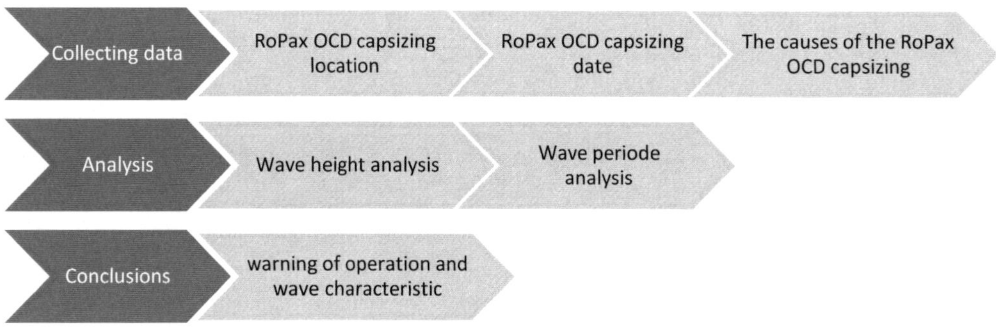

Figure 1. Paper methodology

The first step is to collect ship accident data from the results of ship accident investigations [17]. The body is the official agency of the Indonesian government in charge of investigating accidents on sea, land, and air transportation facilities. The data taken from the NTSC includes the location of the latitude and longitude coordinates where the RoPax OCD sank, data on the date the ship sank, and data on the cause of the ship's sinking due to internal or external factors. Furthermore, a reanalysis of the wave characteristics was carried out by entering the latitude-longitude coordinates. The time of the RoPax accident at Copernicus Marine Service [22]and the results were reanalyzed again using [18] for visualization and reading of the results in more detail, significant wave height, and wave period. From the overall analysis of the significant wave height and wave period, conclusions can be drawn about the dangerous location operation and wave characteristics that cause the risk of ship capsizing.

4. Discussion of RoPax OCD capsizing location

Indonesia was chosen for this case study paper because it is considered ideal for conducting research. This archipelagic country has ± 13,000 islands, has the largest RoPax fleet of ± 300 units, and is the most significant contributor to ship accidents. RoPax OCD location and reverse date data collected from 1996 to 2021 from all parts of Indonesia have been plotted on a map as shown in Figure 1 below:

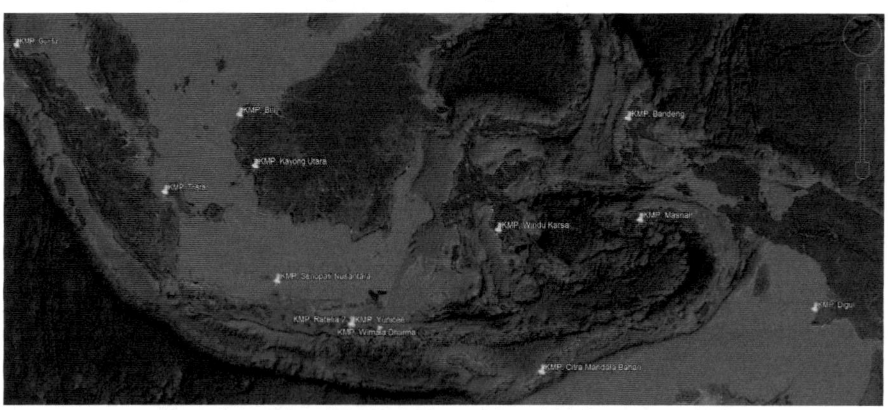

Figure 2. Location of RoPax OCD capsizing which from source [17]

In the map in figure 2 above, which has been recorded for 25 years, it can be seen that there have been 15 cases of ship accidents located almost evenly in Indonesia. In the western region, there are five incidents; in the central region, there are six incidents; in the eastern region, there are three incidents. From plotting the location of the accident onto a map of Indonesia, it was found that 13 cases of the RoPax OCD ship accident occurred not far from the coast near the port, and 2 cases of accidents occurred on the river: KMP Kayong Utara and KMP Bili. Most cases of capsizing locations at sea occurred in the Bali strait. Figure 1 is suitable with data collected by IMO and other researchers that ship accidents mainly occur in nearshore waters [23] [15] [24]. In more detail, the accident location map can be supplemented with the data in table 1 as follows:

Table 1. Data of RoPax OCD Capsiizing with the duration of 25 years

No	Ship name	Incident time	Location / distance from coastline (mile)	Capsizing reason
1	KMP. Gurita	1996/01/19	Sabang / 5.5	Overload, Weather, Stability
2	KMP. Masnait	2000/05/09	Ambon / 0.12	Weather, Stability
3	KMP. Wimala Dharma	2003/09/07	Lembar / 4.5	Overload, WoD, Stability
4	KMP. Digul	2005/07/08	Arafuru	Weather, Stability
5	KMP. Citra Mandala Bahari	2006/02/01	Kupang	Weather, Stability
6	KMP. Tristar	2006/12/28	Bangka	Weather, Stability
7	KMP. Senopati Nusantara	2006/12/28	Kumai / 24	Weather, Stability
8	KMP. Windu Karsa	2011/08/27	Lambasina	WoD, Stability
9	KMP. Munawar	2013/01/03	Selat alas / 4	WoD, Stability
10	KMP. Rafelia 2	2016/03/04	Selat Bali	Overload, Stability
11	KMP. Kayong Utara	2018/02/20	Sungai banyuasin	Overload, Stability
12	KMP. Bandeng	2018/08/19	Maluku Utara	Overload, Stability
13	KMP. Lestari Maju	2018/07/04	Selayar	Overload, Weather, Stability
14	KMP. Bili	2021/02/21	Sungai	Muatan, WoD, Stabilitas
15	KMP. Yunicee	2021/12/31	Selat Bali	Muatan, WoD, Stabilitas

Table 1 above is data on ship accidents taken from sources (KNKT, n.d.). Column 1 shows the number of accidents; the second column of the ship's name; the third column the date of occurrence; the fourth column of the accident location and distance from the beach; and the fifth column is the cause of the accident from the results of the NTSC investigation. Table 1 above shows one accident in 1996, 2000, 2003, 2005, 2011, 2013, and 2016. In 2021, there were two accidents, and in 2006 and 2018, there were three capsizings. From the media information released by the BMKG data at that time, the largest wave height was experienced by KMP Tristas, with a wave height of 3.5 m, followed by KMP. Marsanite, KMP. Wimala Dharma, KMP. Digul, KMP. Citra Mandala Bahari and KMP. Milkfish with a wave height of 2-3 m. Data on the location of the Bali strait occurred two times the incidence of the KMP ship accident. Rafelia 2 and KMP. Yunice. KMP experienced the farthest accident location. Senopati Nusantara had an accident 24 miles from the coastline. From the NTSC investigation report, the cause of most accidents occurred due to stability problems, ocean wave excitation forces, Water on Deck (WoD) problems, and overloading.

5. Discussion of Wave Characteristics

This discussion explores the characteristics of waves globally at the time of the RoPax accident in Indonesian waters, plotted in the form of a scatter map, and then discusses the characteristics of local waves in more detail at the location of the ship accident, which is displayed in graphic form. The wave characteristics discussed include wave height and wave period.

5.1. Characteristics of Global Waves

Indonesia's climatic conditions are generally influenced by the west monsoon from the Indian Ocean and the east monsoon from the Australian continent. The west monsoon brings water vapor that can cause rain, while the east monsoon brings little water vapor that can cause drought. These two strong monsoon winds can potentially affect the characteristics of sea waves in several Indonesian seas. The potential for height characteristics and wave periods change from time to time. Weather and climate conditions worldwide are continually monitored using satellites and coastal radio stations. The characteristics of ocean waves are an essential part of monitoring. Global Ocean Waves Reanalysis Wavers from Copernicus is a site that monitors weather characteristics with a 0.08 x 0.08-degree grid, which can be used to obtain wave conditions at the time of an accident. The plotting of wave height at the time of the accident can be seen in Figure 3 below:

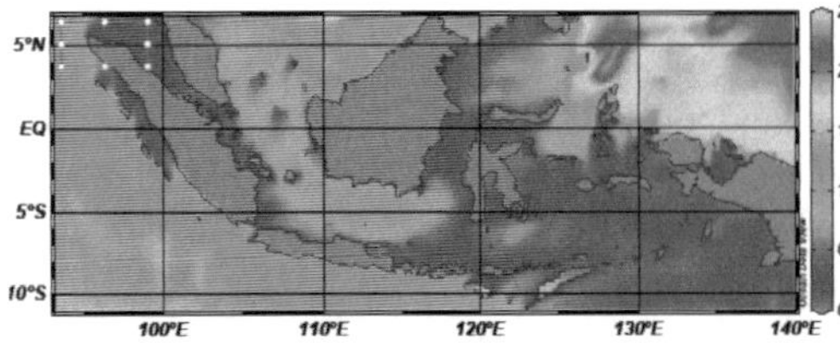

KMP. Gurita

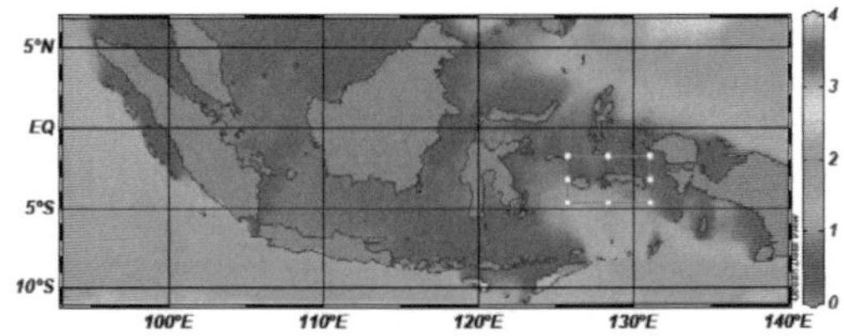

KMP. Masnait

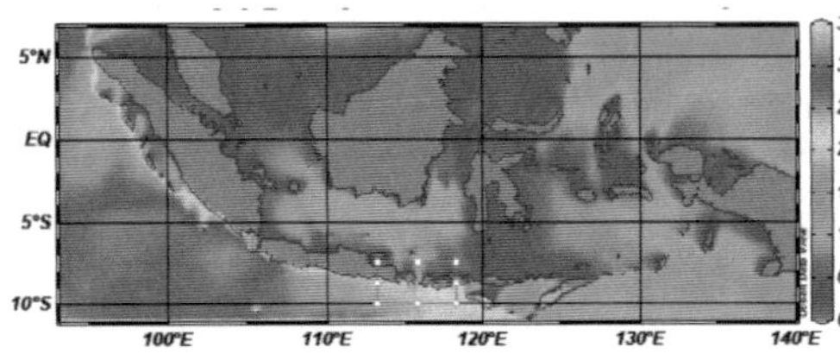

KMP. Wimala Dharma

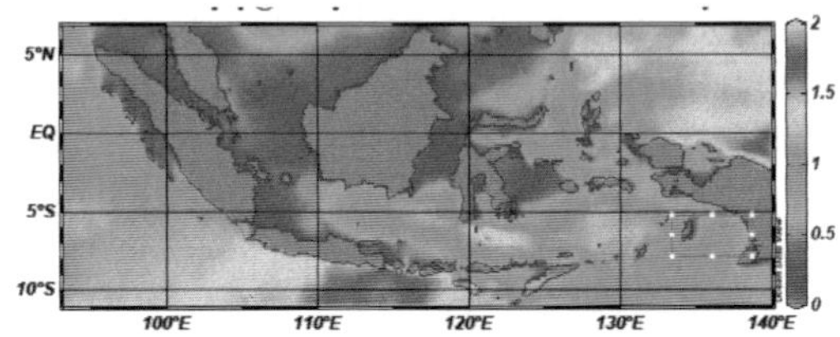

KMP. Digul

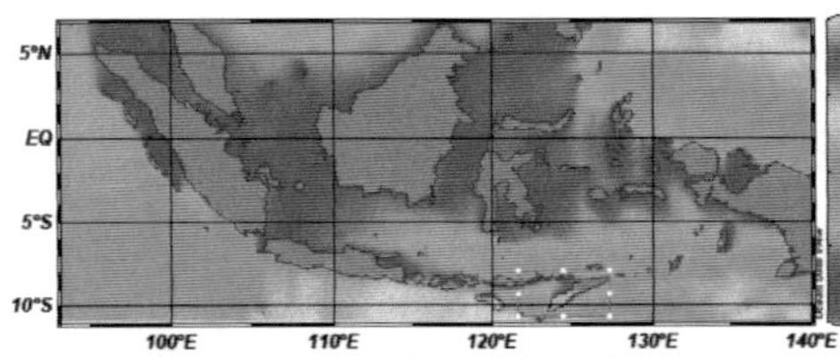

KMP. Citra Mandala Bahari

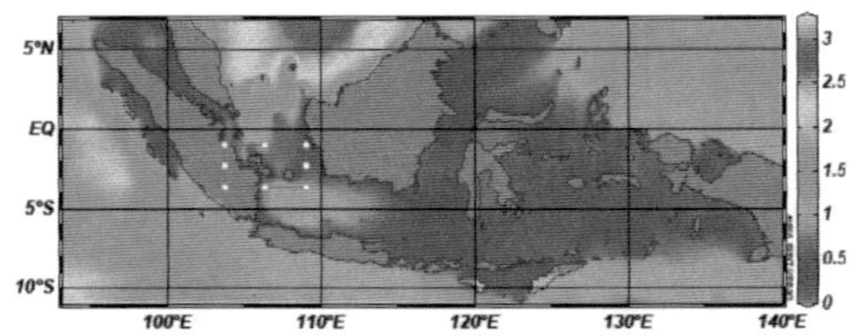

KMP. Tristar

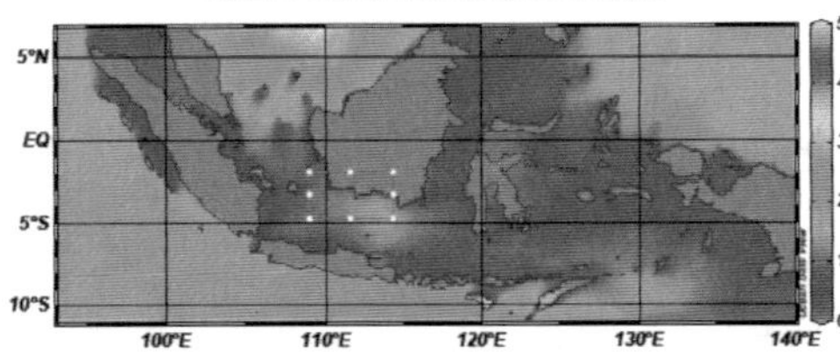

KMP. Senopati Nusantara

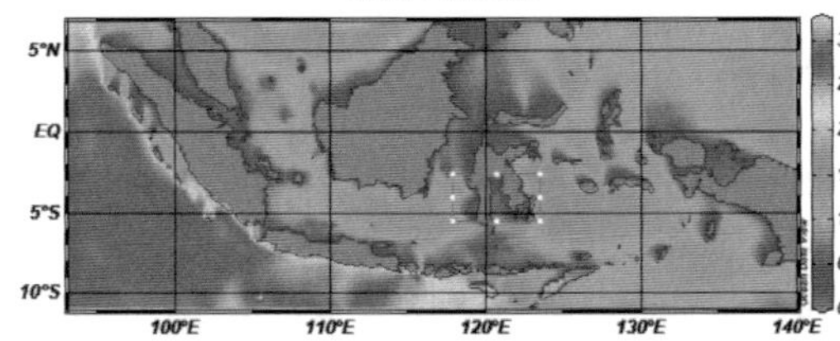

KMP. Windu Karsa

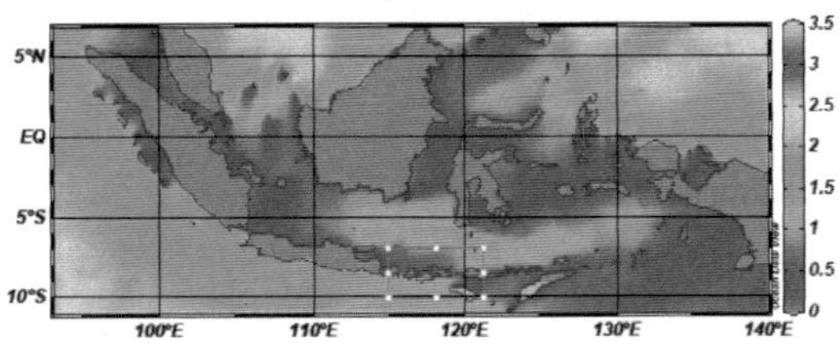

KMP. Munawar

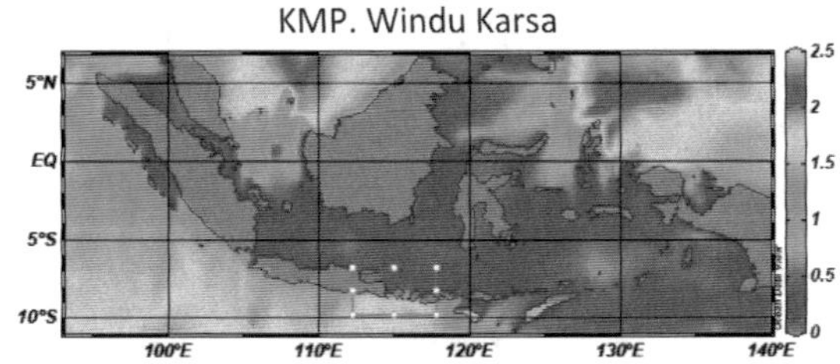

KMP. Rafelia 2

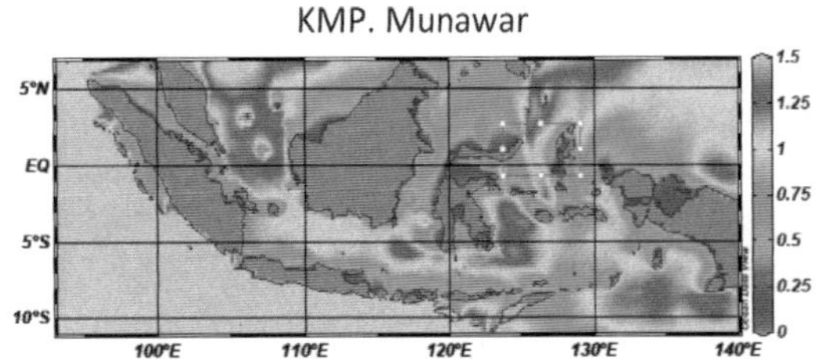

KMP. Bandeng

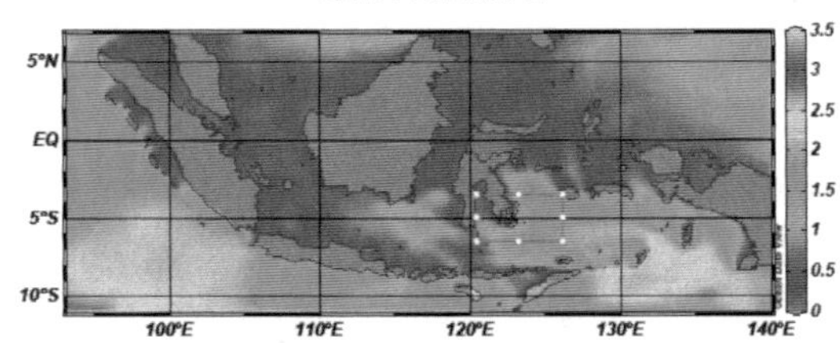

KMP. Lestari Maju

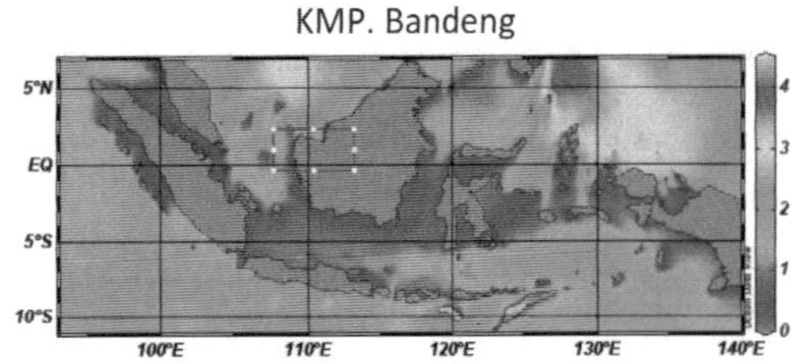

KMP. Bandeng

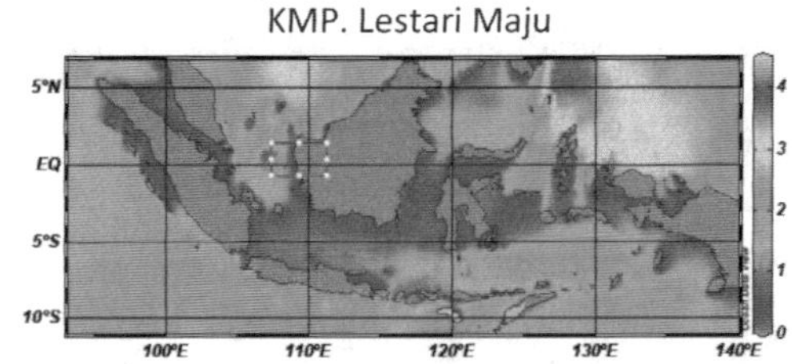

KMP. Bili

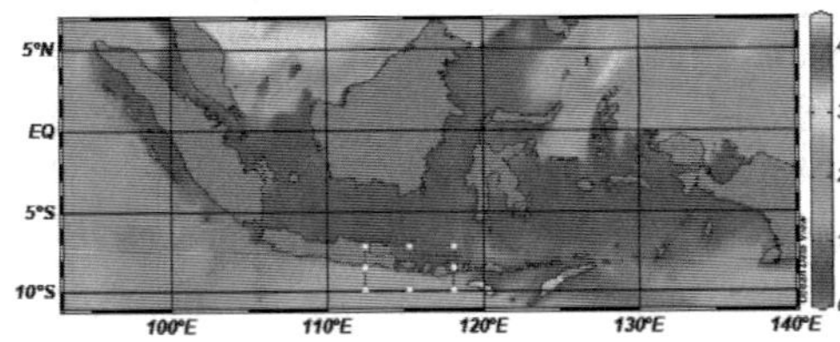

KMP. Yunicee

Figure 3. Characteristics of global waves at RaPax OCD capsizing in Indonesia water

Figure 3 above is a plot of the significant wave height on the global map of Indonesia, and there is a small red box that shows the location of the capsized ship. The x-axis shows the longitude coordinates, the y-axis shows the latitude coordinates, and the scale on the right of the image shows the wave height. From the overall plotting of the map, the area within the big islands of Indonesia is a closed area with a relatively small significant wave height, including the Malacca Strait, Java Sea, Makassar Strait, Maluku Sea, Banda Sea, and Seram Sea. Meanwhile, outside the area, it has significant extreme wave heights in certain months, including the Indian Ocean, Natuna Sea, North Sulawesi Sea, and the Banda Sea. The island of Sumatra protects the waves from the west, Java protects the waves from the south, Papua protects the waves from the east, and Borneo protects the waves from the north. From the data, almost all RoPax cruises only serve in closed sea areas. None of the 15 cases of RoPax inversion occurred at extreme significant wave heights, indicating that the cause of RoPax inversion is due to poor internal stability problems. The incident of ship capsizing occurred in almost all parts of Indonesia with significant variations in small wave heights, indicating the same problem in all operational areas. The location of the RoPax reversal with the highest significant wave occurred in the Sheet Sea with 1 case, and the most occurred in the Bali Strait with 2 cases. When viewed from the global map of Indonesia, the three cases show that the waves in that location were affected by the Indian Ocean.

5.2. Characteristics of Local Significant Waves Height

Significant wave height is one of the main parts of wave characteristics. This significant wave height is caused by wind speed and is influenced by marine morphology. Early warning indicators for ship shipping hazards referred to by BMG use significant wave heights. There are three groups of significant wave heights: significant small waves, significant medium waves, and highly significant waves. In addition to the significant wave height displayed globally for all parts of Indonesia when the RoPax capsized. It can also be displayed in a single coordinate focus point or a local area so that a more accurate significant wave height value is obtained, as shown in the graph as follows:

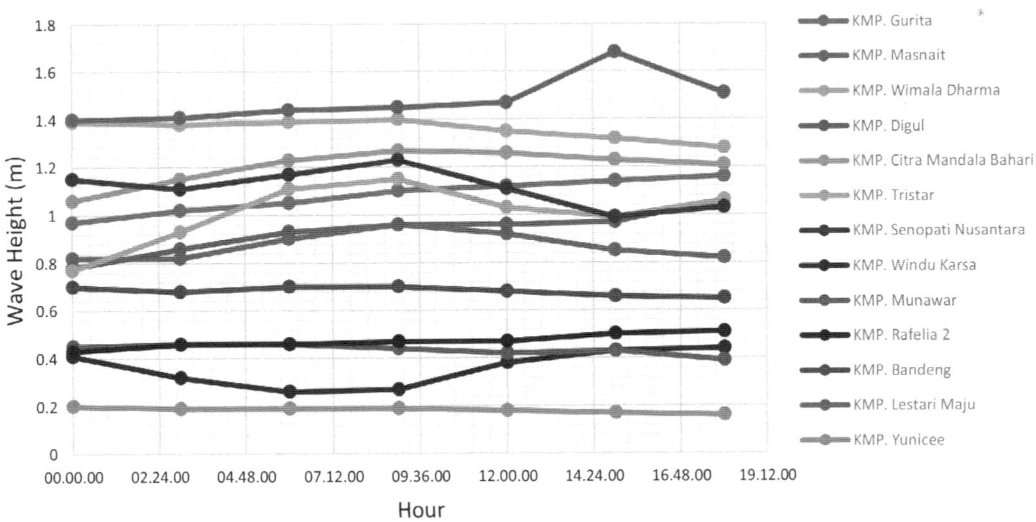

Figure 4. Characteristics of local waves height at RaPax OCD capsizing

Figure 4 or the graph above, the x-axis shows the time coordinates throughout the day from 00:00:00-19:12:00 when the ship capsizes. The y-axis shows the significant wave height. Figure 3 above shows 13 ship accidents at sea, while 2 cases occurred in rivers that were not described because the accident occurred in calm water. In Figure 3 above, the lowest significant wave height is experienced by KMP. Yunicee had an average significant wave height of 0.19 meters, while KMP experienced an enormous wave height. Lestari Maju with a significant wave height of 1.5 meters on average. Meanwhile, the significant wave height in the other RoPax cases occurs at the significant wave height, which varies in the range of the significant wave height if sorted from lowest to largest: KMP. North Kayong, KMP. Bill, KMP. Yunicee, KMP. Windu Karsa, KMP. Munawar, KMP. Rafelia 2, KMP. Bandeng, KMP. Masnait, KMP. Digul, KMP. Tristar, KMP. Senopati Nusantara, KMP. Octopus, KMP. Image of Mandala Bahari, and KMP. Sustainable Forward. The graph shows that the pattern of significant wave heights in one day is relatively unchanged. Some changes in wave patterns that are not too extreme occur in KMP. Wimala Dharma, KMP. Masnait, KMP. Tristar, KMP. Senopati Nusantara and KMP. Sustainable Forward. After analyzing 15 cases of local RoPax overturning, precisely at the coordinates of the sinking ship, it shows that the dominant cause of the RoPax ship is internal ship problems, not harmful weather problems / significant high waves.

5.3. *Characteristics of Local Significant Waves Period*

In addition to the Global Ocean Waves Reanalysis Wavers-Copernicus can record significant wave height data. It can also record wave period, one of the other main parts of wave characteristics. Associated with the wave period with RAO can show the tendency of the cause of the ship capsizing. The restoring force factor is dominant when the ship capsizes in a low period. When a ship capsizes in a high period, the moment of inertia dominates, while when it capsizes in the middle period or at the peak of RAO, the cause of the capsize is the dominant damping moment. In this paper, the local wave period is taken one day at a time, which is shown on the graph as follows:

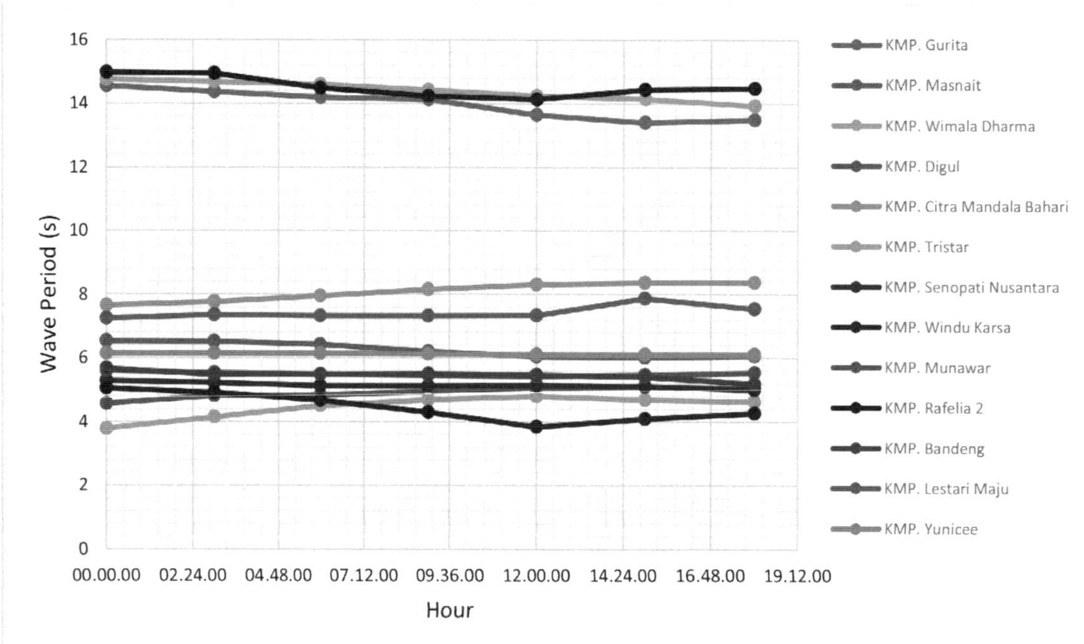

Figure 5. Characteristics of local waves period at RaPax OCD capsizing

In figure 5, the x-axis shows the time coordinates throughout the day from 00:00:00-19:12:00 when the ship capsizes, and the y-axis shows the wave period. The graph shows the wave period pattern in one day with relatively no significant changes. The graph above shows the period pattern divided into two groups. The first group has a period range of 3.9 s - 8.1s. There were 9 cases, from the smallest to the largest: KMP. Windu Karsa, KMP. Tristar, KMP. Senopati Nusantara, KMP. Bandeng, KMP. Yunicee, KMP. Masnait, KMP. Lestari Maju and KMP. Maritime Mandala image. The second group has a period range of 13.9 s -15 s. There were 3 cases, from the smallest to the largest: KMP. Octopus, KMP. Wimala Dharma and KMP. Rafelia 2. These two groups can indicate the cause of the RoPax accident. The low-dominant group indicates the dominant cause due to poor ship stability. Then the second high-period group shows an indication of the accident caused by the size of the ship being less prominent, while the group between the two is not found. Indicates the cause of the accident because the damping moment is not found.

Conclusions

From the results of the analysis and discussion of locations, the characteristics of significant wave heights, and the characteristics of the wave period in 15 cases of ship capsizing that occurred in Indonesia globally and locally, the following conclusions can be drawn:

- The incidence of the sinking of the RoPax almost spread throughout Indonesia's sea areas. Most cases occurred in the Bali Strait, 2 cases of capsizings, and 2 cases occurred in rivers in calm water.
- The waters in Indonesia where RoPax operates are closed waters. Inside, the waters have a low significant wave height, while the waters outside the island bordering the ocean have a high significant wave height,
- The case of the sinking of the RoPax occurred at a relatively low significant wave height with a range of 0.15-1.5 meters. The accident with the highest wave occurred at KMP. Lestari Maju occurs in the Slayar Sea, where most accidents occur in the Bali Strait.
- From the wave period analysis, there are two groups of RoPax ship accidents, namely: the first group has a wave period range of 3.9s - 8.1s, nine ship accidents occur, and the second group has a wave period range of 13.9s -15s, three ship accidents occur,
- In general, from the analysis of significant wave height characteristics and global and local wave periods. It can be seen that the causes of ship accidents are dominated by restoring moment (stability) problems, followed by inertia moment problems (ship size), while accidents due to damping moment (bilge keel) problems are not found.

Acknowledgement

The author would like to thank the Directorate of Research and Community Service - ITS, which has provided funding so that this research can proceed well. The author hopes this research can be used as an operational reference for RoPAx OCD in Indonesia and can be applied for subsequent studies to minimize the risk of ship accidents in Indonesia and the world.

References

[1] T. Dickson, H. Farr, D. Sear, and J. I. R. Blake, "Uncertainty in marine weather routing," *Appl. Ocean Res.*, vol. 88, pp. 138–146, Jul. 2019, doi: 10.1016/j.apor.2019.04.008.

[2] A. Christodoulou, Z. Raza, and J. Woxenius, "The Integration of RoRo Shipping in Sustainable Intermodal Transport Chains: The Case of a North European RoRo Service," *Sustainability*, vol. 11, no. 8, p. 2422, Apr. 2019, doi: 10.3390/su11082422.

[3] A. M. Arof and R. Nair, "The identification of key success factors for interstate Ro-Ro short sea shipping in Brunei-Indonesia-Malaysia-Philippines: a Delphi approach," *Int. J. Shipp. Transp. Logist.*, vol. 9, no. 3, Art. no. 3, 2017, doi: 10.1504/IJSTL.2017.083470.

[4] P. IMO, "Safety of ro-ro ferries." https://www.imo.org/en/OurWork/Safety/Pages/RO-ROFerries.aspx (accessed Aug. 19, 2022).

[5] M. Kaptan, "Analysis of accidents during vehicle stowage on RO-RO vessels by using Fuzzy Bayesian networks," *Ocean Eng.*, vol. 260, p. 111997, Sep. 2022, doi: 10.1016/j.oceaneng.2022.111997.

[6] N. Baird, *Fatal ferry accidents, their causes, and how t y accidents, their causes, and how to prevent them ent them.* University of Wollongong, 2018. [Online]. Available: https://ro.uow.edu.au/cgi/viewcontent.cgi?article=1497&context=theses1

[7] H. Hasanudin, I. KA P Utama, and C. Chen, "Application Side Casing on Open Deck RoRo to Improve Ship Stability," *E&ES*, vol. 135, no. 1, p. 012017, 2018.

[8] A. Coraddu, L. Oneto, B. Navas de Maya, and R. Kurt, "Determining the most influential human factors in maritime accidents: A data-driven approach," *Ocean Eng.*, vol. 211, p. 107588, Sep. 2020, doi: 10.1016/j.oceaneng.2020.107588.

[9] X. Shi, H. Zhuang, and D. Xu, "Structured survey of human factor-related maritime accident research," *Ocean Eng.*, vol. 237, p. 109561, Oct. 2021, doi: 10.1016/j.oceaneng.2021.109561.

[10] N. P. Ventikos, A. D. Papanikolaou, K. Louzis, and A. Koimtzoglou, "Statistical analysis and critical review of navigational accidents in adverse weather conditions," *Ocean Eng.*, vol. 163, pp. 502–517, Sep. 2018, doi: 10.1016/j.oceaneng.2018.06.001.

[11] A. Mrozowska and P. Mrozowski, "The Analysis of Maritime Accidents in 2017," *Annu. Navig.*, vol. 25, no. 1, pp. 233–252, Dec. 2018, doi: 10.1515/aon-2018-0016.

[12] Y. Li and M. Lin, "Regular and irregular wave impacts on floating body," *Ocean Eng.*, vol. 42, pp. 93–101, Mar. 2012, doi: 10.1016/j.oceaneng.2012.01.019.

[13] Z. Zhang and X.-M. Li, "Global ship accidents and ocean swell-related sea states," *Nat. Hazards Earth Syst. Sci.*, vol. 17, no. 11, pp. 2041–2051, Nov. 2017, doi: 10.5194/nhess-17-2041-2017.

[14] F. A. Prasetyoa, M. A. Kurniawanb, and S. Komariyahc, "Indonesian seastate condition and its wave scatter map," 2018.

[15] A. Toffoli, J. M. Lefèvre, E. Bitner-Gregersen, and J. Monbaliu, "Towards the identification of warning criteria: Analysis of a ship accident database," *Appl. Ocean Res.*, vol. 27, no. 6, pp. 281–291, Dec. 2005, doi: 10.1016/j.apor.2006.03.003.

[16] IMO, "Background of criteria regarding righting lever curve properties (part A of the 2008 IS Code)," 2008, [Online]. Available: https://www.imorules.com/GUID-E8AD1425-5B54-4291-AF02-184A91FE1C2B.html

[17] S. K. KNKT, "Accident Reports." http://knkt.dephub.go.id/knkt/ntsc_maritime/maritime_investigation_report1.htm (accessed Aug. 20, 2022).

[18] V. 5. 6. 4 ODV, "Ocean Data View (ODV) software." https://odv.awi.de/ (accessed Aug. 20, 2022).

[19] X. Lv, D. Yuan, X. Ma, and J. Tao, "Wave characteristics analysis in Bohai Sea based on ECMWF wind field," *Ocean Eng.*, vol. 91, pp. 159–171, Nov. 2014, doi: 10.1016/j.oceaneng.2014.09.010.

[20] Ludeno and Serafino, "Estimation of the Significant Wave Height from Marine Radar Images without External Reference," *J. Mar. Sci. Eng.*, vol. 7, no. 12, p. 432, Nov. 2019, doi: 10.3390/jmse7120432.

[21] J.-V. Björkqvist, H. Pettersson, and K. K. Kahma, "The wave spectrum in archipelagos," *Ocean Sci.*, vol. 15, no. 6, pp. 1469–1487, Nov. 2019, doi: 10.5194/os-15-1469-2019.

[22] C. CMEMS, "Copernicus Marine Services," *Global Ocean Waves Reanalysis WAVERYS.* https://marine.copernicus.eu/ (accessed Aug. 20, 2022).

[23] Z. Zhang and X.-M. Li, "Global ship accidents and ocean swell-related sea states.," *Nat. Hazards Earth Syst. Sci.*, vol. 17, no. 11, Art. no. 11, 2017.

[24] IMO, "Background of criteria regarding righting lever curve properties (part A of the 2008 IS Code)," 2008. https://www.imorules.com/GUID-E8AD1425-5B54-4291-AF02-184A91FE1C2B.html (accessed Jul. 19, 2022).

Formal Safety Assessment for determining sensitive area due to traffic of tourism vessels at Lombok Strait

F I Prastyasari, A A B Dinariyana, T F Nugroho and A S Sukmana

Department of Marine Engineering, Institut Teknologi Sepuluh Nopember, Surabaya 60111, Indonesia

fadilla@ne.its.ac.id

Abstract. Lombok Strait has been determined as one of the Indonesia Archipelagic Sea Lane (IASL) passageways. It is also home to many tourism spots, such as Nusa Penida, that result in an increase in traffic density. Nusa Penida has been designated as a Marine Protected Area (MPA) since 2010. However, there is a need to define the sensitive area of Nusa Penida to protect it from the activities of vessels that can further harm the conserved area. Formal Safety Assessment (FSA) is a structured and systematic methodology that aims to improve safety in the maritime, including environmental aspects, using risk analysis and cost-benefit assessment. The result of this study is that ten hazards are categorized as possible hazards and then narrowed into five potential hazards—the one that will be analyzed is tourism activities. The frequency analysis of this study only considers tourism vessels namely Pleasure, Passenger, Fishing, and Sailing Vessels. The consequence analysis provided recommendations for a safe distance of vessels considering the hearing threshold of Sunfish (150 dB) which is 5000 m and coral reef (157 dB) is around 10 m. The proposed design of the sensitive area was based on the Nusa Penida MPA with changes in 2 coordinates considering ship activity as one additional aspect.

1. Introduction

Sea transportation is vital in Indonesia as it is dominated by water with more than two third of its total area [1]. Lombok Strait has been determined as one of the Indonesia Archipelagic Sea Lane (IASL) passageway and it caused this strait to be frequently passed by domestic vessels as well as international vessels. In addition to those vessels, tourist vessels, such as pleasure vessels, passenger vessels, sailing vessels, and fishing vessels, are often found in the vicinity of Lombok Strait due to the tourism attractions; one of which is Nusa Penida area. According to data from the tourism office of Klungkung, Nusa Penida was visited by 722.018 tourists in 2018 to witnessed coral reefs and other marine biota, as shown in Table 1. Consequently, this area could experience a continuous destruction due to tourism activities and international ship activities. Under the Regulation of Government of Klungkung No. 12 of 2010, Nusa Penida has been designated as Marine Protected Area (MPA) [2] to prevent it from further damages. However, MPA does not prohibit the area from international shipping traffic. Hence, Nusa Penida has the urge to design a Sensitive Area (SA) to protect the environmental properties from the potential hazards caused by tourism activities.

Sensitive Area (SA) takes the concept of Particularly Sensitive Sea Area (PSSA) regulated by International Maritime Organization (IMO), as one of the solutions to protect the area because the area has large impact on several aspects, such as ecology, socio-economic, or other scientific aspects, where it is exposed to damage caused by the activities of ocean-going ships. One thing that makes this study different is the shipping activities involved here are made by tourism vessels as their traffic is higher than those bound in international route.

Back in 2016, Lombok Strait, especially in Gili Islands and Nusa Penida, was chosen as a pilot project because it meets the three main elements required by IMO. The first is regional attributes; second is the is prone to damage imposed by international shipping activities; and lastly is availability of Associated Protective Measure (APM) to reduce the risks [3]. This paper evaluates the risk using the Formal Safety Assessment (FSA) to generate recommendations regarding the design of the sea sensitive area in Nusa Penida that aims to improve safety of navigation as well as conserving the marine ecological population.

2. Literature Review

According to Guidelines for the Identification and Designation of Particularly Sensitive Sea Areas in IMO Resolution A.982(24), the definition of PSSA is the area that needs extra care as it has a significance not only for the ecological aspects, but also socio-economic and scientific aspects. It is also potentially threatened by international shipping activities [4]. As the organization that regulate the activities of international shipping, IMO has responded the threat caused by the operation of vessels such as pollutions and/ or physical damages with the possibility to design an area as PSSA and enact the APMs. Currently, there are 17 PSSAs that have been adopted by IMO through the Marine Environment Protection Committee (MEPC). The number is quite small because the area is required to meet one of three criteria at the minimum which is set by IMO. Those criteria are criteria related to ecology; criteria about social, cultural, and economic criteria; and lastly criteria about scientific and educational.

One example of the PSSA that has been adopted in 2017 is the Tubbataha Reefs Natural Park in Philippine that aims to reduce the possibility of environmental damage due to ship groundings and oil spills [5]. The probability of pollution in this particular area has shown an increase during 1999-2007 for about 60% with the associated total loss that could be experienced by a vessel could reach $255,000 [6]. It has a core zone as well as a buffer zone or the so-called Area to be Avoided (ATBA) to be its APM to prevent the international vessels to approach this area.

Indonesia is included in the Coral Triangle together with several other countries: Philippines, Malaysia, Timor Leste, Solomon Islands, and Papua New Guinea, which is a home to 76% species coral reefs in the world [7]. In addition, other fauna are also found in the Nusa Penida waters such as as sea turtles, sharks and cetaceans, as well as rare fishes such as ocean sunfish (*Mola mola*) and manta rays (*Manta birostris*) [8]. One strategy that has been done by the Government of Indonesia to protect the ecosystem of coral reefs is to create a MPA which cause fishing activities to be prohibited in the core zone except in the limited marine tourism zone that is permitted between 4pm to 9am [9]. However, the MPA cannot protect the area from shipping activities performed in the vicinity. Hence, this paper provides a thorough assessment to see how the tourism vessels can endanger the marine population, especially the Sunfish and coral reefs by applying the Formal Safety Assessment to recommend associated protective measures. As the PSSA is applied to reduce the international shipping activities, this study made a slightly different perspective as the tourism activities are heavier compared to the international voyage passing the Traffic Separation Scheme (TSS) Lombok Strait.

3. Formal Safety Assessment (FSA)

Formal Safety Assessment (FSA) is an organized and systematic method that utilizes risk analysis and cost-benefit analysis to improve maritime safety, and to protect life, health, the environment, as well as property [10]. The FSA is performed in the following steps: (1) identification of hazards, (2) risk assessment, (3) identification of risk control options (RCOs), (4) cost and benefit analysis, (5) recommendations for decision making. In its application, FSA is used to help the evaluation of new or existing maritime safety and environmental regulations. Moreover, FSA can be utilized to compare existing regulations to possibly improved regulations to strike a balance between various technical and operational issues, such as maritime safety or marine environment protection and cost.

3.1. Identification of Hazards

Identification of hazards is a step to determine hazards that have influenced the determination of sensitive water areas in the Lombok Strait, especially Nusa Penida. The HAZID generates a list of hazards in tourism activities based on their probability of occurrence. In this study, qualitative hazard identification was carried out using the What-if Analysis method. What-if Analysis is a simple approach for process hazard identification (PHA) by brainstorming in a group with the experts [11]. A list of potential hazards in the Lombok Strait is presented in Table 2.

Table 1. Number of tourists in Nusa Penida.

Year	2014	2015	2016	2017	2018	2019
The num. of tourist	267.971	372.051	378.894	605.423	722.018	839.790

Table 2. List of potential hazards.

No.	Considerations	Potential Hazard
1	Oil spills can cause environmental damage, especially marine biota in MPA specifically to coral reef	Oil Spill (Pollution)
2	Nusa Penida is a popular tourist destination in Indonesia and worldwide. This activity needs to be considered with the high level of tourism activities such as diving, snorkeling, and yachts. Inter-island vessels also have risks in determining this sensitive area.	Tourism Activities
3	Vessels that suffer damage to the steering and propulsion system are very likely to experience accidents such as grounding in MPA that can cause massive damage to ecosystems and marine biota in the area.	Vessel suffered damage
4	The use of trawl can damage fish populations in the vicinity of MPA.	Fisherman Activities
5	This hazard has more potential damage to the environment with the unpredictable movement of anchors from the vessels. The impact could damage coral reefs ecosystem in the MPA permanently	Dragged Anchor

3.2. Risk Assessment

The next step is to analyze and evaluate what factors cause high levels of risk by focusing on potential hazards. Risk assessments consist of two steps: frequency analysis and consequence analysis. This study calculates the frequency of vessels passing through Lombok Strait related to tourism activities, especially in Nusa Penida. Meanwhile, consequence analysis is considering Impacts of underwater noise produced by vessels on marine biota, particularly impact on Sunfish and coral reefs. The recommendation considers the safe distance between ships to marine biota that makes the underwater noise not significantly affect the hearing threshold of marine biota. To determine the level of risk from this maritime activity, there is a recommendation of a safe distance that has adjusted to the hearing threshold of mola-mola (Sunfish) and Coral Reefs. The change in the hearing threshold of Blue Sunfish will happen when there is a change in the underwater sound by 10-20 dB [12]. In comparison, coral reefs will experience an increase of hearing thresholds by 10 dB during the day [13].

3.2.1. Traffic Condition

The traffic condition extracted from marinetraffic.com is shown in Figure 1. Based on this figure, denser vessel activities are monitored in the north part of Nusa Penida Island due to the traffic in Padang Bai and Lembar. Another part with a high density is in the TSS of Lombok Strait. Although the coordinates of MPA have been set as shown in Figure 1, but there are some activities observed because MPA does not have a feature to prohibit the traffic of international vessels, not to mention activities due to tourism vessels.

Despite the fact that PSSA is only applicable to limit the activity of ocean-going vessel, this paper will exclude them from the analysis and only discusses about the activity of tourism vessels. This is due to the higher number of tourism vessel that pass this area compared to ocean-going vessel. The type of vessel that frequently passes inside the MPA is a vessel related to tourism activities. There are four categories, sailing vessels are not included in the analysis as they do not produce significant underwater noise. Table 3 shows the number of vessels that pass the Lombok Strait during May 1st – 31st 2018 and it shows that passenger vessels dominated the traffic due to tourism activities in Lombok Strait with an average of 3-4 ships per day.

3.2.2. Consequence Analysis

Quiet operation condition is a condition of operation when the vessel is having an operation in an area with a sensitive ecosystem. Studies have been conducted in the Salish Sea and it was found that underwater noise for passenger vessel is 175-185 dB and 167 dB for pleasure vessels [14]. A formula

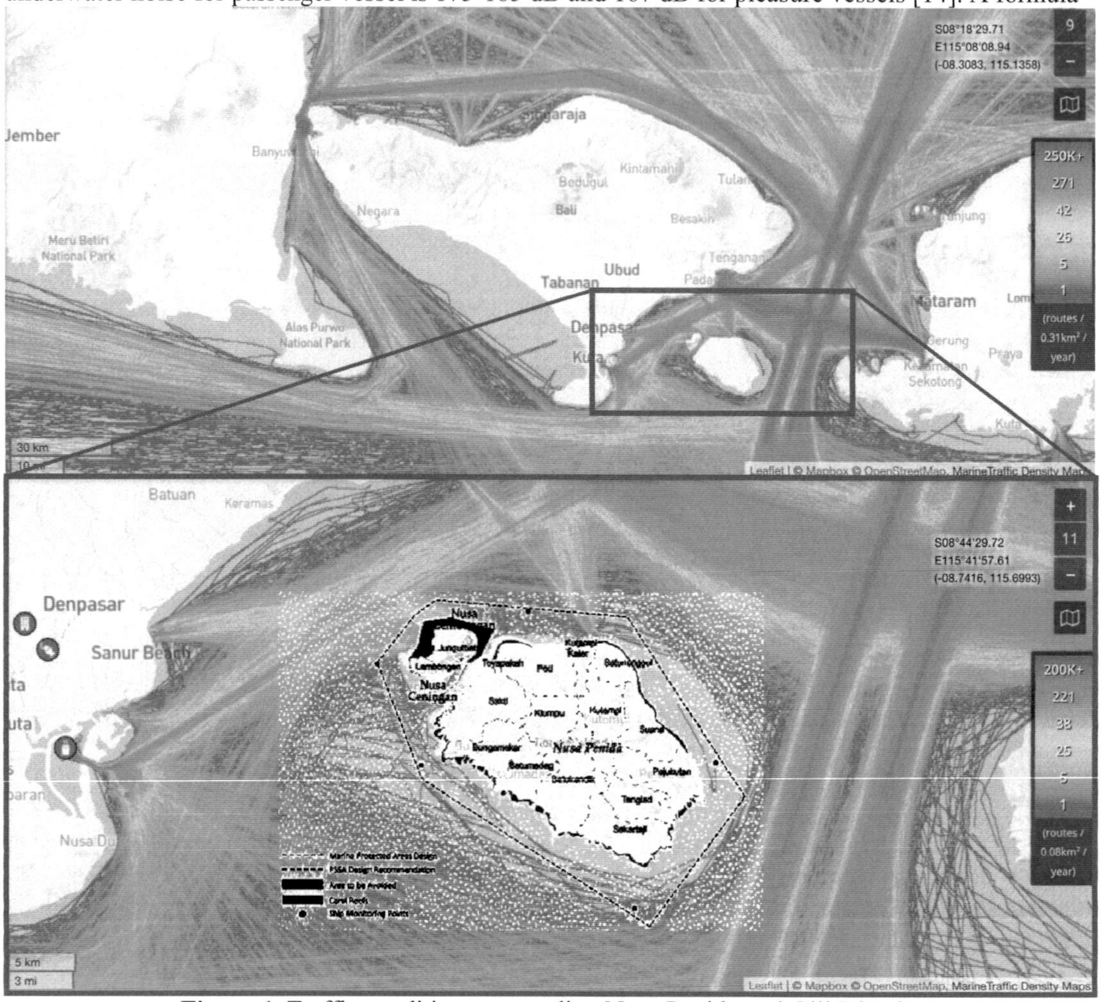

Figure 1. Traffic condition surrounding Nusa Penida and Gili Islands.

taken from American Bureau of Shipping (ABS) is applied to calculate the underwater noise. The calculation of underwater noise in quiet operation conditions for the frequency range between 100 – 1kHz is formulated as follows:

$$UWN = -6 \log f + 179{,}5 \quad (1)$$

Where:

UWN = Underwater Noise
f = Frequency value (100 – 1kHz)

The safe distance is calculated by varying the d value used in the formula below which is issued by the International Council for the Exploration of the Sea (ICES) [10] with consideration of hearing threshold from marine biota nearby Lombok Straits, particularly Sunfish and Coral Reefs.

The Vn or vessel noise is obtained and then the value is compared with the hearing threshold of the Sunfish.

$$Vn = 20 \log d + ht + 30 \tag{2}$$

Where:
d = the adjusted distance until underwater noise produced by vessels is safe
ht = taken as 150 dB, which was categorized into "Noise Level average" [12]. Meanwhile, the coral reef's hearing threshold is 157 dB [13,15].

The results of underwater noise calculation along with the safe distance between ship to Sunfish and coral reefs are presented in Table 4 and Table 5. In Table 4, three types of vessels which contributed to most of the traffic in Lombok Strait were assessed and found that the UWN or underwater noise of those vessels are between 182.68 to 182.84 dB after knowing the frequencies produced by each type of vessel. The value of UWN is then used to the baseline for determining the minimum distance of vessels to the habitat of Sunfish and Coral reefs.

This paper takes the safe distance as the main consideration for defining the risk control option that is discussed in the next chapter. Table 5 shows the vessel noise that can be borne by Sunfish and Coral reefs given the various distances. When the minimum distance is taken as 1000 m, the maximum vessel noise that can be withstood by Sunfish is 182.31 dB or less. Hence, passenger vessels and pleasure vessels that produce UWN around 182.84 dB can have operation within 5000 m from Sunfish habitat. While for the Coral reefs, they can bear vessel noise up to 187.77 dB within 10 m for the closest distance, it means that the vessel operations will not disrupt the population of Coral reefs located in the MPA. Although the safe distance from vessels to the habitats of Sunfish and Coral reefs are different, the farther distance shall be taken to generalize the regulation. This safe distance is used to set up the boundary for area to be avoided (ATBA) as one of the risk control options (RCO) that is discussed in the next subchapter.

Table 3. Dimension of passing vessels according to vessel category.

Vessel Length (m)	Passenger Vessel	Fishing Vessel	Pleasure Vessel	Total
0-10	0	0	0	0
11-20	0	0	3	3
21-30	47	0	2	47
31-40	34	0	1	35
41-50	48	6	1	55
51-60	0	0	0	0
Total	129	6	7	142

Table 4. Result of underwater noise calculations.

Type of Vessel	Frequency	Underwater Noise (dB)
Fishing Vessel	300 Hz	182.68
Passenger Vessel	400 Hz	182.84
Pleasure Vessel	400 Hz	182.84

Table 5. Result of safe distance to sunfish and coral reefs.

Sunfish		Coral Reefs	
d (m)	*Vn* (dB)	d (m)	*Vn* (dB)
1000	182.31	10	187.77
3000	182.72	20	188.00
5000	182.84	40	188.23
6000	182.90	60	188.37

3.3. Risk Control Options

Risk control options (RCO) are proposed following the result of risk assessment which is done in the previous subchapter. This process aimed to find out the right steps to reduce or prevent risks that have been analyzed from the previous stage. According to the risk assessment result, the monthly frequency of passenger vessels, fishing vessels, and pleasure vessels reach 142 and it equals to 1704 vessels annually. Whilst the consequence of the hazard related to the tourism activities showed that the tourism vessels will affect the hearing capacities of Sunfish and Coral reefs when the distance is less than 5000 m and 10 m, respectively. This study proposes two means of RCOs, those are ship monitoring system and Area to be Avoided (ATBA), which is expected to reduce the risk associated with tourism vessel.

3.3.1. Ship Monitoring

Associated Protective Measure (APM) ensures every tourism-related vessel entering the SA is going to be monitored by the local authority which, in this case, is the local representative of the Ministry of Marine Affairs and Fisheries of Republic of Indonesia. Ship Monitoring System requires all tourism vessels that passing Lombok Strait especially near the SA to report to the local authorities. Every vessel going to SA or exit from it, will be monitored by the local authority in 5 predetermined coordinates as shown in Figure 2. Hence, to help local authority some information should be submitted to the local authority, such as vessel name, date and time, location of the vessel, speed of vessel, the destination of vessel, the route will be taken, the draught of vessel, and a total of passengers and crews on board.

3.3.2. Area to be Avoided

Area to be Avoided (ATBA) is one of the proposed APM by IMO to protect the sensitive area by forbidding some vessels or all vessels entering a particular area. The area is where navigation can be so dangerous or important to prevent casualties and environmental damage caused by the vessel. In general, ATBA protects the ecosystem of Sunfish and coral reef, like what the Australian government did to Great Barrier Reef and Torres Strait, and the Coral Sea. Nusa Penida area has a vast coral ecosystem spread from west to east, and it is a perfect match for ATBA itself. To support the recommendation of ATBA, there are several regulations to balance between protecting coral reefs and economic considerations.

- ATBA still can be passed by the vessel with terms and conditions and should be monitored by the local authority. Fishing activities using illegal equipment and diving is prohibited in the area
- Vessel not allowed to stop in ATBA
- As there are many pontoons for tourism purposes that located inside the proposed ATBA, all of them should be relocated into a new area so as to not disturbing the tourism activities as well as the habitat of coral reef.

Pontoon area should be removed to a new area because the early area is placed inside the new ATBA recommendation. This situation makes the vessel more often pass through the ATBA area to reach the pontoon and increases the danger of damaging the coral. New pontoon's area has 500 m length with 50 m width and depth of the area around 5-30 meter. An assumption was made to determine the pontoon's dimension, with each pontoon having 50 m length, 12 m width and the new area should accommodate ten pontoons.

3.4. Cost-Benefit Assessment

3.4.1. Ship Monitoring System

Additional equipment is essential to support the Ship Monitoring System as the report will involve crew on board and the local authority. The proposed reporting system uses Very High Frequency (VHF) radio channel provided by the authorities. Table 6 explains a comparison of three options started from the cheapest tools with limited specifications to the most expensive tools with many specifications to support the monitoring system in the SA area.

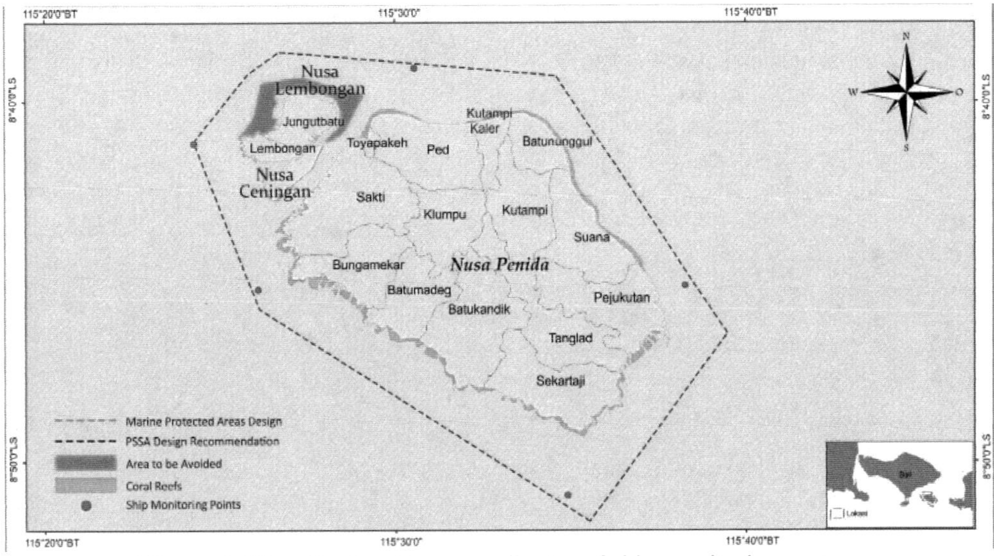

Figure 2. Visualization from coordinates of ship monitoring system.

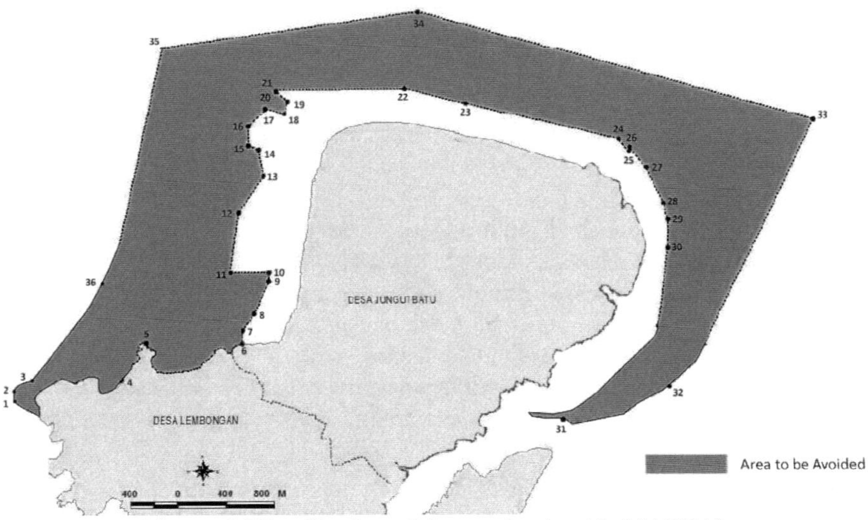

Figure 3 Visualization of Area to be Avoided (ATBA).

Table 6. Comparison between three equipment in terms of cost.

Specifications	POFUNG BF U5R	ALINCO DR-MX15	SAMYUNG STR 600D
Battery Capacity / Power	1800mAh	13.8 V (5.5A)	13.6 V (6 A)
Frequency Range (MHz)	65-108	156.025 – 157.425	156.025 – 157.425
Support Equipment	-	Accumulator	Accumulator
Radio Channel Cap.	128	181	183
Cost of Radio	Rp299,900	Rp2,090,000	Rp7,999,000
Cost of Support Equipment	-	Rp150,000	Rp150,000
Total Cost	Rp299,900 USD21.02	Rp2,240,000 USD157.01	Rp8,149,000 USD571.18

3.4.2. Area to be Avoided

Operational cost changes calculated as the pontoon's area moved up into the new area. Based on marinetraffic.com, Bounty Cruises' gross tonnage is 622 with an average speed of around 13,6 knots. The following calculation presents the initial voyage distance from bounty cruises. Pontoon's new area also put consideration of attractiveness of underwater tourism. The New area is close to Mangrove point, Sakenan point, Blue Corner, and the coral reefs ecosystem in the north of Nusa Lembongan. The depth of the new area is suitable for beginner snorkeling and diving activities so it will be not impacting the income from the tourism sector.

$$C_{ik} = 16.9040 + 0.00198 \times GT \tag{3}$$

Spherical Trigonometry Formula is applied to calculate ship voyage distance, as follows:

$$\cos A = (\cos B \times \cos C + \sin B \times \sin C \times \cos A) \times 60 \text{ Nm} \tag{4}$$

Where:

A = origin angle – destination angle
B = 90° + destination angle
C = 90° + origin angle

Therefore, from the change of the pontoon's location due to ATBA, obtained a change in operational costs of Rp. 1,446,297.96 or $101.37 with a change in the distance is 2.31185 NM and an additional 11 minutes of travel to the new pontoon location.

3.4.3. Benefit Assessment

As a comparison of which risk control option is needed as well as how it affects the coral reefs in the Nusa Penida area, Florida Keys National Marine Sanctuary in 1989 spent US$5,5 million per ha on a 0,3-ha area for coral reefs restoration damaged by ships sinking [16]. Then in Maldives, where restoration used artificial coral and spent almost US$3 million per ha, it still needs 330 workhours for transplantation and 114 work-hours for monitoring [16]. Even so, in the 2,5 years, only 40-60% of coral still alive. Lastly, Tanzania tried a successful low-tech restoration method that only spent US$10,000 per ha that very much needed in developing countries such as Indonesia [17].

The spent cost of coral restoration might not be worth it. For example, in 293 samples that were taken only 64,5% of them were in a good condition and healthy, with money spent around US$ 8,3 million per ha [18]. Even with the restoration process, coral reefs will not back to near-mint condition. However, the restoration process can be 200 times less in developing countries if it involves consensus, the local community, and the stakeholders [19]. The amount of money lost every year on Indonesian tourism if coral was damaged is around US$30-US$5,000 per ha [20].

Tourism activities such as snorkeling and diving in Nusa Penida always increase every year which can endanger the population of Sunfish and affect the income from marine tourism. A study held on Bunaken's Marine Protected Area shows that fish tourism can get up to US$30 million per year [21,22].

3.5. Recommendation

The final stage of the Formal Safety Assessment (FSA) is a recommendation for decision making. In this study, this recommendation generates SA design recommendations in Nusa Penida. The determination of SA uses several considerations, such as ecological aspects, socio-economic and cultural aspects, and traffic density. These considerations based on discussion and survey along with Coral Triangle Centre (CTC).

3.5.1. Ecology Aspect

- The Core zone nearby Toyapakeh Village has good quality coral reefs with a high density between the coral.
- Pejukutan's Village & Jungutbatu's Village core zone is an ecosystem for fish reproduction also good condition on coral reefs.

- Nusa Penida water area and nearby area visited by more or less 2000 tourists.
- Sunfish passing by Sakti village, a cleaning station to clean their body from parasites and fungus, was helped by small fish in the area.
- Manta rays pass south of Nusa Penida Island, near Batumadeg village and Batumandik village on immigration season.
- Many reports show that divers and tourists are careless when visiting nature sites, like stepping on coral reefs and closing with sunfish.

3.5.2. Ecology Aspect

- Nusa Lembongan and Nusa Ceningan citizen has job which depend on water tourism.
- Not every Nusa Penida's citizens are working on water tourism. The majority of Nusa Penida's central region is working as cow breeders.
- In October, there is one day when some places are actively forbidden because of local culture related to religious events.
- When the pandemic hits Indonesia, Nusa Lembongan, Nusa Ceningan, and Nusa Penida's citizens get income from seaweed cultivation and fishing activities.

3.5.3. Vessel Traffic Characteristic Aspect

- In observation by CTC's research team, traffic intensity in one hour can reach around 35-50 small vessels, which carries 4-5 people in each vessel.
- Based on VTS data in May 2018, there are 142 vessels like pleasure, passenger, and fishing vessels.
- Passenger's vessel is the most often pass through in Nusa Penida area because they serve inter-island route from Benoa to Nusa Penida.

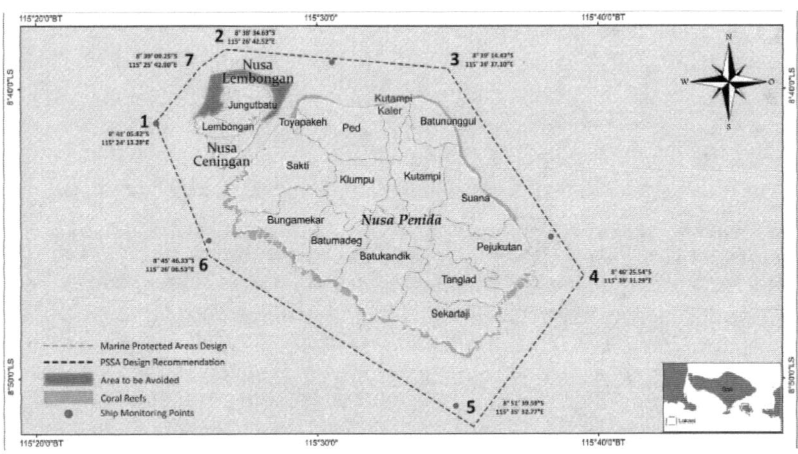

Figure 4. Design of Sensitive Sea Area.

4. Conclusion

This study aimed to make a design of Sensitive Sea Area by mean of formal safety assessment. The results not only presented the design, but also provided a recommendation of the safe distance for ships. The purpose of safe distance is to protect marine life, especially sunfish and coral reefs from underwater noise outcomes from ships. For the safe distance between vessels to sunfish, it is recommended that a distance of 5000 meters or more from the sunfish congregate. Meanwhile, for coral reefs, a recommended distance of at least 5-10 meters is a safe distance between ships and coral reefs.

References

[1] Ministry of Marine Affairs and Fisheries. Laut Masa Depan Bangsa, Mari Jaga Bersama 2019. https://kkp.go.id/artikel/12993-laut-masa-depan-bangsa-mari-jaga-bersama (accessed February 8, 2021).
[2] Marine Protected Area Form District-based MPA Nusa Penida n.d.
[3] Octavian A, Trismadi, Lestari P. The Importance of Establishing Particularly Sensitive Sea Areas in Lombok Strait: Maritime Security Perspective. IOP Conference Series: Earth and Environmental Science 2020;557:012013. https://doi.org/10.1088/1755-1315/557/1/012013.
[4] IMO. Revised Guidelines for the Identification and Designation of Particularly Sensitive Sea Area. 2006.
[5] IMO. Designation of the Tubbataha Reefs Natural Park as a Particularly Sensitive Sea Area 2017.
[6] Heij C, Knapp S, Henderson R, Kleverlaan E. Ship incident risk around the heritage areas of Tubbataha and Banc d'Arguin. Transportation Research Part D: Transport and Environment 2013;25:77–83. https://doi.org/10.1016/j.trd.2013.08.001.
[7] Green AL, Mous PJ. Delineating the Coral Triangle, its ecoregions and functional seascapes. vol. 1, 2008.
[8] Ruchimat T, Basuki R, Welly M. Nusa Penida Marine Protected Area (Mpa) Bali - Indonesia: Why Need to be Protected? Transylvanian Review of Systematical and Ecological Research 2013;15. https://doi.org/10.2478/trser-2013-0016.
[9] Yunitawati D, Clifton J. Governance in the early stages of marine protected area development: A case study of Nusa Penida District Marine Conservation Area, Indonesia. Marine Policy 2021;127:103653. https://doi.org/10.1016/j.marpol.2019.103653.
[10] R. B. Mitson. Underwater noise of research vessels: review and recommendations 2022. https://doi.org/10.17895/ices.pub.5317.
[11] Lyon BK, Popov G. The Power of What-if: Assessing and Understanding Risk 2020.
[12] Scholik A, Yan H. The effects of noise on the auditory sensitivity of the bluegill sunfish, Lepomis macrochirus. Comparative Biochemistry and Physiology Part A, Molecular & Integrative Physiology 2002;133:43–52. https://doi.org/10.1016/S1095-6433(02)00108-3.
[13] Kaplan MB, Mooney TA. Ambient noise and temporal patterns of boat activity in the US Virgin Islands National Park. Marine Pollution Bulletin 2015;98:221–8. https://doi.org/10.1016/j.marpolbul.2015.06.047.
[14] Veirs S, Veirs V, Wood JD. Ship noise extends to frequencies used for echolocation by endangered killer whales. PeerJ 2016;4.
[15] Raick X, Di Iorio L, Gervaise C, Lossent J, Lecchini D, Parmentier E. From the Reef to the Ocean: Revealing the Acoustic Range of the Biophony of a Coral Reef (Moorea Island, French Polynesia). Journal of Marine Science and Engineering 2021;9:420. https://doi.org/10.3390/jmse9040420.
[16] Spurgeon J. Improving the economic effectiveness of coral reef restoration. Bulletin of Marine Science 2001;69:1031–45.
[17] Bowden-Kerby A. Low-tech coral reef restoration methods modeled after fragmentation process. Bulletin of Marine Science 2001;69:915–31.
[18] Bayraktarov E, Saunders MI, Abdullah S, Mills M, Beher J, Possingham HP, et al. The cost and feasibility of marine coastal restoration. Ecological Applications 2016;26:1055–74. https://doi.org/10.1890/15-1077.
[19] Brown B, Fadillah R, Nurdin Y, Soulsby I, Ahmad R. CASE STUDY: Community based ecological mangrove rehabilitation (CBEMR) in Indonesia from small (12-33 ha) to medium scales (400 ha) with pathways for adoption at larger scales (> 5000 ha). Sapiens 2014;7.
[20] Cesar HSJ. Economic Analysis of Indonesian Coral Reefs. The World Bank; 1996.
[21] Davis T. Local and Semi-Local Economic Impacts of Dive Tourism in Bunaken National Park, North Sulawesi, Indonesia. WWF Indonesia; 2005.
[22] Bessa E, Silva F, Sabino J. Impacts of Fish Tourism, 2017, p. 59–72. https://doi.org/10.1007/978-3-319-58331-0_5.

Implementation of Metaverse for Modeling The Ship Bridge Simulator Based on Virtual Reality As Educational Purposes, Case Study: Tanjung Priok Port Area

N Siswantoro[1], D Haryanto[2], M A Hikmahwan[1], T Pitana[1]

[1]Department of Marine Engineering, Institut Teknologi Sepuluh Nopember (ITS), Surabaya, Indonesia
[2]Politeknik Pelayaran Surabaya, Indonesia

corresponding author's e-mail: nurhadi@ne.its.ac.id

Abstract. The development of virtual reality technology has increased in recent years with the characteristics and benefits of virtual reality that can help solve complex problems that exist in the real world. Virtual reality technology has been widely applied as a means of education. The ship bridge simulator is one of the supporting tools for training shipping cadets, but this tool has a relatively high cost. Therefore, this research aims to model the ship bridge simulator. The initial stage in this research is to model the ship bridge simulator belonging to the Surabaya Shipping Polytechnic (Politeknik Pelayaran Surabaya) with a case study of Tanjung Priok Port, referred to as the metaverse area. The process is to convert real objects into 3D assets using Blender software and develop them into a virtual reality application package designed using Unity software. The results obtained at the initial stage of this research are to provide a virtual forward and backward feature of the ship in the metaverse area. This virtual reality application has achieved its goals based on user ratings involving about 30 respondents. It is evidenced by its rating score of 908 out of 1050, which shows that this virtual reality application is included in the excellent category.

Keywords: 3D modeling, metaverse, virtual reality, ship bridge simulator.

1. Introduction

Indonesia is a country that has excellent marine potential, with a coastline of 99,083 lengths making Indonesia one of the countries that have extraordinary potential in the marine industry. The tremendous marine potential has boosted maritime education in Indonesia. There are around 80 shipping schools spread throughout Indonesia, with the potential in the field of shipping education being quite large. It is necessary to support educational equipment that can develop the quality of education that already exists. The shipping school needs a tool to help with the practical training of its students to have readiness before field practice in actual conditions. Learning with simulators is required for shipping schools. At the end of 2017, the Ministry of Transportation's Seafaring Expertise Examination Board conducted a comprehensive seafaring skill test using a simulator. Ship bridge simulators are one of the educational tools already existing in various shipping schools in Indonesia. Figure 1 shows the example of a ship bridge simulator at Surabaya Shipping Polytechnic (Politeknik Pelayaran Surabaya).

Figure 1. Ship Bridge Simulator

To improve the competence of students or cadets, the owned simulator equipment has a function in the scientific field. However, the current ship bridge simulator, owned by several shipping schools, has a very high cost and requires equipment at a much lower price with the same function. Virtual Reality applications have continued to increase over the last few years because they can be used in many fields. This trend is expected to continue in the future with technological advances in technology, such as computer graphics, control, image processing, and other technology-related components [1]. Virtual reality is a technology that allows users to interact with the virtual world environment so users feel like they are in that environment [2]. The main advantage of virtual reality is an experience that makes users feel the sensation of the natural world in the virtual world. With virtual reality, we are taken to another dimension where the depiction of the situation resembles the original shape of the object when in reality, we are still in the same place [3]. Virtual reality has several vital elements. The first element is the virtual world, a three-dimensional environment often realised through media, rendering, display, and others. The second element is immersion, namely the perception of being physically present in the non-physical world. It is a sensation created by virtual reality technology for users to feel a real environment when it is fictitious [4][12]. Immersion is divided into three types, namely mental immersion, the user is made to feel like he is in a natural environment; physical immersion makes the user physically feel the atmosphere around the climate created by the virtual reality; and mentally immersed the sensation felt by the user to dissolve in the environment generated virtual reality [5-7].

Therefore in conducting this research, the author proposes to make virtual reality an alternative means of learning using the ship bridge simulator, which shipping school cadets will later use. Virtual reality allows users to do activities anywhere so that shipping school cadets can easily access educational learning utilising the ship bridge simulator. In addition, the bridge simulator implemented into virtual reality technology requires a more affordable cost than the existing simulators.

2. Methods

The research method is a scientific method that is used to obtain data for later use as research objectives along with other steps such as literature review, problem statements, data collection, analysis, creation and development of virtual reality objects, testing, validation and results. This step is expected to create research results that can provide real solutions.

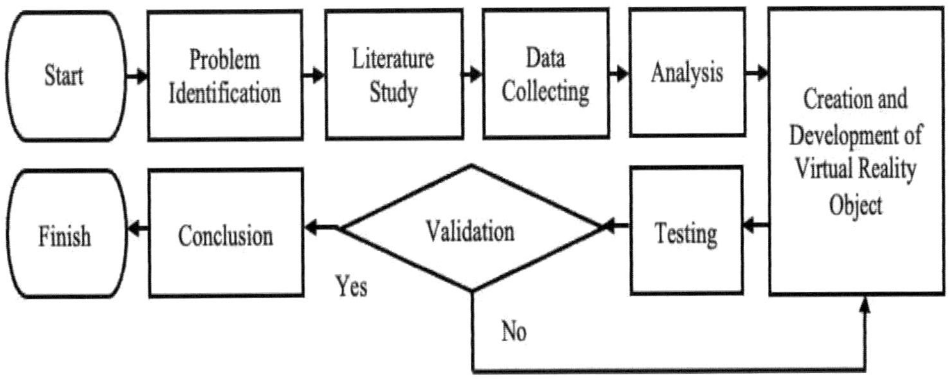

Figure 2. Research Methodology Flowchart

2.1. Problem Statement

The first stage of the methodology used in this research is formulating the problem and then developing the problem to be the research object. The issue raised in this final project is how to apply Virtual Reality technology as an educational medium for ship bridge simulations by students or school cadets to help to learn before carrying out practicals in real conditions.

2.2 Literature Study

The purpose of the literature study is to summarize the fundamental theories and supporting information that will be discussed and then become the basis of this research. This literature review is carried out by summarizing reference sources related to this research. This literature review is obtained from books, journals, papers, and others.

2.3. Data Collecting

This stage is carried out in parallel with the literature study. This stage is the process of collecting data and technical information related to the object of the problem that is used as the basis for research. The data that will be collected for this research is the components in the ship bridge simulator belonging to Surabaya Shipping Polytechnic (Politeknik Pelayaran Surabaya). Other data, such as the features included in the ship bridge simulator, are also needed in this research. The following data needed is the visualization of the Tanjung Priok Port area.

2.4. Analysis

This stage begins with making detailed scenarios in the virtual reality program by defining each object built in the virtual reality simulation. Then the output of this phase needs to be analyzed to get the foundation for building the entire application. This must be done carefully because a small error in this phase will result in the direction of the progress being off track.

2.5. Creation and Development of Virtual Reality Object

This stage focuses on creating every aspect of the virtual environment and the small components that make up the entire virtual world. The requirements for some assets can be seen from the previous analysis phase or the draft scenario. All objects should be tested, and documentation should be created accordingly. Many types of things, scripts, text, graphics, animations, sounds, and hardware, will be developed in this phase. In making a 3D environment for the Tanjung Priok Port area and the canal platform using the blender application. Meanwhile, creating virtual reality applications using the Unity 3D application.

2.6. Testing
The testing phase will be thorough and cover every virtual reality project aspect. Testing should focus not only on unhandled errors and exceptions in the code but also on the overall feel of the virtual environment. The purpose of this phase is to confirm that the progress that has been made is to the requirements.

2.7. Validation
At this stage, a questionnaire regarding user satisfaction with the virtual reality application will be distributed to 30 cadets of Surabaya Shipping Polytechnic (Politeknik Pelayaran Surabaya). Then the results of the questionnaire that have been obtained will be analyzed to get the value of satisfaction with this application. If the assessment results are as desired, it can proceed to the conclusion stage. If it is not appropriate, then return to the location of making and developing virtual reality objects.

2.8. Conclusion
The final step of this research is to conclude the entire process that has been carried out, accompanied by providing answers to some of the existing problems. Then proceed with delivering recommendations based on the results of this further research.

3. Result and Discussion

3.1. Data Collecting
In making this Virtual Reality application, the primary and supporting data were collected from the following sources:

1) General Information of Ship Bridge Simulator
   General information is mainly used to determine the dimensions of the bridge simulator, which will assist in the development of a 3D modelling environment of the bridge simulator, where the dimensions of the virtual bridge simulator are on a scale comparable to the real thing as well as information about the components that are in the ship bridge simulator is also needed because it is an integrated 3D environment of the ship bridge simulator.

2) Bridge Simulator Visual Data
   Visual data was collected from the ship's bridge simulator in the form of photos and videos of the actual conditions based on the survey that has been carried out. The data is needed to support modelling the 3D environment from the ship's bridge simulator.

3) Visual Data of Tanjung Priok Port
   In developing a 3D environment modelling of the Tanjung Priok port, visualization of the natural conditions of the port is very necessary. The collection of visualization data from the Tanjung Priok port area is carried out using the Google Earth application, which will then be developed into a 3D model of the Tanjung Priok port area environment.

A. Research Object
It takes a real ship bridge simulator to be used as an object of research which will later be analyzed and converted into a 3D model. This research used a ship bridge simulator belonging to Surabaya Shipping Polytechnic (Politeknik Pelayaran Surabaya).

Figure 3. Full Bridge Simulator at Surabaya Shipping Polytechnic (Politeknik Pelayaran Surabaya)

B. Equipment List

The following summarizes the list of equipment contained in the ship's bridge simulator:

1) Telephone
2) Navtex Receiver
3) Navigation Aid
4) ECDYSIS
5) ConningRadar
6) VHF
7) Steering
8) Compass

C. Type of Vessel

One of the virtual reality application's later components is the vessel used in simulating the ship bridge simulator. The type of vessel will be modelled later using blender software by adjusting the type of ship that docks in the Tanjung Priok port area.

The type of ship used in this research is IFAMA MAS, with general cargo type and sailing serving the Tanjung Priok port route. The following are the specifications of the IFAMA MAS ship,

1) IMO : 9164251
2) Type : General Cargo
3) L x B : 155m x 25m
4) DWT : 17375 ton

D. Tanjung Priok Port Specification

Tanjung Priok Port specifications can be used as supporting data in 3D environment modelling, which will later be implemented in virtual reality applications. The following are port specifications:

1) Water area: 424 ha
2) Land Area: 605

Table 1. Tanjung Priok Port Spesification

No.	Jetty Type	Total	Length	Depth
1	General Cargo	42	6.597,70	5-11
2	Multipurpose	5	914	8-11
3	Container	13	2.800	9-14
4	Passenger	3	450	9
5	Dry Bulk	8	1.242	4-10
8	Beaching Point	1	66	6

3.2. 3D Modelling

At this stage, each piece of equipment that has been previously registered is modelled into a 3D object. The process of modelling equipment and other 3D assets (environments, etc.) [8] is done using open-source software called Blender. 3D assets for this V.R. application mainly include a 3D ship platform and a 3D environment of the Tanjung Priok Port area.

The process of making 3D models has several ways. Making ports in V.R. Bridge Simulator is simply modelling. This technique is a technique whose initial creation uses objects that the Blender software has provided [9]. In making this port using a cube object. The use of cube objects is done because the cube has a basic shape that is easy to form and reassemble in making 3D objects.

A. Ship Bridge 3D Modelling

Figure 4-5 show the 3D Model of equipment that will be used in a Virtual Reality application.

Figure 4. 3D Model of Ship Bridge Simulator (Front View)

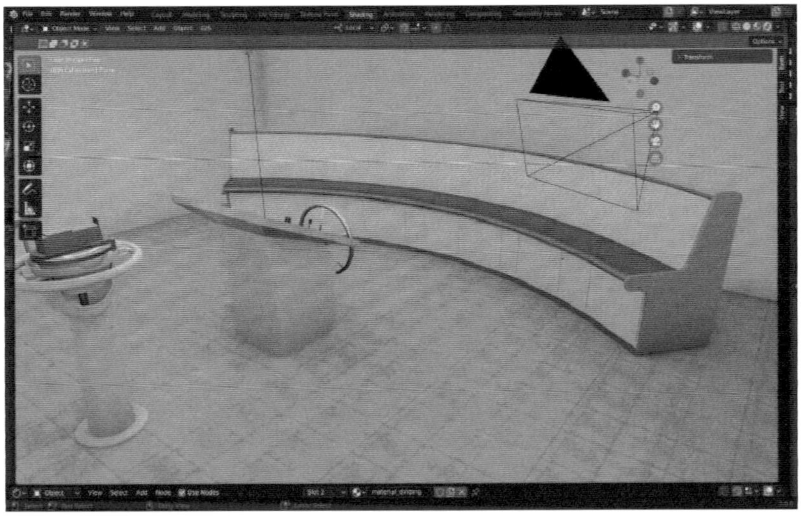

Figure 5. 3D Model of Ship Bridge Simulator (Back View)

B. 3D Modeling of Tanjung Priok Port Area

The Model of the ship bridge simulator and its components can be converted into virtual objects [10], and the Tanjung Priok Port Area will later be used as the primary virtual environment in virtual reality applications. In the application scenario, users can experience driving a ship in a virtual world where the ship they move will sail around the Tanjung Priok Port area. Modelling the port area is essential because the arrangement and dimensions of the port are made as accurate as possible to provide the best experience for users.

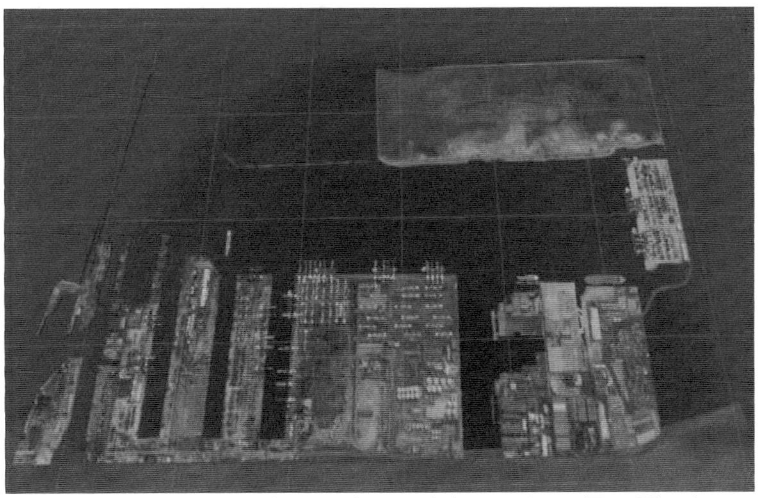

Figure 6. 3D Model of Tanjung Priok Port Area (Top View)

Figure 7. 3D Model of Tanjung Priok Port Area (Front View)

Figure 8. 3D Model of Tanjung Priok Port Area (Side View)

3.3. Virtual Reality Application

In this phase, all 3D and other visual assets will be used and assembled in one virtual environment, eventually becoming one Virtual Reality world. This Virtual Reality application is created using an open source real-time development platform called Unity3D.

A. User Interface

The user interface is the visible part of the virtual reality application that determines how the user interacts with the V.R. application and how information is displayed on the screen. Another way is a communication mechanism between the user and the application system. User interface design is essential because it will determine how a person interacts with the application [11][13-14].

Figure 9. User View using Steering

Figure 10. User View using throttle

3.4. User Assessment

Once developed, V.R. applications need to undergo thorough testing. The function of this test is to find technical errors, bugs (or deficiencies) and to add suggested features to the software. At the same time, user trials will also be conducted to determine whether the application has met the criteria based on its development goals. Supporting questionnaires will also be used to help users rate the software. Questionnaires from this application were distributed physically to the cadets as testers after the trial was carried out at the Surabaya Shipping Polytechnic (Politeknik Pelayaran Surabaya).

A. User Rating Questionnaire Response and Analysis

The data collected from user rating responses can be analyzed to obtain a quantitative score for V.R. applications. The percentage value of each question answer can be done using the Likert scale formula as follows:

$$P = \frac{S}{Ideal\ Score}\ x\ 100\% \tag{1}$$

Table 2 shows summarizes data collection for each response from thirty Surabaya Shipping Polytechnic cadets as respondents who have filled out paper-based questionnaires.

Table 2. Total questionnaire data

No.	**Question**	**Score**					**Total**
		1	2	3	4	5	
1	Is the simulation display in V.R. easy to understand?				7	23	30
2	Is learning to use V.R. more attractive, easier to understand, and practical?			2	13	15	30
3	Can learning using V.R. simulations fulfil functions such as direct simulations in the PoltekPel bridge simulator?			7	18	5	30
4	Does the 3D Model of the ship's bridge match the original?			1	12	17	30
5	Does the 3D Model of the Port match the original?			5	15	10	30
6	Can learning through 3D simulations help indirect (do-it-yourself) learning?			1	11	18	30
7	Do the features contained in the V.R. simulation application meet expectations?		2	7	14	7	30

IOP Conf. Series: Earth and Environmental Science 1166 (2023) 012053 doi:10.1088/1755-1315/1166/1/012053

Table 3. A total score of the questionnaire

Question Number	Percentage value	Total Score
1	95,30%	143
2	88,60%	133
3	78,60%	118
4	90,60%	136
5	83,30%	125
6	91,30%	137
7	77,30%	116
Total Score		908

To get a V.R. app quantitative score based on user ratings, the steps are as follows:

1) Determine the Maximum Score by multiplying (the most significant answer score) by (total questions) and (total respondents).
   Max = 5 x 7 x (30 Respondent) = 1050

2) Determine the Minimum Score by multiplying (the minor answer score) by (the number of questions) and (the number of respondents).
   Min = 1 x 7 x (30 Respondent) = 210

3) Determine the Median value by dividing the sum (maximum total score) and (minimum total score) into two.
$$Med = \frac{(1050 + 210)}{2} = 630$$

4) Determine the value of Quartile I by dividing the sum (median) and (minimum total score) into two.
$$Q1 = \frac{(630 + 210)}{2} = 420$$

5) Determine the value of the third quartile by dividing the sum (maximum score) and (median) into two.
$$Q3 = \frac{(1050 + 630)}{2} = 840$$

The following table shows the quantitative assessment rubrics categorized based on the Maximum Score, Quartile III, Median, Quartile I and Minimum Score values obtained previously.

Table 4. Table of a quantitative assessment rubric

Category	Description	Score
Excellent	Quartile III (Q3) ≤ x ≤ Max. Score	840 - 1050
Good	Median ≤ x ≤ Quartile III (Q3)	630 - 840
Fair	Quartile I ≤ x ≤ Median	420 - 630
Bad	Min. Score ≤ x ≤ Quartile I (Q1)	210 - 420

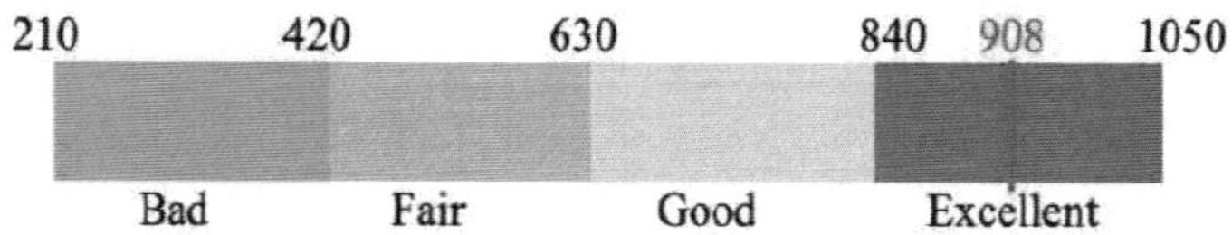

Based on user ratings, namely Surabaya Shipping Polytechnic cadets, it can be concluded that the Virtual Reality Simulator application has a score of 908 out of 1050, which is included in the excellent category.

4. Conclusion

The preliminary research of virtual reality of ship bridge simulator has resulted in a virtual model ship bridge simulator with features steering on the ship has the function of turning the ship to the starboard side or port side of the vessel, speed and 3D model environment of Tanjung Priok Port Area. This application also has been tried by cadets as users. Based on user ratings analyzed by the Likert Scale, the preliminary virtual ship bridge simulator application has achieved its goals. The result showed a 908 out of 1050, which showed that this virtual reality application is included in the excellent category. However, for future development, this application needs improved features, such as weather, wave, maps etc.

Acknowledgements

The authors thank to Digital Marine Operation and Maintenance Laboratory, Department of Marine Engineering, and Institut Teknologi Sepuluh Nopember for funding and supporting facilities this research.

References

[1] Frydenberg, S.G., Nordby, K. 2022 Virtual fieldwork on a ship's bridge: virtual reality-reconstructed operation scenarios as contextual substitutes for fieldwork in design education *Virtual Reality*.

[2] Aukstakalnis, S. & Blatner, D. 1992 *Silicon mirage: The art and science of virtual reality*, Peachpit Press, Berkley.

[3] Fellner, D.W., Behr, J., Bockholt, U. 2009 *Instantreality - a framework for industrial augmented and virtual reality applications. 2nd Sino-German Workshop, "Virtual Reality & Augmented Reality in Industry" 2009*. Invited Paper Proceedings. Participants Edition: 16.-17. April 2009, Shanghai, P.R. China. Shanghai: Shanghai Jiao Tong University.

[4] Ausburn, L. J. & Ausburn, F. B. 2004 Desktop virtual reality: A powerful new technology for teaching and research in industrial teacher education *Journal of Industrial Technical Education*, **41** 4.

[5] Zhang, Rukai. 2018 Construction of virtual ship simulation practical training platform *Advances in Engineering Research*, **166** 269-274.

[6] Hu X.M. 2016 Application of virtual reality technology in classroom teaching reform for navigation majors *Journal of Zhejiang Vocational College of Transportation*, **4** 49-52.

[7] Lauronen J., Ravyse W., Salokorpi M., Luimula M. 2020 validation of virtual command bridge training environment comparing the VR-training with ship bridge simulation *Advances in Intelligent Systems and Computing*, **1212**.

[8] Bown, J., White, E., & Boopalan, A. 2017 Looking for the ultimate display: A brief history of virtual reality. In J. Gackenbach & J. Bown (Eds.), *Boundaries of self and reality online: Implications of digitally constructed realities* 239–259.

[9] Bamodu, O., Ye, X.M., 2013 Virtual Manufacturing and Components of Virtual Reality *Applied Mechanics and Materials*, **318** 83-86.

[10] Huang, C., Zhang, W., & Xue, L. 2022 Virtual reality scene modeling in the context of the Internet of Things *Alexandria Engineering Journal*, **61** 8 5949–5958.

[11] Jong-Heon Kim, Sang-Tae Park, H. Lee, Keun-Cheol Yuk 2001 Virtual reality simulations in physics education *Interactive Multimedia Electronic Journal of Computer-Enhanced Learning,* **3** 2.

[12] Loup G., Serna A., Iksal S., George S., 2016 Immersion and persistence: Improving learners' engagement in authentic learning situations. In Adaptive and Adaptable Learning *Lecture Notes in Computer Science*, **9891** 410–415

[13] De Oliveira R.P., Carim Junior G., Pereira B., Hunter D., Drummond J., Andre M., 2022 Systematic Literature Review on the Fidelity of Maritime Simulator Training *Educ. Sci.*, **12** 817.

[14] Katie A., Joakim D., Kjetil N., Monica L., 2021 Using Operational Scenarios in a Virtual Reality Enhanced Design Process. *Educ. Sci.*, **11** 448.

Approximation method of the axial forces on network of connectors for modular hexagonal floating structures

Nur Hanani[1], Nik Mohd Ridzuan Shaharuddin[1], Arifah Ali[1], Sunarsih[2*]

[1]Faculty of Engineering, School of Mechanical Engineering, Universiti Teknologi Malaysia, Skudai, Malaysia
[2]Department of Marine Engineering, Institut Teknologi Sepuluh Nopember, Surabaya, Indonesia

*sunarsihits@gmail.com

Abstract. Modular floating structures are normally made up of modules which are connected using network of connectors. These connectors are prone to heavy load due to facing the open sea wave. Previous studies have examined simple module configuration such as two rectangular form arranged linearly or three or seven hexagonal form arranged in an oblique pattern experimentally or using CFD and FE packages. Two challenges persist afterward; firstly, on applying the current findings to topology made up of a bigger number of modules and secondly on estimating the axial load values of connector at the preliminary design stage. This study developed an approximation method of axial force at the connector based on basic module configuration of hexagonal modular floating structure and axial load values determined by earlier studies. Four approximation methods vary with number of modules, arrangement and connectors facing 90^0 wave direction were proposed. Early analysis of the proposed method revealed that the axial force of the hexagonal modular connector is affected by the overall structure type and configuration such that hexagonal module-to-module connection requires higher axial load as compared to hexagonal modular cluster-to-cluster linkage due to load distribution amongst the connectors.

1. Introduction

Very large floating structure (VLFS) is applied in coastal and ocean engineering that expected rapid expansion function for the purpose of ocean space utilization. Presently, VLFS function as a floating airport, mobile offshore base, floating storage base, floating pier, floating entertainment base, etc., that gives benefit to the economy, safety and environment friendliness [17]. Therefore, the shape, size and flexibility have the potential to address special requirements in terms of design, analysis, construction, assembly and operation in achieving its various functions of VLFS. The early study dealt with the floating structure of VLFS as a single continuum structure and a variety of VLFS shape design have been proposed. Japanese Society of Steel Construction [1999] has suggested that hexagonal-shaped VLFSs be constructed to allow for easy expansion of the floating structure [7]. Tukuji et al. [2002] treated L-shaped, T-shaped, C-shaped and X-shaped VLFSs. Circular pontoon-type VLFSs are considered by Tukuji et al. (2002), Malte et al. [2003] and Watanabe et al. [2003]. In addition, Tay et el. [2012] proposed shape design of VLFS that grouping by two categories are comprised longish VLFSs with different fore/aft end shapes and comprises various polygonal VLFS plan shapes that are confined within a square boundary or a circle. However, there are disadvantages implicit in the use of single continuum structure, for instance, the huge bending moments which affected the safety of structures and the difficulty in manufacture [18]. Judging from the performance and the construction of floating structures, a multi-module of floating structure has been proposed that solved the problem being the easiness of changes of scale, structure and function and the feasibility of construction in a small-scale shipyard [11]. The multi- module of rectangular floating structure has been proposed by Fu [2007], Zhang [2015], and Eva et al [2012]. Currently, many of researcher proposes the multi–module hexagonal pontoon shape of VLFS [10,11,14,27] which allows for ease in term of expansion and mobility.

Multi–module floating structure consists of individual modules that are detached by connectors. By that, the connector becomes a key element of the multi-module of floating structure [31] has considering serious threats to the entire structure if connector failure happens. The four most common types of connector use by multi-module floating structure are fully flexible, vertical free, hinged connection and fully rigid connection [10]. Each type has its own advantages and disadvantages while it chose by implement with the VLFS function. In addition, Shixiao [2007] proposed four type model of multi-module VLFS that consider both modules and connectors may be modelled as either rigid or flexible are Rigid Module and Rigid Connector (RMRC), Rigid Module and Flexible Connector (RMFC), Flexible Module and Rigid Connector (FMRC) and Flexible Module and Flexible Connector (FMFC). Nowadays, many researchers applied flexible connector in designing their floating structure [3,23] to achieve a greater reduction in the hydroelastic response of the VLFS under wave action. Furthermore, the arrangement of multi-module floating structure also been considered in designing multi-module floating structure. Michailides et al. [2013] is studied deformable floating modules connected with flexible connectors in longitudinal and transverse directions arrangement. Kim [2007] also considers the arrangement of two floating structure unit such as tandem arranged units and side-by-side arranged units.

The study of the performance of multi-module floating structure requires the involvement of hydro elasticity in the corresponding numerical analysis that take into account the effect of both the flexibility of the structure and the presence of the connectors. The connection performs an important role in dynamic responses [31] that applied various method of numerical method calculation implemented by many researchers. Hisaaki et al. [1979] calculated the one-dimensional behavior of floating structures consisting of rigid modules with rigid or pin connectors for regular waves applied a strip method. Direct methods that solve directly the equation of motion using conceptually all nodes of the discretized system of the floating structure applied by Shixiao [2007], Kunihiro [2007], and Tay [2002]. The mode superposition methods, where the hydrodynamic analysis is separated from the eigenvalue analysis of the structure by propose the generalized modes in addition to the six rigid body modes applied by Eva et al [2012] and Michaielis [2013]. Presently, Zhang [2015] proposed a network modeling method that introduced to model the multi-module floating structures of arbitrary topological form where each module under the excitation of waves can be viewed as an oscillator and a connector is viewed as a coupling between the oscillators. While many of researcher focus in hydro elastic of multi-module floating structure, thus, there are limited researcher that study about connector internal load effect [2, 3,16, 21].

Tukuji et al. [2002] stated that the numerical computation of the hydroelastic behavior of VLFS precondition of very time consuming after finalize the estimation of elastic motion of a VLFS for the Mega- Float project in Japan, which adjusted into 5000-m length and 1500-m width within 6s of wave period. As a result, Riggs [1993] introduced approximate methods of hydroelastic analysis theories are applied to predict the dynamic response of 5- and 16-module very large floating structures in regular waves as an option to the more computationally demanding, fully three-dimensional flexible module, flexible connector model in which linear potential theory is adopted. Abd- Azm [2000] applied approximation method for the calculation hydrodynamics of floating pontoons under oblique waves through three equations of motion such as surge, heave and pitch. Xiantao et al [2017] introduced the stiffness approximation method for determining the natural frequency and the motion mode of multi-module floating structure. As concern by Linjian et al. [2017], the determination of stiffness connectors is also the key component influencing the hydrodynamic response of modules and the dynamic constraint force of connector.

2. Mathematical Proposed Method

2.1 *Arrangement of hexagonal modular floating structure*

Circular modular floating structure topology made up of one clusters of hexagonal modules have been selected and each cluster is made up of seven modules. Three stages of the approximation method that defines its mathematical framework are, firstly, identification of Endo's [1997] linear configurations on the given modular floating structure topology, secondly, the determination of connectors' network for the configurations and, thirdly, the determination of axial load on the connectors.

As for Stage 1, the possible overall configuration of a modular floating structure modules depends on the shape of its module. Common shapes are square, rectangular, hexagonal and circular. Modules will be connected using connectors, flexible or otherwise, to form the preferred cluster. Square and rectangular modules are normally connected in matrix form; 2X2, 3X3, 6X6, etc. as a cluster. Hexagonal modules can be connected linearly in a matrix form or obliquely to form a circular topology. A circular module is normally proposed as a large stand-alone but connected to other stand-alone circular using means other than the conventional connectors. Figure 1 show hexagonal modules arranged in oblique configuration by Kunihiro [2007]. Takujumi et al. [2016] propose modular hexagonal module arranged in linear and centralized configuration.

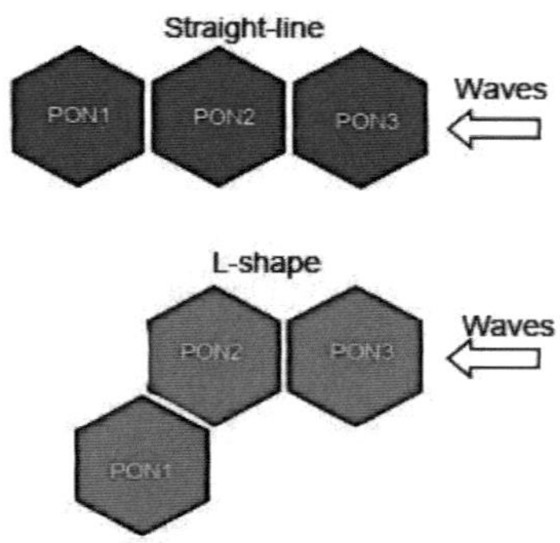

Figure 1. Hexagonal module arranged in straight line and oblique configuration by Kunihiro [2007]

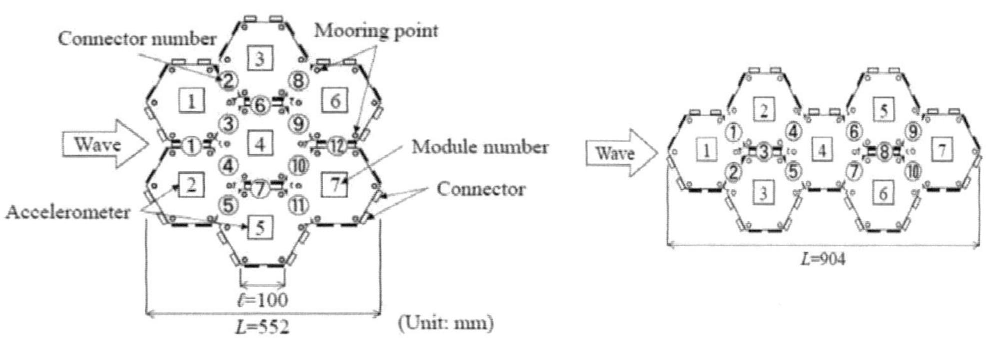

Figure 2. Hexagonal module arranged in linear and centralized configuration by Takujumi [2016]

Basic linear configuration by Endo [1997] adopted by Eva et al. [2015], which consists of two rectangular pontoon modules, are later mapped on the overall hexagonal modular floating structure configuration. Figure 3, Figure 4, Figure 5 and Figure 6 illustrate the 4 mapping strategies for a hexagonal modular floating structure cluster. The two rectangular pontoon configurations refer to Eva [2015] because later the secondary data experiment of axial load calculated by Eva adopted in this mathematical formulation.

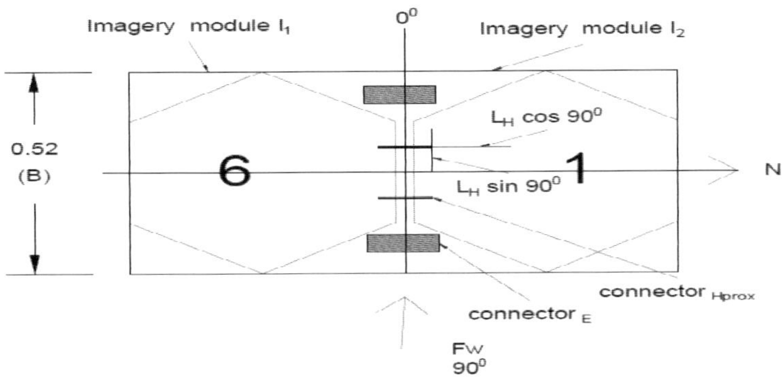

Figure 3. Endo 1 mapping two hexagonal modular with two rectangular modular by Eva et al. [2015]

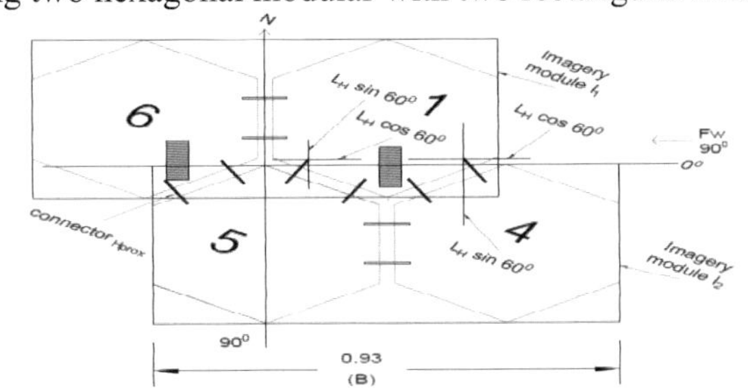

Figure 4. Endo 2 mapping four hexagonal modular with two rectangular modular by Eva et al. [2015]

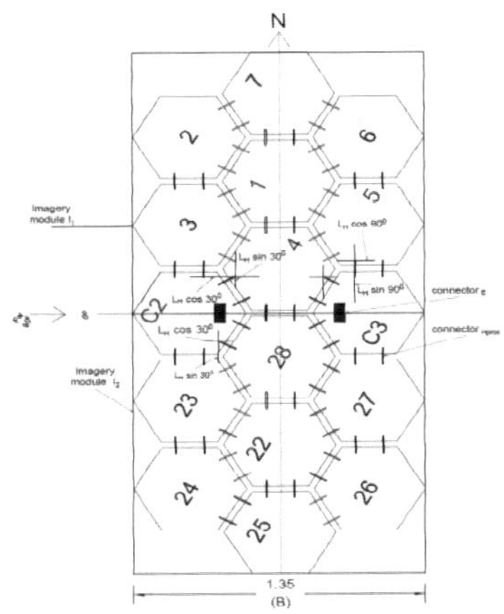

Figure 5. Endo 3 mapping two cluster hexagonal modular with two rectangular modular by Eva et al. [2015]

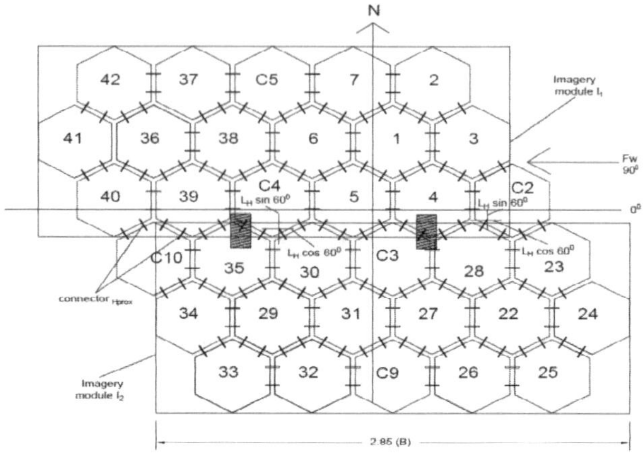

Figure 6. Endo 4 mapping four cluster hexagonal modular with two rectangular modular by Eva et al.[2015]

2.2 *The Algorithm*

The first mapping, here in after referred to as Endo1, is for two hexagonal modules, module 1 and module 6 which are in a linear configuration similar to Endo's [1997]. That also matches the configuration used by Eva's [2015] for the experimental determination of the axial load on the two connectors holding two rectangular modules with wave coming from 90° direction as shown. As such, for a similar set of environmental conditions, it is safe to assume that the load on connectors holding the hexagonal modules is approximately the same as the loads holding Eva's [2015] rectangular modules. Correction for physical difference between the two shapes is assumed handled by the value B/λ ; *B* being the breadth of the module. The value B for B/λ for the hexagonal module is the projected breadth on which the connectors are attached. λ is wavelength impacting the hexagonal modular floating structure. Plots of axial load against B/λ by Eva [2015] are for 3 wave directions (θ); 90°, 60° and 45° with respect to the axial line drawn along the length of the connector. It is to be highlighted that, theoretically, there should be no axial load on the connectors when the wave is impacting the connector perpendicularly. However, Eva [2015] detected connector loadings for that wave since the net effect of the wave on the VLFS is kiosk, 3D motion.

For $Endo1_{90}°$, the two connectors holding the hexagonal modules should experience the same load experiences by the two Eva's [2015] connectors show in Figure 3. Hence,

$$2L_{C/E/90/new\ B/\lambda} = 2L_{C/H/90/(m6-m1)} \quad (1)$$

i.e $$L_{C/H/90//(m6-m1)} = L_{C/E/90//new\ B/\lambda} \quad (2)$$

Where $L_{C/E//90/new\ B/\lambda}$ is the load experiences by the two connectors holding Eva's [2015] rectangular modules, and to be read from the plot of load against B/λ experimentally determined by Eva [2015], and $L_{C/H/90//(m6-m1)}$ is the load experiences by the two connectors holding the hexagonal modules.

For $Endo2_{90}°$ mapping (Figure 4) the loads on the network of six connectors holding the four hexagonal modules are equivalent to two connectors holding two Eva's [2015] rectangular modules each of which represent two hexagonal modules. The general expression for the axial load on connectors for $Endo2_{90}°$ mapping is derived as follows.

The resolved component of $L_{C/H}$ in the axial (north-south) direction (south being +ve), is:F_x

$$2L_{C/H/90/(m1-m50)} \cos 60 + 2L_{C/H/90/(m1-m4)} \cos 30 + 2L_{C/H/90/(m6-m5)} \cos 30 \quad (3)$$

Likewise, the resolved components of $L_{C/H}$ in the east-west direction (east being +ve), is:

$$2\ L_{C/H/90/(m1-m5)} \sin 60 - 2L_{C/H/90/(m1-m4)} \sin 30 - 2L_{C/H/90/(m6-m5)} \sin 30 \quad (4)$$

Assuming that all connectors experience the same load, i.e

$$L_{C/H/90/(m1-m50)} = L_{C/H/90/(m1-m4)} = L_{C/H/90/(m6-m5)} = L_{C/H/90} \quad (5)$$

Hence, the resolved component of $L_{C/H}$ in the axial (north-south) direction

$$F_X = 2L_{C/H/90} \cos 60 + 4L_{C/H/90} \cos 30$$
$$= (1+2\sqrt{3})\ L_{C/H/90} \quad (6)$$

and the resolved component of $L_{C/H}$ in the east-west direction

$$F_Y = 2L_{C/H/90} \sin 60 - 4L_{C/H/90} \sin 30$$
$$= (-2+\sqrt{3}\ )\ L_{C/H/90} \quad (7)$$

Based on the argument given above that the connectors will experience some load even though the wave is impacting perpendicular to the connectors, equation (6) should not, in any case, equate to zero. The $L_{C/H/90}{}^{o}$ as the axial force is the force that acts in the axis direction of connector facing to wave direction. The $L_{C/E//90/new\ B/\lambda}$ is the new axial force was exacted by extrapolation the secondary data of axial load from Eva [2015] using exponential trend graph equation. The axial force connector was calculated and this shall be:

$$(\ 1+2\sqrt{3}\ )\ L_{C/H/90} = 2\ L_{C/E//90/new\ B/\lambda} \quad (8)$$

For Endo3$_{90}$^{o} the mapping is to cater for cluster-to-cluster connection; each cluster comprises of seven hexagonal modules and one additional modules as in Figure 5. Take note of the thirty-four connecting required to establish a perfect connection between cluster. Each cluster is assumed as one Eva's (2015) rectangular module and as such the load experience by the network of eighteen connectors holding the hexagonal modules is equivalent to two Eva's [2015] connectors. The general expression for the axial load on connectors for Endo3$_{90}$^{o} mapping is derived as follows.

The resolved component of $L_{C/H/90}$ in the axial (north-south) direction (south being +ve), is:F_x

$$2L_{C/H/90/(m23-mc2)} \cos 90 + 2L_{C/H/90/(mc2-m3)} \cos 90\ + 2L_{C/H/90/(m28-mc2)} \cos 30 + 2L_{C/H/90/(mc2-m4)} \cos 30$$
$$+ 2L_{C/H/90/(m28-m4)} \cos 90\ + 2L_{C/H/90/(m28-mc3)} \cos 30 + 2L_{C/H/90/(mc3-m4)} \cos 30 + 2L_{C/H/90/(m27-mc3)} \cos 90$$
$$+ 2L_{C/H/90/(mc3-m5)} \cos 90 \quad (9)$$

Likewise, the resolved components of $L_{C/H/90}$ in the east-west direction (east being +ve), are:

$$2L_{C/H/90/(m23-mc2)}\sin 90 + 2\ L_{C/H/90/(mc2-m3)} \sin 90 + 2L_{C/H/90/(m28-mc2)}\ \sin 30 - 2L_{C/H/90/(mc2-m4)}\ \sin 30 +$$
$$2L_{C/H/90/(m28-m4)} \sin 90 - 2L_{C/H/90/(m28-mc3)} \sin 30 + 2L_{C/H/90/(mc3-m4)}\sin 30 + 2L_{C/H/90/(m27-mc3)} \sin 90\ +$$
$$2L_{C/H/90/(mc3-m5)}\sin 90 \quad (10)$$

With the same assumption that all connectors of modules experience the same load, equation 9 becomes

$$F_X = 10L_{C/H/90} \cos 90 + 8L_{C/H/90} \cos 30$$
$$=4\sqrt{3}\ L_{C/H/90} \quad (11)$$

The $L_{C/H/90}{}^{o}$ for Endo3$_{90}$^{o} as the axial force is calculated and this shall equate with load value from Eva (2015) as expressed below.

$$4\sqrt{3}\ L_{C/H/90} = 2\ L_{C/E/90/new\ B/\lambda} \quad (12)$$

Endo 4$_{90}$^{o} maps two clusters connected together and assumed as similar to one of Eva's [2015] rectangular pontoon with another two connected clusters as shown in Figure 6. Two modules have been introduced for the purpose of perfecting the circular topology and making the connection better. In such arrangement the load on the pair of Eva's [2015] connector is sustained by a network of eleven pairs of connectors holding the hexagonal modules together. The general expression for the axial load on connectors for Endo 4$_{90}$^{o} mapping is derived as follow.

The resolved component of $L_{C/H/\theta}$ in the axial (north-south) direction (south being +ve), is:F_x

$$2L_{C/H/90/(m23-mc2)} \cos 60 + 2L_{C/H/90/(m23-m4)}\cos 60 + 2L_{C/H/90/(m28-m4)} \cos 60 + 2L_{C/H/90/(m28-m5)} \cos 60\ + \quad (13)$$
$$2L_{C/H/90/(mc3-m5)} \cos 60\ + 2L_{C/H/90/(mc3-mc4)} \cos 60 + 2L_{C/H/90/(m30-m4)}\ \cos 60 + 2L_{C/H/90/(m30-m39)} \cos 60$$
$$+ 2L_{C/H/90/(m35-m39)} \cos 60 + 2L_{C/H/90/(m35-m40)}\ \cos 60 + 2L_{C/H/90/(mc10-m40)}\ \cos 60$$

The assumption that all connectors experience the same value of load thus allowing (13) to be simplified as

$$F_X = 22\ L_{C/H/90} \cos 60$$
$$=11\ L_{C/H/90} \quad (14)$$

Likewise, the resolved components of $L_{C/H}$ in the east-west direction (east being +ve), are:

$$2L_{C/H/90/(m23-mc2)} \sin 60 - 2L_{C/H/90/(m23-m4)} \sin 60 + 2L_{C/H/90/(m28-m4)} \sin 60 - 2L_{C/H/90/(m28-m5)} \sin 60 + 2L_{C/H/90/(mc3-m5)} \sin 60 - 2L_{C/H/90/(mc3-mc4)} \sin 60 + 2L_{C/H/90/(m30-mc4)} \sin 60 - 2L_{C/H/90/(m20-m39)} \sin 60 + 2L_{C/H/90/(m35-m39)} \sin 60 - 2L_{C/H/90/(m35-m40)} \sin 60 + 2L_{C/H/90/(m10-m40)} \sin 60 \quad (15)$$

And the same assumption allows (15) to be simplified as

$$F_Y = 12\, L_{C/H/90} \sin 60 - 10\, L_{C/H/90} \sin 60 = \sqrt{3}\ L_{C/H/90} \quad (16)$$

Likewise, the $L_{C/H/90}{}^{o}$ for $Endo4_{90}{}^{o}$ as the axial force is calculated and this shall equate with load value from Eva (2015) as expressed in Equation (16). Table 1 organized the final axial force expressions derived.

$$11\ L_{C/H/90} = 2\ _{LC/E/90/new\ B/\lambda} \quad (17)$$

Table 1. The final axial force expressions derived according to each circular configuration.

TOPOLOGY MAPPING	NORTH-SOUTH COMPONENT	EAST-WEST COMPONENT
$Endo1_{90}{}^{o}$	$L_{C/H/90}$	$L_{C/H/90}$
$Endo2_{90}{}^{o}$	$(1+2\sqrt{3})L_{C/H/90}$	$(-2+2\sqrt{3})L_{C/H/90}$
$Endo3_{90}{}^{o}$	$4\sqrt{3}L_{C/H/90}$	$10\ L_{C/H/90}$
$Endo4_{90}{}^{o}$	$11L_{C/H/90}$	$\sqrt{3}\ L_{C/H/90}$

3. Axial Load Connector Analysis

Normally the length of a floating structure is a concern in determining of hydroelastic of floating structure, however, in this research, the breadth of floating structure is taken into charge. This is because the breadth of floating structure is linked to the connector directly facing the wave direction. The breadth of floating increases due to the expansion of modular floating structures from one module linked with one module (Endo 1 and Endo 2) toward a linked cluster with other clusters (Endo 3 and Endo 4). Furthermore, before this modular floating only deals with linear configuration and side-by-side configuration as the function of floating airport and oil storage. Currently, the habitable modular floating structure as an artificial island demands enlarged modules linked as clusters that form a circular configuration.

The increasing number of modules linked together causes the raised number of network connectors. The network of connectors configuration in rectangular and square floating structure layout did not complex as in the hexagonal modular floating structure. Thus, the approximation method explores to development of an easy way for a designer to determine an estimation of axial force for one connector due to complicated module configuration.

The results of the present axial connector load for hexagonal modular floating structure calculated using the approximation method are shown in Figure 7. The breadth of Endo 1, Endo 2, Endo 3 and Endo 4 are 0.52m, 0.93m, 1.35m and 2.85m. From this result, the increasing breadth of the floating structure will decrease the axial force of the connector due to distributed load via many numbers of connectors. Thus, one module linked to one module as Endo 1 needs a higher axial load compared to linked cluster by cluster as Endo 4.

However, this research only takes into account the stiffness same as Eva et al [2015] paper is a concern as an extrapolation of secondary data axial force is extracted to apply in this approximation method. The approximation method load analysis influenced by stiffness has been done by other researchers.

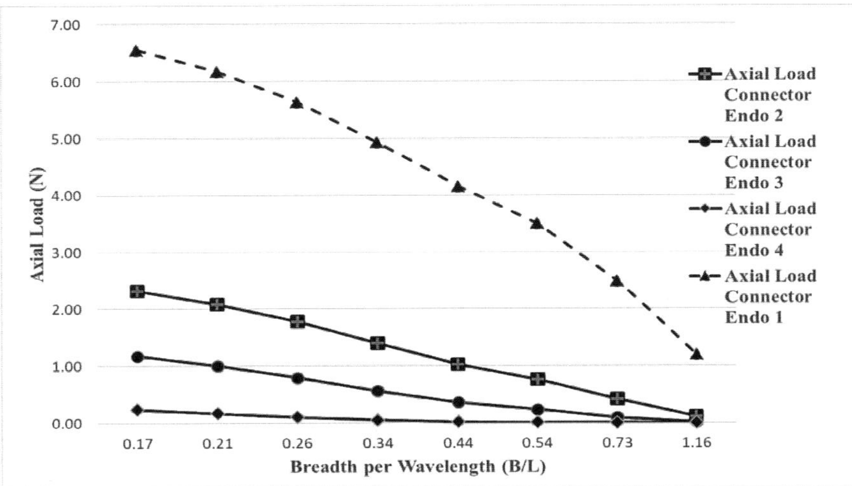

Figure 7. Axial connector force for hexagonal modular floating structure facing 90^0 wave direction

4. Discussion

4.1 *Advantages and strength of the proposed method*

The whole idea of this research is to establish a reference on an estimation of axial load for flexible connector for modular hexagonal floating structures. It has been considered an issue because the connector system is related to the relative motion between modular floating structure modules and is able to sustain forces as a result of wave motion. Furthermore, the connector system is also a key component of a modular floating structure that influences the system safety therefore, the connection load and the dynamic behavior of the system in waves are seriously concerned in the conceptual design of a modular floating structure. Additionally, current methods of load analysis involved complicated mathematical models and techniques that take time to solve then simplified methods of analysis and models are a critical need for design practice.

Limited research has been carried out on the numerical evaluation of internal forces that affect various design parameters compared to hydroelastic behavior for modular floating structures [9]. Riggs and Ertekin [1993] state that the fundamental theories method shows the most accurate analysis technique however, the performance of theory to a larger size causes a large computational problem where the solution will pressure the state-of-the-art in computer technology. The approximation model technology provides a solution that balances the computation time and accuracy [28]. The approximation method for floating structure in terms of hydroelastic behavior [1,20,29] and load analysis in mooring [6, 25, 28] have been done by another researcher. However, the approximation method in connectors for floating structures has not been explored yet. On the other hand, the approximation method for axial load connectors in the building has been done by other researcher [12, 19, 30]. Thus, the idea of an approximation method for axial load analysis for connectors in modular floating structures has been proposed for easier and faster load analysis, especially for complex connector configurations such as in Endo 3 and Endo 4.

4.2 *Limitation and constraints of the proposed method*

Mostly, many researchers consider the rectangular shape of floating structures gradually design in charge of service applications such as floating airport, bridges, and oil storage facilities. Nowadays, the hexagonal shape is the dominant idea of floating structure functional for habitable as the floating artificial island. The shape of floating structure affected the arrangement layout of the modular floating structure which are in linear, side by side, or circular configurations. Thus, a modular hexagonal produces a circular arrangement layout and applied a network connector configuration because a hexagonal has six sides to connect with other modules rather than a rectangular or square shape that has four sides.

The connector load analysis resolved is often applied in a rectangular shape and it is difficult to find secondary data load analysis for other shapes such as hexagonal. In addition, the load analysis studied that splitting the internal connector force into axial force and shear force is a challenge to find it. Meanwhile, Tay et al [2012] have studied altered inscribed rectangular and square shapes with polygons such as hexagon, octagonal and etc to determine which shape that reducing hydroelastic behavior. Hence, the approximation method adopted the secondary data of axial load analysis done by Eva [2015] mapping inscribed rectangular shapes into hexagonal. However, there is some extra space while inscribed rectangular shape into hexagonal that cannot be avoided as shown in each Endo's at Figure 3, Figure 4, Figure 5, and Figure 6.

The internal forces of connectors can be studied with numerous design parameters such as the stiffness of connectors, geometrical dimensions, module layout, and wave field characteristics. Two types of secondary data commonly found in papers by other researchers are wave frequency versus load connector and wave direction versus load. Meanwhile, the expected result of the approximation method for axial load at hexagonal modular floating structure is produced in form of the breadth per wavelength versus axial as pursued by Eva et al. [2015]. Thus, it takes time to analyze the two common secondary data to make a comparison and continues with validating the data of the approximation method that had been determined for the hexagonal modular floating structure. This is because each researcher has their own geometrical dimension of floating structure, stiffness of connector, and wave direction facing the connector when they are conducting their numerical calculation, simulation software and also the experimental.

5. Conclusion

This paper presents a new method proposed as approximation method in calculating axial load for hexagonal modular floating structure. The approximation method concerns about axial load connector of hexagonal modular floating structure has been affected by module arrangement as take into account the linear and circular cluster configuration which also increase the number of connectors applied facing 90^0 wave direction. The increasing number of modules will increase number of connectors, however, the axial load decrease as the estimation one axial load connector in Endo 4 is lower than axial force in Endo 2 and Endo 3. The connector in Endo 1 is facing the same as wave direction 90^0 presumed did not to have any axial load however because this is secondary data from the experimental their have determined the value on it. The approximation method axial load will later continue to develop the mathematical formula axial load in facing 45^0 and 60^0 wave direction. In the future, the calculation axial load of the network connector for hexagonal floating structure by approximation method will be compared with other existing calculation methods and considering the new method in view of other shape of floating structure by exploiting numerical simulations using software such as ANSYS AQWA and ANSYS STATIC.

References

[1] Abul-Azm AG, Gesraha MR. Approximation to the hydrodynamics of floating pontoons under oblique waves. Ocean Engineering. 2000 Apr 1;27(4):365-84.
[2] Kim BW, Hong SY, Kyoung JH, Cho SK. Evaluation of bending moments and shear forces at unit connections of very large floating structures using hydroelastic and rigid body analyses. Ocean engineering. 2007 Aug 1;34(11-12):1668-79.
[3] Michailides C, Loukogeorgaki E, Angelides DC. Response analysis and optimum configuration of a modular floating structure with flexible connectors. Applied Ocean Research. 2013 Oct 1;43:112-30.
[4] Loukogeorgaki E, Michailides C, Angelides DC. Hydroelastic analysis of a flexible mat-shaped floating breakwater under oblique wave action. Journal of Fluids and structures. 2012 May 1;31:103-24.
[5] Loukogeorgaki E, Vasileiou M, Rapanta E. 3D experimental and numerical investigation of the performance of a modular floating structure. InThe Twenty-fifth International Ocean and Polar Engineering Conference 2015 Jun 21. OnePetro.

[6] Kara F. Multibody interactions of floating bodies with time domain predictions. Journal of Waterway, Port, Coastal and Ocean Engineering. 2020 Jun 10.
[7] Japanese Society of Steel Construction. Technical Research Association of Megafloat. Summary of practical research on Megafloat Airport in 1998. TRAM 1999 [in Japanese].
[8] Maeda H, Maruyama S, Inoue R, Watanabe K, Togawa S, Suzuki F. On the motions of a floating structure which consists of two or three blocks with rigid or pin joints. Journal of the Society of Naval Architects of Japan. 1979;1979(145):71-8.
[9] Jiang D, Tan KH, Dai J, Ang KK, Nguyen HP. Behavior of concrete modular multi-purpose floating structures. Ocean Engineering. 2021 Jun 1;229:108971.
[10] Ko KK. Realising a floating city. Ko KK. Releasing a floating city a feasibility study of the construction of a floating city. Delft University of Technology, Civil engineering and Geosciences, Specialisation Hydraulic Structures. 2015.
[11] Ikegami K, Matsuura M, Wada Y, Zhang G. Response characteristics of a multi-connected floating body system in waves. InInternational Conference on Offshore Mechanics and Arctic Engineering 2007 Jan 1 (Vol. 42711, pp. 105-114).
[12] Yang L, Luo Y, Qiu T, Qiu Y, Zheng H. An analytical solution to the buckling of cylindrical shells under arbitrarily distributed axial loads. International Journal of Pressure Vessels and Piping. 2020 Nov 1;187:104169.
[13] Wu L, Wang Y, Li Y, Xiao Z, Li Q. Simplified algorithm for evaluating the hydrodynamic performance of very large modular semi-submersible structures. Ocean Engineering. 2017 Aug 1;140:105-24.
[14] Lister N, Muk-Pavic E. Sustainable artificial island concept for the Republic of Kiribati. Ocean Engineering. 2015 Apr 1;98:78-87.
[15] Pete MA, Meylan MH, Chung H. Wave scattering by a circular elastic plate in water of finite depth: a closed form solution. International Journal of Offshore and Polar Engineering. 2004 Jun 1;14(02).
[16] Newman JN. Wave effects on hinged bodies. Part II—hinge loads. 1997
[17] Park SW, Lee TK, Hong SY. A study of preliminary structural design of pontoon type VLFS. InThe Fourteenth International Offshore and Polar Engineering Conference 2004 May 23. OnePetro.
[18] Shi Q, Xu D, Zhang H. Design of a flexible-base hinged connector for very large floating structures. InInternational Conference on Offshore Mechanics and Arctic Engineering 2018 Jun 17 (Vol. 51203, p. V001T01A015). American Society of Mechanical Engineers.
[19] Cao Q, Xiao L, Guo X, Liu M. Second-order responses of a conceptual semi-submersible 10 MW wind turbine using full quadratic transfer functions. Renewable Energy. 2020 Jun 1;153:653-68.
[20] Riggs HR, Ertekin RC. Approximate methods for dynamic response of multi-module floating structures. Marine Structures. 1993 Jan 1;6(2-3):117-41.
[21] Huang LL, Riggs HR. The hydrostatic stiffness of flexible floating structures for linear hydroelasticity. Marine Structures. 2000 Mar 4;13(2):91-106.
[22] Endo R, Kawakami Y, Imai T, Tosaka N. Identification analysis of structural damage on unit-linked offshore floating models. International Journal of Offshore and Polar Engineering. 1999 Sep 1;9(03).
[23] Fu S, Moan T, Chen X, Cui W. Hydroelastic analysis of flexible floating interconnected structures. Ocean engineering. 2007 Aug 1;34(11-12):1516-31.
[24] Tay ZY, Wang CM. Reducing hydroelastic response of very large floating structures by altering their plan shapes. Ocean Systems Engineering. 2012 Mar 11;2(1):69-81.
[25] Martin T, Kamath A, Bihs H. Modeling and simulation of moored-floating structures using the tension element method. Journal of Offshore Mechanics and Arctic Engineering. 2020 Feb 1;142(1).
[26] Humamoto T, Fujita K. Wet-mode superposition for evaluating the hydroelastic response of floating structures with arbitrary shape. InThe twelfth international offshore and polar engineering conference 2002 May 26. OnePetro.

[27] Humamoto T, Fujita K.. Water Tank Experiment Of Shape-Variable Floating Structures. J. Struct. Constr. Eng. 2016;81(724):1039-49.

[28] Zhu X, Sun L, Li B. Dynamic analysis of vessel/riser/equipment system for deep-sea mining with RBF neural network approximations. Marine Georesources & Geotechnology. 2020 Feb 7;38(2):174-92.

[29] Zhang X, Lu D, Gao Y, Chen L. A time domain discrete-module-beam-bending-based hydroelasticity method for the transient response of very large floating structures under unsteady external loads. Ocean Engineering. 2018 Sep 15;164:332-49.

[30] Yu Y, Chen L, Yu P, Sun W. Analytical approximate solution for large post-buckling behavior of a fixed-pinned beam subjected to terminal force with shear force effect. Iranian Journal of Science and Technology, Transactions of Civil Engineering. 2021 Mar;45(1):159-64.

[31] Zhang HC, Xu DL, Xia SY, Lu C, Qi ER, Tian C, Wu YS. Nonlinear network modeling of multi-module floating structures with arbitrary flexible connections. Journal of Fluids and Structures. 2015 Nov 1;59:270-84.

AUTHOR INDEX